METHODS IN MOLECULAR BIOLOGY™

Series Editor
John M. Walker
School of Life Sciences
University of Hertfordshire
Hatfield, Hertfordshire, AL10 9AB, UK

For further volumes:
http://www.springer.com/series/7651

Protein Electrophoresis

Methods and Protocols

Edited by

Biji T. Kurien and R. Hal Scofield

University of Oklahoma Health Sciences Center
Oklahoma Medical Research Foundation
Department of Veterans Affairs Medical Center
Oklahoma City, OK, USA

Humana Press

Editors
Biji T. Kurien
University of Oklahoma Health Sciences Center
Oklahoma Medical Research Foundation
Department of Veterans Affairs
Oklahoma City, OK, USA

R. Hal Scofield
University of Oklahoma Health Sciences Center
Oklahoma Medical Research Foundation
Department of Veterans Affairs Medical Center
Oklahoma City, OK, USA

ISSN 1064-3745 e-ISSN 1940-6029
ISBN 978-1-4939-6241-9 ISBN 978-1-61779-821-4 (eBook)
DOI 10.1007/978-1-61779-821-4
Springer New York Dordrecht Heidelberg London

Softcover re-print of the Hardcover 1st edition 2012

Cover illustration: A vertical air cooled starch gel analyzing haptoglobin polymers (Figure 13 in Chapter 1; gel run started at 3:50 PM on Tuesday, June 3rd and stopped at 10:50 AM on Wednesday, June 4th, 1958)

Printed on acid-free paper

Humana Press is part of Springer Science+Business Media (www.springer.com)

Preface

We previously edited a book similar to this one in which methods and protocols were presented for protein blotting. The present work concentrates upon methods for electrophoresis. Again, each chapter outlines a specific electrophoretic variant and gives step-by-step instructions. Most often, the instructions found in a typical paper do not allow a laboratory scientist to easily reproduce a technique. Sometimes it seems to us that technique protocols are like food recipes—if one does not already know how to cook the dish, or a very similar dish, then recipes are not very helpful. The instructions given in this textbook are much more detailed than those typically found in research papers (or recipes). We imagine that one will be able to open this textbook in one's lab, and by following the steps given, perform a technique new to that lab without too much difficulty. We hope that investigators will find this is the case. We have elected to open this textbook with a historical account of how gel electrophoresis was developed in the early 1950s by Oliver Smithies, who was then a Research Associate in Connaught Medical Research Laboratory at the University of Toronto. This chapter includes photographs of his laboratory notebooks as well as figures showing some previously unpublished original starch gels from the period of development of the technique. Smithies, who was able to lay his hands on these notebooks with ease despite the fact that the experiments were performed nearly 60 years ago, was elected to the United States National Academy of Sciences in 1971, presumably in part because of his development of gel electrophoresis. Later, he won the Nobel Prize for the development of homologous recombination, which is the basis for gene targeting and gene knockout techniques. One day we called him to discuss about his chapter and he could not come to the phone because he was loading a gel. We only hope that we will be enthusiastic and still going strong, as is Oliver Smithies, at age 86.

Oklahoma City, OK, USA

Biji T. Kurien
R. Hal Scofield

Contents

Contributors

FABIOLA AMAIR-PINEDO • *Universitat de Barcelona, L'Hospitalet de Llobregat, Barcelona, Spain*
CLAUDIA ARNDT • *Carl Gustav Carus University TU Dresden, Dresden, Germany*
MICHAEL BACHMANN • *Carl Gustav Carus University TU Dresden, Dresden, Germany*
NATARAJAN BALASUBRAMANIYAN • *University of Colorado Denver, Aurora, CO, USA*
RAMON BARTRONS • *Universitat de Barcelona, L'Hospitalet de Llobregat, Barcelona, Spain*
HOLGER BARTSCH • *Carl Gustav Carus University TU Dresden, Dresden, Germany*
SUNIL K. BERWAL • *University of Pune, Pune, India*
VLADIMIR BESADA • *Centre for Genetic Engineering and Biotechnology, La Habana, Cuba*
MARK D. BIGGIN • *Lawrence Berkeley National Laboratory, Berkeley, CA, USA*
ENGELBERT BUXBAUM • *Ross University School of Medicine, Roseau, Commonwealth of Dominica, West Indies*
CARMEN CALINESCU • *Université du Québec à Montréal, Montréal, QC, Canada*
MARC CARTELLIERI • *Carl Gustav Carus University TU Dresden, Dresden, Germany*
LILA CASTELLANOS-SERRA • *Centre for Genetic Engineering and Biotechnology, La Habana, Cuba*
MUSTAFA ÇELEBIER • *Laboratory of Foodomics, CIAL, CSIC, Madrid, Spain; Faculty of Pharmacy, Hacettepe University, Ankara, Turkey*
HAN-MIN CHEN • *Catholic Fu-Jen University, New Taipei City, Taiwan*
YASMIN CHISHTI • *Biosciences Division, Advanced Photon Source, Argonne National Laboratory, Argonne, IL, USA*
JUNG-KAP CHOI • *Chonnam National University, Kwangju, South Korea*
MEGAN CHOI • *Lawrence Berkeley National Laboratory, Berkeley, CA, USA*
DAVID CHOVEAUX • *University of KwaZulu-Natal, Pietermaritzburg, Scottsville, South Africa*
ALEJANDRO CIFUENTES • *Laboratory of Foodomics, CIAL, CSIC, Madrid, Spain*
WEI-TAO CONG • *School of Pharmacy, Wenzhou Medical College, Wenzhou, Zhejiang, China*
LEAH COSGROVE • *CSIRO, Preventative Health Flagship, Adelaide BC, SA, Australia; CSIRO, Food and Nutritional Sciences, Adelaide BC, SA, Australia*
GYORGY CSAKO • *Department of Laboratory Medicine, Clinical Center, National Institutes of Health, Bethesda, MD, USA*
MONICA CUBILLOS-ROJAS • *Universitat de Barcelona, L'Hospitalet de Llobregat, Barcelona, Spain*
CHRIS CURSARO • *School of Molecular and Biomedical Science, Adelaide Proteomics Centre, University of Adelaide, Adelaide, Australia*
G. NANDHINI DEVI • *Centre for Biotechnology, Anna University, Chennai, India*

K. Divakar • *Centre for Biotechnology, Anna University, Chennai, India*
Ming Dong • *Lawrence Berkeley National Laboratory, Berkeley, CA, USA*
Yaser Dorri • *University of Oklahoma Health Sciences Center, Oklahoma City, OK, USA*
Giuliano Elia • *University College Dublin, Belfield, Dublin, Ireland*
Rodolfo Federico • *Department of Biology, 3rd University of Rome, Rome, Italy*
Anja Feldmann • *Carl Gustav Carus University TU Dresden, Dresden, Germany*
James D. Fesmire • *Oklahoma Medical Research Foundation, Oklahoma City, OK, USA*
Lydia A. Finney • *Biosciences and X-ray Science Divisions, Advanced Photon Source, Argonne National Laboratory, Argonne, IL, USA*
John S. Fossey • *School of Chemistry, University of Birmingham, Edgbaston, Birmingham, UK*
Kim Y.C. Fung • *CSIRO, Preventative Health Flagship, Australia, Adelaide BC, SA, Australia; CSIRO, Food and Nutritional Sciences, Adelaide BC, SA, Australia*
Lijun Gao • *Baotou Medical College, Baotou, Inner Mongolia, People's Republic of China*
Pennathur Gautam • *Centre for Biotechnology, Anna University, Chennai, India*
J.P. Dean Goldring • *Biochemistry, University of KwaZulu-Natal, Scottsville, South Africa*
Marion L. Greaser • *University of Wisconsin-Madison, Madison, WI, USA*
Shalini Gupta • *University of Illinois at Chicago, Chicago, IL, USA*
Syed R. Haider • *Loughborough University, Loughborough, UK*
Hisao Haniu • *Shinshu University, Nagano, Japan*
Peter Hoffmann • *School of Molecular and Biomedical Science, Adelaide Proteomics Centre, University of Adelaide, Adelaide BC, SA, Australia*
Jin-Sheng Huang • *University of Illinois at Chicago, Chicago, IL, USA*
Sun-Young Hwang • *College of Pharmacy & Research Institute of Drug Development, Chonnam National University, Gwangju, South Korea*
Clara Ibáñez • *Laboratory of Foodomics, CIAL, CSIC, Madrid, Spain*
Tony D. James • *Department of Chemistry, University of Bath, Bath, UK*
P. Jayathilaka • *University of Illinois at Chicago, Chicago, IL, USA*
Li-Tai Jin • *School of Pharmacy, Wenzhou Medical College, Wenzhou, Zhejiang, China*
Jian Jin • *Lawrence Berkeley National Laboratory, Berkeley, CA, USA*
Tripti Khare • *Biosciences Division, Advanced Photon Source, Argonne National Laboratory, Argonne, IL, USA*
Naoka Komori • *University of Oklahoma Health Sciences Center, Oklahoma City, OK, USA*
Stefanie Koristka • *Carl Gustav Carus University TU Dresden, Dresden, Germany*
Robert G.E. Krause • *University of KwaZulu-Natal, Pietermaritzburg, Scottsville, South Africa*
Biji T. Kurien • *University of Oklahoma Health Sciences Center, Oklahoma Medical Research Foundation, Department of Veterans Affairs Oklahoma City, OK, USA*
G.D. Lasanthi • *University of Illinois at Chicago, Chicago, IL, USA*
Jean Paul Lasserre • *Laboratoire de Bactériologie, INSERM U853, Université Victor Segalen Bordeaux 2, Bordeaux Cedex, France; Pôle Protéomique, Plateforme*

Génomique Fonctionnelle, Université Victor Segalen Bordeaux 2, Bordeaux, France; Institut de Biochimie et Génétique Cellulaires, CNRS UMR 5095, Université Victor Segalen/Bordeaux II, Bordeaux Cedex, France
TIEN CANH LE • *Université du Québec à Montréal, Montréal, QC, Canada*
BAO-SHIANG LEE • *University of Illinois at Chicago, Chicago, IL, USA*
HON-CHIU EASTWOOD LEUNG • *Baylor College of Medicine, Houston, TX, USA*
TANYA LEWANOWITSCH • *CSIRO, Preventative Health Flagship, Adelaide BC, SA, Australia; CSIRO, Food and Nutritional Sciences, Adelaide BC, SA, Australia*
SONGPING LIANG • *Hunan Normal University, Changsha, Hunan, People's Republic of China*
JENNIE R. LILL • *Genentech, Inc., South San Francisco, CA, USA*
DANIEL J. MACPHEE • *Memorial University of Newfoundland, St. John's, NL, Canada*
EMILIO MARENGO • *Department of Sciences and Technological Innovation, University of Eastern Piedmont, Alessandria, Italy*
NOELLE MARSH • *Memorial University of Newfoundland, St. John's, NL, Canada*
MIRCEA ALEXANDRU MATEESCU • *Université du Québec à Montréal, Montréal, QC, Canada*
SURESH T. MATHEWS • *Boshell Diabetes and Metabolic Diseases Research Program, Auburn University, Auburn, AL, USA*
HIROYUKI MATSUMOTO • *University of Oklahoma Health Sciences Center, Oklahoma City, OK, USA*
BRIAN MCDONAGH • *Departamento de Bioquímica y Biología Molecular, Universidad de Córdoba, Córdoba, Spain*
PADMAJA MEHTA-D'SOUZA • *Oklahoma Medical Research Foundation, Oklahoma City, OK, USA*
ARMELLE MÉNARD • *Laboratoire de Bactériologie, INSERM U853, Université Victor Segalen Bordeaux 2, Bordeaux Cedex, France*
IRENE MICHALK • *Carl Gustav Carus University TU Dresden, Dresden, Germany*
JONATHAN S. MINDEN • *Mellon Institute, Carnegie Mellon University, Pittsburgh, PA, USA*
MOHAMAD REZA MOHAMADI • *UMR 168, Curie Institute/CNRS/Universite´ Pierre et Marie Curie, Paris, France*
BRUNO MONDOVI • *Department of Biochemical Sciences "Rossi-Fanelli", University of Rome; "La Sapienza", Rome, Italy*
MARTA P. PEREIRA MORAIS • *Department of Biology and Biochemistry, University of Bath, Bath, UK*
NA NA • *Beijing Normal University, Beijing, People's Republic of China*
VICTOR J. NESATYY • *Biomolecular Mass Spectrometry Laboratory, EPFL-FSB-ISIC LSMB, Lausanne, Switzerland*
CARL K. OKERBERG • *Boshell Diabetes and Metabolic Diseases Research Program, Auburn University, Auburn, AL, USA*
MARKUS OTTO • *Department of Neurology, University of Ulm, Ulm, Germany*
JIN OUYANG • *Beijing Normal University, Beijing, People's Republic of China*
JAYANTA K. PAL • *University of Pune, Pune, India*
MANDY PEACH • *Memorial University of Newfoundland, St. John's, NL, Canada*
WENBIN QIN • *Baotou Medical College, Baotou, Inner Mongolia, People's Republic of China*

YASSEL RAMOS • *Centre for Genetic Engineering and Biotechnology, La Habana, Cuba*
CHRISTIAN REICHEL • *AIT Seibersdorf Laboratories, Seibersdorf, Austria*
HELEN J. REID • *Loughborough University, Loughborough, UK*
GUANG REN • *Boshell Diabetes and Metabolic Diseases Research Program, Auburn University, Auburn, AL, USA*
ELISA ROBOTTI • *Department of Sciences and Technological Innovation, University of Eastern Piedmont, Alessandria, Italy*
JOSE LUIS ROSA • *Universitat de Barcelona, L'Hospitalet de Llobregat, Barcelona, Spain*
R. HAL SCOFIELD • *University of Oklahoma Health Sciences Center, Oklahoma Medical Research Foundation, Department of Veterans Affairs Medical Center, Oklahoma City, OK, USA*
BARRY L. SHARP • *Loughborough University, Loughborough, UK*
CAROLINA SIMÓ • *Laboratory of Foodomics, CIAL, CSIC, Madrid, Spain*
OLIVER SMITHIES • *University of North Carolina School of Medicine, Chapel Hill, NC, USA*
RUPALI N. SONI • *University of Pune, Pune, India*
POTHUR R. SRINIVAS • *National Heart, Lung, and Blood Institute, National Institutes of Health Bethesda, Bethesda, MD, USA*
SLAVA STAMOVA • *Carl Gustav Carus University TU Dresden, Dresden, Germany*
YAN SU • *Baotou Medical College, Baotou, Inner Mongolia, People's Republic of China*
PAZHANI SUNDARAM • *Recombinant Technologies LLC, Cheshire, CT, USA*
RANJINI K. SUNDARAM • *Recombinant Technologies LLC, Cheshire, CT, USA*
NOBUAKI TAKEMORI • *Ehime University, Ehime, Japan*
IRA N. TARGOFF • *University of Oklahoma Health Sciences Center; Department of Veterans Affairs Medical Center; Oklahoma Medical Research Foundation, Oklahoma City, OK, USA*
IRANTZU TATO • *Universitat de Barcelona, L'Hospitalet de Llobregat, Barcelona, Spain*
MYRIAM TAVERNA • *University of Paris sud 11, Chatenay Malabry, France*
EDWARD P. TRIEU • *Oklahoma Medical Research Foundation, Oklahoma City, USA*
HENGHANG TSAI • *Genome institute of Singapore, Singapore, Singapore*
JEAN M.H. VAN DEN ELSEN • *Department of Biology and Biochemistry, University of Bath, Bath, UK*
FRANCESC VENTURA • *Universitat de Barcelona, L'Hospitalet de Llobregat, Barcelona, Spain*
ROMAIN VERPILLOT • *University of Paris sud 11, Chatenay Malabry, France*
JEAN-LOUIS VIOVY • *UMR 168, Curie Institute/CNRS/Universite´ Pierre et Marie Curie, Paris, France*
XIANCHUN WANG • *Hunan Normal University, Changsha, Hunan, People's Republic of China*
XU WANG • *College of Pharmacy & Research Institute of Drug Development, Chonnam National University, Gwangju, South Korea*
CHAD M. WARREN • *University of Illinois at Chicago, Chicago, IL, USA*

Chapter 1

How It All Began: A Personal History of Gel Electrophoresis

Oliver Smithies

Abstract

Arne Tiselius' moving boundary electrophoresis method was still in general use in 1951 when this personal history begins, although zonal electrophoresis with a variety of supporting media (e.g., filter paper or starch grains) was beginning to replace it. This chapter is an account of 10 years of experiments carried out by the author during which molecular sieving gel electrophoresis was developed and common genetic variants of two proteins, haptoglobin and transferrin, were discovered in normal individuals. Most of the figures are images of pages from the author's laboratory notebooks, which are still available, so that some of the excitement of the time and the humorous moments are perhaps apparent. Alkaline gels, acidic gels with and without denaturants, vertical gels, two-dimensional gels, and gels with differences in starch concentration are presented. The subtle details that can be discerned in these various gels played an indispensable role in determining the nature of the change in the haptoglobin gene (*Hp*) that leads to the polymeric series characteristic of Hp^2/Hp^2 homozygotes. Where possible, the names of scientific friends who made this saga of gel electrophoresis so memorable and enjoyable are gratefully included.

Key words: Insulin, Filter paper, Starch gel, Serum proteins, Haptoglobin, Transferrin, Vertical gel, Two-dimensional gel, Denaturing gel

Life with electrophoresis began for me almost 60 years ago as a postdoctoral fellow in the laboratory of Dr. Jack Williams, a physical chemist at the University of Wisconsin, Madison. His lab had a state-of-the-art Tiselius apparatus in which the electrophoretic migration of proteins was followed at the two boundaries formed by a U-tube of protein solution with buffer above the two vertical arms (Fig. 1, blue inset box). Arne Tiselius' development of the method (1) and his use of it to show four components in serum (albumin, α-, β-, and γ-globulin) earned him a Nobel Prize in 1948. A particularly exciting example of the power of the method was its use by Linus Pauling and his colleagues (2) in defining sickle cell anemia as a

Biji T. Kurien and R. Hal Scofield (eds.), *Protein Electrophoresis: Methods and Protocols*, Methods in Molecular Biology, vol. 869, DOI 10.1007/978-1-61779-821-4_1, © Springer Science+Business Media, LLC 2012

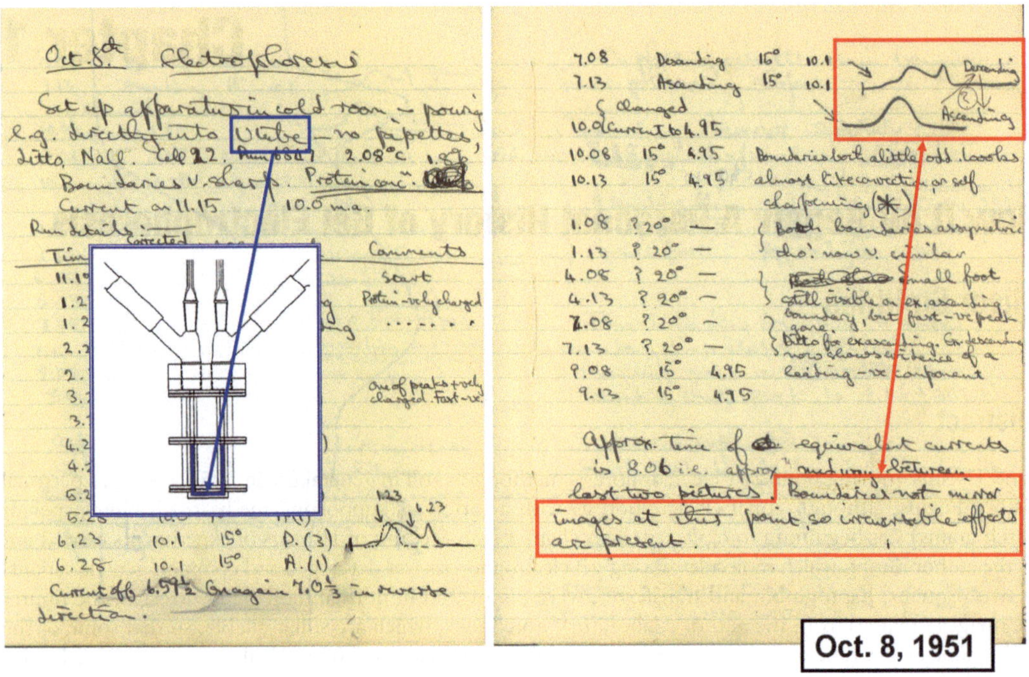

Fig. 1.

molecular disease. The method, however, had two substantial disadvantages. First, because it was a boundary method, separations were overlapping rather than complete:

$$[A\cdots>[A+B\cdots>[A+B+C\cdots \text{U-tube}\cdots A+B+C]\cdots>B+C]\cdots>C].$$

Secondly, because the boundaries were detected by refractive index changes, the method required a cumbersome optical system, and relatively large amounts of protein, of the order of half a gram or more.

The entry in my laboratory notebook, made with a fountain pen in 1951 before the invention of ballpoint pens, records that my first use of the Tiselius electrophoresis method was not very encouraging. I was trying to determine the purity of a recrystallized preparation of a protein, β-lactoglobulin, which I had made from cow's milk, but the ascending and descending boundaries were not mirror images, as they should be (Fig. 1, red boxes). I struggled with the boundary anomaly, off and on, for 2 years but never could get understandable results (3).

My next electrophoresis problem arose during the early days of my first job, which was with Dr. David A. Scott at the Connaught Laboratories, Toronto, Canada. Scott was the first person to crystallize insulin, as a zinc salt, and show its value as a slow release form of the hormone. Not surprisingly, he suggested that I should pick a research task related to insulin. After learning that insulin was typically prepared by extracting pancreases with 70% alcohol containing 2%

concentrated hydrochloric acid, I thought that insulin prepared with such a harsh procedure might be a breakdown product. I therefore chose, as my task, to look for a precursor.

In preparation for an expected need to distinguish the postulated precursor from insulin as normally prepared, I tried a new form of electrophoresis that had recently been championed by Kunkel and Tiselius (4). The method separated proteins on buffer-soaked filter paper. It was zonal and therefore could separate proteins completely:

$$[A+B+C]\cdots>[A]\cdots>[B]\cdots>[C].$$

It also had the advantage of requiring very little protein (<1 mg). But when I used it, my New Year's Day experiment (Fig. 2, left panel) revealed one of its disadvantages—some proteins (including insulin) stick to filter paper and unroll like a carpet.

Kunkel was well aware of the absorption and trailing problem, and he and Slater (5) sought and found a supporting matrix that was free from it. Their preferred matrix was a barely moist bed of potato starch grains. Proteins migrated in the buffer *surrounding* the impervious starch grains. With their method, they could easily separate the five proteins of normal serum (by this time α-globulin had been subdivided into α_1 and α_2).

I heard that Dr. Andrew Sass-Kortsak, a scientific neighbor working at the Toronto Hospital for Sick Children, was using the method, and I visited his lab to see it in action on Saturday, Jan. 23, 1954. The freedom from absorption was spectacular, but the labor required for each run was over-facing; Folin protein determinations of 40 segments cut from the moist starch-grain bed were needed to detect the migrating proteins and develop an electrophoretic pattern (Fig. 2, center panel). Impossible for me; I did not even have someone to wash my lab glassware! But

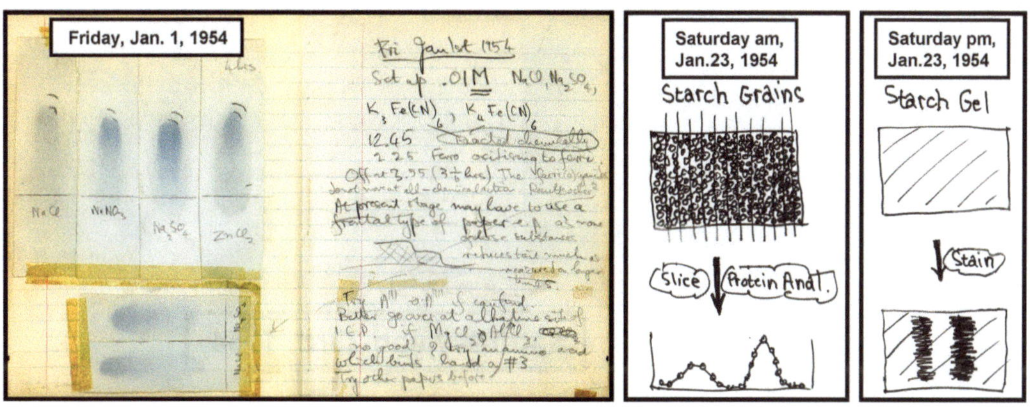

Fig. 2. Reproduced with permission from the Nobel Foundation.

I remembered "helping" my mother do the laundry as a child and noticing that the hot gooey liquid she made to starch the collars of my father's shirts set into a gel when left standing. Perhaps, if I boiled a starch suspension and allowed it to cool, proteins could migrate *through* the resulting jelly and could be detected by staining, in the way that was used to detect proteins in filter-paper electrophoresis (Fig. 2, right panel).

Back at Connaught that Saturday afternoon, I found a bottle of "starch according to Lintner." It made a gel ("jelly"); insulin migrated through the gel without trailing; it could be detected by simply staining the gel, and no protein determinations were needed (Fig. 3). I had invented starch-gel electrophoresis!

Over the next few months, I worked at improving my experiments with insulin and helped my friend Gordon H. Dixon (a graduate student in the Department of Biochemistry, University of Toronto) use the starch-gel system to separate cabbage enzymes (6). Then one day, "just for rough test," I tried electrophoresing serum (Fig. 4, left page). The results were promising, and after only a few more days of work, I was finding at least nine components in serum (Fig. 4, right page), although I was unsure how to label them (Fig. 4, red box and arrow, indicating an erasure).

This was new territory, and (with Dr. Scott's approval) I stopped working on insulin and settled down to making the method reproducible for separating serum proteins. Within a few months, I was ready to publish and had two pretty gels as examples with serum samples from Gordon (GD) and another graduate-student friend, George Connell (Fig. 5, left panel).

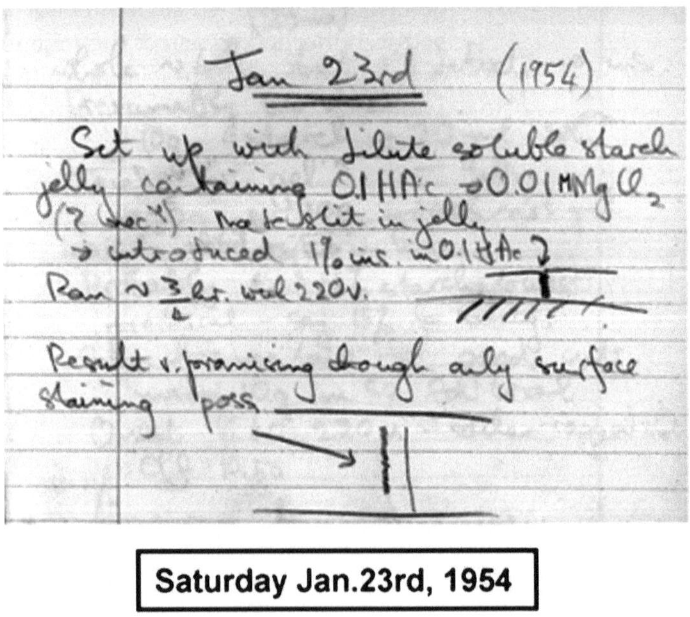

Fig. 3.

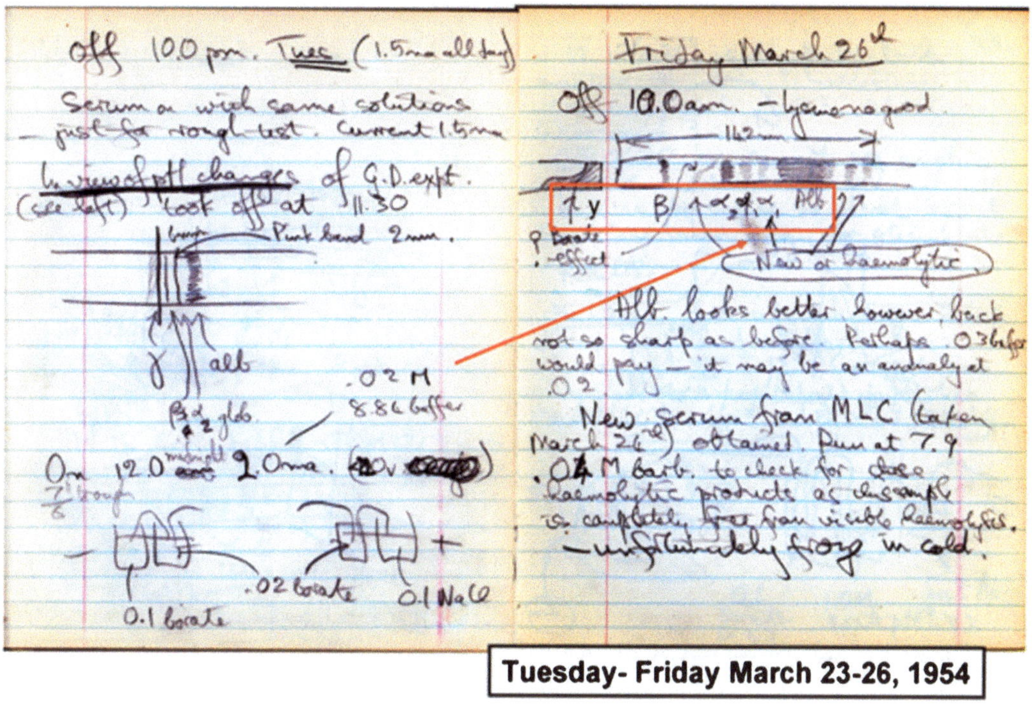

Fig. 4.

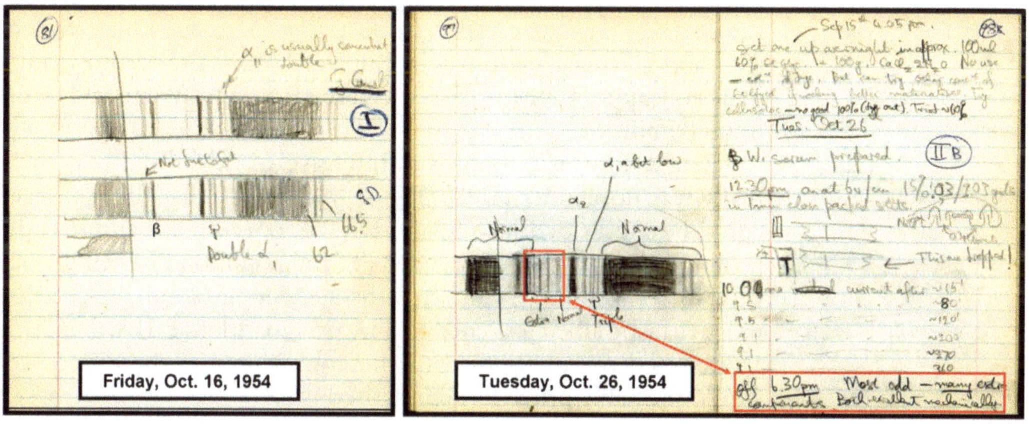

Fig. 5. Reproduced with permission from the Nobel Foundation.

However, for no special reason, I ran another sample before having photographs made (not easy because I had no camera). The serum was from B.W. (Beth Wade), who was a young woman working as a technician in Charles Hanes' lab where Gordon and George were students. The results (Fig. 5, right panel) were "most odd—*many* extra components." I thought I had found a male/female difference and a new way of telling men from women!

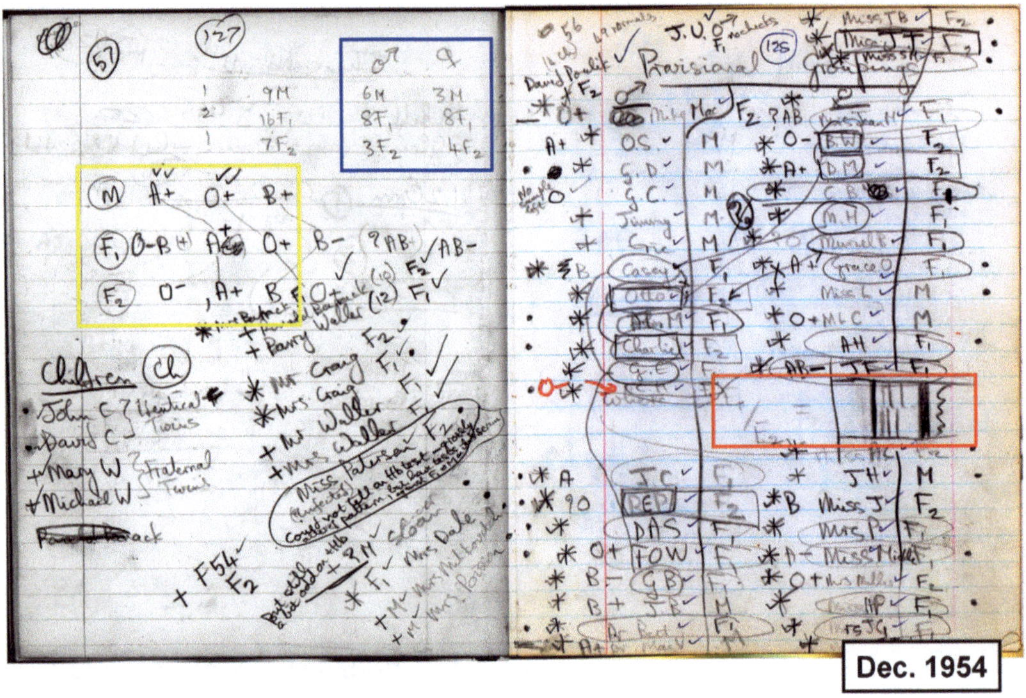

Fig. 6.

Testing serum from another male and another female for each of the next few days appeared to confirm that there was a male pattern (M) and a female pattern (F). But very soon, the serum from a male colleague (Casey Kok) showed the F pattern. We teased Casey unmercifully.

The following 2 months of intensive work showed that there really was no gender effect (Fig. 6, blue box), that the F pattern could be subdivided into F_1 and F_2 (Fig. 6, red box), and that the protein groups were independent of the common red cell groups (Fig. 6, yellow box). So, with the gender notation replaced by more realistic albeit less entertaining designations (I, IIA, and IIB), I published my method, with the comment that the differences would probably prove to be inherited (7, 8).

With help and guidance from Dr. Norma Ford-Walker (Genetics Department, Hospital for Sick Children, Toronto, Canada), a delightful collaborator and teacher, this comment was quickly confirmed—the differences were indeed genetically determined (9). Shortly thereafter, Helge Laurell, University of Uppsala, pointed out to us that the protein with the inherited differences, which by then we knew was a hemoglobin-binding protein, was probably haptoglobin, a hemoglobin-binding protein already described by Jayle et al. (10). Norma and I agreed, and we renamed the three serum groups Hp1-1, Hp2-1, and Hp2-2, with the genotypes being Hp^1/Hp^1, Hp^2/Hp^1, and Hp^2/Hp^2 (11).

Following this introduction to genetics, on the urging of another Connaught scientist, David M. Poulik, who subsequently became a lifelong friend, I took up the task of correlating the serum components seen in starch gels with those seen on filter paper. Dave had been looking at serum samples from hospital patients using filter-paper electrophoresis and had begun switching to starch gel. As a result, he was particularly keen to compare the two systems. He suggested that I should try a test with alpha-naphthyl acetate, a chromogenic substrate which he had been using to detect esterases on filter paper. When I did, the results confirmed his finding that esterase activity was detectable on filter paper as a single zone between α_2- and β-globulin (Fig. 7, filter-paper nomenclature blue). In the starch gel, several enzyme bands were detectable, with the main one migrating between the bands in serum which I thought were α_2- and β-globulins (Fig. 7, provisional starch-gel nomenclature red). I concluded that the two methods agreed (Fig. 7, green box).

But my next test quickly demonstrated that this conclusion was unlikely to be correct. For this test, I cut pieces expected to include α_1-, α_2-, and β-globulin from two filter-paper electrophoresis strips that had been run with Dave's serum, inserted each piece into a starch gel alongside a second piece of filter paper soaked in serum, and ran the gel electrophoresis. My July 19, 1955, notebook

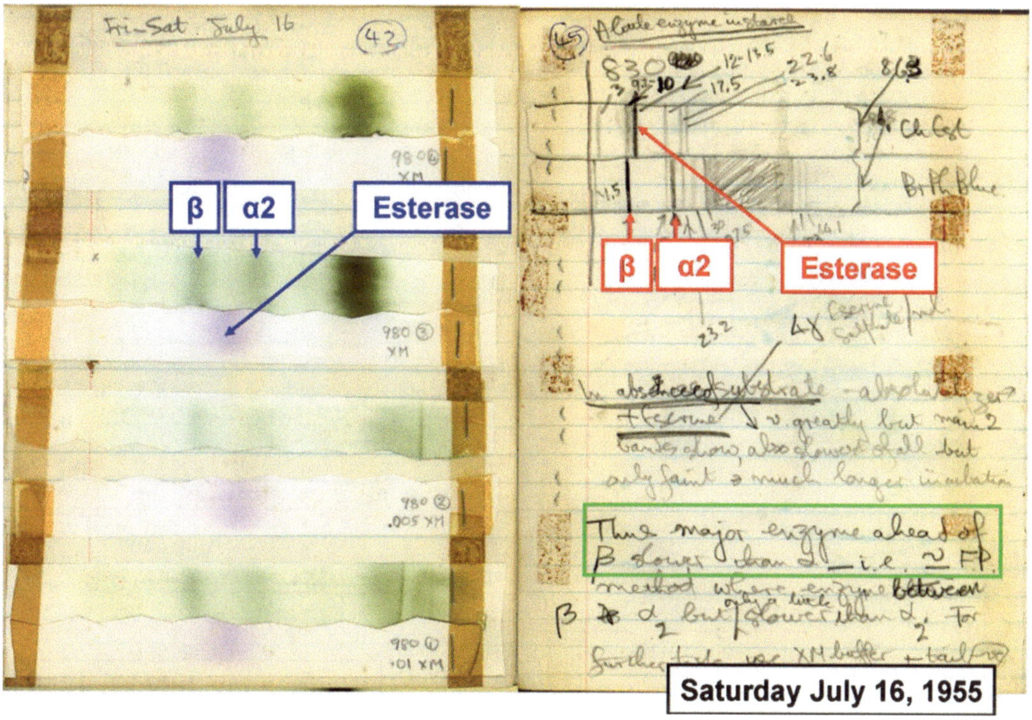

Fig. 7.

entry indicates my surprise—"β (fp)—lin[es] up with $\underline{\alpha_2}$." This was the beginning of a realization that gel separations differ qualitatively from separations in free solution or in buffer surrounding an impervious matrix.

Within a few days, I expanded the comparison experiments to cover all parts of the filter-paper electrophoresis strip. This involved cutting out 20 consecutive transverse pieces from the filter-paper electrophoresis strip (Fig. 8, left page) and running 20 gels to compare the proteins in each piece with whole serum inserted alongside in the same slot in the gel using a second piece of filter paper soaked in serum (Fig. 8, right page). The result confirmed that the gel band which I had provisionally labeled α_2 was in fact a β-globulin and showed that the gel band which I had provisionally labeled β was an α_2 globulin. Dave and I repeated these laborious multi-gel experiments many times. All led to the same conclusion—the two electrophoresis systems were qualitatively different. Filter-paper mobilities approximate those in free solution; the high concentration of starch used in the gels markedly enhances the retarding effects on mobility of increasing molecular size.

However, we were still having a problem in identifying the filter-paper equivalents of the starch-gel pre-albumins. In attempting to overcome this problem, I cut the filter paper *longitudinally* instead of transversely, inserted the forward part of the filter-paper strip into a wide starch gel, and ran the second electrophoresis *at*

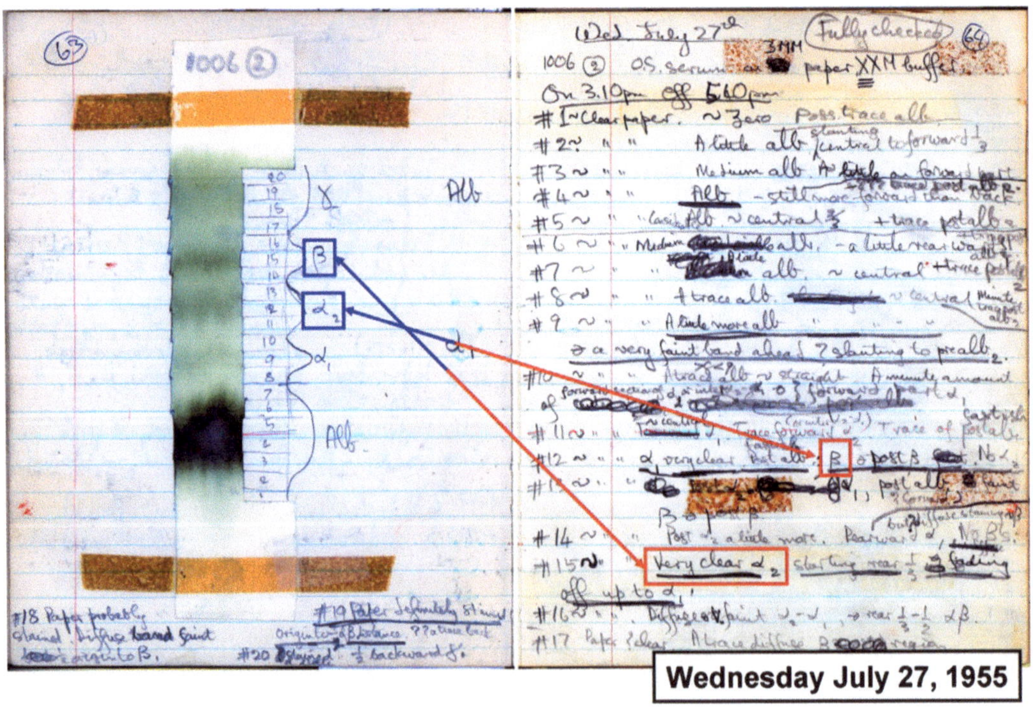

Fig. 8.

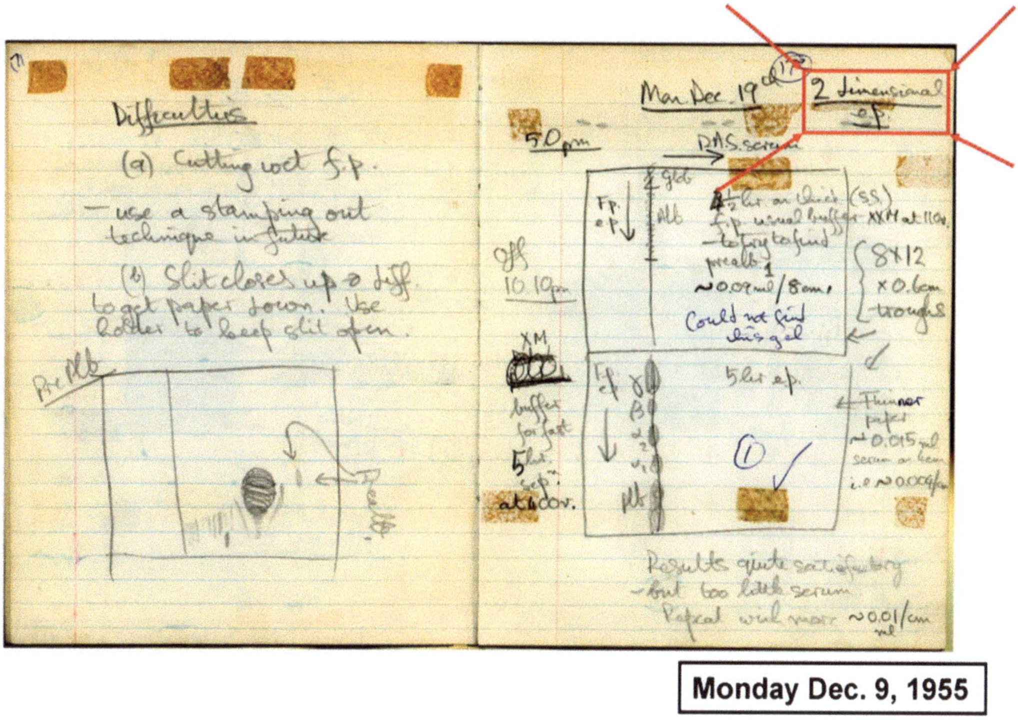

Fig. 9.

right angles to the first (Fig. 9, lower part of left page and upper part of right page). This 90° turn was, I realized, all that was required to have a two-dimensional gel electrophoresis system (Fig. 9, red box), and I tried it immediately with a complete longitudinal strip cut from a still wet filter-paper run (Fig. 9, lower part of right page). (Later, I used a narrow strip of filter paper for the first electrophoresis to avoid one of the difficulties that I noted—"(a) Cutting wet f.p.")

The moment when I thought of the 90° turn is one that remains vivid in my memory; I can still remember where I was standing! The result was that, instead of five proteins in serum, Dave and I could find 20 (12, 13), at that time a very pleasing number, although it pales in comparison to the several hundred protein spots that are now detected in serum with the two-dimensional O'Farrell-type system (14) which combines isoelectric focusing with Laemmli's sodium lauryl sulfate (SDS) electrophoresis (15).

The filter-paper/starch-gel two-dimensional system proved very useful for studying an unusual variant first found by Ralph Engle and Jim Pert (New York Hospital, New York) in serum from an African-American which they sent to me in 1956 for characterization. My Nov. 1, 1956, two-dimensional gel (Fig. 10), with the first electrophoresis now on a narrow filter-paper strip, gave a "V[ery] good result. Shows that the extra zone is a slightly slow β."

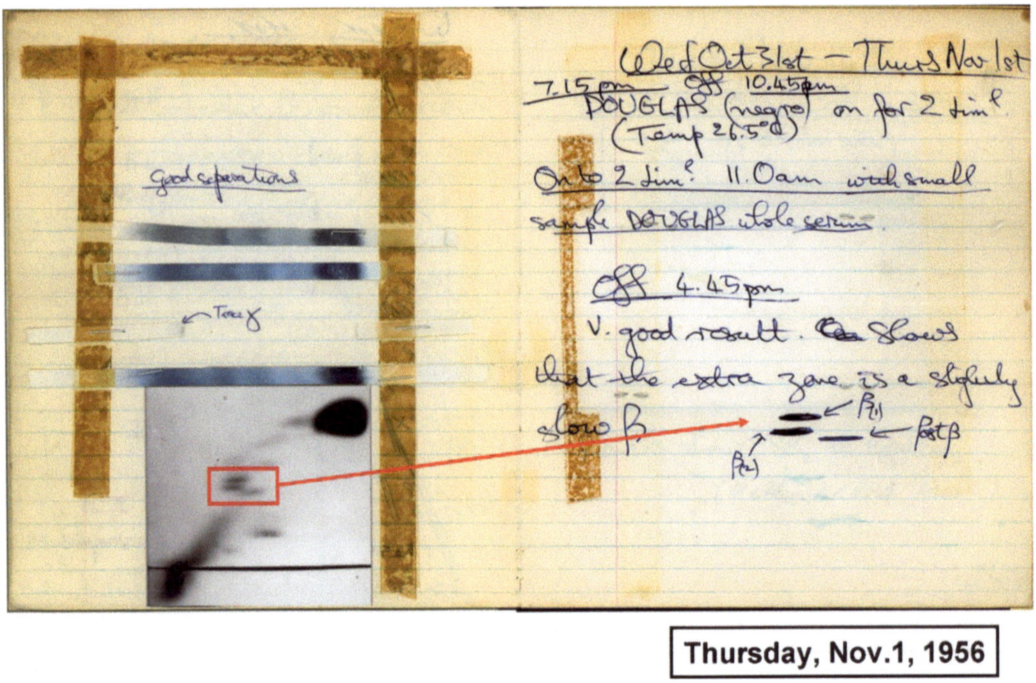

Fig. 10.

Elo Giblett, Charlie Hickman, and I subsequently identified the variable β-globulin as transferrin (16). Elo, who ran the King County Central Blood Bank, Seattle was an ardent blood grouper and became one of the most successful discoverers of new haptoglobin and transferrin variants. Charlie was as keen when it came to cattle (17).

The β-globulin variant first seen in an African-American turned up again in some samples from Australian aborigines that were sent to me by Bill Horsfall, Commonwealth Health Laboratories, Queensland, Australia (18). However, the full blown two-dimensional method was cumbersome for the family studies needed to establish the mode of inheritance. So, in its place, my technician (I now had one, Otto Hiller) and I used a cut-down version of the two-dimensional system. Only the β-globulin-containing region of the narrow filter-paper electrophoresis strip was used, and it could be inserted into a normal-sized one-dimensional starch gel for the second electrophoresis at right angles. With this method, we could test 16 samples a day (Fig. 11, white boxes indicate β-globulin variants) and we used it to study the many samples that Bill collected from his aborigine friends. They were, he explained, very interested and willing to take part in family studies but only occasionally came “out of the bush.” Together, we proved that the β-globulin variants were inherited (19).

Meanwhile, I was unhappy with any of the several methods that I had tried for inserting samples into the gels, which were all

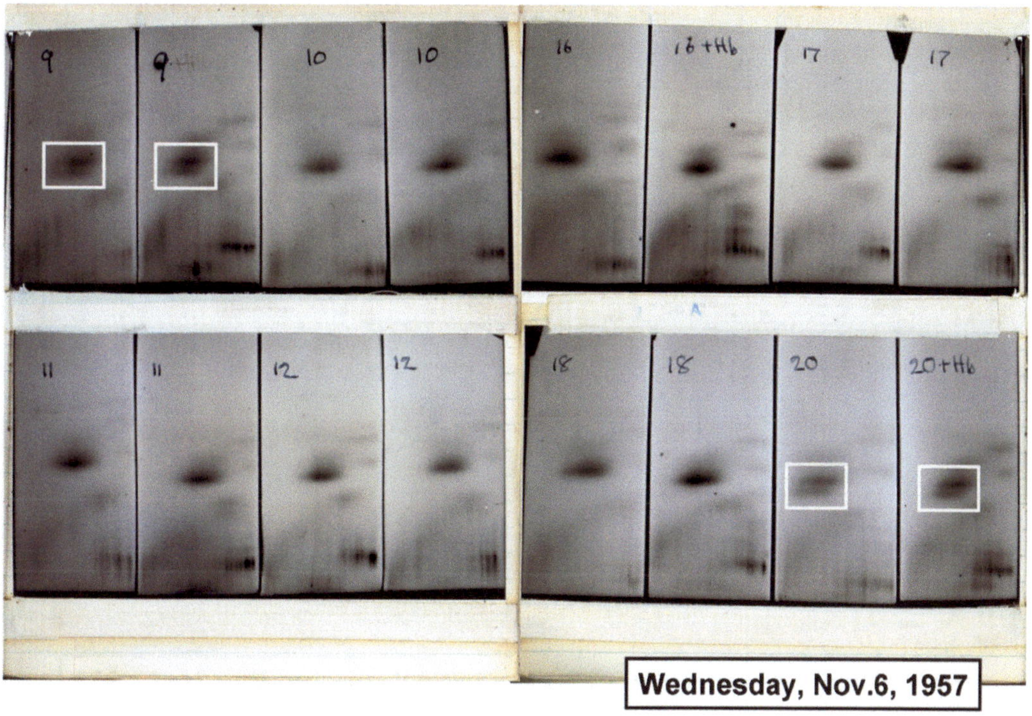

Fig. 11.

at that time being run horizontally. A piece of filter paper soaked in the sample was adequate for many purposes, but absorption was often a problem. A suspension of starch grains was good, but cumbersome. A simple opening in the gel was plagued with edge effects, and electro-decantation occurred as the proteins migrated across the opening in the gel. Eventually, I realized the now obvious solution—run the gel *vertically* with open slots (20). The problems of absorption and electro-decantation completely disappeared, and the full resolving power of the gels was achieved. I illustrate the beauty of these early vertical starch gels with two runs. The first (Fig. 12, left page) is a vertical air cooled (VAC) gel, run downwards on May 31, 1958. This gel compares serum samples having the previously described β-globulin variants with a serum sample having a new faster-migrating variant that Otto had found after running over 400 repetitions of the cut-down version of the two-dimensional system (21, 22). The white box in the figure outlines the β-globulin variants. The resolving power of the vertical starch gel is well illustrated by the easily seen polymers of haptoglobin apparent in this air cooled gel (Fig. 12, blue box). (I also tried water cooling, but the improvement was not worth the added complexity for pursuing the studies that most interested me.)

The second gel (Fig. 13), run a few days later on June 4, 1958, compares George Connell's purified samples of the four inherited types of haptoglobin known at that time (23). At least ten increasingly

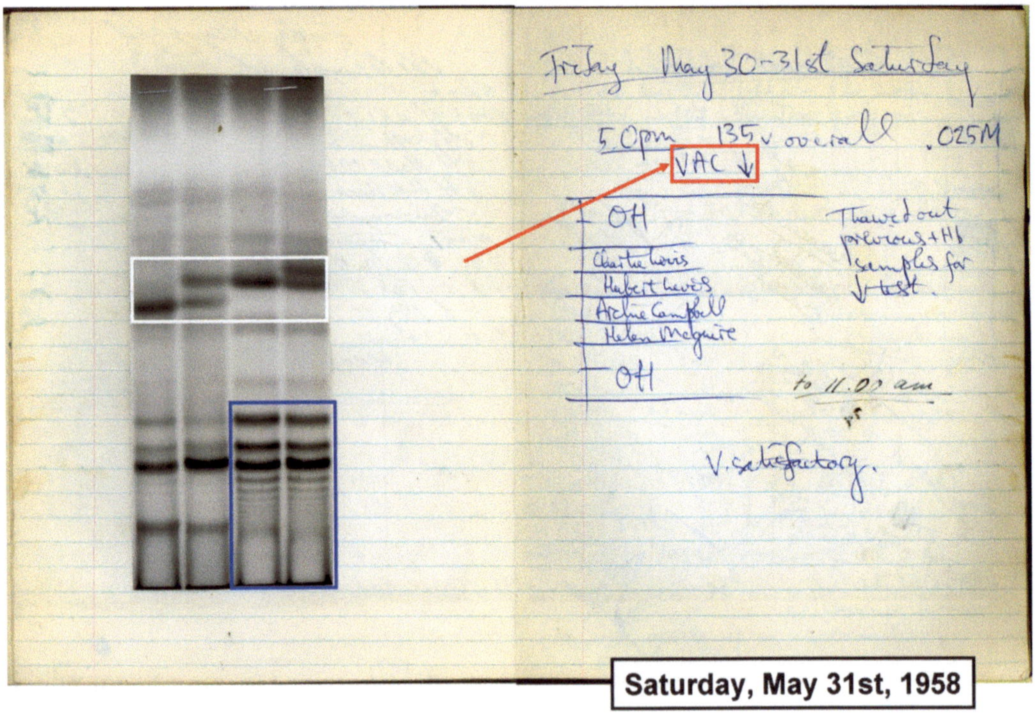
Friday May 30-31st Saturday
5.0pm 135v overall .025M
VAC ↓
OH
Charlie Lewis
Hubert Lewis
Archie Campbell
Helen Maguire
OH
Thawed out previous + Hb samples for test.
to 11.00 am
V. satisfactory.
Saturday, May 31st, 1958

Fig. 12.

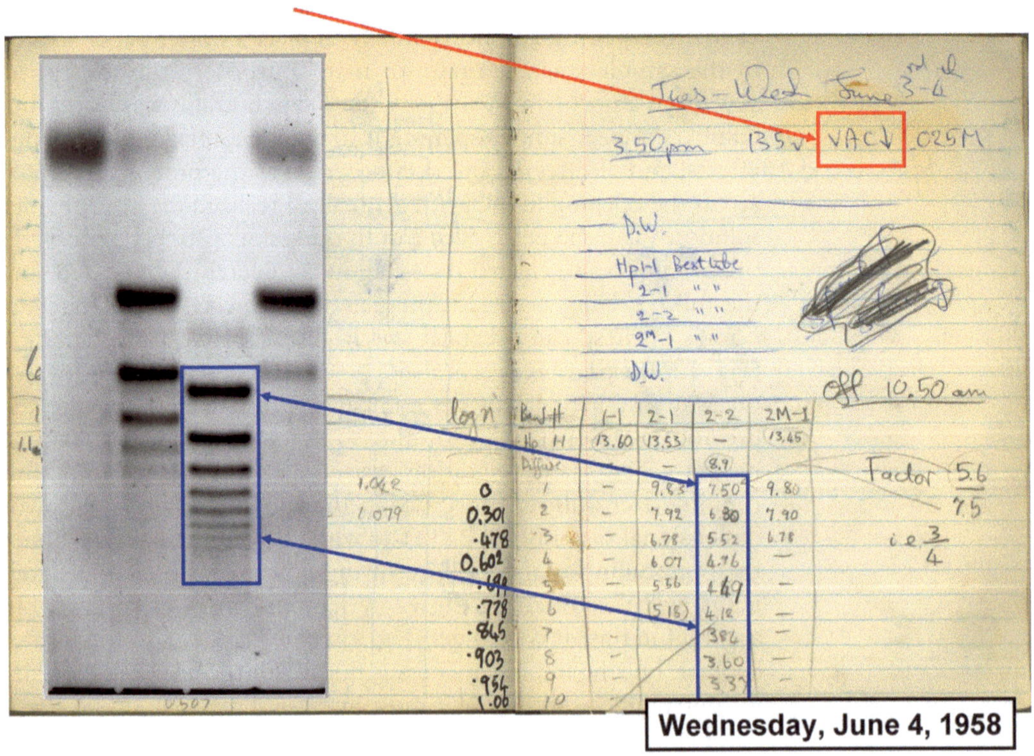
Tues-Wed June 3rd-4th
3.50pm 135v VAC↓ .025M
D.W.
Hp1-1 Best tube
2-1 " "
2-2 " "
2M-1 " "
D.W.
Off 10.50 am
Factor 5.6 / 7.5
i.e. 3/4
Wednesday, June 4, 1958

Fig. 13.

more polymeric forms of haptoglobin are resolved in this vertical starch gel. The right page shows my attempts to find a relationship between their mobilities and the band number (n).

The molecular basis underlying the differences between the haptoglobin variants was of considerable interest because a single protein species in the Hp^1/Hp^1 homozygotes was replaced by a polymeric series in the Hp^2/Hp^2 homozygotes and the heterozygotes had haptoglobins that were not the same as a mixture of the haptoglobins from the homozygotes. Despite this complexity, the problem seemed likely to be approachable using the methods that Vernon Ingram had developed in working out the difference between normal and sickle cell hemoglobin—a glutamic acid residue in the normal β-globin polypeptide was replaced by valine in the sickle cell β-globin (24). My graduate-student friends, George E. Connell and Gordon H. Dixon, were now back at the University of Toronto as Assistant Professors, and together we decided to tackle the haptoglobin problem. My task was to try to find some procedure that would convert the complex polymeric haptoglobins into simpler products that would better mirror the simplicity of their inheritance (25). George's tasks included scaling up production of the purified proteins, developing an assay, and investigating the general physical properties of the products that I was hoping to make (26). Gordon's task was to apply Ingram's methods to the hoped for simpler products (27).

Starch forms gels under a wide range of conditions. This versatility proved of enormous help in carrying out my task, which I approached by testing the electrophoretic properties of purified haptoglobins in gels of many types. For example, I found that the patterns were not affected by running the proteins in a gel made with 0.05 M HCl (pH <2), so the polymers were unlikely to be due to simple ionic bands. The proteins of all three types did not survive 0.05 M NaOH (pH >12) in any form that migrated in gels, not very informative. The usual patterns were seen in the presence of 0.05 M EDTA, so the polymers were unlikely to be due to calcium or similar cross-links. Neither alkaline nor acidic 6 or 8 M urea disassociated the polymers, nor did non-denaturing reduction with thioglycollate. On the other hand, alkaline 8 M urea plus thioglycollate caused depolymerization, and I could see the beginnings of a simpler pattern, although I could not completely understand it (my Wednesday, April 15, 1959, notebook comment was "Splitting has occurred!") (28).

Over the next few months, I tried numerous variations of the urea/thiol procedure and tested whether the treated proteins would still bind hemoglobin. They did not. I also found that a stringy precipitate developed when the urea and the thiol (now β-mercaptoethanol) were dialyzed away. Adding alkali did not affect the precipitate, nor did acid. But adding urea to the acidified material dissolved the precipitate. My conclusion, noted on Wednesday,

July 22, 1959, was that "aggregation is trouble. Therefore run in 4 M gels and try acid gels."

Next day, I ran the treated samples overnight at an acid pH, which proved to be a breakthrough—the products obtained after the urea/thiol treatment at last bore a clear relationship to the genetics. The treated haptoglobins gave markedly different migrating components that were distinguishable in a pH 3.15 formate buffer gel. "Hp1-1 has fast component" (Fig. 14, blue box). "Hp2-2 has diffuse component ~2/3–1/2 mobility" (Fig. 14, red box). "Hp2-1 has both." There was also material that did not migrate in the acid gel without urea, but migrated to some extent in the pH 3.7 gel containing 4 M urea (Fig. 14, green boxes).

After experimenting with different concentrations of urea, different acids, different disulfide-reducing agents, and different ways of protecting the thiol groups generated by the reductant, I finally got a procedure which gave stable reproducible products that migrated in 8 M urea acid gels and mirrored the genetics. The procedure was to reduce the disulfide cross-links with mercaptoethanol at an alkaline pH in the presence of 8 M urea, alkylate the exposed thiol groups with iodoacetamide, and electrophorese the products in 8 M urea gels containing 20% neutralized formic acid. This procedure was very powerful and was a critical element in solving the haptoglobin problem. It showed that the haptoglobins had two polypeptide chains, an invariant slow-migrating β-chain

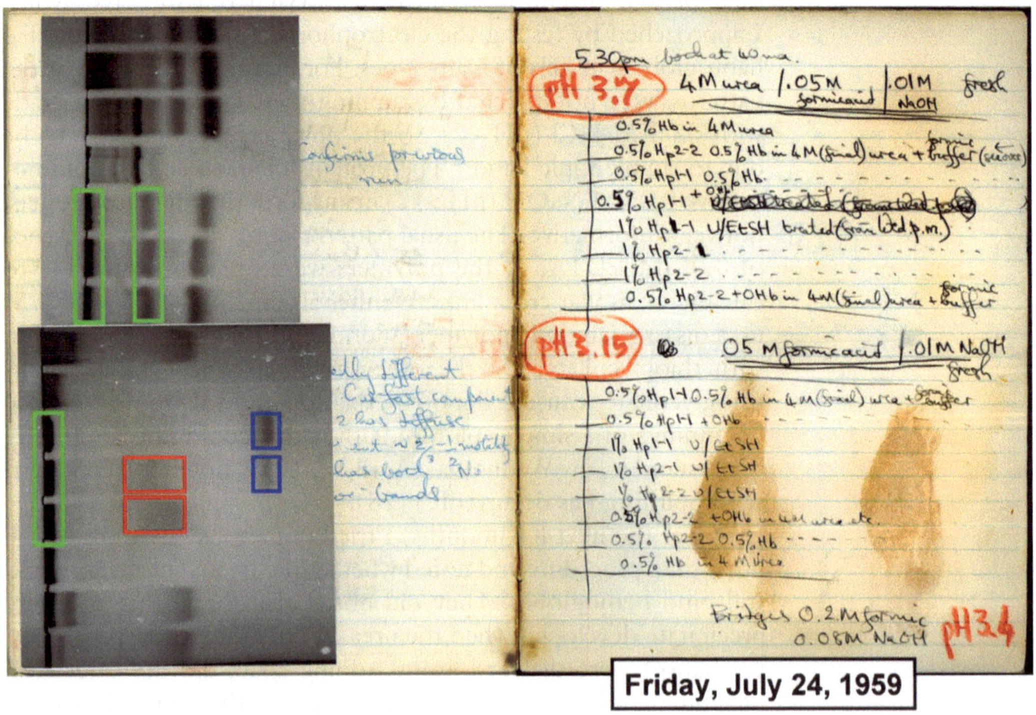

Fig. 14.

and a genetically variable faster-migrating α-chain. And it proved equally powerful in Edelman's use of it to show that urinary Bence Jones proteins were the light chains of the unique immunoglobulins present in the plasma of patients with multiple myeloma (29, 30).

Despite the power of the procedure, the results obtained when we used it to study the haptoglobins were at first somewhat confusing. A gel run on Feb. 18, 1960, to compare the merits of alkylating the urea/thiol-treated products with iodoacetamide or iodoacetic acid illustrates one problem.

We could not understand why the fast component of reduced and alkylated Hp1-1 gave *two* fast-migrating bands in the 8 M urea/formate gels (Fig. 15, red box), while Hp2-2 gave only a *single* band (Fig. 15, blue box). One possibility was that the Hp1-1 product was susceptible to proteolysis during its preparation; and this gel included a small amount of a haptoglobin sample (#3) designed to test this idea—the sample was prepared from plasma that had been treated with a protease inhibitor. It gave a single Hp1α band (Fig. 15, yellow box) and so appeared to support the proteolysis idea (Fig. 15, green box). But the answer turned out to be different—George Connell discovered that the common Hp1-1 type could be divided into three subtypes, as we reported in 1962 (31), and that there were consequently two forms of the Hp^1 gene (32).

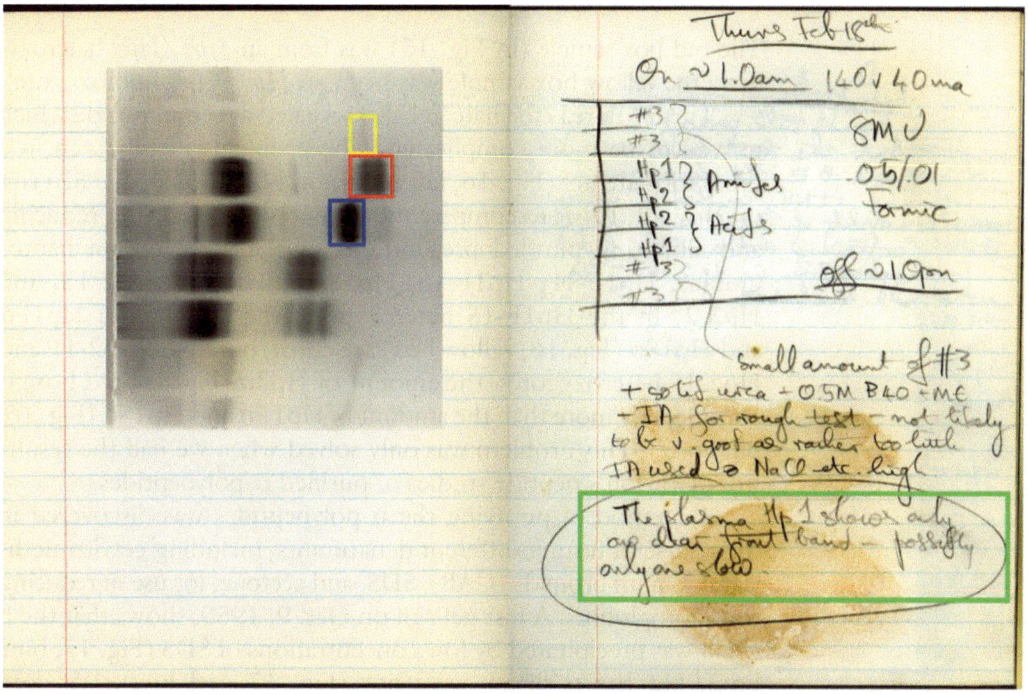

Thursday, Feb.18, 1960

Fig. 15.

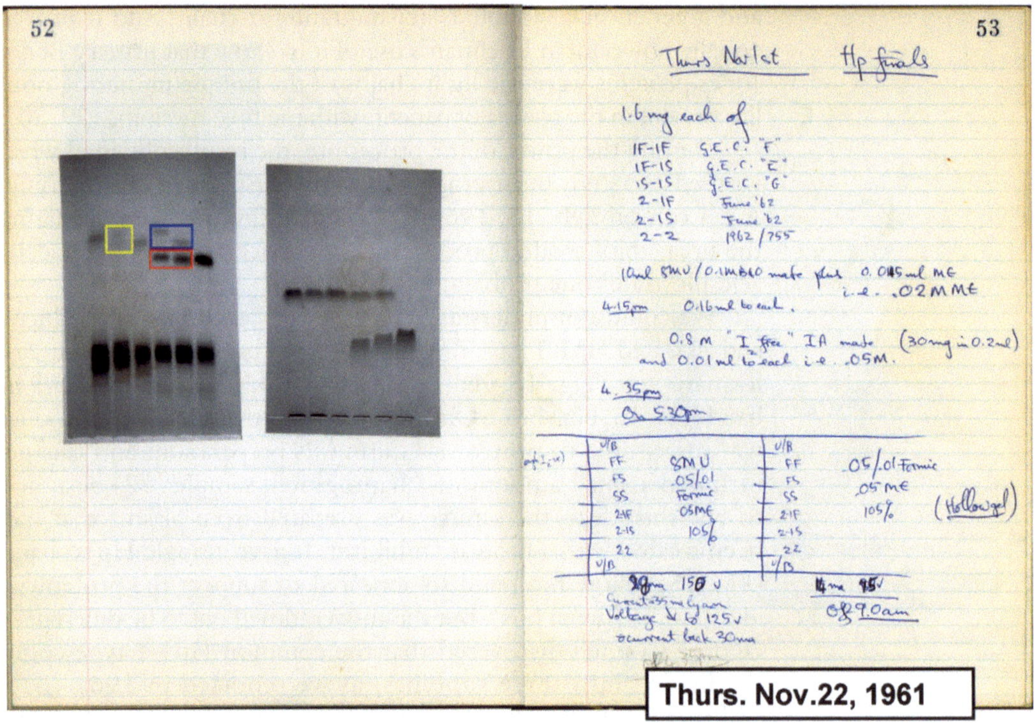

Fig. 16.

The red box sample (in Fig. 15) was from an Hp^{1F}/Hp^{1S} heterozygote; the yellow box sample was from an Hp^{1S}/Hp^{1S} homozygote!

The 8 M urea/formate gel also revealed a second puzzle which proved to be more complicated. The problem is easily seen in a urea/formate gel (Fig. 16, left image on Note Book page 52) run on Nov. 22, 1961, to compare the products obtained after reducing and alkylating purified preparations of the six common haptoglobins, Hp1F-1F, Hp1F-1S, Hp1S-1S, Hp2-1F, Hp2-1S, and Hp2-2. In the Hp1F-1S heterozygote, the amounts of Hp1Fα and Hp1Sα (Fig. 16, yellow box) are equal, but in the Hp2-1F and Hp2-1S heterozygotes, the amount of Hp2α (Fig. 16, red box) is considerably more than the amount of Hp1Fα and Hp1Sα (Fig. 16, blue box). This problem was only solved when we had the results from Gordon's peptide studies of purified α-polypeptides.

A method of purifying the α-polypeptides was discovered in the course of testing different denaturants, including cetyltrimethylammonium bromide (CAB), SDS, and acetone, for use in reducing the haptoglobins. A test gel run on Oct. 9, 1959, shows that the β chain was precipitated by low concentrations of SDS (Fig. 17, blue box), but the α-chains were not (Fig. 17, red box). We used this feature to prepare the variable polypeptides for the next task—Gordon's analysis of the peptides resulting from digesting the purified α-chains with various proteolytic enzymes.

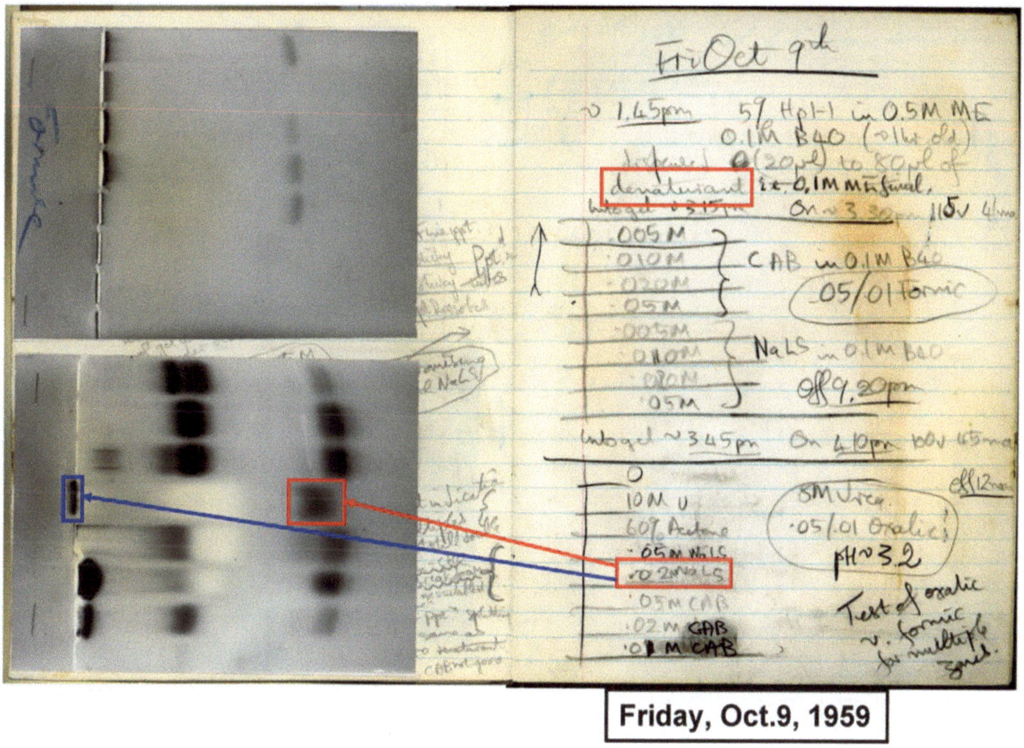

Fig. 17.

Gordon's results, combined with the gel data, led us to the conclusion that almost the whole of the amino-acid sequence of Hp1Fα (ABCDE**F**GHI) together with most of Hp1Sα (ABCDE**S**GHI) occurs in a single molecule of Hp2α (ABCDE**F**GCDE**S**GHI), making it nearly twice as long. We hypothesized that the Hp^2 gene must have been formed by a unique nonhomologous crossing-over event between the Hp^{1F} and Hp^{1S} genes, and we decided to present our data and hypothesis at the 1961 Second International Conference of Human Genetics in Rome. Meanwhile, before the meeting, George would test the hypothesis using the ultracentrifuge, with the expectation that the Hp2α polypeptide would sediment faster than the Hp1Fα and Hp1Sα polypeptides.

We met in Rome the evening before the meeting, and George gave us the news—their sedimentations were identical! Nevertheless, we decided that we would present our data, our hypothesis, and the apparently contradicting ultracentrifuge result!! But we would end our presentation with the statement that we did not believe the ultracentrifuge result and would invent a new way of comparing molecular sizes!!! This led to the final adventure in this saga of gel electrophoresis—the usefulness of varying gel concentrations.

On my return from Rome, I varied the concentration of the starch used in the gels, predicting (Fig. 18, left page) that the *relative* mobilities of Hp1Fα and Hp1Sα would remain constant, as the

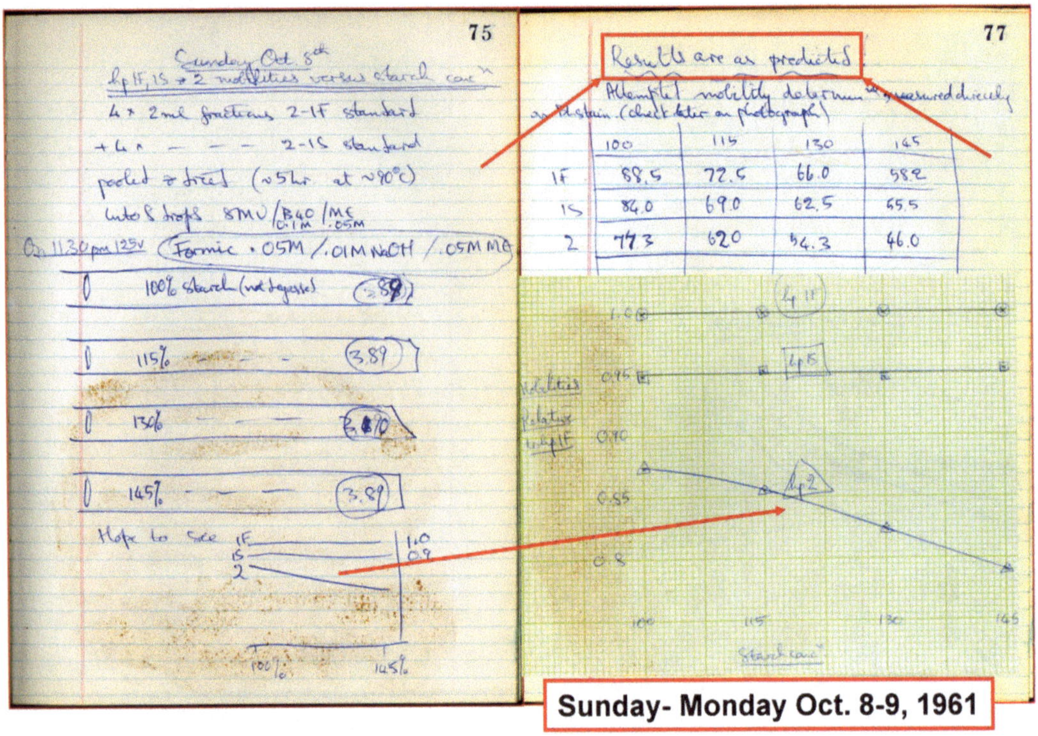

Fig. 18.

starch concentration was increased, since they only differed in charge, but that the relative mobility of the larger sized Hp2α would decrease. The right page shows that the mobilities changed as predicted and therefore that Hp2α was indeed larger than Hp1Fα and Hp1Sα, which were equal in size. "Results are as predicted" (Fig. 18, red box).

George subsequently showed that, in the presence of urea, the ultracentrifuge gave the same result (26). So together, the three of us published our hypothesis (33). And I published my new method (Fig. 19) under the title "Molecular Size and Starch-Gel Electrophoresis" (34). The paper was part of a tributary volume celebrating Tiselius' 60th birthday. It felt particularly appropriate because in 1953 Tiselius had written that "there is a great need for zone separations according to differences in molecular size."

Gradient gels are the modern variants of this principle, and I use them (the only gels that I buy!) enjoyably. The gel in Fig. 20 is a 4–20% polyacrylamide gradient gel (run, like many other of my gels, on a Saturday). It tells me that the gold nanoparticles which I have been making to test a hypothesis related to kidney function (35) have beautiful colors related to their sizes.

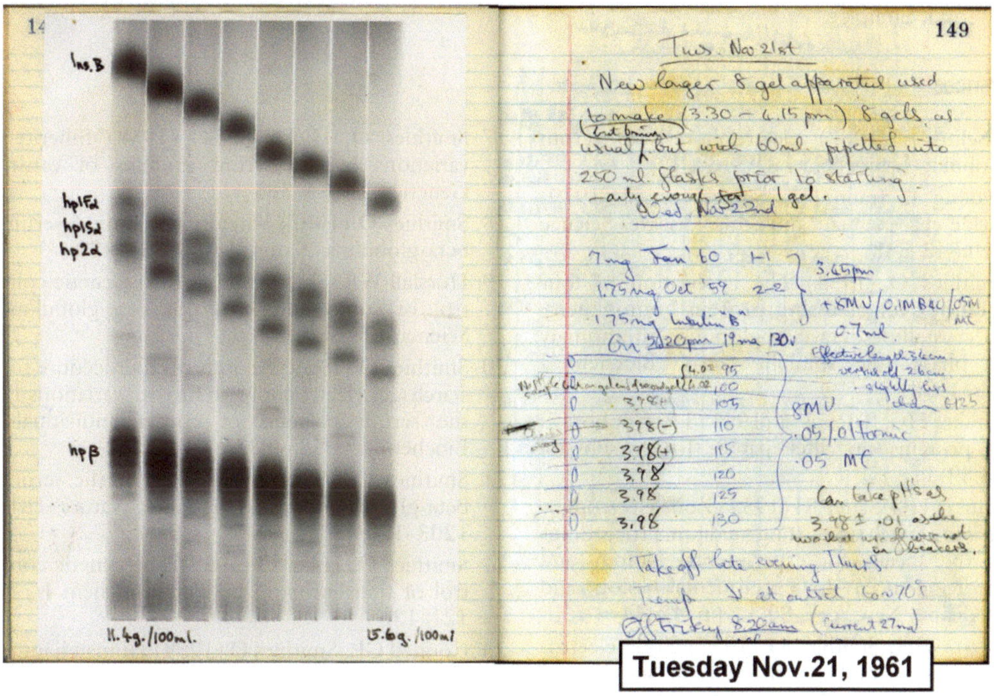

Fig. 19.

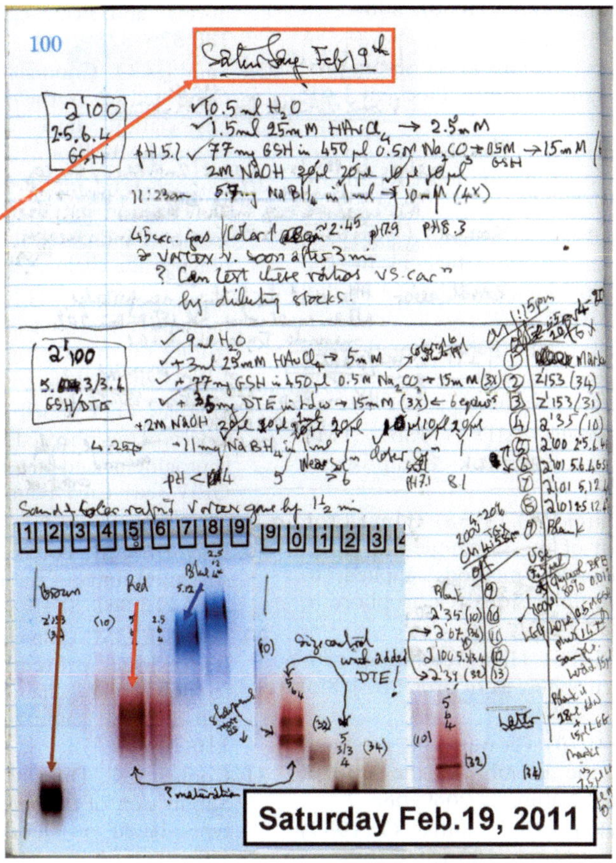

Fig. 20.

References

1. Tiselius A (1937) Electrophoresis of serum globulin. I. Biochem J 31:313–317
2. Pauling L, Itano HA, Singer J et al (1949) Sickle cell anemia a molecular disease. Science 110:543–548
3. Smithies O (1954) The application of four methods for assessing protein homogeneity to crystalline beta-lactoglobulin: an anomaly in phase rule solubility tests. Biochem J 58:31–38
4. Kunkel HG, Tiselius A (1951) Electrophoresis of proteins on filter paper. J Gen Physiol 35:89–118
5. Kunkel HG, Slater RJ (1952) Zone electrophoresis in a starch supporting medium. Proceedings of the society for experimental biology and medicine society for experimental biology and medicine. New York, 80(1), pp. 42–44
6. Dixon GH, Smithies O (1957) Zone electrophoresis of cabbage enzymes in starch gels. Biochim Biophys Acta 23:198–199
7. Smithies O (1955) Grouped variations in the occurrence of new protein components in normal human serum. Nature 175:307–308
8. Smithies O (1955) Zone electrophoresis in starch gels: group variations in the serum proteins of normal human adults. Biochem J 61: 629–641
9. Smithies O, Walker NF (1955) Genetic control of some serum proteins in normal humans. Nature 176:1265–1266
10. Jayle MF, Boussier G, Badin J (1952) Electrophorese de l'haptoglobine et de son complexe hemoglobinique. Bull Soc Chim Biol (Paris) 34:1063–1069
11. Smithies O, Walker NF (1956) Notation for serum-protein groups and the genes controlling their inheritance. Nature 178:694–695
12. Smithies O, Poulik MD (1956) Two-dimensional electrophoresis of serum proteins. Nature 177:1033
13. Poulik MD, Smithies O (1958) Comparison and combination of the starch-gel and filter-paper electrophoretic methods applied to human sera: two-dimensional electrophoresis. Biochem J 68:636–643
14. O'Farrell PH (1975) High resolution two-dimensional electrophoresis of proteins. J Biol Chem 250:4007–4021
15. Laemmli UK (1970) Cleavage of structural proteins during the assembly of the head of bacteriophage T4. Nature 227:680–685
16. Giblett ER, Hickman CG, Smithies O (1959) Serum transferrins. Nature 183:1589–1590
17. Smithies O, Hickman CG (1958) Inherited variations in the serum proteins of cattle. Genetics 43:374–385
18. Smithies O (1957) Variations in human serum beta-globulins. Nature 180:1482–1483
19. Horsfall WR, Smithies O (1958) Genetic control of some human serum beta-globulins. Science 128:35
20. Smithies O (1959) An improved procedure for starch-gel electrophoresis: further variations in the serum proteins of normal individuals. Biochem J 71:585–587
21. Smithies O (1958) Third allele at the serum beta-globulin locus in humans. Nature 181: 1203–1204
22. Smithies O, Hiller O (1959) The genetic control of transferrins in humans. Biochem J 72: 121–126
23. Connell GE, Smithies O (1959) Human haptoglobins: estimation and purification. Biochem J 72:115–121
24. Ingram VM (1957) Gene mutations in human haemoglobin: the chemical difference between normal and sickle cell haemoglobin. Nature 180:326–328
25. Smithies O, Connell GE, Dixon GH (1966) Gene action in the human haptoglobins. I. Dissociation into constituent polypeptide chains. J Mol Biol 21:213–224
26. Connell GE, Smithies O, Dixon GH (1966) Gene action in the human haptoglobins. II. Isolation and physical characterization of alpha polypeptide chains. J Mol Biol 21:225–229
27. Black JA, Dixon GH (1970) Gene action in the human haptoglobins. IV. Amino acid sequence studies on the haptoglobin alpha chains. Can J Biochem 48:133–146
28. Smithies O, Connell GE (1959) Biochemical aspects of the inherited variations in human serum haptoglobins and transferrins. In: Wostenholme GEW (ed) Ciba Foundation Symposium, biochemistry of human genetics pp. 178–193.
29. Edelman GM, Poulik MD (1961) Studies on structural units of the gamma-globulins. J Exp Med 113:861–884
30. Edelman GM, Gally JA (1962) The nature of Bence-Jones proteins. Chemical similarities to polypeptide chains of myeloma globulins and normal gamma-globulins. J Exp Med 116:207–227
31. Connell GE, Dixon GH, Smithies O (1962) Subdivision of the three common haptoglobin types based on 'hidden' differences. Nature 193:505–506

32. Smithies O, Connell GE, Dixon GH (1962) Inheritance of haptoglobin subtypes. Am J Hum Genet 14:14–21
33. Smithies O, Connell GE, Dixon GH (1962) Chromosomal rearrangements and the evolution of haptoglobin genes. Nature 196: 232–236
34. Smithies O (1962) Molecular size and starch-gel electrophoresis. Arch Biochem Biophys Suppl 1:125–131
35. Smithies O (2003) Why the kidney glomerulus does not clog: a gel permeation/diffusion hypothesis of renal function. Proc Natl Acad Sci USA 100:4108–4113

Chapter 2

Introduction to Protein Electrophoresis

Pothur R. Srinivas

Abstract

This chapter discusses, briefly, the developments in electrophoresis of proteins from Tiselius' moving boundary electrophoresis to the modern day two-dimensional polyacrylamide gel electrophoresis. It also touches upon the staining methods used to visualize total proteins postelectrophoresis.

Key words: Electrophoresis, Polyacrylamide gel electrophoresis, Protein staining

1. Introduction

Since proteins are the functional units of the cellular machinery they provide significant information regarding the molecular basis of health and disease. Hence techniques to separate and isolate the various proteins are critical to studying and understanding their functional characteristics. One of the widely used techniques for this purpose is electrophoresis which predates many of the modern day proteomic technologies. Protein electrophoresis involves the migration of charged protein molecules in the context of an applied electric field. The electrophoretic mobility of an individual protein depends on a variety of factors including the aggregate charge on its surface, size, shape, and strength of the applied electric field. Further, the pH of the electrophoresis buffer also influences protein migration since it affects the net charge on the protein surface.

A typical electrophoretic run today involves separation of samples on a gel support that is immersed in buffer with the gel sandwich spanning both electrodes and the proteins migrating as anions toward the cathode. Once separation is achieved, the gel is (a) stained by a variety of methods to visualize the separated proteins, (b) further processed to detect incorporated radioactive or fluorescent tags,

Biji T. Kurien and R. Hal Scofield (eds.), *Protein Electrophoresis: Methods and Protocols*, Methods in Molecular Biology, vol. 869,
DOI 10.1007/978-1-61779-821-4_2, © Springer Science+Business Media, LLC 2012

(c) processed for immunoblotting, (d) characterized for separated polypeptides, or (e) processed for other procedures.

Electrophoretic techniques have progressed from separation in free solution originally described in the moving boundary method by Tiselius (1), to separation on support-based systems that include paper and gel electrophoresis. Tiselius initially used electrophoresis to separate proteins in human plasma in free solution without the use of any support medium. One drawback of this "moving boundary" method was that the heat generated by the current caused the protein bands to become diffuse with only the heads and tails being separated as clean zones (2). This drawback led to the introduction and use of solid support mediums such as paper, starch, and polyacrylamide that helped to prevent the separated proteins from sedimenting. The first of the support mediums to be used was paper.

2. Electrophoresis on a Solid Support

2.1. Paper Electrophoresis

Paper was used as a support medium initially for the separation of snake venom (3). Cellulose-based filter paper was one of the first support media to be used in electrophoresis for the separation of small molecular weight molecules such as amino acids (4). The biggest improvement that this brought about was that all of the sample could now be loaded at the same starting point or line rather than being equally distributed in one part of the electrophoresis apparatus. This allowed for much better separation of the proteins in the sample. Further, the migration of the proteins was not prone to heat-generated convection currents prevalent in free solution electrophoresis. One of the features of apparatus used in paper electrophoresis was the physical separation of the electrode chambers from the electrophoresis compartment to help prevent contamination by products of oxidation and reduction formed at the electrodes (2). Further, a nonconducting liquid, lighter than water was used to dissipate heat generated during the run as well as to keep the filter paper moist. It is interesting to note that paper electrophoresis coupled with chromatography was instrumental in discovering that sickle cell anemia was caused by a single amino acid substitution in the hemoglobin protein (5).

2.2. Gel Electrophoresis

2.2.1. Polyacrylamide Gel Electrophoresis

The demonstration and use of starch gels for the separation of serum haptoglobins was one of the biggest advances in gel-based electrophoresis (6) that also brought with it the concept of sieving. Another significant advance of the starch gel electrophoresis system was the use of discontinuous buffers (7) that was used to great advantage in the PAGE system. The success of the starch gel was instrumental in the development of PAGE. PAGE has been one of the most widely used techniques for protein separation and characterization since the 1970s. Two significant factors contributing to the separation of proteins in PAGE are molecular weight and surface charge. The polyacrylamide support matrix was initially described in 1959 (8) and its use in electrophoresis using a vertical slab gel platform and the disk gel system was demonstrated in 1964 (9, 10). The polyacrylamide gel matrix is formed through copolymerization of acrylamide and a bifunctional cross-linking agent (11). One of the most commonly used cross linking agents is *N*,*N*′-methylenebisacrylamide (Bis) (11, 12). The cross-linking reaction is catalyzed in the presence of *N*,*N*,*N*,*N*-tetramethylethylenediamine (TEMED) and ammonium persulfate (13). One big advantage of the polymerization process is that the pore size of the gel can be varied by changing the ratios of acrylamide and the cross linker, thus enabling the separation of proteins with a wide range of molecular weights. Generally, the pore size is one of the factors that in addition to the electrical field, size, shape, and surface charge determine the distance that a protein migrates and is a function of the molecular sieving capacity of the gel matrix (14). The pore size of a polyacrylamide gel is inversely proportional to the concentration of acrylamide. Higher percentages of acrylamide generally result in better separation of low molecular weight proteins. Hence, the concentrations of acrylamide can be tailored for the desired polyacrylamide pore size based on the proteins that need to be separated. While the introduction of porosity gradients for resolving complex protein samples into sharp separated bands enhanced the versatility of the PAGE platform (15, 16), the use of the anionic detergent sodium dodecyl sulfate (SDS) was a significant step in improving the resolution of the separation (17). SDS was initially used in electrophoresis for the separation and identification of polypeptide chains and subsequently for molecular weight determination which was facilitated by minimizing the effects of the charge differences and ensuring that all the proteins in the sample migrate as anions (17). SDS helps denature and unfold most protein structures thus eliminating differences in secondary and tertiary structures while rendering the proteins as polypeptides.

2.2.2. Buffers

PAGE can be performed at any pH ranging from 2.0 to 10.0 without adversely affecting the proteins being separated (13). In general, buffers with low-ionic strengths facilitate faster rates of protein migration, while buffers with high-ionic strengths reduce migration rates but allow for clearer separation. Concentrated buffers with high-ionic strengths usually result in sharper protein separations since diffusion coefficients are generally lower at high-ionic strengths (13). However, systems should be in place to dissipate the excess heat that is generated under conditions of high-ionic strengths. The discontinuous gel and buffer system first demonstrated by Ornstein (18) and Davis (19) is more widely used today than the uniform gel and buffer phase system since it helps increase the resolution of protein separation. The Laemmli SDS–PAGE system (20) is the most popular version of the discontinuous system. The discontinuous system employs different pH and voltage gradients in different gel compartments. In this system, a gel is polymerized as an upper stacking gel and a lower resolving gel. The stacking gel generally has a lower pH and is of a larger pore size (typically 4% acrylamide) than the resolving gel. The pH and ionic strengths of the buffer used in the stacking gel, the large pore size, and the sample preparation buffer all help concentrate the proteins into a sharp and thin starting zone before they enter the resolving gel (13). The sample preparation buffer usually contains SDS and beta-mercaptoethanol to solubilize, denature, and break up the disulfide bonds in proteins. In addition, they include a dye such as Bromophenol Blue to help visualize migration. The Tris/HCl buffers used in the stacking gel have a lower pH (at least 2 pH units) than the Tris/glycine buffer used for the electrophoresis run and the protein molecules are entrapped as sharp bands between these leading chloride ions and the trailing glycine ions. The resolving gel has a higher percentage of acrylamide (typically 10% acrylamide) and hence a smaller pore size, a higher pH, and salt concentration. Once the protein bands enter the resolving gel, the glycine is ionized, the voltage gradient disappears and the individual proteins are separated based on their sizes (13).

Once the proteins have been separated by SDS–PAGE electrophoresis, they can be visualized through noncovalent chemical staining or metal deposition methods through a series of steps that include (a) fixing to remove substances that would interfere with the stain and to limit background staining, (b) staining with appropriate reagents, and (c) destaining to stop the staining process, remove the excess stain, and prepare the gel for further processing (21). Staining with Coomassie Brilliant Blue is one of the most widely used methods for visualizing total proteins in the microgram range. Coomassie dyes are disulfonated triphenylmethane dyes which stain protein bands blue through interaction with aminoacids (22). Silver staining provides better and enhanced sensitivity than Coomassie dyes for detecting proteins usually at the nanogram range (23). Stains such as Nile red, a phenoxazone dye, SYPRO Ruby, a organometallic ruthenium ion-based stain, and Epicocconone, a fungal derivative are a few of the fluorescent agents used for ultra-sensitive detection of proteins (24–26). In addition, specialized staining products such as Pro-Q Diamond phosphoprotein stains and Phos-tag phosphoprotein stains are available to view posttranslationally modified proteins (27, 28).

2.2.3. Two-Dimensional Gel Electrophoresis

Two-dimensional gel electrophoresis (2-DE) was the main electrophoretic technology used for separating proteins before high throughput mass spectrometry methods matured and were more widely adopted. Initially described in the mid-1970s (29), this method involves the separation of proteins in two dimensions. Separation in the first dimension is based on the respective isoelectric points and the separation in the second dimension is based on molecular weight. In addition to separating proteins from blood and other fluids, 2-DE has been used to evaluate whole cell as well as tissue proteins. Using narrow, immobilized pH gradients in the first dimension increases the resolving power which in turn can help detect low-abundant proteins. A number of techniques that include radioactive or fluorescent labeling and silver staining allow for visualization of the separated proteins. Comparisons between samples can be achieved by quantifying the ratios of spot intensities and a number of computer-assisted image processing tools and software packages have been developed recently to help automate this process and also facilitate data analysis. Spots can be excised and further processed through mass spectrometry platforms to infer qualitative information about the individual proteins. One of the important uses of 2-DE is in the study of posttranslational modifications and hence it remains an important technique for the study of proteins even in the face of recent high throughput proteomics technologies.

References

1. Tiselius A (1938) Elektrophoretische Messungen am Eiweiss. Kolloid Z 85:129
2. Righetti PG (2005) Electrophoresis: the march of pennies, the march of dimes. J Chromatogr A 1079:24–40
3. Von Klobusitzky P, Konig P (1939) Biochemische studien uber die gifte der schlan gengattung. Arch Exp Path Pharmakol 192:271–275
4. Evered DF (1959) Ionophoresis of acidic and basic amino acids on filter paper using low voltages. Biochim Biophys Acta 36:14–19
5. Ingram VM (1958) The comparison of normal human and sicke-cell haemoglobins by fingerprinting. Biochim Biophys Acta 28:539–545
6. Smithies O (1955) Zone electrophoresis in starch gels: group variations in serum proteins of normal human adults. Biochem J 61:629–641
7. Poulik MD (1957) Starch gel electrophoresis in a discontinuous system of buffers. Nature 180:1477–1479
8. Raymond S, Weintraub L (1959) Acrylamide gel as a supporting medium for zone electrophoresis. Science 1959(130):711–712
9. Raymond S (1964) Acrylamide gel electrophoresis. Ann N Y Acad Sci 121:35–65
10. Ornstein L (1964) Disc electrophoresis. I. Background and theory. Ann N Y Acad Sci 28:32–49
11. Gelfi C, Righetti PG (1981) Polymerization kinetics of polyacrylamide gels I. Effect of different cross-linkers. Electrophoresis 2:213–219
12. Righetti PG, Gelfi C, Bosisio AB (1981) Polymerization kinetics of polyacrylamide gels III. Effect of catalysis. Electrophoresis 2: 291–295
13. Andrews AT (1986) Polyacrylamide gel electrophoresis (PAGE): homogeneous gel and buffer systems. In: Electrophoresis: theory, techniques, and biochemical and clinical applications. Oxford University Press, New York
14. Maurer HR (1971) Disc electrophoresis and related techniques of polyacrylamide gel electrophoresis, 2nd edn. Walter de Gruyter, Berlin
15. Margolis J, Kenrick KG (1967) Polyacrylamide gel-electrophoresis across a molecular sieve gradient. Nature 214:1334–1336
16. Margolis J, Kenrick KG (1967) Electrophoresis in polyacrylamide concentration gradient. Biochem Biophys Res Commun 27:68–73
17. Shapiro AL, Vinuela E, Maizel JV (1967) Molecular weight estimation of polypeptide chains by electrophoresis in SDS-polyacrylamide gels. Biochem Biophys Res Commun 28: 815–820
18. Ornstein L (1964) Disc electrophoresis. I. Background and theory. Ann N Y Acad Sci 121:321–349

19. Davis BJ (1964) Disc electrophoresis. II. Method and application to human serum proteins. Ann N Y Acad Sci 121:404–427
20. Laemmli UK (1970) Cleavage of structural proteins during the assembly of the head of bacteriophage T4. Nature 227:680–685
21. Steinberg TH (2009) Protein gel staining methods: an introduction and overview. Methods Enzymol 463:541–563
22. Neuhoff V, Stamm R, Hansjorg E (1985) Clear background and highly sensitive protein staining with coomassie blue dyes in polyacrylamide gels: a systematic analysis. Electrophoresis 6:427–488
23. Switzer RC, Merril CR, Shifrin S (1979) A highly sensitive silver stain for detecting proteins and peptides in polyacrylamide gels. Anal Biochem 98:231–237
24. Daban J-R, Bartolome S, Samso M (1991) Use of the hydrophobic probe Nile red for the fluorescent staining of protein bands in sodium dodecyl sulfate-polyacrylamide gels. Anal Biochem 199:169–174
25. Berggren K, Chernokalskaya E, Steinberg TH et al (2000) Background-free, high sensitivity staining of proteins in one- and two-dimensional sodium dodecyl sulfate-polyacrylamide gels using a luminescent ruthenium complex. Electrophoresis 21:2509–2521
26. Mackintosh JA, Choi H-Y, Bae S-H et al (2003) A fluorescent natural product for ultrasensitive detection of proteins in one-dimensional and two-dimensional gel electrophoresis. Proteomics 3:2273–2288
27. Steinberg TH, Agnew BJ, Gee KR et al (2003) Global quantitative phosphoprotein analysis using multiplexed proteomics technology. Proteomics 3:1128–1144
28. Kinoshita E, Kinoshita-Kikuta E, Takiyama K, Koike T (2006) Phosphate binding tag, a new tool to visualize phosphorylated proteins. Mol Cell Proteomics 5:749–757
29. O'Farrell PH (1975) High resolution two-dimensional electrophoresis of proteins. J Biol Chem 250:4007–4021

Chapter 3

Protein Quantification Methods to Determine Protein Concentration Prior to Electrophoresis

J.P. Dean Goldring

Abstract

During each step of a protein isolation technique, if enzyme activity is to be determined and before a protein mixture is separated on a polyacrylamide electrophoresis gel, it is important to determine the concentration of the protein(s) in solution. Measuring protein concentration involves absorbance in the UV range or staining the protein with dyes or copper. This review describes the various protein determination methods that can be employed to measure protein concentration in solution.

Key words: Protein staining, Coomassie, Bicinchoninic acid, Lowry, Absorbance, Bradford

1. Introduction

The detection and quantification of protein is essential for studying enzyme kinetics and for monitoring protein purity at each step of a protein purification procedure. Methods to determine protein concentration detect particular amino acids by absorbance or interactions of dyes or copper with amino acids and peptide bonds. The choice of method will be influenced by the amount of protein available, the detection limit of the method, the ease of use, the time required to complete the method, and whether one can afford to lose some protein for the measurement. The methods described here are limited to those that detect 10–50 μg protein per sample.

1.1. Detecting Proteins with Absorbance at 280 nm

At 280 nm, amino acid residues with aromatic rings in a protein primary sequence absorb light. The amino acids tryptophan (λ max 279.8 nm) and tyrosine (λ max 274.6 nm) have extinction coefficients, ε, of 5.6 and 1.42 M^{-1} cm^{-1}, respectively, and thus contribute most to the absorbance at this wavelength while

Biji T. Kurien and R. Hal Scofield (eds.), *Protein Electrophoresis: Methods and Protocols*, Methods in Molecular Biology, vol. 869, DOI 10.1007/978-1-61779-821-4_3, © Springer Science+Business Media, LLC 2012

phenylalanine (λ max 279 nm, ε 0.197 M^{-1} cm^{-1}) makes a minor contribution (1). The method follows the Beer–Lambert law, i.e., absorbance is proportional to concentration and path length. It should be noted that the method depends on the relative number of tryptophan and tyrosine residues present in a protein (see Note 1). The method can be used to detect protein in the 20–3,000 μg range. The method is particularly useful for the detection of proteins eluting from chromatography columns as there is no loss of protein (see Note 2). The possibility of nucleic acid contamination should be considered, and compensation can be made (see Note 3).

1.2. Detecting Proteins with the Lowry Method

The Lowry protein assay was based on the biuret reaction with additional steps and reagents to increase the sensitivity of detection. In the biuret reaction, copper interacts with four nitrogen atoms of peptides to form a cuprous complex. Lowry (2) added phosphomolybdic/phosphotungstic acid (Folin–Ciocalteu reagent) which interacts with the cuprous ions and the side chains of tyrosine, tryptophan, and cysteine producing a blue green color that can be detected between 650 and 750 nm (3). The protein detection range is 2–100 μg. The assay was the method of choice to quantify proteins until the development of the Bradford (Coomassie G-250) assay in 1976 and the modified approach to the biuret reaction utilizing bicinchoninic acid introduced in 1985 by Smith (4,5). The Lowry assay has a number of steps, and many common reagents interfere with the method (see Notes 1, 4, 5, 10–12).

1.3. Detecting Proteins with the Smith Bicinchoninic Acid Method

Smith developed the bicinchoninic acid assay (BCA) where the sodium salt of bicinchoninic acid reacts with the cuprous ion generated by the biuret reaction and is thus similar to the Lowry assay (5). The bicinchoninic acid cuprous complex forms a deep blue color which is read at 562 nm, and the detection range is 0.2–50 μg (5). The BCA reagent is stable under alkali conditions so it can be included in the biuret alkaline copper solution. The assay can be run at room temperature or sensitivity can be increased at 60°C. The assay is compatible with detergents, giving it an advantage over both the Lowry and Coomassie dye assays (see Notes 1, 6, 7, 10–12).

1.4. Detecting Proteins with the Bradford Coomassie Blue G-250 Method

The Bradford method of protein determination uses Coomassie Blue G-250 dye which is reddish/brown with an absorbance maximum of 465 nm (4). Under acidic conditions, the dye reacts primarily with arginine and to a lesser extent with lysine, histidine, tyrosine, tryptophan, and phenylalanine residues in proteins producing a blue color with an absorbance maximum at 595 nm (the absorbtion range is between 575 and 615 nm) (4,6), and 0.2–20 μg of protein can be detected. The method is the easiest and fastest of the protein determination methods currently in use (see Notes 1, 8–12).

2. Materials

2.1. Detecting Proteins with Absorbance at 280 nm

1. 1-mL quartz cuvettes are recommended. Some plastic cuvettes are now available where the plastic does not absorb light at 280 nm.
2. A UV spectrophotometer.

2.2. Detecting Proteins with the Lowry Method

1. Spectrophotometer absorbing light in the visible range and 1-mL plastic or glass cuvettes.
2. Copper reagent: Dissolve 20 g of Na_2CO_3 in 260 mL water, 0.4 g $CuSO_4{\cdot}5H_2O$ in 20 mL water, and 0.2 g sodium potassium tartrate in 20 mL water and mix.
3. SDS solution: 1.0 g sodium dodecyl sulfate in 100 mL water (1% w/v).
4. NaOH solution: Dissolve 4 g NaOH in 100 mL water to make a 1 M solution.
5. 2× Lowry working reagent: Mix 3 parts of copper reagent with 1 part SDS solution and 1 part NaOH solution. This reagent is stable for 3 weeks.
6. 0.2 N Folin reagent: Mix 10 mL Folin reagent with 90 mL water. The reagent is stable for several months if stored in an amber bottle.

2.3. Detecting Proteins with the Smith Bicinchoninic Acid Method

1. Spectrophotometer absorbing in the visible range and 1-mL plastic or glass cuvettes.
2. Reagent A: 1 g sodium bicinchoninate, 2 g Na_2CO_3, 0.16 g sodium tartrate, 0.4 g NaOH, and 0.95 g $NaHCO_3$ made up to 100 mL with water. Adjust the pH to 11.25 with NaOH.
3. Reagent B: 0.4 g $CuSO_4{\cdot}5H_2O$ in 10 mL water. Reagents A and B are stable.
4. Mix 100 volumes of reagent A with 2 volumes reagent B for a working solution. The solution is stable for a week at room temperature.

2.4. Detecting Proteins with the Bradford Coomassie Blue G-250 Method

1. Spectrophotometer absorbing in the visible range and 1-mL plastic or glass cuvettes.
2. Dissolve 100 mg Coomassie Brilliant Blue G-250 in 100 mL of 85% (v/v) phosphoric acid and 50 mL 95% (v/v) ethanol. After the dye is dissolved, adjust the volume to 1 L with water. The reagent is stable for several months. Should a precipitate form, filter through filter paper and determine the standard curve again.

3. Methods

3.1. Detecting Proteins with Absorbance at 280 nm

1. Set the UV spectrophotometer to read at 280 nm allowing 15 min for the instrument to equilibrate.
2. Set the absorbance reading to zero with a solution of the buffer and all components except the protein present. Alternately take a reading of dye without protein and subtract the value from each reading of dye with protein.
3. Place the protein solution in the 1-mL cuvette and determine the absorbance. This step should be repeated to obtain duplicate readings. If a reading is obtained with an absorbance value greater than 2, then dilute the protein sample with the parent buffer and determine the absorbance value. An initial one in ten dilution is suggested. The dilution and readings of samples should be performed in duplicate.
4. If the extinction coefficient of the protein is known, then the following equation can be employed. Absorbance = Extinction coefficient × concentration of protein × path length (1 cm) to determine the concentration of the protein.

3.2. Detecting Proteins with the Lowry Method

1. Set the spectrophotometer to read at 750 nm allowing 15 min for the instrument to equilibrate.
2. Prepare a BSA protein standard at 1 mg/mL concentration in duplicate. Dilute the protein standard in a volume of 0.4 mL to give five concentrations over the range of 10–100 μg protein. Perform the dilution and readings for both standard solutions to give four readings at each protein concentration (see Note 1).
3. To 0.4 mL protein sample, add 0.4 mL of the Lowry working reagent and incubate for 10 min at room temperature.
4. Add 0.2 mL of the 0.2 N Folin reagent, vortex and leave for 30 min, and read the absorbance values at 750 nm in the spectrophotometer. Readings can be done between 650 and 750 nm but are more sensitive at 750 nm.

3.3. Detecting Proteins with the Smith Bicinchoninic Acid Method

1. Set the spectrophotometer to read at 562 nm, allowing 15 min for the instrument to equilibrate.
2. Prepare a BSA protein standard at 1 mg/mL concentration in duplicate. Dilute the protein standard in a volume of 20 μL to give five concentrations over a range of 10–50 μg protein. Each dilution of each of the two standards should be done in duplicate to produce four readings per protein concentration.
3. Add 20 μL of protein sample to 1 mL of working solution and incubate for 30 min at 37 or 60°C for increased sensitivity. Cool the sample and place in a 1-mL cuvette and read the absorbance.

3.4. Detecting Proteins with the Bradford Coomassie Blue G-250 Method

1. Set the spectrophotometer to read at 595 nm, allowing 15 min for the instrument to equilibrate.
2. Prepare a BSA protein standard at 1 mg/mL concentration in duplicate. Dilute the protein standard in a volume of 20 μL to give five concentrations over a range of 10–50 μg protein. Each dilution of each of the two standards should be done in duplicate to produce four readings per protein concentration.
3. Add 20 μL of protein solution to 1 mL of dye reagent, mix, and measure the absorbance in a cuvette.

4. Notes

1. The recommended standard protein for protein determination is BSA due to its availability and common use. One should be aware that each protein contains unique amounts of each amino acid. For example, BSA (66.3 kDa) contains 3 Trp, 21 Tyr, 26 Arg, and 60 Lys residues while ovalbumin (45 kDa) contains 3 Trp, 10 Tyr, 15 Arg, and 20 Lys residues. The two proteins will thus provide different standard curves in all the assays described above.
2. Advantages of using absorbance at 280 nm: The method is (a) fast, (b) easily automated, (c) reasonably sensitive, (d) does not destroy protein, and (e) most proteins contain tryptophan and tyrosine residues.
3. Disadvantages of using the absorbance at 280 nm to detect proteins: (a) It should be noted that DNA and RNA have an absorbance maxima at 260 nm but still absorb at 280 nm and have tenfold higher absorbance values at 280 nm compared to the equivalent concentration of protein. Groves et al. developed a method to correct for the presence of nucleic acid in a protein solution (7). A pure protein solution has a ratio of absorbance values (A_{280}/A_{260}) of greater than 1.7 and pure nucleic acid of less than 0.5. (b) Cuvettes should be handled carefully as fingerprints on the cuvette will distort readings.
4. Advantages of the Lowry method: (a) The method is a 100-fold more sensitive than the original biuret reagent. (b) The method is more sensitive than determining absorbance at 280 nm.
5. Disadvantages of the Lowry method: (a) Many common substances (K^+, Mg^{2+}, NH_4^+, EDTA, Tris–HCl, carbohydrates, and reducing agents) interfere with the method. (b) The Folin reagent is reactive for only a short period of time after addition. (c) The method is complicated and requires more steps and reagents than the BCA or Bradford assays. (d) The method is destructive to proteins, i.e., once the protein sample has reacted with the dye, the protein cannot be used for other assays.

6. Advantages of the Smith BCA: (a) The method is less complicated than the Lowry assay. (b) The method is more sensitive than the Lowry assay. (c) The method has less protein/protein variability than the Bradford assay. (d) The volume of reagents can be reduced, and a micro-assay can be performed in 96-well plates. (e) Detergents are compatible with the assay (up to 5%).
7. Disadvantages of the Smith BCA: (a) Reducing reagents (e.g., ascorbic acid) that reduce copper interfere with the assay. (b) Chelating agents (e.g., EDTA) that chelate copper interfere with the assay. (c) The method is destructive to proteins, i.e., once the protein sample has reacted with the copper, the protein cannot be used for other assays.
8. Advantages of the Bradford Coomassie Blue G-250 assay: (a) The method is fast. (b) The method is simple. (c) The reagents are stable for long periods of time. (d) The volume of reagents can be reduced and the assay performed in a 96-well plate. (e) The method is sensitive.
9. Disadvantages of the Bradford Coomassie Blue G-250 assay: (a) The dye stains cuvettes. Cuvettes can be cleaned by washing with a dilute sodium dodecyl sulfate solution. (b) Dye binding depends on the basic amino acid content that can vary between proteins. (c) Concentrated protein solutions can form a precipitate upon contact with the dye reagent. If this is observed, the protein solution should be diluted to determine protein concentration. (d) The method is destructive to proteins, i.e., once the protein sample has reacted with the dye, the protein cannot be used for other assays.
10. It is advisable to dissolve the protein standard in duplicate. If a protein standard is weighed or solubilized incorrectly in a single preparation, then all protein determinations will be inaccurate.
11. Protein determinations with copper or Coomassie Blue become saturated with increasing concentrations of protein.
12. Protein determinations with copper or Coomassie Blue reagents should always be considered as a relative concentration of protein. It is accepted practice to express the relative concentrations of proteins during protein purification steps, provided all protein concentrations are determined with the same method.

References

1. Fasman GD (1990) Practical handbook of biochemistry and molecular biology. CRC, Boston, MA
2. Lowry OH, Rosebrough NJ, Farr AL et al (1951) Protein measurement with the folin phenol reagent. J Biol Chem 193:265–275
3. Stoscheck CM (1990) Quantitation of protein. Methods Enzymol 182:50–68
4. Bradford MM (1976) Rapid and sensitive method for quantitation of microgram quantities of protein utilizing principle of protein-dye binding. Anal Biochem 72: 248–254

5. Smith PK, Krohn RI, Hermanson GT et al (1985) Measurement of protein using bicinchoninic acid. Anal Biochem 150: 76–85
6. Compton SJ, Jones CG (1985) Mechanism of dye response and interference in the Bradford protein assay. Anal Biochem 151:369–374
7. Groves WE, Davis FC, Sells BH (1968) Spectrophotometric determination of microgram quantities of protein without nucleic acid interference. Anal Biochem 22: 195–210

Chapter 4

Protein Solubilization: Attend to the Choice of Lysis Buffer

Mandy Peach, Noelle Marsh, and Daniel J. MacPhee

Abstract

The efficient extraction of proteins of interest from cells and tissues is not always straightforward. In this process, the use of the optimal lysis buffer for protein solubilization should be considered. Here we demonstrate the use of a urea/thiourea lysis buffer, based on O'Farrell's buffer, and compare its effectiveness for solubilization of proteins from smooth muscle with the often utilized RIPA lysis buffer.

Key words: Lysis buffer, RIPA lysis buffer, Urea/thiourea lysis buffer, Protein extraction, Protein solubilization, Small stress proteins

1. Introduction

There are many obstacles to the appropriate extraction of proteins from cells and tissues. For example, it has been estimated that intracellular proteins may represent only a tiny fraction of total cellular protein levels which would make them difficult to extract in reasonable quantities without proper lysis and extraction protocols (1). The hydrophobicity of membrane proteins is also a significant stumbling block, particularly to two-dimensional gel electrophoresis (2). The advent of proteomics has accelerated the development and modification of strategies for cell lysis as well as protein extraction at both small and large scale levels. The result is that despite more traditionally known cell and tissue lysis buffers (3, 4), there are now quite a number of commercially available reagents for use in such procedures (1). Nonetheless, one must consider if a particular lysis buffer is optimally designed to extract the protein(s) of interest, particularly the use of the appropriate denaturants and detergent(s) for protein extraction and solubilization.

Several basic criteria are considered for the production of an effective lysis buffer for protein extraction: (1) pH, (2) ionic strength,

Biji T. Kurien and R. Hal Scofield (eds.), *Protein Electrophoresis: Methods and Protocols*, Methods in Molecular Biology, vol. 869, DOI 10.1007/978-1-61779-821-4_4, © Springer Science+Business Media, LLC 2012

(3) choice of detergents and denaturants, and (4) constituents to improve protein stability or prevent proteolysis. Specific details on these parameters are described in detail elsewhere (see Note 1) (1, 5–8). Although Gromov et al. (9) have reported the development of a lysis buffer that could be used for several different proteomic strategies, it has so far proven difficult to design a lysis buffer effective for isolation of a broad array of proteins. This has necessitated the tailoring of buffers and protocols for individual use.

The *s*mall *h*eat *s*hock *p*roteins (sHSPs) are a family of highly conserved stress proteins ranging in size from 12 to 43 kDa in mass (10, 11). These proteins have the ability to form low and high molecular mass (>300 kDa) complexes with other heat shock proteins (12–14) and possess a dynamic quaternary structure. In periods of stress, sHSP can bind denatured proteins and prevent their irreversible aggregation (15), thus aiding in the assembly, disassembly, stabilization, and the internal transport of intracellular proteins (16). Some sHSP, such as HSPB1, can be found in high quantities (e.g., 2 mg/g of tissue protein) within specific tissues such as different types of muscle (15). These characteristics present challenges to the effective extraction and solubilization of sHSP for SDS-PAGE and immunoblot analysis. Tyson and colleagues (17) demonstrated the very effective extraction of HSPB6 from uterine smooth muscle with a urea/thiourea lysis buffer (see Note 2) and subsequent SDS-PAGE and immunoblot analysis. Clearly, the determination of optimal lysis buffer conditions would be especially important if relative abundance of a specific protein(s) was going to be calculated between control and experimental conditions with downstream immunoblot analysis. Efficient immunoprecipitation of proteins could also be affected by suboptimal lysis conditions. The use of more standard tissue lysis buffers such as RIPA lysis buffers is quite common for extraction of proteins such as uterine smooth muscle proteins (18–20). Thus, using sHSP as an example, the following protocol was used to assess the ability of RIPA lysis buffer and urea/thiourea lysis buffer to solubilize such proteins and demonstrate the importance of lysis buffer choice.

2. Materials

All aqueous solutions utilized deionized ddH_2O or Milli-Q grade water (ultrapure; ~18 MΩ.cm), and all reagents used were electrophoresis or analytical grade.

1. Phosphate-buffered saline (10×): Dissolve 40 g NaCl, 1 g KCL, 7.2 g Na_2HPO_4, and 1.2 g KH_2PO_4 in 500 mL of ultrapure water and adjust the pH to 7.4 with HCl. Filter sterilize and store at room temperature. Create 1× working solutions by diluting with ultrapure water.

2. Modified RIPA lysis buffer: 50 mM Tris–HCl (pH 7.5), 150 mM NaCl, 1% (vol/vol) Triton X-100, 1% (wt/vol) sodium deoxycholate, and 0.1% (wt/vol) SDS. Dissolve 3.02 g of Tris base, 4.38 g of NaCl, 5 g of deoxycholic acid, and 0.5 g of SDS in 400 mL of ultrapure water. Add 5 mL of Triton X-100, mix and bring up the volume to 500 mL following adjustment of pH to 7.5 with HCl. Filter sterilize and store at 4°C (see Note 3).
3. Urea/thiourea lysis buffer: 7 M urea, 2 M thiourea, and 4% (w/v) CHAPS, in 30 mM Tris–HCl (pH 8.5) (17). Combine 4.2 g of urea, 1.52 g of thiourea, and 0.4 g CHAPS in 10 mL of 30 mM Tris–Cl (pH 8.5). Dissolve one tablet each of Mini EDTA-free Protease and PhosSTOP phosphatase inhibitor tablets (Roche Applied Science, Indianapolis, IN, USA) and store the buffer in 1 mL aliquots at −80°C (see Note 4).
4. 1.5 M Tris–HCl resolving gel buffer: Dissolve 18.2 g of Tris base in ultrapure water to a final volume of 100 mL following adjustment of pH to 8.8 with HCl. Store at 4°C.
5. 0.5 M Tris–HCl stacking gel buffer: Dissolve 6.1 g of Tris base in ultrapure water to a final volume of 100 mL following adjustment of pH to 6.8 using HCl. Store at 4°C.
6. SDS-PAGE 15% resolving gel composition: Combine 7.5 mL 30% acrylamide mix (29:1; Sigma Chemical Co), 3.8 mL of 1.5 M Tris–HCl (pH 8.8), 0.15 mL of 10% SDS, 0.15 mL of freshly made 10% ammonium persulfate (w/v; see Note 5), and 0.006 mL of TEMED in ultrapure water to a final volume of 15 mL.
7. SDS-PAGE 10% resolving gel composition: Combine 5 mL of 30% acrylamide mix (29:1), 3.8 mL of 1.5 M Tris–Cl (pH 8.8), 0.15 mL of 10% SDS, 0.15 mL of freshly made 10% ammonium persulfate (w/v), and 0.006 mL of TEMED in ultrapure water to a final volume of 15 mL.
8. SDS-PAGE 4% stacking gel composition: Combine 1 mL 30% acrylamide mix (29:1), 2 mL of 0.5 M Tris–HCl (pH 6.8), 0.08 mL of 10% SDS, 0.04 mL of freshly prepared 10% ammonium persulfate (see Note 5), and 0.008 mL of TEMED in ultrapure water to a final volume of 8 mL.
9. SDS-PAGE running buffer (5×): 15.1 g of Tris base, 94 g glycine, and 50 mL of 10% SDS made up to 1 L in deionized ddH_2O water with a final pH of 8.3. Filter sterilize and refrigerate until use. Dilute to 1× with deionized ddH_2O when required.
10. SDS-PAGE loading dye (2×): 4% (w/v) SDS, 0.2% (w/v) bromophenol blue, 20% (v/v) glycerol in 100 mM Tris–Cl, pH 8.0. Dissolve 0.4 g of SDS, 0.02 g of bromophenol blue in 8 mL of 100 mM Tris–Cl, pH 8.0, and add 2 mL of glycerol (20% final concentration v/v). Divide into 1 mL aliquots and

add β-mercaptoethanol, to a final concentration of 10% (v/v), to a 1 mL working solution. Store at room temperature.

11. Transfer membranes: 0.22 μm nitrocellulose membranes (ThermoFisher Scientific, Nepean, ON, Canada).
12. Gel transfer buffer (10× stock): Dissolve 29 g glycine, 58 g Tris base, and 3.7 g SDS in 1 L of deionized ddH_2O water. Adjust pH to 8.3 and filter sterilize. Store in the refrigerator and produce 1× working solutions by diluting with deionized ddH_2O water.
13. TBST: Dissolve 2.42 g Tris base, 8.0 g of NaCl, 3.8 mL 1 M HCl, and 1.0 mL of Tween-20 in 1 L of deionized ddH_2O water and adjust the pH to 7.6 with HCl. Buffer can be stored in the refrigerator.
14. Immunoblot blocking buffer: Dissolve 5 g of fat-free skim milk powder in TBST buffer and mix by vigorous shaking.
15. Coomassie Brilliant Blue protein stain: Add 0.25 g of Coomassie Brilliant Blue R-250 to 100 ml of methanol:acetic acid solution (45% methanol v/v, 10% acetic acid (v/v) in deionized water). Filter through Whatman paper to remove particulates.
16. Methanol:acetic acid destaining solution: Mix 25 mL of methanol and 7.5 mL of acetic acid in deionized water.
17. Electrophoresis and transfer system: Mini-PROTEAN 3 system (Bio-Rad Laboratories, Hercules, CA, USA).
18. Whatman paper: 3-mm chromatography paper (Schleicher and Schuell Bioscience, Keane, NH, USA).
19. Molecular weight protein markers: Precision Plus Dual Color Protein Standards (Bio-Rad Laboratories, Hercules, CA, USA).
20. Antisera: Rabbit polyclonal anti-HspB8 or anti-HspB6 (Abcam, Inc., Cambridge, MA, USA). Mouse monoclonal anti-smooth muscle calponin (Sigma Chemical Co, St. Louis, MO, USA).
21. HRP-conjugated goat anti-rabbit IgG (H+L) or HRP-conjugated goat anti-mouse IgG (H+L) antisera (Promega Corporation, Madison, WI, USA).
22. Pierce SuperSignal West Pico Chemiluminescent Substrate (MJS BioLynx, Inc., Brockville, Ontario, Canada).
23. Tissue: Rat uterine smooth muscle tissue from d15 of pregnancy.

3. Methods

3.1. Tissue Collection

Isolate all tissue samples rapidly (see Note 6) and place in ice-cooled PBS for rapid washing. Subsequently, place tissues in polypropylene vials and freeze in liquid nitrogen.

3.2. Protein Solubilization

1. In a fume hood, chip off pieces of frozen tissues, quickly weigh the fragments (~100–250 mg) in precooled weigh boats, and place in a precooled mortar on dry ice. Grind the samples into a fine powder with a pestle under liquid nitrogen. The use of a fume hood is recommended to avoid inhalation of vapors from the liquid nitrogen.
2. Transfer the powdered samples to 1 mL of urea/thiourea lysis buffer (see Notes 2 and 3) or 1 mL RIPA lysis buffer in 15-mL polypropylene tubes and homogenize up to 1 min on ice with a Polytron PT10-35 homogenizer (see Note 7).
3. For samples homogenized in urea/thiourea lysis buffer, allow the lysates to settle at room temperature for 30 min while RIPA tissue lysates are kept on ice. Subsequently, transfer all sample lysates to appropriately labeled microcentrifuge tubes, centrifuge at full speed for 15 min in a Beckman Microfuge-18 microcentrifuge, and collect supernatants for protein analysis.
4. Determine sample protein concentrations using the Bradford Assay (21).

3.3. SDS-PAGE and Electroblotting

1. Prepare the polyacrylamide gel casting module according to instructions provided by the appropriate manufacturer (e.g., Mini Protein 3, Bio-Rad Laboratories, Hercules, CA, USA) (see Note 8).
2. Immediately after the addition of TEMED to the resolving gel mixture, add the mixture to the prepared gel cassette with a Pasteur pipet (see Notes 9 and 10). Add isopropanol over the top of the resolving gel to ensure that gel polymerization is not inhibited. After 45 min, remove the overlaid isopropanol by tipping the gel molds to pour off the solvent and soak up residual isopropanol with Kimwipes. Pour the stacking gel in the same manner as the resolving gel. Insert the appropriate gel comb into the stacking gels ensuring that no air bubbles are trapped under the teeth of the comb.
3. Once gel polymerization is complete and the gel assembled in the electrophoresis tank, incubate protein samples (e.g., 50 μg volumes) with 2× SDS-PAGE loading dye at 95°C for 5 min prior to gel loading (see Note 11). Run the gel at 60 V until samples and the prestained molecular mass standards enter the resolving gel and then separate proteins at 100 V until the dye front reaches the bottom of the gel (see Note 12).
4. Following electrophoresis, gently pry open the gel plates with a plastic wedge to recover the gel.
5. To help assess the effective solubilization of sample proteins with the different lysis buffers, stain the polyacrylamide gel with Coomasie Blue for at least 0.5 h with gentle agitation on a shaker. Destain the gel with methanol:acetic acid solution using gentle agitation until bands are finely resolved

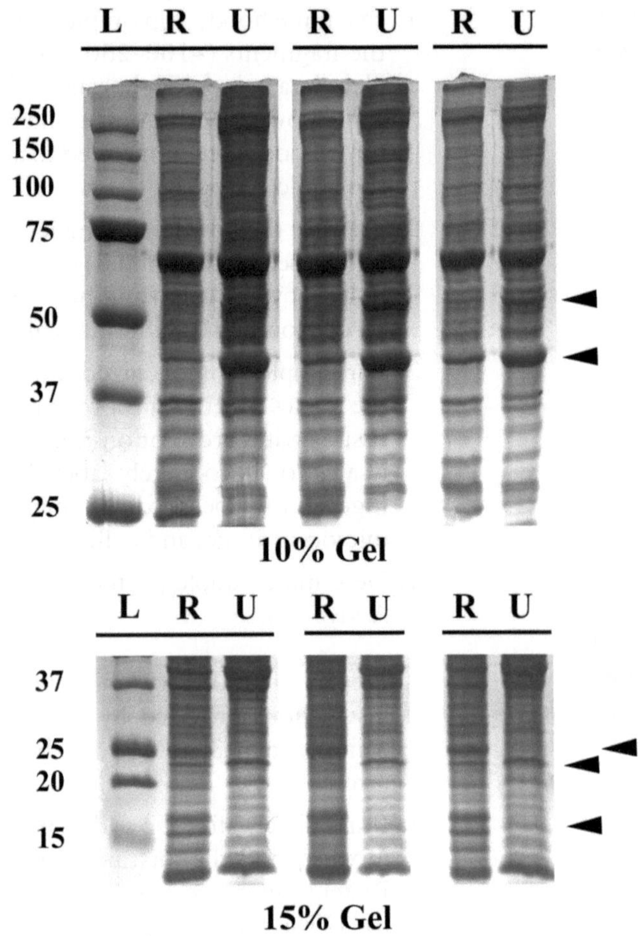

Fig. 1. Proteins were extracted from pregnant rat uterine smooth muscle with RIPA (*R*) or urea/thiourea lysis buffer (*U*) and loaded on 10 or 15% polyacrylamide gels for SDS-PAGE. Gels were stained with Coomassie Blue and destained. Wedges indicate examples of protein species that appeared to solubilize to different extents in the two lysis buffers. *L* protein molecular mass ladder listed in kDa. Paired lanes represent tissue lysates obtained from different animals ($n = 3$).

(~4–8 h; Fig. 1). The use of small pieces of foam added to the destain solution aids the wicking up of dye from the destain solution. Photograph the gel with a gel documentation system.

6. For immunoblot analysis, skip step 5 and place the gel in transfer buffer. Cut a nitrocellulose membrane to the same size of the gel and also place it in transfer buffer. Assemble the gel for electroblotting as has been described in detail elsewhere (22, 23) (see Note 13) and conduct electroblotting for 1 h at 300 mA in transfer buffer with constant buffer stirring in an ice bucket or use an ice pack in the transfer apparatus.

3.4. Immunoblot Analysis

Unless otherwise stated, all incubations should be conducted at room temperature and with constant agitation.

1. Rinse the membrane with TBST for 5 min.
2. Block the blot in 5% skim milk powder/TBST for 30 min (see Note 14).
3. Incubate the membrane in appropriate antisera, diluted in blocking solution, for 1 h.
4. Rinse the blot 1×15 min in TBST, followed by 2×5 min in TBST.
5. Incubate membranes for 1 h in HRP-conjugated goat anti-rabbit IgG (H+L) or HRP-conjugated goat anti-mouse IgG (H+L) antisera (1:10,000 and 1:150,000 dilutions, respectively) diluted in blocking solution.
6. Wash the blot 1×15 min in TBST and then 4×5 min in TBST.
7. Detect proteins on the immunoblot using the Pierce SuperSignal West Pico Chemiluminescent Substrate detection system (Fig. 2). Generate multiple exposures on ECL X-ray films to ensure the linearity of the film exposures.

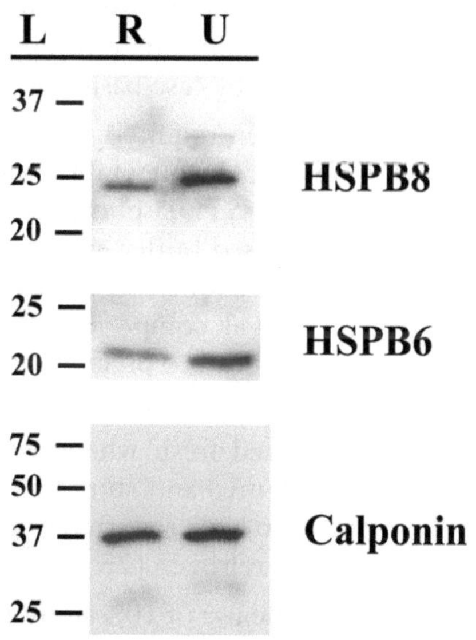

Fig. 2. Representative immunoblot analyses of HSPB6 and HSPB8. Proteins extracted with RIPA (*R*) or urea/thiourea (*U*) lysis buffer were separated by SDS-PAGE and electroblotted. HSPB6 or HSPB8-specific antisera were then used for immunoblot analysis of the two small stress proteins obtained from the different lysis conditions. h-Calponin-specific antiserum was also used to assess calponin expression in the same extraction conditions. Both HSPB6 and HSPB8 were more readily detected from smooth muscle tissue lysates prepared in the urea/thiourea lysis buffer indicating it may be a more optimal lysis buffer for solubilization of these proteins of interest.

4. Notes

1. Some researchers like to modify or develop their own lysis buffers in which case Harlow and Lane (3) have recommended considering the following range of variables to optimize the lysis buffer for downstream western blot analyses: salt concentrations 0–1 M, nonionic detergents 0.1–2%, ionic detergents 0.01–0.5%, divalent cation concentrations 0–10 mM, EDTA concentrations 0–5 mM, and pH 6–9.
2. The urea/thiourea lysis buffer is based on a buffer originally designed by O'Farrell (24). The use of this type of buffer, particularly in 2D gel electrophoresis, has been described in detail elsewhere (25, 26). The combined use of urea and thiourea increases protein solubility since urea is effective at disrupting hydrogen bonds aiding protein unfolding and denaturation while thiourea is much better at reducing hydrophobic interactions between proteins (27). The volumes of lysis buffers utilized are also very important for lysis efficiency. Gorg and colleagues (26) have previously reported the use of 1 mL of this lysis buffer with ~50–100 mg of mammalian tissue. In our hands, it has proven reasonable to use up to a maximum of 250 mg of tissue with a 1 mL volume. It is recommended that the appropriate volume be determined by the investigator on a case by case basis.
3. When required for lysis, take 10 mL aliquots and completely dissolve one tablet each of Mini EDTA-free Protease and PhosSTOP phosphatase inhibitor tablets prior to use. Any unused buffer can be frozen at −20°C for future use. As with just about every lysis buffer, care should be taken to ensure that all components are in solution prior to use as cold storage can lead to precipitation of some constituents (e.g., SDS).
4. It is recommended that the urea/thiourea lysis buffer be prepared fresh whenever possible, but it can also be aliquoted (1 mL) and stored at −80°C for up to several months. It has also been reported that once the buffer is thawed, it should not be refrozen (25, 26).
5. The use of 10% ammonium persulfate in the preparation of the gel mixtures eliminates the need to degas the solution and remove dissolved oxygen.
6. Cells and tissues should be frozen rapidly with liquid nitrogen to avoid protease degradation of proteins in the sample or collected and lysed quickly, preferably while chilled. Since proteases as well as phosphatases can be released during lysis and act on your target protein(s), protease and phosphatase inhibitors should be included in the lysis buffers. Many of these are

produced as cocktails in tablet form for easy purchase, and their use is as simple as dissolving the tablet in the lysis buffer prior to utilization.

7. Methods utilized for tissue disruption clearly depend on the tissue origin. There are a large number of other means to lyse cells/tissues, and readers are directed to Simpson (28) for specific details and discussion of these protocols. When using mechanical homogenization, it is important to avoid the production of excessive amounts of foam as this could decrease your recovery volume (i.e., becomes difficult to recover from the homogenizer). Short bursts of mechanical homogenization, while the sample(s) is cooled with ice, are usually best.
8. Place two to three folded Kimwipes under the thermoplastic rubber gaskets of the casting module and a strip of Parafilm on top of each gasket. This prevents leaks by increasing the thickness of the rubber gaskets upon which the glass plates are held against with a spring loaded lever. Leakage can be a problem as the gaskets age and lose their flexibility and overall thickness from constant use.
9. It is critical to use high-quality SDS from a single source and polyacrylamide that is free of contaminating metal ions. Sambrook and Russell (22) have reported that the migration pattern of polypeptides can change significantly when SDS from different manufacturers are interchanged. Purchase of premade acrylamide from reputable companies is becoming the normal procedure. Acrylamide solutions with a 1:29 bisacrylamide:acrylamide ratio are usually employed as they are capable of resolving polypeptides differing in size as little as 3% (22), but solutions can be purchased with different ratios if necessary to vary the pore size of the gel (23). In addition, Tris base should always be used for the preparation of gel buffers to avoid production of diffuse protein bands and even improper polypeptide migration (23).
10. Leave approximately 1 cm of space below the eventual bottom of the combs for the later addition of the stacking gel mixture.
11. It has been noted that heating of samples containing urea for 2D gel electrophoresis can result in some decomposition of urea and release of isocyanate leading to protein carbamylation and charge heterogeneities of the samples. However, in this instance, there is no need to worry about protein carbamylation during heating of the samples in 2× SDS loading dye at 95°C as the samples are not being used for isoelectric focusing. This heating step is necessary to produce SDS–polypeptide complexes for subsequent SDS–polyacrylamide electrophoresis.

12. The system used here is a discontinuous buffer system. As a result, the SDS–polypeptide complexes in the 4% stacking gel become deposited and concentrated on the surface of the resolving gel. The SDS–polypeptide complexes are then separated in the resolving gel according to size by molecular sieving in a zone of uniform voltage and pH. Greater details on the mechanism of polyacrylamide gel electrophoresis are found elsewhere (22).
13. There are now many options for transfer of polypeptides to membranes, and the reader is directed to a review of these procedures (23). It is also imperative that no air bubbles be trapped between the nitrocellulose membrane and the polyacrylamide gel as this will result in the lack of polypeptide transfer to the membrane in these regions. Use a pipet to roll out any bubbles between the gel and membrane.
14. The researcher should consider the blocking buffer that is most appropriate for the specific antiserum (e.g., skim milk powder vs. BSA). Blocking a blot serves two important purposes. The first is well known in that it can help mask any potential nonspecific binding sites on the membrane itself. The second purpose, being less known and even less understood, is that blocking a membrane can promote renaturation of antigenic sites (29). However, it has been reported that prolonged blocking times (>24 h) can actually remove antigens (30).

Acknowledgments

This work was supported by a Natural Sciences and Engineering Research Council Discovery Grant (#250218) to DJM. MP was a holder of an Alexander Graham Bell NSERC postgraduate fellowship.

References

1. Grabski AC (2009) Advances in preparation of biological extracts for protein purification. Methods Enzymol 463:285–305
2. Cordwell SJ (2008) Sequential extraction of proteins by chemical reagents. 2D PAGE: sample preparation and fractionation. In: Posch A (ed) Methods in molecular biology, vol 424. Humana Press, Totowa, NJ, pp 139–146
3. Harlow E, Lane E (1988) Immunoprecipitation. In: Harlow E, Lane D (eds) Antibodies: a laboratory manual. Cold Spring Harbor Laboratory, New York, p 231
4. Rosenberg IM (2005) Protein analysis and purification, 2nd edn. Birkhauser, Boston, p 37
5. Helenius A, Simons K (1975) Solubilization of membranes by detergents. Biochim Biophys Acta 415:29–79
6. Helenius A, McCaslin DR, Fries E et al (1979) Properties of detergents. Methods Enzymol 56:734–749
7. Dawson RMC, Elliot DC, Elliot WH et al (1986) pH, buffers, and physiological media. In: Dawson RMC et al (eds) Data for biochemical

research. Oxford University Press, New York, pp 417–448
8. Linke D (2009) Detergents: an overview. Methods Enzymol 463:603–617
9. Gromov P, Celis JE, Gromova I et al (2008) A single lysis solution for the analysis of tissue samples by different proteomic technologies. Mol Oncol 2:368–379
10. Salinthone S, Gerhoffer WT (2008) Small heat shock proteins in smooth muscle. Pharmacol Ther 119:44–54
11. Laskowska E, Matuszewska E, Kuczynska-Wisnik D (2010) Small heat shock proteins and protein misfolding diseases. Curr Pharm Biotechnol 11:146–157
12. Kato K, Goto S, Inaguma Y et al (1994) Purification and characterization of a 20 kDa protein that is highly homologous to alpha B crystallin. J Biol Chem 269:15302–15309
13. Pipkin W, Johnson JA, Creazzo TL et al (2003) Localization, macromolecular associations, and function of the small heat shock related protein HSP20 in rat heart. Circulation 107: 469–476
14. Sun X, Fontaine J-M, Rest JS et al (2004) Interaction of human hsp22 (HSPB8) with other small heat shock proteins. J Biol Chem 279:2394–2402
15. Gusev NB, Bogatcheva NV, Marston SB (2002) Structure and properties of small heat shock proteins and their interaction with cytoskeleton proteins. Biochemistry (Mosc) 67:511–519
16. Drieza CM, Komalavilas P, Furnish EJ et al (2010) The small heat shock protein, HSPB6, in muscle function and disease. Cell Stress Chaperones 15:1–11
17. Tyson EK, MacIntyre DA, Smith R et al (2008) Evidence that a protein kinase A substrate, small heat shock protein 20, modulates myometrial relaxation in human pregnancy. Endocrinology 149:6157–6165
18. Williams SJ, Shynlova O, Lye SJ et al (2009) Spatiotemporal expression of α1, α 3, and β1 integrin subunits is altered in rat myometrium during pregnancy and labour. Reprod Fertil Dev 22:718–732
19. Palliser HK, Zakar T, Symonds IM et al (2010) Progesterone receptor isoform expression in the guinea pig myometrium from normal and growth restricted pregnancies. Reprod Sci 17:776–782
20. Shynlova O, Dorogin A, Lye SJ (2010) Stretch-induced uterine myocyte differentiation during rat pregnancy: involvement of caspase activation. Biol Reprod 82:1248–1255
21. Bradford MM (1976) A rapid and sensitive method for the quantification of microgram quantities of protein utilizing the principle of protein-dye binding. Anal Biochem 72:248–254
22. Sambrook J, Russell D (2001) Molecular cloning: a laboratory manual. Cold Spring Harbor Laboratory Press, Cold Spring Harbor, NY, p 2344
23. MacPhee DJ (2010) Methodological considerations for improving western blot analysis. J Pharmacol Toxicol Methods 61:171–177
24. O'Farrell PJ (1975) High resolution two-dimensional electrophoresis of proteins. J Biol Chem 250:4007–4021
25. Weiss W, Gorg A (2008) Sample solubilization buffers for two-dimensional electrophoresis. 2D PAGE: sample preparation and fractionation. In: Posch A (ed) Methods in molecular biology, vol 424. Humana Press, Totowa, NJ, pp 35–42
26. Gorg A, Drews O, Weiss W (2006) Extraction and solubilization of total protein from mammalian tissue samples. Cold Spring Harb Protoc. doi:10.1101/pdb.prot4226. pp 3
27. Rabilloud T (1998) Use of thiourea to increase the solubility of membrane proteins in two dimensional electrophoresis. Electrophoresis 19:755–760
28. Simpson RJ (2009) Preparation of cellular and subcellular extracts. In: Simpson RJ, Adams PD, Golemis EA (eds) Basic methods in protein purification and analysis. Cold Spring Harbor Laboratory Press, New York, pp 39–78
29. Towbin H, Gordon J (1984) Immunoblotting and dot immunobinding—current status and outlook. J Immunol Methods 72:313–340
30. DenHollander N, Befus D (1989) Loss of antigens from immunoblotting membranes. J Immunol Methods 122:129–135

Chapter 5

Native Polyacrylamide Gels

Claudia Arndt, Stefanie Koristka, Holger Bartsch, and Michael Bachmann

Abstract

Usually proteins are separated by polyacrylamide gel electrophoresis (PAGE) in the presence of a detergent and under (heat-) denaturing and (non- or) reducing conditions. The most commonly used detergent is sodium dodecyl sulfate (SDS). The major function of SDS is to shield the respective charge of the proteins present in the mixture to be analyzed and to provide all proteins with a negative charge. As a consequence, the proteins will be separated according to their molecular weight. Electrophoresis of proteins can also be performed in the absence of SDS. Using such "native" conditions, the charge of each of the proteins, which will depend on the primary amino acid sequence of the protein (isoelectric point) and the pH during electrophoresis, will mainly influence the mobility of the respective protein during electrophoresis. Here we describe a starting protocol for "native" PAGE.

Key words: "Native" polyacrylamide gels, "Blue native" polyacrylamide gels, Proteins

1. Introduction

A first hallmark in protein electrophoresis was the introduction of sodium dodecyl sulfate (SDS) (1, 2). In the presence of SDS, the individual charge of a protein is no more relevant for its mobility in a polyacrylamide gel. The detergent forms micelles around the protein molecule. The surface of the micelle is negatively charged. The size of the protein determines the size of the micelle and thereby the mobility. Similar to gel filtration the separation of a sample during electrophoresis depends on the starting volume. During electrophoresis this initial protein "band" gets broader. Thus, the "sharper" the protein band at the beginning the sharper the band after electrophoresis. Therefore, a second hallmark in protein electrophoresis was the introduction of discontinuous gels. A discontinuous gel consists of a stacking gel and a separation gel. The function of the stacking gel is to concentrate the protein sample at the beginning of the electrophoresis. For this purpose, discontinuous electrophoresis systems

Biji T. Kurien and R. Hal Scofield (eds.), *Protein Electrophoresis: Methods and Protocols*, Methods in Molecular Biology, vol. 869,
DOI 10.1007/978-1-61779-821-4_5, © Springer Science+Business Media, LLC 2012

use two different buffer systems. At the beginning of the electrophoresis, the pH in the stacking gel is adjusted in a way that it results in an acceleration of the proteins present in the sample. This is mainly due to the fact that the proteins are the only charged ions in the sample at the pH in the stacking gel. During further electrophoresis, the pH in the gel changes [usually from an acidic pH around 6 to the pH of the electrophoresis buffer and the separation gel (usually around pH 8)]. At this pH, amino acids (such as glycine) which are not charged at the beginning of the electrophoresis will now become charged. As a result, the electric field breaks down and during further electrophoresis the proteins are now separated according to their individual mobility. Commonly the proteins will be heat denatured prior to electrophoresis to separate complexes. In parallel, thiol group containing chemicals such as β-mercaptoethanol or dithiothreitol (DTT) can be added resulting in the reduction of disulfide bridges if present either within the protein or between protein subunits. In the latter case, subunits of complexes will be separated. All these developments and modifications dramatically improved the resolution and separation of proteins during electrophoresis. The major obvious disadvantage of SDS–polyacrylamide gel electrophoresis (PAGE) is that the separated proteins will be heat denatured and complexed by the detergent SDS.

Thus, under certain circumstances electrophoresis using nondenaturing conditions may be of interest: For example, it is possible to recover proteins in their native state after the separation. However, such "native" gel electrophoresis attempts usually are limited due to the lack of all the above described advantages of SDS–PAGE. Especially protein isoforms with only slightly different isoelectric points will not efficiently be separated. Posttranslational modifications such as a varying phosphorylation pattern, methylation, acetylation, various glycosylations will more or less alter the mobility of the protein and influence its separation. In contrast to simple and robust SDS–PAGE protocols, the use of native gel electrophoresis usually requires an optimization of the separation conditions for the respective sample. Although discontinuous native gel electrophoresis systems has also been described for native gel electrophoresis (3) and the use of gradient gels may also be helpful to improve the separation, the easiest way to establish a native gel electrophoresis approach is to start by using a continuous PAGE protocol consisting of a separation gel which can be prepared by just leaving the SDS in the gel, the sample solution, and the gel electrophoresis buffer. Such a system can already result in a good separation (see Fig. 1). As the stacking gel is missing, the starting volume of the sample should be as small as possible and the protein concentration as high as possible. Aside such simple native gel electrophoresis protocols a protocol termed as "blue native" PAGE has also been described. Blue native electrophoresis is based on the ability of Coomassie Brilliant Blue to form stable complexes with proteins. The resulting complexes are negatively charged. Thus, in

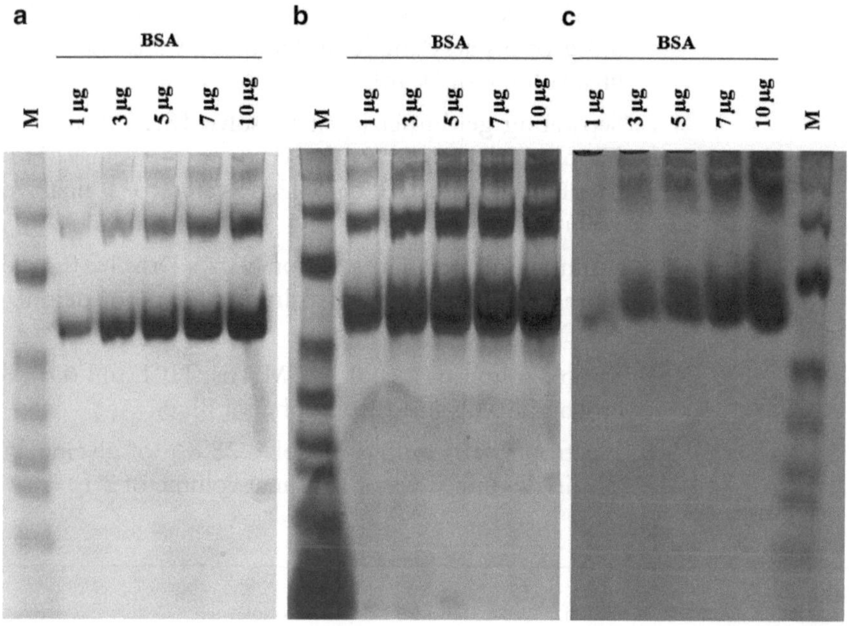

Fig. 1. One dimensional eletrophoresis of BSA. (**a**, **b**) Separation of the BSA sample was performed using native polyacrylamide gels in the absence of SDS. After gel electrophoresis the polyacrylamide gel was stained with Coomassie brilliant blue and documented (**a**). After documentation the same gel was stained using the silver staining technique (**b**). Alternatively BSA was stained with Coomassie brliiant blue prior to the separation by gel electrophoresis. This technique is known as "blue native polyacrylamide gel electrophoresis". The result of this technique is shown in (**c**). Please note: Originally the protein bands in the images a and c were stained with Coomassie brilliant blue and presented originally in blue. In the printed black and white version of these originally blue coloured images the protein bands appear in black.

the case of blue native gel electrophoresis, Coomassie Brilliant Blue plays a similar role as SDS in case of SDS–PAGE. Although Coomassie Brilliant Blue seems to be less critical compared to SDS, protein complexes may also be disrupted by Coomassie Brilliant Blue. Here we describe simple protocols for a native and a blue native PAGE which can be used as an experimental starting protocol for further improvements according to the respective requirements. It should be mentioned that all downstream processing steps as staining of the gels with Coomassie Brilliant Blue, silver, transfer by electroblotting, etc. can be performed as described for SDS–PAGE.

2. Materials

Coomassie Brilliant Blue G-250 solution.

1. Dissolve 2.5 mg of Coomassie Brilliant Blue in 0.45 mL methanol.
2. Add 0.45 mL of deionized water (see Note 1).
3. Add 0.1 mL of glacial acetic acid.

4. Acrylamide solution (40%): Dissolve 20 g of acrylamide and 5.3 g of bis-acrylamide in deionized water (see Note 1) to a final volume of 50 mL.
5. Separating gel buffer (4×): Dissolve 18.15 g of Tris in about 75 mL of deionized water (see Note 1). Adjust pH with HCl to pH 8.8 and add deionized water to a final volume of 100 mL.
6. Ammonium persulfate solution (10%): Dissolve 0.1 g of ammonium persulfate in a final volume of 1 ml of deionized water (see Note 2).
7. Sample solution (2×): 0.187 M Tris/HCl (pH 6.8), 30% glycerol and 80 μg/mL Bromophenol Blue.
8. Electrophoresis buffer: Dissolve 28.8 g of glycine and 6 g of Tris in deionized water to a final volume of 2 L.

3. Methods

3.1. Preparation of a Continuous (10%) Polyacrylamide Gel for Native and Blue Native Gel Electrophoresis

1. Thoroughly clean and dry glass plates, suitable spacers, and comb. Assemble glass plates, spacers, and the comb as described by the manufacturer.
2. Mix 2.5 mL of acrylamide solution (40%), 2.5 mL of separating gel buffer (4×), and 5 mL of deionized water (see Note 3).
3. Degas the mixture to avoid air bubbles in the gel after polymerization and to remove oxygen in the gel solution which otherwise accelerates the polymerization process.
4. Add 50 μL of ammonium persulfate solution (10%) and 10 μL of TEMED which will start the polymerization process.
5. Pour the mixed solution between the glass plates and add the comb.
6. Polymerize the acrylamide for 1 h.
7. Remove the comb carefully. The gel is ready to use.

3.2. Sample Preparation

3.2.1. Sample Preparation for Native PAGE

1. For native gel electrophoresis, mix 10 μl of protein solution with 10 μl sample solution (2×) containing glycerol and the dye Bromophenol Blue.

3.2.2. Sample Preparation for Blue Native PAGE

1. For blue native gel electrophoresis, mix 10 μL of protein solution in sample solution (2×) containing glycerol and the dye Bromophenol Blue with 10 μL of Coomassie Brilliant Blue solution.

3.2.3. Marker Sample

1. To follow the electrophoresis, you can run in parallel prestained marker proteins as commonly used for SDS–PAGE.

3.3. (Blue) Native PAGE Conditions

1. Fill apparatus with gel electrophoresis buffer.
2. Start electrophoresis immediately. [For a gel of 1 mm thickness and 15 cm length, apply about 150 V (constant voltage) which will result in about 20 mA of current] (see Notes 4 and 5).
3. Remove the gel from between the glass plates.
4. After native gel electrophoresis, stain the gel with either Coomassie Brilliant Blue (Fig. 1a) or silver (Fig. 1b).
5. Destain blue native gel with deionized water (the results of blue native gel electrophoresis is shown in Fig. 1c) (see Note 6).

4. Notes

1. Unless stated otherwise, all solutions should be prepared in water that has a resistance greater than 18 MΩ and total organic content of less than five parts per billion. This standard is referred to as "water" in this text.
2. Prepare this solution freshly before preparation of the gel.
3. If necessary, adjust the volume of the gel solution according to your gel equipment.
4. Current will fall during electrophoresis (when constant voltage is applied).
5. The Bromophenol Blue dye front takes about 3 h to reach the bottom of the gel. Greater voltage speeds up electrophoresis, but generates more heat in the gel.
6. Native bovine serum albumin has a heart-like structure, that is stabilized by a series of disulfide bridges. BSA can partially unfold and the different forms can be separated by native PAGE (3, 4).

Acknowledgments

Special thanks for excellent technical support to Christin Gräfe, Livia Schulze, und Barbara Uteß (Inst. Immunology, Dresden, Germany).

References

1. Laemmli UK (1970) Cleavage of structural proteins during the assembly of the head of bacteriophage T4. Nature 227:680–685
2. Studier FW (1973) Analysis of bacteriophage T7 early RNAs and proteins on slab gels. J Mol Biol 79:237–248
3. Niepmann M, Zheng J (2006) Discontinuous native protein gel electrophoresis. Electrophoresis 27:3949–3951
4. He XM, Carter DC (1992) Atomic structure and chemistry of human serum albumin. Nature 358:209–215

Chapter 6

Cationic Electrophoresis

Engelbert Buxbaum

Abstract

Denaturing, discontinuous electrophoresis in the presence of SDS has become a standard method for the protein scientist. However, there are situations where this method produces suboptimal results. In these cases, electrophoresis in the presence of positively charged detergents such as cetyltrimethylammonium bromide (CTAB) may work considerably better. Methods for electrophoresis and staining of such gels are presented.

Key words: Disk electrophoresis, Cationic electrophoresis, CTAB electrophoresis, Detergent, CTAB

1. Introduction

Sodium dodecyl sulfate polyacrylamide gel electrophoresis (SDS-PAGE) (1) and western blotting (2, 3) have become indispensable in the protein laboratory to separate and detect proteins with high resolution. Separation is usually performed on an analytical scale, but preparative equipment is available commercially.

With most proteins the R_f value in SDS-PAGE is proportional to size because proteins bind about 1.4 g SDS per gram of protein, equivalent to one molecule of SDS per three amino acids (4). The negative charges introduced by the detergent far outweigh the charges on the protein itself, thus the charge/mass ratio—and hence the acceleration in an electrical field—is identical for all proteins. The restriction by the gel matrix, however, increases with the size of the protein.

There are, however, some situations where SDS-PAGE performs less well:

- Very hydrophobic proteins (i.e., transmembrane proteins) bind more than the usual 1.4 g SDS per gram of protein, increasing the charge/mass ratio of the protein. Thus, these proteins run faster in SDS-PAGE than expected for their molecular mass.

Biji T. Kurien and R. Hal Scofield (eds.), *Protein Electrophoresis: Methods and Protocols*, Methods in Molecular Biology, vol. 869,
DOI 10.1007/978-1-61779-821-4_6,

- Proteins with a large number of charged amino acids (e.g., histones (5)) also run faster than expected for their molecular mass.
- Glycoproteins contain a highly variable number of negative charges in their sugar side chains. Thus, they do not run as crisp bands in SDS-PAGE, but as broad smears, reducing the achievable resolution of the method.

In the latter two cases, it should be possible to circumvent the problems by using a positively charged detergent in lieu of the negatively charged SDS. Of course, the proteins then run from the positive to the negative electrode. Such attempts have been reported several times in the literature (6–12), but resolution of the gels was usually low, caused by relatively broad protein bands.

The high resolution of SDS-PAGE is the result of band stacking in a discontinuous (multiphasic) buffer system, an effect first described by Ornstein (13) and later theoretically elaborated by Jovin and others (14–18). While the protein moves through the stacking gel, it is electrophoretically concentrated from below 1 mg/mL (as present in usual samples) to several 100 mg/mL. Since the amount of protein cannot change due to mass conservation, the only way in which this can happen is by reducing the volume of the protein band, i.e., its height. Changing from an anionic to a cationic detergent requires a change in buffer composition so that stacking is still possible. Suitable buffers fulfilling the theoretical requirements for stacking are listed on http://www.buffers.nichd.nih.gov. With such selected buffer systems high-resolution electrophoresis of proteins in cationic detergents is possible (19). As with SDS-PAGE, separation in CTAB (cetyl-tri-methyl-am-monium-bromide)–PAGE is based on protein size, as noted also by others (6, 7, 10, 11).

An additional advantage of CTAB compared to SDS is that it efficiently solubilizes membrane proteins, often without damaging their structure (8, 19–22). Thus, one can use the same detergent for electrophoresis that is also used for solubilization and purification, an advantage since extraneous detergent can interfere in PAGE.

2. Materials

All chemicals were of the highest purity available and were obtained mostly from Fluka (Buchs, Switzerland). Antibodies were from Accurate (Westbury, NY). Water came from a Milli-Q system (Millipore, Billerica, MA).

2.1. Casting of CTAB Gels

1. 40% Acrylamide/Bis (37:1): 1.08 g bisacrylamide and 38.9 g acrylamide made to 100 mL with water (see Note 1). Stable for months at 4°C, especially when stored over an anion exchanger.
2. 40% Acrylamide/Bis (19:1): 2.11 g bisacrylamide and 37.9 g acrylamide made to 100 mL with water. Stable for months at 4°C, especially when stored over an anion exchanger.
3. Potassium hydroxide (KOH; 1 M): 5.611 g KOH made to 100 mL with water. Stable at room temperature (RT), if protected from air.
4. 16.6 M acetic acid (commercial 99.5% glacial acetic acid): Stable at RT. Exact molarity is determined once by titration and noted onto the bottle.
5. 10% CTAB: 10 g CTAB made to 100 mL with water. Store at 37°C to increase solubility.
6. Malachite green (1%): 10 mg/mL malachite green in water. Stable at 4°C.
7. Water saturated butanol: *n*-butanol shaken with some water, after phase separation the upper, organic phase is used. Stable at RT.

2.2. For Photo-Polymerization

1. 100 mM Methylene blue: 780 mg methylene blue made to 20 mL with water. Stable for months at 4°C.
2. 100 mM Sodium toluene 4-sulfinate (T4S): 356 mg T4S (anhydrous) made to 20 mL with water. Stable for months at 4°C.
3. 1 mM Diphenyl iodonium chloride (DPIC): 6.3 mg DPIC made to 20 mL with water. Stable for 1 week at 4°C.

2.3. For Fenton System

1. 10 mM Ferrous sulfate ($FeSO_4$): 27.8 mg $FeSO_4 \times 7H_2O$ made to 10 mL with water, make fresh daily.
2. 40 mM Ascorbic acid: 70.5 mg ascorbic acid made to 10 mL with water, make fresh daily.
3. 30% Hydrogen peroxide: commercially available, store at 4°C (see Note 2).

The recipes for both photo-polymerization (5) and a Fenton system (8) are given, obviously only one needs to be prepared.

2.4. Running of CTAB Gels

1. Upper tank buffer: 40 mM (3.56 g/L) β-alanine, 70 mM (2.29 mL/L) acetic acid, 0.1% CTAB. Make fresh each time.
2. Lower tank buffer: 50 mM KOH, 187 mM (3.18 mL/L) acetic acid, 0.1% CTAB. Make fresh each time.

3. Sample buffer (2×): 1.27 mL 1 M KOH (127 mM final), 107 μL acetic acid (187 mM final), 2 mL 10% CTAB (2% final), 100 μL β-mercaptoethanol (β-ME) (1% final), 1 mL glycerol (10% final), 7.21 g urea (12 M final), and 50 μL 1% basic fuchsin (0.005% final), make to 10 mL with water. Stable at RT for a week.

2.5. Staining of CTAB Gels with Ponceau S

1. Fixative: 100 mL glacial acetic acid and 400 mL methanol made to 1 L with water. Stable at RT.
2. Ponceau S solution: 0.1 g Ponceau S and 10 mL glacial acetic acid made to 1 L with water. The solution is stable at RT and may be reused several times.

3. Methods

3.1. Casting of CTAB Gels

1. Mixing table for CTAB–PAGE gels (all volumes in mL). Note that either the reagents for photopolymerization *or* those for the Fenton system should be used.

Solution	5%	7.5%	10%	12.5%	15%	17.5%	20%	Stack
Acrylamide 19:1	–	–	–	–	–	–	–	3.0
Acrylamide 37:1	7.5	11.3	15.0	18.8	22.5	26.3	30.0	–
1 M KOH	2.6	2.6	2.6	2.6	2.6	2.6	2.6.	1.91
Glacial acetic acid	0.962	0.962	0.962	0.962	0.962	0.962	0.962	0.161
Urea	10.8 g	10.8 g	10.8 g	10.8 g	10.8 g	10.8 g	10.8 g	5.4 g
10% CTAB	0.6	0.6	0.6	0.6	0.6	0.6	0.6	0.3
100 mM T4S	0.3	0.3	0.3	0.3	0.3	0.3	0.3	0.15
1 mM DPIC	1.5	1.5	1.5	1.5	1.5	1.5	1.5	0.75
10 mM $FeSO_4$	0.048	0.048	0.048	0.048	0.048	0.048	0.048	0.024
40 mM ascorbate	6.0	6.0	6.0	6.0	6.0	6.0	6.0	3.0
Water to	60	60	60	60	60	60	60	30

2. Pour the mixture, containing either the reagents for photo-crosslinking or the Fenton system, into an Erlenmeyer flask with magnetic stirrer. The stirrer is adjusted for vigorous movement without foam production. Then apply a vacuum for at least 10 min to remove dissolved oxygen. This is essential as oxygen inhibits the polymerization reaction. Do not apply the vacuum before starting the stirrer.
3. Add the catalyst: 18 μL 10 mM methylene blue or 2 μL hydrogen peroxide depending on whether the photo-polymerization or the Fenton system is used. Mix by gentle inversion; do not

shake oxygen into the solution again. At this stage, do *not* add the catalyst to the stacking gel.

4. Pour the separating gel into the casting sandwich and overlay with water-saturated *n*-butanol. The Fenton system will polymerize by itself, the photo-polymerization system requires exposure to a strong source of white light. A sun-exposed window or a halogen lamp may be used. Polymerization is finished when the interface between gel and butanol becomes prominent. Methylene blue will become colorless during polymerization. Normally polymerization should be complete within 15 min.
5. Pour the butanol from the top of the gel, add catalyst to the stacking gel mixture and cast the stacking gel. Insert combs immediately and allow the gel to polymerize. Leave the gel in the fridge overnight (humid chamber), so that reactive intermediates of the polymerization chemistry can disintegrate.

3.2. Running of CTAB Gels

1. Mount the gel in the running chamber according to manufacturers instruction, add upper and lower tank buffer. Sometimes air bubbles get trapped in the wells, these can be rinsed out using a tuberculin syringe with 27-G needle. This process and the loading of samples are aided by malachite green in the stacking gel, which makes the wells easier to see.
2. Load the sample with a 25-μL Hamilton syringe; the needle should have a flat point. Between samples rinse the syringe with upper tank buffer, finally with water.
3. For a standard minigel, electrophoresis is performed at 20 mA per gel (10 mA during stacking), with a maximum voltage of 200 V. Do not forget to *reverse the electrode polarity* compared to SDS-PAGE (see Notes 3–8).

3.3. Staining of CTAB Gels with Ponceau S

1. Fix the gel for 2× 10–15 min on an orbital shaker, then replace the fixative with dye solution for 5 min. The gel can be differentiated by incubation with several changes of fixative (in the same way as is commonly done with CBB-R250), but this procedure is time consuming (see Notes 9–17).

4. Notes

1. Acrylamide is neurotoxic, handle with great care. Ready-made acrylamide/bisacrylamide solutions are commercially available; these avoid the development of dust during weighing. Store the solution over an anion exchanger to remove any acrylic acid which may form to prevent electroendosmosis during runs.

2. Store hydrogen peroxide solutions in the bottles supplied by the manufacturer which have release valves to prevent the build-up of pressure from decomposition.
3. Fenton-polymerization results in gels which are somewhat more brittle than those produced by photo-polymerization. However, if the Hoefer multi-casting stand is used, the non-transparent aluminia back-plates prevent the use of a photo-polymerization system. In that case, the Fenton system must be used. The separation achievable does not depend on the polymerization reaction used, however.
4. Both systems are even more sensitive to the presence of oxygen than the TEMED/APS system used for Laemmli gels. Proper degassing of the gel mixture is essential. In addition, it is important that the combs prevent access of air to the polymerizing stacking gel (19).
5. Gel concentration depends on the molecular mass range of the proteins of interest (23, 24). I have found 5–15% gradient gels most convenient for proteins of 10–200 kDa. If the 10-gel multicaster is used, the gradient can be easily formed with a gradient maker, but for single gels the volume required is too small. In that case, cast a step gradient by mixing:

% desired	5% (μL)	15% (μL)
15.0	0	392
14.0	36	356
13.1	71	320
12.3	107	285
11.3	142	249
10.5	178	214
9.5	214	178
8.6	249	142
7.7	285	107
6.8	320	71
5.9	356	36
5.0	392	0

These solutions are carefully layered on top of each other; this is made easier when the heavy solution contains 10% glycerol (which does not otherwise interfere with electrophoresis). The total volume of that gradient is 2.35 mL, enough for one mini-gel. With the photo-polymerization chemistry you can start polymerization once the gel is cast.

6. Proteins with very high molecular mass (>200 kDa) require low acrylamide concentration that result in very soft, difficult to handle gels. Add 0.5% agarose to stabilize them without any effect on separation.
7. CTAB has a relatively high Krafft point and precipitates if the temperature drops below 18°C. If lower temperatures are desired during electrophoresis consider replacing CTAB with a detergent with lower cmc and/or krafft point like 16-BAC. Note that the stacking properties of buffer systems are temperature-dependent (on http://www.buffers.nichd.nih.gov systems for use at room temperature and at 0°C are available).
8. Prestained molecular weight markers, available for SDS-PAGE from several manufacturers, are unsuitable for CTAB–PAGE even after detergent exchange. The bound dye influences the R_f value and separation is no longer according to molecular mass. It can only be hoped that such standards will also become available for CTAB–PAGE.
9. The staining of gels with Ponceau S is sensitive and fast, but staining with CBB-R250 is also possible. Phenol red can be used in the same way as Ponceau S, the same is probably true for other acidic dyes. Phenol red is fluorescent under acidic conditions, making very sensitive detection of proteins possible.
10. Alternatively destain gels by putting them between filter paper (three sheets of Whatman No. 3 on both sides of the gel) and place it in a tank blotter with blotting buffer. 30 min at 40 V (~200 mA) removes the background stain, while the protein/stain complex is immobile. If need be, proteins become mobile again after incubating the gel in 1 mM KOH for 30 min.
11. Silver staining of CTAB gels is possible, the method of Heukeshoven and Dernick (25) achieves a higher sensitivity than that of Merril et al. (26). You can execute the former method at a constant temperature of 37°C for all steps; replace the glutaraldehyde with formaldehyde for maximum sensitivity.
12. Blotting of CTAB gels ("eastern blotting") is possible with 10 mM KOH and 11 mM acetic acid as buffer. The procedure is similar to western blotting, except for the reversed polarity (19). If the blots are to be immuno-stained, incubate them first with 0.05% SDS in methanol, to prevent nonspecific binding of antibodies to CTAB/protein complexes.
13. After careful removal of CTAB with several changes of fixative, gels may also be stained with "stains all" (3,3′diethyl-9-methyl-4,5,4′,5′-dibenzothiacarbocyanine (27)). This is interesting as different types of protein (acidic, phosphor- and Ca^{2+}-binding) give different colors with "stains all," and because this dye increases the sensitivity of silver staining.

14. PAS-staining of CTAB gels (28) works, but considerable savings in time and chemicals are possible if the staining is performed on eastern blots rather than on gels (29).
15. Gels can also be stained with fluorescent transition metal complexes like "dye hard" (RuBPS). See ref. (30) for details.
16. CTAB is a very mild detergent, which can retain the enzymatic activity of enzymes solubilized with it. Try zymograms (11), proteins in gels specifically stained by their enzymatic activity, at least with monomeric enzymes.
17. Gels can be dried after incubation with 1% glycerol between two sheets of cellophane.

Acknowledgments

This work was supported in part by Kuwait University grant MPB029 and by Ross University.

References

1. Laemmli U (1970) Cleavage of structural proteins during the assembly of the head of bacteriophage T4. Nature 227:680–685
2. Towbin H, Staehelin T, Gordon J (1979) Electrophoretic transfer of proteins from polyacrylamide gels to nitrocellulose sheets: procedure and some applications. Proc Natl Acad Sci USA 76:4350–4354
3. Dunn S (1986) Effects of the modification of transfer buffer composition and the renaturation of proteins in gels on the recognition of proteins on western blots by monoclonal antibodies. Anal Biochem 157:144–153
4. Shirahama K, Tsujii K, Takagi T (1974) Free boundary electrophoresis of sodium dodecyl sulphate-protein polypeptide complexes with special reference to SDS-polyacrylamide gel electrophoresis. J Biochem 75:309–319
5. Rabilloud T, Girardot V, Lawrence J-J (1996) One-and two dimensional histone separations in acidic gels: usefulness of methylene blue-driven photopolymerisation. Electrophoresis 17:67–73
6. Panyim S, Thitipongpanich R, Supatimusro D (1977) A simplified gel electrophoretic system and its validity for molecular weight determinations of protein-cetyltrimethylammonium complexes. Anal Biochem 81:320–327
7. Eley M, Burns P, Kannapell C et al (1979) Cetyltrimethylammonium bromide polyacrylamide gel electrophoresis: estimation of protein subunit molecular weights using cationic detergents. Anal Biochem 92:411–419
8. MacFarlane D (1983) Use of benzyldimethyl-*n*-hexadecylammonium chloride ("16-BAC"), a cationic detergent, in an acidic polyacrylamide gel electrophoresis system to detect base labile protein methylation in intact cells. Anal Biochem 132:231–235
9. Mócz G, Bálint M (1984) Use of cationic detergents for polyacrylamide gel electrophoresis in multiphasic buffer systems. Anal Biochem 143:283–292
10. Akin D, Shapira R, Kinkade J Jr (1985) The determination of molecular weights of biologically active proteins by cetyltrimethylammonium bromide-polyacrylamide gel electrophoresis. Anal Biochem 145:170–176
11. Akins R, Levin P, Tuan R (1992) Cetyltrimethylammonium bromide discontinuous electrophoresis: Mr-based separation of proteins with retention of enzymatic activity. Anal Biochem 202:172–178
12. Hartinger J, Stenius K, Högemann D, Jahn R (1996) 16-BAC/SDS-PAGE: a two-dimensional gel electrophoresis system suitable for the separation of integral membrane proteins. Anal Biochem 240:126–133
13. Ornstein L (1962) Disk electrophoresis: I. Background and theory. Ann N Y Acad Sci 121:321–351
14. Jovin T (1973) Multiphasic zone electrophoresis. I. Steady-state moving-boundary systems formed by different electrolyte combinations. Biochemistry 12:871–879

15. Jovin T (1973) Multiphasic zone electrophoresis. II. Design of integrated discontinuous buffer systems for analytical and preparative fractionation. Biochemistry 12:879–890
16. Jovin T (1973) Multiphasic zone electrophoresis. III. Further analysis and new forms of discontinuous buffer systems. Biochemistry 12: 890–898
17. Jovin T (1973) Multiphasic zone electrophoresis. IV design and analysis of discontinuous buffer systems with a digital computer. Ann N Y Acad Sci 209:477–496
18. Chrambach A (1985) The practice of quantitative gel electrophoresis. VCH, Weinheim
19. Buxbaum E (2003) Cationic electrophoresis and electrotransfer of membrane glycoproteins. Anal Biochem 314:70–76
20. Pritchard D, Crawford C, Duce I et al (1985) Antigen stripping from the nematode epicuticle using the cationic detergent cetyltrimethylammonium bromide (CTAB). Parasite Immunol 7:575–585
21. Freedman D, Nutman T, Ottesen E (1988) Enhanced solubilization of immunoreactive proteins from *Brugia malayi* adult parasites using cetyltrimethylammonium bromide. Exp Parasitol 65:244–250
22. Maki K, Sagara J, Kawai A (1991) A cationic detergent, cetyltrimethylammonium bromide (CTAB), selectively dissociates the intermediate filament of the fibroblast. Biochem Biophys Res Commun 175:768–774
23. Rodbard D, Chrambach A (1971) Estimation of molecular radius, free mobility, and valence using polyacrylamide gel electrophoresis. Anal Biochem 40:95–134
24. Stellwagen N (1998) Apparent pore size of polyacrylamide gels. Electrophoresis 19:1542–1547
25. Heukeshoven J, Dernick R (1988) Improved silver staining procedure for fast staining in phastsystem development unit I. Staining of sodium dodecyl sulfate gels. Electrophoresis 9:28–32
26. Merril C, Goldman D, Sedman S, Ebert M (1981) Ultrasensitive stain for proteins in polyacrylamide gels shows regional variations in cerebrospinal fluid proteins. Science 211:1437–1438
27. Goldberg HA, Warner KJ (1997) The staining of acidic proteins on polyacrylamide gels: enhanced sensitivity and stability of "stains all" staining in combination with silver nitrate. Anal Biochem 251:227–233
28. Segrest J, Jackson R (1972) Molecular weight determination of glycoproteins by polyacrylamide gel electrophoresis in sodium dodecyl sulfate. Methods Enzymol 28:54–63
29. Thornton D, Holmes D, Sheehan J et al (1989) Quantitation of mucus glycoproteins blotted onto nitrocellulose membranes. Anal Biochem 182:160–164
30. Buxbaum E (2012) Fluorescent staining of gels. In: Kurien BT, Hal Scofield R (eds) Protein electrophoresis: methods and protocols. Springer, New York

Chapter 7

SARCOSYL-PAGE: A New Electrophoretic Method for the Separation and Immunological Detection of PEGylated Proteins

Christian Reichel

Abstract

PEGylation of recombinant proteins and synthetic peptides aims to generate biopharmaceuticals with altered physical properties. The modification may lead to a prolonged serum half-life caused by a decreased receptor-mediated endocytosis and/or a delay in renal clearance caused by the increased hydrodynamic volume of the pharmaceutical. MIRCERA, a PEGylated recombinant erythropoietin (rhEpo) frequently used in the treatment of anemia due to chronic kidney disease, has been also abused by athletes as a performance-enhancing drug. While it can be detected by sodium dodecyl sulfate polyacrylamide gel electrophoresis (SDS-PAGE) and immunoblotting, the sensitivity of the test is significantly lower compared to other epoetins. By replacing SDS with sarcosyl in the sample and running buffers, the interaction between SDS and the PEG group of the protein no longer reduces the affinity of the monoclonal anti-Epo antibody (*clone AE7A5*) to the protein chain. Contrary to SDS, sarcosyl only binds to the amino acid chain of the PEGylated protein, thus leading to enhanced antibody binding and a sharper electrophoretic band. While the method was originally developed for anti-doping purposes, it may be also useful for other PEGylated proteins and their electrophoretic separation and immunological detection.

Key words: PEGylated proteins, SDS-PAGE, SAR-PAGE, Sarcosyl, Immunoblotting, Erythropoietin (Epo), Doping control

1. Introduction

Due to the prolonged serum half-life of MIRCERA (ca. 130 h) and its decreased excretion in urine, the abuse of MIRCERA by athletes is preferably detected in serum or plasma samples. MIRCERA, a PEGylated epoetin beta, contains one methoxy polyethylene glycol group (ca. 30 kDa) in covalent linkage to either the N-terminal amino group or the ε-amino group of Lys 52 or Lys 45 of the 165 amino acid chain of human recombinant erythropoietin (rhEpo) beta (1). The resulting molecule has an average molecular mass of ca.

Biji T. Kurien and R. Hal Scofield (eds.), *Protein Electrophoresis: Methods and Protocols*, Methods in Molecular Biology, vol. 869, DOI 10.1007/978-1-61779-821-4_7,

60 kDa and was mainly developed for treating anemia-related diseases [e.g., chronic kidney disease (CKD)]. The PEGylation not only leads to an increase in serum half-life but also to a repeated interaction of each molecule with the Epo receptor, since the interaction is weakened. Therefore, MIRCERA is also known as "continuous erythropoietin receptor activator (CERA)." Since erythropoietin stimulates the body's red blood cell (RBC) production, Epo pharmaceuticals have been mostly abused by athletes in endurance sports. The detection of doping with rhEpo—including analogs, biosimilar epoetins, and MIRCERA—has been regulated by the World Anti-Doping Agency (WADA) in a technical document (2). Since the molecular mass of MIRCERA is profoundly different from the mass of endogenous and recombinant Epo (ca. 30 kDa mass difference), sodium dodecyl sulfate polyacrylamide gel electrophoresis (SDS-PAGE) in combination with western blotting appeared to be an ideal detection method. However, it turned out that SDS-PAGE is less sensitive for the PEGylated analog due to the interaction of SDS with both the PEG and amino acid chain of MIRCERA. By replacing SDS with *N*-lauroylsarcosinate (sarcosyl)—an anionic methyl glycine-based detergent—a significant increase in immunoblotting sensitivity was obtained (3). It was shown that sarcosyl is not able to solubilize PEG the way SDS does. Contrary to SDS, sarcosyl does only bind to the protein part of PEGylated Epo, which leads to a sharper electrophoretic band and an enhanced antibody interaction than on SDS-PAGE. The increased sensitivity of SARCOSYL-PAGE (SAR-PAGE) was explained by the inability of the monoclonal anti-Epo antibody to bind to the fully (i.e., PEG and amino acid chain) solubilized PEGylated protein. Instead, the antibody only interacted with molecules with SDS mainly bound to the protein chain. The result was a "pseudo-sharp" band on western blot but with low sensitivity, since the majority of the MIRCERA molecules was fully solubilized by SDS. Since sarcosyl bound only to the protein chain, no interference with the antibody interaction occurred. This explained the appearance of a "truly sharp" *and* high sensitivity band of the PEGylated protein on western blot (Fig. 1).

2. Materials

Use Milli-Q (MQ) water for preparing all buffers and solutions. Chemicals should be analytical or of electrophoresis grade. Wear non-powdered gloves during the entire procedure (see Note 1).

2.1. Immunoaffinity Purification

1. Epo purification kit (MAIIA Diagnostics, Uppsala, Sweden).
2. Serum samples: MIRCERA-negative and -positive control samples, test samples (typically 200 μL).

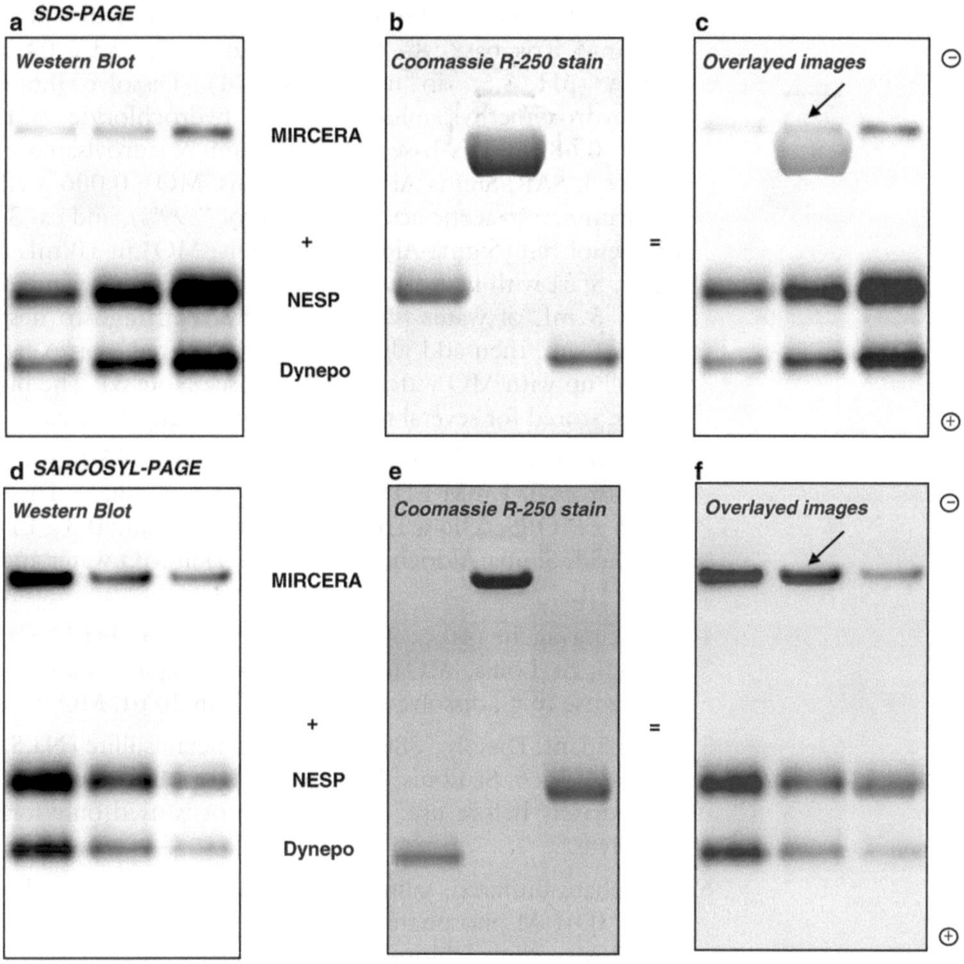

Fig. 1. Relationship between "true band sharpness" on gel and "pseudo-band sharpness" on western blots for MIRCERA. SDS-PAGE (**a–c**) and SAR-PAGE (**d–f**): Immunoblot with monoclonal anti-Epo antibody (*clone AE7A5*), Coomassie R-250 stained identical second gel, and virtual overlay of both images. Note that the majority of the PEGylated protein molecules was not recognized by the antibody on SDS-PAGE (*arrows*). Reproduced from ref. 3 with permission from Wiley.

3. Tris-buffered saline solution (TBS, Sigma-Aldrich, St Louis, MO): Dissolve one pack in MQ water and fill up to 1 L. The final concentration of the buffer is 0.05 M Tris(hydroxymethyl) aminomethane (Tris base), 0.138 M NaCl, 0.0027 M KCl, pH 8.0 at 25°C.
4. Centrifugal microfilters (0.2 μm; Nanosep MF, PALL, Ann Arbor, MI).
5. QIAvac 24 Plus system with vacuum pump, connecting system, and VacValves (QIAGEN, Hilden, Germany).
6. Microcentrifuge for 1.5-mL sample tubes.

2.2. SARCOSYL-PAGE

1. Polyacrylamide gels: Precast Bis–Tris gels (NuPAGE, Invitrogen, Carlsbad, CA), e.g., 10% T, 1.5 mm, 10 wells (see Note 2).

2. Sarcosyl (SAR) sample buffer (4×): 424 mM Tris–hydrochloride, 564 mM Tris base, 8% SAR, 40% glycerol, and 2.04 mM EDTA (pH 8.5; do not adjust) (4). Dissolve 0.666 g Tris(hydroxymethyl)aminomethane hydrochloride (Tris–HCl), 0.682 g Tris base, 0.8 g sodium *N*-lauroylsarcosinate (sarcosyl, SAR, Sigma-Aldrich, St Louis, MO), 0.006 g ethylenediaminetetraacetic acid, 4 g glycerol (≥99%), and ca. 3 mg of phenol red (Sigma-Aldrich, St Louis, MO) in 10 mL MQ water. Start with first dissolving Tris–HCl, Tris base, and SAR in ca. 5 mL of water (vortex), briefly centrifuge to destroy excess foam, then add glycerol and phenol red, vortex again, and fill up with MQ water to 10 mL (see Note 3). The buffer can be stored for several months at room temperature.
3. Sarcosyl (SAR) running buffer: 50 mM MOPS, 50 mM Tris, 0.1% sarcosyl, 1 mM EDTA (pH 7.7; do not adjust). Dissolve 10.46 g MOPS, 6.06 g Tris base, 1 g sarcosyl, and 0.3 g EDTA (free acid, Sigma-Aldrich, St Louis, MO) in MQ water and fill up to 1 L.
4. Reducing agent (40×): 4 M DL-dithiothreitol (DTT, Sigma-Aldrich, St Louis, MO) in MQ water. Prepare immediately before use (e.g., dissolve 12.3 mg DTT in 20 μL MQ water).
5. Antioxidant: Dissolve 380 mg sodium metabisulfite ($Na_2S_2O_5$, Sigma-Aldrich, St Louis, MO) in 1 mL of MQ water. Prepare immediately before use. The antioxidant is used only for the catholyte.
6. Phosphate-buffered saline (PBS, Sigma-Aldrich, St Louis, MO): 0.01 M phosphate, 0.0027 M potassium chloride, and 0.137 M sodium chloride. Dissolve five PBS tablets in MQ water and fill up to 1 L (pH 7.4 at 25°C).
7. Dilution buffer: 0.05% (w/v) bovine serum albumin (ELISA grade, Sigma-Aldrich, St Louis, MO). Dissolve BSA in PBS.
8. PEGylated protein standard: MIRCERA (50 μg/0.3 mL, Roche, Mannheim, Germany). For sensitivity testing prepare a serial dilution down to low pg/medium fg level. Dilute with 0.05% BSA/PBS (see Note 4).
9. Recombinant human Epo standards: e.g., Erypo (Janssen-Cilag, Vienna, Austria), NeoRecormon (Roche, Mannheim, Germany), Dynepo (Shire, Hampshire, UK), Aranesp (NESP, Amgen, Thousand Oaks, CA). For sensitivity testing prepare a serial dilution down to medium/low fg level (absolute amount on gel). Dilute with 0.05% BSA/PBS.
10. Thermomixer comfort with heating block for 1.5 mL sample tubes (Eppendorf, Hamburg, Germany).
11. Sample tubes (1.5 mL, Eppendorf, Hamburg, Germany).
12. GELoader tips (Eppendorf, Hamburg, Germany).
13. Electrophoresis cell (Xcell SureLock Mini-Cell, Invitrogen, Carlsbad, CA) (see Note 5).

14. Power supply (e.g., PowerEase 500, Invitrogen, Carlsbad, CA).
15. Gel knife (Invitrogen, Carlsbad, CA).

2.3. Immunoblotting

1. Transfer membranes: Immobilon-P (0.45 μm) and Durapore (0.65 μm) polyvinylidene fluoride (PVDF) membranes (Millipore, Billerica, MA).
2. Methanol (Merck, Darmstadt, Germany).
3. Western blot transfer buffer I: 48 mM Tris base, 39 mM glycine, 1.3 mM SDS, and 20% methanol (5). Dissolve 5.81 g Tris base, 2.93 g glycine, 0.375 g sodium dodecyl sulfate (Sigma-Aldrich, St Louis, MO) in MQ water, add 200 mL of methanol and fill up to 1 L with MQ water (see Note 6).
4. Western blot transfer buffer II: 0.7% acetic acid. Dilute 7 mL of glacial acetic acid (Merck, Darmstadt, Germany) with MQ water to 1 L.
5. Reducing buffer: 5 mM DTT. Dissolve 77 mg DTT in 100 mL PBS. Prepare immediately before use.
6. Blocking buffer: 5% (w/v) nonfat milk (NFM). Dissolve 5 g NFM (BioRad, Hercules, CA) in 100 mL PBS.
7. Washing buffer: 0.5% (w/v) NFM in PBS.
8. Incubation buffer: 1% (w/v) NFM in PBS.
9. Incubation vessels: Polypropylene and/or glass containers of appropriate size for performing reducing, blocking, washing, and incubation steps.
10. Primary antibody: Monoclonal mouse anti-Epo antibody (*clone AE7A5*, R&D Systems, Minneapolis, MN).
11. Secondary antibody: Biotinylated polyclonal goat anti-mouse IgG (H + L) antibody (ImmunoPure, Pierce, Rockford, IL).
12. Streptavidin–horseradish peroxidase (HRP) complex (Biospa, Milano, Italy).
13. Blotting paper: Extra thick blot paper (7 × 8.4 cm, BioRad, Hercules, CA).
14. Blotter for semidry transfer: Trans-Blot SD Semi-Dry Electrophoretic Transfer Cell (BioRad, Hercules, CA).
15. Power supply: e.g., EPS 3501 XL with adaptor A2/4 (GE Healthcare, Uppsala, Sweden).
16. Incubator (37°C).
17. Shaker (e.g., Stuart see-saw rocker SSL4, Bibby Scientific, Stone Staffordshire, UK).
18. Tweezers (flat tips).
19. Rubber roller.
20. Parafilm.
21. Sponge for cleaning the blotter.

2.4. Chemiluminescent Detection

1. Enhanced chemiluminescent (ECL) substrate: SuperSignal West Pico or West Femto (Thermo Scientific, Rockford, IL; see Note 7).
2. CCD camera: LAS-4000 (GE Healthcare, Uppsala, Sweden) or equivalent.
3. Glass plate.
4. Ethanol (absolute, Merck, Darmstadt, Germany).
5. Pipette tips (5 mL).
6. Image analysis software, e.g., GASepo *v2.1* (6) (Seibersdorf Laboratories, Seibersdorf, Austria), Multi-Gauge *v3.0* (Fujifilm, Tokyo, Japan), or Aida Biopackage (Raytest, Straubenhardt, Germany).

3. Methods

Due to the high protein content of human serum (ca. 60–80 μg/μL) (7) and the low amount of Epo (ca. 26 pg/mL to 1 ng/mL), serum samples cannot be directly applied on electrophoresis gels for detecting Epo by western blotting. Immunoaffinity purification is the most efficient way to remove the majority of these otherwise interfering proteins (see Note 8). The purified samples are then separated with SAR-PAGE, immunoblotted, and detected via chemiluminescence.

3.1. Immunoaffinity Purification

1. Dilute 200 μL of serum sample with 1,800 μL TBS containing 0.1% detergent aid from the kit. The final serum concentration is 10% (8).
2. Filter the diluted sample through a 0.2-μm centrifugal filter in 500 μL aliquots (14,000 RCF/2 min).
3. Mount one dummy column of the kit on the QIAvac system with the VacValve, close valve, fill in 1 mL of TBS, and adjust flow rate to ca. 0.5 mL/min on the vacuum pump.
4. Prepare desorption buffer by mixing 25 μL of desorption buffer component A (supplied with the kit) with 0.25 μL detergent aid.
5. Remove the dummy column and mount the Epo immunoaffinity purification column on the QIAvac system using the VacValve, close the valve, add 1 mL of washing buffer from the kit to the column, open the valve, and let the buffer pass through at ca. 0.5 mL/min, and close the valve.
6. Add the diluted serum sample (2 × 1 mL) to the column, open the valve and let the first milliliter pass through, then close the valve, refill, and open the valve again. The flow rate should be 0.5 mL/min or lower.

7. Wash the column with 1 mL of washing buffer (ca. 0.5 mL/min).
8. Remove the column from the vacuum system, put it in a 1.5-mL tube, and spin in the microcentrifuge at 1,000 RCF for 1 min.
9. Place the column in a new 1.5-mL tube, add 25 μL of desorption buffer, insert the assembly in the microcentrifuge, and spin again for 1 min at 1,000 RCF.
10. Immediately neutralize the eluate with 2.5 μL of adjustment buffer (provided with the kit).
11. Proceed with SARCOSYL-PAGE or store sample at −20°C or below.

3.2. SARCOSYL-PAGE

1. Mix 25 μL of sample or standard (appropriately diluted) with 9.25 μL of SAR running buffer (4×) and 0.93 μL of reducing agent (40×) in 1.5-mL sample tubes. Fill up with MQ water to 37 μL (see Note 9).
2. Heat samples and standards for 5 min at 95°C under shaking (800 rpm, Thermomixer). Cool down on ice and briefly spin in a centrifuge to remove lid condensate.
3. Take one NuPAGE Bis–Tris gel (10% T, 1.5 mm) out of its plastic bag, remove the white tape on the foot of the mold, remove the comb, and briefly rinse the wells with SAR running buffer.
4. Prepare the catholyte by mixing 200 mL of SAR running buffer with 500 μL of antioxidant.
5. Put the gel into the Xcell SureLock Mini-Cell, insert the buffer dam (or a second gel), and lock the sandwich consisting of gel, buffer core, and buffer dam (or second gel) with the gel tension wedge. This assembly builds the inner buffer chamber holding the catholyte.
6. Fill the catholyte in the inner chamber until the wells of the gel are completely covered with liquid. Check if the inner buffer chamber is leaking.
7. Fill samples and standards with GELoader tips (fits, e.g., on an Eppendorf Reference pipette, 2–20 μL) into the wells of the gel. Each well of a 10-well/1.5 mm gel holds about 37 μL of sample.
8. Fill the anolyte (SAR running buffer without antioxidant) in the outer buffer chamber until ca. 1 cm below the upper edge of the cell (this rather large liquid volume serves as a heat sink). Close the cell with the lid and connect it to the power supply.
9. Run for ca. 55 min at constant voltage (limits per gel: 200 V/120 mA/25 W).
10. Stop electrophoresis as soon as the leading phenol red dye front has reached the bottom of the gel.

3.3. Immunoblotting

1. After electrophoresis, release the gel from the mold using gel knife.
2. Place the gel in a vessel containing western blot transfer buffer I.
3. Equilibrate the gel for 3×5 min in blotting buffer under shaking (change the buffer in between).
4. Cut one sheet of Immobilon-P membrane slightly larger than the size of the blotting paper (ca. 9×7 cm). Activate it in methanol (ca. 10 s), then wash in MQ water (ca. 2 min), and finally equilibrate it in western blot transfer buffer I for ca. 5 min.
5. Soak two sheets of extra thick blotting paper with transfer buffer I while the gel is equilibrating.
6. Clean anode and cathode of the Trans-Blot SD Semi-Dry Transfer Cell first with methanol or ethanol, and then with MQ water.
7. After the final gel equilibration step, assemble the western blot sandwich consisting of one sheet of extra thick blotting paper (roll out air bubbles with a rubber roller; protect surface with parafilm during rolling), the PVDF membrane, the gel, and the second sheet of extra thick blotting paper (roll out air bubbles again; assemble directly on the anode of the blotter). Close the blotter with the cathode and lid. Blotting direction is from cathode to anode.
8. Connect the blotter with the power supply and blot for 60 min at 1.0 mA/cm^2 constant current.
9. After blotting is finished, disassemble the blotting sandwich and place the membrane with the side which was next to the gel facing up in PBS and shake for ca. 5 min. Clean the blotter with a new sponge and hot water, and rinse with MQ water.
10. Transfer the membrane into reducing buffer (5 mM DDT in PBS) and incubate it for 45 min at 37°C (make sure that the membrane does not swim up during the incubation period, e.g., by slow shaking).
11. Wash the membrane in PBS and block it in ca. 100 mL 5% NFM/PBS solution for 60 min at room temperature under slow shaking.
12. Discard the blocking solution, wash the membrane briefly in PBS to remove excess blocking solution, and incubate it in primary antibody solution (1:1,000 dilution in 1% NFM/PBS). Slowly shake for 60 min at room temperature or overnight in a cold room or refrigerator.
13. Wash the membrane 3×10 min in 0.5% NFM/PBS (shake faster than before). Briefly rinse membrane in PBS in order to remove milk foam generated during washing.

14. Perform a western double blot in case you cannot completely rule out that your secondary antibody is binding nonspecifically to proteins other than Epo contained in your sample matrix (steps 15–21). Otherwise, directly proceed with step 22 (see Note 10).
15. Cut another sheet of Immobilon-P membrane (ca. 9×7 cm) and activate it in methanol for ca. 10 s while the first membrane is washing in 0.5% NFM/PBS. Rinse the membrane in MQ water and equilibrate it in western blot transfer buffer II (0.7% acetic acid) for ca. 5 min. Cut also one sheet of Durapore membrane (ca. 9×7 cm) and soak directly in transfer buffer II (see Note 10).
16. Soak two sheets of blotting paper in transfer buffer II for ca. 10–15 min.
17. Clean anode and cathode of the Trans-Blot SD Semi-Dry Transfer Cell first with methanol or ethanol, and then with MQ water or transfer buffer II.
18. Assemble the double-blotting sandwich after rinsing the first membrane with PBS (step 13). The sandwich consists of one sheet of extra thick blotting paper, the first Immobilon-P membrane, the Durapore membrane, the second Immobilon-P membrane, and the second sheet of extra thick blotting paper (assemble the sandwich directly on the anode, roll out air bubbles as described above). Close the blotter with the cathode and lid.
19. Connect the blotter with the power supply and blot for 10 min at 0.8–1.0 mA/cm^2 constant current (see Note 11).
20. Remove the second membrane immediately after the blot and rinse ca. 5 min under shaking in PBS to neutralize the acid. Store the first membrane in PBS in the refrigerator or cold room until image analysis is completed (see Note 12). Clean the blotter with a new sponge and hot water, rinse with MQ water.
21. Block the membrane in 5% NFM/PBS solution for 60 min and under slow shaking at room temperature. Rinse in PBS to remove excess blocking solution.
22. Incubate the membrane in secondary antibody solution (diluted 1:2,000 in 1% NFM/PBS) for 60 min at room temperature and under slow shaking (ca. 20–40 mL of solution depending on vessel size).
23. Wash 3×10 min in 0.5% NFM/PBS (shake faster than before), then rinse briefly in PBS to remove milk foam.
24. Incubate the membrane in a dilution of streptavidin–HRP complex in 1% NFM/PBS (1:2,000; 60 min, room temperature, shake slowly).
25. Wash the membrane in PBS (3×10 min) and continue directly with chemiluminescent detection.

3.4. Chemiluminescent Detection

1. Prepare the substrate working solution by mixing both substrate components 1:1 (ca. 0.1 mL/cm^2).
2. Clean the glass plate and tweezers with both ethanol and MQ water, wipe dry (see Note 13).
3. Take the membrane out of the PBS solution with tweezers, briefly remove excess PBS by holding the membrane in a vertical position and with the lower edge touching a sheet of blotting paper.
4. Place the membrane on the glass plate and transfer the plate into the CCD camera.
5. Distribute the substrate evenly on the membrane using the long side of a 5-mL pipette tip (do not directly touch the surface of the membrane).
6. Adjust membrane position and focusing of the camera. Immediately start acquiring images (either by using the automatic mode of the camera or by manually selecting the exposure time) (see Note 14).
7. Evaluate images using image analysis software. The presence of a band matching the position of the MIRCERA standard and the MIRCERA-positive control sample is indicative of the presence of MIRCERA in the test sample (Fig. 2) (3).

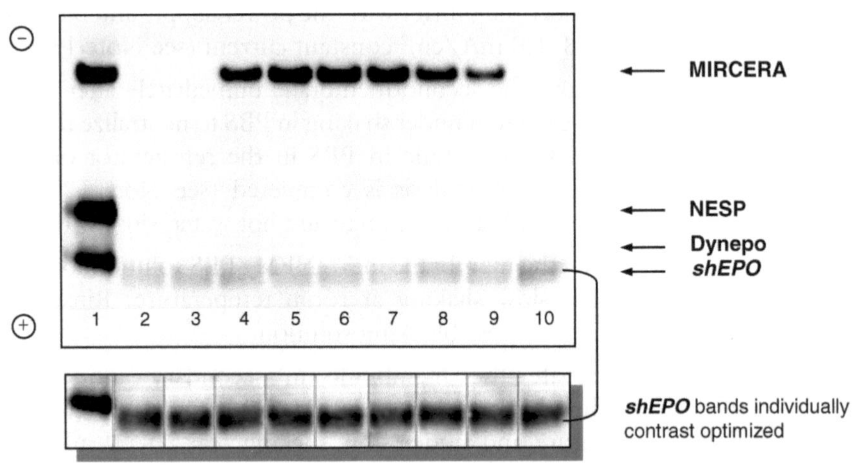

Fig. 2. Detection of MIRCERA in 200 μL of serum by SAR-PAGE. Lane sequence: (*1*) Epo standards (MIRCERA, NESP, Dynepo), (*2–3*) MIRCERA-negative serum samples, (*4–9*) MIRCERA-positive samples, (*10*) negative control serum. Note the lower molecular mass of endogenous serum Epo (shEpo) compared to the other epoetins. Reproduced from ref. 3 with permission from Wiley.

4. Notes

1. The powder of the gloves might lead to interference with the chemiluminescent substrate.
2. Instead of the Bis–Tris buffer system (e.g., MOPS–chloride boundary) the standard Laemmli buffer system (glycine–chloride boundary) (9) can be used. SAR-PAGE was also demonstrated to be compatible with the Neville (borate–sulfate boundary) (10) and Allen-Moore (e.g., borate–citrate boundary) (11) stacking systems. In case you decide to use these systems, make sure that you adapt all gel, sample, and running buffers accordingly by replacing the SDS with sarcosyl.
3. Phenol red is used as front marker dye. The final color of the sample buffer should be reddish.
4. Since the effect of sarcosyl on the separation of PEGylated proteins is based on the noninteraction of sarcosyl with the covalently linked PEG-molecule(s), the negative net charge is primarily dependent on the length of the amino acid chain and its interaction with the negatively charged sarcosyl molecules. Consequently, small PEGylated peptides will migrate much slower on SAR-PAGE than PEGylated proteins, given that the size of the attached PEG group(s) is identical. The effect of sarcosyl on the migration behavior can be visualized by comparing the results obtained on SDS-PAGE and SAR-PAGE after staining the gels with a PEG-specific reagent (Fig. 3). Coomassie R-250 on the other hand does only interact with the amino acid chain—hence does not stain PEG (3). For specifically staining the PEG group(s) first incubate the gel for 15 min in a 5% (w/v) solution of barium chloride in Milli-Q water, then incubate for 15 min in an aqueous solution of 1% iodine/1.5% potassium iodide (w/v), and finally wash the gel until an optimal contrast between the brown PEG-band and the still yellow gel is obtained (ca. 30 min) (12). Avoid prolonged destaining as the PEG-band will then also disappear (if this happens, stain again with the iodine solution). For documentation purposes scan or photograph the destained gel as soon as possible, since the iodine complex is unstable. Proteins without PEG groups are not stained by this method.
5. It is recommended to use a new electrophoresis cell for SAR-PAGE and then use this cell only with sarcosyl. Otherwise, make sure that you extensively clean your usual SDS-PAGE cell in order to completely remove SDS residues in the cell.
6. It is not necessary to replace the SDS in transfer buffer I for sarcosyl. Sarcosyl is only necessary during the electrophoretic

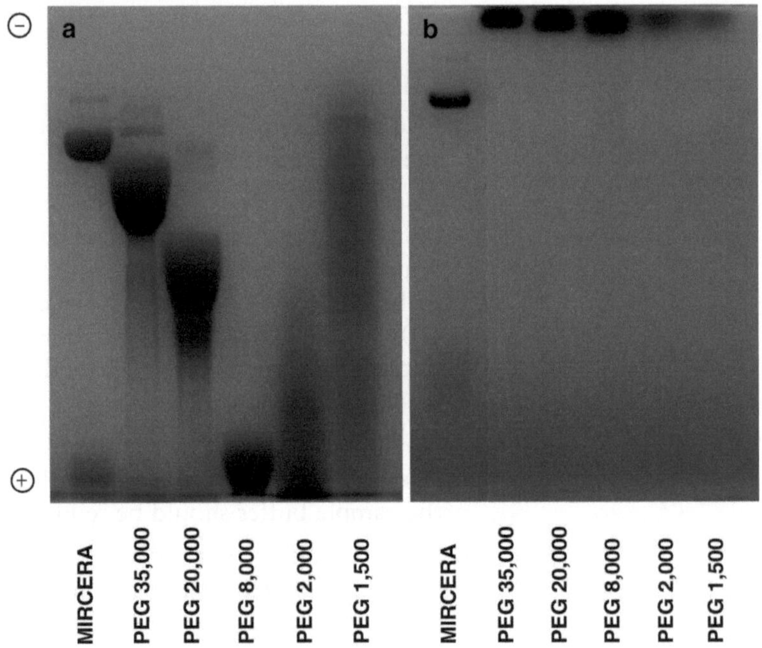

Fig. 3. Comparison of SDS-PAGE (**a**) and SAR-PAGE (**b**) of MIRCERA and polyethylene glycols of different average molecular masses (PEGs) after an iodine-based PEG-specific (Fig. 3b) stain. Applied amounts on gels (Bis–Tris, 10% T, 1.5 mm): MIRCERA (ca. 2.2 μg), PEGs (ca. 5 μg each). Reproduced from ref. 3 with permission from Wiley.

separation for enhanced band sharpness and interaction of MIRCERA with the monoclonal anti-Epo antibody (*clone AE7A5*).

7. For obtaining higher sensitivity use SuperSignal West Femto substrate (Thermo Scientific, Rockford, IL) instead.

8. Aside from the described immunoaffinity purification procedure (13, 14), which uses disposable monolithic columns containing an immobilized monoclonal anti-Epo antibody (*clone 3F6*), alternative procedures were also described (e.g., the removal of high abundant serum proteins by depletion with a multi-affinity column followed by immunoaffinity enrichment of Epo with a commercial Epo-ELISA plate). For additional details see ref. (3).

9. It is recommended to test the sensitivity of your entire system (i.e., immunoblotting, chemiluminescent detection, CCD camera) with a dilution series of MIRCERA and other Epo standards. The typical limit of detection (LOD) for MIRCERA on SAR-PAGE in combination with the described antibodies, streptavidin–HRP complex, and a medium sensitivity ECL substrate (e.g., SuperSignal West Pico) is ca. 3 pg (around 50 amol). With a higher sensitivity substrate (e.g., SuperSignal West Femto) the achievable LOD is in medium fg range (see Fig. 4).

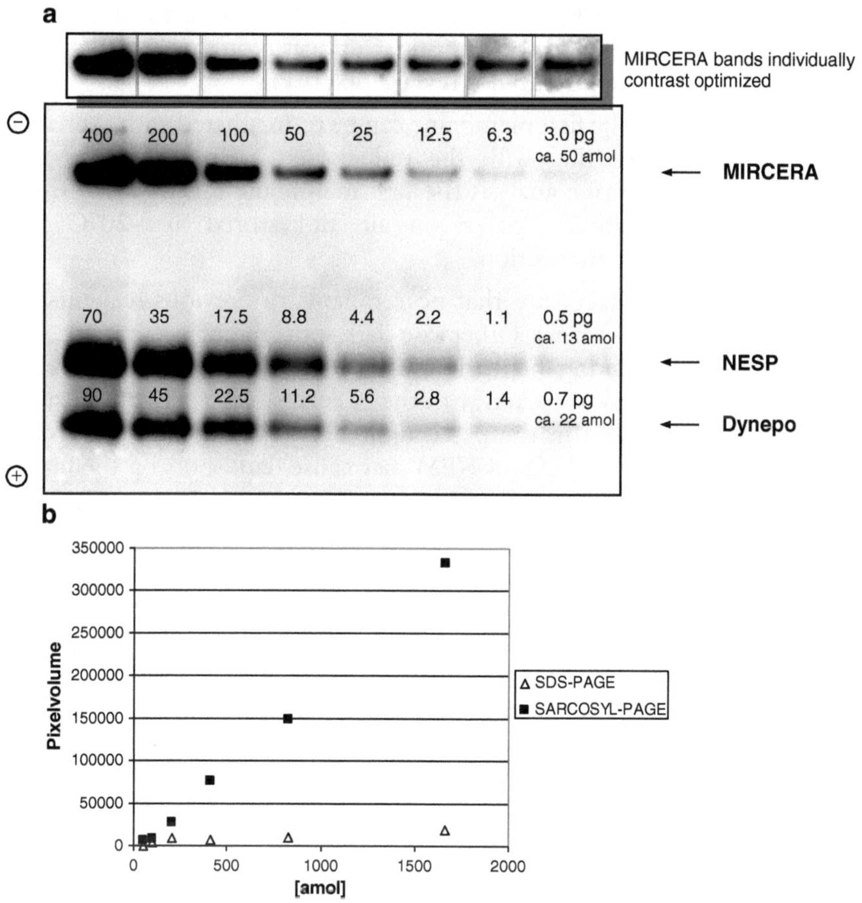

Fig. 4. (**a**) Sensitivity test of SAR-PAGE with a twofold serial dilution of three Epo standards (MIRCERA, NESP, and Dynepo). Typically, low pg to fg sensitivity is obtained with the described detection system (SuperSignal West Pico). (**b**) Densitometric comparison of SAR-PAGE and SDS-PAGE under identical test conditions. Note the significant increase in sensitivity for SAR-PAGE (3). Reproduced from ref. 3 with permission from Wiley.

10. The original procedure of the western double blot was described by Lasne (15, 16) and uses a Durapore membrane between the first and second Immobilon-P membrane. The Durapore membrane is a hydrophilic low protein-binding PVDF membrane in contrast to the hydrophobic high protein-binding PVDF Immobilon-P membrane and acts as a separator membrane between the two Immobilon-P membranes. However, the Durapore membrane can be omitted in this step. The double-blotting technique was originally developed for detecting Epo in urinary retentates by isoelectric focusing in polyacrylamide slab gels and without preceding immunoaffinity purification. An additional explanation for the usefulness of the Durapore membrane during the double-blotting procedure is given elsewhere (see ref. 17).
11. Blotting direction is from anode to cathode this time. Under acidic conditions the interaction between primary antibody

and Epo is disrupted. During blotting the antibody migrates towards the cathode. Connect blotter as usual with the power supply (do not reverse polarity).

12. The first membrane can be re-incubated with primary antibody starting with the reduction (DTT) step. After completing image analysis the first membrane can be dried overnight on blotting paper on air and stored at −20°C for a later re-incubation.
13. Make sure that no remnants of previous reagents are on the tweezers. Otherwise dark spots might occur on the membrane during incubation with the substrate and on the position where it was picked with the tweezers. Also perfectly clean the vessel used in the preceeding PBS washing step of the membrane. Remnants of NFM may cause dark spots on the membrane.
14. Typically, it is not necessary to use a pre-incubation period (e.g., 5 min) after the substrate was put on the membrane and before acquisition of images. Also, it is normally not necessary to remove the substrate and then to cover the membrane with a transparent plastic film (e.g., saran wrap). A pre-incubation period is recommended when high sensitivity is required already for the first image. However, pre-incubation bears the risk that in case of a high antigen concentration the signal is already gone before the first image is acquired due to excessive substrate reaction with HRP (identifiable by the appearance of brown bands on the membrane). Depending on the substrate type (i.e., SuperSignal West Pico or West Femto) typical exposure times range from seconds to 5–10 min.

Acknowledgments

This project has been carried out with the financial support of WADA (grant 08B11CR).

References

1. European Medicines Agency (EMEA) (2007) MIRCERA scientific discussion. http://www.emea.europa.eu/humandocs/PDFs/EPAR/mircera/H-739-en6.pdf. Accessed 27 Nov 2009 (document currently no longer accessible), pp 1–4
2. World Anti-Doping Agency (2009) Harmonization of the method for the identification of recombinant erythropoietins (i.e. epoetins) and analogues (e.g. darbepoetin and methoxypolyethylene glycol-epoetin beta), Montreal. http://www.wada-ama.org/Documents/World_Anti-Doping_Program/WADP-IS-Laboratories/WADA_TD2009EPO_EN.pdf. Accessed 6 Jan 2011
3. Reichel C, Abzieher F, Geisendorfer T (2009) SARCOSYL-PAGE: a new method for the detection of MIRCERA- and EPO-doping in blood. Drug Test Anal 1:494–504
4. Invitrogen (2003) NuPAGE technical guide version E (IM-1001). Invitrogen, Carlsbad, p 49
5. Bjerrum OJ, Schafer-Nielsen C (1986) Buffer systems and transfer parameters for semi-dry electroblotting with horizontal apparatus. In: Dunn MJ (ed) Electrophoresis'86. Wiley-VCH, Weinheim, pp 315–327

6. Bajla I, Holländer I, Minichmayr M, Gmeiner G, Reichel C (2005) GASepo: a software solution for quantitative analysis of digital images in Epo doping control. Comput Methods Programs Biomed 80:246–270
7. Lasne F, Martin L, Martin JA, de Ceaurriz J (2007) Isoelectric profiles of human erythropoietin are different in serum and urine. Int J Biol Macromol 41:354–357
8. MAIIA Diagnostics (2010) Epo purification kit, Upsala. http://www.maiiadiagnostics.com/products/epo_purification_kit/. Accessed 9 Jan 2011
9. Laemmli UK (1970) Cleavage of structural proteins during the assembly of the head of bacteriophage T4. Nature 227:680–685
10. Neville DM Jr (1971) Molecular weight determination of protein-dodecyl sulfate complexes by gel electrophoresis in a discontinuous buffer system. J Biol Chem 246:6328–6334
11. Allen RC, Moore DJ (1971) Electrophoretic system. US Patent 3620947
12. Zheng C, Ma G, Su Z (2007) Native PAGE eliminates the problem of PEG-SDS interaction in SDS-PAGE and provides an alternative to HPLC in characterization of protein PEGylation. Electrophoresis 28:2801–2807
13. Lönnberg M, Dehnes Y, Drevin M, Garle M, Lamon S, Leuenberger N, Quach T, Carlsson J (2010) Rapid affinity purification of erythropoietin from biological samples using disposable monoliths. J Chromatogr A 1217: 7031–7037
14. Dehnes Y, Lamon S, Lönnberg M (2010) Erythropoietin (EPO) immunoaffinity columns: a powerful tool for purifying EPO and its recombinant analogues. J Pharm Biomed Anal 53:1028–1032
15. Lasne F (2001) Double-blotting: a solution to the problem of non-specific binding of secondary antibodies in immunoblotting procedures. J Immunol Methods 253:125–131
16. Lasne F (2003) Double-blotting: a solution to the problem of nonspecific binding of secondary antibodies in immunoblotting procedures. J Immunol Methods 276:223–226
17. Lasne F (2009) Double-blotting: a solution to the problem of nonspecific binding of secondary antibodies in immunoblotting procedures. In: Kurien BT, Scofield RH (eds) Methods in molecular biology, protein blotting and detection, vol 536. Humana Press, New York, p 217

Chapter 8

Tricine-SDS-PAGE

Syed R. Haider, Helen J. Reid, and Barry L. Sharp

Abstract

Tricine-sodium dodecyl sulphate-polyacrylamide gel electrophoresis (tricine-SDS-PAGE) is an efficient way of separating low-molecular-mass proteins. However, the standard system is quite complicated and specifically may not be useful when the separated proteins require to be recovered from the gel for quantitative analysis. Here, we describe a simplified system whereby these smaller proteins can be resolved in comparatively low-percentage gels which have high compatibility with modern detectors such as UV and inductively coupled plasma mass spectrometry (ICP-MS).

Key words: Tricine-SDS-PAGE, Low-molecular-mass proteins, Running buffer, Gel buffer, Gel mixture, Low-percentage gel, Protein standard, Polyacrylamide

1. Introduction

The separation of low-molecular-mass proteins (1–30 kDa), or fragmented peptides derived from enzymatic digestion, is a difficult task. To achieve such separations with high resolution, slab SDS-PAGE-based methods are frequently used and allow several lanes to be run on a single gel under identical conditions. Thus, it is easy to compare and to determine the molecular masses of the separated species by calculating the relative mobility of each band relative to standards run on the same gel. The Laemmli-SDS-PAGE method (1) is widely used for the separation of the proteins; however, for the separation of low-mass proteins, this system requires a very high-percentage gel (>16%) or a gradient gel (2, 3). Such gels have several disadvantages including difficulties in casting, irreproducibility, fragility, and short shelf life (4). Further, it is difficult to quantitatively recover proteins in blotting and electroelution. The high background signals of sulphur and phosphorus in high-percentage gel also make them unsuitable for laser ablation (LA) inductively coupled plasma mass spectrometry (ICP-MS) detection of proteins (5–7).

Biji T. Kurien and R. Hal Scofield (eds.), *Protein Electrophoresis: Methods and Protocols*, Methods in Molecular Biology, vol. 869, DOI 10.1007/978-1-61779-821-4_8, © Springer Science+Business Media, LLC 2012

Tricine-SDS-PAGE is a remarkably simple and efficient alternative way of separating low-molecular-mass proteins/peptides in a single polyacrylamide gel with high resolution. Usually, a 10% gel (with pH 8.45) is employed to separate proteins in the range of 1–100 kDa using two running buffers: a cathode buffer, e.g. 100 mM Tris, 100 mM tricine and 0.1% (w/v) SDS, pH 8.25, and an anode buffer, e.g.100 mM Tris, pH 8.9. However, to obtain highly resolved bands, this method requires three gels (stacking, spacer, and resolving gel) (8–10) and addition of urea (4) in the resolving gel. The use of three gels is tedious, potentially troublesome, and always requires a freshly prepared gel mixture (10). Addition of urea may also create problems in amino acid sequencing (11), and although it is useful for the analysis of low-mass proteins, it crystallizes at low temperature and sometimes decomposes during sample preparation (4). Recently, we have described an improved form of the tricine-SDS-PAGE method employing a simplified procedure which does not require two different electrode buffers, a spacer gel, or addition of urea (12). This modified system is well suited for quantitative analysis and shows excellent compatibility with detectors such as ICP-MS or UV. For such analysis (and where the retention of metals is desired), the softest conditions are preferred, that is, low-percentage gels and low-concentration buffers.

For comparative purposes, the following provides descriptions of separations based on the modified method reagent system and the original method reagent system. These show that using the simplified approach of the modified system, a different buffer and pH are required and that the original reagent system does not work efficiently under these conditions. The user does not need to follow the steps of the original method; they are included only for information.

2. Materials

Prepare all the solutions in ultra-pure water (18 MΩ, e.g. from a Milli-Q water purification system, Millipore Corporation, Bedford, MA). Store at room temperature (if required), unless otherwise stated.

2.1. Tricine-SDS-PAGE Components

1. 2.5 M Tris-HCl buffer, pH 8.8 for modified and pH 8.45 for original system (see Notes 1 and 2): Weigh 302.85 g of Tris base and transfer it to a 1-L graduated cylinder or glass beaker. Dissolve the base by adding 600 mL of deionized water and adjust the pH using HCl. Make up to 1 L with water. This gel buffer can be stored at 4–8°C.
2. Polyacrylamide gel solution: 30% (w/v) solution of acrylamide: bis-acrylamide 29:1 (Sigma-Aldrich, Poole, UK). Store this

polyacrylamide solution at 4–8°C. (It is highly recommended to purchase the ready-made polymer solution from the suppliers instead of preparing it in the laboratory because acrylamide is recognized as a neurotoxin and suspected carcinogen.)

3. Running buffer for modified system: 25 mM Tris, 25 mM tricine, 0.05% (w/v) SDS. Dissolve 3.03 g Tris base, 4.5 g tricine and 0.5 g SDS in 1 L of deionized water (see Note 3). There is no need to adjust the pH. Store at 4–8°C (see Note 4).
4. Running buffer for original system: 100 mM Tris, 100 mM Tricine 0.1% (w/v) SDS. Weigh 12.1 g Tris base, 17.9 g tricine and 1 g SDS and dissolve in 1 L of deionized water (see Notes 3–5). There is no need to adjust the pH. Store at 4–8°C (see Note 4).
5. Sample or loading buffer: 2× sample buffer (Sigma-Aldrich, Poole, UK) containing 100 mM Tris–HCl (pH 6.8) (see Note 6), 1% (w/v) SDS, 4% (v/v) 2-mercaptoethanol, 0.02% (w/v) Coomassie Brilliant Blue (CBB) and 24% (w/v) glycerol. Store at −20°C (see Note 4).
6. Ammonium persulphate (APS) solution: Weigh 0.03 g of APS and dissolve in 1 mL of deionized water (see Note 7).
7. *N*,*N*,*N*′,*N*′-tetramethylethylenediamine (TEMED) (Sigma-Aldrich, Poole, UK) (see Note 8).
8. CBB staining solution: Dissolve 0.025–0.030 g of CBB in 100 mL of 10% (v/v) acetic acid solution (see Note 9).
9. Acetic acid (10% v/v) destaining solution: Prepare 100 mL of 10% (v/v) acetic solution (see Note 9).
10. Glutaraldehyde (5% v/v) fixing solution: Prepare 5% glutaraldehyde solution by diluting five times the 25% (v/v) glutaraldehyde stock solution (Sigma-Aldrich, Poole, UK) (see Note 9). Both stock and diluted solutions should be stored at −20°C (see Note 4).

2.2. Protein Mass Ladder

Protein mass ladder: (a) with a mass range of 2.5–200 kDa (Invitrogen Limited, Paisley, UK) and (b) 1.0–26.6 kDa Ultra Low Molecular Weight Marker™ (Sigma-Aldrich, Poole, UK). Store the protein mass ladder (a) at 4–8°C and protein mass ladder (b) at −20°C (see Note 4). For details of the protein mass ladders (a) and (b), see Table 1.

2.3. Protein Digestion

1. 1, 4-Dithio-DL-threitol (DTT) (100 mM): Weigh 0.015 g DTT and dissolve in 1 mL of deionized water. Store at 4–8°C.
2. Iodoacetamide (IAA) (100 mM): Weigh 0.02 g IAA and dissolve in 1 mL of deionized water. Store at 4–8°C.
3. Ammonium bicarbonate (NH_4HCO_3) (50 mM): Dissolve 0.04 g of NH_4HCO_3 in 10 mL of deionized water (see Note 10).

Table 1
Detail of protein mass ladders (a) and (b)

Protein mass ladder (a)	Protein mass ladder (b)	Molecular mass (kDa)
Myosin	–	200.0
β-galactosidase	–	116.3
Phosphorylase B	–	97.4
Bovine serum albumin (BSA)	–	66.3
Glutamic dehydrogenase	–	55.4
Lactate dehydrogenase	–	36.5
Carbonic anhydrase	–	31.0
–	Triosephosphate isomerase	26.6
Trypsin inhibitor	–	21.5
–	Myoglobin (from horse heart)	17.0
Lysozyme	–	14.4
–	α-lactalbumin (from bovine milk)	14.2
–	Aprotinin (from bovine lung)	6.5
Aprotinin	–	6.0
Insulin B chain	Insulin B chain (oxidized bovine)	3.5
Insulin A chain	–	2.5
–	Bradykinin	1.06

Reproduced from (12) permission of Springer-Verlag © 2010

4. Sequencing grade modified trypsin (0.1 μg/μL) (Promega, Delta House, Southampton, UK): Dissolve 100 μg of the lyophilized powder of modified trypsin in 1 mL of deionized water (see Note 11). Store trypsin solution in aliquots at −20°C. For long-term storage, use −80°C.
5. Prepare sample for protein digestion by dissolving 1 mg of β-casein in 1 mL of deionized water.

3. Methods

All procedures should be carried out at room temperature.

3.1. Protein Digestion

Perform protein digestion in solution for a comparative study according to the following method. Here, β-casein is used as a model protein.

1. Add 10 μL of 1 mg/mL β-casein solution in the mixture of 15 μL of 50 mM NH_4HCO_3 and 1.5 μL of 100 mM DTT solution and incubate at 90°C for 5–10 min.
2. After cooling to room temperature, add 3 μL of 100 mM IAA to this solution and incubate for 25 min in the dark.
3. Add 1 μL of 0.1 μg/μL sequencing grade modified trypsin and incubate at 37°C for 4 h.
4. Finally, add an additional 1 μL of trypsin and incubate overnight at 30°C. Next day, reduce digest mixture to 10 μL using vacuum.

3.2. Gel Casting

1. Prepare all stacking 4% (w/v) and resolving gel mixtures for 7, 9, and 10% (w/v) gels for both the original and modified methods as shown in Table 2 (see Notes 12–14).
2. Cast all the gels in gel cassettes at 7.3-cm height, 8-cm width and 0.75-mm thickness using a mini-gel casting apparatus (Bio-Rad, Hemel Hempstead, UK) (see Notes 15 and 16).
3. Firstly, inject the resolving gel mixture into the gel cassette (see Notes 17 and 18).
4. Do not fill the resolving gel mixture to the top of the gel plates. Leave a space of approximately 1–2 cm from the top. Quickly fill this space with butanol or 70% ethanol (see Notes 19 and 20).
5. The resolving gel usually takes 10–15 min for polymerization. Check for polymerization of the gel in the centrifuge tube and if the gel has been polymerized then pour off the butanol (or 70% ethanol) from the top of the resolving gel. Start injecting the stacking gel mixture onto the polymerized resolving gel in the gel cassette and fill up to the top of the short plates and then quickly insert the gel combs for the loading wells. Check the polymerization of the stacking gel in the centrifuge tube, or by just slightly lifting the combs, and if the stacking gel is polymerized, then remove the gel cassette sandwich from gel casting frame and set this into the electrode assembly. The short plates must be pointing inward in the dual plate electrode assembly (see Fig. 1).

3.3. Sample Preparation

Mix 10 μL of the sample buffer with 10 μL of the digest mixture and heat at 70°C for 5–10 minutes. Dilute 20-fold the protein mass ladder (b) with the 1× sample buffer and heat at 65°C for 2 minutes (see Note 21). There is no need to dilute or heat the protein mass ladder (a).

3.4. Sample Loading and Gel Running

1. Place the electrode assembly containing the two gel cassettes in the running buffer tank.
2. Fill the tank with the running buffer and gently remove the combs. Wash the loading wells with the running buffer using 1-mL pipette tips to remove any un-polymerized acrylamide.

Table 2
Tricine-SDS-PAGE protocol for the preparation of the polyacrylamide gels

Percentage of acrylamide in the gel		Acrylamide/ bis-acrylamide 29:1 30% (w/v) solution (mL)	2.5 M Tris buffer (pH 8.8 for modified and pH 8.45 for original method) (mL)	Deionized water (mL)	TEMED (μL)	APS 30 mg/mL μL	Total volume (mL)
Stacking gel	4	0.66	0.76	3.42	5.0	150	~5
Resolving gel	7	2.33	5.6	1.91	7.0	150	~10
	9	3.00	5.6	1.24	6.0	150	~10
	10	3.33	5.6	0.90	6.0	150	~10
	12	4.00	5.0	0.89	6.0	100	~10
	15	5.00	4.6	0.29	6.0	100	~10
	16	5.33	4.3	0.22	6.0	100	~10

Reproduced from (12) permission of Springer-Verlag © 2010.

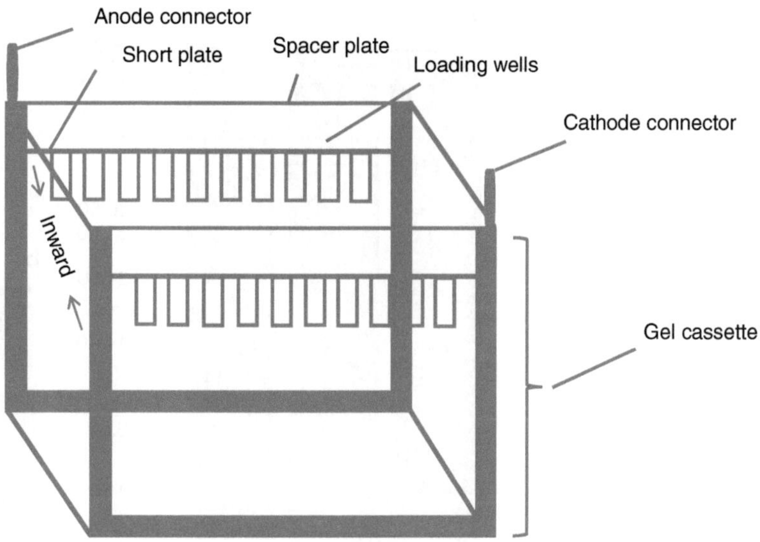

Fig. 1. Dual plate electrode assembly showing the gel cassette sandwich. The short plates must be pointing inwards.

3. Load 7 µL of protein mass ladder (a), 5 µL of protein mass ladder (b) and 10 µL of β-casein digest (as prepared above) in each well of the 7, 9, and 10% (w/v) slab gels for the modified and original tricine-SDS-PAGE methods (see Note 22).
4. Apply a constant voltage of 125–150 V. Do not stop until the dye front touches the bottom (see Note 23).

3.5. Gel Fixing, Staining and Destaining

1. Fix the gels for 25 min (using a shaker) with the fixing solution of 5% (v/v) glutaraldehyde.
2. Stain each gel for 20 min (using a shaker) using CBB staining solution.
3. Perform destaining for 20 min (using a shaker) using 10% (v/v) acetic acid solution (see Notes 24 and 25). See Figs. 2–4 for typical results.

4. Notes

1. The "original method" refers to the previously described (8, 9) tricine-SDS-PAGE, and "modified method" (12) refers to the modification here described.
2. The previously described methods for tricine-SDS-PAGE employed 3.0 M Tris base, whilst the modified method uses 2.5 M—so here, for the comparative study, the concentration of the gel buffer was kept to 2.5 M for both systems.

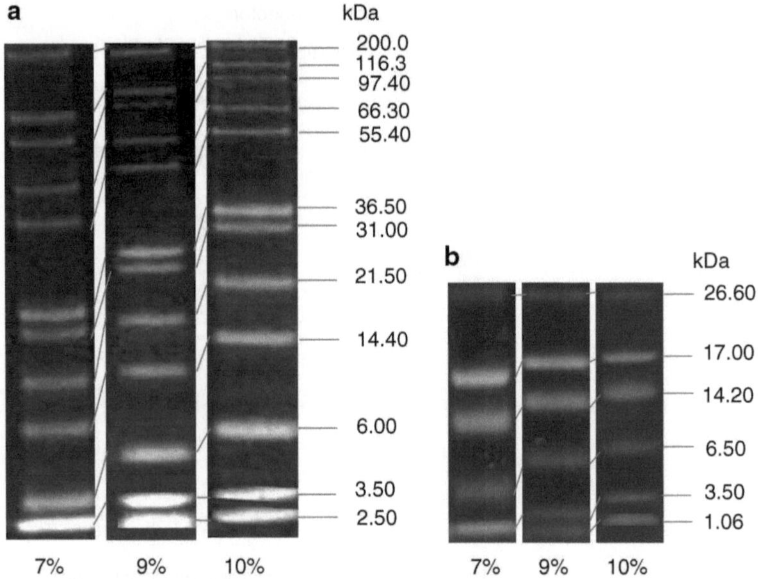

Fig. 2. Modified tricine-SDS-PAGE separation of low-mass proteins in the range of (**a**) 2.5–200 kDa and (**b**) 1.0–26.6 kDa in 7, 9, and 10% gels. For proteins corresponding to the listed masses, see Table 1. (Reproduced from (12) permission of Springer-Verlag © 2010).

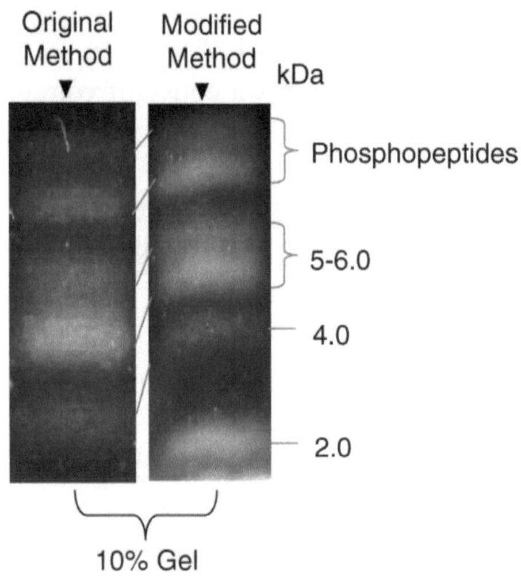

Fig. 3. Head-to-head comparative results for the β-casein digest using a 10% polyacrylamide slab gel compared with the original and modified tricine-SDS-PAGE systems (see Note 26). (Reproduced from (12) permission of Springer-Verlag © 2010).

3. The running buffer can be prepared at 10× concentration and can be diluted accordingly.
4. All the solutions should be used at room temperature.
5. For comparative study between the modified and original tricine-SDS-PAGE system, only one running buffer was used.

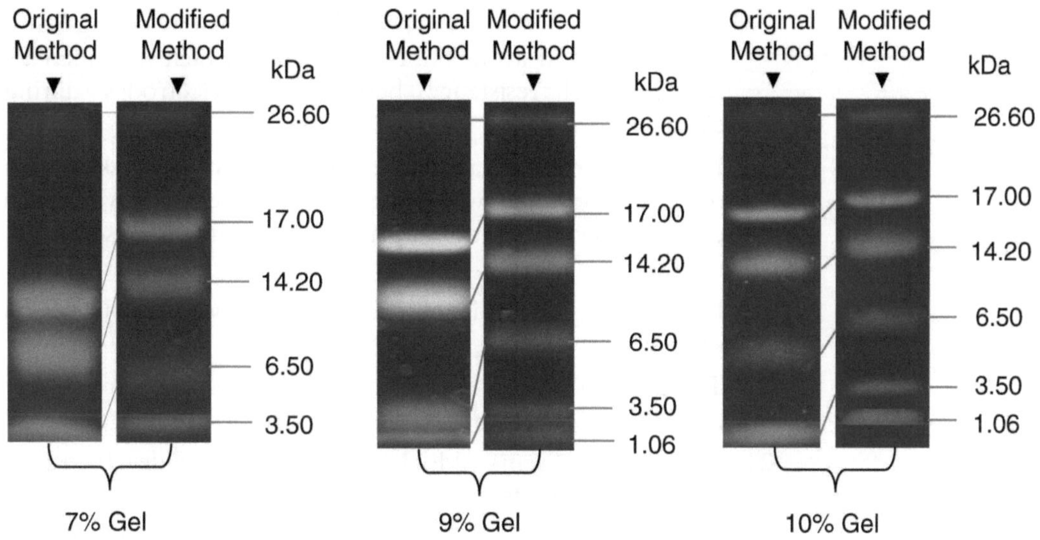

Fig. 4. Effect of increasing the pH of the gel buffer to pH 8.8 showing an additional band in the 7, 9, and 10% modified gel systems. (Reproduced from (12) permission of Springer-Verlag © 2010).

6. Tris base with pH of 7.0 can also be used instead of pH 6.8.
7. APS must always be freshly prepared.
8. TEMED is a very odorous neurotoxin, so extra care is required when handling it.
9. Avoid using methanol in staining, destaining and fixing solutions as this may cause loss of the low-molecular-mass proteins.
10. There is no need to adjust the pH of the ammonium bicarbonate solution.
11. Sequencing grade modified trypsin does not suffer from self-autolysis and minimizes miss-cleavages during the digestion.
12. We prefer a 10–50-mL centrifuge tube for preparing the gel mixture.
13. Do not include SDS, glycerol, or urea in any of the gels.
14. TEMED and APS must be added last in the gel mixture which requires quick injection into the gel cassettes; otherwise, gel will be polymerized in the tube.
15. Acrylamide is recognized as a neurotoxin, so gloves must be worn all the time and the work must be done in a properly ventilated area.
16. For the purpose of comparing the original and modified methods, the same gel casting protocol was used for both the systems.

17. Inject the gel mixture slowly into the gel cassettes using 5-mL pipette tips and avoid bubble formation which may increase the electrical resistance between the electrodes during electrophoresis.
18. In case of direct contact of the gel mixture with the skin, wash the skin thoroughly with water and seek medical advice if the contact is severe because acrylamide can readily be absorbed by the skin.
19. Butanol or 70% ethanol provides a smooth resolving gel surface.
20. The butanol or 70% ethanol must be removed before casting the stacking gel on the top of the resolving gel.
21. For example, Sigma-Aldrich (Poole, UK) provides 10 mL of 2× sample buffer free with the protein mass ladder (b).
22. Do not overfill the wells; otherwise, they will contaminate the neighbouring wells.
23. For rapid separation, a higher voltage of 200–250 can be applied; however, it is not preferred, because it increases the temperature of the buffer which may cause a significant change in the pH of the running buffer and/or resolving gel.
24. Wash the gel with deionized water after each of the fixing, staining and destaining steps.
25. For best results, do not reuse fixing, staining and destaining solutions.
26. Confirmation of the phosphopeptide bands was obtained by whole gel elution and ICP-MS detection of ^{31}P (for details, see ref. (12)).

Acknowledgments

This work was supported by a grant from Loughborough University, Loughborough, Leicestershire, LE11 3TU, UK.

References

1. Laemmli UK (1970) Cleavage of the structural proteins during the assembly of the head of bacteriophage. Nature 227:680–685
2. Hashimoto F, Horigome T, Kanbayashi M, Yoshida K, Sugano H (1983) An improved method for separation of low-molecular-weight polypeptides by electrophoresis in sodium dodecyl sulfate-polyacrylamide gel. Anal Biochem 129:192–199
3. Bothe D, Simonis M, Dohren H (1985) A sodium dodecyl sulphate-gradient gel electrophoresis system that separates polypeptides in the molecular weight range of 1500-100,000. Anal Biochem 151:49–54

4. Okajima T, Tanabe T, Yasuda T (1993) Nonurea sodium dodecyl sulfate-polyacrylamide gel electrophoresis with high-molarity buffers for the separation of proteins and peptides. Anal Biochem 211:293–300
5. Marshall P, Heudi O, Bains S, Freeman HN, Abou-Shakra F, Reardon K (2002) The determination of protein phosphorylation on electrophoresis gel blots by laser ablation inductively coupled plasma-mass spectrometry. Analyst 127:459–461
6. Elliot VL, McLeod C, Marshall P (2005) Combination of gel electrophoresis and ICP-mass spectrometry: novel strategies for phosphoprotein measurement. Anal Bioanal Chem 3:416–423
7. Wind M, Feldmann I, Jakubowski N, Lehmann WD (2003) Spotting and quantification of phosphoproteins purified by gel electrophoresis and laser ablation-element mass spectrometry with phosphorus-31 detection. Electrophoresis 7:1276–1280
8. Schägger H (2006) Tricine-SDS-PAGE. Nature 1:16–22
9. Schägger H, von Jagow G (1987) Tricine sodium dodecyl sulfate-polyacrylamide gel electrophoresis for the separation of proteins in the range from 1 to 100 kDa. Anal Biochem 166:368–379
10. Yim SK, Ahn T, Kim JK, Yun CH (2002) Polyacrylamide gel electrophoresis without a stacking gel: application for separation of peptides. Anal Biochem 305:279–281
11. Hames BD (1998) Gel electrophoresis of proteins. Oxford University Press, New York
12. Haider SR, Reid HJ, Sharp BL (2010) Modification of tricine-SDS-PAGE for online and offline analysis of phosphoproteins by ICP-MS. Anal Bioanal chem 397:655–664

Chapter 9

Analysis of Protein Glycation Using Phenylboronate Acrylamide Gel Electrophoresis

Marta P. Pereira Morais, John S. Fossey, Tony D. James, and Jean M.H. van den Elsen

Abstract

Carbohydrate modification of proteins adds complexity and diversity to the proteome. However, undesired carbohydrate modifications also occur in the form of glycation, resulting in diseases such as diabetes, Alzheimer's disease, autoimmune diseases, and cancer. The analysis of glycated proteins is challenging due to their complexity and variability. Numerous analytical techniques have been developed that require expensive specialised equipment and complex data analysis. In this chapter, we describe a simple electrophoresis-based method that enables users to detect, identify, and analyze these post-translational modifications. This new cost-effective methodology will aid the detection of unwanted glycation products in processed foods and may lead to new diagnostics and therapeutics for age-related chronic diseases and glycosylation disorders.

Key words: Protein glycation, Gel electrophoresis, Boronic acid, Step gel, 2D gel

1. Introduction

1.1. Protein Glycation

Over the past decades, there has been considerable interest in the development of simple and efficient techniques to detect glycated proteins as indicators for chronic age-related diseases such as diabetes, Alzheimer's disease, and cancer (1). Protein glycation is a non-enzymatic post-translational modification whereby saccharides react in a non-random manner with free amino groups at the N-terminus of proteins and with lysine and arginine side chains to form Schiff base or aldimine conjugates. The Schiff base intermediate rearranges to form a more stable ketoamine called the Amadori product, which can undergo a series of reactions including oxidation, dehydration, fragmentation, and degradation to form heterogeneous fluorescent and yellow-brown products collectively

Biji T. Kurien and R. Hal Scofield (eds.), *Protein Electrophoresis: Methods and Protocols*, Methods in Molecular Biology, vol. 869,
DOI 10.1007/978-1-61779-821-4_9, © Springer Science+Business Media, LLC 2012

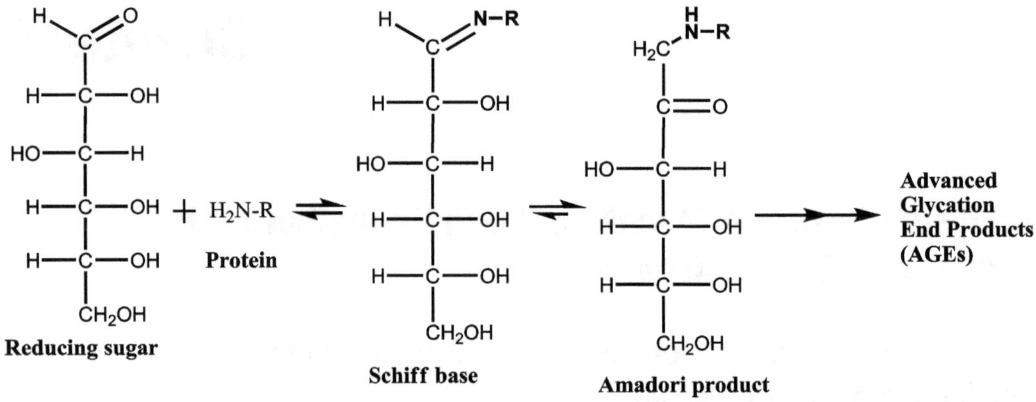

Fig. 1. Schematic diagram of the protein glycation reaction. The condensation reaction between reducing sugar and free amino group of proteins and peptides leads to the formation of a Schiff base, which rearranges to a more stable Amadori product ultimately forming advanced glycation end products (AGEs).

known as the advanced glycation end products or AGEs (2) (Fig. 1). Despite extensive research, the glycation process is still poorly understood due to its complexity and variability (3). Since AGE formation is irreversible, these products accumulate over the lifetime of proteins, thereby affecting their structure and function. Protein glycation is a slow process, and although it takes hours for Schiff bases to form, the formation of the more advanced products is in the order of weeks, depending on physiological conditions such as blood glucose levels (1). The rapid turnover of most proteins, therefore, protects them against potential damage (3). As a result, AGE formation mainly affects long-lived proteins such as albumin, collagen, and lens crystallins (6).

Glycation also occurs during cooking and storage of food, affecting its flavour, colour, texture, digestibility, as well as nutritional value (7). Foerster and Henle have shown that the high temperatures involved in food preparation results in the formation of complexes that are different from those observed in biological systems. Only little is known about the resorption and elimination of these dietary glycated products.

1.2. Analysis of Glycated Proteins

The implication of protein glycation in disease has in recent years led to the development of numerous analytical methods aimed at identifying and quantifying heterogeneous glycated protein products. Glycation analysis techniques include mass spectroscopy (MS) (8), boronate affinity chromatography (BAC) (9), high-performance liquid chromatography (HPLC) (8), and fluorescence spectroscopy (11). Each of these methods has its advantages and limitations. MS, for example, only detects the most abundant proteins, and any information regarding the glycation state is lost when proteins are modified or digested prior to analysis, as part of the sample preparation. Whilst fluorescence spectroscopy, on the other hand, allows monitoring of the course of the glycation reaction, it is biased towards fluorescent AGE products and does not provide any structural information of these adducts. Often the cost and availability of specialised instruments, such as MS, limits their general use. We chose the widely available protein analysis tool, polyacrylamide gel electrophoresis (12), and modified it to facilitate the identification and quantification of glycated proteins (13).

1.3. Uses of Gel Electrophoresis

Polyacrylamide gel electrophoresis was first described by Ornstein and Davis, and Raymond and Weintraub in 1959 (14, 15). This technique utilises a characteristic of proteins that when they are charged at any pH other than their isoelectric point (p*I*), they will migrate in the presence of an electric current. The polyacrylamide gel matrix thereby acts as a molecular sieve to separate proteins, providing information about the molecular weight, charge, subunit composition, and purity of proteins. Polyacrylamide gel properties can be altered by varying the amount of acrylamide and cross-linkers and can be cast into a range of shapes for sensing and detection purposes. Because of its inertness and biocompatibility, polyacrylamide gels can be used in numerous medical applications (16). Gel electrophoresis is simple, highly reproducible (17), cost effective, and sensitive, detecting proteins at picomole levels when using silver staining (18). Multiple samples can be run on a single gel, ensuring identical running conditions and the potential for high throughput analysis. Until recently, this technique could not differentiate between various post-translationally modified proteins due to their similarity in molecular weight. In order to utilise this analytical system for glycated proteins, we incorporated a special class of compounds into polyacrylamide gels that reversibly interact with *cis*-1,2- and 1,3-diol moieties of the carbohydrate protein adducts (19).

1.4. Utilising Boronic Acid–Diol Interactions

The interaction between boronic acids and diols, first described by Lorand and Edwards (20), is unusual because it involves a rapid and reversible covalent interaction (Fig. 2). Boronic ester formation is more favourable at high pH since boronic acid molecules in the tetrahedral anionic state are more reactive than the trigonal analogues. Boronic acids have different affinities for different diols, and this property has been utilised in a number of applications in affinity support from sample separation to sensors (21). Boronic acids were first used in affinity chromatography in 1970 to separate sugars and nucleic acids (22) and were subsequently employed to separate a variety of *cis*-diol compounds including proteins (9, 23). Protein purification (24) and clinical analysis of fructosamine content are some of the reported uses of boronate affinity chromatography (BAC) due to its low cost, speed, and accuracy (25). In this technique, enrichment and isolation of glycated proteins (10) is achieved by passing samples through an agarose matrix containing

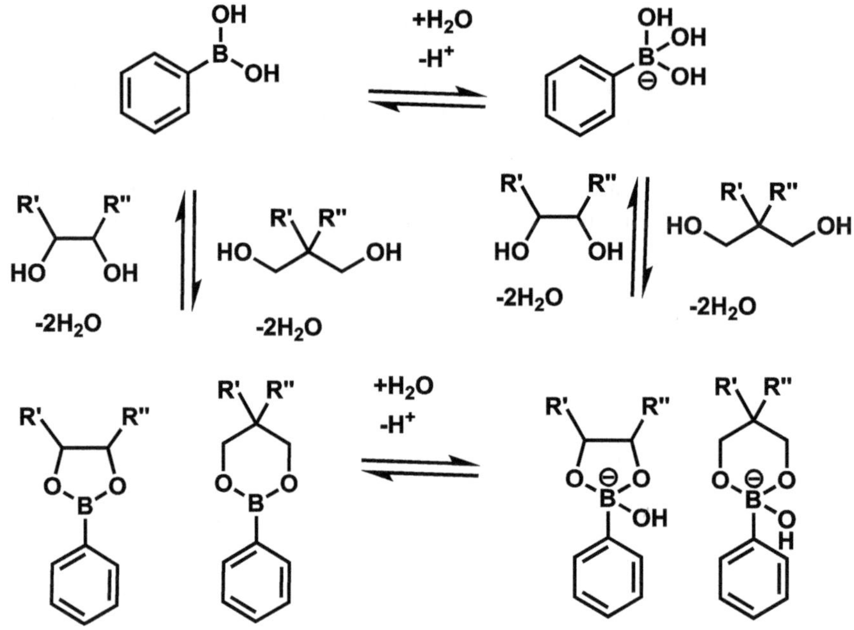

Fig. 2. The interaction between boronic acids and diols. Boronic acids exist in a trigonal state under neutral conditions, or a tetrahedral state under basic conditions. Both states form esters with *cis*-1,2- and 1,3-diols, giving five- and six-membered rings, respectively. Five-membered cyclic boronic esters are more stable than six-membered ones. Furthermore, the condensation reactions are in equilibrium, with the interaction between anionic boronic acids and diols kinetically faster than the neutral boronic acid analogues.

covalently attached *m*-aminophenylboronic acid. Glycated proteins are bound to the matrix whilst unglycated counterparts are removed. The bound fraction of glycated proteins is subsequently eluted using sorbitol-containing buffer, which has a higher affinity for boronic acid. This enrichment of glycated proteins is often used prior to non-chromatographic analyses, such as MS (26).

1.5. The Use of Boronic Acids in Gel Electrophoresis

We first explored the application of boronic acid functionalised polyacrylamide gels for the separation of saccharides (27). Direct analysis of carbohydrates via gel electrophoresis is not possible because these molecules are electroneutral in nature and lack detectable chromophores and fluorophores. To overcome this problem, fluorophore-assisted carbohydrate electrophoresis (FACE) (28) was developed, allowing simple, rapid and sensitive analysis of carbohydrates by labeling them with fluorophores. However, some of the fluorophores used in this method, such as 8-aminonapthalene-1,3,6-trisulphonic acid (ANTS), affect the true nature of carbohydrates by the introduction of a number of negative charges that can result in unexpected migration profiles (28–30).

The incorporation of small quantities (0.5% w/v) of methacylamido phenylboronic acid (MPBA) in boronate-assisted saccharide electrophoresis (BASE) resulted not only in the correction of

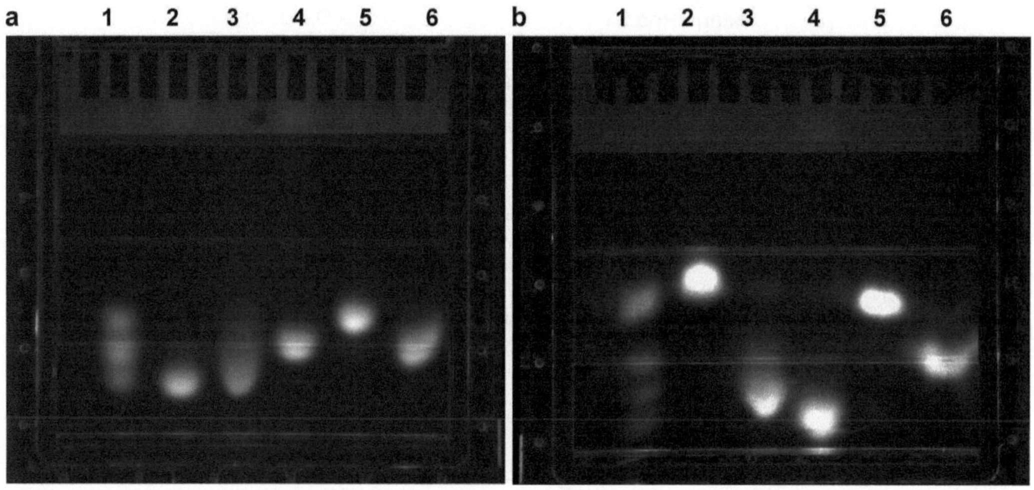

Fig. 3. Boronate-assisted saccharide electrophoresis (BASE). (**a**) FACE separation profile of AMAC-labeled mono- and disaccharides (in the absence of boronate acrylamide) in a 20% polyacrylamide gel at pH 8.3. (**b**) BASE separation profile in the presence of 0.5% w/v MPBA incorporated in a 20% polyacrylamide gel at pH 8.3. Saccharide mixture (*lane 1*), lactose (*lane 2*), galactose (*lane 3*), *N*-acetyl glucosamine (*lane 4*), melibiose (*lane 5*), and glucose (*lane 6*). The incorporation of MPBA in the BASE gels enables the separation of monosaccharides from disaccharides.

aberrant migration profiles of carbohydrates labeled with the neutral fluorophore 2-aminoacridone (AMAC) but also enabled the separation of monosaccharides from disaccharides (Fig. 3) (27). In parallel, we designed a visual detection technique based on a polyacrylamide gel competition assay using the diol-containing dye Alizarin Red S (ARS). We found that ARS was selectively retained in boronic-acid-incorporated gel balls and that it can be displaced by the addition of carbohydrates such as fructose (Fig. 4) (31).

Encouraged by our results from the above-mentioned BASE and gel ball analyses, we hypothesised that sugar-modified proteins, such as glycated proteins, should also interact with gel-incorporated boronic acids, enabling their analysis by gel electrophoresis (13). This concept will be further explored below, using a number of model proteins for glycation analysis.

1.6. Electrophoretic Analysis of Glycated Human Serum Albumin

Human serum albumin (HSA) is the most abundant protein in human plasma and binds a wide variety of compounds, acting as transport proteins for substrates like fatty acids, amino acids, hormones, and drugs (32). HSA is an ideal model system for studying protein glycation because of its abundance (40 mg/mL in blood) and long half-life (approximately 20 days). In healthy adults, 10% of HSA is glycated affecting mainly lysine and arginine residues, with a two- to threefold increase under hyperglycaemic conditions (33). Haemoglobin (Hb) is another protein commonly affected by glycation and is used clinically as a diabetic control index. However, various

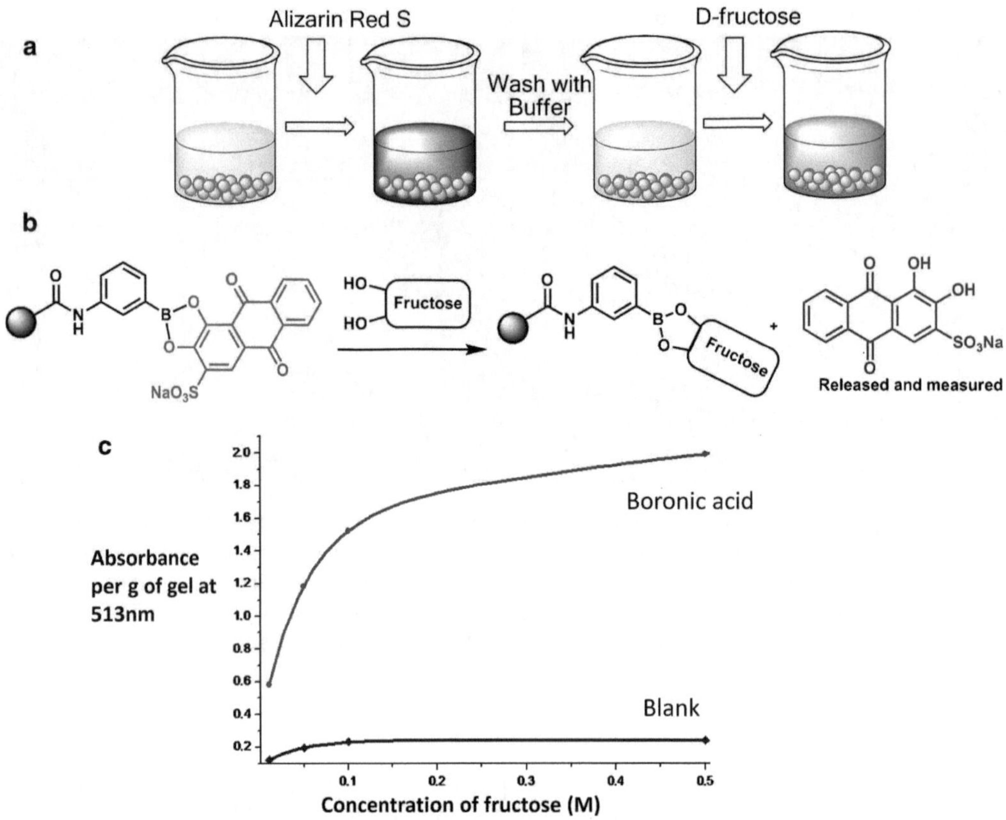

Fig. 4. Alizarin red S (ARS) assay of boronic acid gels. (**a**) A summary of the competition assay procedure. Polyacrylamide gels both with and without boronic acids are cast into gel balls and dyed with 2.5×10^{-4} M ARS, a *cis*-diol containing dye. When the gels are washed with PBS to remove unbound dye molecules, approximately the same amount of dye was removed from both polyacrylamide gel balls with and without incorporated MPBA. The remaining dye molecules in the gel balls are competitively displaced by the addition of carbohydrates such as fructose. More ARS dye is released from the MPBA gel compared to the polyacrylamide control, which can be monitored by measuring the absorbance of solution at 513 nm. These readings indicate that more dye is initially retained in the boronic acid gel. (**b**) The displacement reaction of ARS from MPBA gel on the addition of fructose and (**c**) absorbance at 513 nm of the released dye for both gels during the displacement reaction.

investigations have indicated that glycated HSA is a better glycaemic indicator than glycated Hb (34).

Whilst normal SDS-PAGE analysis cannot distinguish between glycated and unglycated proteins, incorporation of boronic acid (MPBA) in polyacrylamide gel electrophoresis, mP-AGE, results in the retention of glycated proteins whilst the mobility of unmodified proteins remains unchanged. Figure 5 compares the SDS-PAGE profiles of sugar-incubated HSA in gels with (right) and without (left) MPBA. The control polyacrylamide gel shows a typical gel profile of HSA with a single protein band around 67 kDa in all samples. On the other hand, mP-AGE gel containing 0.2% w/v MPBA shows the presence of an additional HSA band at approximately 80–90 kDa. HSA samples that were glycated over a longer period of time with glucose and lactose showed an even higher retention profile in the mP-AGE gel.

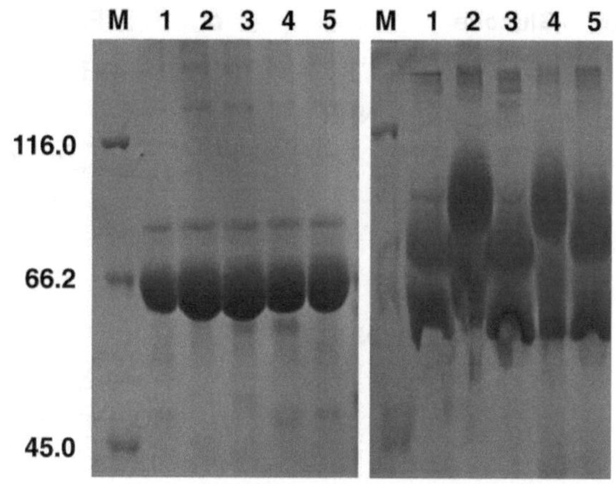

Fig. 5. Methacrylamido phenylboronic acid gel electrophoresis (mP-AGE) analysis of glycated human serum albumin (HSA). SDS-PAGE analysis of glycated HSA on an 8% polyacrylamide (PA) gel with 0.2% w/v MPBA (*right panel*) and without MPBA (*left panel*). HSA samples incubated at 37°C for 21 days in the presence of: PBS buffer (*lane 1*), glucose (*lane 2*), fructose (*lane 3*), lactose (*lane 4*), and glucose-6-phosphate (*lane 5*). Glycated HSA appears as a single band in blank SDS-PAGE gel, whereas additional bands around 80–90 kDa are observed in MPBA gel, revealing different states of protein glycation. HSA glycated with glucose and lactose were retained at a higher position in the gel compared to fructose-incubated HSA. This observation can be explained by differences in affinity of the Amadori adduct with MPBA, as shown in Fig. 6.

1.7. The Molecular Interactions Between Glycated Protein and Boronic Acid in mP-AGE

With our results, we can now provide a molecular model explaining the behaviour of glycated protein in mP-AGE. Boronic acids have different affinities for different carbohydrates, and form the strongest interaction with sugars containing an anomeric *cis*-1,2-diol (Fig. 2). Glucose, the dominant contributor to glycation *in vivo*, contains such an anomeric *cis*-1,2-diol in the protein adduct and an additional 4,5-diol that also interacts with MPBA. The anomeric diol interaction can also be stabilised by electrostatic interaction between the negatively charged boronate and the protonated amino group. In contrast, the pyranose structure from the Heyns rearrangement of the fructose adduct does not contain any *cis*-diols that can interact strongly with the MPBA incorporated in the gel. Although the furanose fructose adduct contains anomeric 3,4- and 4,5-diols, its inability to interact with boronic acid in the gel leads us to conclude that the majority of this adduct exists in the pyranose form in HSA, or the absence of amino group stabilization results in weaker boronic acid binding.

This molecular model can also be used to explain the results obtained with HSA that is glycated with other sugars, such as lactose, glucose-6-phosphate, mannose, and maltose (Figs. 5, 6 and 8). In addition, it elucidates the differences in retention of sugar adducts caused by glycation or from the natural enzymatic process of glycosylation (13). Glycosylated proteins contain *N*- and *O*-linked glycans that do not have anomeric *cis*-diols that can

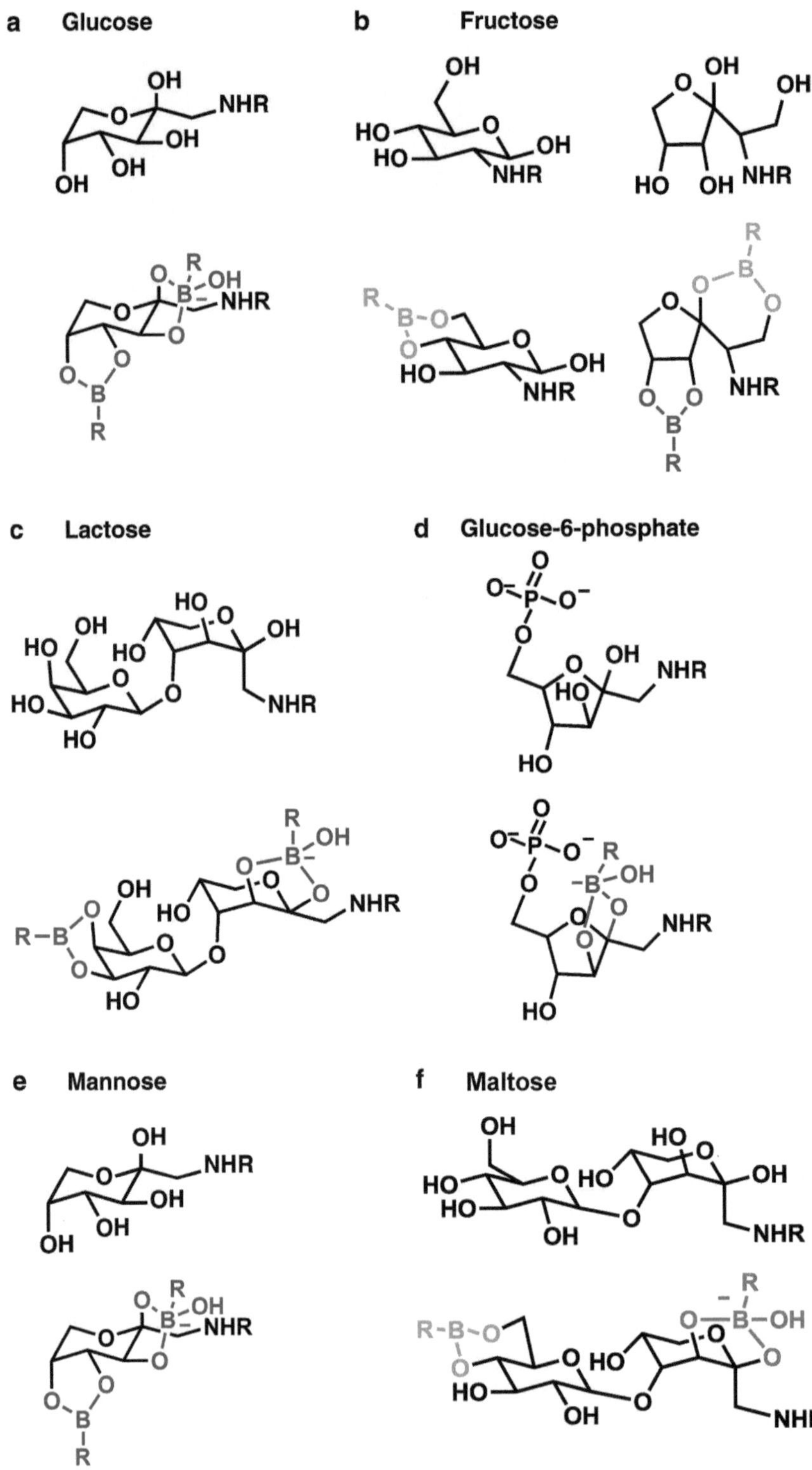

Fig. 6. Molecular model explaining the separation of glycated protein in mP-AGE. This model shows the molecular interactions between phenylboronate and the Amadori product formed when HSA is incubated with (**a**) glucose (fructosamine), (**b**) fructose, (**c**) lactose, (**d**) glucose-6-phosphate, (**e**) mannose, and (**f**) maltose. Structures representing the sugar-modified Amadori adducts are shown at the *top* whilst their boronic acid interactions are shown at the *bottom* respectively. Sugars

interact strongly with boronic acids, and as a result, their electrophoretic mobility is not altered in mP-AGE.

1.8. Detection of Glycation in Recombinant Protein Production

In addition to the analysis of glycated HSA, we would like to mention the potential use of mP-AGE for the detection and separation of glycated protein products that are formed during the production of recombinant proteins in *Escherichia coli* expression systems. Recombinant proteins containing an N-terminal poly-histidine tag are prone to modification by 6-phosphogluconolactone (6PGL), an intermediate of the pentose phosphate pathway. The 6PGL intermediate reacts directly with the N-terminal amino group of proteins resulting in a 259 Da molecular addition. The 6PGL adduct can subsequently be dephosphorylated by a phosphatase to give a single gluconolactone modification of 178 Da (35). Such modifications affect protein purity and activity (4, 5), which are unwanted characteristics of recombinant proteins that are used as therapeutics. We have shown that a single gluconolactone modification can be detected in a recombinant protein using mP-AGE, resulting in a dramatic retention of the carbohydrate modified fraction of this protein in the gel (Fig. 7).

1.9. Recent Improvements of the mP-AGE Technique

1.9.1. Step Gels

Gels containing a gradient of acrylamide concentrations have been used specifically for analysing protein samples containing a mixture of high- and low-molecular-weight species (36). Here, we have used step gels to enhance the separation of the glycated fraction of HSA based on their affinity for MPBA. By stacking three layers of gel with varying boronic acid concentrations, placing the gel with the highest MPBA concentrations at the top, we have been able to maximise the retention of the most heavily glycated protein fractions (Fig. 8).

Fig. 6. (continued) such as glucose and fructose predominantly exist in a pyranose form, but it is the linear form that reacts with amino groups of proteins. Interaction between boronic acid and anomeric *cis*-1,2-diols is preferential (shown in *grey*), as five-membered cyclic boronic esters are more stable than six-membered ones. The weaker 1,3-diol interaction is shown in *light grey*. The Amadori product resulting from glucose incubation contains an anomeric *cis*-1,2- and 4,5-diol available for interaction with MPBA. Additional stabilization results from the electrostatic interaction between the negatively charged boronate and amine group. In contrast, the pyranose configuration of the fructose Amadori product does not have any *cis*-diols for such strong interaction. In its furanose ring structure, the fructose adduct contains an anomeric 3,4,5-diol. The lack of retention in the MPBA gels suggests that the pyranose adduct is predominantly formed in the reaction of HSA with fructose. The lactose adduct contains a *cis*-3′,4′-diol in addition to the anomeric diol. Therefore, it is expected to have similar interactions to the glucose adduct. Although glucose-6-phosphate contains the anomeric *cis*-diol, the proximity of the phosphate group is likely to cause steric hindrance. The Amadori product for mannose is identical to that of glucose and shows a band shift similar to the glucose incubated sample. Maltose is a disaccharide, which has an anomeric diol available for interaction. In the mP-AGE gel, HSA sample glycated with maltose shows the biggest shift (Fig. 8), which could be a result of the flexibility of *cis*-diols occurring in different components of the disaccharide. Molecular carbohydrate structures were produced using ChemDraw (Cambridgesoft).

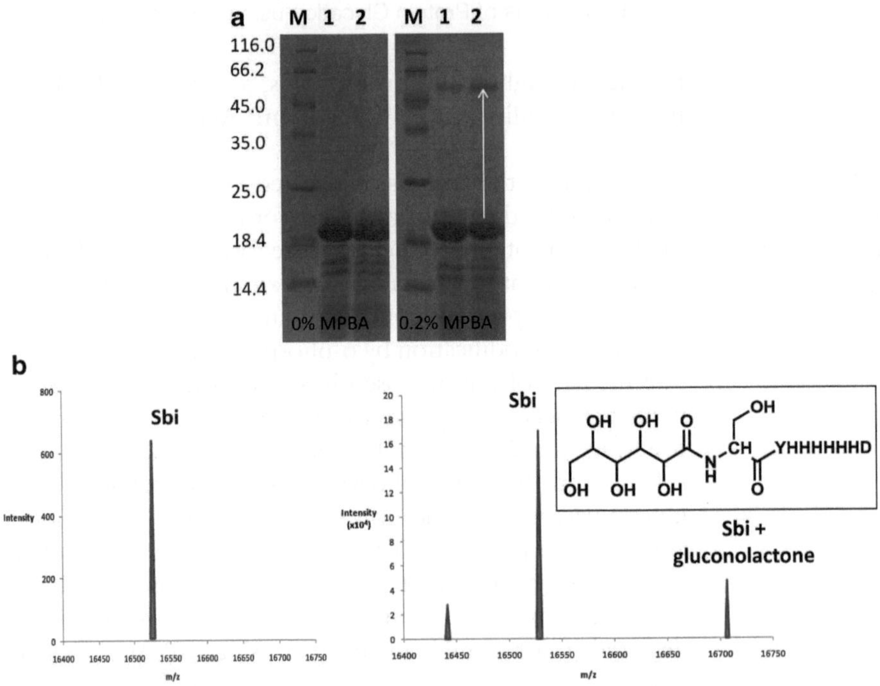

Fig. 7. Analysis of gluconoylated recombinant Sbi-III–IV protein. (**a**) SDS-PAGE analysis of freshly purified Sbi-III–IV protein (*lane 1*) and protein after one hour incubation with 50 mM δ-gluconolactone (*lane 2*) on gels with (*right*) and without (*left*) MPBA. Sbi is gluconoylated at the N-terminal histidine-tag during protein expression in *E. coli*. Normal SDS-PAGE shows identical bands for both freshly purified and gluconoylated protein. The faint band of gluconoylated protein at around 20 kDa is shifted to a position almost three times its molecular weight in the gel containing 0.2% w/v MPBA, whilst the unmodified protein remains unaffected. The dramatic enhancement in protein retention indicates a strong affinity between the gluconoylated protein and the gel-incorporated boronic acid. The linearity of the gluconoyl adduct, and the number of hydroxyl groups present in a resonance structure, makes this adduct very suitable for boronic acid interaction. (**b**) Mass spectra of freshly purified Sbi-III–IV (*left*) and exogenously gluconoylated protein (*right*) with the additional peak corresponding to the addition of a gluconolactone adduct (178 Da). The molecular structure of this adduct is shown in the *inset*.

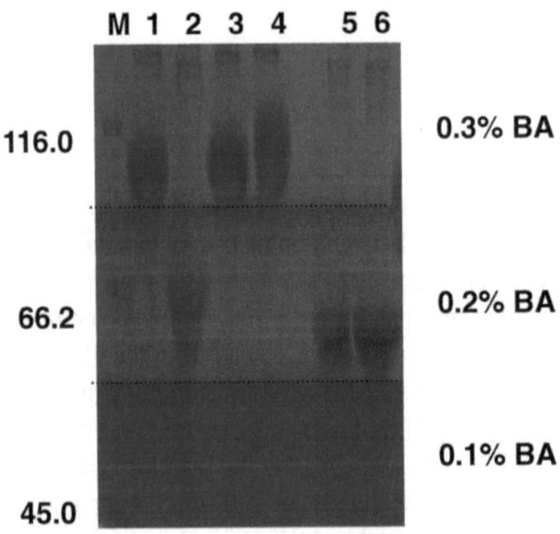

Fig. 8. mP-AGE analysis of glycated HSA using a step gel. SDS-PAGE analysis comparing step gel profiles containing (from *top* to *bottom*) 0.3, 0.2, and 0.1% w/v MPBA. Human serum albumin (HSA) was incubated for 21 days at 37°C with: glucose (*lane 1*), fructose (*lane 2*), mannose (*lane 3*), maltose (*lane 4*), PBS (*lane 6*), and fresh HSA (*lane 5*). Glucose-, mannose-, and maltose-incubated HSA adducts display remarkable gel retention from 66 kDa to around 100–120 kDa due to their boronic acid affinity.

1.9.2. Two-Dimensional Gels

Two-dimensional (2D) polyacrylamide gel electrophoresis can resolve up to 5,000 proteins in complex biological samples (37) by separating proteins according to different properties in each dimension, typically pI and molecular weight. We have developed a 2D gel system where proteins are separated based on their molecular weight (SDS-PAGE) and their affinity for boronic acid (mP-AGE). Unmodified proteins will migrate as a diagonal in the second dimension whilst modified proteins are retained above this diagonal. In Fig. 9, an example of a 2D gel analysis is shown where the gluconoylated fraction of Sbi-III–IV is retained in the gel at a position above the diagonal formed by unaffected protein bands. This method can be a useful tool in proteomic analyses of glycated proteins.

In this chapter, we present novel gel electrophoresis-based methods that can differentiate glycated proteins from unmodified ones with the incorporation of boronic acid (mP-AGE). This cost-effective methodology has the potential to be used in disease diganosis and the screening of glycation inhibitors.

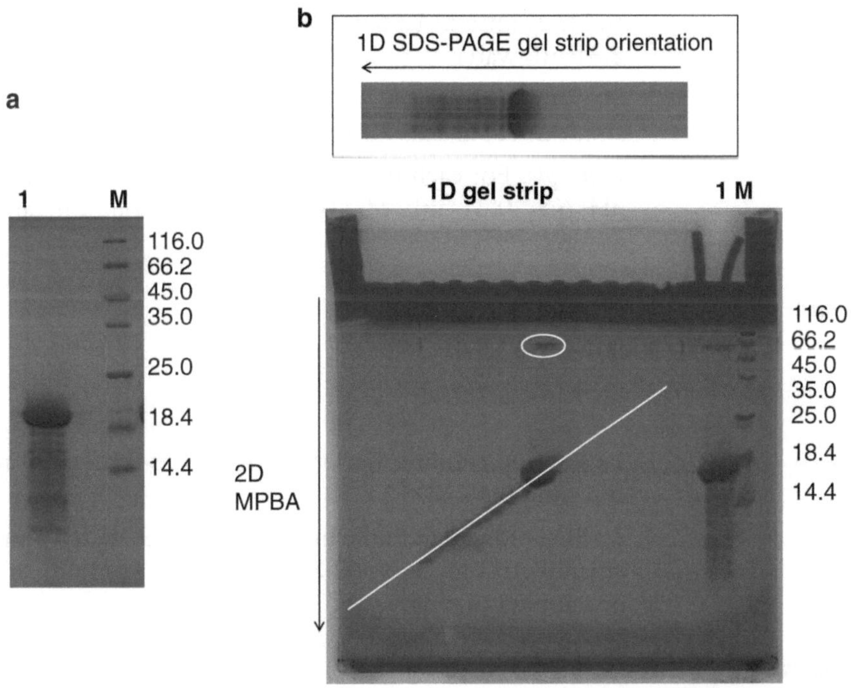

Fig. 9. 2D mP-AGE analysis of gluconoylated Sbi-III-IV (*lane 1*). (**a**) SDS-PAGE gel profile (first dimension) and (**b**) 2D mP-AGE gel profile using 0.5% w/v MPBA (second dimension). The *inset* shows the orientation of the 1D gel strip before electrophoresis in the second dimension. All proteins that do not bind boronic acid appear on the diagonal as they migrate to the same extent in both directions, whilst the gluconoylated protein band is retained above the diagonal in MPBA gel (at around 65 kDa).

2. Materials

2.1. For Casting Gels

1. Gel cassette.
2. Comb.
3. 40% acrylamide solution (29:1 acrylamide: *N,N'*-methylene-bisacrylamide).
4. Tris(hydroxymethyl)aminomethane (Tris).
5. *N,N,N',N'*-tetraethylmethylene diamine (TEMED).
6. 10% ammonium persulphate (APS).
7. Methacrylamido phenylboronic acid (MPBA).

2.1.1. Gel Solutions

1. Resolving gel solution: 8 or 15% polyacrylamide (acrylamide: bisacrylamide 29:1) in 375 mM Tris–HCl pH 8.8 (see Note 1).

   Example: For an 8% polyacrylamide mini gel (dimension height: 100 mm × width: 100 mm × thickness: 1.0 mm), prepare 1.25 mL of 40% acrylamide solution, 1.56 mL 1.5 M Tris–HCl pH 8.8, and 3.40 mL water.

   For 15% polyacrylamide mini gel, use 2.34 mL of 40% acrylamide solution, 1.56 mL of 1.5 M Tris–HCl pH 8.8, and 2.31 mL water.
2. Stacking gel solution: 4% polyacrylamide (acrylamide: bisacrylamide 29:1) in 125 mM Tris–HCl pH 6.8.

   Example: For each mini gel, mix 0.78 mL of 0.5 M Tris–HCl pH 6.8, 0.31 mL of 40% acrylamide solution, and 2 mL water.

2.2. For Electrophoresis

1. Electrophoresis gel tank.
2. Power pack.
3. Protein marker.

2.2.1. Buffers

1. 1× Glycine running buffer: 25 mM Tris–HCl, 190 mM glycine, and 0.1% SDS.
2. 2× SDS reducing sample buffer: 3% w/v SDS, 5% β-mercaptoethanol, 10% glycerol, 62.5 mM Tris–HCl pH 6.8, and 0.01% bromophenol blue.

2.3. For Visualisation

1. Coomassie or silver stain.
2. Light box.

3. Methods

3.1. Sample Preparation

1. Add an equal volume of 2× sample buffer to the protein sample.
2. Heat the mixture at 100°C for 5–10 min for protein denaturation and incorporation of SDS.
3. For Coomassie staining, load approximately 1 μg of protein on the gel depending on sample complexity. In silver stained gels, the sample loading should be decreased by 20-fold to obtain an interpretable gel profile (see Note 2).

3.1.1. Glycation of Human Serum Albumin

1. Incubate HSA (20 mg/mL) with 50 mM saccharides, such as glucose and fructose, at 37°C under aseptic conditions. Azide can be added to prevent bacterial growth.
2. Glycation of HSA can be monitored using fluorescence spectroscopy. Glycated proteins give higher fluorescence intensity between 400 and 550 nm when excited at 370 nm.

3.1.2. Gluconoylation of Sbi-III–IV

1. Incubate Sbi-III–IV (2 mg/mL) with 50 mM δ-gluconolactone for 1 h at 37°C.

3.2. Boronic Acid (MPBA) Synthesis

MPBA can be purchased or synthesised as follows (Fig. 10) (31):

1. Add boronic acid A to diol B to a round-bottomed flask in a 1:1 molar ratio in chloroform.
2. Stir the suspension overnight (~15 h), then filter and evaporate to dryness *in vacuo*, yielding the intermediate Boronic ester C.

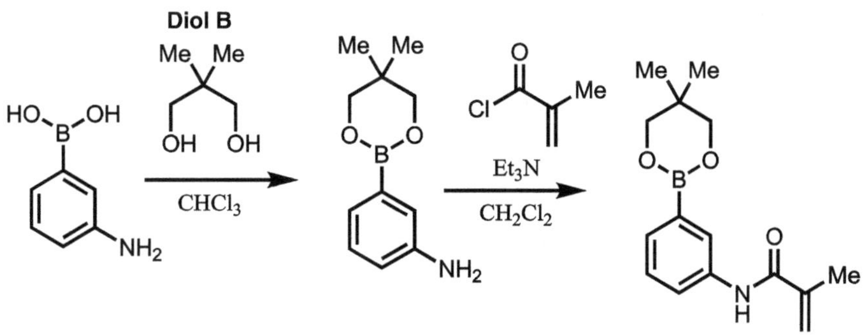

Boronic acid A Boronic ester C MPBA

Fig. 10. Synthesis of methacrylamido phenylboronic acid (MPBA). Synthesis proceeds via two steps, (1) involving the protection of the boronic acid moiety and (2) reaction with acyl chloride to give methacrylamido group. MPBA is soluble in aqueous polyacrylamide solution and can be polymerised into the backbone of the gel matrix aiding sample analysis in electrophoresis.

3. Dissolve boronic ester C in dichloromethane (DCM), then add one molar equivalent of methacryloyl chloride and stir the mixture at 0°C under nitrogen.
4. Add a solution of triethylamine (one molar equivalent) in DCM slowly to the mixture over 2 h and allow to warm to room temperature with stirring for a further 30 min. Wash this organic solution with water, dry over magnesium sulphate, filter, and evaporate to dryness *in vacuo* to give MPBA.

3.3. Casting Gels

3.3.1. Normal Polyacrylamide Gel

1. Polymerize resolving gel solution (see Subheading 2.1.1) with 3 μL TEMED and 31 μL 10% APS. Transfer the solution to an empty gel cassette and add a few drops of saturated butanol or water to the top of the gel solution to level it and prevent atmospheric oxygen from inhibiting the polymerisation process. Gel polymerises in approximately 20 min (see Notes 3–5).
2. Once the gel is set, pour away the butanol, rinse the gel with water and dry the inside of the gel cassette with a piece of filter paper without touching the resolving gel.
3. Polymerize the stacking gel solution (see Subheading 2.1.1) by adding 3 μL TEMED and 16 μL of 10% APS. Pour this solution on top of the resolving gel and insert a gel comb before this gel polymerises to generate sample wells. This gel can be stored at 4°C for about a month (see Note 6).
4. For electrophoresis, remove gel comb and insert the gel cassette into the gel tank apparatus according to the manufacturer's instructions. After adding running buffer, protein samples can be loaded onto the gel and electrophoresis commence. Stop the run when the dye front reaches the bottom of the gel.

3.3.2. MPBA Gel

1. Dissolve the required amount of MPBA in the resolving gel solution prior to polymerisation.
2. Once dissolved, MPBA can be polymerised and will be incorporated into the backbone of the gel matrix ready for use in electrophoresis (see Note 7). Boronic acids are not needed in the stacking gel.

3.3.3. Step Gel

1. Prepare step gels by polymerising resolving gels in layers on top of each other in the gel cassette, varying in polyacrylamide percentage and/or MPBA concentration.
2. Before the addition of every new layer of gel, the previous layer is allowed time to set. In the Fig. 8 example, the same polyacrylamide concentration is used throughout the gel, with each layer varying in boronic acid concentration.

3.3.4. Two-Dimensional Gel

1. 2D gels are made by casting two gels, the first as a normal polyacrylamide gel and the second with MPBA.

2. After running the protein sample in the first dimension SDS-PAGE gel (see Subheading 3.3.1), cut the sample lane out and soak in sample buffer for 10 min.
3. The 1D gel strip is placed horizontally on top of the stacking gel in the gel cassette containing the MPBA gel that is used for the second dimension. The top of the cut strip faces the control lane (Fig. 9) and electrophoresis analysis performed (see Note 8).

4. Notes

1. Gels are often identified with their percentages of acrylamide (%T) and cross-linker (%C) that can be calculated as follows:

   %T = (mass of acrylamide (g) + bisacrylamide (g)) × 100/volume (mL)

   %C = mass of bisacrylamide (g) × 100/(mass of acrylamide (g) + bisacrylamide (g)).
2. Due to the stacking effect, where the rapid migration of proteins in the stacking gel allows them to be stacked into thin zone before resolution, large volumes of 15–20 μL of dilute sample can be loaded onto each lane of the gel.
3. APS and radical initiator TEMED should only be added just before polymerisation is required.
4. Polymerisation is temperature dependent. Therefore, gels should be cast above 20°C. Lower temperatures will result in longer polymerisation time and may cause incomplete polymerisation. Always allow the gels to polymerize at least 2 h prior to electrophoresis to ensure complete polymerisation.
5. To determine if the gel is set, gently tilt the gel cassette to see whether the gel is in solid or liquid state.
6. Mark the bottom of each sample well in the gel cassette with a marker pen to ease sample loading.
7. Maximum dissolution of MPBA in Tris–HCl pH 8.8 gel solution is about 1% w/v. The required amount of MPBA is added to polyacrylamide gel solution in a falcon tube and shaken until MPBA completely dissolves. Do not heat the solution as this might result in unwanted polymerisation. The solubility of MPBA varies with the pH and composition of gel buffer.
8. 2D gel comb should be inserted into the stacking gel. A thicker cassette can be used for the second-dimension electrophoresis gel if inserting the first dimension gel strip proves to be difficult.

Acknowledgements

This work was supported by the Biotechnology and Biological Sciences Research Council (BBSRC) (MPPM). We are grateful for the University of Bath Research Development & Support Office (RDSO) for funding.

References

1. Ulrich P, Cerami A (2001) Protein glycation, diabetes, and aging. Recent Prog Horm Res 56:1–22
2. Lapolla A, Traldi P, Fedele D (2005) Importance of measuring products of non-enzymatic glycation of proteins. Clin Biochem 38:103–115
3. Pokharna H, Pottenger L (1997) Nonenzymatic glycation of cartilage proteoglycans: an in vivo and in vitro study. Glycoconj J 14:917–923
4. Geoghegan KF, Dixon HBF, Rosner PJ et al (1999) Spontaneous (alpha)-N-6-phosphogluconoylation of a "His Tag" in *Escherichia coli*: the cause of extra mass of 258 or 178 Da in fusion proteins. Anal Biochem 267:169–184
5. Yan Z, Caldwell GW, McDonell PA (1999) Identification of a gluconic acid derivative attached to the N-terminus of histidine-tagged proteins expressed in bacteria. Biochem Biophys Res Commun 262:793–800
6. Monnier VM, Cerami A (1981) Nonenzymatic browning in vivo: possible process for aging of long-lived proteins. Science 211:491–493
7. Foerster A, Henle T (2003) Glycation in food and metabolic transit of dietary AGEs (advanced glycation end-products): studies on the urinary excretion of pyrraline. Biochem Soc Trans 31:1383–1385
8. Montgomery H, Tanaka K, Belgacem O (2010) Glycation pattern of peptides condensed with maltose, lactose and glucose determined by ultraviolet matrix-assisted laser desorption/ionization tandem mass spectrometry. Rapid Commun Mass Spectrom 24: 841–848
9. Klenk DC, Hermanson GT, Krohn RI et al (1982) Determination of glycosylated hemoglobin by affinity chromatography: comparison with colorimetric and ion-exchange methods, and effects of common interferences. Clin Chem 28:2088–2094
10. Zhang Q, Ames JM, Smith RD et al (2008) A perspective on the maillard reaction and the analysis of protein glycation by mass spectrometry: probing the pathogenesis of chronic disease. J Proteome Res 8:754–769
11. Wu JT, Tu M-C, Zhung P (1996) Advanced glycation end product (AGE): characterization of the products from the reaction between D-glucose and serum albumin. J Clin Lab Anal 10:21–34
12. Weber K, Osborn M (1969) The reliability of molecular weight determinations by dodecyl sulfate-polyacrylamide gel electrophoresis. J Biol Chem 244:4406–4412
13. Pereira Morais MP, Mackay JD, Bhamra SK et al (2009) Analysis of protein glycation using phenylboronate acrylamide gel electrophoresis. Proteomics 10:48–58
14. Davis BJ, Ornstein L (1959) A new high resolution electrophoresis method. In: Society for the study of blood, New York Academy of Medicine
15. Raymond S, Weintraub L (1959) Acrylamide gel as a supporting medium for zone electrophoresis. Science 130:711
16. Miksík I, Deyl Z (1997) Post-translational non-enzymatic modification of proteins. II. Separation of selected protein species after glycation and other carbonyl-mediated modifications. J Chromatogr B Biomed Sci Appl 699:311–345
17. Garfin DE (2003) Gel electrophoresis of proteins. In: Davey J, Lord MJ (eds) Essential cell biology: a practical approach. Oxford University Press, Oxford
18. Switzer RC, Merril CR, Shifrin S (1979) A highly sensitive silver stain for detecting proteins and peptides in polyacrylamide gels. Anal Biochem 98:231–237
19. Springsteen G, Wang B (2002) A detailed examination of boronic acid-diol complexation. Tetrahedron 58:5291–5300
20. Lorand JP, Edwards JO (1959) Polyol complexes and structure of the benzeneboronate ion. J Org Chem 24:769–774
21. Nishiyabu R, Kubo Y, James TD et al (2010) Boronic acid building blocks: tools for sensing and separation. Chem Commun 47: 1106–1123
22. Weith HL, Wiebers JL, Gilham PT (1970) Synthesis of cellulose derivatives containing the

dihydroxyboryl group and a study of their capacity to form specific complexes with sugars and nucleic acid components. Biochemistry 9:4396–4401
23. Liu X-C (2006) Boronic acids as ligands for affinity chromatography. Chinese J Chromatogr 24:73–80
24. Cartwright SJ, Waley SG (1984) Purification of beta-lactamases by affinity chromatography on phenylboronic acid-agarose. Biochem J 221: 505–512
25. Armbruster DA (1987) Fructosamine: structure, analysis, and clinical usefulness. Clin Chem 33:2153–2163
26. Priego Capote F, Sanchez J-C (2009) Strategies for proteomic analysis of non-enzymatically glycated proteins. Mass Spectrom Rev 28:135–146
27. Jackson TR, Springall JS, Rogalle D et al (2008) Boronate affinity saccharide electrophoresis: a novel carbohydrate analysis tool. Electrophoresis 29:4185–4191
28. Jackson P (1990) The use of polyacrylamide-gel electrophoresis for the high-resolution separation of reducing saccharides labelled with the fluorophore 8-aminonaphthalene-1,3,6-trisulphonic acid. Detection of picomolar quantities by an imaging system based on a cooled charge-coupled device. Biochem J 270:705–713
29. Mahoney DJ, Aplin RT, Calabro A et al (2001) Novel methods for the preparation and characterization of hyaluronan oligosaccharides of defined length. Glycobiology 11:1025–1033
30. Calabro A, Benavides M, Tammi M et al (2000) Microanalysis of enzyme digests of hyaluronan and chondroitin/dermatan sulfate by fluorophore-assisted carbohydrate electrophoresis (FACE). Glycobiology 10:273–281
31. Ma WMJ, Pereira Morais MP, D'Hooge F et al (2009) Dye displacement assay for saccharide detection with boronate hydrogels. Chem Commun 5:532–534
32. Curry S, Mandelkow H, Brick P et al (1998) Crystal structure of human serum albumin complexed with fatty acid reveals an asymmetric distribution of binding sites. Nat Struct Mol Biol 5:827–835
33. Sattarahmady N, Moosavi-Movahedi AA, Ahmad F et al (2007) Formation of the molten globule-like state during prolonged glycation of human serum albumin. Biochim Biophys Acta 1770:933–942
34. Inaba M, Okuno S, Kumeda Y et al, Osaka CKD Expert Research Group (2007) Glycated albumin is a better glycemic indicator than glycated hemoglobin values in hemodialysis patients with diabetes: effect of anemia and erythropoietin injection. J Am Soc Nephrol 18:896–903
35. Burman JD, Leung E, Atkins KL et al (2008) Interaction of human complement with Sbi, a staphylococcal immunoglobulin-binding protein: indications of a novel mechanism of complement evasion by *Staphylococcus aureus*. J Biol Chem 283:17579–17593
36. Margolis J, Kenrick KG (1967) Polyacrylamide gel-electrophoresis across a molecular sieve gradient. Nature 214:1334–1336
37. O'Farrell PH (1975) High resolution two-dimensional electrophoresis of proteins. J Biol Chem 250:4007–4021

Chapter 10

Protein Electrophoresis in Agarose Gels for Separating High Molecular Weight Proteins

Marion L. Greaser and Chad M. Warren

Abstract

Very large proteins (subunit sizes >200 kDa) are difficult to electrophoretically separate on polyacrylamide gels. A SDS vertical agarose gel system has been developed that has vastly improved resolving power for very large proteins. Proteins with molecular masses between 200 and 4,000 kDa can be clearly separated. Inclusion of a reducing agent in the upper reservoir buffer has been found to be a key technical procedure for obtaining optimum resolution.

Key words: SeaKem Gold agarose, Titin, DATD, Large proteins

1. Introduction

Very large molecular weight proteins are difficult to separate by electrophoresis because of their poor penetration into gels using the Laemmli SDS (sodium dodecyl sulfate) polyacrylamide system (1). Protein migration in SDS gels has been found to be linear with the log of the molecular weight (2), so the larger the protein, the more poorly it is resolved from other big proteins. Many workers have attempted to solve this problem by using very low-concentration acrylamide gels (3), acrylamide mixed with agarose (4), or acrylamide gradients (5) to better separate large proteins. Low-concentration acrylamide gels are difficult to use because of their mechanical fragility and distortion during handling. A new electrophoresis system using SDS agarose for protein electrophoresis and blotting has been described (6). Although the system was developed and has been extensively used with muscle samples, it has also been applied in work with other large proteins (such as the von Willebrand factor multimers (7, 8) and analysis of huntingtin

Biji T. Kurien and R. Hal Scofield (eds.), *Protein Electrophoresis: Methods and Protocols*, Methods in Molecular Biology, vol. 869, DOI 10.1007/978-1-61779-821-4_10, © Springer Science+Business Media, LLC 2012

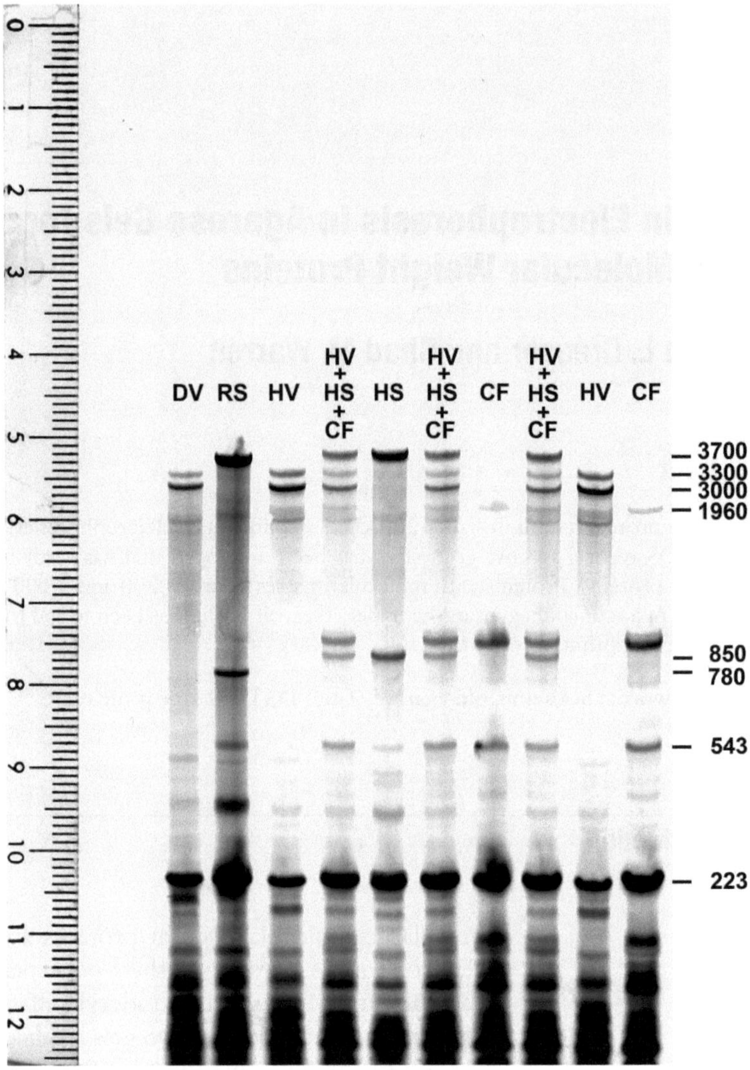

Fig. 1. SDS 1% agarose gel stained with silver. A centimeter ruler is shown on the *left,* and the sizes of the various protein bands in kDa are listed on the *right. DV* dog ventricle, *RS* rat soleus, *HV* human ventricle, *HS* human soleus, *CF* crayfish claw muscle. Human soleus titin is 3,700 kDa, and human ventricle has two titin bands of 3,300 and 3,000 kDa. The bands at 780 and 850 kDa in the skeletal muscle samples are rat and human nebulin, respectively. The myosin heavy chain is 223 kDa.

aggregates (9)), and as a component of a two-dimensional proteomics protocol (10). An example showing the resolution for several muscle samples containing large proteins is shown in Fig. 1. Migration distance shows a linear relationship with the log of the molecular weight (6). This system also allows more quantitative transfer of proteins from the gel for western blots and achieves much higher reproducibility than can be obtained with methods using low-percentage acrylamide.

2. Materials

2.1. Apparatus

1. SE 600 Slab Gel Unit with 16×18-cm glass plates (Hoefer, Holliston, MA) or a similar commercial gel unit (see Note 1).
2. 65°C oven.
3. A constant current power supply.
4. Circulating cooler.

2.2. Stock Solutions

1. Acrylamide gel for plug: 38.5% acrylamide. Weigh 37.5 g of acrylamide and 1 g DATD (*N,N′*-diallyltartardiamide) (Bio-Rad, Hercules, CA) into a beaker, add about 50 mL of water, stir till dissolved, and dilute to 100 mL. Filter through a 0.45-μm filter (such as a Millex-HA, Millipore Corporation, Billerica, MA). Store in a brown bottle in the cold room (4°C). Danger! Avoid skin contact.
2. Reservoir and agarose gel buffer concentrate (5×): 0.25 M Trizma base, 1.92 M glycine, and 0.5% SDS. Store at room temperature.
3. Ammonium persulfate: Prepare a 100 mg/mL solution in water; store frozen in 0.5 mL aliquots (stable indefinitely at -20°C).
4. Sample buffer: 8 M urea, 2 M thiourea, 0.05 M Tris–HCl (pH 6.8), 75 mM DTT, 3% SDS, and 0.05% bromophenol blue (adapted from ref. 11). (Dissolve urea and thiourea and treat with mixed bed resin to remove ionic constituents; then add remaining ingredients. Store at –20°C).
5. 50% v/v glycerol.
6. Acrylamide plug solution: In a 15-mL plastic beaker, add 1.924 mL deionized water, 1.7 mL 50% glycerol, 2.12 mL 3 M Tris–HCl (pH 9.3), 2.72 mL acrylamide plug solution (38.5%), 24 μL 10% ammonium persulfate, and 13 μL TEMED (tetramethylethylenediamine). Mix by gently pipetting a few times. Prepare complete solution only after gel plates are assembled.
7. Agarose gel solution: Weigh 0.8 g of SeaKem Gold agarose (Lonza Group Ltd, Basel, Switzerland) into a 600-mL beaker (see Note 2). To a 100-mL graduated cylinder, add 48 mL of 50% v/v glycerol (see Note 3), 16 mL 5× electrophoresis buffer, and bring volume up to 80 mL with deionized water. Place parafilm over top of the graduated cylinder, mix by inverting a few times, and pour solution into the 600-mL beaker containing the agarose. Place saran wrap over top of beaker and poke a few holes in the saran wrap. Weigh beaker with contents. Place beaker in a microwave oven along with a separate beaker of deionized water. Heat for a total of 2 min (stop every 30 s to swirl—protect hand with an insulated glove) (see Note 4).

Prepare agarose solution after the polyacrylamide plug has been inserted.

8. Coomassie blue protein stain:
   (a) Stain: Dissolve 0.05% (w/v) Coomassie blue R-250 in 50% (v/v) methanol and 10% (v/v) acetic acid. Store at room temperature in a closed container (see Note 5).
   (b) Destain: 10% (v/v) methanol and 7.5%(v/v) acetic acid.
9. Silver protein stain:
   (a) Fixing solution: 50% v/v methanol, 5% v/v glycerol, and 12% v/v acetic acid. Prepare fixing solution by adding, in the following order, 500 mL methanol, 50 mL glycerol, and water to 880-mL mark in a graduated cylinder, mix, then add 120 mL acetic acid and mix. This makes enough for two agarose gels.
   (b) Potassium ferrocyanide solution: potassium ferrocyanide (20 g/L) (500 mL per gel).
   (c) Staining solutions: Solution A—dissolve 25 g sodium carbonate in deionized water and bring up to 500 mL (prepare solution A in 1,000-mL beaker or larger). Solution B—dissolve 5 g silicotungstic acid, 1 g ammonium nitrate, 1 g silver nitrate, and 3.35 mL 37% formaldehyde in deionized water; bring final volume to 500 mL.

3. Methods

3.1. Gel Preparation

1. Volumes listed will provide enough solution for two 16 × 18-cm gels with 1.5-mm spacers.
2. Clean plates and spacers with soap, rinse with distilled water, and finally with ethanol.
3. Assemble gel plates. Place plate on clean bench top. Place spacers hanging half the way off each side of plate. Place second plate on top. Stand up plates and place one side into the clamp. Align spacer with side of plates and clamp and push spacer down so that bottom is flush with the glass plates (top buffer will leak if spacers are not flush with plates).
4. Pour acrylamide plugs in bottom of gel plate assembly (see Note 6). Mix by gently pipetting a few times. Immediately add 2.5 mL to each gel assembly. Add a small amount of water on top of each plug to level the upper surface and provide an oxygen barrier. Allow gel to polymerize for 20–30 min. Drain off water layer by inverting gel plate assembly on a paper towel.
5. Place assembly, 20-lane sample combs, and 60-mL plastic syringe in a 65°C oven for 10 min (see Note 7).

6. Allow freshly prepared agarose to cool for a few minutes at room temperature. Reweigh and add sufficient heated deionized water to replace that lost by evaporation.
7. Draw up about 40 mL of agarose in the pre-warmed 60-mL Luer-Lock syringe and pour each gel slowly until it just overflows the top of the plates. Try to avoid formation of bubbles (if bubbles are present, allow them to migrate to the top of the gel and remove them with the back of the sample comb by pinching off the top bead of agarose). Insert sample combs and allow unit to cool at room temperature for about 45 min (see Note 8).

3.2. Electrophoresis Setup and Sample Loading

1. Add 4 L of buffer to lower chamber (3,200 mL deionized water plus 800 mL 5× electrophoresis buffer). Start cooling unit and stir bar (gels run at 6°C).
2. Prepare 600 mL upper chamber electrophoresis buffer (same concentration as lower chamber buffer). Add 2-mercaptoethanol (final concentration of 10 mM) (12) (see Note 9). Buffer will be poured into top chamber after samples are loaded and assembly placed in unit.
3. Take combs out of gels by bending them back and forth to detach from gel and slowly pull them up. Pour a small amount of upper chamber buffer into a 15-mL beaker and pipette buffer into first and last wells (the rest will fill over). Add buffer to remove any trapped bubbles. Insert pipette tip to deposit sample in bottom of the sample well. Skip the first and last lanes (see Note 10).
4. Running gels. Once samples are loaded, put upper chamber on the assembly. Pour upper chamber buffer into upper chamber from corners (do not pour buffer directly over wells). Place lid on unit and connect to power supply. Turn electrophoresis unit on and run at 30 mA (two gels) for 3 h.

3.3. Staining and Destaining

1. After the tracking dye reaches the bottom of the acrylamide plug, turn off the power and disassemble the plates. Cut off sample wells and acrylamide plug and discard.
2. Gels can be stained with Coomassie blue R-250 or silver (see steps 3–11 below) depending on the sensitivity needed. Place agarose gel in a plastic or glass dish with the Coomassie stain and gently agitate for 1 h. Discard stain and soak gel in methanol–acetic acid destain with gentle shaking until the background clears (usually overnight). Include a Chemwipe in the destain to trap the Coomassie.
3. For more sensitive protein detection, use a silver staining procedure adapted from ref. (13).

4. Add 500 mL of 50% v/v methanol and 10% v/v acetic acid fixing solution to a pre-cleaned polypropylene bin (see Note 11) and slide agarose gel into the solution. Cover and gently agitate on a shaker for 1 h. Discard fixing solution, then transfer bin with gel to a 37°C oven overnight.
5. Add 500 mL deionized water to gel, agitate gently for 20 min, and discard solution.
6. Repeat step 6 two more times (three washes, 20 min each).
7. Add 500 mL potassium ferrocyanide solution to each gel and shake for 5 min, discard solution.
8. Add 500 mL of deionized water to each gel and shake for 5 min, discard water.
9. Repeat gel wash two more times with 500 mL deionized water for 5 min (total of three washes, 5 min each).
10. Slowly add solution B to a stirring solution A just prior to staining. Shake gels in staining solution (500 mL/gel) until bands appear (5–10 min. Do not overstain). Decant staining solution and add 500 mL of 1% v/v acetic acid to each gel and shake for 5 min to stop staining. Decant acetic acid and wash gels with 500 mL deionized water for 5 min.
11. Dry gels between two sheets of wet Mylar and add a couple milliliter of glycerol to make the sandwich less brittle.

4. Notes

1. The agarose gel procedure works equally well with small format gels (i.e., 8 × 10 cm).
2. It is essential to use SeaKem Gold agarose for optimal migration of high molecular weight proteins. This type has large pore size and excellent mechanical stability. Other types of agarose may be used, but the protein mobility will be significantly reduced.
3. Glycerol is included in the mixture to increase the solution viscosity inside the gel and thus sharpen the protein bands.
4. Periodic swirling during the heating step eliminates nonhydrated agarose granules in the final gel.
5. Dissolving the Coomassie in methanol first and then adding the water and acetic acid reduces problems with dye aggregates.
6. The acrylamide plug is used to prevent the agarose from slipping out of the vertical gel plate assembly. Use of DATD as the cross-linker provides an acrylamide that sticks better to the glass plates than if a conventional bisacrylamide

cross-linker is used. Plugs can be poured a day before making the agarose gel (place tape or parafilm over the top of the plates to prevent drying and store in cold room).

7. Preheating the glass plate assembly, well comb, and syringe prevents premature agarose gelling when the solution touches the colder surfaces. In addition, the plates are less likely to crack during pouring if they are closer to the temperature of the hot agarose.
8. Sample combs should extend no longer than 1 cm into agarose; otherwise, they may be difficult to remove. Gels can be used right away or stored overnight in a cold room.
9. The disulfide bond formation of large proteins during electrophoresis also retards their migration and may lead to smearing (see ref. 12). Thus, inclusion of 2-mercaptoethanol in the upper buffer improves resolution of high molecular weight proteins. Alternatively, protein can be alkylated to prevent disulfide bond formation (14).
10. Conventional sample buffers may not be dense enough for the sample to stay at the bottom of the well. If necessary, add additional glycerol (up to 30% v/v final concentration) to increase sample density.
11. Bins may be cleaned with a 50% nitric acid solution to remove silver deposits and then given a final rinse with water.

Acknowledgments

This work was supported by the College of Agricultural and Life Sciences, University of Wisconsin-Madison, and from grants (MLG-NIH HL77196 and Hatch NC1131).

References

1. Laemmli UK (1970) Cleavage of structural proteins during the assembly of the head of bacteriophage T4. Nature 227:680–685
2. Weber K, Osborn M (1969) The reliability of molecular weight determinations by dodecyl sulfate-polyacrylamide gel electrophoresis. J Biol Chem 244:4406–4412
3. Granzier H, Wang K (1993) Gel electrophoresis of giant proteins: solubilization and silver-staining of titin and nebulin from single muscle fiber segments. Electrophoresis 14:56–64
4. Tatsumi R, Hattori A (1995) Detection of giant myofibrillar proteins connectin and nebulin by electrophoresis in 2% polyacrylamide slab gels strengthened with agarose. Anal Biochem 224: 28–31
5. Cazorla O, Freiburg A, Helmes M et al (2000) Differential expression of cardiac titin isoforms and modulation of cellular stiffness. Circ Res 86:59–67
6. Warren CM, Krzesinski PR, Greaser ML (2003) Vertical agarose gel electrophoresis and electroblotting of high-molecular-weight proteins. Electrophoresis 24:1695–1702
7. Wu JJ, Fujikawa K, McMullen BA et al (2006) Characterization of a core binding site for ADAMTS-13 in the A2 domain of von Willebrand factor. Proc Natl Acad Sci USA 103:18470–18474

8. Ott HW, Griesmacher A, Schnapka-Koepf M et al (2010) Analysis of von Willebrand factor multimers by simultaneous high- and low-resolution vertical SDS-agarose gel electrophoresis and Cy5-labeled antibody high-sensitivity fluorescence detection. Am J Clin Pathol 133:322–330
9. Hoffner G, Island ML, Djian P (2005) Purification of neuronal inclusions of patients with Huntington's disease reveals a broad range of N-terminal fragments of expanded huntingtin and insoluble polymers. J Neurochem 95:125–136
10. Oh-Ishi M, Maeda T (2007) Disease proteomics of high-molecular-mass proteins by two-dimensional gel electrophoresis with agarose gels in the first dimension (agarose 2-DE). J Chromatogr B Analyt Technol Biomed Life Sci 849:211–222
11. Yates LD, Greaser ML (1983) Quantitative determination of myosin and actin in rabbit skeletal muscle. J Mol Biol 168:123–141
12. Fritz JD, Swartz DR, Greaser ML (1989) Factors affecting polyacrylamide gel electrophoresis and electroblotting of high-molecular-weight myofibrillar proteins. Anal Biochem 180:205–210
13. Peats S (1984) Quantitation of protein and DNA in silver-stained agarose gels. Anal Biochem 140:178–182
14. Sechi S, Chait BT (1998) Modification of cysteine residues by alkylation. A tool in peptide mapping and protein identification. Anal Chem 70:5150–5158

Chapter 11

Increase in Local Protein Concentration by Field-Inversion Gel Electrophoresis

Henghang Tsai and Hon-chiu Eastwood Leung

Abstract

Proteins that migrate through cross-linked polyacrylamide gels (PAGs) under the influence of a constant electric field experience negative factors, such as diffusion and nonspecific trapping in the gel matrix. These negative factors reduce protein concentrations within a defined gel volume with increasing migration distance and, therefore, decrease protein recovery efficiency. Here, we describe the enhancement of protein separation efficiency up to twofold in conventional one-dimensional PAG electrophoresis (1D PAGE), two-dimensional (2D) PAGE, and native PAGE by implementing pulses of inverted electric field during gel electrophoresis.

Key words: Field-inversion gel electrophoresis, Pulsed-field gel electrophoresis, Two-dimensional polyacrylamide gel electrophoresis, Forward pulse time, Reverse pulse time, Separation efficiency

1. Introduction

Sodium dodecyl sulfate-polyacrylamide gel electrophoresis (SDS-PAGE) is an indispensable technique in protein separation. This technique has only changed marginally over the past three decades (1). Despite its popularity, SDS PAGE, as well as native PAGE for protein separation, suffers the basic limitation of band broadening by diffusion and trapping of biomolecules in gel matrices. Nevertheless, protein separation by SDS-PAGE that interfaces with mass spectrometry (MS) has currently emerged as the method of choice in the forefront of proteomics. Thus, new tools for upstream gel electrophoresis that can improve protein separation efficiency and recovery will possibly lead to new discoveries in downstream processes.

Pulsed-field gel electrophoresis (PFGE) is an elegant, simple, and universally accepted technique for the separation of large DNA molecules (2). Several modifications of PFGE with different

Biji T. Kurien and R. Hal Scofield (eds.), *Protein Electrophoresis: Methods and Protocols*, Methods in Molecular Biology, vol. 869, DOI 10.1007/978-1-61779-821-4_11, © Springer Science+Business Media, LLC 2012

electrophoretic configurations exist (3–5). One modification is the field-inversion gel electrophoresis (FIGE) in which net forward molecular migration is achieved by either employing a longer net forward field time or a higher forward field strength compared with the reverse direction (4). Among the various PFGE techniques, FIGE is likely the easiest to perform with minimal special equipment that generates a highly uniform electric field across the gel (4). Systematic investigations of FIGE on the separation of DNA molecules in agarose gels were previously reported (6–10). However, published studies of the use of FIGE for protein separation are still lacking. Another modification is the orthogonal field alternation gel electrophoresis (OFAGE) techniques (3). Both of these methods were optimized to maximize separation efficiency (band width in the dimension of separation) and selectivity (distance between the center of two bands) for large DNA molecules.

Attempts were made to enhance the separation of proteins by means of PFGE (11). Subsequent applications of PFGE were used to resolve either specific model protein species (12, 13) or high molecular weight muscle myosin heavy chain isoforms (14). However, these approaches were limited to the application of alternating cycles of on-and-off electric fields across a slab gel. This approach inevitably allowed diffusion to occur during the off times. Unwanted band broadening as a result of diffusion compromises general separation efficiency.

There are currently three models to explain molecular migration during PAGE: (1) The extended Ogston (EO) model assumes an overall sphere-like conformation for native or small proteins in which mobility is a function of available gel pores in a regular lattice fashion (15). (2) The reptation model assumes that molecules go through a rather disordered matrix, such as polyacrylamide (16), accounting for the snakelike movement of pearl necklace-shaped polymers, such as protein–SDS complexes. These two models only apply to polypeptides within a certain molecular weight (MW) range at a given cross-linked polyacrylamide concentration. Any deviation from this linearity implies a change or transition in molecular shape (i.e., the radius) and net charge. (3) The door-corridor (DC) model explains the behavior of polypeptides above a critical MW where the mobility of a protein becomes independent of the cross-linked polyacrylamide concentration (17). Effective trapping of migrating molecules by the matrix predominates in this model, and electrokinetic energy is required to overcome the trapping effect (17).

We report here the engineering of a simple field-inversion device (Fig. 1) and an extensive study of protein recovery enhancement using FIGE. Thus, in general, FIGE enhances protein separation by improving local protein concentrations during SDS-PAGE or under native gel conditions. The increased local protein concentration thus improves the observable intensity of protein species in PAGs and also improves the success of downstream peptide sequencing using MS.

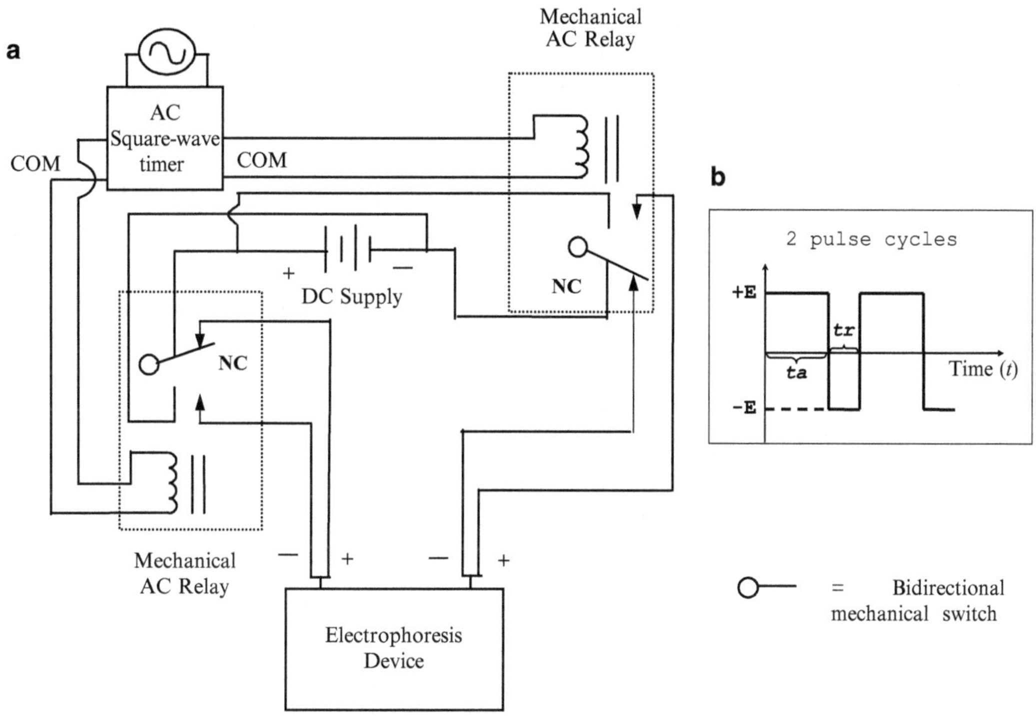

Fig. 1. Electric circuit of pulse generator and field diagram. (**a**) Electric circuit diagram for generating positive (+E) and negative (−E) square wave electric fields during field-inversion experiments. Direct current (DC) supply is from an external source. *NC* normally closed switch, *AC relay* alternate current relay; *Com* common outlet. (**b**) A profile of the electric field during a typical FIGE experiment where +E = −E and t_a (forward field time) is longer than t_r (reverse field time). Reproduced from (18) with permission from BioMed Central.

Taken together, FIGE can be used to complement constant field gel electrophoresis for better protein separation and recovery.

2. Materials

All solutions used should be made by using Milli-Q water; if possible, filter all solution and store in appropriate DURAN® bottles (Schott, Elmsford, NY, USA) unless otherwise mentioned. It is also important to ensure proper pH prior to gel casting and electrophoresis. If protein sequencing using mass spectrometry is to be conducted after electrophoresis, all solutions used should be in molecular biology grade (DNase-, RNase-, and protease-free). All experiments, if possible, are to be conducted in the laminar flow hood. We preferred the use of precast gels if protein sequencing is involved in subsequent steps. Staining of gels can be done using Coomassie blue staining such as the SimplyBlue™ SafeStain (Invitrogen Inc., Carlsbad, CA, USA) or mass spectrometry compatible silver stain, such as the SilverQuest™ silver staining kit (Invitrogen Inc., Carlsbad, CA, USA).

2.1. 1D SDS Polyacrylamide Gel Components

1. Resolving gel buffer: 1.5 M Tris–HCl, pH 8.8. Weigh 181.7 g Tris–HCl base and transfer to a 1-L glass measuring cylinder. Add Milli-Q water to a volume of 900 mL. Mix and adjust pH to 8.8 with HCl. Make up to 1 L with autoclaved Milli-Q water and store at 4°C.
2. Stacking gel buffer: 0.5 M Tris–HCl, pH 6.8. Weigh 60.6 g Tris–HCl base and prepare to 1 L solution as in item 1 and store at 4°C.
3. Thirty percent acrylamide/bis solution (37.5:1 acrylamide:bis): Weigh 146.1 g acrylamide monomer and 3.9 g bis-acrylamide and transfer to cylinder. Add Milli-Q water to a volume of 500 mL. Mix to completely dissolve. Store at 4°C.
4. Ten percent SDS solution: Dissolve 10 g SDS in 100 mL Milli-Q water. Store at room temperature.
5. Ten percent ammonium persulfate: Add 1 g ammonium persulfate to 10 mL Milli-Q water. Store at −20°C (see Note 1).
6. *N,N,N,N′*-tetramethyl-ethylenediamine (TEMED): Store at 4°C.
7. 5× Running buffer: Weigh 15 g Tris–HCl base; 72 g glycine; 5 g SDS. Add Milli-Q water to 1 L; it is not necessary to adjust the pH of the running buffer. It should have a pH of 8.3. Store at room temperature.
8. Laemmli sample buffer (4×): 0.3 M Tris–HCl (pH 6.8), 8% SDS, 40% glycerol, and trace amount of bromophenol blue (BPB). Store at room temperature.
9. 2 M dithiothreitol (DTT) solution: Dissolve 30.8 mg DTT in 100 μL Milli-Q water (see Note 2).
10. Protein standards (Bio-Rad Laboratories Inc., Hercules, CA, USA).

2.2. 2D PAGE Components

1. Rehydration buffer: 7 M urea, 2 M thiourea, 4% CHAPS, and trace amount of BPB. Weigh 10.5 g of urea, 3.8 g of thiourea, 10 g of CHAPS, 500 μL of immobilized pH gradient (IPG) buffer (ampholytes), and 154 mg of DTT to a 100-mL glass cylinder. Make up to 25 mL with Milli-Q water and dissolve all components thoroughly. Store at −20°C.
2. SDS equilibration buffer solution: Weigh 72.1 g of urea, 69 mL of glycerol, 10 mL of 1.5 M Tris–HCl buffer pH 8.8, 4 g of SDS, and trace amount of BPB to cylinder. Make up to 200 mL. Store at −20°C.
3. Electrophoresis buffer: 1.5 M Tris–HCl buffer pH 8.8: Dissolve 181.7 g of Tris–HCl base in 1 L of Milli-Q water. Adjust pH to 8.8 with HCl. Store at 4°C.
4. Agarose sealing solution: Weigh 0.5 g of agarose in 100 mL electrophoresis buffer (see Note 3).

5. 1× SDS running buffer: Weigh 3.03 g of Tris–HCl base, 14.4 g of glycine, and 1 g of SDS in 1-L cylinder and top up to 1 L with Milli-Q water.
6. Ettan™ IPGphor™ II Isoelectric Focusing System (GE Healthcare, Piscataway, NJ, USA).

2.3. Native PAGE Components

1. 40% Acrylamide:bis solution (37.5:1): Weigh 100 g of acrylamide and 2.65 g of bis-acrylamide in 250 mL of Milli-Q water. Store at 4°C (see Note 4).
2. 4× Resolving gel buffer (200 mL): Weigh 36.3 g of Tris–HCl and add Milli-Q water to dissolve. Adjust pH to 8.8 with HCl. Add Milli-Q water to a final volume of 200 mL. Store at 4°C.
3. 4× Stacking gel buffer (50 mL): Weigh 15.1 g of Tris–HCl and add Milli-Q water to dissolve. Adjust pH to 6.8 with HCl and make up final volume to 50 mL. Store at 4°C.
4. 10% Ammonium persulfate (10 mL): Weigh 1.0 g of ammonium persulfate in a 15-mL Falcon tube and add Milli-Q water to a final volume of 10 mL. Store at 4°C.
5. Electrophoresis buffer (1×): Weigh 28.8 g of glycine and 6.0 g of Tris–HCl in a 2-L cylinder and add Milli-Q water to obtain a final volume of 2 L.
6. Water-saturated *n*-butanol (55 mL): Mix 50 mL of *n*-butanol with 5 mL of Milli-Q water. Store at room temperature (see Note 5).
7. 2× Sample buffer (10 mL): Mix 2.5 mL of 4× stacking gel buffer (pH 6.8) with 0.4 mL of glycerol and trace amount of BPB with 5.5 mL of Milli-Q water. Store 0.5 mL aliquots at −20°C for a maximum period of 6 months.

2.4. FIGE Components

A schematic representation of the pulsing circuitry in conjunction with an external electrophoresis unit is provided in Fig. 1. A picture of the prototype can be found elsewhere (see *supplementary data* in ref. 18). Forward and reverse switching of the electric field supply to the gel was achieved by interfacing the voltage supply (PowerPac 1000 power supply, Bio-Rad, Hercules, CA, USA) with an 240 V alternate current (AC) relay (MY2, OMRON, Japan) that could handle a current of 5 A. The rate of the forward and reverse switching was controlled by a MD4E-AP programmable 110–240 V AC switching device (Fuji Electric, Japan) that was able to deliver pulses as short as 1 ms. This simple instrumentation generates square wave form. Amplitude, length, and stability of the pulses were checked and ascertained by a Tektronix oscilloscope model TDS 1000 (see *supplementary data* in ref. 18). Expensive systems for generating inverse pulsed wave can also be purchased from Bio-Rad or GE Healthcare.

3. Methods

All procedures are done at room temperature unless otherwise stated.

3.1. 1D PAGE

1. Use 70% alcohol and Kimwipes to wipe the glass plates used for casting gel and then set up the rest of the apparatus as per manufacturer's instructions.
2. Mix 1.9 mL of separating gel buffer, 2.5 mL of acrylamide/bis solution, 75 μL of 10% SDS solution, and 3.1 mL of Milli-Q water in a 15-mL conical tube. Add 20 μL of 10% ammonium persulfate, 10 μL of TEMED [10% polyacrylamide gel (PAG) in mini cassette: 7.5 cm × 10 cm × 1.0 mm].
3. Invert the tube gently after adding all components.
4. Pour the gel solution into the glass cassette. Allow space for stacking gel and gently overlay with *n*-butanol or water.
5. Remove *n*-butanol, wash with water, and wipe with filter paper.
6. Add the stacking gel solution and put in comb without introducing air bubbles. Prepare the stacking gel solution by mixing 1.25 mL of stacking gel buffer, 0.5 mL acrylamide/bis solution, 50 μL of 10% SDS, and 3.25 mL Milli-Q water in a 15-mL conical tube. Add 20 μL of 10% ammonium persulfate and 10 μL of TEMED.
7. Set up the gel in the tank and make sure that the running buffer is not leaking.
8. For each sample, use ¼ the volume of 4× Laemmli sample buffer. Heat the sample at 95°C for 5 min. Cool the samples to room temperature. Add DTT solution to 50 mM final concentration.
9. Load the samples into the wells using a pipette, ensuring that each well has been loaded with appropriate volume of sample.
10. Reassemble the apparatus and connect the leads to the power supply. Proceed to pulsing (see Subheading 3.3).
11. Stop gel running when dye front reaches the bottom of the gel. Take out the gel and process to gel staining.

3.2. 2D PAGE

1. Select the strip holder(s) corresponding to the IPG strip length (7, 11, 13, 18, or 24 cm) chosen for the experiment. Wash each holder with strip holder cleaning solution supplied to remove residual protein. Rinse thoroughly with double-distilled water. Use a cotton swab or a lint-free paper to dry the holder or allow it to air-dry.

2. Prepare the required volumes per strip as summarized below:

IPG strip length (cm)	Total volume per strip holder (μL, including any sample volume)
7	125
11	200
13	250
18	350
24	450

3. Deliver the solution slowly with sample at a central point in the strip holder channel away from the sample application wells.
4. Remove the protective cover from the IPG strip. Position it with the gel side down and the pointed (anodic) end of the strip directed toward the pointed end of the strip holder. Align the pointed end first and then lower the strip onto the rehydration solution. To help soaking the entire strip, one can gently lift and lower the strip, and slide it back and forth along the surface of the rehydration solution. Tilt the strip holder slightly as needed to assure complete and even wetting. Finally, lower the cathodic (square) end of the strip into the strip holder, making sure that the IPG gel contacts the strip holder electrodes at each end. (The gel can be visually identified once the rehydration solution begins to enter the gel.) Be careful not to trap bubbles under the strip.
5. Apply Immobiline DryStrip Cover Fluid to minimize evaporation and urea crystallization. Pipette the cover fluid dropwise into one end of the strip holder until one half of the strip is covered. Pipette the fluid dropwise into the other end of the strip holder, adding fluid until the entire IPG strip is covered.
6. Place the cover on the holder. Pressure foams on the underside of the cover are to assure that the strip maintains good contact with the electrodes as the gel rehydrates.
7. A minimum of 10 h is required for rehydration; overnight is recommended.
8. Start focusing by increasing the voltage stepwise, to a maximum of 8,000 V. (Different manufacturers have different focusing protocols for IPGs strips. We used Ettan™ IPGphor™ II Isoelectric Focusing System. Detailed protocol can be found in GE's 2D electrophoresis handbook.) Stop focusing when the voltage hour reaches the desired value.
9. Proceed to second-dimension electrophoresis.
10. The second-dimension vertical gel must be ready for use prior to the equilibration of the Immobiline DryStrip after the first-dimension separation.

11. Equilibration is carried out in a two-step process using capped tubes.
12. Place the IPG strips in individual capped tubes, with the support film toward the tube wall.
13. Add the appropriate volume of SDS equilibration buffer (add DTT to 1% final concentration) to each strip. Cap or seal the tubes with flexible paraffin film and place them on their sides on a rocker for the equilibration process. Equilibrate for 15 min.
14. Pour off buffer from above step and add the appropriate volume of SDS equilibration buffer (add iodoacetamide to 4% final concentration) to each strip. Again cap or seal the tubes with flexible paraffin film and place them on their sides on a rocker for the equilibration process. Equilibrate for an additional 15 min.
15. Apply the Immobiline DryStrip gels on top of SDS-PAG. Push the strips gently down to touch the gel surface.
16. Seal the Immobiline DryStrip gel in place with melted agarose sealing solution. Proceed to Subheading 3.4.

3.3. Native PAGE

1. Thoroughly clean and dry the glass plates and three spacers and then assemble them with bulldog clips. Clamp the chamber in an upright and leveled position.
2. Prepare 10 mL of separating gel mixture as follows:

	4%	5%	6%	8%	10%
40% Acrylamide:bis solution (37.5:1)	1 mL	1.25 mL	1.5 mL	2 mL	2.5 mL
4× Separating gel buffer	2.5 mL	2.5 mL	2.5 mL	2.5 mL	2.5 mL
50% Glycerol	2.5 mL	–	–	–	–
ddH_2O	4 mL	6.25 mL	6 mL	5.5 mL	5 mL

3. Degas the solution and then add 5 μL of 10% ammonium persulfate, and 10 μL of TEMED.
4. Mix gently and use immediately (because polymerization starts when TEMED is added). Carefully pour the freshly mixed solution into the gel chamber without generating air bubbles. Pour to a level about 1 cm below where the bottom of the well-forming comb will come when it is in position.
5. Carefully overlay the acrylamide solution with H_2O-saturated *n*-butanol without mixing to eliminate oxygen and generate a flat top to the gel.
6. Polymerize the acrylamide for 1 h.

7. Prepare the 4-mL stacking gel solution as follows. Mix the following:

40% Acrylamide:bis solution (37.5:1)	0.4 mL
4× Stacking gel buffer	1.0 mL
ddH_2O	2.6 mL

8. Degas the stacking gel solution and then add 20 μL of 10% ammonium persulfate and 5 μL of TEMED. Mix gently and use immediately.
9. Pour off the *n*-butanol from the polymerized resolving gel, wash the gel top with water, and fill the gap remaining in the gel chamber with the stacking gel mixture. Insert the comb at the top of the gel.
10. Polymerize the acrylamide for 1 h.
11. When the stacking gel has polymerized, remove the comb without distorting the shapes of the well. Remove the clips holding the plates together and install the gel in the electrophoresis apparatus.
12. Fill the electrophoresis apparatus with electrophoresis running buffer. Push out the bottom spacer from the gel and remove bubbles from both the top and underneath of the gel. Use the gel immediately.
13. While the gel is being polymerized, prepare samples for loading.
14. Dissolve the protein sample solution in a similar volume of 2× sample buffer or dissolve a dry sample in 1× sample buffer. The concentration of sample in the solution should be such as to give a sufficient amount of protein in a volume not greater than the size of the sample well. (The BPB dye in sample buffer indicates when the sample solution is acidic, by turning yellow. If this hap pens, add a little NaOH for the color to turn blue again.)
15. Load the gel with 10–30 μL (20–50 μg) protein sample solution by pipette.
16. Start electrophoresis immediately by turning on power. Proceed to Subheading 3.4.
17. Remove the gel from between the glass plates once the BPB reaches the bottom of the gel.
18. Stain the gel in the staining solution according to the vendor's instruction.

3.4. FIGE

1. The two main reasons for conducting FIGE are (1) to increase the amount of retrievable proteins within a specific locality from the gel by means of detrapping of the migrating proteins (therefore, higher local concentration of protein species can be achieved) (Fig. 2) and (2) to better separate proteins of similar

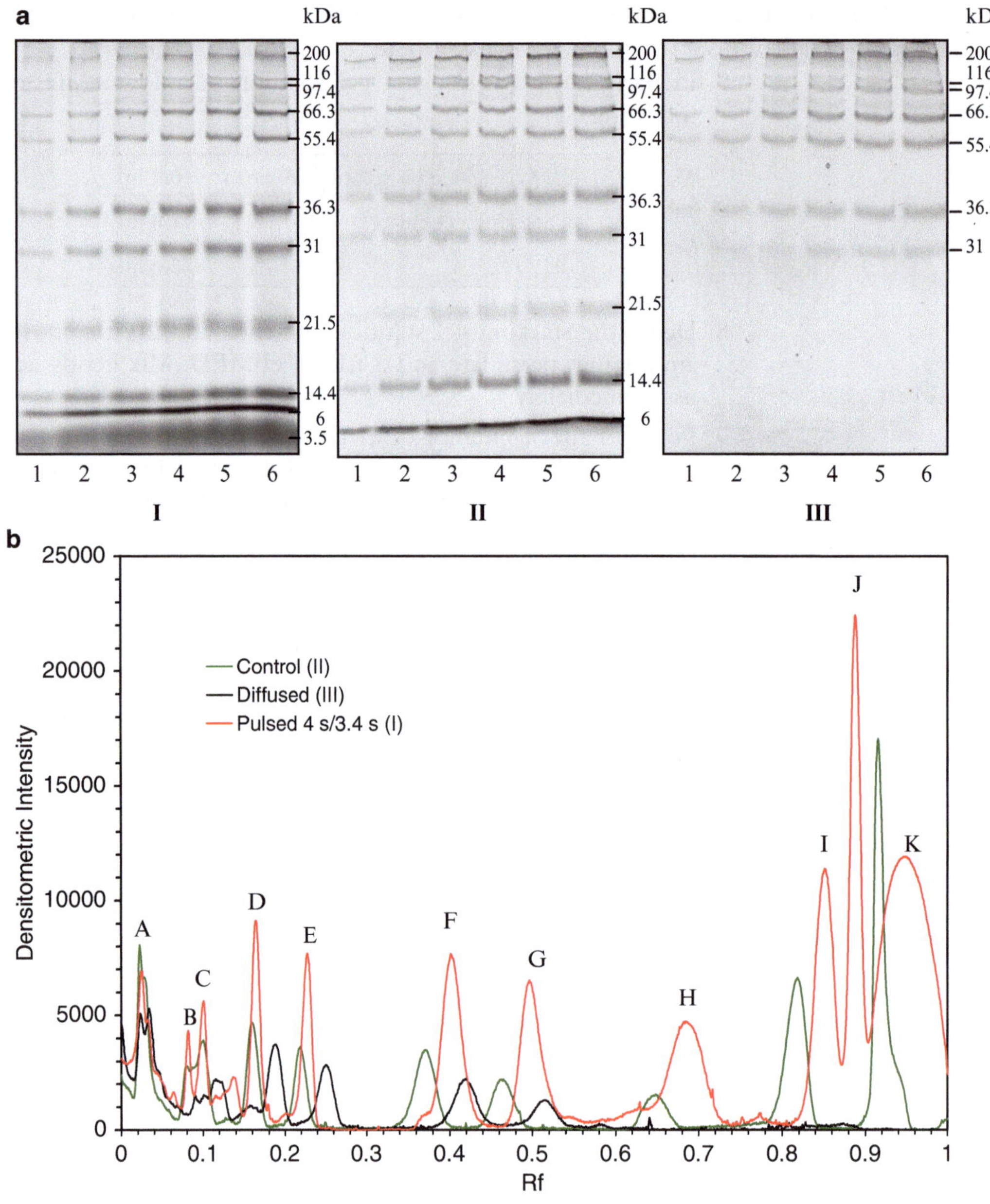

Fig. 2. Increased local concentrations of protein bands upon pulsing. Protein band intensity analyses in FIGE (**a**, I), CFE (**a**, II), and CFE followed by resting within glass plates in room temperature for 12 h (**a**, III). *Lanes 1–6* are 2, 4, 6, 8, 10, and 12 μL of Mark12 protein standards, respectively, in a self-cast Bio-Rad 14% SDS-PAGE 1 mm × 7 cm gel followed by Coomassie blue staining. (**a**, I) Gel was run with a pulsed field at 4/3.4 s at 200 V for 13 h, with an average buffer temperature of 25°C. (**a**, II) Gel was run at a constant field of 200 V for 1 h and an average buffer temperature of 25°C. (**a**, III) Gel was run at a constant field of 200 V for 1 h and left at rest for another 12 h within the glass plates to permit diffusion prior to staining. (**b**) Densitometry analysis of protein bands in the gels of the three conditions tested. Molecular weight was represented by alphabet *A–K*, where *A* = 200 kDa, *B* = 116.3 kDa, *C* = 97.4 kDa, *D* = 66.3 kDa, *E* = 55.4 kDa, *F* = 36.3 kDa, *G* = 31.0 kDa, *H* = 21.5 kDa, *I* = 14.4 kDa, *J* = 6.0 kDa, and *K* = unresolved 3.5/2/0 kDa bands, respectively. Migration distance relative to the dye front (R_f) and intensity of bands from *lane 6* of all three gels were densitometrically analyzed by Bio-Rad Quantity One software. The graph results were the average of two independent experiments. The graph results were subsequently employed in the calculation of peak variance, σ^2, in Table 1. *Red line* represents the sample run in pulsed condition. *Green line* represents the sample run in constant field. *Black line* represents the sample run in constant field and then left for 12 h before staining. Reproduced from (18) with permission from BioMed Central.

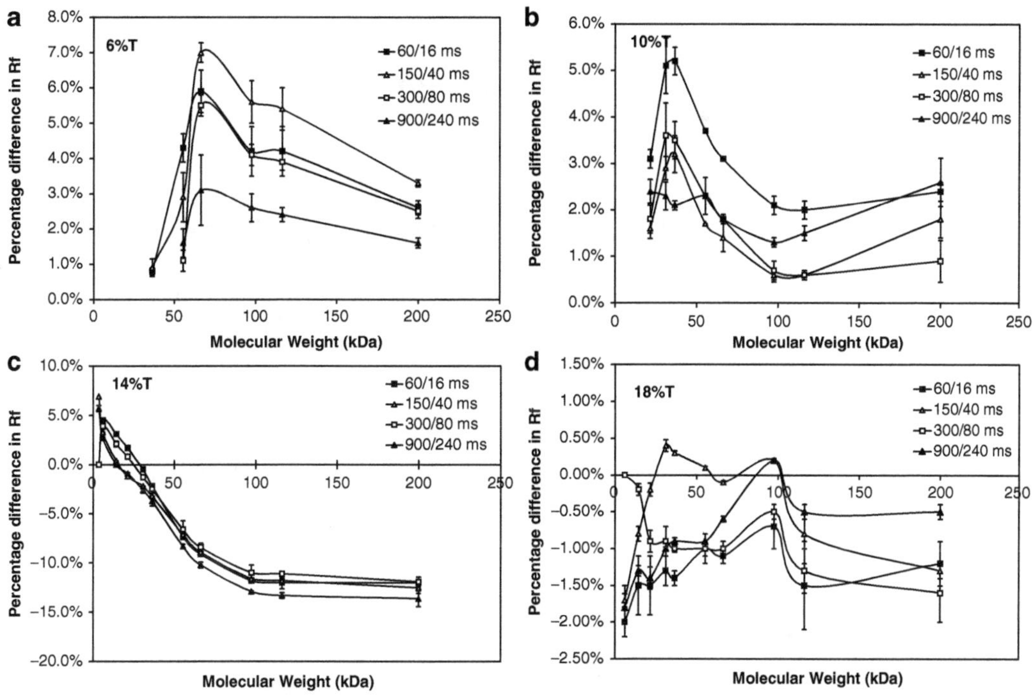

Fig. 3. Changes in mobility upon different pulsing conditions. Comparison of changes in protein mobility between FIGE and CFE conditions in 6% (**a**), 10% (**b**), 14% (**c**), and 18% (**d**) cross-linked polyacrylamide concentration self-cast Bio-Rad SDS-PAG (1 mm × 7 cm). Different concentrations of polyacrylamide were cast in a mini-Protean 3 apparatus. Five microliters of Mark12 protein standards were used. Graphs of mobility were generated based on relative mobilities to the resolving front (R_f) using Quantity One software (Bio-Rad) and expressed as percent differences in R_f with respect to CFE controls. Each data point was the average of two separate experiments. All gels were run at 200 V with the average buffer temperature of 10°C. Positive values denote slower mobility, and negative values denote faster mobility with respect to CFE control. Reproduced from (18) with permission from BioMed Central.

molecular weights, depending on the migrating property of proteins within a particular concentration of PAG (Fig. 3 and spot numbers 3 and 4 in Fig. 4).

2. Before conducting FIGE, it is essential to ensure that the anode and cathode polarities of the power source are connected to the correct electrodes of the timer and the electrophoresis system. A simple voltmeter can be employed to test that forward pulse time (t_a) always give a positive reading and reverse pulse time (t_r) always give a negative reading when the electrodes are connected correctly (see Note 6).

3. Percentages of slab gels used for protein separation should be based on the sizes of the different proteins under investigation with reference to the migration chart for the protein standards obtained from the vendor (see Subheading 2.1, item 10).

4. The tested t_a/t_r ratios effective for one-dimensional PAG electrophoresis (1D PAGE) are in the range of 1.07–3.75. Smaller ratio can be used for separating proteins with closed molecular

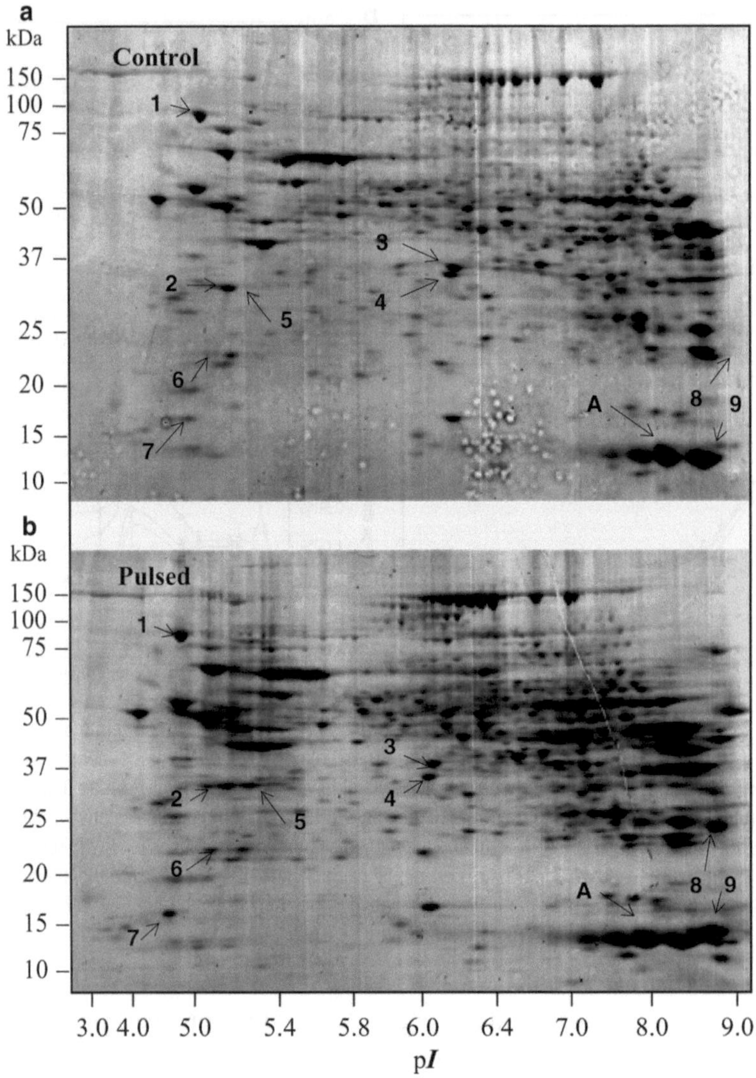

Fig. 4. Effects of FIGE on 2D PAGE analysis of rat liver lysates. Comparison of 2D PAGE images of rat liver lysate under CFE (**a**) and FIGE (**b**) conditions. Each gel represents 100 μg of rat liver lysate separated by isoelectric focusing (IEF) using a nonlinear pH 3–10 IPG strip in the first dimension and a criterion precast SDS 10–20% PAG in the second dimension at room temperature. Control denotes CFE and pulsed denotes FIGE with a t_a/t_r of 400/106 ms in the second-dimension separation. Gels were stained with Coomassie blue. Spots selected for LC-MS/MS analysis are denoted by numbers; see Table 1., and spots IC and IP denote internal controls for equivalent sample loading for control and pulsing, respectively. Reproduced from (18) with permission from BioMed Central.

weight; however, too small a ratio may result in overheating the PAG system. Therefore, smaller ratios are recommended for 1D PAGE when one intends to separate proteins with similar molecular weights (see Fig. 3, Table 1).

5. As for small gels (1 mm × 7 cm gel), we recommend a voltage of 200 V and an average run time of 12 h.

Table 1
Peak variance (σ^2) of proteins separated as in Fig. 2a(I) by FIGE (pulsing) and Fig. 2a(II) by CFE (control)

MW (kDa)	Well	200 (*A*)	116.3 (*B*)	97.4 (*C*)	66.3 (*D*)	55.4 (*E*)	36.5 (*F*)	31.0 (*G*)	21.5 (*H*)	14.4 (*I*)	6.0 (*J*)
σ^2 (peak variance, mm^2)											
Control	–	1.62±0.05[a]	8.83±0.21	6.50±0.18	1.62±0.15	2.63±0.13	5.86±0.28	5.86±0.45	6.94±0.42	4.51±0.20	1.52±0.11
Pulsed	–	1.85±0.07	0.41±0.05	0.85±0.12	0.87±0.07	1.32±0.15	4.51±0.24	4.51±0.38	8.51±1.00	4.51±0.15	1.52±0.05

Peak variance (σ^2) = (full-width half-maximum)^2/5.54. Reproduced from (18) with permission from BioMed Central

[a]Results are the average of two experiments ± SEM

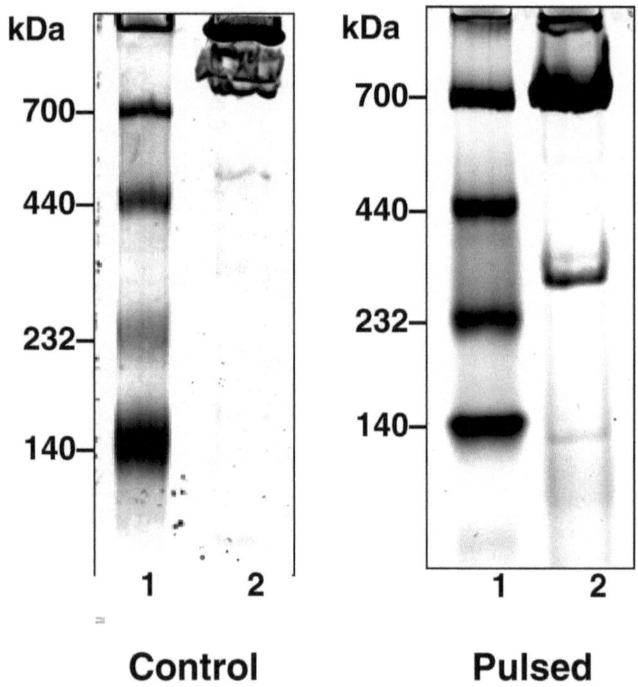

Fig. 5. Effects of FIGE on protein separation under native gel conditions. Comparison of GroEL 14-mer complex (840 kDa) in native PAG between CFE and FIGE conditions. CFE control was the *left panel*. FIGE was at the *right panel*. The gels used were 1 mm × 7 cm native 6% PAG cast in a mini-Protean 3 apparatus. Run time was 2 h 15 min in control condition. Pulsed condition run time was 5 h 30 min with a t_a/t_r of 400/100 ms. *Lane 1*, native MW markers (10 μg total); *Lane 2*, 10 μg purified *E. coli* GroEL native complex (14-mer, 840 kDa). The band at 300 kDa could be a minor cofactor associated with GroEL during purification. The gel was stained with Coomassie blue. Proteins were purposefully overloaded to best represent the effect of "detrapping" of native proteins under pulsing conditions. Reproduced from (18) with permission from BioMed Central.

6. For FIGE in the second dimension in 2D PAGE, we recommend a t_a/t_r of 400/106 ms for a gel with 7–9 cm in width (see Fig. 4).
7. For native gel, we recommend a pulse rate of t_a/t_r at 400/100 ms. If gradient native gel is required for separating large complex native protein mixture, it is better to seek a commercial source. FIGE results are extremely sensitive to the consistency of the gel used (see Fig. 5).

4. Notes

1. Aliquots of 500 μL in Eppendorf tubes are prepared. Tubes are stored in −20°C up to 6 months.
2. It is best to prepare this solution fresh each time.

3. Dissolve 0.5 g agarose in 50 mL Milli-Q water in a 250-mL DURAN® bottle. Put the solution in microwave at high power until it boils. Swirl the solution to ensure complete dissolving. It may take several rounds of microwave boiling and swirling. Add room temperature Milli-Q water to a total volume of 75 mL. Add 25 mL Laemmli sample buffer (4×) to total 100 mL and mix well. Aliquots of 1 mL agarose solution in Eppendorf tubes are prepared. Store solidified agarose solution in −20°C for up to 1 year. Heat the agarose tube to 90°C for 5 min and then keep at 45°C heat block till use.
4. Wear a mask when weighing acrylamide. Put the scale in fume hood when weighing. Mix the solution with a magnetic stir bar on top of a stirrer. Unpolymerized acrylamide is a neurotoxin, and care should be exercised to avoid skin contact. The acrylamide solution can be stored at 4°C for 1 month.
5. Shake the water and *n*-butanol for 1 min. Two layers will be seen. The top is the butanol layer.
6. To ensure a proper t_a/t_r ratio, that is ratio is >1, especially when pulse rate is relatively fast at >5 Hz (number of total t_a and t_r per second), a voltmeter is recommended. When the voltmeter reading becomes unstable, it means that the pulse rate (hertz) is too fast for any accurate detection. In this situation, change the timing to lower hertz (e.g., 0.2 Hz, or $t_a = 4$ s and $t_r = 1$ s) first. Following the proper voltmeter reading, one can increase the pulse rate accordingly. For long period of pulsing, large volume of running buffer should be used to prevent buffer capacity deterioration. At pulse rate of >5 Hz, an electrophoresis system with built-in cooling system is recommended. Alternatively, this can be achieved with large volume of running buffer with the submerged gel during the run. If it is possible, the heat-sinking effect of the running buffer can be enhanced through the use of a magnetic stirrer.

Acknowledgments

This work was supported by the Agency for Science, Technology and Research (ASTAR) of Singapore.

References

1. Laemmli UK (1970) Cleavage of structural proteins during the assembly of the head of bacteriophage T4. Nature 227:680–685
2. Schwartz DC, Cantor CR (1984) Separation of yeast chromosome sized DNA fragments by pulsed field gradient gel electrophoresis. Cell 37:67–75
3. Bancroft I, Wolk CP (1988) Pulsed homogeneous orthogonal field gel electrophoresis (PHOGE). Nucleic Acids Res 16:7405–7418

4. Carle GF, Frank M, Olson MV (1986) Electrophoretic separations of large DNA molecules by periodic inversion of the electric field. Science 232:65–68
5. Chu G, Vollrath D, Davis RW (1986) Separation of large DNA molecules by contour-clamped homogeneous electric fields. Science 234:1582–1585
6. Bostock CJ (1988) Parameters of field inversion gel electrophoresis for the analysis of pox virus genomes. Nucleic Acids Res 16: 4239–4252
7. Heller C, Pohl FM (1989) A systematic study of field inversion gel electrophoresis. Nucleic Acids Res 17:5989–6003
8. Mathew MK, Hui CF, Smith CL et al (1988) High-resolution separation and accurate size determination in pulsed-field gel electrophoresis of DNA. 4. Influence of DNA topology. Biochemistry 27:9222–9226
9. Mathew MK, Smith CL, Cantor CR (1988) High-resolution separation and accurate size determination in pulsed-field gel electrophoresis of DNA. 2. Effect of pulse time and electric field strength and implications for models of the separation process. Biochemistry 27: 9210–9216
10. Mathew MK, Smith CL, Cantor CR (1988) High-resolution separation and accurate size determination in pulsed-field gel electrophoresis of DNA. 1. DNA size standards and the effect of agarose and temperature. Biochemistry 27:9204–9210
11. Tischfield JA, Bernhard HP, Ruddle FH (1973) A new electrophoretic-autoradiographic method for the visual detection of phosphotransferases. Anal Biochem 53:545–554
12. Brassard E, Turmel C, Noolandi J (1991) Observation of orientation and relaxation of protein-sodium dodecyl sulfate complexes during pulsed intermittent field polyacrylamide gel electrophoresis. Electrophoresis 12:373–375
13. Houri A, Starita-Geribaldi M (1994) Pulsed field electrophoresis for the separation of protein-sodium dodecyl sulfate-complexes in polyacrylamide gels. Electrophoresis 15: 1032–1039
14. Sant'Ana Pereira JA, Greaser M, Moss RL (2001) Pulse electrophoresis of muscle myosin heavy chains in sodium dodecyl sulfate-polyacrylamide gels. Anal Biochem 291:229–236
15. Rodbard D, Chrambach A (1970) Unified theory for gel electrophoresis and gel filtration. Proc Natl Acad Sci USA 65:970–977
16. Guo XH, Chen SH (1990) Reptation mechanism in protein-sodium-dodecylsulfate (SDS) polyacrylamide-gel electrophoresis. Phys Rev Lett 64:2579–2582
17. Kozulic B (1994) On the "door-corridor" model of gel electrophoresis. I. Equations describing the relationship between mobility and size of DNA fragments and protein-SDS complexes. Appl Theor Electrophor 4:125–136
18. Tsai H, Low TY, Freeby S et al (2007) Increase in local protein concentration by field-inversion gel electrophoresis. Proteome Sci 5:18

Chapter 12

A Combined Free Flow Electrophoresis and DIGE Approach to Compare Proteins in Complex Biological Samples

Kim Y.C. Fung, Chris Cursaro, Tanya Lewanowitsch, Leah Cosgrove, and Peter Hoffmann

Abstract

Free flow electrophoresis (FFE) has been applied in numerous studies as a protein separation technique due to its multiple advantages such as fast and efficient sample recovery, high resolving power, high reproducibility, and wide applicability to protein classes. As a stand-alone platform however, its utility in comparative proteomic analysis is limited as protein samples must be run sequentially rather than simultaneously which introduces inherent variability when attempting to perform quantitative analysis. Here we describe an approach combining fluorescent CyDye technology (DIGE) with FFE to simultaneously separate and identify differentially expressed proteins in a model cell system.

Key words: Free flow electrophoresis, FFE, DIGE, Proteomics

1. Introduction

Separation of proteins in biological matrices is a major challenge in proteomics and has resulted in the development of numerous technologies primarily involving either electrophoresis or chromatography as the mode of separation. Recent advances in free flow electrophoresis (FFE) have drawn attention to its potential in the broader proteomics community. FFE is a liquid-based separation method amenable to a wide variety of protein classes, and as a result, it is used successfully as a preparative or analytical pre-fractionation tool in proteomics (1–8). The continuous nature of the sample loading allows for relatively high volumes of sample to be analysed which is important for detecting proteins present at low abundance. Separation occurs across an established, well-defined pH gradient and can be performed under native or denaturing conditions. Because the process occurs in aqueous media, sample recovery is highly efficient and potentially quantitative. The versatility of FFE

Biji T. Kurien and R. Hal Scofield (eds.), *Protein Electrophoresis: Methods and Protocols*, Methods in Molecular Biology, vol. 869, DOI 10.1007/978-1-61779-821-4_12, © Springer Science+Business Media, LLC 2012

is also highlighted by its compatibility with many commonly used protein analytical methods such as 1D or 2D gel electrophoresis, chromatography, enzymatic assays, immunoassays, and mass spectrometry.

In its standard format, FFE is not optimal for comparative proteomic studies as protein samples must be run sequentially rather than simultaneously. If an additional separation method is used (e.g. HPLC or 1-DE), the inherent variability associated with comparing multiple images or traces must be taken into consideration when performing comparative or (semi-) quantitative analysis. Here we describe an approach where fluorescent labeling of proteins is performed prior to separation by FFE and 1-DE to allow direct comparison of two protein samples and that also facilitates visualisation of intact proteins and their biologically relevant modified forms (e.g. isoforms, variants, posttranslational modifications). This methodology can be readily applied to measure differential protein expression between any treatment groups.

2. Materials

All solutions are prepared using ultrapure water with a resistance of ≥18 MΩ/cm at 25°C and analytical grade reagents. Use only the highest quality urea and thiourea that have low conductivity. Wear appropriate personal protection when handling potentially harmful reagents and always observe local waste disposal regulations when disposing of waste.

2.1. Cell Lysates and CyDye Labeling Reagents

1. Obtain cell lysate from HT29 cells and HT29 cells treated with butyrate for 48 h as described in (9).
2. Rehydration buffer for solubilization of protein pellet (see Note 1): 7 M urea, 2 M thiourea, 2% CHAPS, 30 mM Tris. Weigh 10.5 g urea, 3.8 g thiourea, 0.5 g CHAPS, and 0.09 g Tris and transfer to 50-mL polypropylene tube (e.g. Falcon™ tube). Add water to 25 mL and mix gently to dissolve.
3. Protein assay kit (e.g. Bradford, bicinchoninic acid, or EZQ).
4. CyDye DIGE Fluor minimal dyes: Cy2, Cy3, and Cy5 (GE Healthcare, Piscataway, NJ, USA). Prepare stock solution by addition of equal volumes of freshly opened, anhydrous *N,N*-dimethylformamide (DMF, >99%) (see Note 2). For example, for each 25-nmol-CyDye vial, add 25 µL DMF (final concentration 1 nmol/µL). Aliquot and store at −80°C wrapped in aluminum foil. These dyes are light sensitive, and exposure to light should be minimized.
5. Quenching solution: 10 mM lysine. Weigh 0.015 g L-lysine, transfer to 15-mL polypropylene tube, and add 10 mL ultrapure water. Aliquots can be stored at −80°C.

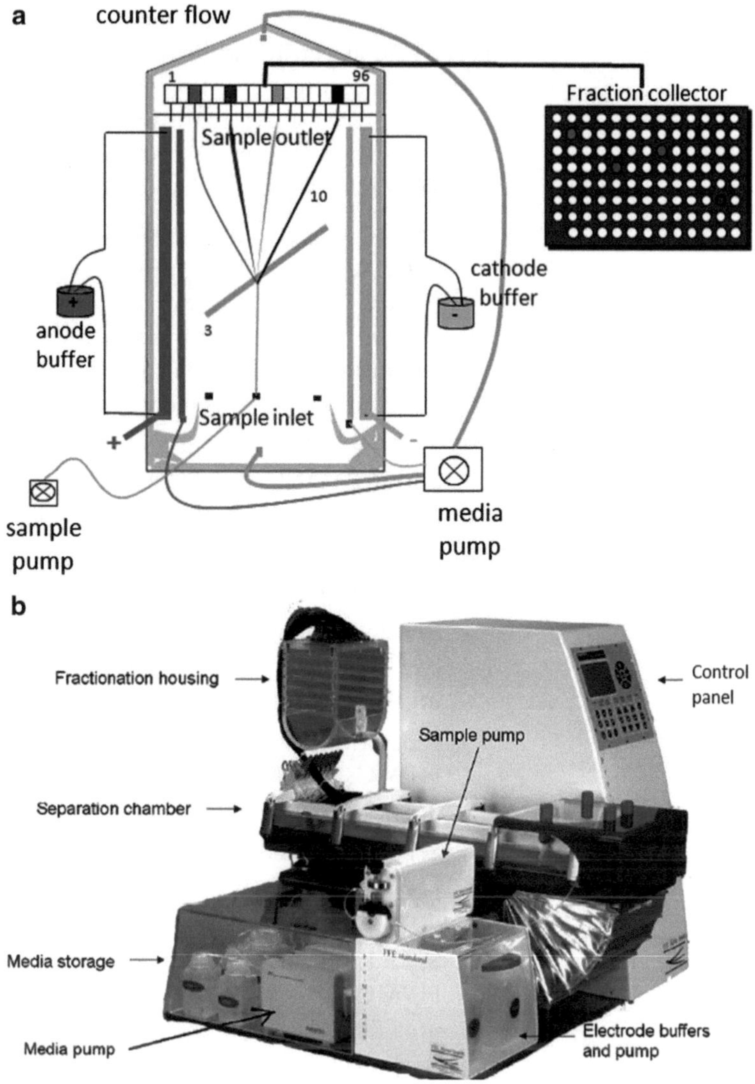

Fig. 1. (**a**) Schematic diagram of FFE apparatus. (**b**) BD Biosciences ProMetHEUS™ FFE apparatus.

2.2. Free Flow Electrophoresis Components

2.2.1. Apparatus

1. FFE apparatus [e.g. ProMetHEUS™ FFE apparatus (BD Biosciences, Bedford, MA, USA)] (see Fig. 1 for a schematic diagram of an FFE apparatus).
2. 96-well microtiter plates for initial performance test (laminar flow) and pI test.
3. 96×2.2-mL polypropylene deep-well plates for collection of FFE fractions.
4. pH meter with micro pH probe (e.g. Thermo™ Ross micro pH probe; Thermo Fisher Scientific, Waltham, MA, USA).
5. Conductivity meter.
6. Microplate reader (capable of absorbance readings at 280, 410, 510 and 600 nm).

2.2.2. Reagents

Weigh the viscous liquids such as hydroxypropyl methylcellulose (HPMC), glycerol, and ProLytes as this is more accurate than volumetric measurement. For each medium, mix thoroughly in a warm (~30°C) water bath, or sonicating water bath if available, until all reagents are dissolved. For best results, the separation media should be freshly made and used within 2 or 3 days:

1. HPMC stock solution (Sigma-Aldrich, St Louis, MO, USA): 0.8% (w/w) in ultrapure water. Prepare by slowly adding 8 g HPMC to 1 L of ultrapure water in a glass beaker with constant agitation, e.g. using a magnetic stirrer. This may take several hours, so is best left overnight. Decant into a 1-L Schott bottle and store in the dark at 4°C.
2. 2-(4-sulfophenylazol)1,8-dihydroxy-3,6-napthalene disulfonic acid (SPADNS; BD Biosciences): 1% (v/v) in ultrapure water (or the appropriate separation medium as required).
3. Carrier ampholytes to establish pH gradient: ProLytes 1, 2, and 3 (BD Biosciences).
4. Anode stabilization medium: 12.5% (w/w) glycerol, 0.05% (w/w) HPMC, 100 mM H_2SO_4, 6 M urea, and 1.7 M thiourea (see Note 3).
5. Separation medium 1: 12.5% (w/w) glycerol, 0.05% (w/w) HPMC, 14.5% (w/w) ProLyte 1, 6 M urea, and 1.7 M thiourea (see Note 4).
6. Separation medium 2: 12.5% (w/w) glycerol, 0.05% (w/w) HPMC, 23.2% (w/w) ProLyte 2, 6 M urea, and 1.7 M thiourea (see Note 5).
7. Separation medium 3: 12.5% (w/w) glycerol, 0.05% (w/w) HPMC, 19.4% (w/w) ProLyte 3, 6 M urea, and 1.7 M thiourea (see Note 6).
8. Cathode stabilization medium: 12.5% (w/w) glycerol, 0.05% (w/w) HPMC, 100 mM NaOH, 6 M urea, and 1.7 M thiourea (see Note 7).
9. Counterflow medium: 12.5% (w/w) glycerol, 0.05% (w/w) HPMC, 6 M urea, and 1.7 M thiourea (see Note 8).
10. Anode electrolyte: 100 mM H_2SO_4 (at least 300 mL).
11. Cathode electrolyte: 100 mM NaOH (at least 300 mL).
12. pI markers (BD Biosciences): diluted 1/20 with separation medium 2.
13. (Optional) Standard protein mixture in water: 2.5 mg/mL stock solutions of myoglobin, cytochrome c, and bovine serum albumin (BSA) with bromophenol blue (0.01%) (bromophenol blue will bind to BSA and stain BSA blue). Combine 100 μL of each and dilute to a final volume of 1 mL with separation medium 2 (final concentration of 0.25 mg/mL for each protein).

2.3. SDS Polyacrylamide Gel Components

2.3.1. Apparatus

1. Precast polyacrylamide gels including appropriate buffer system, sample loading buffer, and electrophoresis apparatus (e.g. Invitrogen NuPAGE® System or Bio-Rad Criterion® System) (see Note 9).
2. Gel imaging system [e.g. Ettan DIGE Imager or Typhoon imager (GE Healthcare)] capable of producing and detecting excitation/emission wavelengths for CyDyes (Cy2: 488/520 nm; Cy3: 532/580 nm; Cy5: 633/670 nm).
3. Image analysis software which enables densitometry analysis (e.g. ImageQuant TL; GE Healthcare).

2.3.2. Reagents

1. Centrifugal molecular weight filters (e.g. Millipore Ultrafree 10 kDa or Vivaspin).
2. Standard molecular weight markers (ECL Plex Fluorescent Rainbow Markers; GE Healthcare).
3. Gel fixing solution: 40% ethanol, 10% acetic acid in ultrapure water.
4. Coomassie Brilliant Blue (CBB) G-250 dye stock: 5% (w/v) in ultrapure water. Store at room temperature.
5. CBB G-250 working solution: For 100 mL, dissolve 8 g ammonium sulphate in 77 mL ultrapure water. Add 1.9 mL 85% phosphoric acid, 20 mL methanol, and 1.6 mL of the 5% CBB dye stock.

3. Methods

3.1. Labeling of Protein Lysate with CyDyes

1. Resolubilize protein pellet in rehydration buffer. Keep volumes to a minimum to obtain a final protein concentration of approximately 8–10 mg/mL. Perform protein assay to determine concentration of each sample. Transfer 650 μg of untreated (sample 1) and treated (sample 2) HT29 cells to clean individual Eppendorf tubes. Prepare the internal standard by combining 325 μg of each sample in a separate Eppendorf tube.
2. To sample 1, add 500 pmol of Cy3 dye (equivalent to 0.5 μL of 1 nmol/μL stock). To sample 2, add 500 pmol of Cy5 dye. To the internal standard, add 500 pmol of Cy2 dye. Incubate samples on ice for 30 min in the dark.
3. Quench labeling reaction with 10 mM L-lysine (1 μL).
4. Combine labeled protein samples and dilute with FFE separation medium 2 to a final concentration of approximately 1.5 mg/mL. Store in the dark at 4°C (see Notes 10 and 11).

3.2. Free Flow Electrophoretic Separation of Samples

Clean and assemble the FFE apparatus as described in the manufacturer's operating manual. Set the temperature of the separation chamber to 15°C. Fill the system with ultrapure water ensuring all air bubbles are removed and leaks are not present. Ensure the 96-fraction capillary tubes are dispensing at an equal rate.

3.2.1. Performance Test for Laminar Flow

This test is performed to ensure that the laminar flow is stable (i.e. the separation media is being delivered at a constant flow into the separation chamber and that turbulence does not exist) and is essential to ensure reproducibility of separations:

1. Move the separation platform into the horizontal position. Set the media pump speed to 80 mL/h (or use specifications provided by manufacturer). Allow the system to equilibrate (10–15 min).
2. Introduce SPADNS (1% solution) via inlet tubes 2, 4, and 6. (Do not introduce SPADNS close to the electrodes as this will lead to contamination of electrode membranes.) Allow the SPADNS to flow through the separation chamber and check that the dye lines are straight and well defined (sharp edges and of a deep red colour).
3. When the SPADNS reaches the end of the separation chamber, begin collecting into a 96-well microtiter plate (approximately 100 μL fractions). The microtiter plate can be visually inspected as the dye should elute as well-defined fractions spanning 12–15 wells (see Fig. 2a). The SPADNS fractions should appear red. Alternatively, the absorbance of the microtiter plate can be read at 510 nm to quantify the SPADNS in the wells.
4. When the results are satisfactory, flush the system with ultrapure water to remove SPADNS (at least 20 min). If the results are unsatisfactory, flush the system with water, and if necessary, disassemble the apparatus and clean the system.

3.2.2. Establish the pH Gradient and Perform the pI Test

These methods have been optimized to establish a linear pH gradient from pH 3 to 9. This can be customized to form a pH gradient suitable for your application:

1. Introduce the separation, stabilization, and counterflow media via the appropriate inlets. Allow the solutions to flow through the system ensuring air bubbles and leaks are not present.
2. Introduce the anode and cathode electrolytes and switch on the electrode pump.
3. Adjust the voltage, current, and power settings to 1,000 V, 50 mA, and 60 W, respectively (or use settings recommended by the manufacturer). Set the media flow rate (80 mL/h or as recommended by manufacturer) and allow the system to equilibrate.

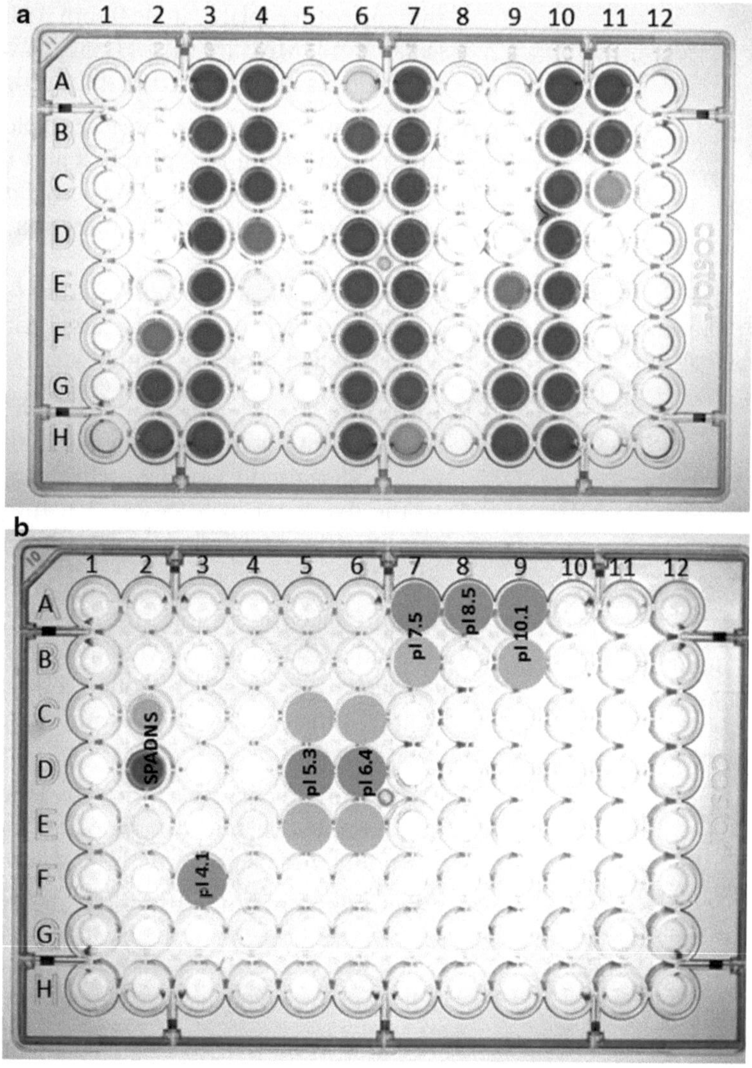

Fig. 2. (**a**) Performance test showing the expected elution pattern of the SPADNS dye. The leftmost fraction from the FFE chamber (anodic side) is delivered to well A1, with the next fraction delivered to B1, etc., down the column and then proceeding to A2. The H12 fraction is collected from the tube closest to the cathodic side of the chamber. (**b**) pI test showing the expected pattern of the markers and SPADNS.

4. Introduce the pI markers at a rate of 2 mL/h (or use settings recommended by the manufacturer). Allow the coloured markers to flow through the separation chamber. A sharp red band (SPADNS) should define the anodic boundary of the separation field. The pI markers should appear as six sharply defined yellow bands within the separation chamber.
5. Collect fractions into a 96-well microtiter plate. The microtiter plate can be visually inspected (see Fig. 2b). The SPADNS should appear in 1–2 fractions, and each individual pI marker should span <4 fractions. Alternatively, read the absorbance of

the microtiter plate at 410 nm (pI markers) and 510 nm (SPADNS). Measure the pH of each well containing pI marker to ensure each has eluted at its appropriate isoelectric point.

6. If the result of this test is acceptable, disconnect the sample tubing from the FFE apparatus and then flush with ultrapure water. Reconnect the tubing and flush with media with the sample pump in the reverse direction.

3.2.3. FFE Separation of CyDye-Labeled Protein Sample

Analysis of the standard protein mixture (steps 1–4) is optional and can be omitted by experienced users. Proceed to step 5 for analysis of CyDye-labeled protein sample:

1. Introduce a small amount of SPADNS (1% w/w in separation medium 2) to help define the beginning of the protein sample. SPADNS is used to help visualize the sample limits (analogous to the dye front in SDS-PAGE). Introduce the standard protein mixture and follow this with a small amount of diluted SPADNS.
2. Begin collecting samples into 96-deep-well plates a few minutes after the initial SPADNS introduced reaches the end of the separation chamber. Collect a small volume (approximately 100 μL) into a 96-well microtiter plate to measure the pH of each fraction.
3. When the end of sample limits is reached (defined by SPADNS), stop sample collection and turn off the voltage. Flush the sample pump and chamber with ultrapure water (at least 20 min).
4. Measure the pH of each well of the 96-well microtiter plate to determine if proteins have eluted at their isoelectric point. Absorbance readings at 410 or 600 nm will identify the wells where the proteins have eluted.
5. Proceed with the CyDye-labeled protein sample using the same process as detailed above, i.e. introduce a small amount of diluted SPADNS (1% w/w in separation medium 2), followed by labeled protein sample, and then followed by diluted SPADNS. Collect fractions into a 96 × 2.2-mL polypropylene deep-well plate and also collect a small volume (approximately 100 μL) into a standard 96-well microtiter plate to measure the pH of each fraction. Store fractions in the dark (e.g. wrapped in aluminum foil) at 4°C (short-term overnight storage) or −80°C (longer-term storage).
6. This process can be repeated to separate multiple protein samples. Ensure sample chamber and pump is flushed with water between each sample and that the separation chamber does not contain any visible precipitation.
7. Once the experiment has been completed, stop the voltage, flush the system with ultrapure water, and follow manufacturer protocols for system shut down.

3.3. SDS Polyacrylamide Gel Electrophoresis

3.3.1. Sample Concentration

The volume of each fraction is approximately 500 μL (or greater depending on the volume collected). A small volume from each fraction can be analysed directly by 1D PAGE (see Subheading 3.3.2), or alternatively, each fraction can be concentrated to increase sensitivity of downstream mass spectrometry identification. Two potential methods for sample concentration are detailed below:

1. Concentration of each sample using centrifugal filters (e.g. 5 or 10 kDa cut-off) (see Note 12):
   (a) Apply each sample to individual centrifugal filters and process as described by manufacturer, ensuring centrifugation limits (*g*-force or rpm) are not exceeded.
   (b) Transfer the retentate to a clean Eppendorf tube ready for SDS-PAGE.
2. Protein precipitation using acetone (see Note 13):
   (a) Transfer each fraction to 15-mL polypropylene tubes and add 5–10 volumes of ice cold acetone to each. Incubate at 4°C overnight in the dark (e.g. wrapped in aluminum foil).
   (b) Spin each sample at 4°C for 15 min using the maximum *g*-force recommended by the manufacturer.
   (c) Decant supernatant and rinse the pellet three times with 100 μL ice cold acetone and gently dispersing the pellet each time to remove residual buffers.

3.3.2. SDS-PAGE, Gel Scanning, and Image Analysis

1. Resuspend fractions to be analyzed in sample loading buffer, using a clean Eppendorf tube for each sample.
2. Assemble and clean the electrophoresis apparatus as per manufacturer operating instructions. Prepare the precast gel cassettes and gel running buffers according to manufacturer instructions.
3. Load the molecular weight marker into the outer most well of the gel and load each fraction to be analysed into the remaining individual wells.
4. Perform electrophoresis using the conditions recommended by the manufacturer.
5. Following electrophoresis, carefully pry the gel cassette open and transfer gel to clean plastic tub. Rinse gently with ultrapure water (2–3 times) and remove to gel scanner.
6. Scan the gel at the appropriate settings for CyDyes (Cy2: 488/520 nm; Cy3: 532/580 nm; Cy5: 633/670 nm).
7. Remove the gel to a clean plastic container, rinse gently with ultrapure water, and apply enough Coomassie fixing solution to just cover the gel. Cover the container with plastic wrap and allow the gel to incubate for 1 h while gently agitating (e.g. using an orbital shaker). Decant the fixing solution and add enough CBB working solution to just cover the gel.

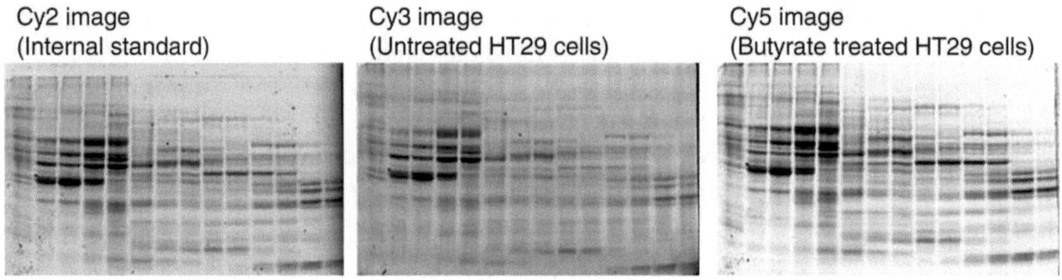

$$\text{Normalised expression for Untreated band} = \frac{\text{Untreated (Cy3)}}{\text{Internal standard (Cy2)}}$$

$$\text{Normalised expression for Treated band} = \frac{\text{Treated (Cy5)}}{\text{Internal standard (Cy2)}}$$

$$\text{Fold change} = \frac{\text{Normalised expression for Treated band}}{\text{Normalised expression for Untreated band}}$$

Fig. 3. 1D gel image of selected FFE fractions scanned at the appropriate settings for each respective CyDye. The formula to calculate the fold change in protein expression of a band of interest using densitometry is shown below.

Cover container with plastic wrap and leave to stain overnight while gently agitating (see Note 14).

8. Gently decant Coomassie stain into waste container and gently rinse gel several times with ultrapure water. Gels can be stored in ultrapure water at 4°C for a maximum 2–3 days or in 1% acetic acid at 4°C for longer periods.
9. Perform image analysis of gels to determine the protein bands that are differentially expressed as a result of treatment. To calculate the fold change in protein expression, potential bands of interest must first be normalised to the internal standard sample (see Fig. 3). Protein bands of interest can be excised, in-gel digestion performed, and the protein can then be identified by mass spectrometry.

4. Notes

1. Protein pellets can be resolubilized in an alternate buffer system, but the compatibility of each buffer reagent with the CyDye labeling procedure should be determined by consulting the product insert.
2. DMF is a hazardous, potentially toxic chemical and should be used in a chemical fume hood. Avoid inhalation and contact with skin. Refer to MSDS and product insert for safety information regarding handling, storage, and disposal.

3. To make 80 g (~70 mL) of anode stabilization medium, weigh 25.20 g urea and 9.05 g thiourea and add to a 100-mL wide-neck Schott bottle or similar, containing 23.25 mL of ultra-pure water and 7 mL of 1 M H_2SO_4. Place the bottle on a balance and, using pipettes, add glycerol (10 g) and 0.8% HPMC stock solution (5 g). Do not heat above 30°C as this decomposes the urea and store the reagents in a cool, dark place until use. The pH and conductivity of the solution should be ~1.7 and ~6.6 mS, respectively.
4. To make 80 g (~70 mL) of separation medium 1, weigh 25.20 g urea and 9.05 g thiourea and add to a 100-mL Schott bottle containing 19.2 mL of ultrapure water. Place the bottle on a balance and, using pipettes, add ProLyte 1 (11.6 g), glycerol (10 g), and 0.8% HPMC stock solution (5 g). The pH and conductivity of the solution should be ~4.5 and ~260 μS, respectively.
5. To make 160 g (~140 mL) of separation medium 2, weigh 50.45 g urea and 18.11 g thiourea and add to a 250-mL Schott bottle containing 24.4 mL of ultrapure water. Place the bottle on a balance and, using pipettes, add ProLyte 2 (37.12 g), glycerol (20 g), and 0.8% HPMC stock solution (10 g). The pH and conductivity of the solution should be ~7.4 and ~430 μS, respectively.
6. To make 80 g (~70 mL) of separation medium 3, weigh 25.20 g urea and 9.05 g thiourea and add to a 100-mL Schott bottle containing 15.3 mL of ultrapure water. Place the bottle on a balance and, using pipettes, add ProLyte 3 (15.52 g), glycerol (10 g), and 0.8% HPMC stock solution (5 g). The pH and conductivity of the solution should be ~10.2 and ~350 μS, respectively.
7. To make 160 g (~140 mL) of cathode stabilization medium, weigh 50.45 g urea and 18.11 g thiourea and add to a 250-mL Schott bottle containing 47.5 mL of ultrapure water and 14 mL of 1 M NaOH. Place the bottle on a balance and, using pipettes, add glycerol (20 g) and 0.8% HPMC stock solution (10 g). The pH and conductivity of the solution should be ~14.0 and ~5.4 mS, respectively.
8. To make 500 g (~435 mL) of counterflow medium, weigh 156.75 g urea and 56.29 g thiourea and add to a 500-mL Schott bottle containing 193.2 mL ultrapure water. Place the bottle on a balance and add glycerol (62.5 g) and 0.8% HPMC stock solution (31.25 g). The pH and conductivity of the solution should be ~9.8 and ~50 μS, respectively.
9. Alternatively, polyacrylamide gels can be handcast. Low-fluorescence glass plates (GE Healthcare, catalogue #80-6442-14) can be used to minimise gel handling during the scanning procedure and minimise potential for contamination

(e.g. contamination with keratin which can interfere with mass spectrometry identification).

10. Ensure that the total salt content of the sample does not exceed 25–50 mM (ideally < 300 uS conductivity) as high salt concentrations will profoundly interfere with the FFE separation process. If required, reduce the salt concentration by diluting further in separation medium 2. Remove any precipitate material by centrifugation.
11. We found that final protein concentrations over 2 mg/mL resulted in precipitation occurring in the separation chamber. Proteins may precipitate when they reach their isoelectric point. If precipitation occurs, dilute the protein sample further in separation medium 2.
12. Concentration of samples using centrifugal filters is limited by the presence of glycerol and HPMC in the sample. These are highly viscous solutions and do not easily pass through the filters, i.e. they will also be concentrated along with protein. Excess glycerol and HPMC may also affect resolution of protein bands by 1D PAGE.
13. Alternate protein precipitation methods can be used, but care must be taken to ensure that the integrity of the CyDye label is maintained. For example, the low pH associated with TCA precipitation methods may reverse the amide linkage between the NHS ester reactive group of the CyDye and the epsilon amino group of lysine in proteins.
14. Alternate staining methods can be used, e.g. silver staining to visualize less abundant proteins. If silver staining the gels, use a mass spectrometry compatible method to reduce interference with downstream analysis.

References

1. Hannig K (1982) New aspects in preparative and analytical continuous free-flow cell electrophoresis. Electrophoresis 3:235–243
2. Hannig K, Heidrich H (1990) Free-flow electrophoresis. GIT Verlag, Darmstadt
3. Hoffmann P, Ji H, Moritz RL et al (2001) Continuous free-flow electrophoresis separation of cytosolic proteins from the human colon carcinoma cell line LIM 1215: a non two-dimensional gel electrophoresis-based proteome analysis strategy. Proteomics 1:807–818
4. Islinger M, Eckerskorn C, Volkl A (2010) Free-flow electrophoresis in the proteomic era: a technique in flux. Electrophoresis 31:1754–1763
5. Krivankova L, Bocek P (1998) Continuous free-flow electrophoresis. Electrophoresis 19: 1064–1074
6. Weber G, Bocek P (1996) Optimized continuous flow electrophoresis. Electrophoresis 17: 1906–1910
7. Weber G, Islinger M, Weber P et al (2004) Efficient separation and analysis of peroxisomal membrane proteins using free-flow isoelectric focusing. Electrophoresis 25: 1735–1747
8. Weber PJ, Weber G, Eckerskorn C (2004) Isolation of organelles and prefractionation of protein extracts using free-flow electrophoresis. Curr Protoc Protein Science 22.5.1–22.5.21
9. Fung KY, Cursaro C, Lewanowitsch T et al (2011) A combined free-flow electrophoresis and DIGE approach to identify proteins regulated by butyrate in HT29 cells. Proteomics 11:964–971

Chapter 13

Immunofixation Electrophoresis for Identification of Proteins and Specific Antibodies

Gyorgy Csako

Abstract

Immunofixation electrophoresis (IFE) is a technique for the identification of proteins within complex mixtures after separation by either conventional zone electrophoresis or isoelectric focusing. Most commonly antigens (which are often immunoglobulins) are separated by electrophoresis followed by precipitation with specific antibodies in situ. However, immunoglobulins with specific reactivity can be also precipitated with the proper antigens after electrophoresis in reverse or reversed IFE. Because of its great versatility, potentially high sensitivity, ease to perform and customize, and relatively low cost with no requirement for expensive instrumentation, manual IFE remains a valuable tool for both clinical diagnostic testing and research. Any low-viscosity body fluid specimen or, possibly, culture fluid could be tested with IFE if proper antibodies (or antigens in reverse[d] IFE) are available. After pretreatment with chaotropic and/or reducing agents, even high-viscosity specimens might be amenable to testing with IFE.

Key words: Amido black, Agarose gel, Immunoglobulins, Immunoprint, Isoelectric focusing, Immunoprecipitation, Lipoproteins, Plasma proteins, Reverse(d) immunofixation electrophoresis, Zone electrophoresis

1. Introduction

Immunofixation electrophoresis (IFE) is an alternative technique to immunoelectrophoresis (IEP) for visualizing specific proteins in situ through an immunologic reaction, namely, antigen–antibody precipitin formation, following separation by electrophoresis in semisolid (gels) or solid media (membranes). Although most commonly antigens (which are often immunoglobulins [Igs]) are separated by electrophoresis followed by precipitation with their corresponding antibodies in situ, immunoglobulins with specific reactivity can be also precipitated with the proper antigens after electrophoresis in situ, and this is called reverse or reversed IFE.

Biji T. Kurien and R. Hal Scofield (eds.), *Protein Electrophoresis: Methods and Protocols*, Methods in Molecular Biology, vol. 869, DOI 10.1007/978-1-61779-821-4_13, © Springer Science+Business Media, LLC 2012

The principle of IFE is simple. The antibodies (or antigens in reverse[d] IFE) are applied to the surface of a gel or onto a cellulose acetate membrane after electrophoresis, and at antigen–antibody equivalence or in moderate antibody excess, they form large precipitates with their counterparts in the gel or membrane. Upon washing, uncomplexed antibodies (or specific antigens in reverse[d] IFE) and other proteins are removed from the gel or membrane, leaving only the immunoprecipitates which are too large and insoluble to be washed out from the pores. The immunoprecipitates remaining in the gel or membrane can be then directly stained or identified by other techniques such as by means of fluorescein-, enzyme-, or radio-labeled primary or secondary antibodies or using a biotin–avidin detection system. In short, IFE permits the identification of proteins within complex mixtures after separation by either conventional zone electrophoresis or isoelectric focusing (IEF) and is suitable for localization of both discrete protein bands and diffuse zones. Further, reverse(d) IFE with (labeled) antigen addition allows for detecting the function of the immunoglobulins separated by electrophoresis.

The concept of IFE was reported independently by Afonso (1) and Wilson (2) as a modified technique of IEP for the quantitation of low protein concentrations in 1964. However, the method became practical only when Alper and Johnson substituted monospecific antisera for polyvalent antisera in 1969 (3, 4). They called this method IFE and used it for improved detection of protein polymorphism. Interestingly, in a commentary published along with reprint of the original article in 1983 (3, 4), the authors recall that their manuscript was rejected by one (unnamed) journal before being accepted for publication in the *Vox Sanguinis*. In addition to protein polymorphisms (5–11), IFE applications also have been reported for the study of protein conversion products of enzymes (3, 4, 12, 13), for the detection and identification of abnormal B-cell clone-produced immunoglobulins (paraproteins) (14, 15), and even for protein quantitation (12). In these studies, IFE was shown to be applicable to a variety of gel types and cellulose acetate membrane as well. In a pivotal series of articles in 1976, Ritchie and Smith (16–18) concluded that "Immunofixation offers the worker an economical means of physically locating a protein in an electrophoretic strip and is ideally suited to forensic medicine, genetic studies, or research. The method is as simple and economical as the commonly used one- or two-dimensional immunoelectrophoresis, yet yields considerably more information." Increasing availability of commercial kits since the mid-1980s greatly contributed to the increasing popularity of IFE for paraprotein detection and identification. IFE in agarose gel gradually replaced IEP and became the new standard for the detection and identification of paraproteins in human body fluids [serum, urine, and cerebrospinal fluid [CSF)] (19–21). In combination with IEF, IFE became a "gold standard" for detecting oligoclonal bands in CSF (22). Finally, zone electrophoresis coupled with IFE is used for distinguishing serum transferrin from CSF asialo-(β_2) transferrin (previously known as β-trace protein) after head trauma or surgery, allowing the diagnosis of CSF leak (23–25). Reverse(d) IFE proved to be a useful tool for identifying the specificity of electrophoretically separated antibodies (14, 26–29). Immunosubtraction methods (also called capillary IFE) in which immunoabsorption of the samples with solid-phase-bound monospecific antibodies (Paragon 2000, Beckman) or formation of slow-migrating immune complexes by mixing the samples with antibodies in liquid phase (CAPILLARYS 2, Sebia) is followed by capillary zone electrophoresis (CZE) became recently available for testing for paraproteins and other proteins (30–32). Although capillary immunosubtraction is relatively fast (no stains and drying) and can be fully automated as a walkaway system, it is less sensitive and generally considered inferior to IFE for paraprotein detection (30–32). The utility of a method proposed for detection of specific antibodies using immunosubtraction and capillary electrophoresis, called immunocapture–immunosubtraction or ICIS (33), requires additional studies. Thus, because of its great versatility, ease to perform and customize, potentially high sensitivity, and relatively low cost with no requirement for expensive instrumentation, manual IFE remains a valuable tool for both basic and clinical research. The following is the description of a simple and practical manual IFE method. Note that, unless stated otherwise, antiserum and antibody variably refer to the same primary reagent of direct immunofixation. Further, depending on the context, specific antibodies may refer to the primary antibodies in direct IFE or to the antibodies identified by their respective antigens in reverse(d) IFE.

2. Materials

Prepare all solutions with distilled or deionized water and high-grade analytical reagents at room temperature (15–30°C). Except stated otherwise, reagents and other materials are stored at room temperature. Unless obtained commercially, prepare and store all reagents and accessories used in these methods without adding sodium azide. Follow all waste disposal regulations for discarding reagents and disposables. As a safe laboratory practice, always wear clean gloves to minimize or eliminate exposure to infectious agents and hazardous chemicals.

2.1. Agarose Gel Plates for Zone Electrophoresis (see Note 1)

1. Hydragel 7 and Hydragel 15/30 Protein(e) alkaline-buffered precast 0.8% agarose gels (Sebia, Norcross, GA, USA). These gels contain Tris–barbital buffer (pH 9.2 ± 0.1) and proprietary additives which are claimed to be “nonhazardous” at the

concentrations used. Store horizontally at room temperature or refrigerated (2–8°C) until stated expiration date (see Note 2).

2. Hydragel 7 and Hydragel 15 Lipoprotein(e) alkaline-buffered precast 0.8% agarose gels (Sebia). These gels contain "buffer" (pH 8.5 ± 0.1) and proprietary additives which are claimed to be "nonhazardous" at the concentrations used. Store horizontally at room temperature or refrigerated until stated expiration date (see Note 2).

2.2. Apparatus and Accessories for Zone Electrophoresis in Agarose Gel

1. Hydrasys® semiautomated electrophoresis system (Sebia) (see Note 3).
2. Buffered, ready to use sponge strips for protein electrophoresis (Sebia). Each contains Tris–barbital buffer pH 9.2 ± 0.3, sodium azide, and proprietary additives which are claimed to be "nonhazardous" at the concentrations used. Store horizontally at room temperature or refrigerated until stated expiration date (see Note 4).
3. Buffered, ready to use sponge strips for lipoprotein electrophoresis (Sebia). Each contains "buffer" pH 8.5 ± 0.1, sodium azide and proprietary additives which are claimed to be "nonhazardous" at the concentrations used. Store horizontally at room temperature or refrigerated until stated expiration date (see Note 4).
4. Precut, single-use applicators (ready to use) (Sebia). Store in a dry place at room temperature or refrigerated.
5. Precut, single-use filter papers (Sebia). Store in a dry place at room temperature or refrigerated (see Note 5).
6. Wet storage chamber supplied with Hydrasys (Sebia).
7. Pipettes or syringe for delivering volumes between 10 and 200 μL.

2.3. Accessories for Immunofixation and Staining in Agarose Gel

1. Thick filter papers (Sebia). Store in a dry plate at room temperature or refrigerated (see Note 6).
2. Lens paper strips for antiserum overlay, 5 × 30 mm each. Cut by means of a scissor from lens paper (VWR Scientific, Radnor, PA, USA) proportionate to the location of the expected plasma protein bands in the electrophoretic lanes (Fig. 1a) (see Note 7).
3. Template for antiserum application. Custom cut from X-ray film with tracks matching the position of the expected plasma protein electrophoretic patterns (Fig. 1b) (see Note 7).
4. Pipettes or syringes for delivering volumes between 0.5 and 100 μL.

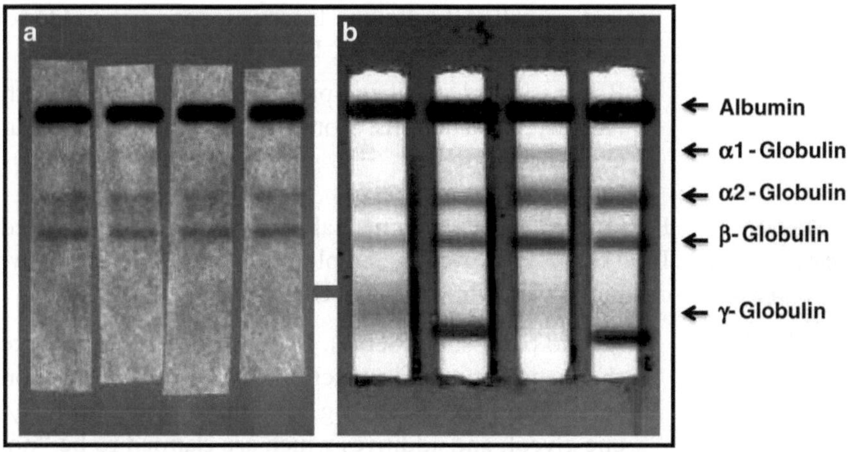

Fig. 1. Two methods of antibody (or in "reverse[d]" IFE: antigen) application for immunofixation following electrophoresis. Applying the antibody (or antigen) solution via cellulose acetate or paper strips (**a**) and placing the antibody (or antigen) solution into the tracks of an antiserum application template (**b**) positioned over the target area of the gel (in this case, electrophoretically separated human serum proteins). For illustration, the methods are shown as overlays on amido black-stained normal and abnormal (monoclonal band present) human serum protein patterns after electrophoresis in agarose gel. In case of (**a**), strips of a simple lens paper were used, while in case of (**b**), the template was prepared by cutting a previously developed X-ray film with a scalpel. *Short horizontal line* in the center indicates approximate point of sample application for electrophoresis.

5. Wet storage chamber supplied with Hydrasys (Sebia) (see Note 8).
6. Staining pans.
7. Paragon dryer (Beckman Coulter, Fullerton, CA, USA).

2.4. Protein Fixative for Reference Lane(s) in Agarose Gel IFE (see Note 9)

For convenience, a commercial protein fixative was used in these IFE runs. However, a simple "homemade" alternative also is described (see Note 10).

1. IFE fixative solution, ready to use (Sebia). This is an acidic fixative of undisclosed composition and contains "nonhazardous" additives at concentrations used. Store at room temperature or refrigerated until stated expiration date.

2.5. Antibodies for Immunofixation (see Notes 11–15)

1. Monospecific animal antisera to human α_1-acid glycoprotein, albumin, α_1-antitrypsin, α_2-macroglobulin, complement factor C3c, and haptoglobin (all from Dade Behring, Deerfield, IL, USA), fibrinogen (Sigma-Aldrich, St. Louis, MO, USA), and transferrin (Behring Diagnostics, Marburg, Germany).
2. Monospecific "mammalian immunoglobulins" (antisera) to human IgG, IgA, and IgM, κ and λ [free and bound] light chains, and free κ and free λ light chains (all from Sebia).
3. Antiwhole human serum (Helena Laboratories, Beaumont, TX, USA), monospecific antisera to human apo-AI and apo-B [Chemicon International Inc.(Millipore), Billerica, MA, USA],

and apo-E [Biodesign International, Inc. (Meridian Life Science, Inc.), Saco, ME, USA].

4. Reconstitute lyophilized preparations according to the manufacturer's instructions. Store all antisera refrigerated until stated expiration date.

2.6. Protein Staining and Destaining Solutions for Agarose Gel IFE (see Note 16)

For convenience, a commercial amido black stain was used in these IFE runs. However, a simple "homemade" alternative also is described (see Notes 17 and 18).

1. Amido black stain (Sebia). Reconstitute and store according to the manufacturer's instructions. The working solution is acidic (pH 2) and contains 0.4 g/dL amido black, 6.7% (v/v) ethylene glycol, and additives which are claimed to be "nonhazardous" at concentrations used. Store working solution in a closed container to prevent evaporation for 1 month.
2. Destaining solution for amido black staining: 5% acetic acid [v/v]: To 950 mL distilled or deionized water, add 50 mL of glacial acetic acid and mix thoroughly using a magnetic stirrer. Store in a closed container for 1 week.

2.7. Sudan Black Staining and Destaining Solutions for Agarose Gel IFE (see Note 19)

For convenience, a commercial Sudan Black stain was used in these IFE runs. However, a simple "homemade" alternative also is described (see Note 20).

1. Sudan Black stain (Sebia). Prepare working solution 30 min before use according to the manufacturer's instructions. Working solution contains ~0.04 g/dL Sudan Black, ~52% (v/v) absolute ethanol, ~48% water, and <0.01% dimethylformamide (v/v). Discard solution after use.
2. Destaining solution and wash solution 1 for Sudan Black staining (Sebia).
3. Wash solution 2 for Sudan Black staining (Sebia).

2.8. Enzymatic Staining for Lipoprotein Cholesterol in Agarose Gel IFE

1. Reagents from an enzymatic cholesterol staining kit: SPIFE® Vis Cholesterol (Helena).
2. Store, reconstitute, and prepare reagents according to the manufacturer's instructions.

2.9. Specimens and Positive Controls for IFE in Agarose Gel (see Note 21)

1. For illustration of the method, use human serum, plasma, and concentrated (30- to 50-fold) urine specimens (see Notes 22 and 23).
2. Collect serum, plasma, and random urine (with no preservatives!) specimens with standard medical procedures. Store refrigerated for up to 1 week (preferably use within 48 h) and

freeze at −20°C for up to 3 months or at −70°C for up to 12 months.

3. For electrophoresis, use serum and plasma samples straight or dilute with 0.85% saline to the concentration range established to be appropriate for target detection in trial runs with different antigen–antibody ratios (i.e., different combinations of sample vs. antiserum dilutions) (see Notes 13, 24, and 25).
4. For electrophoresis, use concentrated urine samples according to the protein concentration of the original (unconcentrated) urine either straight or dilute with 0.85% saline to the concentration range established to be appropriate for target detection in trial runs with different antigen–antibody ratios (i.e., different combinations of sample vs. antiserum dilutions).
5. Positive control specimens for human paraprotein testing: IFE Control Kit (IgG/κ, IgA/λ, IgM) (Helena). Store controls according to the manufacturer's instructions (see Note 26).

3. Methods

Carry out all procedures at room temperature unless otherwise specified. For general guidance and safety issues, consult the operating manual of the electrophoretic system used. Estimated total time needed to complete one gel with any of the electrophoretic and detection methods described below is ~3 h.

3.1. Zone Electrophoresis in Agarose Gel for Proteins and Lipoproteins

Switch on the Hydrasys instrument and proceed according to the Hydragel 7 or 15/30 Protein(e) method or, as appropriate, to the Hydragel 7 or 15 Lipoprotein(e) method for migration. Briefly:

1. Place one applicator for 7 Protein(e) and Lipoprotein(e) plate or two applicators for 15/30 Protein(e) or 15 Lipoprotein(e) plate.
2. Apply 10 μL of sample in each well. Load each applicator within 2 min.
3. Place the applicator(s) into the wet storage chamber with the teeth up. Let the samples diffuse into the teeth for 5 min after the last sample application.
4. Attach buffered strips as appropriate for protein or lipoprotein run to the pins on the uplifted electrode carrier.
5. Unpack the appropriate Hydragel plate and roll quickly and uniformly one thin filter paper onto the gel surface to absorb excess of liquid. Remove and discard the paper immediately.
6. Complete migration setup per manufacturer's instructions, including placement of the gel in the migration module and applications of samples onto the gel.

7. After selecting the proper migration program, carry out electrophoresis.
   (a) Under 10-W constant for serum/plasma and urine proteins using Hydragel 7 Protein(e) or 20-W constant using Hydragel 15/30 Protein(e) at 20°C, controlled by Peltier effect, until 33 Vh have accumulated (~7 min).
   (b) Under 160-V constant voltage for serum lipoproteins using Hydragel 7 and 15/30 Lipoprotein(e) at 20°C, controlled by Peltier effect, until 65 Vh have accumulated (~25 min).

3.2. Immunofixation

1. After stopping electrophoresis, remove gel from the migration module and wipe excess solution from back of gel.
2. Gently blot the gel surface with thin filter paper. Discard paper.
3. When available, apply 0.5 μL of positive control materials as "dots" to designated locations at either end of the electrophoretic tracks (except for the acid-fixative lane) and allow absorption into gel for 2–3 min.
4. For overlay of acid fixative and antisera by means of lens paper strips, carefully place each strip over the protein zones, including the control "spots," using the position numbers of the plate as a guide (Fig. 2a). A 1:1 size print of the stained protein/lipoprotein electrophoretic gel placed under the plate helps in positioning of the strips (Fig. 2a) (see Notes 27–29).
5. Using a micropipette, apply 35 μL of protein fixative onto the strip corresponding to the reference lane and the same amount of appropriate antisera onto each of the other strips. If needed for a protocol, the same amount of antisera can be reapplied after 3–5 min onto the strips.
6. Carefully spread acid fixative and antisera over the entire surface of the strips by means of the tip of the pipette without damaging the gel surface.
7. For using template for antiserum delivery, align the IFE antiserum template with the position of the electrophoretic tracks identified by the printed position numbers of the plate (Fig. 2b). Apply the template to the gel by slightly bending it in a "U-shape." This minimizes or avoids getting air bubbles under the template, especially when very thin plastic templates are used. If bubbles form, gently lift the edge of the template with the edge of a thin spatula, then lay the template back down on the gel to remove the bubble(s).
8. Gently rub around the tracks to ensure complete seal between template and gel surface.
9. Apply 35 μL of IFE protein fixative to the reference track and the same amount of each antiserum to the designated tracks, including the position of the control "spots." Avoid spillage

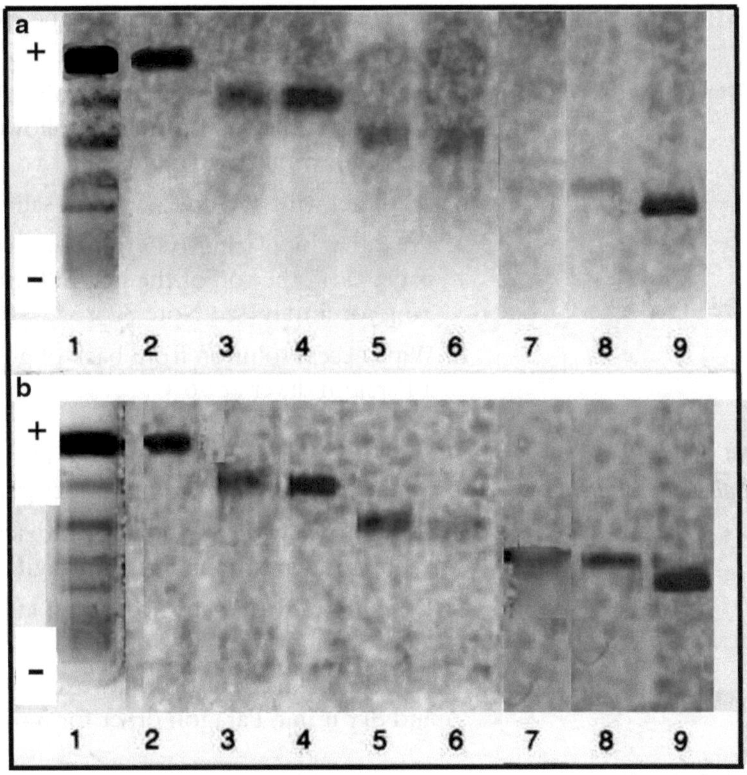

Fig. 2. Detection of various human plasma proteins with IFE in a protein electrophoresis agarose gel using two different methods of immunofixation with monospecific antibodies, followed by protein staining with amido black. Application of antibodies via lens paper (**a**) and antiserum template (**b**) as shown in Fig. 1. *Lane 1*: protein fixation with acid fixative as a reference, while all other lanes represent immunofixation of various plasma proteins with monospecific antibodies as follows: *lane 2*: albumin; *lane 3*: α_1-acid glycoprotein; *lane 4*: α_1-antitrypsin; *lane 5*, haptoglobin; *lane 6*: α_2-macroglobulin; *lane 7*: complement factor C3c; *lane 8*: transferrin; and *lane 9*: fibrinogen.

over tracks. Do not touch gel surface with pipette or syringe tip (see Notes 27–29).

10. Place gel in wet incubation chamber.
11. Cover chamber and incubate gel at room temperature for 30 min (see Note 30).

3.3. Processing of Agarose Gels for Staining

1. Following incubation period, remove gel from wet incubation chamber.
2. In case of overlay with lens paper strips, wash gel with ~75 mL of 0.85% saline with gentle shaking on a platform (or just by hands) for 5 min. Remove and discard strips.
3. In case of cut IFE antiserum template, soak gel in ~75 mL of 0.85% saline with gentle shaking on a platform (or just by hands) for 5 min and gently peel off the template (see Note 31).

4. Remove gel from saline and wipe excess solution from back of gel.
5. Press dry by placing two thick absorbent papers ("thick filter paper") on the top of the gel, followed by adding a weight of ~75 g for 5 min (see Note 32).
6. Wash gel with ~75 mL of 0.85% saline for 5 min.
7. Press dry by placing two thick absorbent papers ("thick filter paper") on the top of the gel, followed by adding a weight of 75 g for 5 min (see Note 32).
8. Wipe excess solution from back of gel and dry gel for 5 min in a Paragon dryer at 56°C.

3.4. Amido Black Staining for Proteins

1. Stain the dried gel by immersing into the amido black stain for 5 min at room temperature under gentle agitation.
2. Destain gel with 5% (v/v) acetic acid solution for 3 min at room temperature under gentle agitation.
3. Destain gel with water for 3 min at room temperature under gentle agitation.
4. Remove gel from water, wipe excess solution from back of gel, and dry it in a Paragon dryer for 5 min at 56°C.

3.5. Sudan Black Staining for Lipoproteins (see Note 33)

1. Stain the dried gel by immersing into the Sudan Black stain for 5 min at room temperature under gentle agitation.
2. Briefly (~1 min) destain gel with water under gentle agitation.
3. Remove gel from water, wipe excess solution from back of gel and dry it in a Paragon dryer for 5 min at 56°C.

3.6. Enzymatic Staining for Lipoprotein Cholesterol (see Note 33)

1. Pour cholesterol substrate into incubation chamber and place agarose gel down.
2. Incubate gel with substrate for 15 min at 30°C.
3. Wash gel briefly (~1 min) with citric acid destaining solution at room temperature.
4. Briefly (~1 min) destain gel in water at room temperature.
5. Remove gel from water, wipe excess solution from back of gel and dry it in a Paragon dryer for 5 min at 56°C.

3.7. Interpretation of IFE Patterns (see Note 34)

General comments:

1. Evaluate gel visually by observing location of specific precipitin bands in relationship to (acid-fixed) reference protein electrophoretic pattern (see Note 35).
2. A stained immunofixed band represents a reaction between a specific antiserum and a protein (antigen) in the sample in both conventional IFE and reverse(d) IFE (see Notes 36 and 37).

3. By comparing the location of a band in the acid-fixed reference pattern with the stained immunofixed band in the same location, determine the identity of a particular protein in the sample (see Notes 38–40).

Specific comments on IFE patterns produced with the manual method on human serum, plasma, and urine specimens:

1. Figure 2: Even in a minimally optimized IFE method with homemade antiserum applicators, various plasma proteins were satisfactorily identified with either method of antibody application for immunofixation after electrophoresis of a human plasma specimen. The "dirty" background apparently is due to inadequate removal of the heavy load of nonreactive proteins from the undiluted or minimally diluted, relatively low-avidity antisera.
2. Figure 3: The same three paraproteins (IgG[κ], IgA[λ], and IgM[κ]), signifying a triclonal gammopathy, were correctly identified by two major semiautomated commercial methods and the manual IFE method in a human serum specimen. Note the clarity of background with the manual IFE method when highly optimized and apparently "prepurified" antibodies (IgG fraction?) were used vs. the "dirty" background in a minimally optimized system in Fig. 2.
3. Figure 4: Based on the IFE patterns, the first urine specimen (A) contains a strong band of free κ light chain, confirmed by testing with anti-free κ antibody, whereas the second urine contains both IgG(λ) and free λ light chain, again confirmed by testing with anti free λ antibody. As in Fig. 3, note the clearness of background when highly optimized and apparently "prepurified" antibodies (IgG fraction?) are used in the manual method. These observations underscore the importance of antiserum quality: a critical need for high-avidity and high-affinity antibodies in the presence of minimal amounts of extraneous proteins to reduce or virtually eliminate background staining, especially when protein stains are used for detection. Obviously, this becomes less of an issue when probes are involved in detection of the immunoprecipitates.
4. Figure 5: The results obtained with plasma lipoproteins demonstrate the versatility of the IFE method. It is interesting to note that more apolipoprotein is carried in the α-lipoprotein fraction than the combined pre-β and β-lipoprotein fraction, while the reverse is true for the transport of both major lipids, triglycerides, and cholesterol. The biochemical results are consistent with these observations.

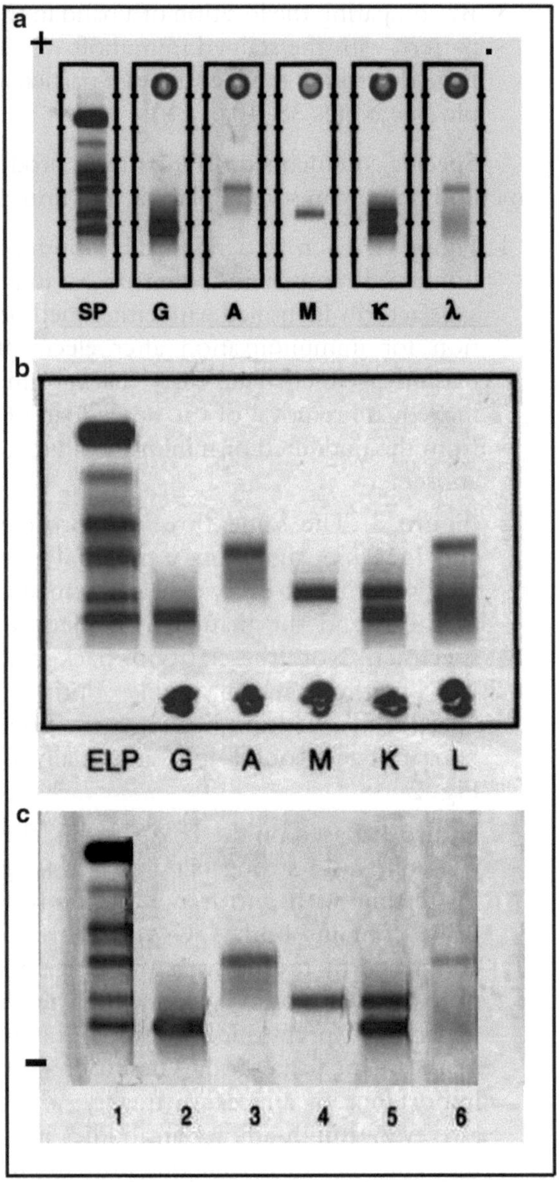

Fig. 3. Comparison of the manual IFE method with two major semiautomated commercial IFE methods for human serum paraprotein detection and identification in agarose gel. (**a**) SPIFE® ImmunoFix kit on the SPIFE 3000 system, acid violet staining (Courtesy of Dr. Rita Ellerbrook of Helena Laboratories, Beaumont, TX, USA). (**b**) Hydragel 2 IF on Hydrasys®, acid violet staining (Sebia Inc., Norcross, GA, USA). (**c**) Manual IFE method on a protein electrophoresis plate and using a hand-cut IFE antiserum template for manually applying acid fixative and antisera, followed by amido black staining. The acid fixative and monospecific antibodies with the manual method are the same as those in the semiautomated Sebia method. Note presence of "dots" of positive controls (Helena) for antibody identification in both commercial IFE's. Codes: *lane 1*: SP and ELP, acid fixative; *lane 2*: G and G, antihuman IgG heavy chain-specific antibody; *lane 3*: A and A, antihuman IgA heavy chain-specific antibody; *lane 4*: M and M, antihuman IgM heavy chain-specific antibody; *lane 5*: κ and K, antihuman κ (free and bound) light chain-specific antibody; and *lane 6*: λ and L, antihuman λ (free and bound) light chain-specific antibody.

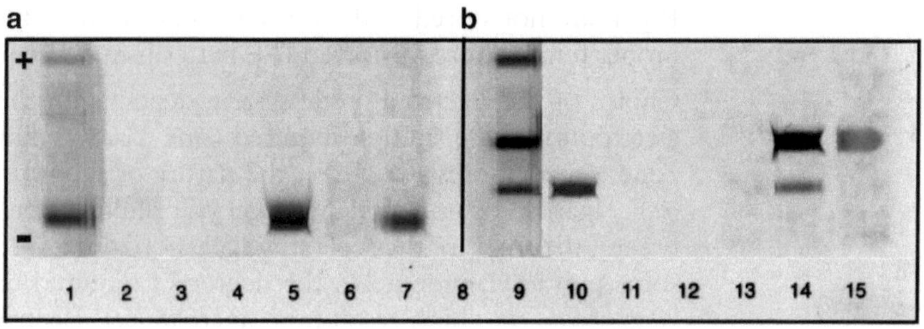

Fig. 4. Detection of paraproteins in two different human urine specimens with the manual IFE method in protein electrophoresis agarose gel using a home-cut IFE antiserum template for acid fixative and antibody application, followed by amido black staining. Both urine specimens were preconcentrated ~50-fold using Amicon Ultra-4 10K microconcentrators (Millipore, Billerica, MA, USA) for electrophoresis. The acid fixative and monospecific antibodies are the same as those in the semiautomated Sebia method. Codes: *lanes 1 and 9*: acid fixative; all other lanes received monospecific antibodies to human Ig heavy and light chains as follows: *lanes 2 and 10*: IgG heavy chain; *lanes 3 and 11*: IgA heavy chain; *lanes 4 and 12*: IgM heavy chain; *lanes 5 and 13*: κ (free and bound) light chain; *lanes 6 and 14*: λ (free and bound) light chain; *lane 7*: free κ light chain; and *lane 15*: free λ light chain.

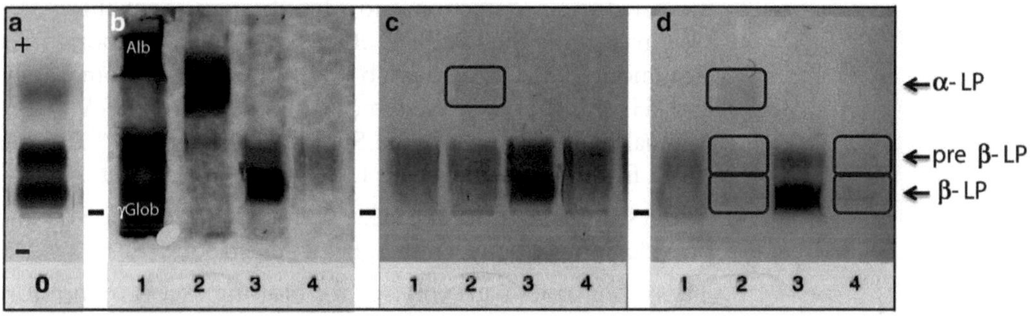

Fig. 5. Manual IFE method used for detecting various lipoprotein constituents via immunofixation of human whole serum and apolipoproteins-AI, apo-B, and apo-E in an agarose gel for lipoprotein electrophoresis. (**a**) Standard Sudan Black staining for lipids (Sebia), (**b**) through (**d**) staining after immunofixation with antibodies against human whole serum (*lane 1*), apo-AI (*lane 2*), apo-B (*lane 3*), and apo-E (*lane 4*), using a hand-cut antiserum template for application: (**b**) Amido black staining for proteins (Sebia method), (**c**) Sudan Black staining for lipids (Sebia method), (**d**) enzymatic staining for cholesterol (Helena method). *Short horizontal lines* indicate point of sample application. The serum specimen tested contained total cholesterol 289 mg/dL (desirable: <200, borderline high cardiovascular risk: 200–239), triglycerides 219 mg/dL (*N*: <150, high: 200–499), high-density lipoprotein-cholesterol (HDL-C) 35 mg/dL (high risk: <40), low-density lipoprotein-cholesterol ($LDL\text{-}C_{calculated}$) 210 mg/dL (optimal: <100, very high cardiovascular risk: ≥190), apolipoprotein A-I 118 mg/dL (*N*: 110–205), apolipoprotein B 146 mg/dL (*N*: 55–140), lipoprotein electrophoresis α-lipoprotein 14% (*N*: 17–46), pre-β-lipoprotein 42% (*N*: 2–32), β-lipoprotein 43% (*N*: 39–66), and chylomicrons 1% (*N*: 0–2).

4. Notes

1. Besides alkaline agarose, applications have been also reported for agar, acid starch, acidic agarose, agarose-acrylamide, polyacrylamide and cellulose acetate (Cellogel) gels (gel thickness commonly ≤1 mm), and dry cellulose acetate membrane (thickness 0.12–0.16 mm).

2. If gels are not stored in the horizontal position and/or at the proper temperature, distorted IFE patterns may occur.
3. Choice of the electrophoretic system depends on the type of electrophoretic separation required (one- or two-dimensional zone electrophoresis vs. IEF), the nature of support medium (e.g., agarose vs. polyacrylamide gel vs. cellulose acetate membrane), the shape of the separating gel (horizontal vs. disk with some polyacrylamide gels), the degree of required resolution [e.g., low- vs. high-resolution agarose gel electrophoresis (HRAGE)], availability of plate cooling (e.g., a must for IEF and HRAGE), and, obviously, the size of the gel or membrane (fit into the electrophoretic chamber). Any electrophoretic system that satisfies these requirements is suitable for generation of a gel or cellulose acetate membrane with the target proteins separated for subsequent immunofixation. Currently, there are at least four major commercial kits with dedicated instrumentation for clinical diagnostic testing of human body fluid specimens (serum, neat and/or concentrated urine, and/or CSF) for paraprotein detection and identification. Semiautomated systems include the SPIFE® 2000/3000, Helena Laboratories, Beaumont, TX, USA; Interlab G26, Interlab Sri, Rome, Italy; and Hydrasys-Hyrys® system, Sebia, Norcross, GA, USA. Manual methods include the Standard and Minifix™ IFE kits (both from The Binding Site Ltd., London, England).
4. These strips constitute the electrophoresis buffer reservoir and provide contact between the gel and electrodes.
5. These thin paper pads are used for blotting excessive moisture off the gel surface before sample application.
6. These single-use, thick absorbent papers are used for removing unprecipitated proteins off the gel surface after immunofixation.
7. Immunofixation of electrophoretic gels or membranes can be carried out by either immersing into an antibody solution (all polyacrylamide disks and often cellulose acetate membranes) or overlaying with an antibody solution (horizontal gels and membranes). In this latter approach, either the entire surface is spread by an antibody solution or different electrophoretic zones are exposed to different antibody solutions by means of semipermeable membranes or (filter) papers (Fig. 1). Application of antiserum directly to the gel surface may be advantageous over application via cellulose acetate and paper strips because (a) much smaller volumes (i.e., less antibody or less antigen in reverse[d] IEF) are needed to develop the same intensity IFE patterns, (b) more rapid diffusion of antiserum into the gel is less likely to be associated with antigen excess in most cases, and (c) the range of antigen concentrations detectable

with a given concentration of antibodies is wider (20). On the other hand, direct application of the antiserum to the surface of a gel requires availability of proper application templates which may be less accessible than cellulose acetate or paper strips. Nonetheless, as shown in Fig. 1, both types of IFE antiserum application devices can be made from materials that either can be purchased or are already available in the laboratory. Proportionate strips of cellulose acetate membrane or filter papers such as Whatman No. 1 or, simply, a lens paper (e.g., VWR Scientific, Radnor, PA, USA) for the area of gel or membrane containing the target protein band(s) can be cut using scissors (Fig. 1a). Likewise, antiserum application templates can be custom cut from thin, pliable plastic sheets or X-ray film using a scalpel (Fig. 1b).

8. Any level humid incubation chamber can substitute this chamber for room temperature incubations. Incubation at higher temperatures (up to 45°C) will shorten the incubation time for immunoprecipitation.
9. A variety of protein fixatives, often with undisclosed composition in commercial IFE kits, have been proposed and used over the years. Apart from the type of electrophoretic medium (e.g., agarose and polyacrylamide gels or cellulose acetate membrane), choice of the fixative often is arbitrary, based on personal experience and practicability.
10. Alternative "homemade" protein fixative (10% acetic acid [v/v] and 5% sulfosalicylic acid [w/v] in water): Weigh out 5.8 g of 5-sulfosalicylic acid dehydrate (MW 254.2), add it to a graded beaker, dissolve in 80 mL of distilled water under continuous mixing using a magnetic stirrer, then add 10 mL of glacial acetic acid, mix again, and add distilled water up to 100 mL. Store in a closed container at room temperature for 1 month.
11. High-affinity and high-avidity antisera with optimized dilutions to avoid prozoning will produce maximal immunoprecipitation. These attributes, along with their relatively low price, make polyclonal antibodies the reagent of choice for most immunoprecipitation-based applications. High titer has the advantage of allowing for higher antiserum dilutions which, in turn, reduce the undesirable background protein "load" in the gel or membrane. Because of their generally low affinity, monoclonal antibodies are rarely suitable for IFE. This is in contrast with their high applicability to and widespread use in the alternative immunodetection method, immunoblotting (western blotting).
12. In order to minimize the protein "load" on the gel or membrane during the immunofixation step, instead of using full

antisera, the immunoglobulin fraction (IgG) of antisera is preferred. This is a common approach in commercial kits for human paraprotein testing to reduce the background due to incomplete removal of nonreactive serum proteins. Alternatively, high dilution of high-titered full antisera helps to minimize the protein "load" in the gel or membrane and, hence, to reduce the background. In reverse(d) IFE, the same considerations apply: use of partially or completely purified antigens will minimize the background staining due to incompletely removed nonreactive proteins.

13. For each antigen–antibody combination, the optimal ratios should be determined in trial runs using several combinations of dilutions.

14. For enhancement of both speed and intensity of immune complex formation, inclusion of 4% polyethylene glycol (PEG, M_r 6,000) in the antiserum has been proposed (34), but it did not gain wide acceptance.

15. While replacing polyvalent antisera for monospecific ones was revolutionary at the time when IFE has been introduced in 1969 (3, 4), polyvalent antisera with well-defined specificities became recently advantageous at least in one application. Several semiautomated commercial methods for paraprotein detection and identification now offer the option of screening serum and/or concentrated urine samples with pentavalent antisera consisting of a pool of monospecific antibodies to IgG, IgA, and IgM heavy chains and κ and λ free and bound light chains (e.g., SPIFE® IFE Pentavalent kit for screening serum, Helena Labs.; Serum and concentrated urine pentavalent immunofixation, Interlab G26, Interlab Sri; and Hydragel 6/12 IF Penta for screening serum, Hydrasys, Sebia). The concept may be applicable to other protein antigen or specific antibody systems as well.

16. Besides fluorescein, radioactive, or enzyme probe-labeled primary or secondary antibodies, a variety of stains have been used for detecting protein in the immunoprecipitates of IFE. Major differences among these stains involve detection sensitivity [ranging from a few ng to 500 ng (often 50–100 ng) of target protein per band], electrophoretic media compatibility (e.g., agarose vs. polyacrylamide gel vs. cellulose acetate membrane), and "reversibility" (e.g., both Ponceau S and Nigrosin[e] stainings are completely reversible). Comprehensive direct comparisons of all of these stains are not available, and band detection sensitivity also is greatly affected by unequal affinity of stains toward various proteins and by factors unrelated to the stain (e.g., optimization of the precipitin reaction). Nevertheless, an approximate order of detection sensitivity for commonly used stains is estimated as follows: Silver > Coomassie

Brilliant Blue R-350 > Coomassie Brilliant Blue R-250 or G-250 > Ponceau S > Nigrosin(e) > Acid Violet > amido black. Note that the use of Ponceau S is limited to cellulose acetate membrane (and nitrocellulose membrane in immunoblotting). Finally, practicability (e.g., stability on storage, speed of staining and destaining, and concern for toxicity) and simplicity of use also play a major role in selection of stains for IFE. These considerations explain why amido black, a relatively insensitive dye, is so widely used.

17. Amido black (MW 616.5), a dark-red to black powder, stains proteins with a blue-black color. Of the many synonyms used for amido black, a short list is as follows: Acid Black 1, Acidal Black 10B, Acidal Navy Blue 3BR, Amidoschwarz, Amido Black 10B, Buffalo Black NBR, Eriosin Blue Black B, Naphthalene Blue Black., Naphthalene (Black) 12B, Naphthol Blue Black, and C.I. 20470. For protein detection in electrophoretic gels and membranes, amido black stains have been prepared in a number of ways, and specific formulations of the stain often are not disclosed in publications and IFE kits for paraproteins. While the nature of the electrophoretic matrix needs to be considered, the choice of a particular solution is often based on personal preference. Note that after staining of lipoproteins with Sudan Black, the proteins no longer take amido black.
18. Alternative "homemade" amido black staining solution: Prepare a saturated solution of the dye (~1 g/dL) in 10% acetic acid in methanol (v/v) by dissolving 1 g of amido black in 450 mL of methanol in a glass beaker, then add 50 mL of glacial acetic acid. Mix well using a magnetic stirrer for at least 30 min then filter. Store in a closed container for up to 2 months.
19. Sudan Black (MW 457), also known as Fat Black HB, Solvent Black, Sudan Black B, and C.I. 26150, is a dark-brown to black powder. It is slightly basic and combines with acidic groups in compound lipids, resulting in a blue-black color. Sudan Black is thought to stain mostly triglycerides in human plasma lipoproteins.
20. Alternative "homemade" Sudan Black staining solution: Prepare a saturated solution of Sudan Black in 60% ethanol by adding an excess of the dye (100 mg) to 100 mL of warm (37°C) 60% ethanol solution under constant mixing with a magnetic stirrer. Keep it at 37°C for 16 hrs, allow to cool, let it stand for few days at room temperature, filter, then store in a dark bottle. Stable for up to 2 months. Alternative "homemade" destaining solutions: 50% (v/v) ethanol then 40% (v/v) ethanol. Store in a closed container for up to 2 months.

21. Any low-viscosity body fluid specimen or culture fluid could be tested with IFE if proper antibodies (or antigens in reverse[d] IFE) are available. High-viscosity specimens (e.g., sera from certain gammopathy cases, most often of the IgM, and, less commonly, of the IgA and IgG type) can be pretreated with special chaotropic agents (such as the Fluidil with undisclosed composition in the Sebia system) and/or with the reducing agent 2-mercaptoethanol (2-ME). Cryoproteins may cause gelation of the serum specimens at room temperature and, hence, interfere with sample application for electrophoresis. Pretreatment with a chaotropic agent such as the Fluidil (Sebia) and/or preheating the specimen at 37°C for 30–60 min would reverse the "gelation" and allow for proper sample application for electrophoresis.
22. In order to get insight into the impact of IFE on clinical practice and research, PubMed searches were performed on July 14, 2011. While these retrievals obviously underestimated the true usage of this method, they still show that IFE is a widely used laboratory technique. Using the search term "immunofixation," a total of 1,128 articles were retrieved, of which 852 or 64% (852/1,128) were diagnosis related based on adding "diagnosis" as a second search term. Using the search terms "immunofixation electrophoresis," a total of 715 articles were retrieved from PubMed, of which 543 or 64% (543/715) were diagnosis related based on adding "diagnosis" as a third search term. These data indicate that most commonly (approximately two-thirds of the cases) IFE is used as a diagnostic test. A quick review of the title (and, occasionally, the abstract) of the articles retrieved by adding the search term "diagnosis" to either retrieval approach revealed that practically all are related to paraprotein detection and identification in hematologic malignancies. Further, the identical proportions of diagnosis-related immunofixation (electrophoresis) articles with the two retrieval approaches suggest that the word "immunofixation" is used virtually identically to IFE in the literature.
23. Most clinical urine specimens (dipstick protein ≤3+ or protein concentration <200 mg/dL), virtually all CSF specimens, and, likely, many culture fluid specimens require concentration for IFE testing after zone electrophoresis. Up to 50-fold enrichments with yields of 60–120 μL of concentrate could be obtained by centrifuging 2-mL samples in Amicon Ultra-4 microconcentrators (Millipore, Billerica, MA) (10 K for urines and 30 K for CSF at 3,100 × *g* for 14 min and 2,600 × *g* for 8 min, respectively, at 20°C in a swing-out rotor with a radius of ~11.5 cm). Retentate recovery is claimed to be high (>90%) with these devices. However, it is noteworthy that concentrating

specimens may lead to artifacts, including protein losses that differ from sample to sample and may be substantial (up to 50%) with some other devices.

24. To avoid antigen excess and minimize unwanted diffusion, samples generally need to be adjusted to contain approximately 5–50 mg/dL or 50–500 ng/μL of the protein under study.
25. Because of the concentrating (focusing) effects of IEF, the protein concentrations required for band detection are only approximately one-tenth of those with conventional zone electrophoresis (20).
26. Use of positive controls ensures that the proper antibodies (or, in case of reverse[d] IFE, antigens) are used in the proper locations in the gel or membrane for immunofixation. Along with a commercial kit for IFE testing of paraproteins, one manufacturer markets such controls (IFE Control Kit [IgG/κ, IgA/λ, IgM], Helena). If commercially not accessible, use solutions of "purified" target antigens (or, if available, specific antibodies in reverse[d] IFE) to confirm proper usage of each antibody (or antigen in reverse[d] IFE). While high purity is desirable to reduce the total protein "load" (and, hence, risk of false positivity), greatly "enriched" preparations might be satisfactory if their reactivity pattern with the corresponding monospecific antibody (or antigen in reverse[d] IFE) has been already established. Commercially available purified or recombinant preparations could be a major source of positive controls. Alternatively, specimens available in the laboratory either as a home-purified product or without further purification from natural sources can be considered for use as positive controls. For instance, for gammopathy testing of any body fluid specimen, urine specimens with a single strong paraprotein band (IgG, IgA, IgM, free κ, or free λ) might be appropriate for this purpose. The controls are applied after electrophoresis either to a preformed well (Helena) or simply to the surface of the gel and allowed to be absorbed into the gel as a "dot" (this manual method and Sebia IFE in Fig. 3). After protein staining, a positive antigen–antibody reaction affirming that the right antibody was used in the right place will be indicated by the development of a ring or dot (Fig. 3). Target concentrations of the positive controls usually are in the range of 5–50 mg/dL (corresponding to 25–250 ng of target protein assuming 0.5 μL is being applied to each control spot). For obvious reasons, performance of controls should be verified before routine use.
27. Since improper antibodies (or antigens in reverse[d] IFE) obviously would cause inaccurate results, great care should be taken to use the "right" antibodies in the "right" position of the gel. In order to minimize or avoid this type of errors, manufacturers of commercial kits for paraprotein detection and identification

use differently colored antibody reagents. Color is removed from the gel during the wash steps.

28. Application of antibodies (or antigens in reverse[d] IFE) should be no later than 10 min after completion of electrophoresis.

29. The volume of antiserum (or antigen in reverse[d] IFE) needed for immunofixation varies with the type and character of electrophoretic matrix (e.g., gel vs. membrane, thin vs. thick gels), mode of antiserum application (immersion into the antiserum vs. direct surface overlay, overlay using cellulose acetate membrane or thick filter paper vs. thin lens paper), and, obviously, with the size of the target area (defined by width and length of the tracks [lanes] and expected distribution of target proteins). Because of the loss due to uptake by the delivery device itself, direct overlay of an electrophoretic track generally requires less antiserum (or antigen in reverse[d] IFE) than overlay via a membrane or filter paper (see Note 7). Probably, searching for paraproteins represents the far extreme in antiserum volume needs because it requires covering virtually the entire area of serum electrophoretic separation from the anodal end (albumin) to the cathodal end of the normal gamma globulin zone. In fact, in order to assure that no paraproteins are missed, immunofixation (~1 cm) beyond the cathodal end of the normal gamma globulin zone is necessary to detect rare cases of far-cathodal migrating paraprotein variants. Most target proteins separated in the electrophoretic matrix are localized in a relatively short segment of a track and, therefore, require smaller volumes for immunofixation. Antiserum (or, in reverse[d] IFE, antigen) volume needed for immunofixation is variably reported as total per gel (e.g., 0.25 mL for a 10-position gel), total per track (lane) (e.g., 25–80 μL/track [lane]), or volume per unit of surface area (e.g., 10.2 μL/cm^2).

30. In order to speed up and enhance the antigen–antibody reaction, higher than room temperature incubations up to 45°C for 10–30 min have been suggested and used for paraprotein IFE . Since high temperatures will increasingly denature proteins, heat stability of both antigens and antibodies needs to be considered for establishing optimal incubation with a given system.

31. Unlike membranes or paper strips, IFE antiserum application templates may be reused after washing with mild detergents and, when indicated, disinfecting.

32. Many protocols call for using multiple layers of filter papers and adding more weight (as much as 1–1.5 kg) for increasing the efficiency of gel preparation for subsequent staining. Increasing the weight is particularly advantageous for "cleaning up" the gel after immunofixation. On the other hand, the

present protocol with a "minimalist" approach shows that even approximately one-tenth of the "optimal" weight load is able to generate usable IFE patterns if IgG fractions of the antisera are being used (Figs. 3 and 4). Additional cycle(s) of wash and press dry can be also considered for improving background "cleanup."

33. In case of complex particles such as lipoproteins, each containing several different apolipoproteins, a series of monospecific antibodies targeting the same immunoprecipitate location can be used to "dissect" the various protein constituents (Fig. 5b). Further, in addition to using stains or probes to detect the protein content of immunoprecipitates formed in the electrophoretic matrix, special stains or enzyme reactions can be applied to detect the nonprotein components. Examples include but are not limited to the detection of various lipid components of lipoproteins (Fig. 5c, d). A common lipid stain, Sudan Black, detects mostly triglycerides (Fig. 5c). For more specific identification and quantification of lipids in electrophoretically separated lipoproteins, enzymatic cholesterol, triglyceride, and phospholipid assays have been developed (35–38). These assays can be readily adapted for the detection of specific lipid components such as cholesterol in immunoprecipitates and allow for the study of colocalization of various apolipoproteins and lipids (Fig. 5d). Using a similar concept, calcium, iron, esterase, oxidase activity, etc., can be also detected in the immunoprecipitates.

34. Although the original intention of IFE was to identify proteins with antigenic property in the electrophoretic media in situ, it was later discovered that diffusion of the immune reactants can generate interpretable patterns in another way as well. When antibodies are applied with a cellulose acetate overlay, protein antigens from the separatory gel also diffuse into the cellulose acetate membrane and form precipitates there. This membrane can be then washed and stained for these immunoprecipitates, resulting in an "immunofixation print" or "immunoprint" (11). However, it was also observed that, while any cellulose acetate membrane is suitable for applying the antiserum for precipitation in the separating gel, those for generating good immunoprints require certain brands of membranes such as the ones used initially (Separphore III, Gelman Sciences, Ann Arbor, MI, USA, or Millipore Corp., Bedford, MA, USA). Details of the immunoprint technique have been summarized by Johnson (20).

35. Since IFE is suitable for localization of both discrete protein bands and diffuse zones, in the common patterns of paraproteinemic cases, the monoclonal or oligoclonal bands are present as discrete band(s) against the background of polyclonal

Igs as a diffuse zone in the gamma region. Detection sensitivity of IFE for paraproteins is generally higher in specimens with low background of polyclonal Igs (e.g., urines with no or low-level proteinuria and most CSF specimens). Because the co-occurrence of antigenically similar diffusely migrating and restricted mobility proteins in the same specimen is rather unique to the combination of polyclonal and monoclonal Igs, similar phenomena are less likely with other proteins.

36. Antigen excess in direct IFE results in increased diffusion and wider band. In extreme cases, this excess is associated with the presence of an unstained area in the center of an immunoprecipitin band, variably described as "central clear spot," "hole," or "doughnut." Rerunning the sample at a higher dilution (i.e., lowering the antigen concentration) will eliminate this prozoning artifact.

37. Because of the increased detectability of immunofixed protein bands with increased protein mass due to the antigen–antibody reaction, proteins occurring at low concentration in the electrophoresed sample may not always be visible in the acid-fixed reference lane. Consequently, if artifactual bands can be ruled out (see Notes 38 and 39), weak IFE bands in the absence of matching bands in the reference lane should be interpreted as being consistent with low target protein concentrations.

38. Multimeric paraproteins, especially IgM, and immune complexes may occasionally adhere to the gel matrix. During IFE testing for paraproteins, these bands will appear in all five antisera reaction areas of the gel at or around the point of application. In most cases, true IgM or, possibly, other monoclonal bands still can be identified with a marked increase in size and staining intensity where they react with the specific antisera for their heavy chain and light chain. Re-running the sample after pretreatment with the reducing agent 2-ME or a chaotropic agent (e.g., Fluidil, Sebia) usually eliminates or markedly weakens the artifactual bands and confirms the initial assessment for the type of the monoclonal band.

39. Care should be taken to avoid overconcentration of urine samples because this can cause protein aggregation and consequent decrease of solubility. Consequently, artifactual IFE bands may appear with all anti-Ig antibodies when tested for paraproteins. Similar phenomena may be also observed in urine with other antibodies and with other type of specimens that require preconcentration. Reelectrophoresis of such samples on dilutions usually eliminates these artifacts.

40. Routine IFE methods, including commercial kits for paraproteins, do not test initially for IgD and IgE heavy chains and free (unbound) κ and λ light chains. However, manufacturers

of commercial IFE kits provide antisera for testing for IgD and IgE heavy chains and free κ and free λ light chains as well. There is even a dedicated kit available for testing with a semi-automated method for free light chains in neat urine (Bence-Jones Immunofixation, Interlab Sri). When light chains without corresponding heavy chains are identified in a serum IFE pattern and inappropriate antigen–antibody ratios, deteriorated or low-avidity antisera, or possible errors in antibody application (antibody omitted or wrong), can be ruled out, additional testing with antisera against free κ and free λ light chains should be carried out. If these are negative, monospecific antibodies to IgD and IgE heavy chains should be used for identification of rare cases of the respective monoclonal Igs. If even these tests are negative, then the possibility of "unreactive" light chains needs to be entertained (39–41). Some IgA and IgD paraproteins have been found to have either a configurational change in their light chain or an enzymic cleavage in their molecule which prevents the precipitation reaction with antilight chain antisera, hence is the name "unreactive" light chains (39–41). In such very rare cases of immunologically inaccessible light chains, it is often necessary to isolate the abnormal Ig and use special techniques for its proper characterization, thereby ruling out true heavy chain disease (39–41). Since light chains without corresponding heavy chains are relatively common in urine IFE, retesting the samples with anti-free κ and anti-free λ usually is sufficient for diagnosis.

References

1. Afonso E (1964) Quantitative immunoelectrophoresis of serum proteins. Clin Chim Acta 10:114–122
2. Wilson AT (1964) Direct immunoelectrophoresis. J Immunol 92:31–34
3. Alper CA, Johnson AM (1969) Immunofixation electrophoresis: a technique for the study of protein polymorphism. Vox Sang 17:445–452
4. Alper CA, Johnson AM (1993) Immunofixation electrophoresis: a technique for the study of protein polymorphism. Vox Sang 65:76
5. Johnson AM, Cleve H, Alper C (1975) Variants of the group-specific component system as demonstrated by immunofixation electrophoresis. Report of a new variant, Gc Boston (Ge B). Am J Hum Genet 27:728–736
6. Johnson AM (1976) Genetic typing of alpha1-antitrypsin by immunofixation electrophoresis, identification of subtypes of Pi M. J Lab Clin Med 87:152–163
7. Lieberman J, Gaidulis L (1976) Simplified alpha1-antitrypsin phenotyping by immunofixation of acid-starch gels. J Lab Clin Med 87:710–716
8. Grunbaum BW, Zajac PL (1977) Rapid phenotyping of the group specific component by immunofixation on cellulose acetate. J Forensic Sci 22:586–589
9. Cleve H, Constans J, Berg S et al (1981) Gc revisited: six further Gc-phenotypes delineated by isoelectric focusing and by polyacrylamide gel electrophoresis. Hum Genet 57:312–316
10. Mauff G, Hummel K, Pulverer G (1975) Properdin factor B (glycine-rich beta-glycoprotein or C3 proactivator)-polymorphism: genetic and biochemical aspects. First application to paternity cases. Z Immunitatsforsch Exp Klin Immunol 150:327–338
11. Arnaud P, Wilson GB, Koistinen J et al (1977) Immunofixation after electrofocusing: improved method for specific detection of serum proteins with determination of isoelectric points. I. Immunofixation print technique for detection of alpha-1-protease inhibitor. J Immunol Methods 16:221–231

12. Baumstark JS (1978) Quantitative immunofixation of proteins following zone electrophoresis in agarose gel: application to the determination of the stoichiometry of the alpha1-antitrypsin-elastase interaction. J Immunol Methods 23:79–89
13. Morrison R, Noppinger K, Brown MG Jr (1985) Immunofixation of complement component C3 phenotypes in bloodstains after cellulose acetate electrophoresis. J Forensic Sci 30:1221–1225
14. Cawley LP, Minard BJ, Tourtellotte WW et al (1976) Immunofixation electrophoretic techniques applied to identification of proteins in serum and cerebrospinal fluid. Clin Chem 22:1262–1268
15. Chang CH, Inglis NR (1979) Convenient immunofixation electrophoresis on cellulose acetate membrane. Clin Chim Acta 65:91–97
16. Ritchie RF, Smith R (1976) Immunofixation. I. General principles and application to agarose gel electrophoresis. Clin Chem 22:497–499
17. Ritchie RF, Smith R (1976) Immunofixation. II. Application to typing of alpha1-antitrypsin at acid pH. Clin Chem 22:1735–1737
18. Ritchie RF, Smith R (1976) Immunofixation. III. Application to the study of monoclonal proteins. Clin Chem 22:1982–1985
19. Kyle RA, Greipp PR (1978) The laboratory investigation of monoclonal gammopathies. Mayo Clin Proc 53:719–739
20. Johnson AM (1982) Immunofixation electrophoresis and electrofocusing. Clin Chem 28:1797–1800
21. Keren DF (1999) Procedures for the evaluation of monoclonal immunoglobulins. Arch Pathol Lab Med 123:126–132
22. Freedman MS, Thompson EJ, Deisenhammer F et al (2005) Recommended standard of cerebrospinal fluid analysis in the diagnosis of multiple sclerosis: a consensus statement. Arch Neurol 62:865–870
23. Laurenzi MA, Link H (1978) Localization of the immunoglobulins G, A and M, beta-trace protein and gamma-trace protein on isoelectric focusing of serum and cerebrospinal fluid by immunofixation. Acta Neurol Scand 58: 141–147
24. Bateman N, Jones NS (2000) Rhinorrhoea feigning cerebrospinal fluid leak: nine illustrative cases. J Laryngol Otol 114:462–464
25. Bleier BS, Debnath I, O'Connell BP et al (2011) Preliminary study on the stability of beta-2 transferrin in extracorporeal cerebrospinal fluid. Otolaryngol Head Neck Surg 144:101–103
26. Ottó S (1982) Reversed immunofixation agar gel electrophoresis. Immunol Lett 4:85–86
27. Mehta PD, Patrick BA, Thormar H et al (1982) Oligoclonal IgG bands with and without measles antibody activity in sera of patients with subacute sclerosing panencephalitis (SSPE). J Immunol 129:1983–1985
28. Storstein A, Monstad SE, Honnorat J et al (2004) Paraneoplastic antibodies detected by isoelectric focusing of cerebrospinal fluid and serum. J Neuroimmunol 155:150–154
29. Stich O, Rauer S (2007) Antigen-specific oligoclonal bands in cerebrospinal fluid and serum from patients with anti-amphiphysin- and anti-CV2/CRMP5 associated paraneoplastic neurological syndromes. Eur J Neurol 14:650–653
30. Bossuyt X, Bogaerts A, Schiettekatte G et al (1998) Detection and classification of paraproteins by capillary immunofixation/subtraction. Clin Chem 44:760–764
31. Litwin CM, Anderson SK, Philipps G et al (1999) Comparison of capillary zone and immunosubtraction with agarose gel and immunofixation electrophoresis for detecting and identifying monoclonal gammopathies. Am J Clin Pathol 112:411–417
32. Yang Z, Harrison K, Park YA et al (2007) Performance of the Sebia CAPILLARYS 2 for detection and immunotyping of serum monoclonal paraproteins. Am J Clin Pathol 128:293–299
33. Paquette DM, Banks PR (2001) Detection of specific antibodies using immunosubtraction and capillary electrophoresis instrumentation. Electrophoresis 22:2391–2397
34. Pascali E, Pezzoli A, Chiarandini A (1982) Immunofixation: application to the identification of "difficult" monoclonal components. Clin Chem 28:1404–1405
35. Leglise D, Menez JF, Person B et al (1982) A detailed lipidograph: enzymatic determination of cholesterol, phospholipids and glycerides in plasma lipoprotein after a cellulose acetate electrophoretic procedure (author's transl) [Article in French]. Clin Chim Acta 118: 265–277
36. Nauck M, Winkler K, März W et al (1995) Quantitative determination of high-, low-, and very-low-density lipoproteins and lipoprotein(a) by agarose gel electrophoresis and enzymatic cholesterol staining. Clin Chem 41:1761–1767
37. Winkler K, Nauck M, Siekmeier R et al (1995) Determination of triglycerides in lipoproteins separated by agarose gel electrophoresis. J Lipid Res 36:839–847

38. García-Sánchez C, Torres-Tamayo M, Juárez-Meavepeña M et al (2011) Lipid plasma concentrations of HDL subclasses determined by enzymatic staining on polyacrylamide electrophoresis gels in children with metabolic syndrome. Clin Chim Acta 412:292–298
39. Cejka J, Kithier K (1979) IgD myeloma protein with "unreactive" light chain determinants. Clin Chem 25:1495–1498
40. Netto D, Vladutiu AO (1981) A simple technique for identification of "unreactive" light chains of immunoglobulins. Clin Chim Acta 116:253–260
41. Rabhi H, Ghaffor M, Abbadi MC (1989) Spontaneous enzymatic cleavage of IgD myeloma protein giving a pattern of delta heavy chain disease. Arch Inst Pasteur Alger 57:135–140

Chapter 14

Microchip Electrophoresis, with Respect to "Profiling of Aβ Peptides in the Cerebrospinal Fluid of Patients with Alzheimer's Disease"

Mohamad Reza Mohamadi, Romain Verpillot, Myriam Taverna, Markus Otto, and Jean-Louis Viovy

Abstract

Aggregation of beta amyloid peptides especially Aβ1-42 in amyloid plaques is one of the major neuropathological events in Alzheimer's disease. This event is normally accompanied by a relative reduction of the concentration of Aβ1-42 in the cerebrospinal fluid (CSF) of patients developing the signs of Alzheimer's disease. Here, we describe a microchip gel electrophoresis method in a polydimethylsiloxane (PDMS) chip that enables rapid profiling of major Aβ peptides. The method was applied to compare the relative concentration of Aβ1-42 with other Aβ peptides, for example, Aβ 1-40 in CSF. In order to increase the sensitivity of detection, Aβ peptides in the CSF samples were first captured and concentrated using magnetic beads coated with specific anti-Aβ antibodies.

Key words: Microchip electrophoresis, Beta amyloid, Cerebrospinal fluid and Alzheimer's disease

1. Introduction

The growing number of Alzheimer's patients, associated with the aging of the population, raises a strong demand for efficient treatment and also for reliable methods for diagnosis of AD in the early stages. It is commonly accepted that AD has a very long (>10 years) asymptomatic evolution period during which irreversible neurological damages are progressively happening. Although there is no short-term hope of treatments able to reverse the existing damages, there are more and more hopes to find neuroprotective treatments able to retard the disease's progression. Due to the slow evolution of the disease, even a moderately efficient treatment can have a strong impact on the quality of life and survival, if diagnosis can be made early enough. Currently, diagnosis of AD

Biji T. Kurien and R. Hal Scofield (eds.), *Protein Electrophoresis: Methods and Protocols*, Methods in Molecular Biology, vol. 869, DOI 10.1007/978-1-61779-821-4_14, © Springer Science+Business Media, LLC 2012

is mainly based on clinical and neurological symptoms. However, at early stages neurological symptoms are less clear and distinguishing AD from other forms of dementia or even normal aging is challenging. Therefore, there is increasing demands for biomarker screening and developing methods for the detection of AD-specific markers at early stages.

The most important and widely accepted biomarkers for AD are beta amyloid peptides. Aβ peptides are formed by proteolytic cleavage of a transmembrane protein called amyloid precursor protein (APP) (1–4). Cleavage at different sites is possible, and several Aβ peptides are found in measurable quantities in biological fluids of healthy individuals. The ratios of concentrations of these different peptides are relatively stable in the CSF of healthy individuals (5). During the progression of AD, Aβ peptides, especially Aβ1-42, are accumulated in amyloid plaques. Therefore, a selective reduction of Aβ1-42 in CSF from AD patients was reported in several studies, while the total level of Aβ peptides was found to be in the normal range (6).

Currently, conventional slab gel followed by western blotting is the principal method for quantitation of Aβ peptides in CSF samples (5). However, slab gel electrophoresis is time consuming, difficult to be automated, and hard to be used for screening the large number of samples that is required for biomarker studies. Recent advances of microchip electrophoresis (MCE) demonstrated the potential of MCE as a reliable analytical method for the profiling of proteins and peptides in biological samples (6). As compared to conventional electrophoresis methods, MCE involves significant advantages, such as low sample consumption and a strong potential for automation and integration with different on-chip sample preparations (7, 8).

Here, we introduce a gel-based MCE method for the analysis of fluorescently labeled Aβ peptides in human CSF. Due to the complexity and variability of real CSF sample, and the low concentration of Aβ peptides as compared to other highly abundant peptides and proteins in CSF, a sample preparation step prior to the MCE analysis is necessary. We used an immunocapture method based on capturing of Aβ peptides using magnetic beads coated with anti-Aβ peptides (9). In the future, the integration of this immunocapture step into microfluidics can be contemplated (unpublished work by Mohamadi et al).

2. Materials

2.1. Standard Peptides, Fluorescent Dye, and Antibodies

1. Synthetic Aβ peptides (Aβ1-37, Aβ1-38, Aβ1-39, Aβ1-40, and Aβ1-42) were purchased from AnaSpec (Fremont, CA, USA). We followed the protocols proposed by the producer for dissolving and aliquoting the original batch of standard Aβ

peptides (see Note 1). Typically, Aβ peptides should be dissolved in alkaline buffer or ddH_2O (depending on the type) at a concentration of 1 mg/mL. 5–10 μL aliquots can be kept at −20°C for a few months or at −80°C for a long time. Fluoprobe-488 NHS ester was purchased from Interchim (Montlucon, France). Fluoprobe-488 NHS is a powder and should be prepared in dimethyl sulfoxide (DMSO) at a concentration of 10 mg/mL. Prepare 5–10 μL aliquots of the fluorescent dye and keep them in dark and at −20°C for a few months.

2. Monoclonal antibody anti-Aβ 6E10 was purchased from Covance (Covance, Emeryville, CA, USA).

2.2. RIPA Buffer

Prepare 50 mM Tris–HCl, pH 7.2 including 150 mM NaCl, 1% Triton X-100, 0.1% SDS, 1 mM sodium orthovanadate (Na_3VO_4), 1 mM phenylmethanesulfonyl fluoride (PMSF), 1 mM ethylenediaminetetraacetic acid (EDTA), 1 mM protease inhibitor cocktail (Roche, Basel, Switzerland) (see Note 2).

2.3. Borate Buffer 2× for Fluorescent Labeling

Prepare 80 mM Borate in ddH_2O and adjust the pH at 10.5 using NaOH.

2.4. Materials for Preparation of SU-8 Master

Silicon wafer, 2 in. (5.1 cm) circular diameter (Neyco, France), SU-8 2050 resin (Microchem Corp, Newton, MA, USA) (see Note 3). MicroChem's SU-8 Developer (Microchem Corp, Newton, MA, USA), UV exposure machine, with wavelength range of 360–400 nm (Speed plate 650, MC10, Marchetti) (see Note 4), photomask and isopropyl alcohol.

2.5. Materials for Preparation of Polydimethylsiloxane (PDMS Mold)

1. PDMS kit containing PDMS elastomer and curing agent (Sylgard184, Dow Corning, Wiesbaden, Germany). Spin coater machine, glass microscope slides (75 × 25 × 1 mm thick).
2. Plasma oxidizer such as PDC-32G plasma cleaner (Harrick Scientific, Ossining, NY, USA).

2.6. Materials for Microchip Capillary Electrophoresis

1. Methylcellulose (MC-400) (Sigma Chemical Co., St. Louis, MO, USA).
2. High-voltage programmable power supply: we used a HVS448 3,000 V power supply (LabSmith, Livermore, CA, USA) controlled by a personal computer, platinum wire electrodes, electric connectors. An inverted fluorescence microscope to detect in real time fluorescently labeled peptides: we used Olympus IX71, equipped with a 10 × 0.3 NA objective lens (Olympus, Tokyo, Japan). A 100-W mercury arc lamp for illumination. A CCD camera connected to the microscope: we used a Nikon digital DS-Qi1 camera (Nikon, Tokyo, Japan); an image grabbing software or an image grabbing/processing software (we used NIS Elements by Nikon) which allows online intensity measurement of the recorded images by CCD.

3. Methods

Carry out all procedures at room temperature unless otherwise specified.

3.1. Preparation of Photomask

1. Design the microchip using designing software such as AutoCad or QCad. A typical design for MCE is shown in Fig. 1a. Set the channel width at 100 μm by choosing the line width in your design. Set the distances from the intersection of injection and separation microchannels to the buffer reservoir (BR), the sample reservoir (SR), and sample waste reservoir (SW) at 5 mm and to the buffer waste reservoir (BW) at 35 mm. Print the designed microchip onto a transparency film using a high-resolution printer (at least 3,000 dpi or higher). Our photomasks were printed by SELBA (Versoix, Switzerland), other local retailer may be available for high-resolution printing (see Note 5).

3.2. Preparation of Master for PDMS Microchip

1. Microchips were fabricated by rapid prototyping and PDMS technology (10), starting with a master composed of a positive relief of SU-8 resin on a silicon wafer made by soft lithography (Fig. 1b) (see Note 6). Generally photolithography consisted of several steps as mentioned below.

3.2.1. Wafer Pretreatment and Spin Coating of Initiator

1. To obtain maximum process reliability, substrates should be clean and dry prior to applying the SU-8 resist. Start with a solvent cleaning, or a rinse with dilute acid, followed by a deionized water rinse. Where applicable, substrates should be subjected to a piranha etch/clean (sulfuric acid and hydrogen peroxide) (see Notes 7 and 8).

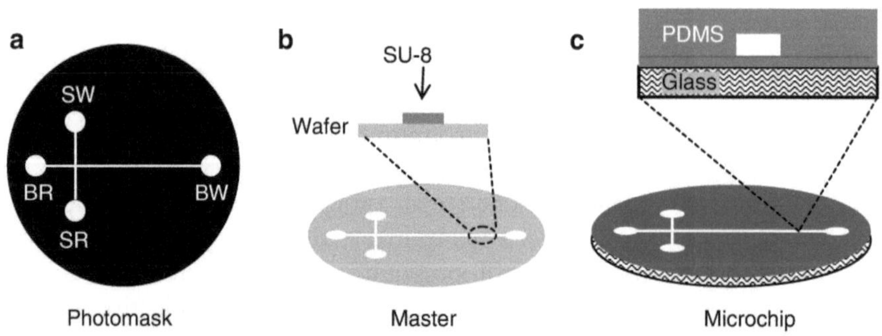

Fig. 1. Schematic depicting three steps for preparing microchip. Figure is not to scale. (**a**) Shows a typical microchip design layout on a photolithography mask, the indications on the figure are *BR* buffer reservoir, *BW* buffer waste reservoir, *SR* sample reservoir, *SW* sample waste reservoir. (**b**) Silicon wafer on which SU-8 resin is photopolymerized. (**c**) Shows the prepared microchip. If you follow the same protocol given in this chapter, channel cross-section will be 100 μm (*width*) by 50 μm (*depth*).

2. To dehydrate the surface, bake at 120°C for at least 5 min on a hot plate. Treat the surface of the wafer using oxygen plasma for 1 min. Put the wafer on a spin coater and pour about 1 mL of MCC Primer 80/20 (HMDS) and spin coat for 20 s at 500 rpm and acceleration of 300 rpm/s (see Note 9).

3.2.2. Su-8 Resin Coating on the Silicon Wafer by Spin Coating

1. Spin coating is a step for spreading photoresist on the surface of wafer evenly. It ensures the consistency in the depth of microchannels of the device. In addition, the thickness and as a result the depth of the microchannel depends on the spinning time and speed. Pour about 1 mL of the photoresist resin per inch of wafer in the center of the wafer and let it spread naturally. Put the wafer at the center of the spinner and first spin the wafer at a low speed of 500 rpm for 10 s with an acceleration of 100 rpm/s and then speed up at 2,000 rpm for 30 s with an accelerating ramp of 300 rpm/s (see Note 10).

3.2.3. Soft Baking or Pre-Exposure Baking of the Coated Wafer

1. This step is used to evaporate the solvent and densify the photoresist film on the surface of the wafer. Place the coated wafer in an oven or hot plate set at 65°C for 3 min and then in an oven or hot plate set at 95°C for 6 min (see Note 11).

3.2.4. Ultraviolet Exposure

1. The photomask controls the UV exposure on the photoresist-covered wafer allowing the transfer of microchannel structures onto the wafer. Place the negative photomask on the coated wafer and fix it with paper tapes. Program the UV exposure machine to 30 s exposure time. Start the UV exposure (see Note 12).

3.2.5. Postexposure Bake

1. Postexposure baking of exposed wafer permits the exposed photoresist to cross-link, thereby allowing it to be more resistant to the developer in the next step. Place the coated wafer in an oven or a hot plate at 65°C for 1 min, and then at 95°C for 7 min. Let the wafer cool to room temperature.

3.2.6. Developing of Exposed Wafer

1. The unexposed photoresist will be removed by SU-8 developer, leaving the cross-linked photoresist which forms inverted microchannels in the wafer. After the developing step, the imprinted microchannels will be obviously visible. Submerge the exposed wafer completely in an undiluted SU-8 developer in a beaker. Shake the beaker vigorously in the developer for at least 1 min or until the unexposed photoresist, as can be signified by white spots on the wafer, is completely washed off (see Note 13). Rinse the wafer thoroughly with isopropyl alcohol in another clean beaker. If the white spots are still observed on the surface of the wafer, repeat the developing step and again clean the surface of wafer with fresh running isopropanol. Dry the wafer using N_2 gas or airstream.

3.3. PDMS Replication of Si Wafer Mold

1. After fabrication of wafer mold, the PDMS replica can be easily replicated by a simple molding procedure.
2. Mix well the PDMS elastomer and curing agent in the ratio of 10:1 (V/V) in a single-use plastic container.
3. Place the silicon wafer mold on a plastic petri dish and pour the mixture into the petri dish to cover the surface of the wafer completely.
4. Place the uncovered petri dish in a vacuum degasser for 15 min to remove the bubbles from the PDMS mixture. When there is no other bubble in the PDMS mixture, take it out. Cover the petri dish and label your name and date of the experiment on it.
5. Leave the petri dish in an oven at 67°C for 2 h to cure the PDMS mixture. Take the PDMS mold out and wait for it to reach room temperature. Peel off the PDMS microchip from the wafer mold with a sharp scalpel. Make reservoirs on the PDMS microchip with a 3 or 5 mm circular punch (see Note 14).

3.4. Preparation of Glass Slide Coated with PDMS

1. Rinse the glass slide with acetone and then with ethanol. Dry the glass slide with N_2 gas or airstream. Put the glass slide in the oxygen plasma source. Set the plasma machine for 1 min. Put the glass slide on the spin coater and pour 2 mL of the prepared PDMS–elastomer mixture on the glass slide. Set the spin coater at 1,000 rpm for 20 s and with an acceleration of 300 rpm/s (see Note 15).

3.5. Adhering PDMS Microchip to PDMS-Coated Glass Slide Using Oxygen Plasma Bonding

1. The PDMS has self-adhesive property, hence allowing reversible sealing with most materials such as glass and polymers. The sealing will be improved if the surfaces of both the PDMS and sealing substrate are free of particles. If there are particles on the PDMS surface wash the surface with ethanol. Dry the microchip by N_2 gas. In order to achieve more robust chips, we recommend inducing permanent bonding, thanks to plasma treatment: Place the microchip and PDMS-coated glass slide in the oxygen plasma source. Set the plasma machine for 1 min. Bring out the glass slide and the PDMS microchip. Locate the PDMS microchip on the PDMS-coated glass and press naturally (see Note 16). Keep the microchip at 67°C in an oven for at least 1 h before use. Store the prepared chip in a dust-free box such as a single-use petri dish before use (Fig. 1c).

3.6. Surface Coating and Microchip Electrophoresis

1. To control the electroosmotic flow in the microchannel, the surface of the microchannel should be coated. Inject a solution of 0.2% of poly(dimethylacrylamide-*co*-allyl glycidyl ether) (PDMA-AGE) in ddH_2O into the microchannels. Keep the chip in the dark for at least 30 min and then rinse out the excess polymer by extensive ddH_2O washing (see Note 17).

3.7. Preparation of 2% Methylcellulose (MC)

Prepare 2% MC in 40 mM borate, pH 9.5. Prepare 25 mL hot water (about 40–45°C). Add 1 g MC to the hot water and gently mix it using a glass rod. Use the glass rod to gently press the big agglomerated MC formed in hot water (see Note 18). Add 25 mL of 80 mM borate buffer, pH 9.5, plus 0.01% Tween-20 which has been kept in refrigerator for at least 1 h before use. Mix the buffer and polymer and keep the mixture at 4°C for a few more hours until the polymer solution is completely clear (11, 12).

3.8. Preparation and Labeling of Standard Aβ Peptides

1. Take an already prepared aliquot of Aβ, add equal volume of borate buffer, pH 10.5 (2×) (13).
2. Add a given volume of fluorescent dye to the peptide solution. If your peptide is prepared in NH_4OH aqueous solution, add equal volume of the borate buffer, pH 10.5 (1×) to your peptide solution.
3. Lyophilize the sample and dissolve the powder in the desired volume of borate buffer solution (pH 10.5 [1×]) (see Note 19). Typically, a tenfold excess (molar ratio) of the fluorescent dye should be used for labeling (see Note 20). Incubate the mixture at 4°C for 1 h in dark (see Note 21).

3.9. Preparation of CSF Samples

1. The CSF samples analyzed in this study were obtained from patients attending the outpatient memory clinic in Ulm (see Note 22). Collection and analysis of the CSF sample were approved by the Ethics Committee in Ulm. All samples were anonymized before transmission to our laboratory (see Note 23). CSF was taken, aliquoted within 2 h, and stored at −80°C (14).
2. CSF can be directly used for the labeling reaction. Add 10 μL of Fluoprobe NHS-488 (10 mg/mL in DMSO) to 250 μL of human CSF.
3. Incubate the mixture at 4°C for 1 h in dark. The labeled CSF was then used for immunocapture of Aβ peptides (see Note 24).

3.10. Preparation of Magnetic Beads

1. Add 80 μL of monoclonal antibody anti-Aβ (1 mg/mL) to 2 mL of magnetic microparticles M280 Dynabeads (10 mg/mL, 6.7×108 beads/mL) coated with sheep antimouse IgG. The activated beads can be stored in 4°C for few months.

3.11. Affinity Immunocapture of Labeled Aβ and CSF

1. Take 25 μL of the activated magnetic beads in 1 mL tube and add 1 mL RIPA buffer to the beads.
2. Let the beads rotate for 5–10 min in the RIPA buffer.
3. Use a magnetic particle separator to keep the beads and remove the buffer.

4. Add 250 μL RIPA buffer to 250 μL labeled CSF sample or to the 25 μL of labeled standard Aβ peptides.
5. Add the mixture to the magnetic beads.
6. Let the mixture rotate at 4°C and in dark overnight.
7. Use a magnetic separator to separate the beads form the mixture.
8. Wash the beads by adding 1 mL RIPA and rotate for 10 min.
9. Use a magnetic particle separator to separate the beads.
10. Repeat the washing step three times.
11. Wash the beads with 1 mL ddH_2O, follow the same procedure for washing the beads with RIPA buffer.
12. Add 200 μL of 0.16% NH_4OH in ddH_2O to the beads and use sonication (see Note 25) for 10 min at room temperature.
13. Use a magnetic particle separator to separate the beads.
14. Use a Speedvac to completely dry the eluted solution and resolve the dried eluted peptides in electrophoresis buffer (borate 40 mM, pH 9.5).
15. Keep the sample for MCE analysis in dark and at 4°C before use.

3.12. Microchip Capillary Electrophoresis

1. Place 10–15 μL of MC 2% in sample waste reservoir. Slight pressure using a syringe is necessary to introduce the MC solution into the microchannel. Make sure that MC solution has filled all the channels and reached all reservoirs.
2. Use microscope to make sure there is no air bubble trapped in the channels. Remove the excess of MC solution in buffer waste reservoir.
3. Place 15 μL electrophoresis buffer (borate 40 mM, pH 9.5) in all the reservoirs. Fix the platinum wire electrodes in the reservoirs (see Note 26).
4. Replace the buffer in the sample reservoir with 15 μL of sample when ready to run the voltage program. Run the recording software before running the voltage program. The voltage program includes two steps, injection and separation.
5. To inject the sample apply +550 V in SW reservoir while keeping other reservoirs at Ground for 60 s.
6. Switch to separation step by applying +250 V in SW and SR reservoirs and +950 V in BW while keeping BR at Ground.
7. Measure the total fluorescent intensity at 5 mm to the BW in the separation microchannel and plot the measured intensity against recording time to achieve electropherograms (Figs. 2 and 3).

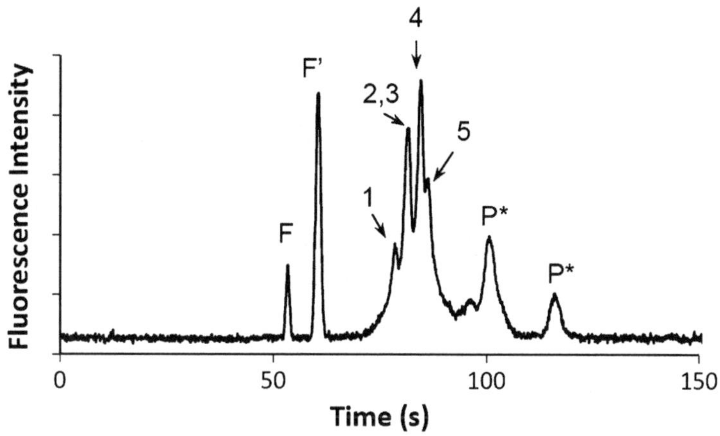

Fig. 2. Electropherogram from the separation of a mixture of five Aβ peptides labeled with Fluoprobe-488 [Aβ1-37 (1), Aβ1-38 (2), Aβ1-39 (3), Aβ1-40 (4), and Aβ1-42 (5)] after immunocapture. Reprinted from Ref. 9 with permission from the American Chemical Society.

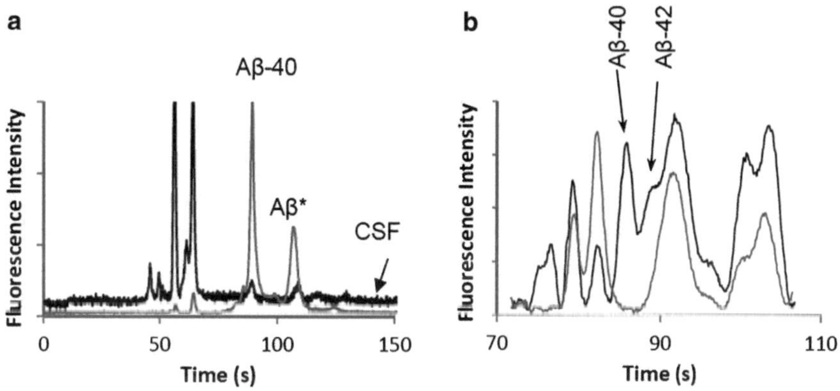

Fig. 3. (**a**) Electropherogram of labeled CSF. CSF (250 μL) was labeled with Fluoprobe-488, and the Aβ peptides in the labeled CSF were isolated using magnetic beads coated with anti-Aβ. An electropherogram of standard Aβ1-40 labeled with Fluoprobe-488 is superimposed to identify the peak from Aβ1-40 in CSF. Aβ* corresponds to the peak for aggregated or modified peptides. (**b**) Electropherogram of Aβ peptides immunocaptured from 250 μL of CSF which were labeled after immunocapture. An electropherogram (dashed electropherogram) of a blank control showing the peaks related with the fluorescent dye is superimposed in this electropherogram. The separation window of Aβ1-40 and Aβ1-42 is out of the zone for strong peaks related with the fluorescent dye. Reprinted from Ref. 9 with permission from the American Chemical Society.

4. Notes

1. Instead of using alkaline buffer which is normally recommended for dissolving Aβ peptides, you can also dissolve and aliquot each Aβ peptide (Aβ1-37, Aβ1-38, Aβ1-39, Aβ1-40, and Aβ1-42) in an aqueous solution of 0.16% ammonium hydroxide (NH_4OH), which will help limit the aggregation of peptides during their storage. Note that NH_4OH is an active compound and may interfere with fluorescent labeling, thus an

additional lyophilization step to remove all traces of NH_4OH prior to the labeling step is necessary.

2. The buffer data are mentioned in molarity. In general, the buffer volume needed for MCE is very low, so the buffers can be prepared in 50 mL volumes. All the buffers should be filtered by 0.2 μm filters and kept in refrigerator before use.
3. Beware that SU-8 resin are light and temperature sensitive.
4. Beware that chronic UV exposure may lead to skin cancer and eye cataract. Wear UV glasses and leave the room during the UV exposure.
5. For printing smaller structures in the photomask, higher resolutions are recommended.
6. Soft lithography normally requires clean room facilities, a clean room (a room away from dusts) for doing photolithography techniques and a constant room temperature for consistency in the coating by SU-8 resin. Instead of using clean room facility, photolithography can be done under a hood with laminar clean air; however, there is higher risk of defects.
7. These steps are crucial if glass slides are used as substrates for lithography and can be avoided when using commercial silicon wafer prepared for photolithography.
8. Safety note: Piranha is a very strong oxidizing agent, and reacts violently with organic materials or solvents, so it must be handled with extreme precaution under chemical hood. Before using piranha, ask your institution about the procedure for piranha waste disposal.
9. Adding MCC Primer 80/20 (HMDS) step is recommended to improve adherence of SU-8 resin to the silicon wafer.
10. During the spin-coating process, a buildup of photoresist may occur on the edge of the substrate. In order to remove this edge a small stream of isopropanol using a syringe at the edge of the wafer can be used. This procedure will help the photomask to be placed in close contact with the wafer, resulting in improved resolution and aspect ratio.
11. Let the wafer cool to room temperature. If you see wrinkles on the SU-8 film repeat the heating procedure until wrinkles disappear.
12. UV exposer can be done using an automated aligner, which is normally part of clean room facilities for lithography. Caution: UV exposure is dangerous for the eyes and potentially may cause skin cancer, use UV protection glasses and protect your skin when using UV.

13. Long developing step may detach the polymerized structure too and should be avoided. Caution: Developer is toxic and should be used under chemical hood.
14. Normal biopsy punches can be used and special punches for cutting PDMS with wide range of diameters are also commercially available.
15. Pay attention that the upper side of the glass, when put in the plasma container, should be used for PDMS coating.
16. Do not overpress the PDMS microchip as the channels may be attached to the surface. Be careful to put the surfaces together which are exposed to the plasma.
17. We used a homemade PDMA-AGE synthesized copolymer kindly provided by Marcella Chiari. The protocol for synthesis of PDMA-AGE is published (see ref. (15)) and therefore excluded from this chapter.
18. Adding cold buffer should be done in 5–10 min for preparation of MC solution in water. During this time keep mixing the MC solution using a glass rod.
19. Aβ peptide mixture can be prepared and lyophilized prior to labeling this way. The labeling conditions for all the peptides in the mixture are similar. Lyophilized peptide samples can be stored several days before performing the labeling reaction.
20. The ratio of added fluorescent dye depends on the concentration of your peptides. If you are preparing peptides with concentration lower than 10 μg/mL, higher ratios (50-fold) of fluorescent dye are recommended.
21. Labeled peptides could be used on the same day if they are kept in a refrigerator and in dark.
22. Obtaining CSF from patients should be done by a specialist and processed in specialized laboratories (e.g., neurochemical labs).
23. Follow the procedure given by your institution, following national and international legislation, for using biological samples from patients.
24. Fluorescent labeling can also be done after immunocapture and on the eluted peptides.
25. Normal ultrasonic bath used in your laboratory with an average power of 100 W can be used.
26. Be aware of electric shock when working with high-voltage power supply, use appropriate tapes to fix the electrode, and avoid any direct contact with the electrode during electrophoresis.

Acknowledgments

This work was supported by European grants LSHB CT 06 037953 (NeuroTAS) and FP7-NMP-09-246513 (NADINE).

References

1. Haass C, Schlossmacher MG, Hung AY et al (1992) Amyloid β-peptide is produced by cultured cells during normal metabolism. Nature 359:322–325
2. Hardy J, Selkoe DJ (2002) The amyloid hypothesis of alzheimer's disease: progress and problems on the road to therapeutics. Science 297:353–356
3. Klein WL, Krafft GA, Finch CE (2001) Targeting small Aβ oligomers: the solution to an Alzheimer's disease conundrum? Trends Neurosci 24:219–224
4. Lambert MP, Barlow AK, Chromy BA et al (1998) Diffusible, nonfibrillar ligands derived from Aβ1–42 are potent central nervous system neurotoxins. Proc Natl Acad Sci USA 95:6448–6453
5. Wiltfang J, Esselmann H, Bibl M et al (2002) Highly conserved and disease-specific patterns of carboxyterminally truncated Aβ peptides 1–37/38/39 in addition to 1–40/42 in Alzheimer's disease and in patients with chronic neuroinflammation. J Neurochem 81: 481–496
6. Lewczuk P, Wiltfang J (2008) Neurochemical dementia diagnostics: state of the art and research perspectives. Proteomics 8:1292–1301
7. Jakeway SC, de Mello AJ, Russell EL (2000) Miniaturized total analysis systems for biological analysis. Fresenius J Anal Chem 366: 525–539
8. Chovan T, Guttman A (2002) Microfabricated devices in biotechnology and biochemical processing. Trends Biotechnol 20: 116–122
9. Mohamadi MR, Svobodova Z, Verpillot R et al (2010) Microchip electrophoresis profiling of Aβ peptides in the cerebrospinal fluid of patients with Alzheimer's disease. Anal Chem 82:7611–7617
10. Xia Y, Whitesides GM (1998) Soft lithography. Angew Chem Int Ed Engl 37:550–575
11. Mohamadi MR, Mahmoudian L, Kaji N et al (2007) Dynamic coating using methylcellulose and polysorbate 20 for nondenaturing electrophoresis of proteins on plastic microchips. Electrophoresis 28:830–836
12. Mohamadi MR, Kaji N, Tokeshi M et al (2007) Online preconcentration by transient isotachophoresis in linear polymer on a poly(methyl methacrylate) microchip for separation of human serum albumin immunoassay mixtures. Anal Chem 79: 3667–3672
13. Verpillot R, Esselmann H, Mohamadi MR et al (2011) Analysis of amyloid-β peptides in cerebrospinal fluid samples by capillary electrophoresis coupled with LIF detection. Anal Chem 83:1696–1703
14. Jesse S, Brettschneider J, Süssmuth SD et al (2011) Summary of cerebrospinal fluid routine parameters in neurodegenerative diseases. J Neurol 258:1034–1041
15. Chiari M, Cretich M, Horvath J (2000) New adsorbed coatings for capillary electrophoresis. Electrophoresis 21:909–916

Chapter 15

A Foodomics Approach: CE-MS for Comparative Metabolomics of Colon Cancer Cells Treated with Dietary Polyphenols

Mustafa Çelebier, Clara Ibáñez, Carolina Simó, and Alejandro Cifuentes

Abstract

The potential of capillary electrophoresis-mass spectrometry (CE-MS) for metabolomics is demonstrated through the analysis of metabolites from human HT29 colon cancer cells treated and non-treated with dietary polyphenols. Prior to CE-MS analysis, four different metabolite purification strategies are investigated. Namely, the results obtained after methanol deproteinization, ultrafiltration, and two solid-phase extraction methods using C18 and polymer-based cartridges are described. These generic methods can have broad applications to analyze metabolites in a large variety of matrices and fields, including the new Foodomics area.

Key words: Foodomics, Metabolomics, Colon cancer, Dietary polyphenols, CE, CE-MS

1. Introduction

One of the main goals in modern food science and nutrition is to improve our limited understanding of the roles of nutritional compounds at molecular level. Dietary polyphenols are currently receiving considerable interest for their presumed role in the prevention of various degenerative diseases such as cancer and cardiovascular diseases (1). However, the molecular mechanisms underlying the biological activity of these compounds and their cellular targets are still unknown. Foodomics has emerged as a new discipline that studies the food and nutrition domains through the application of advanced "omics" technologies in order to improve consumer's well-being, health, and confidence (2). The advent of "omics" techniques—interpreted broadly as a suite of high-throughput technologies for the generation, processing, and

Biji T. Kurien and R. Hal Scofield (eds.), *Protein Electrophoresis: Methods and Protocols*, Methods in Molecular Biology, vol. 869, DOI 10.1007/978-1-61779-821-4_15, © Springer Science+Business Media, LLC 2012

application of scientific information about the composition and functions of genome, proteome, and metabolome—has created unprecedented opportunities for increasing our understanding of how nutrients modulate gene and protein expression and ultimately influence cellular metabolism. In order to gain detailed insight on the molecular mechanisms governing the metabolic pathways, metabolomics is a must in the new Foodomic field. Metabolome can be defined as the full set of endogenous or exogenous low molecular weight metabolic entities of approximately <1,000 Da (metabolites) and the small pathway motifs that are present in a biological system (3). Considering the great diversity and heterogeneity of metabolites, simultaneous purification and analysis of all metabolites from a biological sample is also a challenge. Sample preparation for metabolomics will depend on the type of sample being analyzed, the subsequent method of analysis, and the metabolomic approach (targeted or non-targeted). As a result, at present no single analytical methodology or platform is available to detect, quantify, and identify all metabolites in a certain sample, although it is now generally accepted that mass spectrometry (MS)-based techniques (together with nuclear magnetic resonance) are crucial for metabolomics. However, due to the great complexity of metabolomes in biological samples, an analytical separation technique is usually coupled to MS to avoid signal suppression and subsequent loss of metabolic information. In this regard, capillary electrophoresis (CE) is particularly suited for the rapid separation of ionic and polar compounds with very high resolution using extremely small reagents and sample volumes. The coupling CE-MS is described in this work for metabolite fingerprinting (4). Time of flight (TOF) mass analyzer is used in the CE-MS coupling due to its high scan speed and high mass resolution and accuracy, what perfectly fits with the narrow peaks provided by CE. In summary, four different sample preparation procedures for cytosolic metabolite extraction from human HT29 colon cancer cells are described in this work, together with a complete CE-MS approach for the study of the effect of dietary polyphenols on colorectal cancer cells' metabolome.

2. Materials

2.1. Reagents and Samples

1. Distilled water from Milli-Q system (Millipore, Bedford, MA, USA).
2. MS grade water from Scharlau (Barcelona, Spain).
3. Formic acid (98–100%) from Riedel-de Haën (Seelze, Germany).

4. Phosphate buffered saline (PBS): 138 mM sodium chloride, 2.7 mM potassium chloride, and 10 mM sodium hydrogen phosphate, pH 7.4.
5. Protease inhibitor cocktail containing 104 mM 4-(2-aminoethyl) benzenesulfonyl fluoride (AEBSF), 1.5 mM pepstatin A, 1.4 mM E-64, 4 mM bestatin, 2 mM leupeptin, and 80 μM aprotinin (Sigma-Aldrich).
6. Amino acids (arginine, lysine, histidine, γ-aminobutyric acid (GABA), valine, serine, leucine, threonine, glutamine, proline, aspartic acid, and tyrosine) (Sigma-Aldrich).
7. Tyramine (Sigma-Aldrich).
8. DL-methionine sulfone (Sigma-Aldrich).
9. 1,4-Piperazinediethanesulfonic acid (PIPES) (Sigma-Aldrich).
10. Human HT29 colon cancer cells (Universidad de Miguel Hernandez, Elche, Spain).
11. Dulbecco's modified eagle medium (DMEM).
12. Rosemary polyphenolic extracts were obtained by SFE using CO_2 at 150 bar as solvent with 7% ethanol (5).

2.2. Solutions for Sample Treatment, Metabolite Extraction, and CE-ESI-TOF MS Method

1. Homogenization buffer solution for sample treatment: 10 mM Tris–HCl, 5 mM EDTA, and 120 mM NaCl, pH 7.4. Dissolve 1.21 g of Trizma® base, 1.87 g of EDTA, and 3.51 g of NaCl in 95 mL of Milli-Q water and adjust the pH to 7.4 by adding HCl. Then reach a final volume of 100 mL by adding Milli-Q water using a volumetric flask.
2. Sodium formate (5 mM) for MS calibration: Mix 5 mL of 1 M NaOH with 99.5 mL of H_2O:2-propanol (50:50 v/v) with 0.2% formic acid and make up volume to 100 mL.
3. Sodium hydroxide (1 M): Dissolve 4 g of sodium hydroxide in 95 mL of Milli-Q water. Then reach a final volume of 100 mL using a volumetric flask.
4. Amino acids solution to corroborate metabolite identification. Dissolve amino acids in Milli-Q water at the following concentrations: arginine (0.58 mg/mL), lysine (0.49 mg/mL), histidine (0.52 mg/mL), GABA (0.34 mg/mL), valine (0.39 mg/mL), serine (0.35 mg/mL), leucine (0.44 mg/mL), threonine (0.40 mg/mL), glutamine (0.49 mg/mL), proline (0.38 mg/mL), aspartic acid (0.44 mg/mL), and tyrosine (0.09 mg/mL).
5. 1 M Formic acid, pH 1.8: Mix 3.77 mL of formic acid with 95 mL of Milli-Q water in a volumetric flask, then reach a final volume of 100 mL.
6. Isopropanol–water (50:50 v/v): Mix 20 mL of MS grade 2-propanol with 20 mL of MS grade water.
7. Water–methanol (95:5, v/v). Mix 95 mL of Milli-Q water with 5 mL of methanol.

2.3. Instrumentation for Sample Preparation Extraction Procedures

1. Polytron® homogenizer (Capitol Scientific, Austin, TX, USA)
2. Evolute™ ABN (water-wettable polymer-based sorbent that extracts acidic, basic, and neutral analytes) cartridges (Biotage, Cardiff, Wales, UK).
3. Isolute C18 Endcapped cartridges (Biotage).
4. Amicon Ultra 3 kDa centrifugal device (Millipore, Bedford, MA, USA).
5. Savant SpeedVac SC 200 (Thermo Fisher Scientific, Waltham, MA, USA).

2.4. Instrumentation for CE-ESI-TOF MS Analysis

1. High-performance capillary electrophoresis (CE) system model P/ACE 5500 (Beckman, Fullerton, CA, USA) controlled by a PC equipped with System Gold software (Beckman).
2. Syringe pump (Cole Palmer, Vernon Hills, IL, USA).
3. Uncoated fused-silica capillaries (50 µm id and 90 cm total length) (Composite Metal Services, Worcester, England).
4. Orthogonal electrospray interface model G1607A (Agilent Technologies, Palo Alto, CA, USA).
5. TOF MS instrument (micrOTOF) (Bruker Daltonics, Bremen, Germany) controlled by a PC equipped with micrOTOF control software (Bruker Daltonics).

2.5. Software for Data Processing and Analyzing

1. Data Analysis 3.3 software (Bruker Daltonics).
2. Free available databases used for tentative identification (6–10):

   Database HMDB (http://www.hmdb.ca/)

   Metlin (http://metlin.scripps.edu/)

   KEGG compound (http://www.genome.jp/kegg/compound/)

   KEGG Pathway (http://www.genome.jp/kegg/pathway.html)

   PubChem (http://pubchem.ncbi.nlm.nih.gov/)

3. Methods

3.1. Sample Preparation

1. Grow human HT29 cells in DMEM supplemented with 5% heat-inactivated fetal calf serum, 2 mM L-glutamine, 50 U/mL of penicillin G, and 50 µg/mL of streptomycin, at 37°C in humidified atmosphere and 5% CO_2. Treat the cell culture with 10 µM rosemary polyphenolic extract for 72 h.
2. Wash human HT29 colon cancer cells with PBS solution and then centrifuge at 1,000 × *g* for 10 min to form the pellets.
3. Suspend the pellets with homogenization buffer (5.5 mL) and protease inhibitor cocktail (10 µL/10^6 cells).

4. Disrupt the cells with a Polytron homogenizer (3 cycles of 30 s, in ice) and centrifuge for 14 min at 14,000 × *g* and 4°C.
5. Discard the pellet (nuclear fraction) and take the supernatant (cytosolic fraction).
6. Centrifuge the supernatant for 1 h at 100,000 × *g* and 4°C.
7. Store the supernatant (cytosolic fraction) at −80°C until metabolite purification procedure is carried out. This temperature stops enzyme activity and allows samples to be stored safely without continuing metabolic activity.

3.2. Metabolite Extraction Procedures

3.2.1. Solid-Phase Extraction

1. Use Evolute™ ABN columns and Isolute C18 Endcapped cartridges for SPE. The conditions of elution are identical for both solvents.
2. Pass 1 mL of methanol followed by 1 mL of water through the cartridges.
3. Add 350 μL of Milli-Q water to 150 μL of cytosolic fraction obtained from the cell culture.
4. Load this solution onto the column and wash with 1 mL of water–methanol (95:5, v/v).
5. Use 500 μL of methanol to elute the sample.
6. Aliquot eluted sample in 100 μL volume.
7. Vacuum dry the aliquots and store the dried extracts at −80°C.
8. Prior to CE-MS analysis, dissolve dried extracts in 20 μL of MS grade water.

3.2.2. Protein Precipitation

1. Add 350 μL of methanol to 150 μL of cytosolic fraction obtained from the cell culture.
2. Incubate the solution at −20°C during 2 h to reduce protein solubilization.
3. Centrifuge the suspension for 5 min at 20,000 × *g* and 4°C.
4. Discard the pellet and collect the supernatant fraction.
5. Aliquot supernatant fraction in 100 μL volume.
6. Vacuum dry the aliquots and store the dried extracts at −80°C.
7. Prior to CE-MS analysis, dissolve dried extracts in 20 μL of MS grade water.

3.2.3. Ultracentrifugation

1. Add 350 μL of Milli-Q water to 150 μL of cytosolic fraction obtained from the cell culture.
2. Using the Amicon Ultra 3 kDa centrifugal device, centrifuge the solution for 70 min at 14,000 × *g* and 20°C.
3. Collect with a pipette the fraction with molecular weight lower than 3 kDa and aliquot in 100 μL volume.

4. Vacuum dry the aliquots and store the dried extracts at −80°C.
5. Prior to CE-MS analysis, dissolve dried extracts in 20 μL of MS grade water.

3.3. CE-ESI-TOF MS Metabolic Analysis of Human HT29 Colon Cancer Cells

1. Condition all new capillaries by flushing with 0.1 M NaOH for 20 min, followed by water for 20 min and background electrolyte (BGE) for 5 min (see Note 1). Use the following BGE for CE separation (1 M formic acid, pH 1.8) (see Note 2).
2. After each run, condition the capillary with Milli-Q water for 4 min, followed by BGE for 4 min (see Note 3).
3. Set the interface between CE and TOF MS via a sheath-flow configuration, in which electrical contact at the electrospray needle tip is established via a flow of conductive sheath liquid with the following composition: isopropanol–water (50:50 v/v) (see Note 4), and delivered at a flow rate of 0.24 μL/min by a syringe pump. Use next nebulizer and drying gas conditions: 0.4 bar N_2 and 4 L/min N_2, respectively, and maintain the ESI chamber at 200°C (see Note 5).
4. Operate the micrOTOF to acquire spectra in the *m/z* range of 50–600 every 90 ms. Make external and internal calibration of the TOF MS instrument (see Note 6) by introducing 5 mM sodium formate solution through the separation capillary.
5. Make sample injection at the anodic end using N_2 pressure of 0.5 psi (34.5 mbar) for 80 s (see Note 7).
6. Use +25 kV as running voltage at a constant temperature of 25°C to carry out electrophoretic separation by using 1 M formic acid as BGE.
7. Analyze each sample in triplicate by CE-MS.
8. Depending on metabolite extraction method (see Subheading 3.2), four different base peak electropherograms (BPE) are obtained by CE-MS (Fig. 1).

3.4. Metabolomic Fingerprinting of Human HT29 Colon Cancer Cells Treated with Dietary Polyphenols

Follow the CE-ESI-TOF MS metabolomic approach explained in Subheading 3.3 to obtain the metabolic fingerprinting from human HT29 colon cancer cells treated and non-treated with dietary polyphenols (Fig. 2).

3.5. Data Analysis

1. Using the Data Analysis 3.3 software, annotate the accurate *m/z* value, migration time, and area from each peak.
2. Depending on established *m/z* differences, eliminate the redundant responses from same ion, such as isotopic peaks, fragments, adducts, dimers, etc. Remove the low reliability signals such as non-peak-like shapes or spike noise.

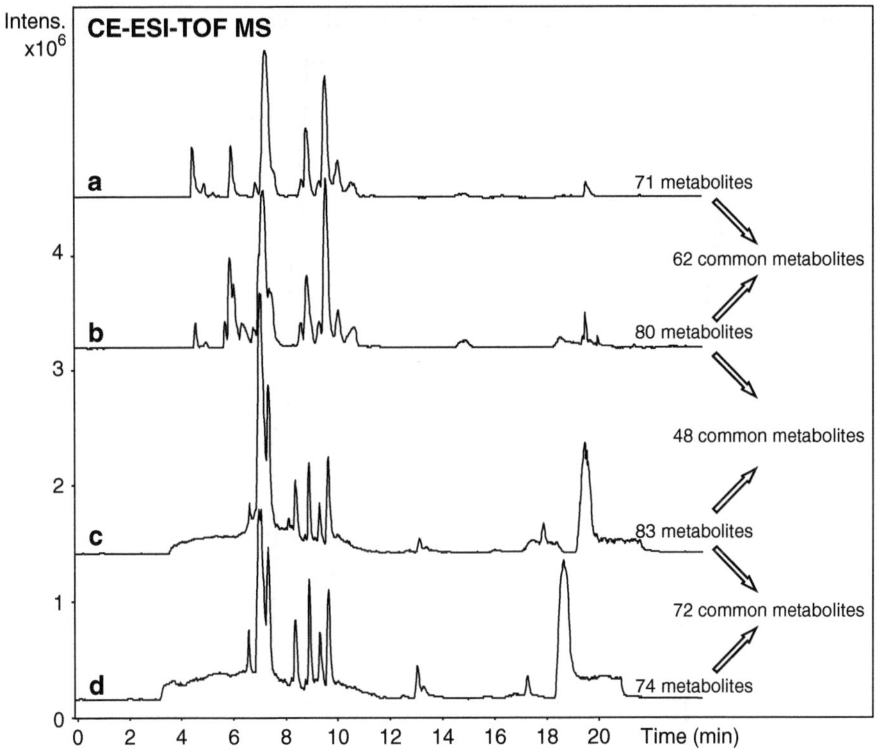

Fig. 1. CE-ESI-TOF MS base peak electropherograms (BPE) of the cytosolic fraction from HT-29 colon cancer cells after four different metabolite extraction procedures. (**a**) SPE extraction using Isolute C18 Endcapped cartridge. (**b**) SPE extraction using Evolute™ ABN cartridge. (**c**) Metabolite extraction using methanol deproteinization. (**d**) Metabolite extraction using ultrafiltration with an Amicon Ultra 3 kDa centrifugal device.

3. As a result of using a high mass accuracy TOF MS instrument, Generate-Molecular Formula editor in Data Analysis 3.3 software is capable of providing a list of possible elemental formulas (with errors below 5–10 ppm).
4. Enlighten standard functionalities such as minimum/maximum elemental range, electron configuration, and ring-plus double-bond equivalents and compare the theoretical and the experimental isotopic pattern (Sigma-Value™) for increased confidence in the theoretical molecular formula.
5. Use some free available databases (see Subheading 2.5) in order to work out tentative identification based on the obtained molecular formula.
6. Confirmation of amino acid identification is carried out by using commercial compounds (see Subheading 2.2).
7. After analysis by CE-ESI-TOF MS of human HT29 colon cancer cells using different metabolite purification protocols, 71 and 80 different metabolites are detected in C18 and ABN extract, respectively. Similar extraction selectivity is obtained since 62 are common metabolites. On the other hand, 83 and

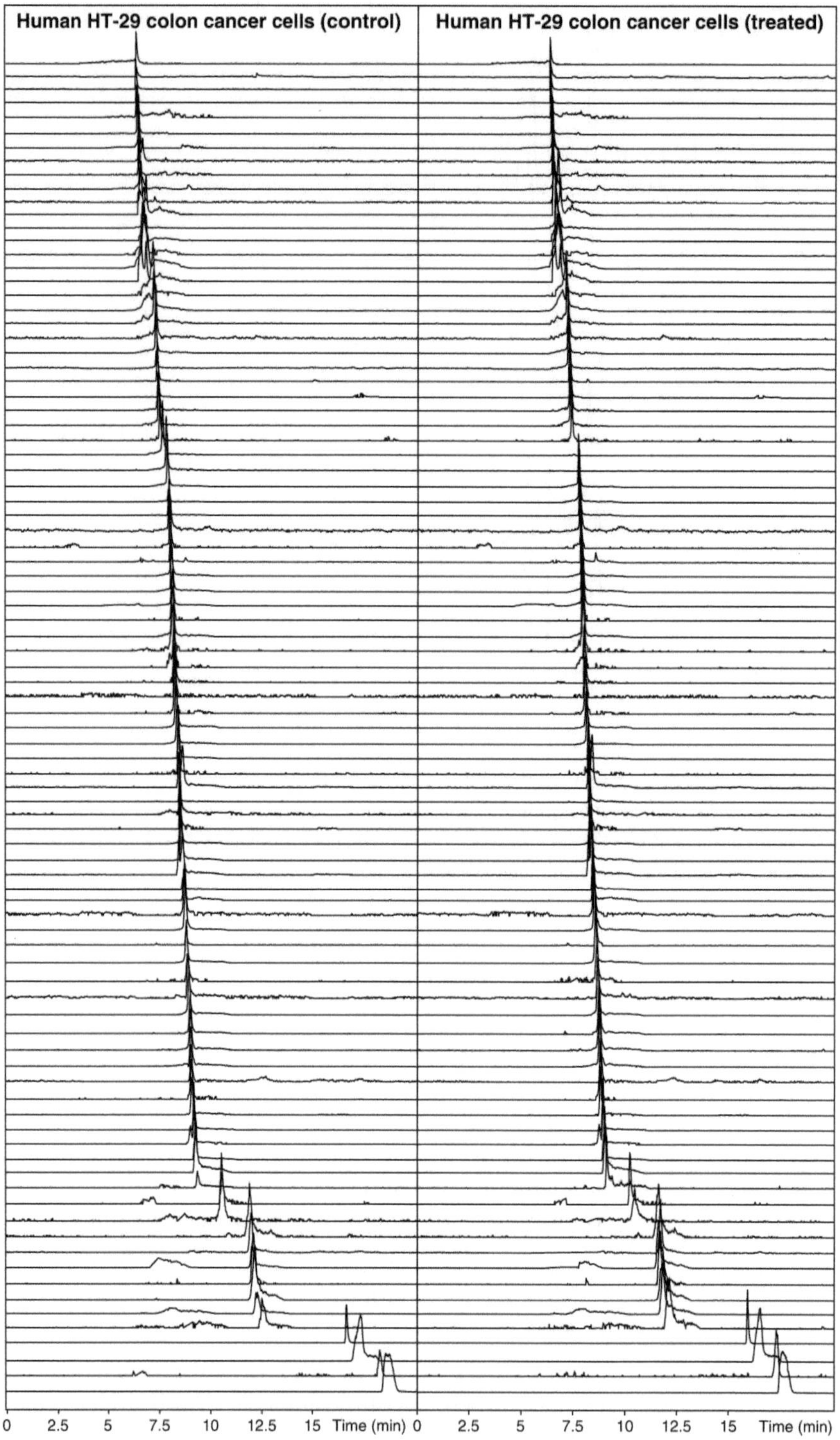

Fig. 2. CE-ESI-TOF MS extracted ion electropherograms (EIE) of the cytosolic fraction from human HT-29 colon cancer cells: control (*left panel*) and treated with dietary polyphenols (*right panel*). Metabolite purification is carried out through methanol deproteinization.

74 metabolites are detected in MeOH and ultrafiltrate extract, respectively. Similar extraction selectivity is obtained since 72 are common metabolites. Comparing ABN (80 metabolites) and methanol (83 metabolites) purification, different extraction selectivity is observed with only 48 common metabolites.

8. After data analysis of ABN and methanol extracts, a total number of 115 different metabolites with molecular masses from 90.0 to 777.2 Da are observed. 44 metabolites are tentatively identified and classified in several groups. Most abundant compounds are amino acids (histidine, arginine, lysine, GABA, valine, serine, leucine (or isoleucine), threonine, glutamine, proline, tyrosine, and aspartic acid) and polypeptides, although other compounds such as lipoic acid derivatives, steroid derivatives, carbohydrates derivatives, amino alcohol, amino acid phosphates, nucleosides, quaternary amines, pterins, acyl phosphates, and nucleotides are also present. Among all tentatively identified compounds, 20 of them are related to the following metabolic pathways: glycan biosynthesis and metabolism, organismal systems, excretory system, neuroactive ligand–receptor interaction, carbohydrate metabolism, lipid metabolism, nucleotide metabolism, energy metabolism, metabolism of cofactors and vitamins, environmental information processing, genetic information processing, and amino acid metabolism.
9. After CE-ESI-TOF MS analysis of metabolite extracts (obtained through methanol deproteinization) human HT29 colon cancer cells treated and non-treated with dietary polyphenols were compared. Ninety different compounds are detected with this method in less than 20 min (see Fig. 2). Significant differences are found for ions: 130.1602, 132.0772, 212.1167, 240.1487, 271.0557, 118.0870, 150.0586, 166.0861, 182.0802, and 307.0840 m/z. Among them, creatine, valine, tyrosine, and glutathione disulfide were identified.

4. Notes

1. Since the presence of nonvolatile components decreases the MS sensitivity, increases the background noise, and, under extreme conditions, can clog the system, volatile sheath liquids and separation BGEs must be used while working with MS.
2. A low pH should be used for the analysis in order to avoid analyte adsorption onto the inner capillary wall and confer positive charge to the analytes to improve their ESI ionization yield. Under these conditions, most amino acids, amines, nucleosides, small peptides, and, in general, basic compounds present overall positive charge and they migrate before EOF

(electroosmotic flow). Moreover, using these electrophoretic conditions allows to detect some acidic compounds (bearing some negative charge in their structure) being carried to the MS by the residual EOF.

3. In order to eliminate adsorbed compounds from the capillary wall, after each analysis, the capillary is rinsed.
4. Even some different mixtures of acetonitrile, methanol, and isopropanol with water could be used, in order to obtain the best response on sensitivity and signal stability of metabolites, 50:50 (v/v) isopropanol–water mixture is recommended as sheath liquid for the electrospray interface.
5. Even the spray is stable between 2 and 6 μL/min, it is recommended to use 4 μL/min as sheath liquid flow rate to reach the best S/N value during CE-ESI-TOF MS analysis of metabolites. Nebulizer pressure (0.2–0.7 bar) has also some effect on MS sensitivity, and a nebulizer pressure of 0.4 bar is advised to obtain the best results.
6. External and internal calibration of the TOF MS instrument should be performed prior to each analysis. Masses for calibration of the TOF MS instrument are next: 90.9766, 158.9641, 226.9515, 294.9389, 362.9263, 430.9138, 498.9012, and 566.8886 *m/z*.
7. Since different salt concentration of the purified samples may change the amount of injected sample, use pressure injection instead of electrokinetic injection. By applying 0.5 psi pressure for 80 s, a sample plug of 2.7 cm (3% of the capillary length) is injected.

Acknowledgments

This work was supported by AGL2008-05108-C03-01 (Ministerio de Ciencia e Innovación, Spain) and CSD2007-00063 FUN-C-FOOD projects (Programa CONSOLIDER, Ministerio de Educacion y Ciencia, Spain). MC thanks The Council of Higher Education, Turkey for his grant. CI thanks Ministerio de Educacion y Ciencia, Spain for her FPI grant.

References

1. Reddy L, Odhav B, Bhoola KD (2003) Natural products for cancer prevention: a global perspective. Pharmacol Ther 99:1–13
2. Cifuentes A (2009) Food analysis and foodomics. J Chromatogr A 1216:7109
3. Garcia-Cañas V, Simó C, Leon C, Cifuentes A (2010) Advances in nutrigenomics research: novel and future analytical approaches to investigate the biological activity of natural compounds and food functions. J Pharm Biomed Anal 51:290–304
4. Ramautar R, Somsen GW, de Jong GJ (2009) CE-MS in metabolomics. Electrophoresis 30:276–291
5. Herrero M, Plaza M, Cifuentes A, Ibáñez E (2010) Green processes for the extraction of bioactives from Rosemary: chemical and functional characterization via ultra-performance liquid chromatography-tandem mass spectrometry and in-vitro assays. J Chromatogr A 1217:2512–2520
6. Wishart DS, Knox C, Guo AC, Eisner R, Young N, Gautam B et al (2009) HMDB: a knowledgebase for the human metabolome. Nucleic Acids Res 37:D603–D610
7. Smith CA, O'Maille G, Want EJ, Qin C, Trauger SA, Brandon TR et al (2005) METLIN: a metabolite mass spectral database. Ther Drug Monit 27:747–751
8. Kanehisa M, Goto S, Furumichi M, Tanabe M, Hirakawa M (2010) KEGG for representation and analysis of molecular networks involving diseases and drugs. Nucleic Acids Res 38:D355–D360
9. Kanehisa M, Goto S, Hattori M, Aoki-Kinoshita KF, Itoh M, Kawashima S et al (2006) From genomics to chemical genomics: new developments in KEGG. Nucleic Acids Res 34:D354–D357
10. Aoki-Kinoshita KF, Kanehisa M (2007) Gene annotation and pathway mapping in KEGG. Methods Mol Biol 396:71–91

Chapter 16

Peptide Fractionation by SDS-Free Polyacrylamide Gel Electrophoresis for Proteomic Analysis via DF-PAGE

Yassel Ramos, Vladimir Besada, and Lila Castellanos-Serra

Abstract

Here we present a procedure for peptide fractionation by SDS-free polyacrylamide gel electrophoresis, based on discontinuous buffer systems. In the absence of SDS, peptide migration depends both on their molecular mass and on their net charge at the electrophoresis pH. By selecting the separation pH, peptide mobility is modulated. In the original discontinuous buffer system (Tris/glycine), peptides that migrate to the anode have p*I* values below 6.8 and distribute along the lane in a p*I* decreasing order, while at acidic pH, as that afforded by histidine/MOPS buffer system, peptides with p*I* below 5.5 are fractionated. Separation at acid pH is particularly useful for recovering phosphopeptides as well as other highly negatively charged peptides, as those containing sialic or sulfate substituents. Both separation conditions in Tris/glycine and in histidine/MOPS are applicable to proteomic studies, by dual-fractionation polyacrylamide gel electrophoresis (DF-PAGE). First, complex protein samples are separated via SDS-PAGE, and after in-gel proteolysis, peptides are loaded on a second SDS-free gel, where they are separated as described here.

Key words: Fractionation, Peptide, Mass spectrometry, Proteomics, Protein, Electrophoresis, PAGE

1. Introduction

SDS-polyacrylamide gel electrophoresis (SDS-PAGE) (1) has become the most popular tool for protein electrophoresis. In this system, all proteins are negatively charged and, in consequence, they migrate towards the anode. This universally known technique was preceded by the seminal contribution of Ornstein (2) that described a discontinuous buffer system for separating negatively charged proteins. Proteins are concentrated in a very thin starting zone before entering the separation gel, generating thin bands that significantly increase the electrophoretic resolution. In this system,

Biji T. Kurien and R. Hal Scofield (eds.), *Protein Electrophoresis: Methods and Protocols*, Methods in Molecular Biology, vol. 869, DOI 10.1007/978-1-61779-821-4_16,

proteins and peptides migrate according to their electrophoretic mobility which is proportional to their charge and inversely proportional to their size. Several discontinuous buffer systems for electrophoresis have since been developed (3) and extensively applied in biochemical research (4, 5).

We have described methods for peptide fractionation using a SDS-free PAGE; the corresponding protocols are presented here (6, 7). About 30 years ago, West et al. described the separation of peptides at acidic or at basic pH in 40–50% polyacrylamide resolving gels in the absence of SDS (8). The procedure was abandoned in the 1980s when reversed-phase chromatography became the method of choice for peptide separation. We have modified this old procedure, demonstrating that efficient peptide fractionation can be accomplished in 15% polyacrylamide gels, and by selecting the composition of the gel and electrophoresis buffer, we influence peptide migration, achieving higher separation efficiency for specific p*I* ranges.

We applied discontinuous buffer electrophoresis systems to select and simultaneously fractionate peptides with either p*I* lower than 5.5 or 6.8. At pH 6.8, as afforded by the Tris/glycine system, peptides with p*I* between 6.8 and 2.9 are fractionated and remain into the gel. By using a new electrophoresis buffer based on MOPS and histidine, the dissociating functions of the acidic amino acids are partially titrated, and both surface charge and electrophoretic mobility of the peptides decrease (Fig. 1). Under these conditions, peptides with p*I* higher than 5.5 (slightly acid, neutral, and basic peptides) are positively charged and would migrate to the cathode, while those with lower p*I*, as peptides with several aspartic, glutamic, sialic, or phosphate groups, penetrate the gel, concentrate, and then separate according to their charge-to-mass ratio. The method has been successfully combined with conventional protein electrophoresis separation and applied to the analysis of complex biological samples as bacterial and human proteomes (6, 7).

Several equipment for the electrophoretic fractionation of peptides have been developed and are commercially available (i.e., "off-gel electrophoresis" (9), "free flow electrophoresis," (10) or Rotofor (11)). However, the procedures described here allow peptide separation by using standard slab gel vertical electrophoresis equipment that are commonly used in biochemical laboratories, with a performance similar or better than those commercial equipment. They can be integrated to conventional SDS-PAGE (6, 7), allowing first a fractionation at protein level, and then at peptide level, a procedure named dual-fractionation polyacrylamide gel electrophoresis (DF-PAGE) (6), which has proved efficiency in proteomic studies of complex natural samples.

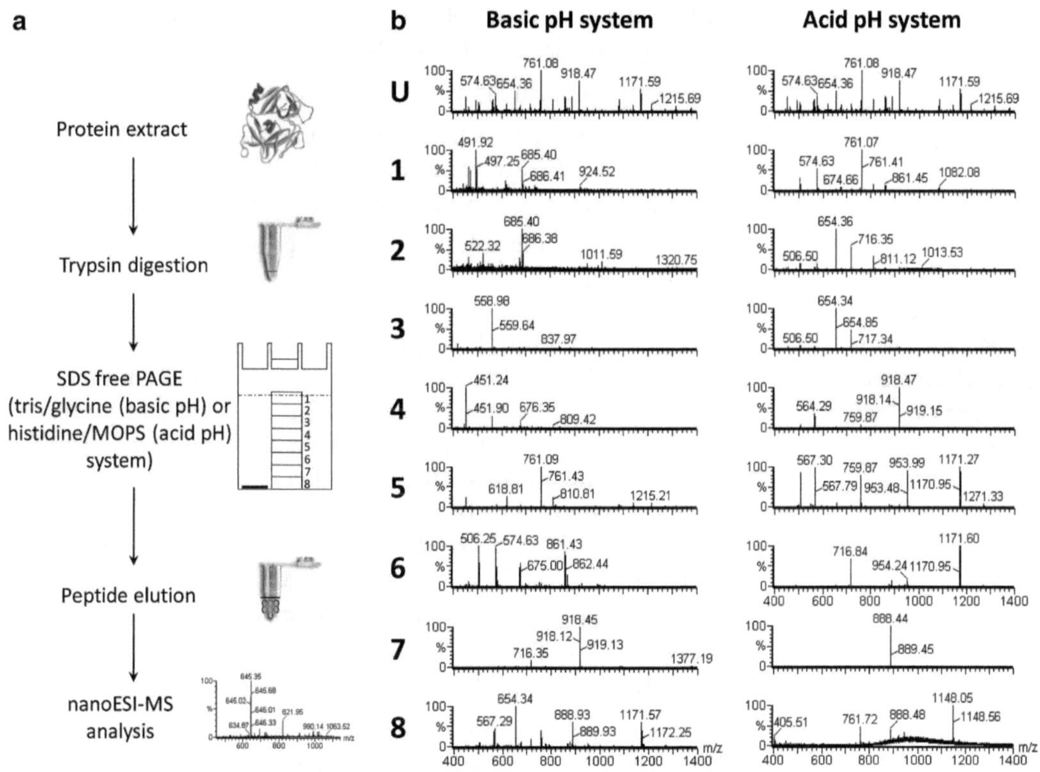

Fig. 1. (**a**) Schematic representation of the experimental workflow for tryptic peptide fractionation by PAGE. (**b**) Mass spectra of tryptic peptides from streptokinase separated in a 15% T SDS-free PAGE at basic and acid pH. The lane was cut in eight slices. U: unfractionated peptide mixture, 1–8: peptides obtained from slices 1 to 8, respectively. Reprinted with permission of Ramos et al. (2011) Electrophoresis 32(11):1323–1326. Reproduced from ref. 7 with permission from John Wiley and Sons.

2. Materials

Reagents are electrophoresis or analytical grade from international suppliers. Use freshly collected ultrapure water, with resistivity of 18 MΩ or higher. Avoid contact with reagents and solutions and use gloves during all manipulations. Acrylamide, bisacrylamide, and their solutions are highly toxic. They should be manipulated with extreme care, solids should be handled only on a hood, and recipients containing the reagents should be kept covered to avoid accidental dispersion. Follow established waste disposal procedures recommended for all materials and solutions.

2.1. Solutions Common to Gel Electrophoresis in Tris/Glycine and Histidine/MOPS

1. Thirty percent acrylamide/bisacrylamide solution (29.2:0.8 acrylamide: bisacrylamide) : Weigh 29.2 g of acrylamide and 0.8 g bisacrylamide, transfer to a 100-mL graduated cylinder containing about 60 mL of water, and mix during 20 min.

Add water to a final volume of 100 mL, transfer to an amber bottle, and store at 4°C (see Note 1).

2. Ammonium persulfate: 10% solution in water (see Note 2).
3. *N, N, N, N′*-tetramethylethylenediamine (TEMED) (Sigma Chemical Company, St. Louis, MO, USA). Store at 4°C.
4. Bromophenol blue solution: Dissolve 0.1 g bromophenol blue in 100 mL loading solution 1× as described below for each pH buffer system.

2.2. Solutions for Gel Electrophoresis of Peptides in Tris/Glycine

1. Electrode solution (10×) pH 8.3: 0.25 M Tris/HCl, 1.92 M glycine, pH 8.3. Prepare this solution as a concentrated 10× stock solution. For this, weigh 30.3 g Tris and 144 g glycine, stir in 700 mL water, transfer it to a 1-L graduated cylinder, and add water to make it 1 L. Label the solution "electrode buffer Tris/glycine 10×, pH 8.3." Just before use, dilute 100 mL of 10× electrode buffer to 1 L with water (final concentration 0.025 M Tris/HCl, 0.192 M glycine).
2. Stacking gel solution (pH 6.8): 0.25 M Tris/HCl, pH 6.8. Weigh 30.3 g Tris and prepare a 1 L solution as in previous steps, but adjusting the pH to 6.8 with HCl. Store at 4°C.
3. Resolving gel solution (pH 8.8): 0.75 M Tris/HCl, pH 8.8. Weigh 90.85 g Tris and transfer it to a 1-L graduated cylinder. Add water to a volume of 900 mL. Stir and adjust pH to 8.8 by adding HCl. Add water to a final volume of 1 L. Store at 4°C.
4. Loading solution (2×): 0.125 M Tris/HCl (pH 6.8), 25% glycerol. Transfer 5 mL of stacking solution (Tris/HCl 0.25 M pH 6.8) to a 10-mL polypropylene conical tube. Add 2.5 g glycerol and water to a final volume of 10 mL and mix. Store at 4°C.

2.3. Solutions for Gel Electrophoresis of Peptides in Histidine/MOPS

1. Cathode solution (10×): Histidine/MOPS 2 M, pH 5.5. Prepare the cathode solution as a concentrated 10× stock solution. For this, weigh 418.52 g MOPS (Sigma Chemical Company, St. Louis, MO, USA) and transfer it to a 1-L glass beaker. Add water to a volume of 900 mL. Stir and adjust pH to 5.5 by adding histidine prepared as a water-saturated solution. Transfer to a 1-L graduated cylinder and add water to a final volume of 1 L. Store at 4°C. Just before use, for using as cathode electrolyte, dilute 100 mL of 10× stock solution to 1 L with water (final concentration 0.2 M).
2. Anode solution (2×): Histidine/HCl 0.4 M pH 6.5. Weigh 62.06 g histidine and transfer it to a 1-L glass beaker. Add water at 37°C (see Note 3) to a volume of 900 mL. Mix under magnetic agitation on a magnetic stirrer provided with a heating plate set at 37°C. After complete dissolution of histidine, remove from the heating plate, let it equilibrate to room

temperature, and adjust its pH to 6.5 by adding concentrated HCl. Transfer to a 1-L graduated cylinder and add water to a final volume of 1 L. Store at 4°C (see Note 4). For using as anode electrolyte, just before use, dilute 500 mL of 2× stock solution to 1 L with water (final concentration 0.2 M) (see Note 5).

3. Stacking gel solution (2×): Histidine/HCl 0.25 M pH 5. Weigh 38.79 g histidine and transfer it to a 1-L glass beaker. Add water to a volume of 900 mL. Stir and adjust pH to 5 by adding concentrated HCl. Transfer to a 1-L graduated cylinder and add water to a final volume of 1 L. Store at 4°C (see Note 6).
4. Loading solution (2×): Histidine/HCl 125 mM, glycerol 25%, pH 5. Transfer 5 mL of stacking solution (histidine/HCl 0.250 M pH 5) to a 10-mL polypropylene conical tube. Add 2.5 g glycerol and water to a final volume of 10 mL and mix. Store at 4°C.

2.4. Solution for Peptide Elution

1. 5% Formic acid: Prepare 10 mL of 5% formic acid in water. Store at 4°C.

3. Methods

We cast the gel at room temperature, generally 24 h before use (see Note 7). The gel is conserved overnight at 4°C for use on next day. Reported gel solutions are for 8 mm × 10 mm × 1 mm vertical minigels. To avoid interference with mass spectrometric analysis of fast moving peptides, the marker dye is not applied with peptide samples. To visually follow the electrophoretic migration, bromophenol blue solution is loaded in free lanes between those carrying the samples.

For peptides from in-gel proteolysis, gel pieces are applied directly into the well (see Subheading 3.3.1), or peptides are passively eluted, concentrated, and applied (see Subheading 3.3.2). Better peptide recovery is obtained when the "in-gel loading" method is used.

After electrophoresis, gels are not stained and, to avoid peptide diffusion, they are immediately processed as described in Subheading 3.3.1. Gel blocks can be conserved indefinitely at −80°C, until peptide elution for analysis.

3.1. 15% Polyacrylamide Gel Electrophoresis for Peptide Separation in Tris/Glycine Gels

1. Prepare a 15% polyacrylamide resolving gel by mixing 4 mL of resolving gel solution (0.75 M Tris/HCl, pH 8.8) and 4 mL of acrylamide/bis solution. Add 40 μL of ammonium persulfate (10%) and 6 μL of TEMED and cast gel within the minigel vertical cassette provided with 1-mm-thick spacers. Allow

about 2.5 cm free space for stacking the gel. Overlay the gel solution with isobutanol.

2. Prepare the stacking gel by mixing 2 mL of stacking gel solution pH 6.8 (0.25 M Tris/HCl, pH 6.8), 0.8 mL of acrylamide/bis solution, and 1.12 mL water. Add 40 μL of ammonium persulfate (10%) and 4 μL of TEMED. Insert a 10-well gel comb 1 mm thick immediately without introducing air bubbles. Let the gel polymerize overnight at 4°C.
3. Mix peptide samples with loading solution 2× (0.125 M Tris/HCl, pH 6.8, 25% glycerol) and apply samples, leaving empty lanes in-between. Apply bromophenol blue dye solution to the empty lanes.
4. Add the electrode solution 1× (0.025 M Tris/HCl, 0.192 M glycine, pH 8.3) to the electrode compartments. Electrophorese at 30 mA till the bromophenol blue dye has reached the bottom of the gel.

3.2. 15% Polyacrylamide Gel Electrophoresis for Peptide Separation in Histidine/MOPS Gels

1. Prepare a 15% polyacrylamide resolving gel by mixing 4 mL of anode solution 2× (histidine/HCl 0.4 M pH 6.5) and 4 mL of acrylamide/bis solution. Add 40 μL of ammonium persulfate (10%) and 6 μL of TEMED and cast the gel within the minigel vertical cassette provided with 1-mm spacers. Allow about 2.5 cm free space for stacking the gel. Overlay the gel solution with isobutanol.
2. Prepare the stacking gel by mixing 2 mL of stacking gel solution 2× (histidine/HCl 0.25 M) pH 5, 0.8 mL of acrylamide/bis solution, and 1.12 mL water. Add 40 μL of ammonium persulfate (10%) and 4 μL of TEMED. Insert a 10-well gel comb 1 mm thick immediately without introducing air bubbles.
3. Mix peptide samples with loading solution 2× (histidine/HCl 0.125 M, glycerol 25%, pH 5) and load samples, leaving empty lanes in-between. Apply bromophenol blue dye solution to the empty lanes.
4. Add the cathode solution 1× (final concentration 0.2 M) to the upper electrode compartment and the anode solution 1× (final concentration 0.2 M) to the lower electrode compartment. Electrophorese at 30 mA till the bromophenol blue dye has reached the bottom of the gel.

3.3. Dual-Fractionation Polyacrylamide Gel Electrophoresis

3.3.1. In-Gel Loading

1. Proteins separated by SDS-PAGE are digested in-gel. Load in-gel-digested gel piece to any of the above described SDS-free systems.
2. "In-gel loading": Add 10–15 μL of loading solution 2× (described above for each pH buffer system) to the gel pieces containing the digested proteins and incubate for 15 min at 25°C.

3. Place the gel blocks as well as any surrounding solution into the well of the gel and cover it with cathode buffer 1×. Apply to the empty lanes the bromophenol blue dye solution.

3.3.2. In-Solution Loading

1. Add 100 μL of 0.025 M ammonium bicarbonate to the gel pieces containing the digested proteins and shake during 20 min. Centrifuge and collect the solution in labeled polypropylene tubes.
2. Repeat the procedure two more times, collecting the peptide eluate in the same corresponding tube.
3. Concentrate the solution volume under centrifugal vacuum evaporator to approximately 10 μL.
4. Mix peptide samples with loading solution 2× (described above for each pH buffer system) and load samples, leaving empty lanes in-between. Apply to the empty lanes the bromophenol blue dye solution.

3.4. Gel Processing and Peptide Elution for Mass Spectrometry Analysis

1. Following electrophoresis, immediately open the gel cassette with the aid of a spatula, the gel remains on one of the glass plates. Cut the gel lanes vertically where samples were applied; remove the accompanying section of the stacking gel. Slice the resolving gel lane in about ten sections of equivalent height and place each section in an independent, adequately labeled polypropylene tube. Gel slices can be conserved indefinitely at −80°C, until processing for peptide elution and analysis.
2. Peptide elution. With the aid of a spatula, take the gel slice out of the tube, place it on a clean glass plate, cut the gel slice into gel blocks of about 1 mm^3, and transfer the cubes back to the polypropylene tube. Add 5% formic acid solution just to cover the gel blocks and close the polypropylene-labeled tubes. Shake the tubes for peptide extraction during about 20 min in a vibrating shaker. Centrifuge and collect the solution in labeled polypropylene tubes.
3. Repeat the procedure two more times, collecting the peptide eluate in the same corresponding tube.
4. Concentrate the solution volume under centrifugal vacuum evaporator to approximately 20 μL.
5. Desalt the samples using reversed-phase C-18 microcolumns (as ZipTips), according to manufacturer's instruction for peptide desalting and recovery. For relatively simple samples, as peptide mixtures of 2–3 proteins, the eluted peptides can be directly analyzed by mass spectrometry, while complex peptide samples, as those derived from cells or subcellular organelles, nano LC-mass spectrometry is recommended.

4. Notes

1. Prepare this solution every month.
2. Use freshly prepared solution.
3. To accelerate dissolution of histidine, use previously warmed water and mix on a controlled temperature heating plate set at 37°C.
4. At 4°C, histidine partially precipitates. To use the solution, stir it until redissolution on a controlled temperature heating plate set at 37°C.
5. The anode solution is also used for preparing the resolving gel, where its final concentration, after mixing with the acrylamide/bis solution, is 0.2 M.
6. The final concentration in the stacking gel, after mixing with the acrylamide/bis solution, is 0.125 M.
7. This facilitates thorough acrylamide polymerization, reducing the concentration of residual free monomers in the matrix.

Acknowledgments

This work was granted by the research program of the Centre for Genetic Engineering and Biotechnology (Havana, Cuba).

References

1. Laemmli UK (1970) Cleavage of structural proteins during the assembly of the head of bacteriophage T4. Nature 227:680–685
2. Ornstein L (1964) Disc electrophoresis. I. Background and theory. Ann N Y Acad Sci 121:321–349
3. Jovin TM (1973) Multiphasic zone electrophoresis. IV. Design and analysis of discontinuous buffer system with a digital computer. Ann N Y Acad Sci 209:477–496
4. Schägger H, von Jagow G (1987) Tricine-sodium dodecyl sulfate-polyacrylamide gel electrophoresis for the separation of proteins in the range from 1 to 100 kDa. Anal Biochem 166:368–379
5. Tastet C, Lescuyer P, Diemer H et al (2003) A versatile electrophoresis system for the analysis of high- and low-molecular-weight proteins. Electrophoresis 24:1787–1794
6. Ramos Y, Gutierrez E, Machado Y et al (2008) Proteomics based on peptide fractionation by SDS-free PAGE. J Proteome Res 7:2427–2434
7. Ramos Y, Garcia Y, Pérez-Riverol Y et al (2011) Peptide fractionation by acid pH SDS-free. Electrophoresis 32:1323–1326
8. West MHP, Wu RS, Bonner WM (1984) Polyacrylamide gel electrophoresis of small peptides. Electrophoresis 5:133–138
9. Heller M, Michel PE, Morier P et al (2005) Two-stage Off-Gel™ isoelectric focusing: protein followed by peptide fractionation and application to proteome analysis of human plasma. Electrophoresis 26:1174–1188
10. Xie H, Rhodus NL, Griffin et al (2005) A catalogue of human saliva proteins identified by free flow electrophoresis-based peptide separation and tandem mass spectrometry. Mol Cell Proteomics 4:1826–1830
11. Xiao Z, Conrads TP, Lucas DA et al (2004) Direct ampholyte-free liquid-phase isoelectric peptide focusing: application to the human serum proteome. Electrophoresis 25: 128–133

Chapter 17

Tris–Acetate Polyacrylamide Gradient Gels for the Simultaneous Electrophoretic Analysis of Proteins of Very High and Low Molecular Mass

Monica Cubillos-Rojas, Fabiola Amair-Pinedo, Irantzu Tato, Ramon Bartrons, Francesc Ventura, and Jose Luis Rosa

Abstract

Polyacrylamide gel electrophoresis (PAGE) is one of the most powerful tools used for protein analysis. We describe the use of Tris–acetate buffer and 3–15% polyacrylamide gradient gels to simultaneously separate proteins in the mass range of 10–500 kDa. We show that this system is highly sensitive, it has good resolution and high reproducibility, and that it can be used for general applications of PAGE such as Coomassie Brilliant Blue staining and immunoblotting. Moreover, we describe how to generate mini Tris–acetate polyacrylamide gels to use them in miniprotein electrophoresis systems. These economical gels are easy to generate and to manipulate and allow a rapid analysis of proteins. All these features make the Tris–acetate–PAGE system a very helpful tool for protein analysis.

Key words: Electrophoresis, Giant proteins, Gradient gel, PAGE, Protein separation, Tris–acetate

1. Introduction

The sodium dodecyl sulfate polyacrylamide gel electrophoresis (SDS-PAGE) is a tool frequently used for protein analysis. Glycine–SDS-PAGE (also known as Laemmli–SDS-PAGE) and tricine–SDS-PAGE, which are based on Tris–glycine and Tris–tricine buffer systems, respectively, are the most commonly used variants to separate proteins by their size (1, 2). Polyacrylamide gels are composed of acrylamide and the cross-linker bisacrylamide. Variations in the concentrations of acrylamide and bisacrylamide into Laemmli–SDS-PAGE and tricine–SDS-PAGE allow separating proteins in the range of 1–500 kDa. Proteins smaller than 30 kDa are well resolved using tricine–SDS-PAGE. Glycine–SDS-PAGE

Biji T. Kurien and R. Hal Scofield (eds.), *Protein Electrophoresis: Methods and Protocols*, Methods in Molecular Biology, vol. 869,
DOI 10.1007/978-1-61779-821-4_17, © Springer Science+Business Media, LLC 2012

system is recommended for proteins larger than 30 kDa (2, 3). The different separation characteristics of these systems are related to the *p*Ka values of glycine and tricine that permit different electrophoretic mobilities and increasing the resolution of proteins of a specific mass range. Thus, to separate proteins in a wide mass range, it would be necessary to use both electrophoresis systems (2). Alternatively, it is also recommended to use a gradient gel with high percentage of acrylamide at the bottom and a lower percentage at the top to resolve giant and small proteins in the same PAGE (3).

In this chapter, we describe an electrophoresis system based on Tris–acetate gels to separate proteins in the mass range of 10–500 kDa (4). In the last few years, several companies have commercialized Tris–acetate gels to separate proteins of different mass range. The composition of these gels is not available for customers and, for this reason, the researcher has no choice but to purchase precast gels and their optimized buffers. These electrophoresis systems are based on a discontinuous buffer system that results in a neutral pH environment during electrophoresis, stabilizing both proteins and gel matrix and providing a better band resolution. The relevance of the environment to which proteins are exposed was previously showed by Hachmann and Amshey using Tris–glycine gels (5). At basic pH, there is a high possibility of protein modification, especially by alkylation of cysteine sulfhydryls, due to the cysteine pKa that ranges from 8 to 9 and because the reducing agent contained in the loading buffer does not co-migrate with the proteins. The use of near-neutral pH gel formulations, electrophoresis buffers at a pH lower than the cysteine pKa, and reducing agents (e.g., sodium bisulfite) that move into the gel ahead of the proteins to maintain a reducing environment reduces this risk significantly (5–7). We have assessed several combinations of these parameters (concentrations, pH, acrylamide:bisacrylamide ratios, gradients) obtaining the best results using a 3–15% polyacrylamide gradient gel made with 0.2 M Tris–acetate (pH 7.0) buffer and a standard acrylamide:bisacrylamide solution (40:1), containing a 2.6% concentration of bisacrylamide, 0.042% ammonium persulfate (APS), and 0.12% *N*,*N*,*N*,*N′*-tetramethylethylenediamine (TEMED) (4).

The composition of the running buffer (50 mM tricine, 50 mM Tris, 0.1% SDS, 1.3 mM sodium bisulfite, pH 8.2) is similar to the buffer used for tricine–SDS-PAGE (2), but the antioxidant sodium bisulfite is added in order to maintain the proteins in a reduced state during the run. The sample of proteins is prepared in a sample buffer that contains lithium dodecyl sulfate (LDS): 250 mM Tris–HCl pH 8.5, 2% (w/v) LDS, 100 mM DTT, 0.4 mM EDTA, 10% (v/v) glycerol, 0.2 mM phenol red, and 0.2 mM Coomassie Brilliant Blue (CBB) G. This sample buffer allows the reduction of disulfides of the proteins under mild heating conditions (10 min at 70°C) avoiding the cleavage of Asp-Pro bonds that occurs when the sample is heated at 100°C in Laemmli sample buffer (8). After the run, proteins in the gel can be stained (e.g., CBB) or transferred to a membrane (e.g., nitrocellulose or polyvinylidene fluoride (PVDF)). To transfer the run to a PVDF membrane, we have used a transfer buffer containing 20% methanol, 25 mM bicine, 25 mM Bis–Tris, 1 mM EDTA, and 1.3 mM sodium bisulfite, pH 7.2. Under these conditions, it is possible to separate and simultaneously analyze proteins in the mass range of 10–500 kDa in a single PAGE. This system also provides high sensibility in detecting small amounts of protein: it is possible to analyze by immunoblotting proteins around 500 kDa, running samples containing only 10 μg of total protein. Interestingly, we have optimized this Tris–acetate system to use small gels (8 cm × 8 cm) in miniprotein electrophoresis systems, which are easier to manipulate and cheaper to fabricate due to reduced reagent volume requirements. Moreover, the time required to run and transfer these gels is also reduced.

2. Materials

Prepare all solutions using ultrapure water (Milli-Q quality) and analytical grade reagents.

2.1. Tris–Acetate Polyacrylamide Gradient Gel Components

1. 3 M Tris–acetate pH 7.0 gel buffer (15×): Dissolve 36.33 g Tris base in 90 mL water. Mix and adjust pH with acetic acid. Make up to 100 mL with water. Store at 4°C.
2. LDS sample buffer (4×): 4 mL of 2.5 M Tris–HCl pH 8.5, 0.8 g LDS (Sigma, St Louis, MO, USA), 0.006 g EDTA, 5 mL glycerol (Roche, Indianapolis, IN, USA) 80%, 0.75 mL of 1% CBB G solution, and 0.25 mL of 1% phenol red solution. Mix well and store at room temperature.
3. 40% Acrylamide solution: 37.5:1 (acrylamide/bisacrylamide mix). Store at 4°C.
4. APS: 10% solution in water.
5. TEMED. Store at 4°C.
6. DTT (1,4-dithiothreitol): 1 M solution in water. Aliquot and store at −20°C. We recommend not to freeze samples again for further uses.
7. Running buffer (1×): Dissolve 8.95 g tricine, 6.06 g Tris base, 1 g dodecyl sulfate sodium salt (SDS), and 0.25 g sodium bisulfite in 1 L water. Mix and store at 4°C. The pH of this solution is 8.24. It is not necessary to adjust it.
8. Gradient Maker (Hoefer Instruments, San Francisco, CA).
9. Stir bar and agitator.

10. Mini-PROTEAN II electrophoresis system (Bio-Rad, Hercules, CA, USA).
11. Bicinchoninic acid (BCA) assay kit (Thermo Scientific, Rockford, Ill, USA).

2.2. Immunoblotting Components

1. PVDF membranes (Millipore, Bedford, MA, USA).
2. Transfer buffer (20×): Dissolve 40.8 g Bicine, 52.4 g Bis–Tris, and 3 g EDTA in 500 mL water. Store at 4°C. For western, dilute this buffer to 1× in water with 20% methanol (e.g., for 1 L, 50 mL of 20× transfer buffer/750 mL of water/200 mL methanol and add 0.25 g of the antioxidant sodium bisulfite). The pH of this solution is 7.2. pH adjustment is not necessary.
3. Whatman chromatography papers (17 CHR, Whatman).
4. Mini Trans-Blot Electrophoretic Transfer Cell (Bio-Rad).

3. Methods

Carry out all procedures at room temperature unless otherwise specified.

3.1. Casting the Gel

1. We use the gel cassette of Mini-PROTEAN II to cast the gel (see Fig. 1).
2. Select and prepare the appropriate gradient solutions to separate the proteins of interest. The volumes of reagents required for a mini 3–15% polyacrylamide gradient gel (dimensions: 8 cm×8 cm; spacers: 1.5 mm) are indicated in Table 1 (see Note 1).
3. Mix solutions before pouring into the chambers of the gradient maker. Add 4 mL of the lower polyacrylamide percentage solution (3%) into the non-outlet chamber (reservoir) (left chamber of gradient maker in Fig. 1a) and 4 mL of the higher polyacrylamide percentage solution (15%) into the outlet chamber that must be equipped with a stir bar (right chamber of gradient maker in Fig. 1a). The valves are closed at this moment. Keep the remaining volume (3 mL) of the lower percentage solution to add later (see Note 2).
4. Start pouring the mixture (or mixed solution) into the glass sandwich. Open the outlet valve. Let the higher percentage solution arrive at the bottom of the glass sandwich and open immediately the other valve. This way, solutions are mixed and form the gradient progressively while pouring. The flow must be constant. Fast flow will cause turbulence that disrupts the gradient, and slow flow rates will cause polymerization of the acrylamide before completing the pouring.

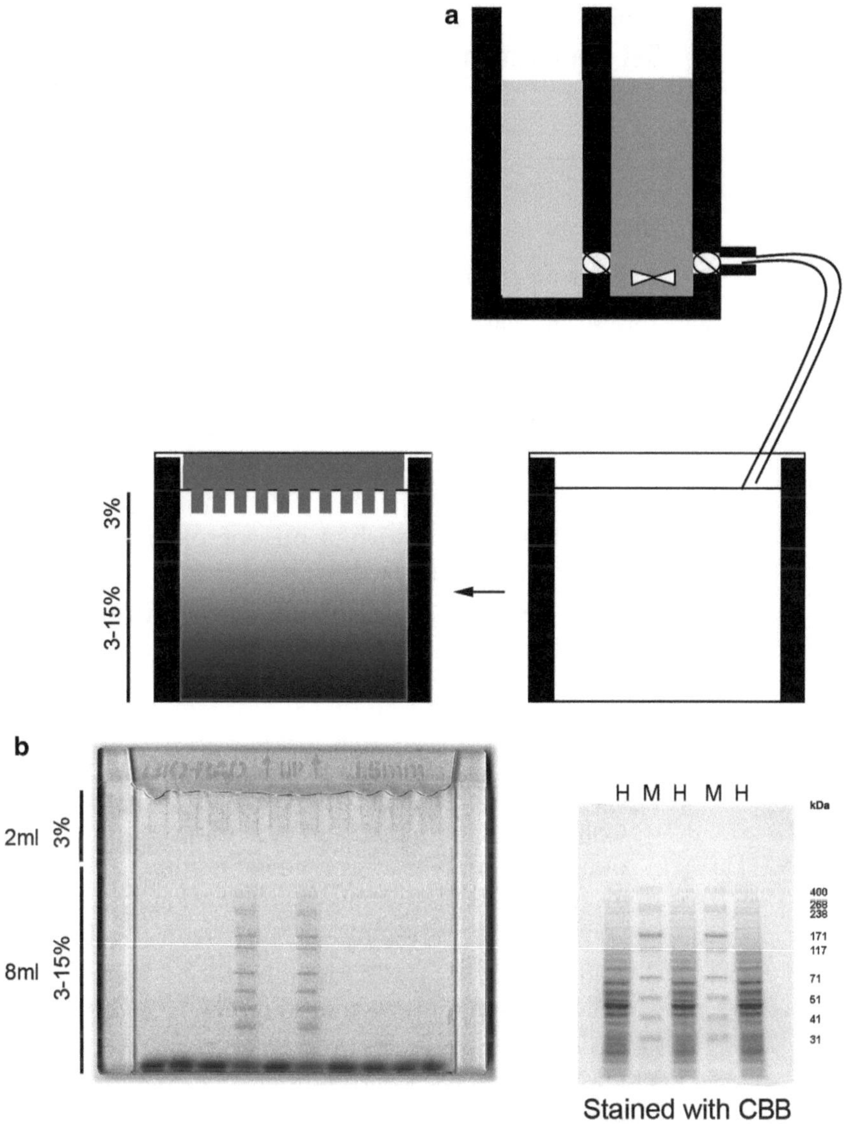

Fig. 1. The Tris–acetate polyacrylamide gel. (**a**) Schematic representation of how to make a 3–15% gradient gel with a gradient maker. The 3 and 15% polyacrylamide solutions [from a 40% acrylamide (37.5:1 acrylamide/bisacrylamide mix) stock solution] are poured in *left* and *right* compartments, respectively, of the gradient maker at the same time. Immediately after the gradient is made, a 3% solution is added before letting the gradient gel solidify between glass plates from a slab apparatus. (**b**) Example of Tris–acetate polyacrylamide gel. Gel was made in glass plates from Bio-Rad (Mini-PROTEAN II system) with 1.5 mm spacers. Volumes used are indicated (left side). Pre-stained protein standards (31, 41, 51, 71, 117, 171, 238, 268, and 460 kDa from Invitrogen) and 50 μg of protein from a lysate of HEK 293 cells were separated in a 3–15% polyacrylamide gradient gel (*left side*) and visualized by CBB staining (*right side*). Standards and lysates are indicated as M and H, respectively. Reproduced from ref. 4 with permission from Elsevier.

5. When solutions inside the chambers are about to consume, add 3 mL of the lower percentage solution into the reservoir chamber and wait until the total pouring of the sandwich is completed. Put the comb. Recover the volume that did not

Table 1
Gel composition to make a 3–15% gradient gel

	3%	15%
3 M Tris–acetate pH 7.0 (15×) (mL)	0.47	0.27
40% Acrylamide solution (mL)	0.53	1.5
Milli-Q water (mL)	6	2.23
Total volume (mL)	7	4
TEMED (μL)	8.75	5
APS 10% (μL)	33.25	19

enter into the sandwich (1 mL approximately) to know when the gel has polymerized.

6. Wash the gradient maker immediately with abundant water to prevent acrylamide polymerization.
7. Leave the gel to polymerize (see Note 3).

3.2. Sample Preparation and Protein Loading

1. Lyse the samples (see Note 4).
2. Centrifuge the samples (e.g., 15,000×*g* for 15 min at 4°C). Collect the supernatant and determine the protein concentration by the BCA assay or a similar method. Add sample buffer 4× without DTT to a final concentration 1×. At this moment, you can use these samples or store them at −20°C.
3. Add DTT (final concentration: 100 mM, e.g., 5 μL of 1 M DTT in 50 μL of total volume) when protein samples are used in PAGE (see Note 5).
4. Heat the samples for 10 min at 70°C and load them into the wells of the gel (see Note 6).

3.3. Electrophoresis Conditions

1. Run at $\Delta V = 130$ V ($I = 115$ mA). The run time estimated for a 3–15% gradient gel is approximately 1 h 15 min. As shown in Fig. 1b and Fig. 2, molecular weight markers ranging from 10 to 460 KDa were separated. These gels can be used for CBB staining (Fig. 1b, right) or to perform immunoblotting (Fig. 2).

3.4. Transfer Conditions

1. Submerge the fiber pads and the Whatman paper in transfer buffer. Place the gel in transfer buffer using a separate tray.
2. Activate the PVDF membrane in methanol for 2 min, wash with deionized water, and submerge in transfer buffer.
3. When the phenol red dye reaches the end of the gel, turn the power supply off. Remove the cast (glasses with the gel).

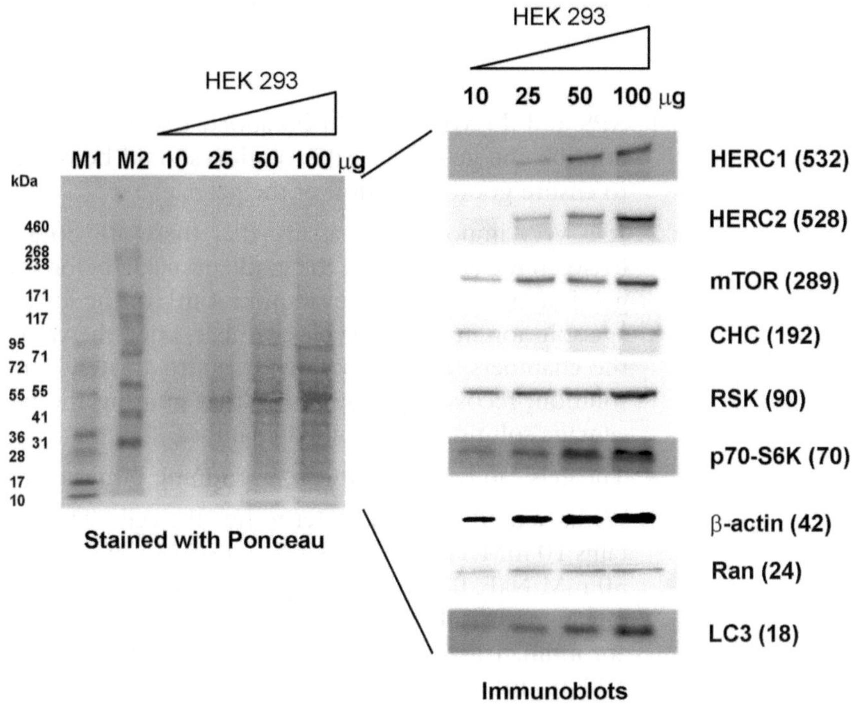

Fig. 2. Applications of the Tris–acetate polyacrylamide gel: analysis by staining and immunoblotting of proteins in the mass range of 10–500 kDa. Lysates from HEK 293 cells were loaded, and then they were run in a 3–15% polyacrylamide gradient gel, stained with S-Ponceau (*left side*), transferred to a PVDF membrane, and analyzed by immunoblot (*right side*) with specific antibodies against the following proteins: HERC1 (532 kDa), HERC2 (528 kDa), mTOR (289 kDa), CHC (192 kDa), RSK (90 kDa), p70-S6K (70 kDa), β-actin (42 kDa), Ran (24 kDa), and LC3 (18 kDa). To check the system sensitivity, 10–100 μg of protein was analyzed. Two different pre-stained protein standards, M1 (10, 17, 28, 36, 55, 72, and 95 kDa from Fermentas) and M2 (31, 41, 51, 71, 117, 171, 238, 268, and 460 kDa from Invitrogen), were used (*left side*). Reproduced from ref. 4 with permission from Elsevier.

Rinse the cast carefully with deionized water and equilibrate it in transfer buffer for few minutes.

4. Separate the thinner glass of the gel carefully with the help of a spatula. Cover the gel with Whatman paper and separate both the gel and the paper from the cast with the help of tweezers or a spatula (see Note 7).
5. Assemble the sandwich as follows (from bottom to top, in the transfer direction): fiber pad, Whatman paper, Whatman paper with the gel, PVDF membrane, two Whatman papers, and fiber pad.
6. Slide the assembled transfer cassette into Mini Trans-Blot Electrophoretic Transfer Cell. Add the transfer buffer and connect the power supply.
7. Perform the transfer for 2 h ($I = 200$ mA) or overnight ($\Delta V = 20$ V).

4. Notes

1. APS and TEMED should be added last, immediately before pouring the gels. The APS solution should be prepared fresh to ensure good resolution of the gel.
2. It is very important to assure that there are no air bubbles between the chambers of the gradient maker before casting the gel. For air bubbles removal, pour 4 mL of the lower percentage solution in the reservoir chamber, open the valve between the chambers, and allow flowing approximately 1 mL of the solution, recovering it and after that pouring the higher percentage solution in the outlet chamber.
3. The gels can be stored at 4°C in running buffer up to 3 days.
4. We have used CHAPS buffer to lyse the cells. This buffer contains 10 mM Tris–HCl pH 7.5, 100 mM NaCl, 0.3% CHAPS, 50 mM NaF, 0.5 mM EDTA, 1 mM sodium vanadate, 1 mM phenylmethylsulfonyl fluoride, 5 μg/mL leupeptin, 5 μg/mL aprotinin, 1 μg/mL pepstatin A, 50 mM β-glycerophosphate, 100 μg/mL benzamidine, and 1 μM E-64. After rocking during 20 min at 4°C, lysates are centrifuged at 15,000 × *g* for 15 min at 4°C. Pellets are discarded and supernatants analyzed.
5. Freeze samples in sample buffer 1×. Do not freeze the samples after thawing. Add fresh DTT just before heating the samples.
6. To eliminate traces of non-polymerized acrylamide, rinse wells with running buffer before applying samples.
7. Because the bottom of the gel is more dense, it is important to start separating the glass from the bottom. Extracting the gel from glasses without the paper implies the risk of distortion of the upper part of the gel; this part is very viscous and must be handled carefully.

Acknowledgments

This study was supported by grants from Spain (BFU2008-02084 and BFU2011-22498), European Union (FEDER), ISCIII (RETICRD06/0020), and Generalitat de Catalunya (2009SGR1059). M.C.R, F. A. P., and I. T. are supported by a doctoral fellowship from IDIBELL, Generalitat de Catalunya, and by the Juan de la Cierva program, respectively.

References

1. Maizel JV (2000) SDS polyacrylamide gel electrophoresis. TIBS 25:590–592
2. Schägger H (2006) Tricine-SDS-PAGE. Nat Protoc 1:16–22
3. Casas-Terradellas E, Garcia-Gonzalo FR, Hadjebi O et al (2006) Simultaneous electrophoretic analysis of proteins of very high and low molecular weights using low-percentage acrylamide gel and a gradient SDS-PAGE gel. Electrophoresis 27:3935–3938
4. Cubillos-Rojas M, Amair-Pinedo F, Tato I et al (2010) Simultaneous electrophoretic analysis of proteins of very high and low molecular mass using Tris-acetate polyacrylamide gels. Electrophoresis 31:1318–1321
5. Hachmann JP, Amshey JW (2005) Models of protein modification in Tris-glycine and neutral pH Bis-Tris gels during electrophoresis: effect of gel pH. Anal Biochem 342:237–245
6. Schägger H, von Jagow G (1987) Tricine-sodium dodecyl sulfate-polyacrylamide gel electrophoresis for the separation of proteins in the range from 1 to 100 kDa. Anal Biochem 166: 368–379
7. Wiltfang J, Arold N, Neuhoff V (1991) A new multiphasic buffer system for sodium dodecyl sulfate-polyacrylamide gel electrophoresis of proteins and peptides with molecular masses 100,000–1000, and their detection with picomolar sensitivity. Electrophoresis 12:352–366
8. Kubo K (1995) Effect of incubation of solutions of proteins containing dodecyl sulfate on the cleavage of peptide bonds by boiling. Anal Biochem 225:351–353

Chapter 18

SDS-PAGE for ^{35}S Immunoprecipitation and Immunoprecipitation Western Blotting

Edward P. Trieu and Ira N. Targoff

Abstract

This report discusses recent methods of sample preparation and gel electrophoresis for ^{35}S immunoprecipitation (IP) and IP western blotting. In both methods, IP is used to obtain purified proteins, and sodium dodecyl sulfate polyacrylamide gel electrophoresis (SDS-PAGE) is used to separate the proteins on a gel. In ^{35}S IP, the proteins are radiolabeled and visualized on film by fluorography; in IP blotting, proteins are transferred onto nitrocellulose paper, and antibodies are used to detect specific proteins. A similar IP and SDS-PAGE method can be used for both procedures, but IP blotting has the potential advantages of improvement in sensitivity for low-abundance proteins and enhanced specificity for identification of proteins from a mixture. Some of the technical adaptations discussed here to facilitate IP blotting and avoid loss of beads or purified proteins may also be useful for ^{35}S IP.

Key words: Immunoprecipitation, S-35 immunoprecipitation, Immunoblot (western blot), Autoantibody, Autoantigen, Dimethyl pimelimidate, Isotachophoresis, SDS-PAGE

1. Introduction

Sodium dodecyl sulfate polyacrylamide gel electrophoresis (SDS-PAGE) of proteins is routinely used in modern laboratories to separate mixtures of proteins into individual bands before applying other methods such as fluorography and western blotting for analysis of the proteins. In 1964, Ornstein designed the polyacrylamide gel with a discontinuous buffer system (1) based on the 1897 isotachophoresis theory of Kohlrausch (2). When he observed proteins as a blue band stacked into a thin band from an isoelectric focusing gel electrophoresis, he knew that pH played an important role in the stacking mechanism and explored the theory of Kohlrausch. Laemmli improved on this system in 1970 by using

Biji T. Kurien and R. Hal Scofield (eds.), *Protein Electrophoresis: Methods and Protocols*, Methods in Molecular Biology, vol. 869, DOI 10.1007/978-1-61779-821-4_18, © Springer Science+Business Media, LLC 2012

SDS and adjusted pH (3). This system of SDS-PAGE continues to be widely used to fractionate different kinds of proteins and analyze different kinds of protein mixtures. For example, SDS-PAGE is used for determining the molecular weight of proteins and for the analysis of proteins isolated by immunoprecipitation from cell extracts. For ^{35}S IP, the cells are labeled by growing in appropriate media with reduced amount of methionine, after addition of (^{35}S)-methionine. The cells incorporate (^{35}S)-methionine to produce ^{35}S-labeled proteins and are used as antigen source for IP. ^{35}S IP sensitivity can be adjusted by changing the amount of methionine used. Beginning with the labeling method of Pollard in 1994, methionine was used at 1 mg/L in the labeling medium (4). However, it was found that this level did not provide strong enough labeling to detect some of the proteins that were present in the cell in low copy numbers. Therefore, the level of cold methionine in the labeling medium was reduced to 0.5 mg/L (see Fig. 1). This decrease in cold methionine resulted in an increase in the signal of ^{35}S in the gel. However, later, the sensitivity level was raised to a higher level by using mainly (^{35}S)-methionine in the labeling medium for pulse chase labeling of the proteins (5).

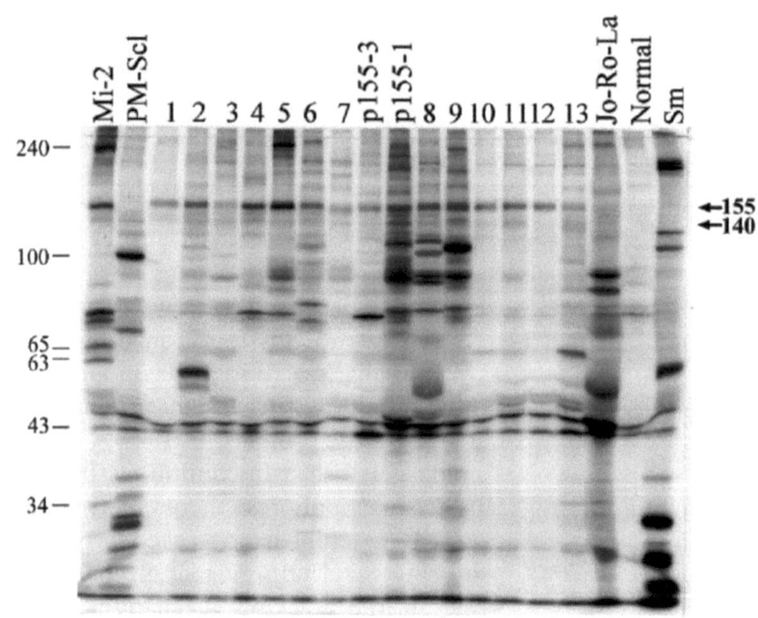

Fig. 1. SDS-PAGE fluorography of ^{35}S IP with 0.5 mg/L of methionine in the labeling medium. 13 anti-p155/140 autoantibody-positive sera (*lanes* 1–13) were screened with an IP of ^{35}S-labeled cell extract. These sera precipitated a similar protein at 155 KDa with several sera showing additional autoantibodies. In the other lanes, control sera with the well-known autoantibodies Mi-2, PM-Scl, Jo-Ro-La, and Sm were used.

SDS-PAGE has been used in clinical testing as part of ^{35}S IP and IP blotting for identification of autoantibodies in patient sera. (6) For example, ^{35}S IP has long been used to detect and identify autoantibodies of significance for autoimmune myositis (7). ^{35}S IP more recently revealed additional autoantibodies, anti-p155/140 and anti-MJ-140, that were more difficult to conclusively identify by simple SDS-PAGE of immunoprecipitates. IP blotting was employed to confirm results when these antibodies were suspected after ^{35}S IP testing (Fig. 2). Initially, the prototype serum was used for IP, followed by separation by SDS-PAGE and transfer, after which the test sera were used for blotting to confirm identity with prototype serum. However, some test sera with ^{35}S IP results suggesting anti-p155/140 were negative for any bands by IP blot, even when the test serum itself was used for both the IP and blotting steps. This suggested exclusive reaction with conformational epitopes. When such IP-positive, blot-negative sera were tested in a reverse manner, using test serum for the IP step and prototype serum for the blotting step, the expected bands were detected. (6, 7) This method has also been successfully applied to detection of other autoantibodies.

The overall procedure has four sections: (1) samples preparation, (2) gel preparation to separate proteins from the samples, (3) gel electrophoresis of protein samples, and (4) fluorography and western blotting.

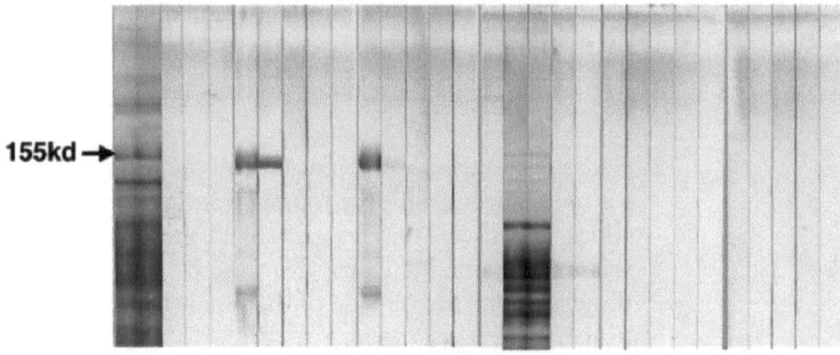

Fig. 2. IP blotting after screening with ^{35}S IP. Six immunoprecipitates shown were prepared using the test sera (horizonatal labels). They were separated by SDS-PAGE and transferred to nitrocellulose. When blotting, these test sera were the reference sera. The three anti-p155/140-positive sera, p155-1, p155-2, and p155-3, immunoprecipitated the p155/140 native antigen from the cell extract and the reference serum, p155-1, detected it on the blot as a denatured protein as seen on every first strip of p155-1, p155-2, and p155-3. Reference serum p155-2 could recognize the denatured form of antigen but p155-3 could not. PM-Scl, NL-1, and NL-2 were the controls and were all negative for p155/140 antigen (reproduced from ref. 6 with permission from John Wiley & Sons).

2. Materials

2.1. Sample Preparation

2.1.1. ^{35}S Immunoprecipitation

1. Protein A-Sepharose-CL-4B (GE Healthcare, Sweden).
2. Immunoprecipitation (IP) buffer: 10 mM Tris–HCl, pH 8.0, 0.5 M NaCl, 0.1% NP-40 (Thermo Scientific, Rockford, IL, USA) or Igepal CA-630 (Sigma-Aldrich, St. Louis, MO, USA). Store at 4°C.
3. Human or animal sera. Remember to use exposure control plan to protect both personnel and environment from blood-borne pathogens.
4. Bovine serum albumin (BSA) blocking solution: 4% BSA in 10 mM Tris–HCl, pH 8.0, 0.5 M NaCl, 0.01% NP-40 or Igepal CA-630. Store at −70°C in a small aliquot of 5–10 mL. A small amount of detergent in the solution aids in even coating of solution inside the tube for blocking.
5. HeLa or K562 cell lines obtained from the American Type Culture Collection.
6. Dialyzed fetal bovine serum (DFBS) (Sigma): Store in small aliquots, 5–10 mL at −70°C not more than 6 months.
7. L-Glutamine–penicillin–streptomycin solution, 100× (Sigma). Store at −20°C.
8. Complete medium: MEM Eagle with Earl's Salts (Sigma-Aldrich) supplemented with 10% bovine calf serum, glutamine, and penicillin/streptomycin. Store at 4°C.
9. Labeling medium: MEM Eagle (Modified) with Earl's Salts without L-Glutamine and L-Methionine (MP Biomedicals LLC, Solon, OH, USA). Store at 4°C.
10. Complete labeling medium: above medium with 0.5 mg/L of methionine.
11. Methionine, cell-labeling grade, L-(^{35}S) (Perkin Elmer, #NEG009L005MC).
12. Radiolabeling box: A box used to contain small flasks of cells for radiolabeling with the top open and was made up of 1/2-in. -thick Plexiglas. The Plexiglas shield the other cell culture that was in the same incubator from the β rays radiation. A typical dimension for the outside of the box is 6 × 9 × 5½ in.
13. Cell culture flasks, 25 cm^2.
14. L-Methionine (Sigma): 1 mg/mL of methionine; sterile filter and store at 4°C.
15. Activated carbon-loaded paper, size 8 × 10 in. (Whatman). Use to absorb the ^{35}S waste product from cell metabolism.

16. Lysis buffer for ^{35}S-IP: 1% NP-40 or Igepal CA-630 in IP buffer supplemented with 100× protease inhibitor cocktail (Sigma). Prepare immediately before use.
17. Whatman Vacu-Guard Filter (Fisher Scientific, Dallas, TX, USA). Put on the vacuum line to avoid contamination of ^{35}S to the vacuum system.
18. Transfer pipette, capillary tip.
19. SDS sample buffer, 1×: 10 mM Tris–HCl, pH 6.8, 10% glycerol, 2% SDS, 0.0025% bromophenol blue, 1% mercaptoethanol. Store at −20°C in 5–10 mL aliquots.
20. Hamilton syringe, 50 μL (Fisher Scientific).

2.1.2. Immunoprecipitation Blotting

1. Micro Bio-Spin Chromatography Columns, caps, and tip enclosures (Bio-Rad, Hercules, CA, USA) or 96-well filter plates with silicone mat closure that seal both the top of the plate and the bottom drain ports (Arctic White LLC, Bethlehem, PA, USA). The alternative 96-well filter plate is Nunc 96-well Filter Plates with PET frit (Fisher Scientific).
2. Costar, V-shaped well untreated, clear polystyrene 96-well plate with silicone plate seal (Fisher Scientific).
3. Antibody-binding buffer (ABB): 0.2 M triethanolamine (Sigma), pH 8.9, 0.5 M NaCl. Store at 4°C.
4. Cross-linking solution: 30 mg/mL of dimethyl pimelimidate (DMP; Sigma) in ABB. Prepare immediately before use. Store DMP at 4°C.
5. Quenching buffer: 0.1 M Tris–HCl, pH 8.5, 0.5 M NaCl. Store at −20°C.
6. Unbound antibody elution solution: 1.0 M acetic acid, 0.15 M NaCl. Store at room temperature (RT).
7. Lysis buffer for IP blotting: 0.3% NP-40 or Igepal CA-630 in IP buffer supplemented with 100× protease inhibitor cocktail.
8. Micro incubator shaker: Model E36 Micro Incubator Shaker (Bionexus Inc., Oakland, CA, USA) holds two 96-well plates, or titer plate shaker (LabLine Instruments, Inc, Melrose Park, ILL, USA) holds four 96-well plates.
9. Sonicator (Ultra Sonic Processor).
10. SDS sample buffer, 2×: 20 mM Tris–HCl, pH 6.8, 20% glycerol, 4% SDS, 0.0025% bromophenol blue, 2% mercaptoethanol. Store at −20°C in 5–10 mL aliquots.
11. Protein-free (PBS) blocking buffer (Thermo Scientific). Add 0.05% Tween-20 for the dilution of the sera.

2.2. Gel Preparation

1. Protean II xi gel apparatus from Bio-Rad, 16-cm glass plates, 25-well combs, and spacers for each of 0.75 mm and 1.5 mm gels.

2. Polyacrylamide solution, 30%: Polyacrylamide is not stable once in solution. Prepare immediately or a few days before use. Store at 4°C in a dark bottle or cover with aluminum foil. Polyacrylamide is a neurotoxin when it is a monomer. Therefore, polymerize excess solution before disposal.
3. Resolving buffer, 4×: 1.5 M Tris–HCl, pH 8.8. Store in 50 mL aliquots at −20°C.
4. Stacking buffer, 4×: 0.5 M Tris–HCl, pH 6.8. Store in 50 mL aliquots at −20°C.
5. Water-saturated butanol: Add 10 mL of water into a 50-mL tube and add 40 mL of butanol. Cap the tube, shake well, and let it sit to separate; use the water-saturated butanol in the top layer. Store at RT.
6. Ammonium persulfate solution, 10%: Store at −20°C in 2 mL aliquots. Reuse the remaining solution if refrozen at −20°C.
7. TEMED.

2.3. Gel Electrophoresis of Protein Samples

1. A small power supply with a current capacity of 500 mA or higher.
2. Refrigerated circulator (Fisher Scientific, Dallas, TX, USA).
3. Running buffer, 5×: 125 mM Tris–HCl (pH 8.6), 960 mM glycine, 0.5% SDS. Store at RT.
4. Molecular weight markers for western blot gel: Fisher BioReagents EZ-Run Rec Protein Ladder (Fisher Scientific).
5. Prestained molecular weight markers for ^{35}S-gel: Precision Plus Protein all Blue Standard (Bio-Rad).

2.4. Fluorography and Western Blotting

2.4.1. Fluorography

1. Gel dryer with vacuum: Savant SG210D SpeedGel (Thermo Savant, Holbrook, NY, USA) (Fisher Scientific).
2. X-ray film: Hyperfilm MP or ECL, size 8×10 in. (GE Healthcare Bioscience Corp., Piscataway, NJ, USA), (VWR, West Chester, PA, USA).
3. X-ray cassette for fluorography, size 10×12 in. or 14×17 in. (Fisher Scientific).
4. X-ray developing machine and a dark room.
5. Whatman filter paper, grade 3, size 46×57 cm (Fisher Scientific).
6. 0.5 M sodium salicylate: Sodium salicylate is light sensitive when in solution. Store the solution in a dark bottle at RT. Over time, the solution turns dark yellow and blue and should be discarded.
7. A roller or a plastic spreader: Plastic spreader is a piece of flexible plastic for spreading soft things such as glue onto a surface.

2.4.2. Western Blotting

1. Nitrocellulose paper (Bio-Rad).
2. Trans-Blot cell (Bio-Rad).
3. Protein transfer buffer: 25 mM Tris–HCl, 192 mM glycine, 20% methanol, 20 mM sodium hydroxide, and 0.005% SDS (pH 8.3). Store at 4°C.
4. Rocking shaker: Labquake Shaker (Spectrum Chemicals and Laboratory Products, New Brunswick, NJ, USA). When shaking, the diluted serum solution becomes a wave over the membrane and thus can use less volume for the incubation.

3. Methods

3.1. Sample Preparation

First, the antibody of interest from animal or human serum binds to the protein A-Sepharose beads during incubation of serum with protein A-Sepharose beads. The wash buffer removes unbound antibodies and other serum proteins. Then, the antibodies on the protein A-Sepharose bind their specific antigen during incubation with a homogenate or cell extract. The immunoprecipitate forms on the Sepharose beads from the reaction of the antibody with the antigen. Then, the wash buffer is used to wash away unbound proteins from the beads. Following the wash, SDS is used to separate the antigen from the beads. At this step, for the ^{35}S IP discussed here, SDS sample buffer releases both the antigen and the antibody from the beads; for the IP-blot procedure, SDS sample buffer releases only the antigen from the beads because the antibody is cross-linked to protein A-Sepharose beads.

3.1.1. ^{35}S Immunoprecipitation

1. *Prepare antibody–protein A-Sepharose:* Label on the cap of 0.5-mL microcentrifuge tubes a number for each serum sample to be used, and add into each tube the following: 400 μL of IP buffer, 10 μL of protein A-Sepharose CL-4B 1:1 with IP buffer, and 10–15 μL of serum.
2. Cap and put tubes into a plastic freezer box that has dividers for 50 tubes to keep the tubes organized and in the same orientation. Let the tubes agitate on the rotator for 2 h to overnight at 4°C.
3. Centrifuge briefly (approximately 30 s) at 5,000 × *g* and aspirate the buffer to above the bead level about 10 μL. Add 400 μL of blocking solution into each tube and place on rotator overnight at 4°C. BSA binding to the beads is necessary to reduce the background from ^{35}S-labeled proteins.
4. Wash the beads thrice with IP buffer and store the beads in a refrigerator until the ^{35}S-labeled protein is ready.

5. *Labeling of HeLa cells for ^{35}S immunoprecipitation*: Grow HeLa cells to near confluence (with a little space between the cells) in a 25-cm^2 flask with complete medium (see Note 1).
6. To prepare 50 mL of complete labeling medium with 0.5 mg/L of methionine, 10% DFBS, and penicillin/streptomycin in labeling medium, add into a 50-mL tube the following: 44.5 mL of labeling medium, 25 μL of 1 mg/mL of methionine solution, 500 μL of L-glutamine–penicillin–streptomycin solution, and 5 mL of dialyzed fetal calf serum. Warm up to 37°C just before use (see Note 2).
7. Wash the flask of cells thrice with 2–2.5 mL of complete labeling medium. The washes remove the complete medium that has 15 mg/L of methionine.
8. Now cells are ready for labeling. Each flask is usually labeled at the "medium" sensitivity level as described in Table 1, and that is assumed in the steps below (see Note 3).
9. After pipetting out the remaining wash medium, add into the flask 1.7 mL of the complete labeling medium and 500 μCi of (^{35}S)-methionine with a filtered pipette tip. Place the flask into the radiolabeling box and cover with activated carbon-loaded paper and place the box into a cell culture incubator (37°C, 5% regulated CO_2 level and humidified).
10. Culture the cells with (^{35}S)-methionine overnight for 16–20 h.
11. *Prepare ^{35}S-labeled cell extract:* In the morning, remove the box containing the (^{35}S)-methionine flask of cells from the incubator and place on an absorbent paper. Remove the used ^{35}S-medium from the flask with a capillary tip transfer pipette or with an aspirator connected to a vacuum flask for collection of all ^{35}S radioisotope waste.
12. Wash the cells thrice with 2–2.5 mL of ice-cold PBS per flask.

Table 1
Sensitivity Levels for ^{35}S-labeling

^{35}S-labeling level	Weak	Medium	Strong	Unit in
Cold methionine	1	0.5	0	mg/L
Volume of media use per flask of cells	2.5	1.7	1.3	mL
^{35}S-Methionine, 0.5 mCi/flask	25–30	25–30	25–30	μL
Specific activity of ^{35}S per mL	200	294	385	μCi/mL
Artifacts or background	Weak	Low	High	–

13. After removing the wash buffer, use the cells immediately or store for a few days at –70°C, but not longer than 7 days because ^{35}S decays, changes its property, and comes off from the protein.
14. Add into the flask 0.5 mL of ice-cold lysis buffer. Turn the flask up and down on each side to distribute the lysis buffer and wait about 5 min or until the cells come off the flask surface. The detergent from the lysis buffer dissolves the cell membrane and the high salt in the buffer disrupts the chromatin structure. Therefore, there is no viscosity from the DNA.
15. Transfer 0.5 mL of ^{35}S-labeled cell extract into a 2.0-mL centrifuge tube with a capillary tip transfer pipette. Rinse the flask surface for the remaining cell extract and cell debris with another 500 μL of lysis buffer. Combine the lysis buffer wash with the cell extract in the 2-mL centrifuge tube. Use one tube of 2.0 mL per flask of cells. Cap the tubes and vortex to release more antigens from the cell debris.
16. Centrifuge for 5 min at 12,000 × *g* at RT or preferably at 4°C.
17. Transfer the ^{35}S-labeled cell extract from each tube with 1-mL filtered pipette tip into a 15-mL conical tube and dispose of the cell debris pellet. Bring the volume of ^{35}S-labeled cell extract from 1 to 4 mL with 2 mL of BSA blocking solution and 1 mL of IP buffer. Each flask of cells will thus provide enough labeled cell extract for ten samples (400 μL of ^{35}S-labeled cell extract per sample). Be sure to measure the volume of the cell extract with a little bit over so that all tubes will get the same volume of cell extract.
18. Cap the conical tube and invert to mix. Keep the ^{35}S-labeled cell extract on ice.
19. *Immunoprecipitation*: Add 400 μL of ^{35}S-labeled cell extract into each tube with antibody–protein A-Sepharose (see Subheading 3.1.1, steps 1–4), cap, and place on the rotator to rotate for 2.5–3 h at 4°C.
20. Centrifuge the tubes to bring the beads down to bottom of the tube. Aspirate the ^{35}S-labeled cell extract into a flask for ^{35}S radioisotope waste container and wash the beads 4× with 500 μL of IPP buffer. Discard the ^{35}S-labeled cell extract and the wash buffer from the flask into a jug for ^{35}S radioisotope waste and log the usage of radioisotope.
21. Wash 1× with 50 mM Tris–HCl, pH 6.8 (1/10 dilution from 4× stacking buffer).
22. Aspirate the Tris–HCl buffer to gel level with a 200-μL pipette tip or with a Hamilton syringe. Add 30 μL of 1× SDS sample buffer into each tube, cap, and vortex. Store the samples at –70°C (see Note 4 for troubleshooting).

3.1.2. Immuno-precipitation Blotting

1. *Prepare the antibody cross-links to protein A-Sepharose*: Add into each well of a 96-well plate with v-shaped well bottom, 20 μL of 1:1 protein A-Sepharose CL-4B beads to ABB, and 20 μL of undiluted serum and 150 μL of ABB. Put a silicone-sealing mat on the 96-well plate to seal the wells. Place the plate on a rotator and let it rotate on the rotator for 2 h to overnight at 4°C. Do not use the Micro Bio-Spin columns in this step because the porous support void volume isolates some of the antibody and protein A-Sepharose from the binding.
2. The next day, use a 200-μL wide-bore pipette tip for each well; transfer the beads from each well of the 96-well plate to each Micro Bio-Spin column. Place the columns on vacuum manifold or tall tubes to collect the flow-through (7).
3. Rinse the wells thrice with 200 μL of ABB to collect the remaining beads.
4. Wash the beads thrice on the Micro Bio-Spin columns with 300 μL of ABB.
5. Plug the bottom of the columns with tip enclosures and push each column onto a well of a deep-well microplate. Micro Bio-Spin columns fit well onto a 96 × 2-mL/square well plate and the micro incubator shaker holds the deep-well microplate for the shaking. Add into each Micro Bio-Spin column 100 μL of cross-link solution. Place on the micro incubator shaker for 15 min at low speed. Do not place the columns on a rotator unless the columns are full with liquid because the beads will stick onto the side of the columns or onto the caps and may be lost.
6. When time is up after 15 min, immediately add 200 μL of quenching solution into the columns and continue to shake for 10 min. The quenching solution neutralizes the remaining binding sites while the beads are still in the suspension to avoid cross-linking between the beads when they settle down to the bottom of the column.
7. Unplug the bottom of the columns and place the columns on vacuum manifold or tall tubes. Wash twice with 300 μL of ABB.
8. Add 200 μL of unbound antibody elution solution into the columns and immediately wash the beads thrice with 300 μL of ABB.
9. Plug the bottom of the columns, add about 200 μL of IP buffer, and put on the cap cover. Keep the columns in the refrigerator overnight or over the weekend.
10. *Prepare K562 cell extract*: Plan ahead for the amount of cell extract to be used. For each column with 10 μL of beads, 100–150 μL of cell extract is needed. K562 cell extract can be prepared in a large

amount for several columns and extra extract can be frozen at –70°C with 10% glycerol for later use.

11. Add 10 μL of 100× protease inhibitor cocktail to a 15-mL conical tube containing 1 mL of K562 frozen cell pellet and let the cells thaw in a beaker containing tap water at RT. Vortexing the conical tube a few times will help to thaw the cells quicker.
12. Add 5 mL of lysis buffer for IP blotting containing 0.3% NP-40 and 1× protease inhibitor cocktail into the conical tube with thawed K562 cell pellet. Cap the tube and mix.
13. Place the conical tube in a beaker containing ice and water and sonicate the cells (power of 30 W, six times with 20 s sonication, and 10-s rest between each sonication).
14. Transfer the cell extract into a centrifuge tube or microcentrifuge tubes and centrifuge at 10,000×*g* for 20 min at 4°C. Pour the cell extract into a new tube and discard the cell debris pellet.
15. Take the columns out from the refrigerator and unplug each column to drain IPP buffer. Plug each column, add 100–150 μL of K562 cell extract, and put on caps. Push the columns onto the wells of 96-deep-well plate that is on the micro incubator shaker and incubate them with shaking for 3 h at 4°C.
16. When incubation is complete, take the columns out from the 96-deep-well plate. Unplug each column to drain the cell extract into a well of the 96-deep-well plate. Wash the beads eight to ten times with 300 μL of ice-cold IP buffer.
17. Wash each column twice with 300 μL of ice-cold 50 mM Tris–HCl, pH 6.8 (from 1/10 dilution of the stacking buffer).
18. Shake off the water from each column into the sink and hit lightly on the bottom of the column onto a piece of sponge or a piece of paper towel to absorb some of the excess water that may still remain inside the column (see Note 5).
19. When no more water issues out from the bottom and the outside of columns are dried, put each column into a 1.5-mL microcentrifuge tube.
20. Add 20 μL of 2× SDS sample buffer into each column. The porous support of the column and the beads could hold up to about 20–25 μL of water. Centrifuge briefly. Take the same sample buffer from each 1.5-mL microcentrifuge tube and put it back into the same column to completely elute the antigen from the column. Centrifuge again, about 30 s to 1 min at 5,000×*g*. Label each 1.5-mL tube with the antigen name and cap the tubes. There will be about 40–50 μL of sample in each tube.
21. Store the samples at –70°C until gel electrophoresis.

3.2. Gel Casting

The proteins from the samples are separated on a polyacrylamide gel using the discontinuous buffer system of Laemmli (3).

3.2.1. Prepare the Gel-Casting Chamber

1. Read manufacturer's manual for instructions to set up the gel apparatus. In short, to set up the gel-casting chamber for one gel with 16-cm glass plates for the Bio-Rad Protean II xi, place the following on a clean surface in this order: longer glass plate, two spacers on each side, and the 16-cm glass plate on top of the two spacers. For ^{35}S IP, use 0.75-mm spacers to avoid cracking of the ^{35}S-gel while drying, but for western blot IP, use 1.5-mm spacers.
2. While holding the glass plate–spacer sandwich upright on a flat surface such as a glass plate, pull two spacers both side outward a little, and then push the left side clamp to the left side of the glass plates. Do the same to the right side of the glass plates with right side clamp. Make sure both glass plates and the spacers are even inside of both clamps by pushing them together and the spacer ends are flat on the bottom of the glass plates by pushing the top of spacers down. Now, while holding one clamp steady, tie up the clamp of the right side of the glass plate and then the left side of the glass plates. Check the bottom of the two glass plates. If they are not smooth to touch, repeat the setup.
3. Attach the gel-casting chamber (glass plate–spacer sandwich) onto the base of the gel-casting stand and mark on the short glass plate 4–4.5 cm from the top of the short glass plate with a Sharpie marker for the resolving gel level.

3.2.2. SDS Polyacrylamide Gel Casting

1. The volume of gel solution for ^{35}S IP samples with Protean II xi gel apparatus is 17 mL while the volume of gel solution for IP-blotting samples is 34 mL.
2. Add into each 125-mL side arm vacuum flask the resolving gel solution and the stacking gel solution, according to Table 2, but leave out TEMED until the last step to polymerize the gel. Use double distilled water for all solutions.
3. Degas the resolving gel solution for at least five min at RT.
4. Add 24 μL of TEMED to the flask, swirl lightly to mix. While holding the gel-casting chamber at about a 45° angle to the bench top, pour the solution into the gel-casting chamber and repeat if there are two gels. Fill the solution to the mark of the resolving gel level.
5. Apply water-saturated butanol across the top and let the solution sit about 5–10 min to polymerize.
6. Prepare stacking gel solution and degas at RT for at least 1 min.

Table 2
Gel Solutions for 8.5% SDS-PAGE (8)

Gel solutions	Resolving		Stacking	
Final percent use	8.5%		3.9%	
Final volume use	34.00 mL		10.00 mL	
0.50 M Tris–HCL pH 6.8	0		2.50 mL	(0.125 M)
1.5 M Tris–HCL pH 8.8	8.50 mL	(0.375 M)	0	
30% Polyacrylamide	9.63 mL		1.30 mL	
Water	14.82 mL		6.00 mL	
5% SDS	0.68 mL	0.1% w/v	0.20 mL	0.1% w/v
10% AP	0.34 mL	1.0% w/v	0.05 mL	0.5% w/v
TEMED	24 μL	0.07% v/v	10 μL	0.1% v/v

7. When the resolving gel solution polymerizes, pour off the butanol from the top of the gel into the sink and use double distilled water to flush out the remaining butanol with a squirt bottle. Lay the gel-casting chamber on one side to let the water come out and blot dry the top of the glass plates with a paper towel. To remove the water inside of the upper gel chamber, lay the gel horizontally and insert into the upper gel chamber two pieces of Kimwipe folded over onto a piece of transparent film with the same width of the gel chamber. Do not let the paper touch the gel.
8. Add TEMED to the stacking gel solution, swirl to mix, and, while holding the gel-casting chamber at about 45° angle to the bench top, pour the stacking gel solution into the gel-casting chamber and insert the comb between the glass plates. Pull the comb up and push down hard to remove bubbles on the comb.
9. Place a piece of Saran wrap on top of the upper gel and place a metal clamp on top to reduce space between the comb and the glass plates. To avoid drying of the stacking gel, place a piece of wet sponge onto the casting stand and cover the whole gel with a plastic bag. Leave the gel overnight or over the weekend at RT.

3.3. Gel Electrophoresis of Protein Samples

3.3.1. Gel Electrophoresis

1. Pull the comb slowly out from the gel (adding water can help to pull the comb out). If the gel fingers are not straight, straighten them with a Hamilton syringe needle. Mark the bottom of each well with a Sharpie marker or use the water to adhere a template (with pre-marks of the bottom of the wells on a strip of transparent tape) to the glass plate.

2. Attach the gel with the clamps to the cooling unit of the gel apparatus to form the upper buffer reservoir. If running only one gel, block the other side of the cooling unit similarly with a piece of plastic to replace the glass plates.
3. Fill up the lower reservoir tank with 2–2.5 L of running buffer. Fill up the upper reservoir with 250–300 mL of running buffer. After the gel is assembled with the cooling unit on the bench top, wait a few minutes to check for any leak; if not, put the gel assembled with the cooling unit and with the running buffer on top into the lower tank buffer.
4. Take the samples out from the −70°C freezer along with the molecular weight markers. Thaw the samples with warm water.
5. To denature the samples, heat at 95°C for 5 min or microwave for 90 s (9).
6. Use a Hamilton syringe to load ^{35}S-IP samples and prestained molecular weight markers. For western blot IP, use 200 μL narrow tips. Wash the Hamilton syringe thrice with distilled water to get rid of remaining sample and the beads. The beads can clog up the syringe if the plunger is pulled out rapidly.
7. When the loading is finished, turn on the cooling unit, put on the lid of gel apparatus, and connect the leads from the lid to a power supply. Turn on the power supply and set the current for gel electrophoresis. For one gel (0.75 mm thickness) of ^{35}S IP, use 13 mA; for one gel (1.5 mm thickness) of western blot IP, use 26 mA. If running 0.75 and 1.5 mm gels together in a single gel apparatus or in separate gel apparatus, use 39 mA—a combined current for two gels.
8. Turn up the current when the blue tracking dye enters the resolving gel. For ^{35}S-gel, use 15 mA; for western blot IP gel, keep at 26 mA until the end.
9. Let the gel run until the blue dye front reaches near the bottom about 1–0.5 cm. It takes about 5½–6 h for ^{35}S-gel and about 6–6½ h for western blot gel.
10. Turn off the power supply and cooling unit and disassemble the gel apparatus. Take the gel with the cooling unit out from the lower tank and place it into a large tray. Release the gel from the cooling unit and remove the clamps from the glass plates. The gel is ready for the next step.

3.4. Fluorography and Western Blotting

When the gel electrophoresis is complete, remove the gel from the gel apparatus and the glass plates and process according to either ^{35}S IP or western blot IP. For ^{35}S IP, the gel is soaked in sodium salicylate, a fluor that absorbs the radiation from ^{35}S and emits light through the gel (10). Dry the gel in a gel dryer after it absorbs the sodium salicylate. The gel side of dried gel is exposed to X-ray film in a cassette to record the image of ^{35}S-labeled proteins in the gel. For western blot IP, the proteins from the gel are electrophoretically transferred onto a membrane according to western blot protocol (7).

3.4.1. Fluorography

1. Separate the glass plates with a gel wedge. Remove the stacking gel with a long blade.
2. Drop the gel into a shallow tray containing 100 mL of 0.5 M sodium salicylate and place on a shaker at low speed for 30 min.
3. Place a piece of saran wrap onto a large piece of plastic or cardboard. Put on gloves to avoid contact with sodium salicylate solution. Pull the gel up with fingers or a gel clip from Bio-Rad and place the gel on the saran wrap. Hold the gel steady with the fingers since it is slippery. However, the gel can be held more securely with a gel clip.
4. Cut a piece of Whatman filter paper # 3, about 1–2 in. larger than the gel all around. Spray 70% ethanol onto the filter paper with a squirt bottle to make it damp and wrinkles start to show up. Use a paper towel to remove the excess ethanol and the paper becomes flat. Place the damp filter paper on the gel.
5. Use a roller or a plastic spreader over the filter paper to drive the excess sodium salicylate solution out from the gel surface. Dried sodium salicylate on the gel surface quenches ^{35}S signal.
6. Place the porous support of the gel dryer on top and turn the whole assembly over. The saran wrap side should be on the top and the paper side is on the porous support of the gel dryer. If the saran wrap is on the bottom, the gel will not dry.
7. Remove the large piece of plastic or cardboard on top and position the gel with the porous support onto the gel dryer. Cover the gel with the silicone cover and turn on the power. Let the gel dryer run at 80°C with vacuum for 1–2 h. Do not take the gel out before it is dried because the gel can break immediately into pieces.
8. When the drying is done, take the gel out from the gel dryer. Trim the excess paper around the gel to fit into the film cassette. Use an autoradiography marking pen or a radioactive pen to mark the molecular weight markers on the gel or the orientation of the gel (see Note 6).
9. In the dark room, position a film inside the cassette (without intensifier screen) and place the gel side, with no saran wrap, on the film and close the cassette (11). If the cassette has an intensifier screen that cannot be removed, then place a piece of paper on the screen and use that side for the film (see Note 7). Place the cassette for a few days in a –70°C freezer.

3.4.2. Western Blotting

1. Set up protein transfer for western blotting. The procedure has been described previously (7). In brief, set up the protein transfer onto a nitrocellulose paper under water to avoid bubbles: fill into a tray the protein transfer buffer and put in a piece of Whatman filter paper #3 and a piece of nitrocellulose paper larger than the gel. Push them down to the bottom of the tray. Drop the gel into the tray and orient the gel to the middle of the nitrocellulose paper. Place a second filter paper into the water and align with the gel. Take the gel sandwich out from the tray and place on a glass plate with the nitrocellulose paper side on top. Roll out the buffer between the gel and the nitrocellulose paper and place it inside a cassette. Close the cassette and, with the nitrocellulose paper side toward the red lead, position the cassette inside the protein transfer tank. Fill up the tank with protein transfer buffer and transfer the proteins at 200 mA overnight at 4°C in a refrigerator.
2. The next day, take the membrane out from the tank, stain with Fast Green and air-dry for at least 15 min. Cut the membrane into strips, and block with 5% nonfat dry milk in PBS-T for 1–3 h at 37°C.
3. Antibody development with 1/100 diluted serum in PBS-T with 1% nonfat dry milk or with protein-free (PBS-T) blocking buffer (see Note 8) is on a rocking shaker overnight at 4°C. The membrane sizes can vary. Hence, place the membrane into a plastic container or a boat made out of aluminum foil, if there is no other way. If turning aluminum foil into a boat, it is a perfect tray for membrane strips because it can vary in size and uses less liquid; however, be aware that leaking may occur because of erosion.
4. Continue with conjugate incubation and development with substrate. After substrate development, rinse the membrane strips thrice with distilled water to remove the substrate and then stop the reaction completely with 80–85°C water. Hot water denatures the enzyme immediately and thus more effectively stops the development. Remove the water and pour into the tray 50–70% ethanol, enough to cover the membranes. Ethanol dehydrates the membrane so that the adhesive glue can bind to the wet membrane. Thus, while the membrane strips are still wet with ethanol, align them together with face downward on a piece of transparency. Secure the strips together by putting on the back of them with a piece of large size packing tape. Alternatively, apply spray glue on a transparency film and apply this to the back of the membrane strips. Air-dry and document the membrane.

4. Notes

1. Each 25-cm^2 flask of cells contains about two million cells when fully confluent. To prepare the flask with cells to use on the same day, use about 1.5–1.8 million cells to seed into the flask. It takes about 5 h for the cells to settle down and attach. HeLa cells attach to the flask surface to grow. Therefore, it is much easier for the washing of the cells. However, when cells in suspension are used, use half of the normal cell density to label the cells (4). Be sure the cells are healthy and alive before using for IP.
2. There is a practice of pre-gassing for ^{35}S-labeling medium. Depending on the depth of media, it takes from 1 to 4 h for the oxygen to diffuse into the media and reach the cells. The cells need the oxygen to grow. Thus, it would help them by preincubating some of the labeling medium in a flask in the CO_2 incubator at least a few hours before use.
3. If the ^{35}S signal for the protein is too weak, the "strong" labeling level can be used; however, remember that the background level would also increase. More often, the IP test fails because of insufficient protein. The low sensitivity level is commonly used to enhance the weak signal of proteins by reducing the background because when the background is high, it is really difficult to see the protein bands among the artifacts.
4. Problems may arise with ^{35}S IP because it is a very sensitive system. First, the whole gel may turn dark suddenly for no apparent reason. Four possible causes include the following: (a) The beads from a bad lot of protein A-Sepharose bind to ^{35}S-labeled proteins. (b) Denatured BSA in the blocking step binds to proteins nonspecifically. To test for this, remove the blocking step. (c) Cell debris is incompletely removed from the ^{35}S-labeled cell extract. (d) Low pH from the buffer denatures ^{35}S-labeled proteins which can bind nonspecifically to the beads. Second, some of the cell material cannot pass through the porous support of the column if using column for the IP. To overcome this problem from the washing step after IP, dilute the cell extract to tenfold and let the beads settle down to the bottom of the column for 20–30 min and then remove the buffer from the top. Aspirate the cell extract or wash buffer down to near the beads level and repeat the same for four washes. Third, signal may be too weak so that it takes a longer time than expected for the film exposure. This problem may be caused by methionine in the dialyzed FCS. Do not store the dialyzed FCS more than 6 months. Fourth, there may be protein overloading and loss of resolution in the area of immunoglobulin. The amount of IgG may be too high for other

proteins to migrate into it (12). Try a few samples with a reduced amount of protein A-Sepharose beads and run the gel at about 1/4 to 1/3 lower current.

5. Do not centrifuge the columns to remove the water inside the beads before elution of the purified protein because this water contains the purified protein.
6. To make a radioactive pen, take a Sharpie Permanent Marker with ultrafine point and pull the tip out and add 2–3 μL of (^{35}S)-methionine into the pen with 25 μL of ethanol and reinsert the pen tip. The pen will be ready to use in a few days.
7. Intensifying screens work well only for high energetic penetrative emission of ^{32}P by absorbing the energy that extend far away from the film back onto the screen and emit out the blue light. The intensifying screens do nothing for less energetic emission from ^{3}H, ^{14}C, or ^{35}S, but may decrease the resolution. Therefore, for fluorography with ^{3}H, ^{14}C, or ^{35}S, do not use an intensifier screen (11). ^{35}S emits β rays 30 cm into air, but only 0.25 mm into water or plastic, a very short distance and require a close contact of the gel side to the film (9). A piece of aluminum foil can stop ^{35}S β rays completely. If there is more than one gel in the cassette, put a piece of aluminum foil between the gels.
8. Sometimes, 1% nonfat dry milk in PBS-T is not a good solution to use for antibody development when precious antibodies need to be saved for later use by freezing at −20°C. Nonfat dry milk usually forms curd during the incubation and precipitates out from the solution after freezing. Therefore, a blocking solution developed by Pierce, using no protein for blocking of the membrane, is a perfect choice to use. Add 5–10% glycerol to the solution to protect the antibodies during freezing at −20°C.

References

1. Ornstein L (1964) Disc electrophoresis. I. Background and theory. Ann N Y Acad Sci 121:321–349
2. Kohlrausch F (1897) Ueber concentrations-verschiebungen durch electrolyse im inneren von lösungen und lösungsgemischen. Ann Phys Chem 62:209–239
3. Laemmli UK (1970) Cleavage of structural proteins during the assembly of the head of bacteriophage T4. Nature 227:680–685
4. Pollard JW (1994) The in vivo isotopic labeling of proteins for polyacrylamide gel electrophoresis. Methods Mol Biol 32:67–72
5. Paulsson KM, Jevon M, Wang JW et al (2006) The double lysine motif of tapasin is a retrieval signal for retention of unstable MHC class I molecules in the endoplasmic reticulum. J Immunol 176:7482–7488
6. Targoff IN, Mamyrova G, Trieu EP et al (2006) A novel autoantibody to a 155-kd protein is associated with dermatomyositis. Arthritis Rheum 54:3682–3689
7. Trieu EP, Gross JK, Targoff IN (2009) Immunoprecipitation-western blot for protein of low abundance. In: Kurien BT, Hal Scofield R (eds) Protein blotting and detection: methods and protocols, vol 536, Methods in molecular biology., pp 259–275
8. Hames BD (ed) (1998) Gel electrophoresis of proteins: a practical approach, 3rd edn, The practical approach series. Oxford University Press, Oxford

9. Horscroft NJ, Roy P (1997) Thermal denaturation of proteins for SDS-PAGE analysis by microwave irradiation. Biotechniques 22: 224–226
10. Bonner WM (1987) Autoradiograms: ^{35}S and ^{32}P. Methods Enzymol 152:55–61
11. Laskey RA (1994) Radiometric methods for detection in blots. In: Dunbar BS (ed) Protein blotting: a practical approach, The practical approach series (Rickwood D and Hames BD, eds). Oxford University Press, Oxford, pp 121–132
12. Kessler SW (1981) Use of protein A-bearing staphylococci for the immunoprecipitation and isolation of antigens from cells. Methods Enzymol 73:442–459

Chapter 19

Two-Dimensional Gel Electrophoresis: Vertical Isoelectric Focusing

Yaser Dorri

Abstract

Two-dimensional gel electrophoresis (2-DE) is one of the most powerful tools for separating proteins based on their size and charge. 2-DE is very useful to separate two proteins with identical molecular weights but different charges, which cannot be achieved with just sodium dodecyl sulfate polyacrylamide gel electrophoresis (SDS-PAGE). Here, a simpler and easier version of 2-DE is presented which is also faster than all the currently available techniques. In this modified version of 2-DE, isoelectric focusing is carried out in the first dimension using a vertical SDS-PAGE apparatus. Following the first-dimensional IEF, each individual lane is excised from the IEF gel and, after a 90° rotation, is inserted into a second-dimensional SDS-PAGE, which can be stained with Coomassie Brilliant Blue for protein analysis or immunoblotted for further analysis. This version of IEF can be run in less than 2 h compared to the overnight run required by O'Farrell's method. Difficult tube gel casting and gel extrusion as well as tube gel distortion are eliminated in our method. This method is simpler, faster, and inexpensive. Both dimensions can be done on the same SDS-PAGE apparatus, and up to ten samples can be run simultaneously using one gel.

Key words: 2D gel electrophoresis, Isoelectric focusing, Sodium dodecyl sulfate polyacrylamide gel electrophoresis, SDS-PAGE, 2-DE, IEF

1. Introduction

Two-dimensional gel electrophoresis (2-DE) is a well-known and powerful technique used for separating proteins based on their size and charge. The usefulness of this method is that proteins that are separated by their molecular mass in sodium dodecyl sulfate polyacrylamide gel electrophoresis (SDS-PAGE) can be further separated based on charge. Era of electrophoresis can be reached back to late 1800, when Lodge introduced this technique by attaching electrodes to jelly-filled tubes and measured ionic velocities (1). However, the

Biji T. Kurien and R. Hal Scofield (eds.), *Protein Electrophoresis: Methods and Protocols*, Methods in Molecular Biology, vol. 869, DOI 10.1007/978-1-61779-821-4_19, © Springer Science+Business Media, LLC 2012

results were not impressive. Despite many efforts to improve this phenomenon, a century later, Laemmli introduced SDS-PAGE in 1970 (2). Followed by that, Margolis and Kenrick introduced a reliable method of 2-DE in 1969 (Cited in 3), and several years later, MacGillivary and Rickwood in 1974 made slight improvements to this technique (Cited in 4). In 1975, O'Farrell was able to introduce tube-based isoelectric focusing, a well-modified version of 2-DE with impressive resolution (5). In O'Farrell's method, the IEF gel is cast within a glass tube and the proteins of interest are electrophoresed through it. However, researchers find number of problems with this method: (a) it is difficult to cast the gel, (b) the tube gel is difficult to extrude following an IEF run, and (c) possibility of gel breakage during gel extrusion and the long run time which can take up to 16 h (6). Ever since, there have been many improvements to 2-DE in terms of resolution and sample solubility; however, almost all the methods require a long run time for the first-dimensional IEF. Among the current methods of 2-DE, immobilized pH gradient (IPG), flatbed IEF, and IPGphor Isoelectric Focusing System strip gel methods are well known (6). Sample leakage and long run time as well as high-voltage requirement for running the gel are the disadvantages of these methods (7–9). Here, a modified version of 2-DE, vertical IEF, is introduced. This method uses the same apparatus as that used in second-dimensional SDS-PAGE (6). In this method, SDS-PAGE Mini-PROTEAN II gel apparatus is used for the first and second-dimensional run. Therefore, this method eliminates the use of expensive equipment required by previous methods. Faster run time and easy gel handling in this method are among advantageous factors, compared to other 2-DE methods (6). In this modified method, each lane of the first-dimensional IEF gel is excised and, after a short wash, it is rotated 90° and inserted onto the second-dimensional SDS-PAGE. Up to ten samples can be analyzed in one IEF run. Total run time for both dimensions in this method is less than 3 h (6). This method is discussed in detail here.

2. Materials

Prepare all solutions with ultrapure water (prepared by purifying deionized water, to attain a sensitivity of 18 MΩ cm at 25°C) and analytical grade reagents. Prepare and store all reagents at room temperature (unless indicated otherwise). Diligently follow all waste disposal regulations when disposing waste materials.

2.1. First-Dimensional IEF Gel Components

1. Acrylamide (30%)/bis-acrylamide (1.5%) solution (see Note 1).
2. *N,N,N′,N′*-tetramethylethylenediamine (TEMED).
3. Riboflavin (Sigma-Aldrich, Inc., St. Louis, MO, USA).

4. Bio-Lyte 5–8 (Amersham Biosciences Co., NJ, USA).
5. Riboflavin (0.004%)/TEMED (0.45%) (see Note 2).
6. Ammonium persulfate (APS): 10% solution in water (see Note 3).
7. Urea.
8. Triton X-100—20% (see Note 4).
9. Triton X-100—10% (see Note 4).
10. Acetic acid.
11. Tris base.
12. IEF lysis buffer: 2 M thiourea, 5 M urea, 0.25% CHAPS, 0.25% Tween-20, 0.25% SB 3–10 (sulfobetaine-10), 0.25% ampholyte (isoelectric point 5–8), 10% 2-propanol, 1 mM sodium vanadate, 12.5% water-saturated isobutanol, 5% glycerol, and 0.79% β-mercaptoethanol (see Note 5).
13. Acetic acid solution: 0.005 M, pH 5.0. Store at 4°C.
14. Tris–HCl: 0.1 M Tris, pH 8.0. Store at 4°C.
15. Overlay buffer: 3.0 g of urea, 2 mL of 10% Triton X-100, and 0.5 mL of 40% Bio-Lyte pH 5–8 and made to 10 mL with distilled water (see Note 6). Store at 4°C.
16. Equilibration buffer (125 mL of 0.5 M Tris–HCl pH 6.8, 20 g of SDS, and 50 mL of β-mercaptoethanol, made up to 1 L) (see Note 7).
17. Plastic piece (2 cm × 2 mm) (6).

2.2. SDS Polyacrylamide Gel Components

1. Acrylamide (30%)/bis-acrylamide (0.8%).
2. Resolving gel buffer: 1.5 M Tris–HCl (see Note 8). Store at 4°C.
3. Stacking gel buffer: 0.5 M Tris–HCl, pH 6.8 (see Note 9). Store at 4°C.
4. APS: 1.5% solution in water (see Note 10).
5. SDS-PAGE running buffer: 0.025 M Tris, 0.192 M glycine, 0.1% SDS, pH 8.3 (see Note 11).
6. SDS lysis buffer (5×): 0.3 M Tris–HCl (pH 6.8), 10% SDS, 25% β-mercaptoethanol, 0.1% bromophenol blue, 45% glycerol. Store at −20°C.

3. Methods

3.1. First-Dimensional IEF Gel (4% IEF)

In this method, the IEF gel is cast just like a SDS-PAGE gel using a Mini-PROTEAN II gel apparatus (used for mini SDS-PAGE gels) (see Fig. 1). Unlike other IEF gel methods, this method suggests an inexpensive apparatus which is cheaper than currently introduced methods. Both dimensions can be done on the same SDS-PAGE

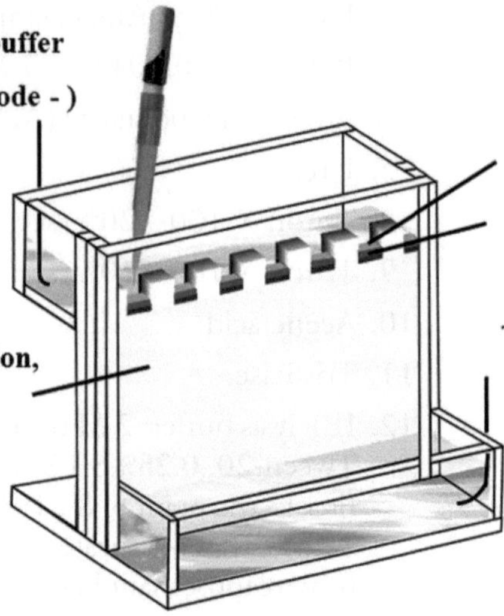

Fig. 1. Schematic representation of first dimension IEF gel. Mini-PROTEAN II SDS-PAGE apparatus is used to run the first-dimensional IEF gel. A thin film of Spinkote from Beckman was coated along the sides of the apparatus to provide a tight seal and thus prevent mixing of buffers between *upper* and *lower* chambers. Adapted from ref. 6.

apparatus, and up to ten samples can be run simultaneously in less than 3 h, using one gel. Here, the first step of this modified version of 2-DE is introduced:

1. Set up the apparatus similar to that in SDS-PAGE (see Fig. 1).
2. Prepare the solution for a 4% IEF gel by adding 6.13 g of urea to 2.5 mL of deionized water and let it mix (see Note 12). Once mixed, add 1.6 mL of acrylamide (30%)/bis-acrylamide (1.5%) solution, 1.5 mL of riboflavin/TEMED solution, 1.2 mL of 20% Triton X-100, and 0.6 mL Bio-Lyte 5–8. Let it mix for 10–15 min.
3. Add 200 μL APS and quickly cast the gel within a 7.25 cm × 10 cm × 1.0-mm gel cassette using 1-mm spacer (see Note 13).
4. Place a 10-well comb at the top of the gel.
5. Place the gel apparatus in front of a UV light overnight at room temperature for polymerization (see Note 14). Once polymerized, either use immediately or cover it with a wet tissue paper and wrap it thoroughly with clear plastic wrap to prevent gel from drying.
6. To run the gel, a thin film of Spinkote from Beckman was coated along the sides of the Mini-PROTEAN II gel apparatus to provide a tight seal. This will create a greasy area along the sides and thus prevent mixing of buffers between upper and lower chambers. Then attach the polymerized IEF gel to the Mini-PROTEAN II gel apparatus using clamps (6) (see Figs. 1 and 2). Then remove the comb.

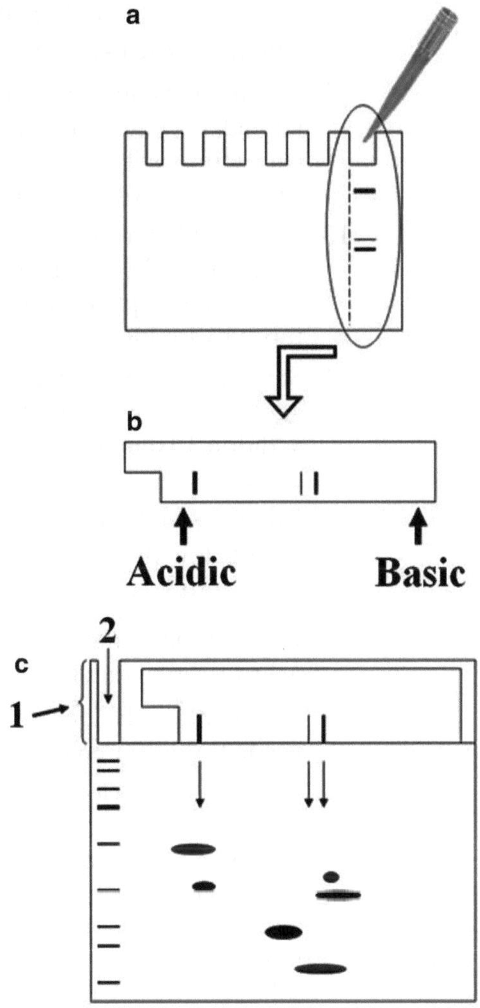

Fig. 2. Schematic representation of 2-DE: (**a**) Each lane of the first-dimensional IEF is excised. (**b**) Excised lane is rotated 90° counterclockwise. (**c**) Excised gel strip is inserted into the empty space (1) of the second-dimensional SDS-PAGE gel. Protein marker can be loaded into the well created by plastic piece (2).

7. Prepare the samples by adding 12.25 mg of urea and 34 μL of IEF lysis buffer to 5 μL of desired sample. Mice liver sample (post-nuclear fraction derived from 10% homogenate) is used here (6) (see Note 15).
8. Fill the top chamber with acetic acid (pH 5) buffer, including inside the wells of IEF gel (see Fig. 1).
9. Pour Tris–HCl (pH 8) in the bottom chamber of apparatus (see Fig. 1).
10. Prefocus the gel by running it at 100 V (constant voltage) for 50 min.
11. To add samples, first remove the acetic acid buffer from the upper chamber and wells using a 10-mL pipette and a 1-mL

syringe (see Note 16). Place 60 μL of samples (the whole amount of each sample) in each well using a P200 pipette (6).

12. Fill the remaining space in each well with overlay buffer. Use enough overlay buffer to cover the samples and fill all the wells (including wells without samples, if any), to prevent samples from mixing with the acetic acid buffer (see Fig. 1).
13. Slowly add more acetic acid (pH 5) to the top of overlay buffer (over the wells) till the top chamber is filled. It is important to follow this step with care and slowly to prevent samples from mixing with buffer.
14. Start the running process by applying 100 V (constant voltage) for 15 min followed by 200 V (constant voltage) for an additional 10 min. Then change the setting to 20 W (constant power).
15. Run for 30 min or until a yellow horizontal line across the gel, right below the wells, is observed (see Note 17).
16. Turn off the power supply.
17. Remove the clamps and glass plate from the gel. Transfer the gel to a clean container and rinse it with deionized water.
18. Incubate the IEF gel with equilibration buffer (125 mL or as needed) for 20 min on a shaker, in a laboratory hood.
19. Wash 2× with deionized water for 5 min.

3.2. Second-Dimensional Sodium Dodecyl Sulfate Polyacrylamide Gel Electrophoresis

1. Prepare a 10% SDS-PAGE resolving gel solution (solution A): Prepare the second-dimensional SDS-PAGE gel solution by mixing 2.5 mL of resolving buffer, 3.33 mL of acrylamide mixture, and 4 mL water in a 50-mL conical flask. Degas the solution with vacuum for 10 min (6) (see Note 18).
2. Prepare another set of 10% SDS-PAGE resolving gel solution (solution B) as stated above in a separate flask. Degas the solution with vacuum for 10 min. Once finished, wrap it with plastic wrap and keep it aside (see Note 18).
3. Add 80 μL of APS and 20 μL of TEMED to solution A and immediately cast gel within a 7.25 cm × 10 cm × 2.0-mm gel cassette using 2-mm spacer. Allow 1.5–2.0 cm space on top, to lay the first-dimensional gel strip (see Figs. 3, 4a, and Note 19). Let the gel polymerize.
4. Excise each lane of the first-dimensional IEF vertically (see Figs. 2 and 3) and, after a 90° counter clockwise rotation, insert it into the top of the SDS-PAGE gel, right inside the 1.5–2.0 empty space (see Note 20 and Figs. 4a and c). By doing so, the left side of the gel strip represents the acidic side, while the right side represents the basic end of the IEF gel strip (see Fig. 2). Keep the gel to the right side of the empty space (see Fig. 4d).
5. Add 180 μL of APS and 40 μL of TEMED to solution B and immediately pour the solution into the empty space (over the

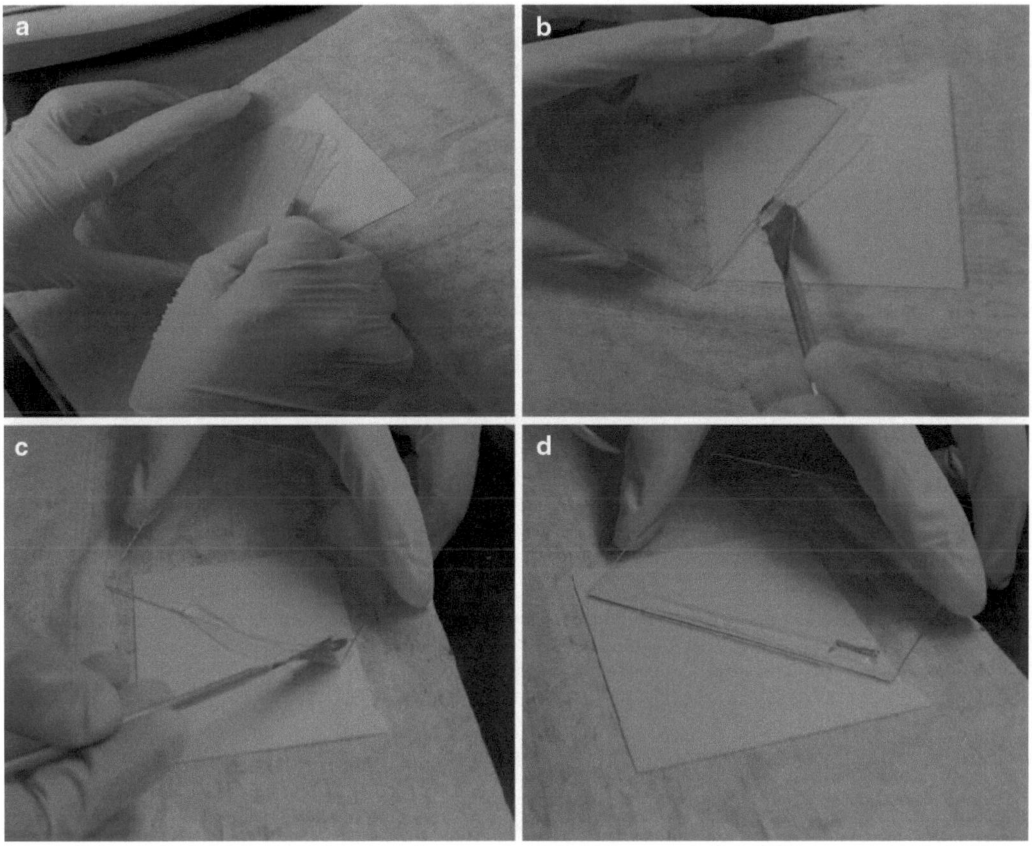

Fig. 3. Transfer of IEF gel strip onto the second-dimensional SDS-PAGE gel: (**a**) Each lane of the first-dimensional IEF is excised using a blade and glass plate. (**b**) IEF gel strip is pushed over a glass plate starting from one end, using a spatula. (**c**) The other end of IEF gel strip is pushed over the glass plate. (**d**) The gel strip is completely moved on to the glass plate and is ready to be inserted on to the second-dimensional SDS-PAGE gel.

IEF gel strip). This solution must fill all the remaining space around the gel strip. Before the polymerization of the top section of the gel, make a well for loading the molecular marker by inserting a 2-cm-long plastic piece (2 mm thick) into the left side of the empty space (now filled with solution B), next to the gel strip (see Fig. 4d). Once the top portion polymerized, remove the 2-cm plastic piece (see Note 21).

6. Once the top section of the SDS-PAGE gel is polymerized, assemble the apparatus and fill the top and bottom chambers with SDS-PAGE running buffer.
7. Load 10–20 μL of molecular marker into the well created by plastic piece (see Fig. 4d).
8. Run the gel at 14 mA (constant current) until the marker passes into the bottom portion of SDS-PAGE gel (see Note 22). Then run at 30 mA (constant current) until the marker reaches the bottom of the gel.

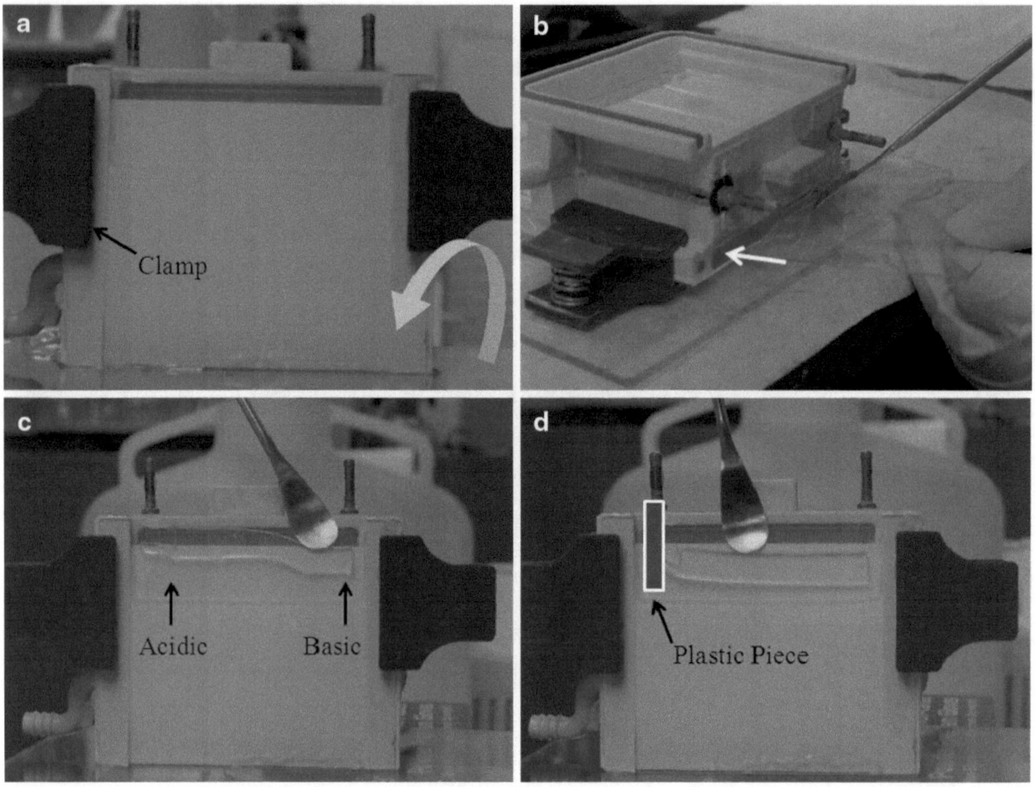

Fig. 4. The insertion of the IEF gel strip on to the second-dimensional SDS-PAGE: (**a**) Flip the Mini-PROTEAN II SDS-PAGE apparatus to sit horizontally. (**b**) Place the glass plate carrying the IEF gel strip (see Fig. 3d) over the glass plate of second-dimensional SDS-PAGE apparatus. Carefully push the gel strip into the empty space. (**c**) Push the gel inside using a spatula. (**d**) Place a 2-cm plastic piece to the left side of the IEF gel strip and pour the solution B over the gel strip.

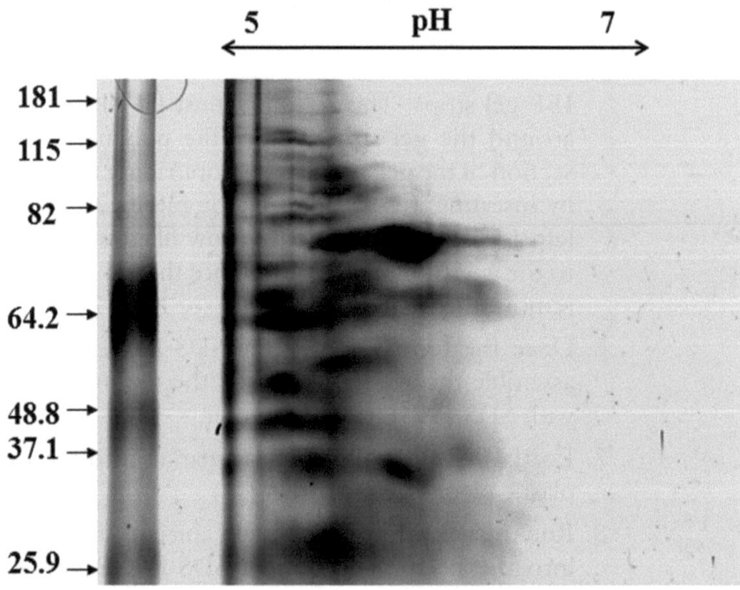

Fig. 5. CBB stained 2-DE gel (10%): Mouse liver post-nuclear fraction analyzed on 2-DE gel.

9. Turn off power supply.
10. Stain the gel using 0.05% Coomassie Brilliant Blue (CBB) for 2 h (for best result, stain overnight) (10) or use the gel for immunoblotting (11, 12).
11. Destain the gel using destaining solution and Kimwipe (see Fig. 5) (13).

4. Notes

1. Weigh 30 g of acrylamide and 1.5 g bis-acrylamide while wearing a mask and transfer them to a 100-mL graduated cylinder containing about 50 mL of water, under a laboratory fume hood. Add a magnetic stir bar and mix it on magnetic stirrer for about 30 min. A low level of heat can speed up the mixing process. Make up to 100 mL. Wrapped the bottle with aluminum foil and store it at 4°C.
2. Weigh 4 mg of riboflavin and transfer it to a 100-mL graduated cylinder containing about 50 mL of water. Add a magnetic stir bar and mix it on a magnetic stirrer for 30 min. Then add 0.45 g of TEMED and mix for 3 min. Make up to 100 mL. Store at 4°C, with bottle wrapped with aluminum foil.
3. Weigh 100 mg of APS and transfer it to 1 mL of water in a 1-mL Eppendorf tube. Mix for 1 min and keep at room temperature for immediate use. It is better to make fresh APS solution when needed.
4. Weigh 20 g of Triton X-100 and transfer it to a 100-mL graduated cylinder containing about 50 ml of water. Mix for 15 min and make up to 100 mL with water. To make Triton X-100—10%, weigh 10 mg of Triton X-100 and mix in water and make to 100 mL with water.
5. To prepare 10 mL of IEF lysis buffer, weigh 1.52 g of thiourea, 3.0 g of urea, 25 mg of CHAPS, 25 mg of SB 3–10, and 1.22 mg of sodium vanadate and transfer them into a 10-mL graduated cylinder containing 4 mL of water. Add 25 mg of Tween-20, 25 mg of ampholyte 5–8, 1 mL of isopropanol, 1.25 mL of water-saturated buffer, 0.5 mL of glycerol, and 79.6 μL β-ME into the solution. Add a magnetic stir bar and mix the solution on a magnetic stirrer for 30 min. Make up to 10 mL. Make 1 mL aliquots and store at 4°C in 1-mL Eppendorf tubes.
6. To make overlay buffer, mix 2 mL of 10% Triton X-100 with 3 g of urea and 0.5 mL of Bio-Lyte (pH 5–8). Make it to 10 mL.

7. Weigh 20 g of SDS and mix it in 500 mL of water. Add 125 mL of 0.5 M Tris–HCl (pH 6.8) and 50 mL of BME. Make up to 1 L with water. Mix for 30 min. Store at room temperature.
8. Resolving gel buffer: 1.5 M Tris–HCl, pH 8.8. Add about 100 mL water to a 1-L graduated cylinder or a glass beaker. Weigh 181.7 g Tris and transfer to the cylinder. Add water to a volume of 900 mL. Mix and adjust pH with HCl. Make up to 1 L with water. Store at room temperature.
9. Stacking gel solution was not used in this method; however, stacking gel can be used to polymerize the IEF gel strip in the empty space above the SDS-PAGE gel (6).
10. APS (1.5%): Weigh 150 mg of APS and mix in 10 mL of water. Mix for 5 min. A low level of heat can speed up the mixing process.
11. Take 100 mL of 10× native buffer (can be prepared by mixing 30.3 g of Tris base with 144 g glycine in 500 mL water for 30 min; then make it to 1 L) and mix in 500 mL of water. Add 10 ml of 10% SDS. Make it to 1 L with water.
12. When making IEF gel solutions, first weigh 6.13 g of urea and mix in 2.5 mL water in a 10-mL conical flask for 10–15 min with medium level of heat. Add a magnetic stir bar and let it mix on magnetic stirrer. Solution should be warmed at all time otherwise it will form into crystals. Once urea is dissolved, add 1.5 mL of riboflavin/TEMED solution, 600 μL of Bio-Lyte, 1.6 mL of acrylamide/bis-acrylamide, and 1.2 mL of Triton X-100. Mix 10 min while being warmed. Add 80 μL of APS right before the use of the solution. IEF gel solution must be immediately used as it may polymerize.
13. The thickness of the first-dimensional IEF gel must be a little smaller than that of SDS-PAGE. It is because each lane of the first-dimensional IEF will be transferred on top of the second-dimensional SDS-PAGE, and since the gels expand during run time, and may not fit onto SDS-PAGE apparatus on the second dimension, it would be useful to have a 1-mm thickness for the first-dimensional IEF gel and 2-mm thickness for the second dimension.
14. IEF gel was placed in front of UV light overnight as a convenience. Our results indicate 1-h UV exposure of gel is sufficient.
15. Heating the samples containing urea can speed up the mixing process.
16. Use a 10-mL pipette to remove most of the buffer solutions over the wells and then use 1-mL syringe to draw the rest of the buffer within the wells. All the buffer solutions must be carefully removed to allow an empty space for the samples.

17. By the end of the first-dimensional run time, we observed a yellowish horizontal line, right below the wells. We noticed that if we increase the run time, proteins will run out of the gel. Therefore, this horizontal line shows the end of the first-dimensional run time.
18. For a better resolution, it is recommended to make 4–20% gradient gel. Here, a 10% gel was made as a convenience. Solution B will be used to fill the top portion of the second-dimensional SDS-PAGE, where the IEF gel strip is placed (see Fig. 2c). A stacking gel solution can also be used to make solution B (here, a regular SDS-PAGE resolving gel solution was used as solution B). This solution will polymerize the IEF strip on top of the SDS-PAGE gel.
19. As mentioned earlier, the thickness of the second-dimensional SDS-PAGE must be bigger than the first-dimensional IEF so that the gel strip can be easily inserted onto the top of the SDS-PAGE (see Fig. 4c). Here, the thickness of the first-dimensional IEF was 1 mm, and for the second-dimensional SDS-PAGE, the thickness is 2 mm.
20. In this step, remove the glass (top plate) and let the gel remain on the back plate (see Fig. 3a). Rinse the gel with water and, after cleaning a glass plate with 70% ethanol, excise each lane of the first-dimensional IEF, vertically as shown in Fig. 3a, using a blade. Once each well is excised, carefully transfer each well on a glass plate using a spatula (see Figs. 3b, c and d). The excised IEF gel strip is flexible enough to be rotated along the sides up to 45° without breaking, so start lifting one side of the IEF strip using a spatula and place it on the glass plate. Then move to the other end of the IEF strip and move this end on the glass plate (see Fig. 3c). Once both ends are on the glass plate, the middle part of the IEF gel strip can be easily lifted and transferred on the glass plate by a little push. In the next step, lay the SDS-PAGE apparatus horizontally (from vertical position) so the IEF gel can easily slide into the empty space (see Fig. 4a). While the SDS-PAGE apparatus is in horizontal position, slowly transfer the IEF gel over the top of SDS-PAGE glass plate. Using a spatula, the IEF gel is placed over the SDS-PAGE glass plate, starting with the right side of the IEF gel (see Fig. 4b). Once the right side of the IEF gel strip is over the SDS-PAGE plate, push the left side of the IEF gel with spatula and then carefully push the whole gel inside the empty space (see Fig. 4c). Do not force the IEF gel strip to go deep inside as the IEF gel strip will nicely fit in the empty space once the SDS-PAGE solution is poured over it. Here, the top of the gel strip (placed in the left side of the empty space of the SDS-PAGE) represents the acidic part, and bottom represents the basic part of the gel (see Fig. 4c).

21. As mentioned earlier, a 2-cm plastic piece is inserted to the left side of the SDS-PAGE empty space to create a well for the protein marker. Once the top gel portion is polymerized, slowly remove the plastic piece.
22. Since in the second-dimensional SDS-PAGE proteins are migrating from the gel strip to the SDS-PAGE gel, it is better to run the gel, starting with a low current (14 mA). Once the protein marker moves through the SDS-PAGE gel, then change the current to 30 mA.

References

1. Lodge O (1886) Brit Assoc Repts #38
2. Laemmli UK (1970) Cleavage of structural proteins during the assembly of the head of bacteriophage T4. Nature 227:680–685
3. Zhen C, Jen-Fu C, Qing-Yu H (2005) Current progress in sample preparation for two-dimensional electrophoresis in proteomics. Curr Proteomics 2:259–268
4. Rabiloud T et al (2010) Two dimensional gel electrophoresis in proteomics: past, present and future. J Proteomics 73:2064–2077
5. O'Farrell PH (1975) High resolution two-dimensional electrophoresis of proteins. J Biol Chem 250:4007–4021
6. Dorri Y, Kurien BT, Scofield RH (2009) A simple and faster version of two-dimensional gel electrophoresis using vertical, mini SDS-PAGE gel apparatus. Iran J Chem Chem Eng 28:51–56
7. Knisley KA, Rodkey LS (1986) Affinity immunoblotting. High resolution isoelectric focusing analysis of antibody clonotype distribution. J Immunol Methods 95:79–87
8. Zuo X, Hembach P, Echan L et al (2002) Enhanced analysis of human breast cancer proteomes using micro-scale solution isoelectrofocusing combined with high resolution 1-D and 2-D gels. J Chromatogr B Analyt Technol Biomed Life Sci 782:253–265
9. Klein J, Harding G, Klein E (2002) A new isoelectric focusing gel for two-dimensional electrophoresis constructed in microporous hollow fiber membranes. J Proteome Res 1:41–45
10. Kurien BT, Scofield RH (1998) Heat mediated quick Coomassie blue protein staining and destaining of SDS-PAGE gels. Indian J Biochem Biophys 35:385–389
11. Dorri Y, Khalili AW, Scofield RH (2009) Whole and strip nitrocellulose membrane as well as a new-line-immunoblotting of antigen using the chemiluminescence technique. Methods Mol Biol 536:417–432
12. Kurien BT, Scofield RH (1997) Multiple immunoblots after non-electrophoretic bidirectional transfer of a single SDS-PAGE gel with multiple antigens. J Immunol Methods 205:91–94
13. Dorri Y, Kurien BT (2010) Environmentally safe removal/disposal of Coomassie Brilliant Blue from gel destain and used gel stain. Anal Biochem 404:193–196

Chapter 20

Isoelectric Focusing in Agarose Gel for Detection of Oligoclonal Bands in Cerebrospinal and Other Biological Fluids

Gyorgy Csako

Abstract

Isoelectric focusing (IEF) coupled with immunodetection (immunofixation or immunoblotting) has become the leading technique for the detection and study of oligoclonal bands (OCBs) in cerebrospinal fluid (CSF) and also is increasingly used in other body fluids such as the tear and serum. Limited commercial availability of precast agarose IEF gels for research and a need for customization prompted reporting a detailed general protocol for the preparation and casting of agarose IEF gel along with sample, control, and isoelectric point marker preparation and carrying out the focusing itself for CSF OCBs. However, the method is readily adaptable to the use of other body fluid specimens and, possibly, research specimens such as culture fluids as well.

Key words: Agarose gel, Cerebrospinal fluid, Control specimen, Gel casting, Immunoglobulins, Isoelectrofocusing, Immunodetection, Oligoclonal bands, p*I* marker, Tear

1. Introduction

Detection of clonally produced immunoglobulins (Igs) in the blood and various body fluids has become a critical diagnostic test in clinical practice and also is of major research interest. One of the most important applications is in the area of neurological diseases. Intrathecal Ig production is detected by the presence of oligoclonal Igs in the cerebrospinal fluid (CSF), and oligoclonal band (OCB) testing for IgG in the CSF is now the primary laboratory test for the diagnosis and follow-up of multiple sclerosis (1–5). As a noninvasive alternative for specimen collection, the tear has been recently proposed to replace CSF for OCB testing (6, 7).

Biji T. Kurien and R. Hal Scofield (eds.), *Protein Electrophoresis: Methods and Protocols*, Methods in Molecular Biology, vol. 869, DOI 10.1007/978-1-61779-821-4_20, © Springer Science+Business Media, LLC 2012

Based on charge differences mostly due to amino acid sequence variations, clonal production of Igs traditionally has been detected by the presence of distinct bands of highly restricted mobility on zone electrophoresis using agar gel, cellulose acetate membrane, agarose, or polyacrylamide gels (3). Posttranslational modifications such as glycosylation may also alter the charge on an Ig molecule. Consequently, Igs originating from the same clone of cells and sharing identical primary structure can migrate in different and limited regions of a gel upon isoelectric focusing (IEF). IEF is electrophoresis in a pH gradient, which concentrates ("focuses") proteins at their isoelectric points (p*I*s) and allows them to be separated on the basis of charge differences potentially as small as a p*I* of 0.01. The resultant patterns reflect the microheterogeneity, also called limited heterogeneity, of proteins such as Igs. In combination with the introduction of highly sensitive detection techniques for OCBs, IEF is now the most widely used laboratory method for detecting intrathecal Ig synthesis. A variety of IEF methods using polyacrylamide or agarose gels with or without immunodetection for human CSF OCB testing have been developed over the past three decades. Agarose IEF methods combined with immunodetection (immunofixation or immunoblotting) for increasing detection sensitivity and providing high specificity have been standardized and now are proposed as the "gold standard" for detecting IgG OCBs in the human CSF in diagnostic laboratories (2). There are commercial versions of this kind of IEF methods, such the Hydragel 3/9 Isofocusing with immunofixation (Sebia Inc., Norcross, GA) and the SPIFE® IgG IEF with immunoblotting (Helena Laboratories, Beaumont, TX), which have been both approved for clinical diagnostic testing by the US Federal Drug Administration (FDA). A polyacrylamide IEF method coupled with immunodetection (CSF Analysis kit with PhastGel™, GE Healthcare, Piscataway, NJ) also is available commercially as a kit. Nevertheless, because detection and study of OCB patterns has been largely limited to plasma (serum) (8, 9) and CSF (1–5, 8, 9), including its proposed surrogate, the tear (7, 8), and research continues even on OCBs in the CSF itself, there is a continued need to have manual versions of these methods that are amenable to modifications for laboratory research. The following is the description of agarose IEF for the study of OCBs in CSF. However, this method is readily adaptable and modifiable to the study of OCBs in other body fluids and in select research specimens such as culture fluids as well.

2. Materials

All solutions are prepared with "distilled water" (refers here to ultrapure water: deionized water further purified to obtain a resistance of 18 MOhm at 25°C) and high-grade analytical reagents at room temperature (15–25°C). Reagents and other materials are stored at room temperature, unless stated otherwise. Unless obtained commercially, all reagents used in these methods are prepared and stored without adding sodium azide. Follow all waste disposal regulations for discarding reagents and disposables. As a safe laboratory practice to minimize or eliminate exposure to hazardous chemicals and infectious agents, always wear clean gloves. If the gels are to be further processed for silver staining, make sure that the gloves are also powder-free when handling the gel. Wash gloves with soap and water and pat dry.

2.1. Components of Agarose IEF Gel Solution

1. Distilled water.
2. IEF Agarose (GE Healthcare [formerly Amersham Biosciences], Piscataway, NJ, USA).
3. Sorbitol.
4. Carrier ampholytes (a mixture of amphoteric substances): Pharmalyte broad range pH 3–10 (GE Healthcare). Store refrigerated at 4–8°C (see Note 1).

2.2. Preparation of Agarose IEF Gel Solution

Prepare an ~1% agarose solution freshly for casting one 125 × 250 × 0.5-mm agarose IEF gel (pH 3–10) essentially according to a procedure reported previously [Instructions 52-1536-00 AF (GE Healthcare Bio-Sciences AB, Uppsala, Sweden, 2006)]:

1. Add 0.2 g dry IEF agarose powder and 2.4 g sorbitol to 18 mL of distilled water.
2. Heat the mixture to boiling in a water bath or by using an electric heater equipped with a magnetic stirrer.
3. Once the solids are dissolved (after ~10 min), cool the solution to 60–70°C before adding 1.3 mL of the carrier ampholytes (Pharmalyte broad range pH 3–10).
4. Mix thoroughly, keep the solution heated at the same temperature, and proceed immediately to cast the gel as in Subheading 2.5 (see Note 2).

2.3. Component and Accessories for Agarose IEF Gel Casting

1. GelBond® Film for agarose, 124 × 258 mm (GE Healthcare).
2. Glass plate, 125 × 260 × 3 mm—support for GelBond® in gel casting (GE Healthcare).
3. Glass plate, 125 × 240 mm, 0.5 mm—parallel spacer for gel casting (GE Healthcare).
4. Plastic syringe, 20 mL with short flexible tubing attached. Alternatively, use a plastic pipette of appropriate caliber.
5. FlexiClamps (GE Healthcare).
6. Roller (GE Healthcare).

2.4. Assembly of Mold for Agarose IEF Gel Casting

Assemble the mold for gel casting essentially according to the instructions in the Multiphor II Electrophoresis System User Manual (Code No. 18-1103-43, Amersham Biosciences [now GE Healthcare], Uppsala, Sweden, 1998):

1. Clean the glass plates with a mild detergent, then rinse them with distilled water, and dry.
2. Pipette 2–3 ml of distilled water (see Note 3) onto the 3-mm thick glass plate and center the GelBond® Film with its hydrophilic side up over the plate (see Note 4).
3. Cover the film with a sheet of blotting paper.
4. Starting longitudinally at one end of the glass plate, begin pressuring the blotting paper with the film beneath it evenly with the roller to eliminate air bubbles and to seal the film to the plate with a minimum of interfacing water. Remove the paper and remove any excess water with a tissue.
5. Place the glass plate with spacer on top of the film, leaving 10 mm free at both ends.
6. Hold the plates securely together with one FlexiClamp on each of the long sides.
7. Place the assembled mold in a heating cabinet at 70°C for 10 min.

2.5. Casting of Agarose IEF Gel (see Note 5)

Cast the gel essentially according to the instructions in the Multiphor II Electrophoresis System User Manual [Code No. 18-1103-43, Amersham Biosciences (now GE Healthcare), Uppsala, Sweden, 1998] (see Note 6):

1. Take the prewarmed mold out of the heating cabinet.
2. Slowly fill the syringe or pipette with the molten agarose solution. Avoid getting air bubbles into the syringe or pipette.
3. Hold the mold by one hand at a 20–45° angle with support plate positioned lower and inject or pipette by the other hand the agarose solution at a moderate, steady rate between the glass plate with spacer and the GelBond Film at the elevated end. The gel solution will fill the space by capillary action.
4. As the mold fills, gradually lower the angle to decrease the flow of the gel solution.
5. When completely filled, position the mold horizontally and allow the agarose solution to gel at room temperature for 15 min.
6. Before opening the mold, carefully remove the excess (10-mm wide) gel from both "open" ends with a scalpel.
7. Remove the two FlexiClamps and position the mold on one long side.

8. Grasp the GelBond® Film and glass plate with spacer with one hand and the 3-mm glass plate with the other and pull them gently apart (see Note 7).
9. First remove the 3-mm glass plate, then the glass plate with spacer from the gel.
10. Place the agarose gel supported by the GelBond® Film in a humidified chamber at 4°C for at least 1 h (see Note 8).
11. The gel is now ready for use. Alternatively, store the gel in a humidified chamber at 4°C for up to 3 days.

2.6. Instrument and Accessories for Agarose IEF Gel Electrophoresis

1. Horizontal IEF system: Multiphor® II Flatbed Electrophoresis System, including electrophoresis unit Multiphor® II, thermostatic circulator MultiTemp® III, and EPS 3501 Power Supply [Pharmacia, then Amersham Biosciences (now GE Healthcare)].
2. IEF electrode strips (GE Healthcare).
3. EPH/IEF sample application foil (agarose), 24 samples, 2–4 μL (GE Healthcare) (see Note 9).
4. Sample application syringe, 1–5 μL, with stoppers (e.g., Hamilton Microliter syringe from Thermo Fisher Scientific).
5. Filter paper: EPH electrode wicks, paper 104×253 mm (GE Healthcare) or Whatman blotting paper (Thermo Fisher Scientific).
6. Paper towels (see Note 10).
7. Anolyte (anode [+] solution): 0.05 M sulfuric acid. Follow safety precautions!
8. Catholyte (cathode [−] solution): 1 M sodium hydroxide. Follow safety precautions!

2.7. Specimens

1. Collect CSF and serum specimens with standard medical procedures. Store at 2–8°C and use within 48 h, freeze at −20°C for up to 3 months or at −70°C for up to 12 months (see Note 11).
2. Use neat or unadulterated CSF (see Note 12).
3. Determine the IgG concentrations in both CSF and serum (see Note 13).
4. Prepare samples for agarose IEF by diluting both CSF and serum specimens with physiologic saline to the same concentration range deemed appropriate for the detection sensitivity of the method to be used (e.g., 1–2 mg total IgG/dL might be appropriate for an immunofixation-based method) (see Note 14).

2.8. Controls

1. OCB-positive and OCB-negative specimens. Store small aliquots (10–20 μL) frozen in 80-μL sealed capillary tubes at -20°C for up to 3 months, at -70 to 80°C for up to a year, and avoid or minimize repeat freezing–thawing (≤3 cycles!) (see Note 15).
2. Bring frozen aliquots of the controls to room temperature [preferably by briefly (~5 min) placing them in a 37°C water bath], mix well with brief (3–5 s) vortexing after thawing, then dilute the same way as the unknown patient samples.
3. Suggest using an OCB-positive CSF in combination with an OCB-negative serum in each gel.

2.9. pI Markers

1. Broad range p*I* marker: GE Healthcare Broad pI kit, pH 3–10, a lyophilized mixture of stable, salt-free, highly purified proteins (Thermo Fisher Scientific Inc.). Store refrigerated at 4–8°C (see Note 16).
2. Reconstitute lyophilized p*I* marker preparation according to the manufacturer's instructions. Store frozen in small aliquots (10–20 μL) in 80-μL sealed capillary tubes at -20°C for 3 months or, preferably, at -70°C for up to 12 months.

3. Methods

Carry out all procedures at room temperature unless otherwise specified. For general guidance and safety issues, consult instrument's operating manual:

3.1. IEF System Setup

1. Connect the electrophoresis unit Multiphor® II to the thermostatic circulator MultiTemp® III.
2. Check the cooling platform with a leveler and, if necessary, adjust the level.
3. Thoroughly clean the cooling surface with distilled water and towel dry.
4. Set the temperature to 10°C and begin equilibration of the electrophoresis chamber at least 15 min before starting IEF.

3.2. Gel Placement

1. Bring the agarose IEF gel to room temperature (10–20 min). Place the gel on a clean, level surface (clean filter paper or clean plastic tray). Do not lay the gel on the bench top!
2. Lightly blot the gel with one filter paper/gel blotter to remove the surface moisture. After 10–30 s, gently peel off the paper from the gel surface and also blot excess water from around the perimeter of the gel.

3. Pipette 1–2 mL of distilled water onto the center of the cooling platform and spread it with a gloved finger.
4. Place the gel with the plastic film facing down in the middle of the cooling surface in the electrophoresis chamber using the screened template on the platform to center the gel. While holding the (opposite) diagonal corners, apply the gel to the platform by bending it in a "U-shape." This minimizes or avoids getting air bubbles under the gel.
5. If bubbles form, with the edge of a thin spatula, gently lift the edge of the gel, then lay the gel back down on the platform to remove the bubble.
6. Blot excess moisture with paper towel from the base of the electrophoresis chamber (around the gel) being careful to not disturb the gel (see Note 17).

3.3. Electrode Strip Placement

1. Soak one electrode strip into 3 mL anolyte (0.05 M sulfuric acid).
2. Blot the strip for ~1 min on a filter paper to remove excess liquid.
3. Place the blotted strip to the anodic side of the gel.
4. Soak a second strip into 3 mL catholyte solution (1 M sodium hydroxide) and repeat step 2.
5. Place this strip 100 mm away from the anodic strip on the gel. The strips should be well aligned (parallel) and completely inside the gel on the respective anodic and cathodic gel side.
6. Make sure that the electrode strips have good, even contact with the gel.

3.4. Sample Application

1. Position the sample application foil (also called application "template" or "mask") across the gel such a way that the samples can be loaded at 20 mm from the anodic strip on the gel. Make sure that no air bubbles become trapped between the template and the gel (see Note 18).
2. With the syringe, apply 2–4 μL (volume dependent on sensitivity of the OCB detection method) of the prediluted controls and paired CSF and serum samples (side by side!), and if applicable, the p*I* marker over each slot, without touching the gel surface.
3. Allow samples to absorb into the gel (usually within 3–5 min).
4. Gently peel off the template when the last sample has been completely absorbed and the sample sites look dry (maximum 20 min).

3.5. Running IEF

1. Place the electrode holder on the electrophoresis unit in the shallow depressions.
2. Carefully center the EPH/IEF electrodes over the strips and connect electrical leads.
3. Lower carefully so that the electrodes rest on and make good contact with the electrode strips.
4. Place the safety cover over the unit and connect the leads into the power supply.
5. Run IEF at 10 W for 30 min and then at 20 W until focusing is completed after 1,500 Vh. Limit voltage to 1,250 V. Total run time is ~90 min at 10°C.

3.6. Ending IEF

1. Upon completion of IEF (~90 min), turn the power supply off (see Note 19).
2. Disconnect the Multiphor II® unit from the power supply.
3. Remove the safety lid from the unit.
4. Carefully remove the electrode holder.
5. Remove the gel from the electrophoresis chamber.
6. Gently pull the electrode strips from the gel and discard them.
7. Process the gel immediately as intended for direct protein staining (first placing it in a fixative) or for immunodetection by methods described elsewhere (see Note 20).
8. Remove the electrodes from the electrode holder.
9. Rinse the electrodes with distilled water to remove the strong acidic and basic solutions.
10. Air-dry or carefully dry with paper towel the electrodes for storage.
11. Check the platinum wire for possible damage.

4. Notes

1. For higher resolution of IgG OCBs with IEF, the pH gradient can expanded in the basic region with Pharmalyte broad range pH 8–10.5 (GE Healthcare). In turn, using ampholytes in the range of pH 3–8 improves the resolution for IgM OCBs (Fig. 1).
2. Agarose solution remains liquid until the temperature drops to about 40°C.
3. As alternatives, 0.1% solution of a nonionic detergent such as Triton X-100® and ethanol/water (95/5 vol.) have been suggested with no clear advantage over water.

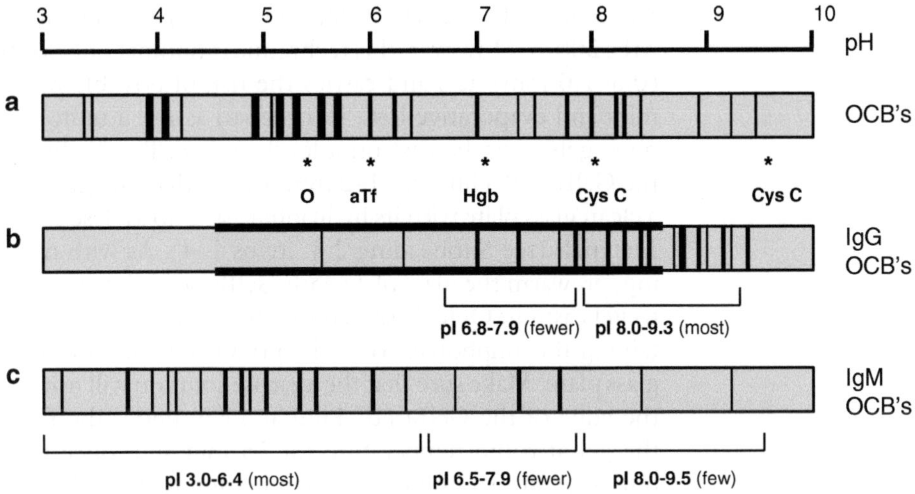

Fig. 1. Scheme for the location and relative frequency of common oligoclonal bands (OCBs) and for the location of common non-Ig bands in IEF patterns of CSF. Based largely on data reported in (3, 11, 13–15). (**a**) Direct protein staining: *Asterisks* show location of O, origo (application point) and common non-Ig bands such as *aTf* asialotransferrin, *Hgb* hemoglobin (occurring in hemolyzed specimens), and *Cys C* cystatin C ("γ-trace protein"). Note that two abundant CSF proteins (albumin, p*I* ~4.8 and prealbumin, p*I* ~4.7) are located near the application point but are not identified in the pattern. (**b**) Immunodetection for IgG: Heavy *horizontal lines* indicate location of polyclonal IgG (p*I* range 4.7–8.6). (**c**) Immunodetection for IgM.

4. The GelBond® film, a transparent, flexible polyester sheet, has a specially treated hydrophilic side that firmly adheres to agarose gels as they form. This property can be tested by observing the behavior of a drop of water on both sides of the film: the water will "spread" on the hydrophobic surface, whereas it will "bead" on the hydrophobic side. Since GelBond® is nonporous, it cannot be used with electroblotting or capillary blotting applications (Document #18652-1007-03, Lonza Rockland, Inc., Rockland, ME, 2007).
5. Precast Agarose IEF gel, supported on GelBond® (Lonza), along with necessary accessories is available commercially and adaptable to OCB testing: (Lonza) IsoGel IEF plate, pH 3–10, 125 × 100 mm (Thermo Fisher Scientific Inc. or Westburg BV, Leusden, The Netherlands). Using the Multiphor® II system, the IEF running conditions reported for protein separation with this gel [2.5 μL sample/site applied, gel prefocused at 1 W for 10 min, then focused at 2,000 V (maximum), 25 mA (maximum), and 25 W (maximum) for 60 min at 10°C] would be applicable to CSF OCB testing as well.
6. If commercial precast agarose IEF gel is not preferred, gel casting mold is not available, and/or GelBond® sizes other than those fitting the commercially available mold are to be used, horizontal agarose IEF gels can be prepared using an open casting procedure outlined previously (Lonza GelBond® Film protocol sheet for Casting Agarose Gels on GelBond Film,

Document #18652-1007-03, Lonza Rockland, Inc., Rockland, ME, 2007). However, this technique is optimal only for thicker (0.5–1.0 mm) gels and carries the risk of possible nonuniformity and evaporative loss. In order to assure a uniform thickness, gels must be cast on a level surface. Briefly, first adhere the GelBond® Film (cut 1–2 mm smaller than the glass plate) to a clean glass plate with its hydrophilic side up as described under Materials (see Subheading 2.4, items 1–4). As with mold casting, prewarm the glass plate with GelBond® Film attached prior to gel casting. Quickly spread the molten agarose solution containing the ampholytes over the prewarmed film on top of the glass plate. Make sure that the agarose solution will not overflow the edge of the GelBond® Film and run under the film. Cool the gel at room temperature for 15 min and, once it has set, refrigerate it in a humidified chamber for at least 1 h. The gel is now ready for use. Alternatively, the gel can be stored in a humidified chamber at 4°C for up to 3 days (see Note 8).

7. Agarose gel poorly adheres to glass.
8. Either commercially available humidity chambers (e.g., Humidity chamber 125 × 260 mm, GE Healthcare) or a tightly closed plastic container with wet towel at the bottom and a leveled glass plate placed over short glass rods (or plastic pipettes) can be used.
9. As an alternative to using sample application foil, IEF sample application pieces (GE Healthcare) can be used. However, these pieces require much larger sample volumes. They are soaked with 15–20 μL of sample solution before placing them on the gel surface for 30–45 min.
10. Use only white paper towels because colored paper towels my exude a dye into the gel and electrode strips.
11. CSF and serum specimens were claimed to be stable for 7 days at "ambient" temperature, 14 days "refrigerated," 3.5 months frozen at −20°C, and for three freeze-thaw cycles (10).
12. Because of the high sensitivity of currently available methods for OCB testing, concentrating CSF is now considered obsolete (2). Further, concentrating CSF leads to artifacts, including protein losses that differ from sample to sample and may reach as much as 50% for IgG (3).
13. Standardized commercial methods currently are available for serum and CSF IgG measurements (e.g., immunonephelometry by Siemens BN-II).
14. Reliable detection of CSF OCBs requires standardization so that identical amounts of a patient's CSF IgG and serum IgG are applied side by side on the gel. In most cases, the CSF IgG concentration requires little adjustment for IgG OCB testing and only the serum requires dilution. Rarely, the CSF IgG

concentration may be too low, and in such cases, despite its recognized shortcomings, concentrating CSF specimens is necessary to attain the detection sensitivity of the particular method used for OCBs. Illustrating the rarity of these cases, in a large study 98.2% (388/395) of the CSF specimens received for OCB testing had an IgG concentration of >1 mg/dL, and the lowest IgG concentration was 0.7 mg/dL (10).

15. Controls help to determine the reliability of any given IEF run. For proper interpretation of the results, recognized control samples (e.g., an "OCB"-positive CSF and an "OCB"-negative serum) should be included in each run. While commercial kit methods for diagnostic IEF testing with immunodetection for CSF IgG OCB include appropriate control specimens, such specimens are not readily obtainable outside of kit users. If commercially not available, collect and save aliquots of single or pooled CSF and serum specimens that have been previously analyzed and confirmed for the presence or absence of OCB for use as positive and negative OCB controls.
16. Although routine use of p*I* markers is not required for diagnostic CSF "OCB" testing, these markers may be useful (a) to confirm the analytical performance of the IEF method, (b) to study the properties of the "oligoclonal bands" for research purposes, and (c) to recognize non-Ig bands in the absence of immunodetection (Fig. 1a).
17. This step is important to avoid fanning.
18. Unlike conventional electrophoresis, the point of application of the sample on the gel is not as critical in IEF. This is because whatever the starting point may have been, the proteins in the sample will concentrate in the pH regions equal to their isoelectric point.
19. Because IEF is an end-point method, the stop time is less important than in conventional electrophoresis. Proteins actually become more focused as time progresses beyond the "stop time."
20. Immunodetection methods for OCBs are advantages over traditional direct protein staining methods both in terms of higher sensitivity and greater specificity (2, 3, 11). Since the overwhelming majority (>99%) of all OCBs are known to be of the IgG type, diagnostic assays most commonly test for IgG OCBs. However, IEF tests followed by immunodetection using monospecific antibodies to IgG subclasses, IgM, IgA, kappa, and lambda chains also have been developed for CSF OCBs (3, 11). Furthermore, the utility of testing for specific antibodies among OCBs by using IEF coupled with "reverse" immunodetection (i.e., with use of antigens) has also been well established. Examples include detection of paraneoplastic antibodies with this approach (9, 12).

References

1. Polman CH, Reingold SC, Edan G et al (2005) Diagnostic criteria for multiple sclerosis: 2005 revisions to the "McDonald Criteria". Ann Neurol 58:840–846
2. Freedman MS, Thompson EJ, Deisenhammer F et al (2005) Recommended standard of cerebrospinal fluid analysis in the diagnosis of multiple sclerosis: a consensus statement. Arch Neurol 62:865–870
3. Link H, Huang Y-M (2006) Oligoclonal bands in multiple sclerosis cerebrospinal fluid: an update on methodology and clinical usefulness. J Neuroimmunol 180:17–28
4. Regeniter A, Kuhle J, Mehling M et al (2009) A modern approach to CSF analysis: pathophysiology, clinical application, proof of concept and laboratory reporting. Clin Neurol Neurosurg 111:313–318
5. Schäffler N, Köpke S, Winkler L et al (2011) Accuracy of diagnostic tests in multiple sclerosis—a systematic review. Acta Neurol Scand 124(3):151–164
6. Devos D, Forzy G, de Seze J et al (2001) Silver stained isoelectrophoresis of tears and cerebrospinal fluid in multiple sclerosis. J Neurol 248:672–675
7. Calais G, Forzy G, Crinquette C et al (2010) Tear analysis in clinically isolated syndrome as new multiple sclerosis criterion. Mult Scler 16:87–92
8. Mehta PD, Patrick BA, Thormar H et al (1982) Oligoclonal IgG bands with and without measles antibody activity in sera of patients with subacute sclerosing panencephalitis (SSPE). J Immunol 129:1983–1985
9. Storstein A, Monstad SE, Honnorat J et al (2004) Paraneoplastic antibodies detected by isoelectric focusing of cerebrospinal fluid and serum. J Neuroimmunol 155:150–154
10. Fortini AS, Sanders EL, Weinshenker BG et al (2003) Cerebrospinal fluid oligoclonal bands in the diagnosis of multiple sclerosis. Isoelectric focusing with IgG immunoblotting compared with high-resolution agarose gel electrophoresis and cerebrospinal fluid IgG index. Am J Clin Pathol 120:672–675
11. Mehta PD (1991) Diagnostic usefulness of cerebrospinal fluid in multiple sclerosis. Crit Rev Clin Lab Sci 28:233–251
12. Stich O, Rauer S (2007) Antigen-specific oligoclonal bands in cerebrospinal fluid and serum from patients with anti-amphiphysin- and anti-CV2/CRMP5 associated paraneoplastic neurological syndromes. Eur J Neurol 14:650–653
13. Laurenzi MA, Link H (1979) Characterisation of the mobility on isoelectric focusing of individual proteins in CSF and serum by immunofixation. J Neurol Neurosurg Psychiatry 42:368–372
14. Sharief MK, Thompson EJ (1992) Distribution of cerebrospinal fluid oligoclonal IgM bands in neurological diseases: a comparison between agarose electrophoresis and isoelectric focusing. J Neurol Sci 109:83–87
15. Nakashima I, Fujihara K, Sato S et al (2005) Oligoclonal IgG bands in Japanese patients with multiple sclerosis. A comparative study between isoelectric focusing with IgG immunofixation and high-resolution agarose gel electrophoresis. J Neuroimmunol 159:133–136

Chapter 21

Native Flatbed Isoelectric Focusing for Determining Antibody Clonotype Distribution

Biji T. Kurien and R. Hal Scofield

Abstract

A sensitive method to analyze specific antibody clonotype changes in a patient with systemic lupus erythematosus, who developed autoantibodies to the Ro 60 autoantigen under observation, is described in this chapter. Patient sera collected over several years were separated by flatbed isoelectric focusing (IEF). Following electrofocusing, the gel is now ready to be analyzed by affinity immunoblotting utilizing Ro 60-coated nitrocellulose membrane to determine oligoclonality of the anti-Ro 60-containing sera.

Key words: Affinity immunoblotting, Clonotype distribution, Systemic lupus erythematosus, Ro 60 autoantigen, Flatbed IEF

1. Introduction

Isoelectric focusing (IEF) serves as a useful procedure for studying the heterogeneity of antibody and immunoglobulin (Ig) clonotypes (1). Patterns of antigen-specific antibody clonotype can tell whether changes in cell population happen during ongoing immune responses as a response to regulatory influences. It can also show whether changes in hybridoma cell lines can occur with time (2). Previous studies investigated these changes by immobilizing the separated antibody clonotypes following IEF and incubating them with radioactive antigen. Radiolabeled hapten was allowed to diffuse into a gel, in one method, prior to precipitation of Ig with sodium sulfate followed by detection of hapten-specific clonotype distribution by autoradiography (3). Ig was precipitated in the gel with sodium sulfate, in another study, immediately following completion of the focusing run. It was then cross-linked with glutaraldehyde followed by the addition of labeled antigen or anti-Ig (4). Subsequent studies showed that fixation with glutaraldehyde could diminish the antigen-binding capability of certain Ig (5). In addition,

Biji T. Kurien and R. Hal Scofield (eds.), *Protein Electrophoresis: Methods and Protocols*, Methods in Molecular Biology, vol. 869, DOI 10.1007/978-1-61779-821-4_21, © Springer Science+Business Media, LLC 2012

it was shown that the earlier study was unable to define optimal cross-linker (glutaraldehyde or suberimidate) concentration since certain antibodies could not be fixed at cross-linker concentrations that substantially inactivated others. The excess time needed to diffuse antigen into the gel and for rinsing the unbound antigen out of the gel is another drawback of these methods, which can take several days especially when using radioactive probes.

One procedure for immobilizing focused antibodies entailed the use of nitrocellulose membranes. Focused antibodies are transferred electrophoretically or non-electrophoretically to nitrocellulose membrane, and labeled antigen is employed to detect antigen-specific clonotypes (6). Yet another procedure involved laying the gel with the focused antibodies with agarose containing antigen-coated sheep erythrocytes (7). In this procedure, antibodies diffuse into the RBC-containing gel, bind the antigen-coated cells, and lyse the cells after complement addition.

In this chapter, we describe a procedure in which a 60,000 molecular weight Ro autoantigen was first passively immobilized on nitrocellulose membrane and placed in contact with an IEF gel-containing autoantibodies (derived from a patient with systemic lupus erythematosus (SLE) who developed antibodies to the Ro 60 autoantigen over time) focused according to its isoelectric point. After diffusion mediated transfer to membrane, the nonspecific antibody clonotypes are removed by washing and the antigen-specific antibody clonotypes are detected using alkaline phosphatase conjugated anti-Ig (8).

SLE is a complex, chronic autoimmune disease in which autoantibodies target self-antigens, including the Ro (or SS-A) ribonucleoprotein complex. Anti-Ro 60 antibodies occur in up to 40% of patients with SLE (8). The epitopes of the Ro 60 autoantigen targeted by SLE patients have been characterized previously (9, 10). Even though anti-Ro 60 sera commonly bound short Ro 60 peptides, it did not bind the denatured Ro 60 antigen well. In addition, the antibodies that bound Ro 60 octapeptides were also found to bind the native protein (10).

Examples of some SLE autoantibodies appearing and disappearing, at times in association with specific disease manifestations, therapy, or generalized clinical disease activity, have been reported. For example, antibodies to native DNA is associated with renal disease, and the detection of this autoantibody may be an indication of disease exacerbation (11). Antibodies to the P autoantigen (ribosomal P antigens) can appear with an increase of neurologic or renal disease. Autoantibodies like anti-Ro 60, on the other hand, occur in some normal subjects as well as in SLE patients prior to onset of disease (8) and develop only rarely during the course of SLE.

This study was carried out after the identification of a patient with SLE who developed antibodies to the Ro 60 autoantigen after about 10 years of illness. Figure 1 shows that anti-Ro 60 clonality increases in complexity, and affinity to the Ro 60 antigen also increases as the response developed.

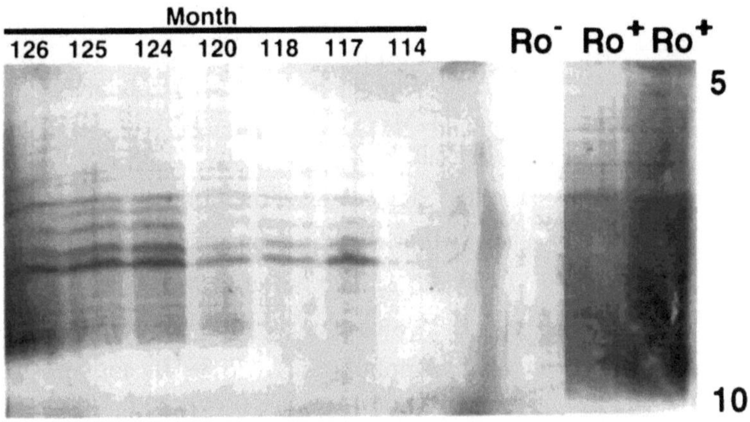

Fig. 1. Affinity immunoblotting of the patient's sera obtained at different months following first observation, showing anti-Ro 60-specific IgG clonotypes. Sera from an anti-Ro 60-negative SLE patient (Ro$^-$) and from two typical anti-Ro 60-positive patients (Ro$^+$) are shown for comparison. The pH range of the IEF gel is shown on the *right*.

2. Materials

Prepare all solutions using ultrapure water (prepared by purifying deionized water, to attain a sensitivity of 18 MΩ cm at 25°C) and analytical grade reagents. Prepare and store all reagents at room temperature (unless indicated otherwise). Diligently follow all waste disposal regulations when disposing waste materials. We do not add sodium azide to the reagents.

1. 25% Glycerol (v/v): Add 25 mL glycerol to 75 mL of distilled water. Mix well.
2. 5× Acrylamide (26.5% T, 3% C): Add about 25 mL water to a 100-mL graduated cylinder or a glass beaker. Weigh 12.84 g acrylamide and 0.4098 g bis acrylamide and transfer to the cylinder (see Note 1). Add a spatula of AG 501-X8 (D) mixed-resin beads (Bio-Rad, Hercules, CA, USA) and stir using a magnetic stir bar on a magnetic plate for about 30 min. Make up to 50 ml (after removing the stir bar) with water and filter through a 0.45-µm Corning filter (see Note 2). Store at 4°C, with bottle wrapped with aluminum foil (see Note 3).
3. 10% Tween-20: Add 90 mL of distilled water into a glass beaker. Add 10 mL Tween-20 and mix.
4. 2% Ammonium persulfate: Weigh 0.02 g ammonium persulfate and dissolve in 1 mL of distilled water (see Note 4).
5. *N.N.N.N′*-Tetramethylethylenediamine. Store at 4°C (see Note 5).
6. Alkaline phosphatase buffer: Weigh 6.1 g of Tris–HCl, 2.9 g sodium chloride, and 0.51 g magnesium chloride.6H_2O and make it to 500 mL with water after adjusting pH to 9.3 with HCl (see Note 1). Store at 4°C.

7. Nitro blue tetrazolium (NBT)/5-bromo-4-chloro-3-indolyl phosphate (BCIP): Dissolve 1 g NBT in 20 mL of 70% dimethylformamide (DMF). Dissolve 1 g BCIP in 20 mL of 100% DMF. Add 33 μL of BCIP and 66 μL of NBT to 10 mL of alkaline phosphatase buffer just before adding to membrane.
8. Nitrocellulose membrane.
9. 0.5 M sodium bicarbonate solution, pH 9.5.
10. Phosphate buffered saline (PBS), pH 7.4.
11. PBS containing 0.05% Tween-20 (PBST).
12. Ampholytes: pH 3–10 and pH 8–10.5 (Amersham/GE Healthcare, Piscataway, NJ, USA).
13. Ro 60 autoantigen (Immunovision, Springdale, AK, USA).
14. Glass plates: Two 5″ by 4″ glass plates.
15. Medium binder clips (1¼ in.).
16. Small binder clips (¾ in.).
17. Gasket with three edges, about 3 mm wide, to serve as spacer between the plates.
18. LKB-2117 Multiphor apparatus for IEF (LKB Instruments, Rockville, MD, USA).
19. Model 3000/300 power supply (Bio-Rad).
20. pH 3 and pH 10 solutions (Serva Electrophoresis GmbH, Heidelberg, Germany).
21. Helium gas.
22. Sample applicator strip (Bio-Rad).
23. Paper wicks.

3. Methods

All procedures are at room temperature unless otherwise specified.

1. Pipette 5.6 mL of distilled water into a conical flask. Add 2 mL of 25% glycerol followed by 2.1 mL of the 5× acrylamide solution. Then add 300 μL of pH 3–10 ampholytes followed by 100 μL of pH 8–10.5 ampholytes.
2. Degas this solution by bubbling helium through it for 15 min. Rinse the metal end of degassing tube first with water and wipe dry with Kimwipes.
3. While the solution is degassing, set up the gel apparatus. Soak the gasket in water for few min. Mop dry with Kimwipes.
4. Take one glass plate and lay the gasket on top of the glass plate around the edges so that it will seal the bottom and two sides

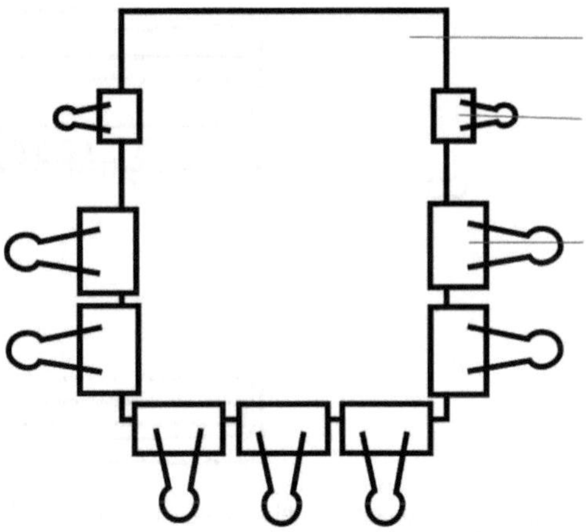

Fig. 2. Gel assembly for isoelectric focusing.

of the plates. Lay the other glass plate on top of the gasket. Clamp the clips around the edges of the plates (bottom, the left side, and the right side) (see Fig. 2). Stand the gel upright using the base of the clips (see Fig. 2) to pour the gel. Prepare a 2% ammonium persulfate solution fresh.

5. After degassing is complete, the metal end of the degassing tube is cleaned with water. To the degassed solution, add 100 μL of 10% Tween-20 and mix gently. Then add 100 μL of 2% APS. Have a Pasteur pipette ready for pouring the gel. Add 10 μL of TEMED and mix gently. Pipette the gel mixture into the Pasteur pipette and transfer into the gel apparatus quickly. Attempts should be made to avoid bubbles. Fill up the gel apparatus to the top. Polymerization should begin within minutes. However, let the assembly stand for 2 h without disturbance.
6. Turn cooling unit on and set it on 4°C in preparation for focusing.
7. After 2 h, carefully remove one of the glass plates and gasket. The gel will remain on one of the glass plates.
8. Lay the glass plate on top of the IEF unit, with the gel side facing up (wipe of water on top of the unit beforehand). Place the smaller cover in place and press down slightly so as to make imprints for the wicks. Cut two wicks to the size of the gel (be as close as possible). Soak the top wick in Serva pH 3 solution and the bottom wick in Serva pH 10 solution. Dab off excess solution and place where imprints were made by cover (see Fig. 3).
9. Put smaller cover back on, making sure connection is made with both wicks. Connect red and black wires. Put on larger cover and make connections to power supply (red = positive; black = negative).

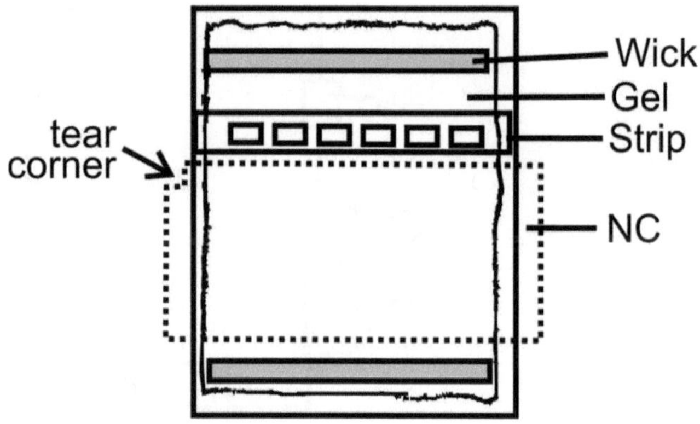

Fig. 3. The membrane–gel assembly following flat bed IEF. The gel bond is shown in this figure. However, we did not use the gel bond to support the gel. The gel was directly in contact with the glass plate.

10. Prefocus by setting constant voltage 200 V for 20 min, then increase voltage to 400 V for another 20 min. Prepare samples for application.
11. Turn of power supply, disconnect wires, and remove covers. Take applicator strip and lay on top of gel 1–2 cm below top wick. Make sure strip is stuck to the gel well. Strip can hang over gel a little (see Fig. 3; see Note 6).
12. Apply the samples, being careful not to spill over into other wells. Replace the covers and make the connections. Turn the power supply on to 12 W constant power. Focus for approximately 1–2 h. When focusing, the voltage will rise, and the current will drop. The rate at which these two parameters change is much faster in the beginning than the end. The run is complete when the voltage is between 1,800 and 2,000 V and the current 3–5 A. When the change appears to be very slow or not at all, turn off the unit (see Note 7).
13. After the run is complete, transfer the focused protein from the gel to membrane. Take gel off the flatbed and remove applicator strip.
14. Stain gel with appropriate stain or carry out transfer to membrane (12, 13).

4. Notes

1. Having water at the bottom of the cylinder helps to dissolve the Tris–HCl relatively easily, allowing the magnetic stir bar to go to work immediately. If using a glass beaker, the Tris–HCl can be dissolved faster if the water is warmed to about 37°C.

However, the downside is that care should be taken to bring the solution to room temperature before adjusting pH.

2. Wear a mask when weighing acrylamide. To avoid exposing acrylamide to coworkers, cover the weigh boat containing the weighed acrylamide with another weigh boat (similar size to the original weigh boat containing the weighed acrylamide) when transporting it to the fume hood. Transfer the weighed acrylamide to the cylinder inside the fume hood and mix on a stirrer placed inside the hood. Unpolymerized acrylamide is a neurotoxin, and care should be exercised to avoid skin contact. Mixed resin AG 501-X8 (D) (anion and cation exchange resin) is used when acrylamide solution is made since it removes charged ions (e.g., free radicals) and allows longer storage. Some investigators store the prepared acrylamide along with this resin in the refrigerator. However, we filter them out before storage. The used mixed resin should be disposed of as hazardous waste. Manufacturer's warning states that this resin is explosive when mixed with oxidizing substances. The resin contains a dye that changes from blue-green to gold when the exchange capacity is exhausted.
3. The acrylamide solution can be stored at 4°C for 1 month. Acrylamide hydrolyzes to acrylic acid and ammonia. The acrylamide mixture, buffer, and water can be prepared in large batches, frozen in aliquots (for greater day-to-day reproducibility) and used indefinitely (see ref. 12). Remove the required amount, bring to room temperature, and add the other ingredients for polymerization. However, in our laboratory, we make the acrylamide solution fresh about every month when we cast our own gels.
4. We find it is best to prepare this fresh each time.
5. We find that storing at 4°C reduces its pungent smell.
6. Large well = 10 all, medium wells = 5 all, and small well = 1 all. Only lay one size of wells on to the gel. Cut if necessary. Strips may be used again.
7. During the end of the run, the gel must be watched carefully in case a fire starts. Many times, the gel will burn near the applicator strip. If this happens, turn off the unit. The gel can still be used if it had been focused for a long time. The bands are usually below the strip.

Acknowledgments

This work was supported by NIH grant ARO1844 and Oklahoma Center for the Advancement of Science and Technology to RHS.

References

1. Awed ZL, Williamson AR, Askonas BA (1968) Isoelectric focusing in polyacrylamide gel and its application to immunoglobulins. Nature 219: 66–67
2. Knisley KA, Rodkey LS (1986) Affinity immunoblotting. High resolution isoelectric focusing analysis of antibody clonotype distribution. J Immunol Methods 95:79–87
3. Askonas BA, Williamson AR, Wright BE (1970) Selection of a single antibody-forming cell clone and its propagation in syngeneic mice. Proc Natl Acad Sci USA 67:1398–1403
4. Keck K, Grossberg AL, Pressmann D (1973) Specific characterization of isoelectrofocused immunoglobulins in polyacrylamide gel by reaction with 125 I-labeled protein antigens or antibodies. Eur J Immunol 3:99–102
5. Nicolotti RA, Briles DE, Schroer JA, Davie JM (1980) Isoelectric focusing of immunoglobulins: improved methodology. J Immunol Methods 33: 101–115
6. Friedenson B, Soong CJ (1984) A simple general method of exactly comparing different binding activities and antigenic properties of antibodies by obtaining duplicate copies from a single isoelectric focusing gel. J Immunol Methods 67:235–242
7. Phillips JM, Dresser DW (1973) Antibody isoelectric spectra visualized by antigen-coated erythrocytes. Eur J Immunol 3:738–740
8. Scofield RH, Zhang F, Kurien BT et al (1996) Development of the anti-Ro autoantibody response in a patient with systemic lupus erythematosus. Arthritis Rheum 39:1664–1668
9. Scofield RH, Harley JB (1991) Autoantigenicity of Ro/SSA antigen is related to a nucleocapsid protein of vesicular stomatitis virus. Proc Natl Acad Sci USA 88:3343–3347
10. Huang S-C, Yu H, Scofield RH, Harley JB (1995) Human anti-Ro autoantibodies bind peptides accessible to the surface of the native Ro autoantigen. Scand J Immunol 41:220–228
11. Lloyd W, Schur PH (1981) Immune complexes, complement, and anti-DNA in exacerbations of systemic lupus erythematosus (SLE). Medicine (Baltimore) 60:208–217
12. Harlow E, Lane D (1988) Electrophoresis. Appendix I. In: Harlow E, Lane D (eds) Antibodies. A laboratory manual. Cold Spring Harbor Laboratory, New York, p 638
13. Wu M, Stockley PG, Martin WJ II (2002) An improved western blotting technique effectively reduces background. Electrophoresis 23: 2373–2376

Chapter 22

Two-Dimensional Gel Electrophoresis: Glass Tube-Based IEF Followed by SDS-PAGE

Hiroyuki Matsumoto, Hisao Haniu, Biji T. Kurien, and Naoka Komori

Abstract

The genome information combined with data derived from modern mass spectrometry enables us to determine the identity of a protein once it is isolated from a complex mixture. Two-dimensional gel electrophoresis established more than three decades ago serves as a powerful protocol to isolate many proteins at once for such protein analysis. In the first two decades, the original procedure to use a glass tube-based isoelectric focusing (IEF) had been commonly used. Since an IEF in glass tubes is rather difficult to maneuver, a new method to use an IEF on a thin agarose slab backed by a plastic film (IPG Dry Strip) has been invented and is now widely used. In this chapter, we describe the original protocol that uses a glass tube-based IEF because, the capacity of protein loading and resolving power of this type of classic two-dimensional gel is still indispensible.

Key words: 2D-gel, Glass tube-IEF

1. Introduction

Identification of isolated proteins by peptide mass fingerprinting (1, 2) became simplified because of the availability of genome information. This makes isolation of proteins of interest from a complex mixture an important protocol in biomedical sciences. Among various protocols of two-dimensional (2D) gel electrophoresis (3–7) published more than 35 years ago, those described by O'Farrell (5) and Ames and Nikaido (7) have been used frequently. Although these 2D gel protocols are powerful, their disadvantages are that the entire procedures are rather tedious and require skills. Especially, somewhat difficult steps involve (1) to prepare reliable isoelectric focusing (IEF) gel in a glass tube and (2) to focus proteins in the disc IEF gels in the first dimension. Because of technical difficulty in these steps, a simpler and user-friendly protocol for the IEF step using a thin layer of agarose IEF made on a plastic

Biji T. Kurien and R. Hal Scofield (eds.), *Protein Electrophoresis: Methods and Protocols*, Methods in Molecular Biology, vol. 869, DOI 10.1007/978-1-61779-821-4_22,

backing (i.e., IPG Dry Strip) has been developed recently. One clear difference between original glass tube IEF and the IPG Dry Strip IEF is the loading capacity of protein mixture; several hundred micrograms of proteins can be loaded on an IEF gel made in a glass tube, whereas an IPG Dry Strip IEF gel can take only ~150 μg of proteins. There appear to be some differences in focusing patterns of proteins on the final 2D gels on a slab SDS-PAGE. However, since the IEF gradients made on a glass tube and on an IPG dry strip are usually different, it is impossible to draw any scientific conclusion on the difference of 2D gel patterns between these two systems without doing detailed experiments. In this chapter, we will describe a 2D gel electrophoresis based on a glass tube IEF gel that our group has been using for more than three decades. The protocol described in this paper is based on O'Farrell (5) and its modification by Miyazaki et al. (8–10).

2. Materials

1. An IEF apparatus with a jacketed beaker (10.5 cm i.d., 12.5 cm inner height) used as a lower-buffer chamber (Buchler Instruments—Labconco, Kansas City, MO, USA) or a similar tube gel apparatus such as one from Bio-Rad (Hercules, CA, USA).
2. An SDS-PAGE slab gel apparatus such as a vertical slab gel apparatus Model V16 (GIBCO-BRL, Gainsburg, MD, USA) or equivalent.
3. Borosilicate glass tubes (14–15 cm in length with 3.0 mm i.d. and 5.0 mm o.d.).
4. Power supply appropriate for IEF focusing and SDS-PAGE.
5. Stock solutions for IEF Gel:

   Solution A: Acrylamide (30%, w/v)-bis-Acrylamide (1.5%, w/v).

   Solution B: Riboflavin (0.004%, w/v)-*N*,*N*,*N*′, *N*′-tetramethyethylendiamine (TEMED; 0.45%, v/v).

   Solution C: Triton X-100 or NP-40 (20%, w/v).

   Solution D: Ammonium persulfate (1.5%, w/v).

   Ampholyte: Bio-Lyte 3/10 (Bio-Rad) or equivalent.
6. IEF lysis buffer consisting of:

   Urea, 9.5 M.

   Triton X-100 or NP-40 (2.0%, w/v).

   Bio-Lyte 3/10 (2.0%, w/v).

   2-Mercaptoethanol (5%, v/v).

High-quality water.

To make this solution, weigh urea in a small gradient cylinder. Add other ingredients and dissolve the urea and adjust the final volume.

7. IEF overlay buffer consisting of:

   Urea, 5.0 M.

   Triton X-100 or NP-40 (2.0%, w/v).

   Bio-Lyte 3/10 (1.0%, w/v).

   Use the same procedure as described for IEF lysis buffer. IEF lysis and IEF overlay buffers can be stored in aliquots at −20°C.

8. Electrode solutions:

   Top chamber (positive polarity): 20 mM H_3PO_4.

   Lower chamber (negative polarity): 1 M NaOH.

   These solutions are designed to run IEF from the acidic side to the basic side (8, 9). To run an IEF gel from the basic side, one should refer to O'Farrell (5) for the concentration of buffers. We reuse 1 M NaOH solution in the lower chamber several times. Store the solutions at room temperature.

9. Equilibration buffer:

   Tris–HCl (pH 6.8), 62.5 mM.

   SDS (2%, w/v).

   2-Mercaptoethanol (5%, v/v).

   It is convenient to make 1 L of this solution. Store the solution at room temperature.

10. Agarose stock solution:

    Agarose, 1% (w/v) in equilibration buffer.

    Microwave to dissolve the agarose, make 2- to 3-mL aliquots in 15-mL tubes, and store at room temperature. One should liquefy agarose by boiling prior to use. This agarose solution is used for overlaying the first-dimension IEF gel on top of the second-dimension SDS-polyacrylamide slab gel.

11. Prestained molecular weight markers (GIBCO-BRL).

12. Chromic–sulfuric acid cleaning solution.

3. Methods (see Note 1)

3.1. Sample Preparation

IEF samples can be prepared either from a tissue block, culture cells, or a tissue/cell homogenate solution (see Note 2).

1. A tissue block weighing 1–5 mg can be homogenized in 90 μL of IEF lysis buffer. In this case, the dilution of the lysis buffer by the tissue fluid will not affect the performance of IEF significantly.
2. After homogenization, the sample is homogenized further by sonication in a sonication bath for at least 20 min to break down chromosomal DNAs.
3. The sample is then centrifuged at 15,000 × *g* for 5 min at 25°C to precipitate insoluble debris or aggregated DNAs.
4. The recovered supernatant (usually 50–100 μL) is loaded on top of the IEF gel.

3.2. Isoelectric-Focusing Gel Electrophoresis (First Dimension)

1. It appears that preparation of high-quality IEF gels is crucial for a successful IEF. By "high quality," we mean homogeneously polymerized IEF gel in clean glass tubes. The method developed by Miyazaki et al. (8) utilizes a combination of photoinduced polymerization by riboflavin and light-independent polymerization by ammonium persulfate of acrylamide. Such a combination appears to make high-quality IEF gels routinely. IEF tubes must be cleaned thoroughly by soaking overnight in chromic–sulfuric acid cleaning solution (see Note 3) and by rinsing with hot water followed by deionized water. The tubes are dried in a vacuum oven at 80°C for 1 h. We have observed that, as long as IEF tubes are kept wet, that is, not allowed to dry after each electrophoresis, cleaning them with a common laboratory detergent solution is sufficient. When tubes are clean and dry, mark them 1.5–2.0 cm from the top, seal the bottom end of each with Parafilm, and set up the tubes on a tube stand.
2. IEF gel solution (8, 9) made in a 125-mL filtering flask and consists of:

   Urea, 8.5 M.

   Acrylamide (4.0%, w/v) bisacrylamide (0.2%, w/v).

   Riboflavin (0.0005%, w/v) TEMED (0.056%, v/v).

   Triton X-100 or NP-40 (2.0%, w/v).

   Bio-Lyte 3/10 (2.0%, w/v).

   Ammonium persulfate (0.01%, w/v).

   To make IEF gel solution, the following steps should be performed.

   This recipe makes 12 mL of IEF gel mixture.

3.3. Casting IEF Gels in Glass Tubes

1. Weigh 6.13 g of urea in a 125-mL filtering flask.
2. Add the following ingredients in order: 1.6 mL of stock solution A, 1.5 mL of stock solution B, 1.2 mL of stock solution

C, 0.6 mL of Bio-Lyte 3/10 (40%, w/v), and 2.5 mL of high-quality water.

3. Dissolve the urea completely and degas the mixture for 5 min. During the process, we usually cover the flask with aluminum foil in order to keep riboflavin from light. It is also recommended that the ambient light be kept minimal by turning off some of the room lights until the polymerization is started at step 7.
4. Add 0.08 mL of stock solution D and mix quickly by swirling the flask.
5. By using a 1-mL syringe with appropriate tubing attached to the needle, fill the glass tubes with the IEF gel mixture up to the 1.5- to 2.0-cm mark.
6. Overlay water to about 5 mm on top of the IEF gel solution.
7. Start polymerization by illuminating the tubes with a fluorescent lamp.
8. Fifteen minutes after the start of polymerization, seal the top of each tube with Parafilm.
9. Continue illumination for at least 4 h or longer. Alternatively, an overnight illumination under normal room light will also polymerize IEF gels efficiently.

3.4. Loading the IEF Sample Solution

1. Set up the tubes on a tube-gel apparatus, load samples, and slowly overlay IEF overlay buffer on the top of each sample (see Note 4).
2. Perform electrophoresis first at 100 V for 1 h and then at 300 V for 15–18 h. At the end of the run, we often increase the voltage to 500 V and run the apparatus for 1–2 h. Consequently, the total voltage-hours is somewhere between 5,000 and 7,000. It is necessary to perform preliminary studies to determine the optimum voltage-hours for each system.
3. After a run, push the IEF gel out of the tube, using a syringe filled with water, and shake the gel in 20 mL of equilibration buffer for 15 min in order to equilibrate the gel in the SDS-PAGE environment. Change the equilibration buffer three times before proceeding to SDS-PAGE. If necessary, freeze the gels in the same buffer at −20°C until use; however, it should be kept in mind that some low molecular mass proteins (smaller than 15 kDa) may diffuse out by freezing and thawing. When we freeze IEF gels, we usually perform equilibration only twice.

3.5. Sodium Dodecyl Sulfate Polyacrylamide Gel Electrophoresis (Second Dimension)

1. Prepare separating and stacking gels according to a conventional SDS-polyacrylamide gel recipe (11). The length of our stacking gel is somewhere between 1.5 and 2.0 cm. To make a stacking gel, seal its top by means of a glass rod that is slightly

shorter than the opening width of the glass plate and pour stacking gel solution until it touches the glass rod. The glass rod will prevent acrylamide from being exposed to air and allow the gel to polymerize quickly and evenly.

2. After the gel is polymerized, remove the glass rod and load the IEF tube gel equilibrated with equilibration buffer. Seal the space between the tube gel and the stacking gel with 1% (w/v) agarose.
3. At this time, a well for loading molecular weight markers can be made by placing a piece of spacer ($10 \times 5 \times 0.8$ mm) over a stacking gel.
4. After the agarose solidifies, remove the spacer and load prestained molecular weight markers.
5. Carry out SDS-PAGE according to Laemmli (11). The running front can be tracked by dropping 0.1% (w/v) bromophenol blue into the upper chamber or by using equilibration buffer containing bromophenol blue (0.01%, w/v).
6. The resulting 2D gels can be stained by Coomassie blue and dried.

4. Notes

1. It is recommended that the 2D gel protocol described in this chapter should be modified for each project in order to optimize the results. We have applied the protocol to mouse retinas, human vitreous samples, and rat dorsal root ganglion samples (12–14).
2. For cultured cells, the procedures described by Steinberg and Coffino (15) are used with some modifications. Briefly, $\sim 1.5 \times 10^6$ cells are spun down, and the cell pellet is dissolved in IEF lysis buffer to make the final volume less than 90 μL. In the case of bacterial cells, a 0.5 $A_{590-600}$ equivalent is usually adequate to start with. Again, in these two cases, the dilution of lysis buffer can be disregarded. If tissue homogenates are the starting material, mix homogenate (equivalent to 100–500 μg of total protein) with IEF lysis buffer to make the final volume less than 90 μL. One hundred micrograms of total protein is usually enough to see many protein spots by silver staining, but may not be enough for Coomassie blue staining if the protein of interest is in minor abundance. With our system, we can observe a linear increase in the size and intensity of protein spots up to 500 μg of total protein without detectable loss of small protein spots. In preparing IEF samples from culture cells or tissue homogenates, do not dilute the IEF lysis

buffer too much; we have observed that a dilution of up to 30% (at least) of lysis buffer, caused by an aqueous sample, does not affect the IEF. Thus, the volume of IEF buffer should be decided on the basis of the sample volume.

3. Chromic–sulfuric acid solution can be reused many times, until its brown color becomes greenish. Special care needs to be taken for the disposal of the used chromic acid solution, which makes the use of chromic acid inconvenient.
4. IEF gels can be kept at 25°C for at least 1 week or as long as the bottom and top ends of the gels do not dry out. Before loading IEF gels onto a tube-gel apparatus, remove the water layer over the gels, using a Kimwipe, and also remove the Parafilm from the bottom ends (although we have never had an IEF gel come out of a tube during electrophoresis, it could happen; if this appears to be the case, dialysis membrane can be used to seal the bottom end in order to prevent the gel from slipping out).

References

1. Matsumoto H, Komori N (2000) Ocular proteomics: cataloging photoreceptor proteins by two-dimensional gel electrophoresis and mass spectrometry. Methods Enzymol 316: 492–511
2. Matsumoto H, Kurono S, Matsumoto M et al (2005) Mass spectrometry of biomolecules in proteomics. In: Meyers RA (ed) Encyclopedia of molecular and cell biology and molecular medicine. Wiley-VCH, Germany, pp 557–585
3. Kenrick KG, Margolis J (1970) Isoelectric focusing and gradient gel electrophoresis: a two-dimensional technique. Anal Biochem 33:204–207
4. Klose J (1975) Protein mapping by combined isoelectric focusing and electrophoresis of mouse tissues. A novel approach to testing for induced point mutations in mammals. Humangenetik 26:231–243
5. O'Farrell PH (1975) High resolution two-dimensional electrophoresis of proteins. J Biol Chem 250:4007–4021
6. Scheele GA (1975) Two-dimensional gel analysis of soluble proteins. Characterization of guinea pig exocrine pancreatic proteins. J Biol Chem 250:5375–5385
7. Ames GF, Nikaido K (1976) Two-dimensional gel electrophoresis of membrane proteins. Biochemistry 15:616–623
8. Miyazaki K, Hagiwara H, Yokota M, et al (1978) Two-dimensional electrophoresis: Basic methods of experiment In: Ui N, Horio T (eds) Isoelectric focusing and isotachophoresis. Kyoritsu Shuppan, Tokyo, Japan [In Japanese], p 183
9. Matsumoto H, Pak WL (1984) Light-induced phosphorylation of retina-specific polypeptides of *Drosophila in vivo*. Science 223:184–186
10. Matsumoto H, Kurien B, Takagi Y et al (1994) Phosrestin I undergoes the earliest light-induced phosphorylation by a calcium/calmodulin-dependent protein kinase in *Drosophila* photoreceptors. Neuron 12:997–1010
11. Laemmli UK (1970) Cleavage of structural proteins during the assembly of the head of bacteriophage T4. Nature 227:680–685
12. Haniu H, Komori N, Takemori N et al (2006) Proteomic trajectory mapping of biological transformation: application to developmental mouse retina. Proteomics 6:3251–3261
13. Komori N, Takemori N, Kim HK et al (2007) Proteomics study of neuropathic and non-neuropathic dorsal root ganglia: altered protein regulation following segmental spinal nerve ligation injury. Physiol Genomics 29: 215–230
14. Shitama T, Hayashi H, Noge S et al (2008) Proteome profiling of vitreoretinal diseases by cluster analysis. Proteomics Clin Appl 2:1265–1280
15. Steinberg RA, Coffino P (1979) Two-dimensional gel analysis of cyclic AMP effects in cultured S49 mouse lymphoma cells: protein modifications, inductions and repressions. Cell 18:719–733

Chapter 23

One-Dimensional and Two-Dimensional Immobilized Metal Affinity Electrophoresis

Bao-Shiang Lee, G.D. Lasanthi, P. Jayathilaka, Jin-Sheng Huang, and Shalini Gupta

Abstract

Immobilized metal affinity electrophoresis (IMAEP) is a straightforward method in which metal ions are embedded in a polyacrylamide gel strip with a negligible electrophoretic migration. Due to the preferential binding between metal ions and the phosphate group, this method uses immobilized metal ions like iron, manganese, aluminum, or titanium to capture phosphoproteins from a mixture of phosphoprotein and nonphosphoproteins. IMAEP has also been incorporated into a traditional two-dimensional (2D) sodium dodecyl sulfate polyacrylamide gel electrophoresis (SDS-PAGE) system (isoelectric focusing-PAGE) to increase its resolving power. In 2D IMAEP, the metal ions in polyacrylamide gel strip are overlaid on top of the second dimensional polyacrylamide gel to stop electrophoretic migration of phosphoproteins. Data shows that there is no detrimental effect of SDS in IMAEP on the extraction of phosphoproteins from a mixture of proteins. In addition, SDS exposes phosphate groups by unfolding the phosphoproteins to facilitate metal ion–phosphate binding while supplying the protein with negative charges.

Key words: Immobilized metal affinity electrophoresis, Immobilized metal affinity column, Phosphoproteins, 2D SDS-PAGE, Immobilized pH gradient, Pro-Q diamond phosphoprotein gel stain, β-casein, Mass spectrometry

1. Introduction

A novel immobilized metal affinity electrophoresis (IMAEP) technique for the extraction of phosphoproteins from a protein mixture by embedding metal ions in a polyacrylamide gel strip has been reported by us (1–3). IMAEP is similar in principle to immobilized metal affinity column (IMAC) which is one of the most popular methods used to enrich phosphoproteins or phosphopeptides (4–8). In IMAEP, the metal ion is incorporated in the gel matrix by simply adding the metal ion into acrylamide gel solution before polymerization. For the entrapment of the metals, we have experimented with a wide variety of materials (1), including iron

Biji T. Kurien and R. Hal Scofield (eds.), *Protein Electrophoresis: Methods and Protocols*, Methods in Molecular Biology, vol. 869, DOI 10.1007/978-1-61779-821-4_23, © Springer Science+Business Media, LLC 2012

(III) chloride, iron (II) chloride, nickel (II) chloride, manganese (II) chloride, magnesium (II) chloride, silver chloride, tin (II) chloride, scandium (III) oxide, aluminum (III) chloride, vanadium (III) chloride, gallium (III) nitrate, cupric sulfate, chromium (III) chloride, zirconium (IV) hydrogen phosphate, sodium chloride, potassium chloride, titanium (III) acetate, calcium (II) chloride, and titanium (IV) oxide nanopowder. The capture of phosphoproteins using IMAEP is demonstrated by analyzing a protein mixture of nonphosphoproteins such as human albumin, lysozyme, amyloglucosidase, carbonic anhydrase, dephosphorylated α-casein, and human hemoglobin and phosphoproteins such as α-casein, β-casein, ovalbumin, and phosvitin. Data shows that electrophoretic migration of phosphoproteins ceases upon encounter with metal ions like Al^{+3}, Ti^{+3}, Fe^{+3}, Fe^{+2}, and Mn^{+2} whereas non-phosphorylated proteins migrate normally. The same behavior is observed with highly complex protein cell extracts such as protein extract from MCF-7 human breast carcinoma cells. In a subsequent study (2), protein samples are heated with 2% w/v SDS for 15 min at 85°C such that they become negatively charged and migrate electrophoretically through the embedded metal ions toward the anode. In addition, SDS exposes phosphate groups by unfolding the phosphoproteins to facilitate metal ion–phosphate interaction. To extend the usefulness of the one-dimensional (1D) IMAEP method, the IMAEP technique is built into the high-resolution two-dimensional (2D) sodium dodecyl sulfate polyacrylamide gel electrophoresis (SDS-PAGE) (9–12) which allows for better separation, identification, and quantification of larger number of proteins. In 2D SDS-PAGE, proteins are resolved by isoelectric focusing (IEF) using immobilized pH gradient (IPG) gel strip in the first dimension, which differentiates proteins according to their isoelectric points (p*I*s), followed by PAGE in a second dimension in the presence of SDS, which separates proteins according to their molecular weight. In 2D IMAEP (3), the metal ions containing gel strip is overlaid on top of the second dimensional polyacrylamide gel to extract phosphoproteins. These captured phosphoproteins are readily identifiable by a downstream procedure using standard proteomic techniques such as in-gel trypsin digestion and mass spectrometry. One of the advantages of 2D IMAEP over IMAC is that besides enrichment of phosphoproteins, it also separates the phosphoproteins based on their p*I*s.

2. Materials

Prepare all solutions in 18 MOhm water with clean containers and use ultrapure reagents. Store all solutions and reagents at room temperature unless mentioned otherwise. Always keep safety considerations in mind.

2.1. 1D Native Immobilized Metal Affinity Electrophoresis

1. 30% Stock monomer acrylamide gel solution. Store at 4°C.
2. 4× Resolving gel buffer: 1.5 M Tris–HCl, pH 8.8. Store at 4°C.
3. Native gel running buffer: 25 mM Tris–HCl, 192 mM glycine, pH 8.3. Store at 4°C.
4. 10% (w/v) Ammonium persulfate solution in water (prepare fresh).
5. *N*, *N*, *N′*, *N′*-tetramethylethylenediamine (TEMED).
6. 2× Native sample buffer: 125 mM Tris–HCl, pH 8.8, 20% glycerol, 0.005% bromophenol blue. Store at 4°C.
7. 10 × 8 × 1-mm plastic strips (homemade).
8. Metal ion solution: 1 M iron (III) chloride, manganese (II) chloride, cupric sulfate, aluminum (III) chloride, and titanium (III) acetate solution in water. Store at 4°C.
9. 100 × 100 × 1-mm empty gel cassette (Invitrogen, Carlsbad, CA).
10. Power supply: PowerEase 500 programmable power supply (Invitrogen).

2.2. SDS 1D Immobilized Metal Affinity Electrophoresis

1. 10% Sodium dodecyl sulfate (SDS) (see Note 1).
2. SDS gel running buffer: 25 mM Tris–HCl, 192 mM glycine, 0.1% SDS, pH 8.3. Store at 4°C.
3. 2× SDS sample buffer: 125 mM Tris–HCl, pH 6.8, 20% v/v glycerol, 0.2 M DTT, 4% SDS, and 0.005% bromophenol blue. Store at 4°C.
4. Gradient maker (Hoefer, Holliston, MA).
5. Light acrylamide gel solution: 7.5% (w/v) acrylamide gel solution, containing resolving gel buffer, APS, and TEMED (prepare fresh).
6. Heavy acrylamide gel solution: 15% (w/v) acrylamide gel solution, containing resolving gel buffer, APS, and TEMED (prepare fresh).
7. MCF-7 human breast carcinoma cells (American Type Culture Collection, Rockville, MD).
8. Lysis buffer: 50 mM HEPES, pH 7.5, 150 mM NaCl, 10% glycerol, 1% Triton X-100, 1 mM EGTA, 1.5 mM $MgCl_2$, 100 mM serine/threonine phosphatase inhibitor sodium fluoride (NaF), and 1 mM tyrosine phosphatase inhibitor sodium orthovanadate (Na_3VO_4).

2.3. 2D Immobilized Metal Affinity Electrophoresis

1. Rehydration buffer: 8 M urea, 2 M thiourea, 4% CHAPS, 2 mM EDTA, pH 8.0, 250 mM DTT, 2 mM tributyl phosphine, 0.5% ampholytes, and 0.005% bromophenol blue (prepare fresh).

2. Precast 7 cm IPG pH 3–10 dry gel strip (GE Bioscience, Piscataway, NJ, USA). Store at −20°C.
3. PROTEAN IEF cell (Bio-Rad Laboratories, Hercules, CA).
4. Equilibrium solution I: 50 mM Tris–HCl (pH 8.8), 6 M urea, 30% glycerol, 2% SDS, 0.005% bromophenol blue, and DTT (1% w/v) (prepare fresh).
5. Equilibration buffer II: 50 mM Tris–HCl (pH 8.8), 6 M urea, 30% glycerol, 2% SDS, 0.005% bromophenol blue, and iodoacetamide (2.5% w/v) (prepare fresh).
6. Pro-Q diamond phosphoprotein gel stain (Molecular Probes, Eugene, OR).
7. PageRuler™ Prestained Protein Ladder (Fermenta, Burlington, ON, Canada).
8. FX fluorescent imager (Bio-Rad).

3. Methods

Always manipulate gels and solutions with gloved hands. Perform all procedures at room temperature unless indicated otherwise. Strictly follow all safety regulations.

3.1. 1D Native Immobilized Metal Affinity Electrophoresis

1. Prepare the native 7.5% acrylamide gel solution by mixing 7.5 mL of 30% stock monomer acrylamide gel solution, 7.5 mL 4× resolving gel buffer, 14.8 mL deionized water, and 10 μL TEMED (see Note 2). Add 190 μL 10% APS to the gel solution immediately before casting the gel.
2. Each sample well (18×8×1 mm) is made by inserting one plastic strip at a correct depth in a 100×100×1-mm empty gel cassette, and 8 mL of the gel solution is poured into the cassette to cast the gel (see Note 3).
3. Following gel formation and removal of the plastic strips, 75 μL of the native 7.5% acrylamide gel solution containing APS, TEMED, and 1.5 μL of 1 M metal ion solution is added to the wells and allowed to polymerize (see Note 4).
4. The gel is run using native gel running buffer at ~3 W with a current of ~22 mA and a voltage of ~125 V for ~1.5 h with PowerEase 500 programmable power supply. Fe^{+3}, Mn^{+2}, and Cu^{+2} IMAEPs are run without loading the protein sample before and after electrophoresis (Fig. 1a) (see Note 5). In addition, Al^{+3}, Ti^{+3}, Fe^{+3}, Mn^{+2}, and Cu^{+2} IMAEPs are run with human albumin, carbonic anhydrase, dephosphorylated a-casein, human hemoglobin, a-casein, b-casein, and phosvitin (Fig. 1b–d) (see Note 6). Gels are stained with Coomassie blue

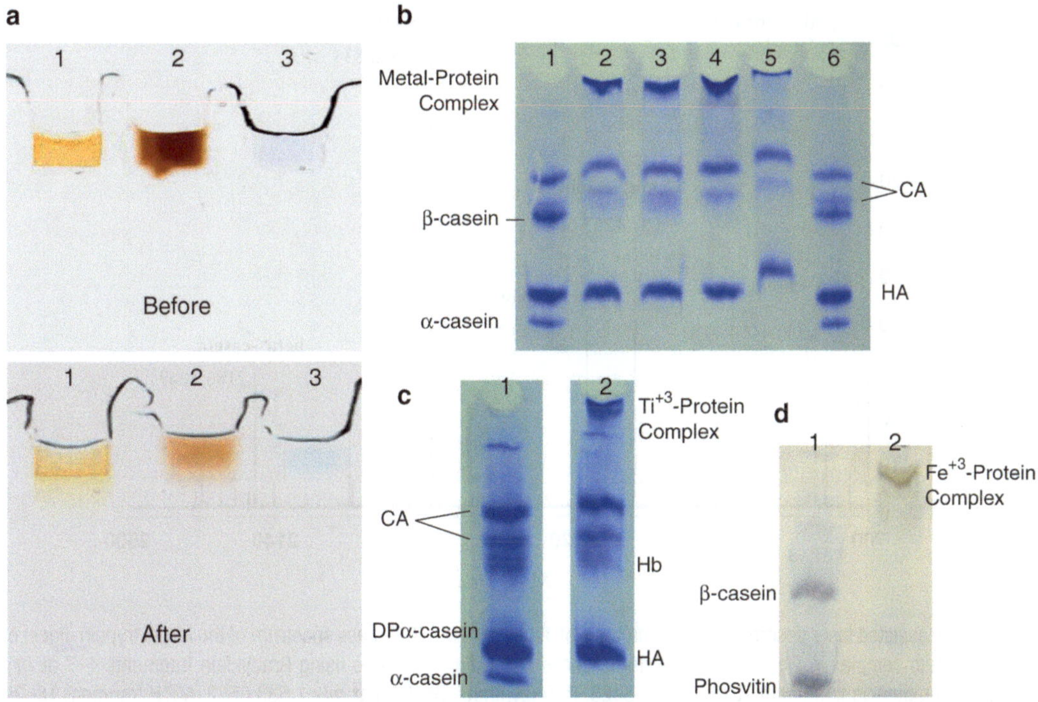

Fig. 1. (**a**) IMAEP without protein sample of a 7.5% native polyacrylamide gel before and after gel run with 1.5 μL of 1 M Fe^{+3} (*lane 1*), Mn^{+2} (*lane 2*), or Cu^{+2} (*lane 3*) incorporation. (**b**) IMAEP of human albumin (HA), carbonic anhydrase (CA), α-casein, and β-casein with no metal (*lane 1*), 1.5 μL of 1 M Al^{+3} (*lane 2*), Ti^{+3} (*lane 3*), Fe^{+3} (*lane 4*), Mn^{+2} (*lane 5*), or Cu^{+2} (*lane 6*) incorporation. (**c**) IMAEP of hemoglobin (Hb), HA, CA, α-casein, dephosphorylated α-casein (DP α-casein) with Ti^{+3} (*lane 2*) or no metal incorporation (*lane 1*). (**d**) IMAEP of phosvitin and β-casein with ferric ion (*lane 2*) or no metal incorporation (*lane 1*). Zn^{+2}-reverse staining was used to visualize phosvitin. All other proteins were visualized with Coomassie blue G-250 staining. Proteins (10 μg each) are dissolved in native sample buffer. Electrophoresis was done at ~3 W, 125 V, and 22 mA (reproduced with permission from the Association of Biomolecular Research Facilities).

G-250 or Zn^{+2}-reverse staining and documented using a scanner or a digital camera and a light box.

5. The metal–protein complex is excised from the IMAEP gel and the captured phosphoproteins are identified by standard proteomic techniques using in-gel trypsin digestion and mass spectrometry (Fig. 2).

3.2. 1D SDS Immobilized Metal Affinity Electrophoresis

1. The procedure of performing 1D SDS IMAEP is the same as in 1D native IMAEP except that 0.1% SDS is added to the native acrylamide gel solution and/or native gel running buffer. In addition, 4% SDS is added to the 2× native sample buffer and protein sample is heated at 80°C for 15 min (see Note 7).

2. MCF-7 cell is maintained in Modified Eagle's Medium supplemented with 10% fetal bovine serum and incubated at 37°C with 5% CO_2 until 100% confluent. Harvested cells are lysed

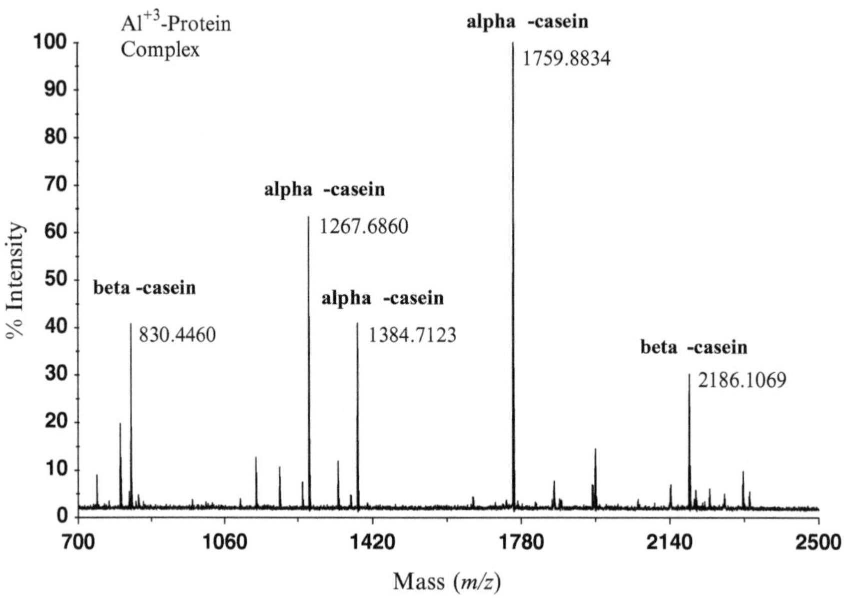

Fig. 2. Matrix-assisted laser desorption/ionization time-of-flight (MALDI–TOF) mass spectrum of the in-gel trypsin digest of the Al^{+3}–protein complex in Fig. 1b, *lane 2*. External mass calibration was done using Bradykinin fragments 1–7 at *m*/*z* 757.3997, Angiotensin II (human) at *m*/*z* 1,046.5423, $P_{14}R$ (synthetic peptide) at *m*/*z* 1,533.8582, ACTH fragment 18–39 (human) at *m*/*z* 2,465.1989, and insulin-oxidized B (bovine) at *m*/*z* 3,494.6513. Internal mass calibration was done using trypsin autodigestion peaks at *m*/*z* 842.5100 and 2,211.1046 (reproduced with permission from the Association of Biomolecular Research Facilities).

with lysis buffer with 10% proteinase inhibitor cocktail. Debris is removed by centrifugation at 14,000 × *g* for 20 min at 4°C. The lysates with protein concentration 17 mg/mL (Bradford protein assay) were stored at −80°C until used.

3. The 1D native and SDS IMAEP is run at ~3 W with a current of ~22 mA and a voltage of ~125 V for ~1.5 h (Fig. 3) (see Note 8). Gels are stained with Coomassie blue G-250 and documented using a scanner or a digital camera and a light box.
4. The metal–protein complex is excised from the IMAEP gel and the captured phosphoproteins are identified by standard proteomic techniques using in-gel trypsin digestion and mass spectrometry (Fig. 4) (see Note 9).

3.3. 2D Immobilized Metal Affinity Electrophoresis

1. A protein mixture (10 g each protein) of amyloglucosidase from *Aspergillus niger* (p*I* 3.6, MW 100 kDa), β-casein from bovine milk (p*I* 4.6, MW 24 kDa), human albumin (p*I* 5.2, MW 66 kDa), carbonic anhydrase from bovine erythrocytes (p*I* 6.6, MW 30 kDa), and lysozyme from chicken egg white (p*I* 9.3, MW 14 kDa) is dissolved in 120 μL rehydration buffer.

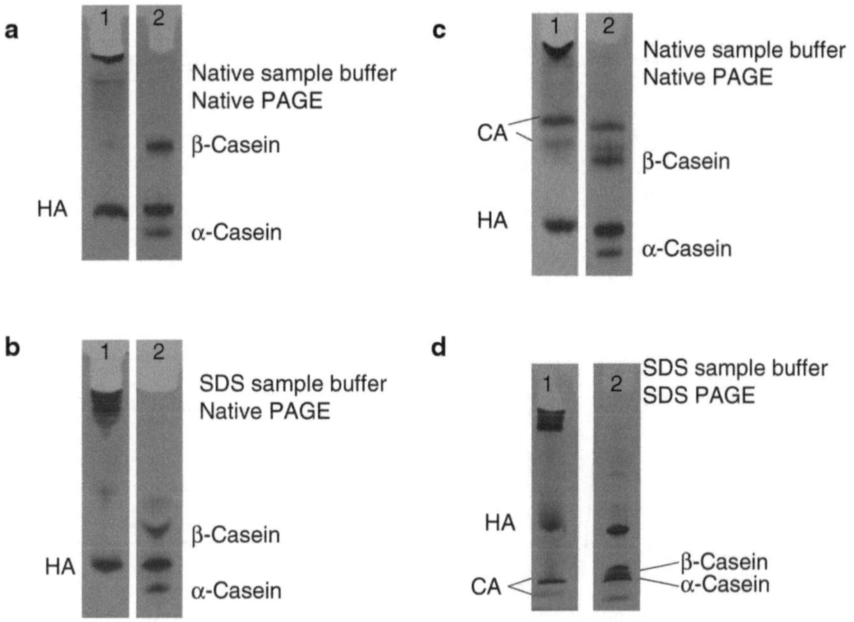

Fig. 3. (**a**) Native sample buffer and native IMAEP gel of human albumin (HA), α-casein, and β-casein with 2 μL of 1 M Fe^{+3} (*lane 1*) or no metal incorporation (*lane 2*). (**b**) SDS sample buffer and native IMAEP gel of human albumin (HA), α-casein, and β-casein with 2 μL of 1 M Fe^{+3} (*lane 1*) or no metal incorporation (*lane 2*). (**c**) Native sample buffer and native IMAEP gel of human albumin (HA), carbonic anhydrase (CA), α-casein, and β-casein with 2 μL of 1 M Fe^{+3} (*lane 1*) or no metal incorporation (*lane 2*). (**d**) SDS sample buffer and SDS IMAEP gel of human albumin (HA), carbonic anhydrase (CA), α-casein, and β-casein with 2 μL of 1 M Fe^{+3} (*lane 1*) or no metal incorporation (*lane 2*). All proteins were visualized with Coomassie blue G-250 staining. 10 μg each protein is used. Electrophoresis was done at ~3 W, 125 V, and 22 mA (reprinted with permission from Wiley-VCH).

This solution is used to rehydrate a 7-cm IPG pH 3–10 IPG dry gel strip overnight.

2. The IPG gel strip is run with the PROTEAN IEF cell for a total of 50,000 Vh.
3. After the IEF run, the IPG gel is equilibrated with the equilibrium solution I for 20 min, followed by equilibrating with equilibrium solution II for another 20 min. The IPG gel is then overlaid on top of a second dimension 7.5–15% gradient IMAEP gel.
4. The second dimension 7.5–15% gradient IMAEP gel is generated according to the following procedure: native 7.5 and 15% acrylamide gel solutions (3.9 mL) are poured into light and heavy solution reservoir chambers of a gradient maker and are used to fill an empty gel cassette. A 2D sample comb is inserted on top of the gel (see Note 10).
5. Following gel formation and removal of the 2D sample comb, one larger well of 70 × 8 × 1 mm for the IMAEP gel strip and IPG strip and one smaller well of 2 × 8 × 1 mm for the protein

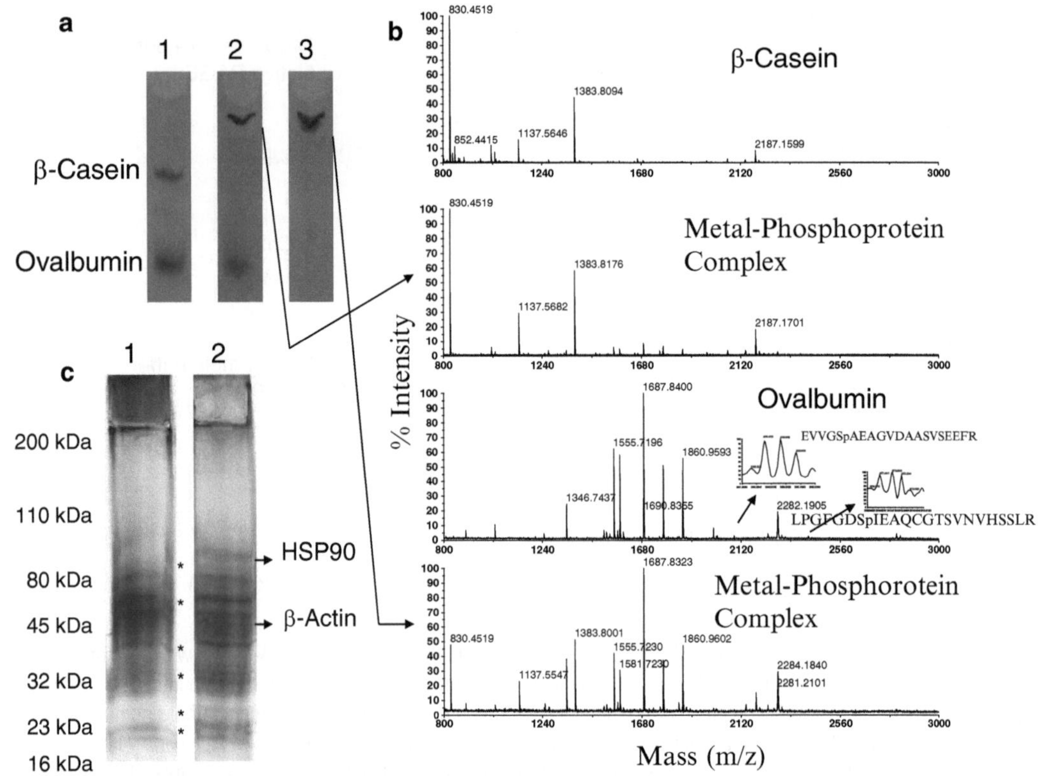

Fig. 4. (**a**) Native sample buffer and native IMAEP gel of ovalbumin and β-casein with no metal (*lane 1*) or 2 μL of 1 M Fe^{+3} incoporation (*lane 2*) and SDS sample buffer and native IMAEP gel of with 2 μL 1 M Fe^{+3} (*lane 3*). (**b**) Matrix-assisted laser desorption/ionization time-of-flight (MALDI–TOF) mass spectra of the in-gel trypsin digest of β-casein, the Fe^{+3}–phosphoprotein complex in *lane 2*, ovalbumin, and the Fe^{+3}–phosphoprotein complex in *lane 3*. External mass calibration was done using Bradykinin fragments 1–7 at *m/z* 757.3997, Angiotensin II (human) at *m/z* 1,046.5423, $P_{14}R$ (synthetic peptide) at *m/z* 1,533.8582, ACTH fragment 18–39 (human) at *m/z* 2,465.1989, and insulin-oxidized B (bovine) at *m/z* 3,494.6513. Internal mass calibration was done using trypsin autodigestion peaks at *m/z* 842.5100 and 2,211.1046. (**c**) SDS sample buffer and SDS IMAEP gradient gel (7.5–15%) of total protein extracts of MCF-7 human breast carcinoma cells with 2 μL of 1 M Fe^{+3} (*lane 1*) or no metal incorporation (*lane 2*). All proteins were visualized with Coomassie blue G-250 staining. 100 μg of total protein is loaded onto each lane. Electrophoresis was done at ~3 W, 125 V, and 22 mA for 2 h. *Asterisks* indicate possible phosphoproteins. HSP90 stands for heat shock protein 90 (reprinted with permission from Wiley-VCH).

marker are created. 525 μL native 7.5% acrylamide gel solution with APS, TEMED, and 7 μL 1 M Fe^{+3} solution was added to the larger well and allowed to polymerize.

6. The first-dimension IPG strip gel is overlaid on top of the IMEAP gel and then the second dimension of the 2D IMAEP is run at ~3 W with a current of ~22 mA and a voltage of ~125 V for ~1.5 h (Fig. 5). Gels are stained with Coomassie Blue G-250 or Pro-Q diamond phosphoprotein gel stain and documented using a scanner or a FX fluorescent imager.

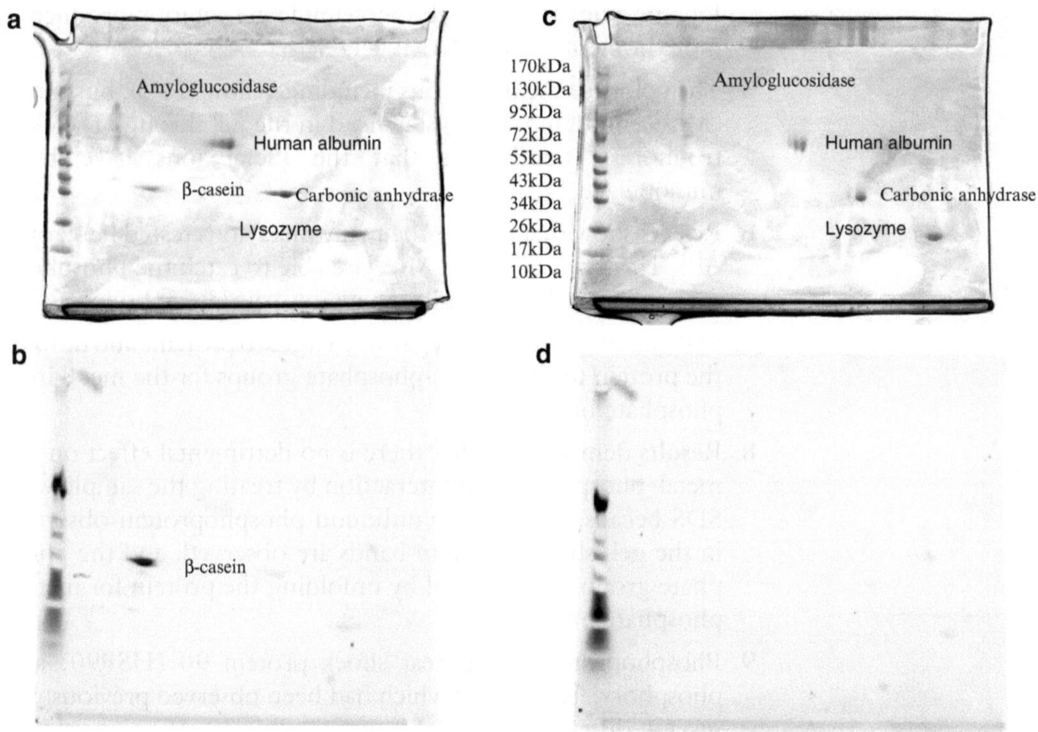

Fig. 5. Two-dimensional polyacrylamide gel electrophoresis of a protein mixture of amyloglucosidase, β-casein, human albumin, carbonic anhydrase, and lysozyme. (**a**, **b**) Normal 2D-PAGE (**c** and **d**) Fe^{+3} 2D IMAEP. A 7.5–15% gradient native Tris–glycine polyacrylamide gel with SDS Tris–glycine gel running buffer was used. All proteins were visualized with Coomassie blue G-250 staining for (**a**, **c**). All proteins were visualized with Pro-Q diamond phosphoprotein staining for (**b**, **d**). Proteins (10 μg each) are dissolved in 2D sample buffer. Electrophoresis was done at ~3 W, 125 V, and 22 mA. PageRuler™ Prestained Protein Ladder (Fermenta, Burlington, ON, Canada) was used to estimate the molecular masses of the proteins (reproduced by permission of the Association of Biomolecular Research Facilities).

4. Notes

1. 10% SDS can be prepared by dissolving 10 g of ultrahigh pure SDS with gentle warming. Precipitation will form if the solution is at lower temperature. SDS undergoes hydrolysis at elevated temperatures causing decomposition of alkyl sulfates into fatty alcohols and sodium sulfate.
2. The gel solution is degassed under vacuum in a 50-mL BD Falcon conical tube sealed with a rubber septum.
3. Scotch tape is used to hold the plastic strips during the gel forming process.
4. The amount of the metal ion should be adjusted to optimize the IMAEP performance. Polymerization time for the gel can be controlled by changing the amount of the 10% APS.

Pipetting up and down the acrylamide gel solution produces a more homogenized IMAEP gel.

5. The colors of the metal ions including yellow (Fe^{+3}), burgundy (Mn^{+2}), and blue (Cu^{+2}) stay fixed in the gel through the electrophoresis confirming that the metal ions have been immobilized.
6. Results show that among the many metal ions tested here, only Al^{+3}, Ti^{+3}, Fe^{+3}, Fe^{+2}, and Mn^{+2} are able to catch the phosphoproteins through metal–phosphate ion-pair interaction.
7. SDS treatment imparts negative charges to proteins and unfolds the protein to display the phosphate groups for the metal ion–phosphate binding.
8. Results demonstrate that there is no detrimental effect on the metal–phosphoprotein interaction by treating the sample with SDS because there is no unbound phosphoprotein observed in the gel, sharper protein bands are observed, and the phosphate groups are exposed by unfolding the protein for metal–phosphate interaction.
9. Phosphoproteins, like heat shock protein 90 (HSP90) and phosphorylated β-actin which had been observed previously in several different types of cancer cells (13–15), are clearly retained by the immobilized Fe^{+3} ion with MCF-7 protein extract. Protein extracts from HEK293 human embryonal kidney cells, *Saccharomyces cerevisiae*, Chinese Hamster Ovary cells, and mouse brain all show similar results with either Fe^{+3} or Al^{+3}.
10. We simply reused the comb that come with precast 2D Gel from Invitrogen.

Acknowledgments

We thank the support of the Research Resources Center at the University of Illinois at Chicago.

References

1. Lee BS, Jayathilaka GDL, Huang JS, Gupta S (2008) Immobilized metal affinity electrophoresis: a novel method of capturing phosphoproteins by electrophoresis. J Biomol Tech 19: 106–108
2. Lee BS, Jayathilaka GDL, Huang JS, Decresce D, Borgia JA, Zhou X, Gupta S (2008) Modification of the immobilized metal affinity electrophoresis using sodium dodecyl sulfate polyacrylamide gel electrophoresis. Electrophoresis 29:3160–3163
3. Gupta S, Jayathilaka GDL, Huang JS, Lee BS (2010) Two-dimensional immobilized metal affinity electrophoresis for capturing a phosphoprotein. J Biomol Tech 16:160–162
4. Anderson L, Porath J (1986) Isolation of phosphoproteins by immobilized metal (Fe^{+3}) affinity chromatography. Anal Biochem 154:250–254

5. Witze ES, Old WM, Resing KA, Ahn GN (2007) Mapping protein post-translational modifications with mass spectrometry. Nat Methods 4:798–806
6. Gaberc-Porekar V, Menart V (2001) Perspectives of immobilized-metal affinity chromatography. J Biochem Biophys Methods 49:335–360
7. Neville DC, Rozanas CR, Price EM, Gruis DB, Verkman AS, Townsend RR (1997) Evidence for phosphorylation of serine 753 in CFTR using a novel metal-ion affinity resin and matrix-assisted laser desorption mass spectrometry. Protein Sci 6:2436–2445
8. Zachariou M (2004) Immobilized metal ion affinity chromatography. In: Aguilar M (ed) HPLC of peptides and proteins. Humana Press, Totowa, NJ, p 89
9. O'Farrell PH (1975) High resolution two-dimensional electrophoresis of proteins. J Biol Chem 250:4007–4021
10. Bjellqvist B, Ek K, Righetti PG, Gianazza E, Gorg R, Westermeier R, Postel W (1982) Isoelectric focusing in immobilized pH gradients: principle, methodology and applications. J Biochem Biophys Methods 6:317–339
11. Hames BD (1998) Electrophoresis of protein. Oxford University Press, Oxford
12. Issaq HJ, Veenstra TD (2008) Two-dimensional polyacrylamide gel electrophoresis (2D-PAGE): advances and perspectives. Biotechniques 44:697–700
13. Rush J, Moritz A, Lee KA, Gao A, Goss VL, Speck EJ (2005) Immunoaffinity profiling of tyrosine phosphorylation in cancer cells. Nat Biotechnol 23:94–101
14. Giorgianni F, Zhao Y, Desiderio DM, Beranova-Giorgianni S (2007) Toward a global characterization of the phosphoproteome in prostate cancer cells: identification of phosphoproteins in the Lncap cell line. Electrophoresis 28: 2027–2034
15. Wang S, Zheng Y, Yu Y, Xia L, Chen G, Yang Y, Wang L (2008) Phosphorylation of beta-actin by protein kinase C-delta in camptothecin analog-induced leukemic cell apoptosis. Acta Pharmacol Sin 29:135–142

Chapter 24

Two-Dimensional Difference Gel Electrophoresis

Jonathan S. Minden

Abstract

Two-dimensional difference gel electrophoresis (2D DIGE) is a modified form of 2D electrophoresis (2DE) that allows one to compare two or three protein samples simultaneously on the same gel. The proteins in each sample are covalently tagged with different color fluorescent dyes that are designed to have no effect on the relative migration of proteins during electrophoresis. Proteins that are common to the samples appear as "spots" with a fixed ratio of fluorescent signals, whereas proteins that differ between the samples have different fluorescence ratios. With the appropriate imaging system, difference gel electrophoresis (DIGE) is capable of reliably detecting as little as 0.2 fmol of protein, and protein differences down to ±15%, over a ~20,000-fold protein concentration range. DIGE combined with digital image analysis therefore greatly improves the statistical assessment of proteome variation. Here we describe a protocol for conducting DIGE experiments, which takes 2–3 days to complete.

Key words: Proteomics, Difference gel electrophoresis, DIGE, Digital fluorescent gel imaging, IPG strips

1. Introduction

The central goals of proteomics include identifying protein changes that differentiate normal and diseased states in cells, tissues, or organisms and examining how protein changes correlate with developmental age and environment. The first stage in comparative proteomics is to separate complex mixtures of protein into individual components; this is typically done using gel electrophoresis (at the whole protein level) or column chromatography (at the peptide level). Both of these separation schemes have advantages and disadvantages. We have focused on 2D electrophoresis (2DE) because of its accessibility to most laboratories. This approach was described simultaneously by several groups in 1975 (1–3).

Biji T. Kurien and R. Hal Scofield (eds.), *Protein Electrophoresis: Methods and Protocols*, Methods in Molecular Biology, vol. 869, DOI 10.1007/978-1-61779-821-4_24, © Springer Science+Business Media, LLC 2012

Despite the substantial advances in the technology since its launch—the most notable of which was the introduction of immobilized pH gradients in the first dimension (4, 5)—some of the more significant systemic shortcomings have remained unsolved. The most troublesome of these is the inherent lack of reproducibility between gels. Efforts to surmount this limitation have mostly focused on developing computational methods for gel matching. These approaches have had limited success because the sources of gel-to-gel variation are numerous, complex, and difficult to model.

Difference gel electrophoresis (DIGE) was developed to overcome the irreproducibility problem in the 2DE methodology by labeling two samples each with a different fluorescent dye prior to running them on the same gel (Fig. 1) (6, 7). The fluorescent dyes

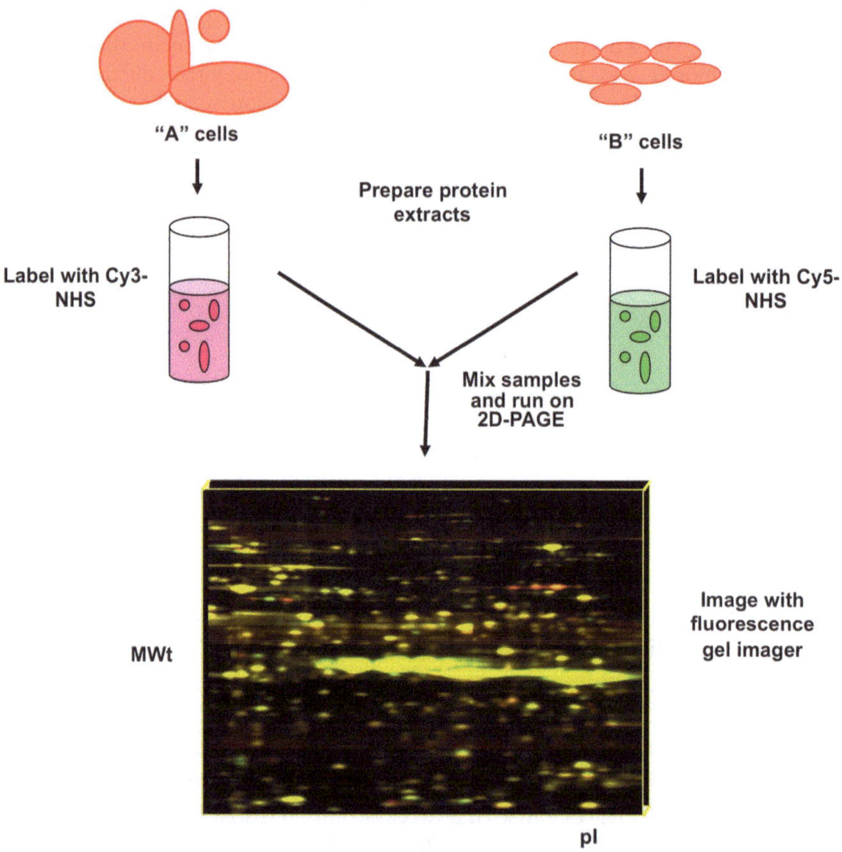

Fig. 1. Schematic of DIGE analysis. Extracts are made of two cell samples, denoted "A" and "B." These extracts are separately labeled with Cy3-NHS and Cy5-NHS, which covalently link to lysine residues. A low stoichiometry of labeling is used, where ~5% of all proteins carry a single dye molecule. The labeled protein extracts are then combined and co-electrophoresed on a 2DE gel. The gel is then imaged on a fluorescent gel imager at the Cy3 and Cy5 wavelengths. Shown here is a color overlay of Cy3 (*green*) and Cy5 (*red*) images of *Drosophila* embryo extracts. Regions of equal Cy3 and Cy5 signals appear yellow. *MWt* molecular weight, *pI* isoelectric point.

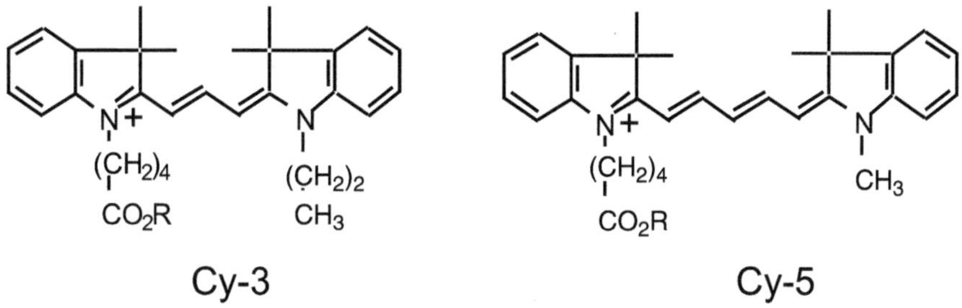

R = NHS

Fig. 2. Chemical structure of DIGE dyes. Propyl-Cy3-NHS and methyl-Cy5-NHS are shown here. These compounds are charge and mass matched. The fluorescent characteristics of Cy3 and Cy5 are dictated by the three-carbon and five-carbon polyene chains, respectively, linking the two indoline rings.

used in DIGE, Cy3-NHS, and Cy5-NHS (Fig. 2) are cyanine-based, molecular weight-matched, amine reactive, and positively charged. These characteristics, coupled with sub-stoichiometric labeling, result in no electrophoretic mobility shifts arising between the two differentially labeled samples when they are co-electrophoresed. Therefore, in DIGE, every identical protein in one sample superimposes with its differentially labeled counterpart in the other sample, allowing for more reproducible and facile detection of differences. Furthermore, DIGE is a sensitive technique, capable of detecting as little as 0.2 fmol of protein, and this detection system is linear over a ~20,000-fold concentration range (6, 8, 9).

The most important considerations in performing DIGE experiments are experimental design and sample preparation (10). DIGE has been used to analyze proteome changes from a wide variety of cell types and bodily fluids, including serum (11–14). The sample-preparation protocol depends on the cell type. Most samples require mild homogenization in lysis buffer to extract protein. DIGE is an extremely sensitive method, in which a 15% change in protein abundance is more than two standard deviations (SDs) above the normal variation (9). One must take great care in deciding which samples to compare, while bearing in mind the sources of variation. If specific tissues are to be compared, one must carefully dissect the tissue to avoid variation. Hypothetically, if one has 10% contamination of neighboring tissue in one sample and 5% variation in another, it might lead to artifactual protein differences. One way to avoid tissue contamination is to use laser microdissection, which provides a precise method for capturing specific populations of cells (15–17). Another source of variation can arise during sample cleanup or fractionation. If sample cleanup or fractionation is planned, it is best to label the samples independently and then combine them prior to cleanup or fractionation. This will alleviate variation due to handling. Simply measuring total protein after sample cleanup or fractionation will not guard against variation in loss of specific proteins during processing.

The protocol described here outlines the steps we use in performing two-dimensional difference gel electrophoresis (2D DIGE) experiments. We typically use a two-dye approach with Cy3 and Cy5. For comparing model systems, such as *Drosophila* or yeast, the two-dye approach is more reproducible than the three-dye method, which utilizes a Cy2-labeled pooled sample for normalization (18). The three-dye method is better suited to multiple sample comparisons where there is considerable genetic variation, as in the analysis of human samples (19). The precursor to the three-dye method was a two-dye method comparing a Cy3-labeled pooled control + test sample and Cy5-labeled control or test samples (20). We always run two gels for each comparison, in which the order of labeling is reversed (referred to as reciprocal labeling). This allows one to differentiate between sample-dependent differences and rare dye-dependent differences. The latter are presumably due to incomplete solubilization of proteins, allowing the two dyes to associate differently with the protein. Loading equal amounts of each protein sample is advisable, but high precision is not required as slight load differences are normalized during image analysis. All steps should be performed on ice or in a cold room if necessary. As soon as lysis is complete, samples can be stored at −80°C in aliquots of 100–250 μg at concentrations >1 mg/mL. When comparing whole-cell extracts, we generally load 80–200 μg of each sample (160–400 μg total protein).

High-quality digital imaging of DIGE gels is essential for detecting proteome changes. To peer as deeply into the proteome as possible, it is important to use an imager that is capable of true 16-bit data collection. Fluorescent gel imagers come in two formats: scanner based or CCD camera based. There are several commercial gel imagers, such as the Typhoon imager (Amersham Biosciences/GE Healthcare), which is commonly used for DIGE gel imaging. The Typhoon imager is based on a scanning laser-illumination system and photomultiplier detector. The gel-scanning systems that are suitable for DIGE analysis require a separate device for spot picking. We use a homemade CCD-based fluorescence imager that has an integrated spot-picking robot (Fig. 3).

The main issue to consider in gel imaging is that one should try to utilize the full dynamic range of the detector. The dynamic range of the detector is measured in terms of the number of discrete gray levels or bits of information that each pixel can accommodate. The CCD camera in our imager has a 16-bit dynamic range or 65,536 gray levels. Ideally, one would like to load enough fluorescently labeled protein and set the exposure parameters so that the brightest protein spot nearly saturates the full dynamic range of the detector. Another important issue is the fluorescent background. Many of the materials and substances that are placed

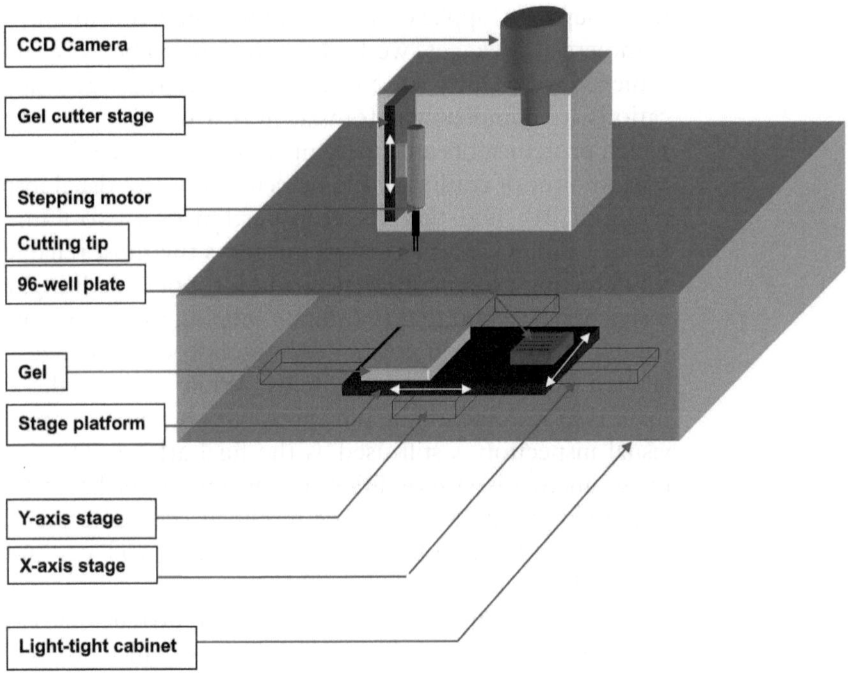

Fig. 3. Fluorescence gel imager/spot picker. This diagram illustrates the components of the DIGE gel imager. This device comprises the following components: a scientific-grade Peltier-cooled 16-bit CCD camera (Photometrics/Roper Scientific), an 105-mm macro lens (Nikon), a computer-driven stage (New England Affiliated Technology), and a spot-picker drive (Applied Precision). All electronic components are controlled by a computer workstation running Windows and custom software. Not shown is the illumination system, which is composed of two 250 W quartz-tungsten-halogen light sources (Oriel) that direct their light through motorized filter wheels (Ludl). The light is directed onto the gel via fiber-optic light guides (Oriel), and fluorescence wavelengths are selected by band-pass filters (Chroma Technology).

in the imager can have intrinsic fluorescence. For example, borosilicate glass is highly fluorescent so any glass used to hold the gels in the imager must be made of quartz or fused silica. Although background fluorescent signals can be removed computationally, they limit the dynamic range of the gel images and mask signals from low-abundance proteins. Our gel-imaging system is capable of detecting as little as 0.2 fmol of protein and can detect proteins over a ~20,000-fold concentration range. As there are a limited number of such gel-imaging systems currently in use, we will not describe our imaging protocol here in detail.

Image analysis in our laboratory is done using several software applications, including ImageJ, QuickTime, and SExtractor. There are integrated image-analysis packages for DIGE gels, the most commonly used of which is DeCyder (Amersham Biosciences/GE Healthcare). We use ImageJ and QuickTime for image visualization and annotation. SExtractor is an astrophysics freeware application that has been adapted to quantify protein-spot intensities and ratios (https://sourceforge.net/projects/sextractor) (21). Although these separate applications are more time-consuming than the commercial packages, we find the analysis to be as reliable as that achieved with DeCyder. None of the current image-analysis applications are completely automatic in that the computer determines which protein spots are changing and the degree of change. A certain amount of editing of the detected spots by visual inspection is required. We find that visual inspection of a two-frame looping movie of the Cy3 and Cy5 gel images is the most reliable method for detecting protein differences and for spot-list editing. For visual inspection, we find that two-frame looping movies are much more robust than superimposing two pseudo-color images of Cy3-labeled and Cy5-labeled proteins. While state-of-the-art, automated spot-detection applications are becoming more robust, visual inspection is still used as the final arbiter. The outcome of image analysis is a list of difference-protein spots that indicate putative differentially expressed or modified candidate proteins. This spot list can be used for large-scale protein-profiling studies using standard bioinformatics tools (22).

Despite DIGE being sensitive and reproducible, two caveats need to be mentioned. First, 2DE does not efficiently resolve integral membrane proteins. This is due to their hydrophobic domains causing precipitation during isoelectric focusing (IEF). Other laboratories are working to solve this problem. Second, labeling with the amine-reactive DIGE dyes limits one to sub-stoichiometric labeling (also known as minimal labeling), where less than 5% of all proteins carry a single bound dye molecule, and the rest have no bound dye. For proteins that are >25 kDa, there is no appreciable molecular weight shift between labeled and unlabeled protein, while there is a slight but predictable shift for smaller proteins in which unlabeled proteins run about half a spot diameter faster than their labeled counterparts. This shift problem has been addressed with the development of cysteine-reactive dyes. These dyes allow one to saturation label all available cysteines, which eliminates the shift between labeled and unlabeled proteins, as all proteins are maximally labeled (15, 16, 23). Regardless of these limitations, DIGE combined with MS is a sensitive, robust, and useful approach for comparative proteomics.

2. Materials

2.1. Reagents

Use double-distilled H_2O unless otherwise stated. The source and purity of ingredients is critical, especially for solutions used for labeling and IEF:

1. SDS: 20% w/v stock solution.
2. HEPES: 100 mM Na-HEPES, pH 8.0, stock.

3. Acrylamide: 30% T, 2.6% C stock solution.
4. Ammonium persulfate (APS): 10% w/v stock kept at −20°C.
5. IEF strips (e.g., pH 3–10 NL, 18-cm Immobiline DryStrips, Amersham Biosciences/GE Healthcare, Piscataway, NJ).
6. Whatman 3-mm filter paper.
7. Nylon strips (30–130 mesh Nitex® screen, Sefar America Inc).
8. IEF strip-cleaning detergent (Amersham Biosciences/GE Healthcare).
9. RBS 35 detergent (concentrate; Pierce, Thermo Scientific Rockford, IL).

2.2. Equipment

1. IPGphor IEF apparatus (Amersham Biosciences/GE Healthcare).
2. SE660 vertical gel-electrophoresis apparatus (Hoeffer, Inc., Holliston, MA).
3. Fluorescent gel imager/spot picker (custom made, drawings are available upon request).
4. Model 485 Gradient Former gradient maker (Bio-Rad, Hercules, CA).
5. DryStrip rehydration tray (Amersham Biosciences/GE Healthcare).
6. Image-analysis software applications (e.g., ImageJ, QuickTime, SExtractor).
7. Protein-digestion robot (e.g., ProGest, Genomics Solutions).

2.3. Reagent Setup (see Note 1)

1. Lysis buffer: 7 M urea, 2 M thiourea, 4% (wt/vol) CHAPS, 10 mM DTT, and 10 mM Na-HEPES (pH 8.0) (see Note 2).
2. Dye solutions: Add 25 μL DMF to each 25-nmol tube of Cy3-NHS (Amersham Biosciences/GE Healthcare). As the extinction coefficient of Cy5 is greater than that of Cy3, add 37.5 μL DMF to each 25-nmol tube of Cy5-NHS (Amersham Biosciences/GE Healthcare) (see Note 3).
3. Quencher: 5 M methylamine-HCl and 10 mM HEPES (pH 8.0). Dissolve 2.38 g HEPES in 38.8 mL of 40% vol/vol methylamine aqueous solution (see Note 4).
4. Rehydration buffer: 7 M urea, 2 M thiourea, 4% (wt/vol) CHAPS, 10 mM DTT, 2 mM acetic acid, 0.002% (wt/vol) bromophenol blue, and 1% (wt/vol) IPG buffer. Use the appropriate IPG buffer (Amersham Biosciences/GE Healthcare) that corresponds to the pH range of the IEF strips in the experiment (see Note 5).
5. Equilibration stock: 50 mM Tris–HCl (pH 8.8), 6 M urea, 30% (vol/vol) glycerol, 2% (wt/vol) SDS, and 0.002% (wt/vol) bromophenol blue (see Note 6).

Table 1
Composition of acrylamide solutions needed to form a 10–15% gradient gel

Stock solution	Light	Heavy	Stack
Acrylamide (30% T, 2.6% C)	8.25 mL	12.25 mL	400 μL
1.5 M Tris–HCl, pH 8.8	6.25 mL	6.25 mL	
0.5 M Tris–HCl, pH 6.8			800 μL
Sucrose		3.75 g	
SDS (20% w/v)	125 μL	125 μL	16.5 μL
H_2O	10.375 mL	4.175 mL	2.05 mL
APS (10% w/v)	82.5 μL	82.5 μL	16.7 μL
TEMED	8.25 μL	8.25 μL	1.7 μL

6. Equilibration buffers I and II: Prepare equilibration buffers I and II as needed using 10 mL per IEF strip: equilibration buffer I comprises 1% (wt/vol) DTT in equilibration stock; equilibration buffer II comprises 2.5% (wt/vol) iodoacetamide in equilibration stock.
7. Gradient gel solutions: See Table 1 for the composition of the three acrylamide solutions needed to form a 10–15% gradient gel that is 14×24 cm (see Note 7).
8. Tank buffer: 25 mM Tris, 190 mM glycine, and 0.1% (wt/vol) SDS. Can be made or purchased as a 10× stock.
9. Agarose-sealing solution: 125 mM Tris–HCl (pH 6.8), 0.1% (wt/vol) SDS, 1% (wt/vol) agarose, and 0.002% (wt/vol) bromophenol blue. Heat in a microwave until the agarose dissolves.
10. Fixative: 40% (vol/vol) methanol (HPLC grade) and 1% (vol/vol) acetic acid (glacial).
11. IEF strips: IEF strips are available in different pH ranges and sizes. pH 3–10 NL strips are a good starting point for whole-cell extracts; "NL" indicates that the strip has a nonlinear pH gradient, with increased resolution between pH 5 and 7. The 13-cm strips fit well into the Hoeffer SE 660 system that we use for SDS-PAGE. If using 18 cm strips, the ends will have to be cut to fit. This does not lead to appreciable loss of proteins when using pH 3–10 NL strips, as most samples do not have many proteins focused with pI values near the extremes. In practice, using 18 cm pH 3–10 NL strips improves the resolution

of proteins in the region where most proteins lie. pH 4–7 strips might be used instead of pH 3–10 NL if most proteins are within this pI range, which is the case for bacterial samples. Narrow pH range strips (1 pH unit) are available for closer study of proteins within a region of interest.

3. Methods

3.1. Casting the Gradient Gel

1. Assemble the gel cassette using cleaned and dried plates. We use homemade 10–15% gradient gels. The glass plates used for the SE660 vertical gel-electrophoresis apparatus are 24 × 18 cm. The spacers used are 1.5 mm in thickness (see Note 8).
2. Set up a standard two-chamber gradient maker so that the outlet tube directs the acrylamide solution into the top of the gel cassette (see Note 9).
3. Add SDS to the heavy and light acrylamide solutions and mix well. Then add 10% (wt/vol) APS and TEMED when ready to pour the gel.
4. Pour heavy and light solutions simultaneously into the front and back chambers of the gradient maker, respectively. Open the mixing channel and the front pinchcock simultaneously to start pouring the gel. Overlay with ~750 μL water-saturated *n*-butanol (top layer) and allow the gel to polymerize for ≥1 h (see Note 10).
5. Drain off the *n*-butanol and rinse with water to remove. Pour the stacking gel and overlay with *n*-butanol (see Note 11).

3.2. Rehydrating the IEF Strip

1. Rehydrate Immobiline DryStrips in a DryStrip rehydration tray according to the manufacturer's instructions (see Note 12).

3.3. Sample Preparation

1. Place the cells or tissue sample in a 1.5-mL centrifuge tube that has a fitted pestle. Sufficient material to yield 100–250 μg protein is needed. Rinse the cells with an ice-cold low-salt Na-HEPES (pH 8.0) buffer that does not contain primary amines. Remove excess liquid and add 100 μL lysis buffer. Homogenize the cells with a few passes or turns of the pestle. Centrifuge the sample in a microcentrifuge for 5–15 min at 15,000 × *g* at 4°C to remove unbroken cells and debris (see Note 13).
2. Measure protein concentration using a Bradford assay. As urea, CHAPS and DTT affect the Bradford assay, the standards and samples should all be made up in lysis buffer. Add 2 μL sample/standard to 1,600 μL water in a plastic test tube and mix. Add 400 μL Bradford reagent, mix, and measure the OD at 595 nm

within 1 h. If the sample is too concentrated, dilute with lysis buffer. A standard curve should be made with BSA dissolved in lysis buffer at a concentration range of 0.5–2.5 mg/mL. The blank should be made with 2 μL lysis buffer (see Note 14).

3.4. Protein Labeling

1. The following four steps describe a typical DIGE experiment in which protein samples denoted "A" and "B" are to be compared. Set up four 1.5-mL microcentrifuge tubes, numbered 1–4. Pipette 80 μL protein sample A, which contains 80–200 μg of protein in lysis buffer, into tubes 1 and 3. Pipette 80 μL protein sample B, which contains 80–200 μg protein in lysis buffer, into tubes 2 and 4 (see Note 15).
2. Add 1 μL Cy3-NHS stock solution to tubes 1 and 4. Add 1 μL Cy5-NHS stock solution to tubes 2 and 3. Mix by vortexing. Incubate in the dark on ice for 15 min (see Note 16).
3. To each tube, add 1 μL quenching solution (see Note 17).
4. In a new tube labeled "gel 1," combine the entire contents of tubes 1 and 2. In a new tube labeled "gel 2," combine the contents of tubes 3 and 4. Add 1.6 μL of the appropriate IPG buffer solution (1 μL for every 100 μL of sample) to the gel 1 and 2 tubes. Mix by vortexing and briefly spin in a microcentrifuge at top speed for a few seconds at 4°C to consolidate all of the liquid (see Note 18).

3.5. First-Dimension Electrophoresis: IEF

1. IEF should be set up according to the manufacturer's instructions with a few modifications. Refer to the user manual for more details and precautions to be taken while setting up IEF. Turn on the IPGphor. Place ceramic strip holders on the IPGphor platform, with the pointed end toward the anode.
2. Cut two 3-mm filter paper wicks for each gel and immerse in HPLC water. The wicks should be approximately the width of the gel (see Note 19).
3. Open the rehydration cassette assembly. Remove a rehydrated IEF strip and rinse it briefly with HPLC water. Blot excess water. Place the IEF strip in the holder with the acidic end toward the anode. Place wicks at both ends of the strip, making sure to blot excess water from the wicks before placing on the gel. Place an electrode on each of the wicks. Place a sample cup near the electrode on the acidic end of the gel. Cover the surface of the gel with ~1 mL of DryStrip Cover Fluid (see Note 20).
4. Pipette the sample into the sample cup, taking care to avoid bubbles. Place the lid over the strip holder. Start the IEF program shown in Table 2, using 50 μA per strip at 18°C (see Note 21).

Table 2
Recommended program for running 18 cm, pH 3–10 IEF strips

Step	Voltage	Voltage gradient type	Duration (h)	Duration (kVh)
1	500	Step-n-hold	1	
2	4,000	Step-n-hold	1	
3	8,000	Step-n-hold	2–4	
4	8,000	Step-n-hold		30–40

3.6. Equilibration

Remove strips and place in 100-mm Petri dishes curled along the inside of the dishes and the gel surfaces facing inward. Equilibrate in 10 mL room temperature equilibration buffer I for 15 min with swirling. Rinse briefly with water. Equilibrate in 10 mL room temperature equilibration buffer II for 15 min with swirling. Drain buffer. Equilibrated IEF strips can be run on the second dimension immediately (see Note 22).

3.7. Second-Dimension Electrophoresis: SDS-PAGE

1. Melt agarose-sealing solution in a microwave and place in a beaker of hot water to keep melted. Drain the *n*-butanol from the top of the second-dimension gel and rinse well with water.
2. Remove the IEF strip from the equilibration II solution or thaw if previously equilibrated and frozen. Place IEF strip on top of the second-dimension gel with the plastic touching the back glass and the acidic end of the strip toward the left (see Note 23). Gently push the IEF strip down until it contacts the stacking gel. Cover the IEF strip with melted agarose until it just covers the IEF strip (see Note 24).
3. Place the gel in the electrophoresis unit and fill the upper and lower chambers with tank buffer. The lower tank should be filled with 3.5 L tank buffer and stirred constantly at 4°C in a cold room.
4. Electrophorese at a constant current with a maximum voltage set at 500. Typically, we set the current at 10–25 mA per gel depending on how long we want the run to take (e.g., 25 mA per gel requires ~10 h to complete and 15 mA per gel takes ~16 h).
5. At the completion of electrophoresis, remove the gels from the glass plates. The stacking gel, dye front, and IEF strip should be removed, and the gel soaked in fixative for ≥2 h with gentle swirling (see Note 25).

3.8. Image Acquisition

Acquire images using your chosen imaging system. The DIGE experiment described will yield four images: two each from gel 1 and reciprocal gel 2. Each image is 1,024 × 1,280 pixels, with a resolution of 135 μm per pixel. These images should be stored as raw unsigned 16-bit data.

3.9. Image Analysis

Perform image analysis. Image analysis in our laboratory is done using several software applications: ImageJ, QuickTime, and SExtractor. Visually inspect a two-frame looping QuickTime movie of the Cy3 and Cy5 gel images to detect protein differences. The outcome of image analysis is a list of difference-protein spots that indicate significant differences between the two protein samples being compared.

In general, we rely on the two-frame looping movies to identify the significant protein changes we want to identify by MS. The exogenously added BSA serves as a loading control to balance the image-display parameters for making two-frame looping movies. More precise quantification is done later using SExtractor. To accurately assess the level of protein change, the Cy3 and Cy5 images are summed and submitted to SExtractor for spot detection, which generates a list of elliptical objects that specify individual protein spots. This list is used as a mask to determine the protein spot intensities in background-subtracted Cy3 and Cy5 images. This approach typically yields an SD of 5–7% for unchanging spots. Visual difference-protein detection requires <1 h for an experienced user to complete. Computational spot-detection methods are becoming more reliable, but require manual editing. The computational methods are particularly useful when comparing multiple gels.

3.10. Spot Picking

Use the integrated spot-cutting tool to "click" on spots of interest, cut a 1.8-mm diameter plug from the gel, and deposit it in a 96-well plate designed for a protein-digestion robot (ProGest). The wells of the receiving plate should be filled with 1% (vol/vol) acetic acid. The imager software allows one to create a list of protein spots to be retrieved. After all the desired spots are cut, the acetic acid solution should be removed and the plate moved to the digestion robot (see Note 26).

3.11. Protein Identification

Identify the difference proteins using MS. This can be done by routine methods using a variety of instruments. DIGE labeling with Cy3-NHS or Cy5-NHS labels only a small fraction of proteins in the lysate: ~5% of all proteins have a single dye molecule bound, whereas the rest have none. Hence, DIGE labeling does not affect the MS identification of proteins. It is important to note, however, that the DIGE imaging system is capable of detecting <0.2 fmol of protein. To identify low-abundance difference proteins, one must either enrich for the proteins of interest or pool protein from multiple DIGE gels (see Note 27).

3.12. Timing

1. Gel setup: 1 h.
2. Rehydrating the IEF strip: 1 h for set up and 12 h for rehydration.
3. Sample preparation: 1 h.
4. Protein labeling: 1 h.
5. First dimension IEF: 6–12 h.
6. Equilibration of IEF strip for second dimension: 30 min.
7. Second dimension SDS-PAGE: 6–12 h.
8. Image acquisition: 30 min per gel.
9. Image analysis: 1 h per 24 × 18 cm gel with ~1,500 protein spots.
10. Spot picking: 1–2 h.

4. Notes

1. Wear gloves at all times to avoid keratin contamination, which is a key obstacle to further analysis by MS.
2. Never heat urea solutions above room temperature (18–21°C) as this will cause the formation of cyanate, which carbamylates protein and produces charge trains in 2DE gels (23). The pH of this solution is critical for the labeling reaction. We typically store the lysis buffer in 1 mL aliquots at −80°C.
3. Use fresh, dry DMF of the best quality. Store dye solutions at −80°C and allow them to equilibrate to ice temperature before opening the tube.
4. Slowly add concentrated HCl with stirring, until the pH reaches 8.0. This step should be done on ice in a fume hood. You will need ~60 mL acid to reach the desired pH. The total volume should be 100 mL. Aliquot and store at −80°C.
5. Store rehydration buffer at 4°C. Use within 7 days.
6. Can be stored at 4°C for 4–6 weeks.
7. These solutions can be stored without SDS, APS, and TEMED for 2–3 weeks at 4°C.
8. One can bypass this step by purchasing premade gradient gels for 2DE.
9. Make sure the opening of the outlet tube is sufficiently wide that the pouring process takes <2–3 min; otherwise, the gel might begin to polymerize during the pouring process. We use a disposable 200-μL pipette tip, cut at an angle, which fits snuggly between the glass plates and allows for rapid casting of the gradient gel.

10. If pouring several gels at a time, rinse the gradient maker with water between gels. Use a magnetic stirrer and stir bar in the front chamber to ensure sufficient mixing of the heavy and light solutions.
11. Acrylamide is a neurotoxin; wear gloves and use appropriate handling precautions. Gels should be allowed to polymerize for ≥8 h if one intends cutting protein spots for MS. Unpolymerized acrylamide causes side-chain and amino-terminal modification of proteins, and these might pose problems for MS. Gels can be allowed to polymerize overnight. They should be layered with sufficient water-saturated *n*-butanol that the gels do not dry out. It is alright to pour the stacking gel just 4–6 h before the electrophoresis is set up. Commercial gels might be suitable, provided they do not contain any fluorescent contaminants. Gels might be stored at 4°C for 2 days, provided care is taken to prevent drying.
12. We prefer to use sample cups to load protein samples. We do not recommend including the protein sample in the rehydration solution as this leads to vertical and horizontal streaks and protein loss (24). Make sure the IPG buffer matches the pH profile of the IEF strip. The Immobiline DryStrips need to be rehydrated in rehydration buffer for 8–22 h.
13. The majority of our samples are *Drosophila* embryos, which lyse easily by this method. Keep the centrifuge tube on ice during homogenization or work in a cold room. Some cell types might require more vigorous homogenization, such as *Saccharomyces cerevisiae*, which requires vortexing at high speed in the presence of glass beads after addition of lysis buffer. A glass homogenizer may be used for larger samples, but the lysate needs to be centrifuged at top speed in a microcentrifuge for 5–15 min at 4°C to remove debris.
14. Ideally, the protein concentration of the lysate should be 1–2 mg/mL. If the lysate concentration is too high, dilute with lysis buffer. If the concentration is too low, the sample may be concentrated by precipitation, but this could create problems with protein loss during precipitation, re-solubilization, and pH adjustment. It is preferable to start with sufficient material to yield a suitably concentrated lysate. Another important consideration is the removal of contaminating factors that affect IEF, such as salt, lipids, detergents, and nucleic acids. There are a number of commercial protein cleanup kits that are worth testing; however, one must be cautious about protein loss and pH effects. We recommend labeling the protein samples first and then combining the samples before using a protein cleanup protocol.

15. Try to use equal amounts of protein in each sample. Optionally, you can also add 1 μL of 1 mg/mL BSA in lysis buffer to each tube as a loading control. Mix by vortexing.
16. To synchronize the labeling time, add the dye to the walls of all reaction tubes. The Cy3 and Cy5 reaction tubes containing samples to be loaded on the same gel are briefly spun, vortexed, and spun again in a microcentrifuge at top speed for a few seconds at 4°C. Allow the dye stocks to warm up to ice temperature before opening.
17. Again, the solution is dispensed onto the walls of the tubes, and the reaction is started as described in the previous step. Incubate in the dark on ice for 30 min. The duration of quenching can be extended up to 2 h if necessary. After labeling, the samples can be stored at −80°C. Thaw on ice prior to applying to the IEF strips.
18. The samples should be loaded onto IEF strips immediately after combining.
19. The wicks may be made a little longer than they are wide.
20. You can also cover the entire length of the gel between the cathodic electrode and the sample cup holder with nylon-mesh strips cut to the width of the opening in the ceramic strip holder and the length of the exposed IEF strip. Gently press down with tweezers to make sure the strips are contacting the surface of the gel. The nylon mesh helps retain a thin film of cover fluid over the surface of the gel, thereby preventing it from drying out. Avoid unnecessary delays during this step to prevent the IEF strip from drying. The IEF gel is fragile and requires careful handling.
21. We cover the IEF machine with an opaque cover to limit photobleaching. The time taken to reach 30–40 kVh using the above protocol is between 6 and 12 h. The duration depends on the sample and can vary from run to run depending on the amount of salt and other contaminants in the sample. Gels should be equilibrated in equilibration buffer immediately after the run.
22. Wash all IEF accessories on finishing IEF. Use double-distilled water for washing. Wash strip holders with the IEF strip-cleaning detergent. Use Q-Tips. Avoid scratching. Rinse thoroughly with water and then absolute ethanol. Wash electrodes with water and ethanol. Nitex strips should be washed with RBS detergent, water, and finally ethanol. Clean strip-holder lids and the IPGphor platform. After equilibration, the IEF strips can be stored at −80°C for several weeks.

23. Make sure the gel side is not touching the front glass.
24. Make sure not to damage the IEF gel. Dispense some from each end so that it covers evenly. Make sure there are no bubbles. If there are bubbles, remove them once the agarose solidifies.
25. Gels can be stored in fixative for several months without significant loss of fluorescent signal. For long-term storage, refrigeration is recommended. However, protein spots to be identified by MS should be cut out and frozen at –80°C within 1 week.
26. A key difference between the gel-imaging system we use and the Typhoon imaging system is that we use an open-gel format that is amenable to imaging and spot picking within the same device, whereas the Typhoon imager is best suited to imaging the gel within the electrophoresis plates, so spot picking is done on a different device. The open-gel format does not require the gel to adhere to one of the two electrophoresis plates. The cut gel fragments can be stored at –80°C for at least several weeks.
27. There are two main reasons for poor results: insufficient protein labeling and incomplete IEF. Insufficient labeling is commonly due to the pH of the lysate being <8.0 or the presence of excess primary amines, such as Tris. If there is sufficient sample, one can test the pH using a small piece of pH paper. Tris <50 mM does not appear to interfere with protein labeling, and 30 mM is recommended. Incomplete focusing is usually the result of interfering contaminants, such as salts and nucleic acids. Salts, which can be removed using a protein cleanup kit, prevent the IEF from reaching the desired voltage. Nucleic acids make the sample viscous and difficult to handle and also interfere with focusing. Nucleic acids can be removed by commercially available cleanup kits or can be broken up by sonication or the addition of nuclease (1).

    There will always be gel-to-gel variation for 2DE gels. This is due to a myriad of uncontrollable features in manufacturing, sample preparation, and electrophoretic conditions. DIGE was developed to circumvent some of this variability. The digital-imaging system and fluorescence labeling also provide a three or four orders of magnitude linear response, which provides much more accurate estimation of protein abundance than silver or Coomassie blue staining of proteins. DIGE protein differences can be visualized in a variety of ways, including color overlay (Fig. 1) and two-frame looping movies.

Acknowledgments

The development of DIGE would not have been possible without the efforts and dedication of my students and staff (Mustafa Ünlü, Liz Morgan, Chris Lacenere, Surya Viswanathan, Lei Gong, Mamta Puri, Anupam Goyal, and Susan Down).

References

1. O'Farrell PH (1975) High resolution two-dimensional electrophoresis of proteins. J Biol Chem 250:4007–4021
2. Scheele GA (1975) Two-dimensional gel analysis of soluble proteins. Characterization of guinea pig exocrine pancreatic proteins. J Biol Chem 250:5375–5385
3. Klose J (1975) Protein mapping by combined isoelectric focusing and electrophoresis of mouse tissues. A novel approach to testing for induced point mutations in mammals. Humangenetik 26:231–243
4. Gorg A, Postel W, Gunther S (1988) The current state of two-dimensional electrophoresis with immobilized pH gradients. Electrophoresis 9:531–546
5. Hamdan M, Righetti PG (2005) Proteomics today: protein assessment and biomarkers using mass spectrometry, 2D electrophoresis, and microarray technology. Wiley, Hoboken, NJ
6. Ünlü M, Morgan ME, Minden JS (1997) Difference gel electrophoresis: a single gel method for detecting changes in protein extracts. Electrophoresis 18:2071–2077
7. Viswanathan S, Ünlü M, Minden JS (2006) Two-dimensional difference gel electrophoresis. Nat Protocols 1:1351–1358
8. Lilley KS, Friedman DB (2004) All about DIGE: quantification technology for differential-display 2D-gel proteomics. Expert Rev Proteomics 1:401–409
9. Gong L, Puri M, Unlu M et al (2004) Drosophila ventral furrow morphogenesis: a proteomic analysis. Development 131:643–656
10. Shaw MM, Riederer BM (2003) Sample preparation for two-dimensional gel electrophoresis. Proteomics 3:1408–1417
11. Lilley K (2002) Protein profiling using two-dimensional difference gel electrophoresis (2-D DIGE). In: Coligan JE, Dunn BM, Speicher DW, Wingfield PT (eds) Current protocols in protein science. Wiley, New York, pp 22.22.21–22.22.14
12. Marouga R, David S, Hawkins E (2005) The development of the DIGE system: 2D fluorescence difference gel analysis technology. Anal Bioanal Chem 382:669–678
13. Wang H, Clouthier SG, Galchev V et al (2005) Intact-protein-based high-resolution three-dimensional quantitative analysis system for proteome profiling of biological fluids. Mol Cell Proteomics 4:618–625
14. Okano T, Kondo T, Kakisaka T et al (2006) Plasma proteomics of lung cancer by a linkage of multi-dimensional liquid chromatography and two-dimensional difference gel electrophoresis. Proteomics 6:3938–3948
15. Kondo T, Seike M, Mori Y et al (2003) Application of sensitive fluorescent dyes in linkage of laser microdissection and two-dimensional gel electrophoresis as a cancer proteomic study tool. Proteomics 3:1758–1766
16. Sitek B, Potthoff S, Schulenborg T et al (2006) Novel approaches to analyse glomerular proteins from smallest scale murine and human samples using DIGE saturation labelling. Proteomics 6:4506–4513
17. Zhou G, Li H, DeCamp D et al (2002) 2D differential in-gel electrophoresis for the identification of esophageal scans cell cancer-specific protein markers. Mol Cell Proteomics 1:117–124
18. Karp NA, Lilley KS (2005) Maximising sensitivity for detecting changes in protein expression: experimental design using minimal CyDyes. Proteomics 5:3105–3115
19. Tonge R, Shaw J, Middleton B et al (2001) Validation and development of fluorescence two-dimensional differential gel electrophoresis proteomics technology. Proteomics 1:377–396
20. Yan JX, Devenish AT, Wait R et al (2002) Fluorescence two-dimensional difference gel electrophoresis and mass spectrometry based

proteomic analysis of *Escherichia coli.* Proteomics 2:1682–1698

21. Bertin E, Arnouts S (1996) SExtractor: software for source extraction. Astron Astrophys 117:393–404
22. Suehara Y, Kondo T, Fujii K et al (2006) Proteomic signatures corresponding to histological classification and grading of soft-tissue sarcomas. Proteomics 6:4402–4409
23. Shaw J, Rowlinson R, Nickson J et al (2003) Evaluation of saturation labelling two-dimensional difference gel electrophoresis fluorescent dyes. Proteomics 3:1181–1195
24. Zhou S, Bailey MJ, Dunn MJ et al (2005) A quantitative investigation into the losses of proteins at different stages of a two-dimensional gel electrophoresis procedure. Proteomics 5:2739–2747

Chapter 25

Protein Identification on Archived 2-D Gels

Hiroyuki Matsumoto, Nobuaki Takemori, and Naoka Komori

Abstract

Because of the availability of genome information combined with proteomics techniques, it is possible to determine the identity of a protein which had been isolated many years ago on a two-dimensional gel electrophoresis and stored in a dry state as a data archive. The protocol described in this chapter will assist researchers who want to know the identity of a protein separated decades ago when no techniques were available to determine the identity of the protein.

Key words: Protein identification, Archived gels

1. Introduction

Identification of isolated proteins by peptide mass fingerprinting (PMF) based on genome information (1, 2) has been established and has been routinely used in the past decade. Quite often, both one-dimensional and two-dimensional (2-D) gels run on polyacrylamide slabs have been dried on a filter paper or in between cellophane films and stored at room temperature. We discovered that such archived gels, stored over two decades, still serve as a source for protein identification by PMF (3). This could serve as a useful technique to confirm or even discover protein identity from dried gels stored in data archive. In our original paper (3), the in-gel digestion by TPCK–trypsin was performed by the method reported by Rosenfeld et al. (4) and adapted to mass spectrometry (5, 6). Later we modified the protocol to optimize the efficiency of PMF (7–9). The protocol described in this chapter is based on the recent modifications (7–9).

Biji T. Kurien and R. Hal Scofield (eds.), *Protein Electrophoresis: Methods and Protocols*, Methods in Molecular Biology, vol. 869,
DOI 10.1007/978-1-61779-821-4_25, © Springer Science+Business Media, LLC 2012

2. Materials

1. Sample protein spots separated by (2-D) gel electrophoresis, which had been stained by Coomassie Blue, and dried either on a filter paper or in between cellophane papers (see Note 1). Typically, a small protein spot (approximately 2×2 mm) is excised.
2. Stock solutions for in-gel digestion:
   (a) 50% (v/v) Acetonitrile/100 mM ammonium bicarbonate (pH 8.9).
   (b) 50% (v/v) Acetonitrile/5% (v/v) trifluoroacetic acid (TFA).
   (c) 0.2% (v/v) TFA solution.
   (d) 2,5 Dihydroxybenzoic acid (DHB) solution for mass spectrometry (MS) analysis: 2% (w/v) DHB in 50% (v/v) acetonitrile/0.1% (v/v) TFA.
   (e) DHB solution for MS analysis: 2% (w/v) DHB in 50% (v/v) acetonitrile/0.1% (v/v) TFA.
3. MASCOT (PMF) search program (Matrix Science, London, UK).

3. Methods

1. Soak protein spot overnight at 4°C in 50% (v/v) acetonitrile/100 mM ammonium bicarbonate adjusted to pH 8.9.
2. Incubate destained gel pieces with 1 mL of acetonitrile for 30 min to dehydrate the gels.
3. After removing acetonitrile, rehydrate the gel pieces with 2 μL of trypsin solution for 30 min at 4°C.
4. Add 50 μL of 10 mM ammonium bicarbonate and incubate at 37°C for 5–10 h (see Note 2).
5. Transfer the solution containing tryptic digests into a new 0.65-mL polypropylene tube.
6. Vigorously shake the residual gel in 50 μL of 50% (v/v) acetonitrile/5% (v/v) TFA for 20 min to extract the rest of the tryptic digests from the gel.
7. After a brief centrifugation step, collect and transfer the supernatant into the 0.65-mL tube containing the original digest solution (see Step 5).
8. Concentrate the collected peptide solution down to near dryness using vacuum centrifuge.

3.1. Peptide Mass Fingerprinting by Matrix-Assisted Laser Desorption Ionization Time-of-Flight Mass Spectrometry

1. Reconstitute peptide samples with 5 μL of 0.2% (v/v) TFA solution.
2. Load the peptide solution (0.5 μL) onto a MALDI-TOF sample target plate with a fresh DHB solution (0.5 μL) and dry the mixture completely at room temperature.
3. Subject the peptide samples on the MALDI sample target to PMF analysis using MALDI-TOF MS.
4. After measuring the masses of the tryptic peptides by MALDI-TOF MS, submit the observed masses to MASCOT PMF search program at http://matrixscience.com.
5. We routinely perform MASCOT search against the National Center for Biotechnology Information nonredundant database using the following parameters: (a) unlimited protein molecular weight and pI ranges; (b) presence of protein modifications including acrylamide modification of cysteine, methionine oxidation, protein N terminus acetylation, and pyroglutamic acid; and (c) mass tolerance of ± 0.25 Da. We consider the confirmation to be positive when a significant MOWSE score ($p < 0.05$) is generated.

4. Notes

1. Proteins can be identified by this method from 2-D gels dried over 20 years ago and stored at room temperature.
2. During the incubation, trypsin continues to digest a protein in a gel piece and some of the tryptic peptides come out of the gel piece.

References

1. Matsumoto H, Komori N (2000) Ocular proteomics: cataloging photoreceptor proteins by two-dimensional gel electrophoresis and mass spectrometry. Methods Enzymol 316:492–511
2. Matsumoto H, Kurono S, Matsumoto M et al (2005) Mass spectrometry of biomolecules in proteomics. In: Meyers RA (ed) Encyclopedia of molecular and cell biology and molecular medicine. Wiley-VCH Verlag GmbH, Germany, pp 557–585
3. Matsumoto H, Komori N (1999) Protein identification on two-dimensional gels archived nearly two decades ago by in-gel digestion and matrix-assisted laser desorption ionization time-of-flight mass spectrometry. Anal Biochem 270:176–179
4. Rosenfeld J, Capdevielle J, Guillemot JC et al (1992) In-gel digestion of proteins for internal sequence analysis after one- or two-dimensional gel electrophoresis. Anal Biochem 203: 173–179
5. Matsumoto H, Kurien BT, Takagi Y et al (1994) Phosrestin I undergoes the earliest light-induced phosphorylation by a calcium/calmodulin-dependent protein kinase in *Drosophila* photoreceptors. Neuron 12: 997–1010
6. Kinumi T, Tobin SL, Matsumoto H et al (1997) The phosphorylation site and desmethionyl N-terminus of *drosophila* phosrestin I *in vivo* determined by mass spectrometric analysis of proteins separated on two-dimensional gel electrophoresis. Eur Mass Spectrom 3:367–378

7. Takemori N, Komori N, Matsumoto H (2006) Highly sensitive multistage mass spectrometry enables small scale analysis of protein glycosylation from two-dimensional polyacrylamide gel. Electrophoresis 27:1394–1406
8. Takemori N, Komori N, Thompson J Jr et al (2007) Novel eye-specific calmodulin methylation characterized by protein mapping in *Drosophila melanogaster*. Proteomics 7:2651–2658
9. Takemori N, Komori N, Matsumoto H (2009) MS analysis of protein glycosylation, Chapter 150. In: Walker JM (ed) The protein protocols handbook, 3rd edn. Humana, New York, pp 1381–1388

Chapter 26

Diagonal Electrophoresis for the Detection of Protein Disulfides

Brian McDonagh

Abstract

Cysteines are one of the most rarely used amino acids in proteins, therefore when conserved in proteins they usually play critical roles in structure, function, or regulation of the protein. These cysteines or thiols can be reversibly oxidised to sulfenic acid (–SOH), thiyl radicals (–S•) or nitrosothiols (–SNO) or form both inter-and intra-disulfide bridges (PSSP). The protein thiol groups PSSPs, represent a larger active redox pool than glutathione and are likely to be directly involved in cellular defence against oxidative stress. Diagonal electrophoresis is a relatively simple technique to analyze the formation of protein disulfides by sequential non-reducing/reducing electrophoresis. Proteins that do not form disulfides, electrophorese identically in both dimensions and form a diagonal after the second dimension, proteins that contained intra-chain disulfides lie above this diagonal, while those that formed inter-disulfides fall below the diagonal. This technique therefore allows for the detection and identification of protein disulfides.

Key words: Redox proteomics, Diagonal eletrophoresis, Disulfide bonds, Thiols, Oxidative stress

1. Introduction

Although cysteines are one of the most rarely used amino acids within proteins, they can play a crucial role in structure and function, largely due to the wide range of oxidation states that sulphur can occupy and are often referred to as sulphur switches (1). These sulphur switches provide an important means of controlling protein function and allow the proteome to respond to environmental, genetic, biochemical, and pathological perturbations. The proteome response is of considerable interest as it can relate to the cell stress response and can take the form of post-translational modifications and inter-protein interactions. The formation of reactive oxygen species (ROS) by endogenous processes such as

Biji T. Kurien and R. Hal Scofield (eds.), *Protein Electrophoresis: Methods and Protocols*, Methods in Molecular Biology, vol. 869, DOI 10.1007/978-1-61779-821-4_26, © Springer Science+Business Media, LLC 2012

electron transport, enzyme mechanisms, signal transduction as well as pro-oxidants can initiate such a redox proteomic response. Oxidative stress (OS) has been defined as an imbalance between the generation of ROS and their neutralisation by antioxidants, caused by excessive production of ROS, loss of antioxidant defences, or both (2). This definition has been further refined to account for two different mechanistic outcomes, macromolecular damage, and the disruption of redox signaling and control (3). ROS are essential in the physiological control of cell function in biological systems and are continuously produced in living cells. Excess levels of ROS can lead to irreversible modifications on proteins such as carbonylation; however, under normal conditions it can lead to reversible redox proteomic responses on protein thiols that can include, inter-/intra-disulfide bond formation, glutathionylation, nitrosylation, and sulfenic acid formation. Many of these reversible modifications play a protective role preventing further oxidative damage, and as they are reversible, protein function may be restored when normal oxidative conditions are restored. In unstressed mammalian cells it was shown that disulfides in proteins (PSSP) account for 6 and 9.5% of protein sulfhydryls in HEK and HeLa cells, respectively, and after treatment with the thiol-specific oxidant diamide this increased to 24 and 25%, while the steady state level of glutathione–protein mixed disulfides (PSSG) is less than 1% increasing to 15% after treatment (4). Therefore, it would appear that protein thiols represent an important and significant redox buffer within the cell.

Within any individual protein there may be a number of cysteine residues and could allow for multiple modifications of a given protein. There are a number of redox sensitive proteins with a cysteine residue as part of the active site, allowing for the protein to be switched on or off depending on its redox state, one of the most widely identified is glyceraldehyde 3-phosphate dehydrogenase (5). There are a number of proteins where the cysteine is not within the active site but modulates its activity by changing its conformation or influencing its regulatory role; iron sulphur complexes (ISC) that are required for its activity, e.g. aconitase (6). Interactions with other proteins or molecules are another feature of cysteine residues that can affect protein activity. Allosterically regulated proteins that require an activator are often based on a thiol exchange interaction involving cysteines, e.g. pyruvate kinase uses fructose bisphosphate (fbp) as a heterotrophic activator and it contains a cysteine residue in its fbp-binding site (7).

A relatively simple method developed for the analysis of both inter- and intra-chain disulfide bonds, originally developed in the 1970s, is the use of diagonal electrophoresis (8). Diagonal electrophoresis involves sequential non-reducing/reducing electrophoresis, essentially manipulating the redox state of thiols within proteins (see Fig. 1). Initially, a protein mixture is electrophoresed under

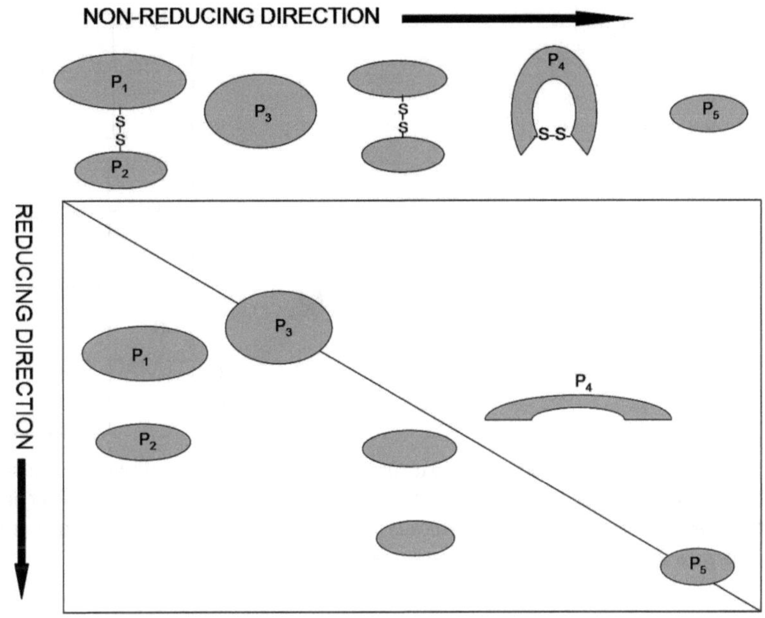

Fig. 1. Schematic of principles of diagonal electrophoresis. Protein samples are initially electrophoresed under non-reducing conditions; a gel track is excised, exposed to reducing agent, and placed orthogonally on a SDS-PAGE gel. Proteins lacking disulfides migrate forming a diagonal as they electrophorese identically in reducing and non-reducing conditions. Proteins with inter-chain disulfides (*P1* and *P2*) migrate as separate polypeptides below the diagonal while those with intra-chain disulfides (*P4*) migrate slower under reducing conditions and thus are above the diagonal.

non-reducing conditions in an alkylated buffer to prevent thiol disulfide exchange. The entire lane is then excised from the one-dimensional electrophoresis gel, treated with a reducing agent, placed orthogonally on top of a second gel and electrophoresed as normal under reducing conditions. After staining, proteins that do not contain disulfide bonds congregate along a diagonal since they electrophorese identically in both reducing and non-reducing conditions. Proteins involved in inter-chain disulfide bonds will fall below the diagonal as they have a lower apparent molecular mass in reducing conditions. Proteins that contain intra-chain disulfide bonds will run above the diagonal as they have a larger apparent molecular mass (see Fig. 2). Protein spots can then be excised, tryptic digested, and identified by mass spectrometry. This technique has led to the identification of proteins previously not known to form disulfide bonds (9, 10). It has also been used in combination with other techniques, such as affinity purification of redox targets before separation (11) and to identify proteins by immunoblotting (12).

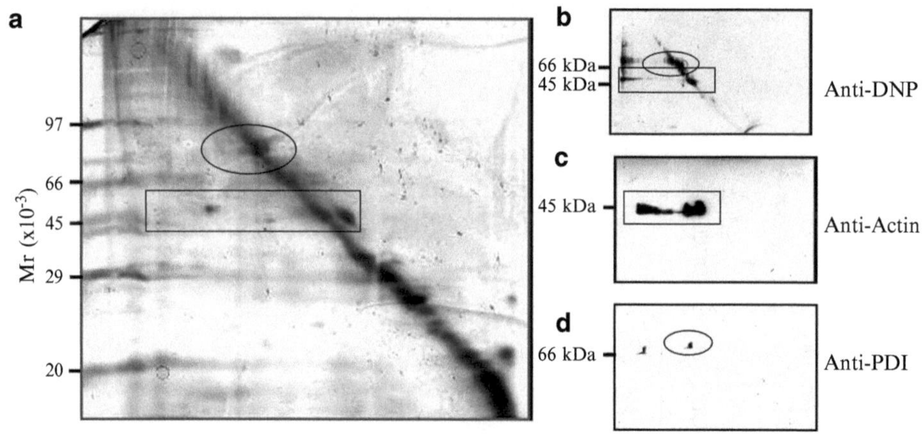

Fig. 2. (**a**) Diagonal electrophoresis of 50 μg of a derivatised protein sample with dinitrophenylhydrazine (for carbonylation). Diagonal gel was subsequently immunoblotted with various antibodies: (**b**) Anti-DNP for detection of carbonylated proteins, (**c**) anti-actin—*rectangular area* corresponds to actin-associated proteins, and (**d**) anti-protein disulfide isomerase (PDI)—*the circular area* corresponds to thiol exchange protein PDI.

2. Materials

1. An example of an alkylating buffer solution, 50 mM Tris–HCl, pH 7.5, 2 mM EDTA, and 40 mM iodoacetamide (IAM) or similar alkylating agents such as *N*-ethylmaleimide.
2. Sample buffer (4×): 1.25 mL 0.5 M Tris–HCl, pH 6.8, 3.0 mL glycerol, 1 mL 20% SDS, and bromophenol blue.
3. 20% Trichloroacetic acid (TCA).
4. Resolving acrylamide gel solution.
5. Stacking acrylamide gel solution.
6. Running buffer: Make 1 L containing 14.4 g glycine, 3.03 g Tris base, and 1 g SDS.
7. Equilibration buffer: 6 M urea, 0.375 M Tris–HCl, pH 8.8, 2% SDS, 20% glycerol, and either 2% dithiothreitiol (DTT) or 2.5% IAM.
8. Agarose solution: 0.1% agarose in running buffer plus bromophenol blue.
9. Two sets of slab-gel apparatus, e.g. ATTO and BioRad mini Protean system or alternatively slab-gel apparatus with two sets of spacers of different thickness (see Note 1).
10. Power supply.
11. Rocking platform.

3. Methods

3.1. Sample Preparation

1. Protein extracts are prepared using an alkylating buffer solution.
2. Proteins are precipitated in 20% TCA and resuspended in non-reducing sample buffer (4×) to give a final protein concentration of 50 μg/15 μL (see Note 2).

3.2. Electrophoresis

1. Prepare a 12% polyacrylamide resolving gel with 4% stacking gel (see Note 3).
2. Add 15 μL non-reducing sample buffer containing 50 μg of protein to a well. Use alternate wells in the non-reducing first dimension so that it will be easier to excise the lanes.
3. One-dimensional electrophoresis (1-DE) is carried out at 60 V until dye reaches the end of the gel (see Note 4). The entire lane is excised using a sharp fine scalpel (see Notes 5 and 6).
4. The entire lane is soaked for 20 min in equilibration buffer containing 2% DTT at room temperature on a rocker, followed by 20 min in equilibration buffer containing 2.5% IAM (see Note 7).
5. The gel lane is rinsed in running buffer and placed horizontally on a 12% resolving gel (see Note 3). Carefully slot excised lane onto slab gel using the blunt end of small spatula. It helps if the excised lane is kept moist using running buffer; gel lanes can become very fragile when dry (see Note 8).
6. A protein standard mixture can be electrophoresed alongside diagonal gel by pipetting 5 μL of mixture onto a small cube of blotting paper (see Note 9).
7. Layer warm agarose containing bromophenol blue on top of the slab gel. Ensure there are no air bubbles between gel lane and slab gel as this will interfere with protein separation.
8. Carry out electrophoresis as per normal until dye front has reached the bottom of the gel. Proteins may be detected by any of the staining techniques applicable to 2DE. Proteins may also be electroblotted for immunoblot analysis. This technique may also be applied to radio- or fluorescently labeled proteins (see Note 10).

4. Notes

1. In the second dimension it is important to use a gel with greater thickness than first dimension. For instance, in the first dimension a gel with 0.75- or 1.0-mm spacers could be used followed

by a gel of 1.5 mm thickness in the second dimension. Alternatively in larger gels, spacers with different thicknesses can be used. Also large gels can give option of running two samples from small first-dimension rigs alongside each other (9).

2. If using 50 μg of protein/well in the first dimension it is better to precipitate 200 μg of protein and resuspend in 60 μL of sample buffer overnight. This ensures complete and homogenous resuspension and also allows at least three replicates of the sample to be used. 50 μg of protein is a guideline as the precise amount will depend on the abundance of proteins of interest and size of the gel rigs. Samples can be resuspended in any sample buffer so long as there are no reducing agents present.
3. Percentages of acrylamide gels can be varied to optimize resolution for proteins of interest.
4. Electrophoresis in the first dimension is carried out slowly. If voltage or current is too high, separation of proteins in the non-reducing dimension will result in streaking of the final gel.
5. The lane width should not be exceeded as it will result in poor resolution in second dimension. Also lane edges should be kept as smooth as possible by using sharp scalpel.
6. The gel lane should not be allowed to dry during excision as it will become very brittle; use running buffer to keep moist during excision.
7. Alternative reducing and alkylating agents may be used, for instance it is reported that NEM has a higher reactivity than IAM.
8. Excised lanes can be stored at –70°C at this stage.
9. This allows easier detection of standards; however, it is not as accurate as for determining the molecular weight of proteins that have been electrophoresed in both dimensions.
10. The thiol-specific fluorescent reagent 5- Iodoacetamidofluorescein (5-IAF) could be included after reduction with DTT. This would specifically label newly reduced thiols and offer superior sensitivity for detection of proteins.

Acknowledgements

This work was supported by Grants P06-CVI-01611 from the Andalusian Government and BFU2006-02990 and BFU2009-08004 from the Spanish Government.

References

1. Jones DP (2008) Radical-free biology of oxidative stress. Am J Physiol 295:C849–C868
2. Giustarini D, Milzani A, Aldini G, Carini M, Rossi R, Dalle-Donne I (2005) S-nitrosation versus S-glutathionylation of protein sulfhydryl groups by S-nitrosoglutathione. Antioxid Redox Signal 7:930–939
3. Jones DP (2006) Redefining oxidative stress. Antioxid Redox Signal 8:1865–1879
4. Hansen RE, Roth D, Winther JR (2009) Quantifying the global cellular thiol-disulfide status. Proc Natl Acad Sci USA 106:422–427
5. Mohr S, Hallak H, de Boitte A, Lapetina EG, Brune B (1999) Nitric oxide-induced S-glutathionylation and inactivation of glyceraldehyde-3-phosphate dehydrogenase. J Biol Chem 274:9427–9430
6. Chen XJ, Wang X, Butow RA (2007) Yeast aconitase binds and provides metabolically coupled protection to mitochondrial DNA. Proc Natl Acad Sci USA 104:13738–13743
7. Susan-Resiga D, Nowak T (2004) Proton donor in yeast pyruvate kinase: chemical and kinetic properties of the active site Thr 298 to Cys mutant. Biochemistry 43:15230–15245
8. Sommer A, Traut RR (1974) Diagonal polyacrylamide-dodecyl sulfate gel electrophoresis for the identification of ribosomal proteins crosslinked with methyl-4-mercaptobutyrimidate. Proc Natl Acad Sci USA 71: 3946–3950
9. Brennan JP, Wait R, Begum S, Bell JR, Dunn MJ, Eaton P (2004) Detection and mapping of widespread intermolecular protein disulfide formation during cardiac oxidative stress using proteomics with diagonal electrophoresis. J Biol Chem 279:41352–41360
10. Cumming RC, Andon NL, Haynes PA, Park M, Fischer WH, Schubert D (2004) Protein disulfide bond formation in the cytoplasm during oxidative stress. J Biol Chem 279:21749–21758
11. Mata-Cabana A, Florencio FJ, Lindahl M (2007) Membrane proteins from the cyanobacterium Synechocystis sp. PCC 6803 interacting with thioredoxin. Proteomics 7:3953–3963
12. McDonagh B, Sheehan D (2007) Effect of oxidative stress on protein thiols in the blue mussel *Mytilus edulis*: proteomic identification of target proteins. Proteomics 7:3395–3403

Chapter 27

Two-Dimensional Blue Native/SDS Gel Electrophoresis of Multiprotein Complexes

Jean Paul Lasserre and Armelle Ménard

Abstract

The two-dimensional blue native/sodium dodecyl sulfate polyacrylamide gel electrophoresis (2D BN/SDS-PAGE) is a method of choice for the investigation of protein complexes. This highly resolvent separation method is unique in that it facilitates the identification of many protein complexes simultaneously. Because of its simplicity and suitability, the 2D BN/SDS-PAGE can be now applied to a wide range of organisms, including bacteria, viruses, yeasts, animals, and plants. Moreover, recent modifications have made it possible to apply this method to the study of whole protein complexes of an organism. Here, we propose protocols for the investigation of the whole complexome of bacteria as well as eukaryotic cells.

Key words: Complexome, Interactome, Protein complex, Two-dimensional blue native/SDS-PAGE, Electrophoresis, Coomassie Brilliant Blue

1. Introduction

"Tell me who your friends are and I'll tell you who you are!" An expression perfectly suited to the two-dimensional blue native/sodium dodecyl sulfate polyacrylamide gel electrophoresis (2D BN/SDS-PAGE). Indeed, this method allows the investigation of protein–protein interaction and can be applied to the global and simultaneous identification of complexes in an organism, providing insight into protein function and cell behavior. Intact protein complexes can be separated under native conditions by their apparent molecular mass (M_r) using a standard PAGE system and the dye Coomassie Brilliant Blue G-250 (CBB G-250): the BN-PAGE (first dimension). This system combined with SDS-PAGE (second dimension) was first described by Schägger et al. (1) and allowed the separation of the proteins of each complex according to their M_r (Fig. 1). Since the invention and development of many mass

Biji T. Kurien and R. Hal Scofield (eds.), *Protein Electrophoresis: Methods and Protocols*, Methods in Molecular Biology, vol. 869, DOI 10.1007/978-1-61779-821-4_27, © Springer Science+Business Media, LLC 2012

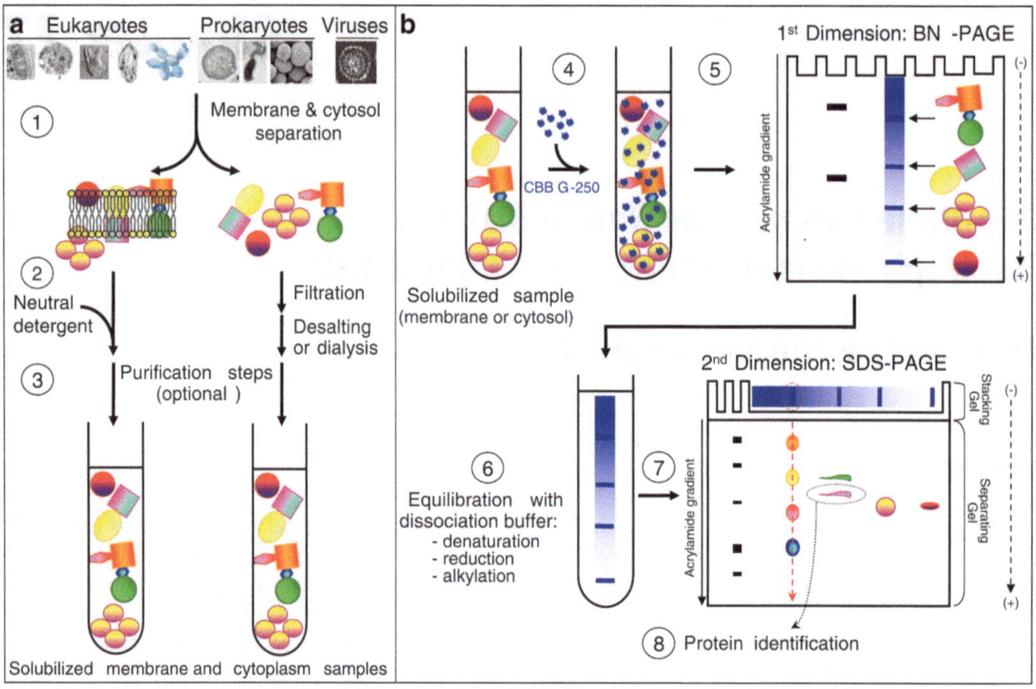

Fig. 1. Sample preparation (**a**) and principles of the 2D BN/SDS-PAGE (**b**). (1) The membrane and cytoplasmic samples are solubilized and separated into two different fractions. (2) Solubilization of protein complexes belonging to the membrane using neutral detergent. Removal of "small molecules" from cytoplasmic samples using filtration, desalting, or dialysis. (3) Optional purification steps (see Note 21). (4) After solubilization and prior to the BN-PAGE, CBB G-250 dye is added to the sample to facilitate the migration. (5) The different protein complexes present in this mixture are separated by BN-PAGE (migration under native conditions), according to their molecular mass. (6) Following the electrophoresis, an individual unstained lane is excised from the gel and subjected to dissociation: denaturation (SDS), reduction (β-mercaptoethanol), and alkylation (iodoacetamide). As a result, the native protein complexes dissociate to their proteins which are uniformly negatively charged by SDS. (7) The individual lane with dissociated proteins is loaded onto an SDS-PAGE (migration under denaturing conditions) which allows the separation of the proteins according to their molecular mass. Following this second electrophoresis, the gel proteins are stained with CBB G-250 or silver stain. Subunits of a protein complex are found aligned with a similar shape in the second dimension gel: the set of spots indicated by the *red dotted arrow* represents subunits attributed to one protein complex. (8) Each spot of interest is excised; the proteins are destained and submitted to trypsin digestion, the resulting peptides being analyzed by mass spectrometry.

spectrometry techniques and given the ease and suitability of the 2D BN/SDS-PAGE, this approach has generated growing interest and is now widely used for the study of complexomes of many organisms.

We applied the 2D BN/SDS-PAGE to analyze the whole complexome of Gram-negative bacteria such as *Escherichia coli* (2) and *Helicobacter pylori* (3, 4), as well as the complexome from eukaryotes such as the yeast *Yarrowia lipolytica* (5) and human colon adenocarcinoma Caco-2 cells (6). With regard to *E. coli*, 306 protein complexes were identified, of which 109 had never been reported (2). In *H. pylori*, 13 multiprotein complexes could be initially described (3), and more recently, the complexome of two different *H. pylori* strains was analyzed leading to the recovery of

90 different complexes (4). A study of hydrophobic binding protein complexes in *Y. lipolytica* allowed the identification of 40 different alkane-binding protein complexes, many of them being comprised of hypothetical proteins (5). The 2D BN/SDS-PAGE, when applied to Caco-2 cells, allowed the identification of 58 organellar protein complexes, completing the complexomic network in which 32 complexes had never been reported (6). In these studies, purification steps using physical or chemical properties of the multiprotein complexes were used in order to confirm protein–protein interactions and to highlight complexes hardly visible on crude extracts. Using the protocols reported here, it was possible to identify complexes in the range of 20–1,300 kDa.

2. Materials

All solutions must be prepared using ultrapure water (prepared by purifying deionized water to attain a sensitivity of 18 MΩ cm at 25°C). All reagents must be prepared and stored at 4°C (unless otherwise indicated). Follow all appropriate safety precautions diligently when handling products and organisms, as well as all waste disposal recommendations. (*Read carefully the recommendations in* see Notes 1 and 2 *before starting*).

2.1. Cells and Culture

1. Culture *H. pylori* B38 and J99 strains simultaneously under identical conditions on agar plates for 48 h at 37°C (4).
2. Culture *E. coli* DH5a strain in liquid medium for 6 h at 37°C (2).
3. Culture *Y. lipolytica* strain H222 for 24 h in liquid medium at 28°C (5).
4. Culture adherent human colorectal cancer Caco-2 cells to confluence (2–3 days) at 37°C (6).

2.2. Extraction of Protein Complexes from H. pylori and E. coli

1. Sterile phosphate-buffered saline (PBS).
2. Native extraction buffer (NEB): 750 mM 6-amino-*n*-caproic acid (Sigma-Aldrich, Saint-Louis, MO, USA) and 50 mM Tris. Weigh 9.84 g of 6-amino-*n*-caproic acid and 605.7 mg Tris. Adjust volume to 100 mL with water (see Note 3).
3. One Shot disruptor (Constant Systems Ltd, Northants, UK). Optima TLX ultracentrifuge (Beckman Coulter, Brea, CA, USA).
4. Phenylmethanesulphonyl fluoride (PMSF), DNase I, *n*-dodecyl-β-D-maltoside (DDM) (see Notes 4–7).
5. Miracloth membrane (22–25 μm) for quick filtration of gelatinous homogenates (Calbiochem, San Diego, CA, USA).
6. 5 mL HiTrap™ Desalting Column prepacked with Sephadex™ G-25 Superfine (GE Healthcare).

7. Membrane dialysis (cut-off: 14,000 Da, cut-off depending on the mass of the complexes of interest) (Medicell International Ltd, London, UK).
8. PMSF 0.5 M stock solution: Dissolve 0.87 g PMSF in dimethyl sulfoxide and bring up to a final volume of 10 mL. Then, use this stock solution to resuspend the bacteria in NEB to obtain a final concentration of 1 mM (see Note 7).
9. DDM 10%: Dissolve 10 g of DDM in water to equal a final volume of 100 mL.

2.3. Extraction of Protein Complexes from *Y. lipolytica* and Caco-2 Cells

1. H.A 0.45-μm pore filters (Millipore, St-Quentin-en-Yvelines, France) for *Y. lipolytica*. Sterile plastic cell scraper (Greiner Bio-One France SAS, Courtaboeuf, France) for Caco-2 cells.
2. Sterile PBS.
3. NEB and DDM (see Subheading 2.2).
4. EDTA-free protease inhibitor cocktail and PMSF (see Note 7).
5. CellD cell disruptor (CellD, Roquemaure, France) for *Y. lipolytica* only.
6. Alkane (decane, C10, Sigma-Aldrich) for *Y. lipolytica* and Dounce glass homogenizer (Sigma-Aldrich) for Caco-2 cells.
7. Thomas Teflon pestle homogenizer (Thomas Scientific, Swedesboro, NJ, USA) for *Y. lipolytica* only.
8. Spermine (Sigma-Aldrich) and Miracloth membrane (22–25 μm) for Caco-2 cells (see Note 4).

2.4. Treatment of the Extracts Before BN-PAGE

1. Loading buffer: 0.5 M 6-amino-*n*-caproic acid, 5% CBB G-250. Weigh 65.6 mg of 6-amino-*n*-caproic acid and 50 mg of CBB G-250. Adjust to a final volume of 1 mL with water (see Notes 3 and 8).

2.5. Blue Native: Polyacrylamide Gel

1. Gradient casting apparatus (Sigma-Aldrich) for BN-PAGE (Fig. 2).
2. The gel dimensions are $20 \times 14.5 \times 0.1$ cm.
3. Gel buffer 3× stock solution: 1.5 M 6-amino-*n*-caproic acid, 150 mM Tris. Add approximately 100 mL of water to a 1-L graduated cylinder (see Note 9). Weigh 196.8 g 6-amino-*n*-caproic acid and 18.17 g Tris. Transfer to the cylinder and adjust to a final volume of 1 L with water (see Note 3).
4. Filter PlusOne™ 40% acrylamide solution and PlusOne™ 2% *N*,*N*′-methylene-bisacrylamide solution (GE Healthcare) on a 0.2-μm membrane prior to use. Acrylamide powder ultrapure (Euromedex, Souffelweyersheim, France), only if necessary (see Note 10).
5. PlusOne™ ammonium persulfate (APS): 10% solution in water. Weigh 50 mg of APS and add 500 μL of water. Store at −20°C.

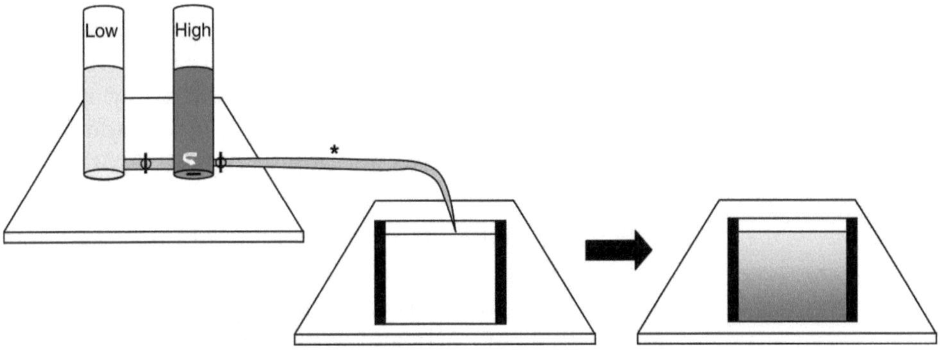

Fig. 2. Gradient casting apparatus for BN-PAGE. Gradient makers consist of two containers filled with high and low % acrylamide solutions. These containers are joined by a narrow connector at their bases, with the container having higher % acrylamide also having an additional outlet at its base. As the highly concentrated acrylamide solution is drained from the higher acrylamide container, it is replaced from below due to the equilibration of hydrodynamic pressure which keeps the levels in the high and low acrylamide containers equal. The higher acrylamide container is constantly stirred, which causes the solution draining from the higher part to be progressively diluted with low concentrated acrylamide solution from the lower part. When the gradient maker is empty, the outflowing material is essentially 100% low. (*Asterisk*) No peristaltic pump is represented, but it is possible to use one to regulate the gradient maker flow. These steps must be carried out rapidly to avoid polymerization of the solutions before the gel is fully cast.

6. PlusOne™ TEMED, PlusOne™ glycine, PlusOne™ glycerol, and Tris. A 20% SDS solution is available from Eurobio (Courtaboeuf, France) (see Notes 11 and 12).
7. Water-saturated isopropanol. Mix 50% water and 50% isopropanol (see Note 13).
8. Anode buffer: 75 M PlusOne™ glycine, 50 mM Tris. Weigh 5.6 g of glycine and 6 g of Tris. Adjust with water to a final volume of 1 L (see Note 14).
9. Cathode buffer: 75 M glycine, 50 mM Tris, 0.004% CBB G-250. Weigh 5.6 g glycine, 6 g Tris and 0.04 g CBB G-250. Adjust with water to a volume of 1 L (see Notes 9 and 15).
10. Prepare thyroglobulin (669 kDa) and bovine serum albumin (66 kDa) (Sigma-Aldrich) (see Note 16) in NEB to obtain a final concentration of 1 mg/mL for each protein.

2.6. Equilibration of Strips

1. 70% Ethanol-washed ruler.
2. Tris and iodoacetamide, 20% SDS. β-mercaptoethanol (see Note 12).
3. Equilibration buffer: 125 mM Tris–HCl, pH 6.8, and 1% SDS. Add 25 mL of stacking gel buffer (see Subheading 2.5), 5 mL of 20% SDS, and bring up to a final volume of 100 mL with water. Mix and adjust pH with 37% HCl solution.
4. 1 M dithiothreitol (DTT). Weigh 1.54 g of DTT and adjust with water to a final volume of 10 mL. Mix and prepare 650 μL aliquots; store at −20°C (see Note 17).

2.7. Sodium Dodecyl Sulfate: Polyacrylamide Gel

1. The gel dimensions are 22 × 16.5 × 0.1 cm.
2. Resolving gel buffer: 1.5 M Tris–HCl, pH 8.8. Add approximately 100 mL of water to a 1-L graduated cylinder. Weigh 181.7 g of Tris and transfer to the cylinder. Add water to a volume of 900 mL. Mix and adjust to pH 8.8 with 37% HCl. Adjust to 1 L with water.
3. Stacking gel buffer: 0.5 M Tris–HCl, pH 6.8. Add approximately 100 mL of water to a 1-L graduated cylinder. Weigh 60.6 g of Tris and transfer to the cylinder. Add water to a volume of 900 mL. Mix and adjust to pH 6.8 with 37% HCl. Adjust to 1 L with water.
4. Use the solutions listed in four and five from Subheading 2.3.
5. SDS-PAGE 1× running buffer: 0.025 M Tris–HCl, pH 8.3, 0.192 M glycine, 0.1% SDS. Prepare a 10× native buffer solution stock (0.25 M Tris, 1.92 M glycine) prior to the dilution. Weigh 30.3 g of Tris and 144 g of glycine, mix, and adjust the volume to 1 L with water. Dilute 100 mL of 10× native buffer solution to 990 mL with water and add 10 mL of 10% SDS. Gently add the SDS solution last, since it makes bubbles.
6. Acrylamide (purity 99.4%)/40% bis-acrylamide sol ratio 37.5/1.1 L (see Note 10).

2.8. Protein Staining

2.8.1. Silver Staining

PROTSIL1 Silver Staining Kit (Sigma-Aldrich).

2.8.2. Colloidal Blue Staining

1. Acetic acid.
2. Incubation solution: 10% ammonium sulfate, 0.1% CBB G-250, 3% orthophosphoric acid, 20% ethanol (see Notes 18 and 19).

2.9. Protein Destaining and Excision from Gel

1. Sterile scalpel (Swann-Morton, Sheffield, UK) or 70% ethanol-washed die cutter (Dominique Dutscher SAS, Brumath, France).
2. Acetonitrile, trypsin solution, ammonium bicarbonate, and acetic acid.

3. Methods

Carry out all procedures at 4°C unless otherwise specified (see Note 1). Gloves should be worn for all the steps (extract preparations, BN-PAGE, SDS-PAGE, staining, protein excision). (*Read carefully the recommendations in* see Notes 1 and 2 *before starting*.) The sample preparation and principle of the 2D BN/SDS-PAGE are described in Fig. 1. The protocols presented here refer to published BN-PAGE experiments (2–6) and are not exhaustive.

Some details (such as prevention of protein phosphorylation or protein dephosphorylation) are available in the relevant publications.

3.1. Membrane and Cytosolic Complex Extraction from Cells

After culture, harvest cells and treat them as described below. Extraction of membrane and cytosolic complexes can be performed simultaneously. It is important to keep the extracts cool on ice and to perform all the steps described below at 4°C (unless otherwise indicated) (see Notes 1 and 2). Extracted samples can be used immediately or kept frozen at −80°C until further analyses. Decontaminate all organic waste before disposal.

An overview regarding the extraction and analysis of various bacterial protein complexes using BN-PAGE (from *Archaea*, through Gram-positive and Gram-negative bacteria to *Chlamydia*) was recently published (see ref. 7 and references that are cited therein).

3.1.1. Extraction of Protein Complexes from H. pylori and E. coli

1. Harvest bacteria by a centrifugation at 6,000 × *g* for 10 min. Discard the supernatant and wash the pellet twice at 2,500 × *g* for 10 min in sterile ice-cold PBS buffer or water for *H. pylori* and *E. coli*, respectively.
2. Suspend bacteria recovered from the pellet in NEB supplemented with PMSF (1 mM final concentration) and pass them through a One Shot disruptor at 2 kb (one shot).
3. Centrifuge the lysate at 6,000 × *g* for 20 min.
4. Discard the pellet containing unbroken cells. Add a 0.2 mg/mL final concentration of DNase I to the supernatant and incubate for 1 h at 25°C (see Notes 4 and 6).
5. Centrifuge this extract at 100,000 × *g* for 30 min. The pellet contains membrane multiprotein complexes and supernatant cytosolic multiprotein complexes.

3.1.2. Membrane Extraction

1. Suspend the pellet (obtained in Sub Heading 3.1.1. Step 5) in 8 mL of NEB with 1 mM PMSF and break again through a One Shot disruptor at 2 kb (one shot) (see Note 20).
2. Centrifuge the resulting lysate at 6,000 × *g* for 30 min.
3. Discard the pellet and centrifuge the supernatant at 100,000 × *g* for 30 min.
4. Resuspend the resulting pellet with 1 and 3 mL of NEB supplemented with 2% DDM for *H. pylori* and E. *coli*, respectively.
5. Centrifuge this sample at 100,000 × *g* for 30 min. The resulting supernatant contains the membrane multiprotein complexes.
6. After solubilization of the complexes, different protocols can be used for the purification of the samples before the BN-PAGE in order to obtain a better result after the electrophoresis (see Note 21).

3.1.3. Cytosol Extraction

1. After steps 1–5 (see Subheading 3.1.1), filter the supernatant containing cytosolic multiprotein complexes (obtained in Subheading 3.1.1, step 5) through a Miracloth membrane.
2. For high-resolvent protein complex separation, a desalting or dialysis step is required. Load the filtrate onto a 5-mL HiTrap™ desalting column equilibrated with NEB, according to the manufacturer's recommendations. A dialysis in NEB buffer with 30% glycerol is sometimes also used to eliminate salts and small molecules (4).
3. After the extraction of cytosolic complexes, different protocols can be used for purification of the samples before the BN-PAGE in order to obtain better results after the electrophoresis or to target specific complexes (see Note 21).

3.1.4. Extraction of Alkane-Binding Protein Complexes from Y. lipolytica

1. Filter five cultures (200 mL each) through an H.A. filter according to the manufacturer's recommendations.
2. Recover the cells on the filters and wash with 50 mL ice-cold sterile PBS.
3. Resuspend cells in 5 mL of NEB supplemented with 1 mM PMSF and 50 μL of an EDTA-free protease inhibitor cocktail.
4. Centrifuge the extracts at 3,000 × *g* for 5 min.
5. Resuspend the resulting pellet in 7 mL of NEB until reaching the OD of 2,000 at 600 nm (approximately 5 g of wet weight of the cells) and cool it on ice.
6. Disrupt cells with a CellD cell disruptor.
7. Centrifuge disrupted cells at 3,000 × *g* for 5 min.
8. Collect and centrifuge supernatant for 10 min at 6,000 × *g*.
9. Extract the supernatant with 4 mL of alkane (decane, C10) for 30 min and freeze it at −80°C.
10. Centrifuge again the upper organic fraction (alkane phase) at 100,000 × *g* for 45 min.
11. Recover and extract again the pellet in 4 mL of alkane (decane, C10) for 30 min and dry under nitrogen flux.
12. Resuspend this fraction in 200 μL of NEB and add DDM to a final concentration of 2% (w/v).
13. Centrifuge directly the solubilized samples with DDM at 100,000 × *g* for 30 min. Discard the pellet and use the supernatant containing the alkane-binding protein complexes.

3.1.5. Extraction of Organellar Protein Complexes from Caco-2 Cells

Organelles correspond to subcellular structures having a specialized function (like a mitochondrion, a Golgi apparatus, an endoplasmic reticulum, etc).

1. Wash the Caco-2 cells three times with 10 mL of ice-cold sterile PBS.

2. Add again 10 mL of PBS and scrape cells using a sterile plastic cell scraper and rinse with 5 mL of PBS.
3. Centrifuge the cells (15 mL) at 800 × *g* for 10 min.
4. Wash twice abundantly the resulting pellet with ice-cold PBS (800 × *g* for 10 min).
5. Suspend the final pellet in NEB supplemented with 1 mM PMSF and a protease inhibitor cocktail according to the manufacturer's instructions.
6. Break manually cells on ice using a Dounce glass homogenizer.
7. Centrifuge the lysate at 15,000 × *g* for 15 min in order to recover the organellar fraction in the pellet.
8. Suspend the pellet in NEB and wash three times by centrifugation at 15,000 × *g* for 15 min.
9. Suspend the pellet containing the organellar fraction in NEB and break again manually on ice using a Thomas Teflon pestle tissue homogenizer.
10. Centrifuge the lysate at 100,000 × *g* for 45 min at 4°C. Discard the pellet and add spermine to the supernatant to achieve a final concentration of 1 mM (see Note 4).
11. Centrifuge this lysate at 100,000 × *g* for 45 min at 4°C.
12. Discard the pellet and filter the supernatant containing the organellar components through a Miracloth membrane.

3.2. Blue Native: Polyacrylamide Gel Electrophoresis

1. Different linear acrylamide gradient gels were used for the first dimension (see Note 22).
2. Handcast gels: All glass plates and spacers should be clean and completely dry. Assemble the glass plate on a clean surface. Lay the shorter glass plate down first, then place the joint around the plate for sealing and two spacers of equal thickness. Then place the longer glass plate on top of the spacers so that the bottom ends of the spacers and glass plates are aligned. Clamp the cassette sandwich and drop vertically. Check that the plates are level on the bottom.
3. Place the gradient maker on a stirring stand so that the bottom of the chambers is higher than the top of the gel cassette sandwich. Place a stir bar in chamber A. Attach a narrow-bore tygon tube to the outflow. Attach a pipette tip to the other end of the tube and clamp into position so that the outflow is directed into the cassette from the top center. Angle the cassette slightly to allow the solution to run down the tall plate. Pour the gels with the aid of a gradient maker.
4. Place a mark on the glass plate 1 cm below the teeth of the comb. This indicates the level to which the separating gel is poured. Remove the comb.

Table 1
Blue native-PAGE gel composition

Acrylamide (%)	3%	4%	7%	8%	9%	12%	13%	18%
PlusOne™ acrylamide 40% solution (mL)	1.125	1.5	2.625	3	3.375	4.5	4.875	–
PlusOne™ *N*,*N*′-methylene-bisacrylamide 2% solution (mL)	0.703	0.937	1.641	1.875	2.109	2.812	3.046	4.219
Acrylamide powder ultra pure (g)[a]	–	–	–	–	–	–		2.7
3× Stock solution (mL)[b]	5	5	5	5	5	5	5	5
PlusOne™ glycerol (mL)	–	–	–	–	1.5	1.5	1.5	1.5
Water (mL)	8.084	7.475	5.646	5.037	2.928	1.100	0.491	1.5
10% APS (μL)	80	80	80	80	80	80	80	80
TEMED (μL)	8	8	8	8	8	8	8	8

The volumes of the two gel solutions combined are for a 30 mL gel casting
[a]Acrylamide powder should be used for concentrations higher than 14%
[b]See Subheading 2.5

5. Prepare the high and low % gel solutions by combining all reagents according to Table 1 without APS and TEMED (see Note 23).
6. Close the stopcock between the gradient maker chambers and the outlet from the gradient maker.
7. Add the specified quantities of APS and TEMED to the low and high % acrylamide gel solutions according the gradient required detailed in Table 1 and mix well by swirling gently. Then, pour the acrylamide separating gel mixture.
8. Place the low % gel solution into the non-outlet side of the gradient maker (chamber B). Open the stopcock between the two chambers and allow 0.1–0.3 mL of the solution to flow through to clear any bubbles. Close the stopcock between the gradient maker chambers and transfer the excess low % gel solution from the non-outlet side (chamber B) to the outlet side of the gradient maker (chamber A) (see Note 24).
9. Place the high % gel solution into the outlet side (chamber A) of the gradient chamber. The next steps must be carried out rapidly to avoid polymerization of the solutions before the gel is fully cast.
10. Start mixing in the outlet side (chamber A). Open the outlet, and then open the stopcock between the chambers. Open at a flow rate which will drain the solutions in 8–10 min.

11. When all of the solution is dispensed, remove the pipette tip from the top of the cassette. Overlay the gel immediately with water-saturated isopropanol (see Note 13). Flush the gradient apparatus immediately with water to prevent polymerization within the system.
12. Allow the gel to polymerize for 8 h.
13. After polymerization of the gel gradient, pour the stacking gel mixture. Remove the water/isopropanol overlaying the gel, wash with water, and then drain the excess water with strips of filter paper.
14. Prepare the stacking gel solution. Combine all reagents except APS and TEMED. Prepare the stacking gel by mixing 1.33 mL of 3× stock solution, 0.3 mL of 40% acrylamide, 0.187 mL of 2% acrylamide, and 2.173 mL of water.
15. Add 32 μL of 10% APS and 3.2 μL of TEMED to the stacking gel solution.
16. Place the comb in the gel sandwich.
17. Pour the stacking gel solution and allow it to polymerize for 1 h. Remove the comb slowly.
18. Place the gel in the buffer chamber. Add the anode buffer into the chamber at the bottom. With a syringe, remove the bubble at the bottom of gel.
19. Remove the comb and wash the wells with anode buffer (see Note 25).
20. Before loading the sample, add 1 μL of loading buffer to 20 μL of sample while mixing back and forth gently with a pipette. Use molecular mass size standards for BN-PAGE analysis (see Note 16). Load each sample mixture in a well and fill each well with cathode buffer. Then add the cathode buffer into the chamber at the top.
21. Connect the electrical leads to a suitable power pack with the proper polarity (black to black and red to red). Run the gel at 4°C, 1 W, 100 V, and 28 mA during approximately 36 h, depending on the gradient used.
22. Stop the electrophoresis when the tracker dye is approximately 1 cm above the end of the glass plates. Protein complexes are generally not visible during the electrophoresis and will become visible following Coomassie or silver staining. Since the gels stained (Fig. 3a) after the first dimension cannot be used for a second dimension, it is recommended to perform each loading twice: stain the first lane to verify the correct migration and to determine the M_r of the complexes and use the second unstained lane for the second dimension.

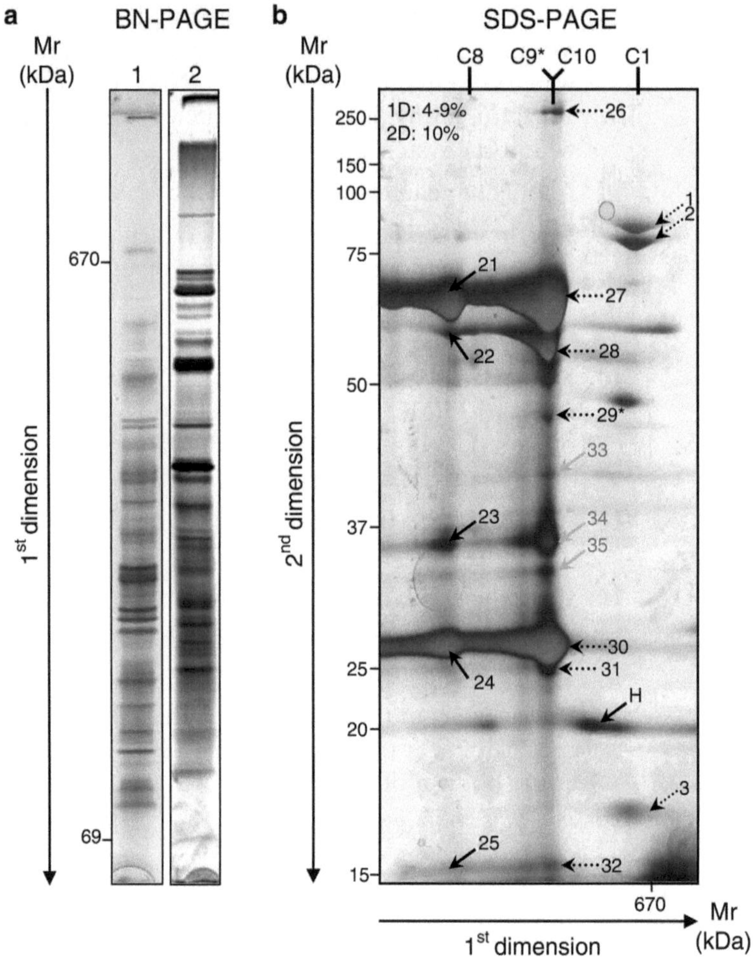

Fig. 3. First and second dimensions of 2D BN/SDS-PAGE technology after staining. (**a**) BN-PAGE (3–18% linear acrylamide gradient gel) of an *E. coli* (strain DH5a) crude cytoplasmic sample. Proteins were visualized using Colloidal Blue or silver staining (*lanes 1* and *2*, respectively). Note that the gels stained after the first dimension cannot be used for the second dimension (see Subheading 3.2). (**b**) SDS-PAGE (10% acrylamide gel) of *H. pylori* (strain J99) cytoplasmic sample purified using the liquid isoelectrofocalization method (Rotofor system, $pI = 5$) prior to applying the 2D BN/SDS-PAGE (BN-PAGE: 4–9% linear acrylamide gradient gel). Enlargement of a part of Fig. 2 of a research originally published in (3). Complex C1: spots #1–3 (JHP0632, JHP0633, and JHP0631). Complex C8: spots #21–25 (UreB, GroEL, IlvC, UreA, Tpx). Complex C9: spots #26–31 (RpoB, UreB, GroEL, UreA, spot #29*, DeoD and Tpx, where spot # contains a mixture of three different proteins: JHP0411, GlyA, and GdhA). Complex C10 spots #33–34 (AmiE, Ggt, and PorB). H: putative multihomooligomeric complex.

3.3. Equilibration of Strips

Perform this equilibration at room temperature.

1. Following electrophoresis, pry the gel plates open with the use of a spatula. The gel will remain on one of the glass plates. Excise individual lanes from the BN-PAGE gel with ruler and equilibrate in 10 mL of equilibration buffer for 5 min.
2. Immerse each individual lane in 9.35 mL of equilibration buffer supplemented with 0.65 mL of 1 M DTT (65 mM final) for 20 min.

Table 2
SDS-PAGE gel composition

	Resolving gel			Stacking gel
	10%	12%	15%	4%
Acrylamide/bis-acrylamide 40% 37.5 (mL)	5.32	6.38	7.98	800
Resolving gel buffer (mL)[a]	5.32	5.32	5.32	–
Stacking gel buffer (mL)[a]	–	–	–	2
Water (mL)	10.419	9.359	7.759	5.112
SDS 20% (μL)	106.4	106.4	106.4	40
APS10% (μL)	106.4	106.4	106.4	40
TEMED (μL)	8	8	8	8

[a]See Subheading 2.5

3. Soak then each individual lane for 20 min in 10 mL equilibrating buffer supplemented with 0.25 g of iodoacetamide (135 mM final).
4. Wash then the gel strip with 10 mL of equilibrating buffer for 5 min.

3.4. Sodium Dodecyl Sulfate-Polyacrylamide Gel Electrophoresis

1. Different % acrylamide gels can be used for the second dimension according to the M_r of targeted complexes.
2. Handcast gels as in Subheading 3.2.
3. Draw a line on the glass plate 3 cm below the top of the plate. This indicates the level to which the separating gel will be poured.
4. Prepare the resolving gel solutions at room temperature by combining all reagents according to Table 2 with the exception of APS and TEMED.
5. Add the specified amounts of APS and TEMED to the resolving gel solution according to Table 2 and mix well by swirling gently. Then pour the acrylamide separating gel mixture (see Note 23).
6. Immediately overlay the gel with water-saturated isopropanol (see Note 13).
7. Allow the gel to polymerize for 2 h.
8. Remove the water/isopropanol overlaying the gel, wash with water, and then drain the excess water with strips of filter paper.
9. Pour the stacking gel in two steps. For this, draw a line on the glass plate 1 cm above the separating gel. Prepare the stacking

gel solution. Combine all reagents (adding APS and TEMED at the end according to Table 2) and mix well by swirling gently and separate in two vials.

10. Pour the acrylamide stacking gel mixture from the first vial while the second vial is kept on ice.
11. Repeat steps 6–8.
12. Place the lane previously equilibrated (see Subheading 3.3) on the top of this poured stacking gel. Pour then additional stacking gel mixture contained in the second vial around it.
13. Add APS and TEMED (Table 2) to the second stacking gel solution and pour directly onto the lane previously equilibrated.
14. Allow the stacking gel to polymerize for 1 h.
15. Place the glass plates containing the gel in the buffer chamber. Add the running buffer into the chambers at the top and bottom. Remove the air bubbles at the bottom of gel with a syringe containing migration buffer.
16. Connect the electrical leads to a suitable power pack with the proper polarity (black to black and red to red). Run the gel at room temperature, 120 V, and 28 mA during approximately 2–4 h, depending on the percentage of acrylamide gradient used.
17. Stop the electrophoresis when the tracker dye is approximately 1 cm above the end of the glass plates.

3.5. Visualization of Proteins After Electrophoresis

1. Push one of the spacers out to the side of the plates without removing them.
2. Gently twist the spacer so that the upper glass plate pulls away from the gel.
3. Cut the gel on one side (to orientate the gel).
4. Remove the gel by gently grasping two corners of the gel and immerse it completely in the specific solution for silver or blue Colloidal staining (see Subheadings 3.5.1 and 3.5.2).

3.5.1. Silver Staining

Proteins are visualized after staining of the gels using the PROTSIL1 silver staining kit, according to the manufacturer's instructions (Fig. 3a and see Note 18).

3.5.2. Colloidal Blue Staining

CBB G-250 can also be used for gel staining. After two ultrapure water washings lasting 3 min each, place the gels overnight in the incubation solution with constant stirring. Then wash twice the gels for 1 min in ultrapure water and twice for 1 h in 5% acetic acid. After staining, estimate the approximate molecular mass of the visualized protein bands by comparing them with the molecular mass markers (see Note 18).

3.6. Definitions, Problems, and Limits Encountered for Certain Identifications

3.6.1. Definitions

1. Multiprotein complexes are separated according to their size and shape during the first dimension: the BN-PAGE. Each multiprotein complex is then denatured in a second dimension electrophoresis (SDS-PAGE) and the protein alignment allows the identification of interactive proteins. The basic condition for the identification of a multiprotein complex is that proteins of the same multiprotein complex co-migrate in the first dimension and the corresponding spots are found aligned with a similar shape in the second dimension (1).
2. It should be noted that different migrations of some proteins are often observed during the separation in the second dimension electrophoresis, due to multiple isoforms of these proteins. Their occurrence is explained by posttranslational modifications changing their physicochemical criteria (pI, M_r, binding affinity) (3, 4).
3. Different oligomerization states of some complexes also exist as previously pointed out (2, 3, 6).

3.6.2. Problems and Limits Encountered for Certain Identifications

Various difficulties of interpretation of the results obtained by the 2D BN/SDS-PAGE technique can be encountered. Consequently, the assignment of spots to a complex becomes difficult.

1. The shape of the spot present in the lower M_r range is sometimes rather diffuse and difficult to interpret. A good example is spot #3 in complex C1 (Fig. 3b). In comparison to spots #1 and 2, spot #3 has a more rounded and more diffuse shape. This is due to a more diffuse gel migration in this M_r range. In the present case, the problem of spot #3 was solved with other migrations and other acrylamide gradients, allowing the attribution of this spot to complex C1 (3).
2. A mixture of proteins can be identified in certain spots, and as a result, the same shape criterion is more difficult to apply. (1) Different complexes can indeed co-migrate in the first dimension, and therefore, some of their proteins can be present in the same spot in the second dimension, or (2) proteins with similar M_r can also belong to the same complex. Purification, other migrations, or other acrylamide gradients can be useful in such cases. For example, see spot #29 in complex C9 (Fig. 3b).
3. It is sometimes difficult to assign weakly visible proteins to a complex because the same shape criterion becomes difficult to apply. For example, spots #26–32, found several times to be aligned with the same shape (not shown) in the crude extract, were also identified with the additional spots #33–35 (Fig. 3b) after a purification. Thus, the difficulty was to attribute spots #26–32 and 33–35 either to a single complex or to two different complexes. Indeed, the absence of spots #33–35 on the gel

from the crude extract can be explained in two ways: (1) either the proteins present in spots #26–32 and 33–35 are all part of the same complex, and spots #33–35 which were less intense were not identified in the crude extract but revealed after a purification, or (2) spots #26–32 belong to a different complex than that of spots #33–35 but with the same apparent pl which would explain why they were not found in the crude extract except after the purification. As a precautionary measure to avoid describing a false complex, two different complexes are described.

4. Homooligomeric complexes can be also described with this method although the same shape criterion is no longer applicable (2, 3). Thus, particular care should be taken regarding the description of homooligomeric complexes. Indeed, as recently shown (4), some partners, hardly or not visible during the staining, can be present, leading to the description of false complexes or to homooligomeric complexes with an inadequate number of subunits.
5. The resolution of the gels must be of high quality to avoid incoherent migrations in the first and second dimensions. For example, the M_r of a complex is sometimes lower than the M_r sum of the subunits, suggesting the presence of at least two complexes which co-migrate in the first dimension. This highlights the usefulness of sometimes performing a purification step prior to 2D BN/SDS-PAGE.
6. The 2D BN/SDS-PAGE is of great interest to compare the complexes of an organism in two different physiological states. However, this method has limitations when comparing organisms with a high variability, as it was demonstrated for *H. pylori* (4).

3.7. Protein Excision from Gel and Trypsin Digestion

1. Excise each spot of interest with a sterile scalpel or a die cutter. Concerning the coloration with silver staining, destain the pieces of gel using the PROTSIL2 silver staining kit. With regard to the CBB G-250 staining, rinse spots in 150 µL of buffer containing 25 mM ammonium bicarbonate and 50% acetonitrile until complete destaining.
2. Remove the solvent mixture and replace it by 100% acetonitrile for 10 min.
3. After shrinking of the gel pieces, remove acetonitrile and dry the gel pieces in a vacuum centrifuge at room temperature.
4. Cover the gel pieces with 10 ng/µL trypsin solution and 50 mM ammonium bicarbonate, rehydrate at 4°C for 10 min, and finally incubate overnight at 37°C.
5. Add ammonium bicarbonate (50 mM) to the gel pieces and incubate for 15 min at room temperature under rotary shaking.

6. Collect the supernatant and add an ultrapure water/acetonitrile/acetic acid (47.5/47.5/5) solution onto the gel slices for 15 min. Repeat this step twice.
7. Pool supernatants and concentrate them in a vacuum centrifuge to a final volume of 25 μL.
8. Finally, acidify digested products by the addition of 1.5 μL of acetic acid and store them at −20°C until identification by mass spectrometry.

4. Notes

1. Keep nondenaturing conditions during the sample extraction and BN-PAGE. Ideally, perform manipulations at 4°C and place organisms directly on ice without heating or repeated freezing in order to avoid artificial aggregate formation, disruption, or loss of subunits from complexes. Avoid the use of nondenaturating or reducing agents.
2. Wear gloves when preparing the extracts, solutions, and gels as well as when handling gels or excising proteins from a gel for safety reasons and also in order to protect proteins from degradation by proteases and to avoid the presence of human keratin which would interfere with mass spectrometry analyses.
3. The 6-amino-*n*-caproic acid reagent/solution is an irritant and should be handled with gloves.
4. The negatively charged nucleic acids bind to proteins, which is worrisome during the preparation of the extract since it is the main interfering factor during the migration of proteins. Thus, effective elimination of nucleic acids from protein samples is highly recommended for optimizing BN-PAGE electrophoresis. Here, we propose the elimination of DNA by different treatments with DNase or spermine. Alternatively, nucleic acids can be efficiently removed by enrichment of the nucleic acid-binding proteins by precipitation with streptomycin sulfate, with subsequent elimination of DNA by precipitation with polyethyleneimine (8).
5. For BN-PAGE, DDM, digitonin, and Triton X-100 have also proven to be useful. However, DDM and digitonin are the most commonly used (9). DDM resulted in the improvement of membrane protein solubilization. Detergents are toxic, and gloves should be worn when handling buffers or samples containing these detergents.
6. Store resuspended DNase I solution at −20°C. DNase I is sensitive to physical denaturation. Do not vortex the resuspended DNase I solution.

7. Prepare PMSF in anhydrous dimethyl sulfoxide as a stock solution because it is rapidly degraded in water.
8. The anionic dye CBB G-250 used during the BN-PAGE electrophoresis binds nonspecifically to the surface of all proteins, particularly on aromatic residues and on arginines. This binding of a large number of negatively charged dye molecules to proteins facilitates the multiprotein complex migration toward the anode, and the tendency for protein aggregation is thus reduced considerably (10).
9. Having water at the bottom of the cylinder helps to dissolve components (Tris, CBB G-250) relatively easily and avoids lumps, allowing the magnetic stir bar to go to work immediately.
10. Wear gloves and glasses to avoid skin and eye contact when using acrylamide and bisacrylamide which are highly neurotoxic. The manufacturer's warning states that this resin is explosive when mixed with oxidizing substances. Use preferentially liquid acrylamide. However, powder should be used for concentrations higher than 14% when performing the BN-PAGE. The acrylamide solution can be stored at 4°C for 1 month and at –20°C for longer periods.
11. Storing TEMED at 4°C reduces its pungent smell.
12. SDS as a powder is an irritant; use preferentially liquid SDS. β-Mercaptoethanol is toxic; use gloves and work under a hood.
13. Before use, mix the water/isopropanol solution vigorously in order to saturate the isopropanol.
14. This buffer can be kept for many months at 4°C.
15. Add the CBB G-250 at the end. This buffer can be kept for many months at 4°C.
16. Other molecular weight markers such as aldolase (158 kDa), catalase (232 kDa), and ferritin (440 and 880 kDa) can also be added. It was recently pointed out that soluble marker proteins should not be used for mass estimations of membrane proteins, and optimal sets of membrane protein markers were recently reported (11).
17. Store this buffer at –20°C in aliquots. When an aliquot is unfrozen, it must not be frozen again.
18. Silver staining is a highly sensitive method but not a quantitative method, unlike the CBB G-250 staining which is quantitative but less sensitive.
19. It is preferable to dissolve CBB G-250 in a small volume of ultrapure water prior to the final dilution.
20. Break the membrane extract a second time in order to avoid contamination of the membrane extracts by cytosolic components. Indeed, during the first bacterial breaking, membranes

tend to close in on them even with proteins from the cytosol. To prevent this, dilute the membrane pellet in a high volume and break a second time.

21. Purification steps using physical or chemical properties of the multiprotein complexes constitute an important aspect to confirm protein–protein interactions and to increase the recovery number of multiprotein complexes. Thus, after solubilization of the complexes, different protocols can be used for the purification of the samples before the BN-PAGE: filtration, desalting column, liquid isoelectrofocalization, gel filtration, ionic column, deoxycholate extraction, chloroform/methanol extraction, sucrose gradient, affinity purification, density centrifugation, and ammonium sulfate precipitation have proven to be effective (3, 4, 12–22). Moreover, different amounts of proteins can be loaded onto the gels, and various acrylamide concentrations in first and second dimensions can be used. After the purification step, the sample can be desalted and/or concentrated using Vivaspin concentrator (Vivascience, Aubagne, France) or Centricon Centrifugal Filter Units (Amicon, Inc., Beverly, MA, USA) with the adapted cut-off.
22. An acrylamide gradient may not be necessary. When necessary, it must be defined to allow optimization of complex focalization and separation in a given size range. Precast BN gels and buffers are also commercially available.
23. Add APS and TEMED immediately before pouring gel, as these reagents promote polymerization.
24. Some backflow into the low % reservoir may occur, due to the density variation between the solutions. It will not substantially alter the final gradient. Faster flow will cause turbulence in the gel which will disrupt the gradient. Slower flow rates will allow polymerization to occur before pouring is complete. A peristaltic pump may be used to regulate the flow from the gradient maker. These steps must be carried out rapidly to avoid polymerization of the solutions before the gel is fully cast.
25. Wash the wells with anode buffer since this buffer does not contain CBB G-250, allowing a better visualization of the experiment.

Acknowledgments

Particular thanks are extended to Prof. Marc Bonneu for his innovative and brilliant ideas. The study on *Helicobacter pylori* was financially supported by INSERM, the University of Bordeaux 2, the European consortium INCA (Infection and Cancer) from the sixth PCRD of the EU, the Conseil Régional d'Aquitaine, and the Association pour la Recherche sur le Cancer (ARC). The study on *Escherichia coli* was financially supported by the University of Bordeaux 2 and the Conseil Régional d'Aquitaine. The study on *Yarrowia lipolytica* was financially supported by INRA and CNRS and Oxyrane (grant no. B 05344). Predoctoral and postdoctoral students were supported in part by fellowships from the Fondation pour la Recherche Médicale (*H. pylori*), by EU grant number QLRICT-2001-00015 under the RTD program "Quality of Life and Management of Living Resources" (*E. coli*). We thank Lindsay Megraud for proofreading the manuscript.

References

1. Schagger H, von Jagow G (1991) Blue native electrophoresis for isolation of membrane protein complexes in enzymatically active form. Anal Biochem 199:223–231
2. Lasserre JP, Beyne E, Pyndiah S et al (2006) A complexomic study of *Escherichia coli* using two-dimensional blue native/SDS polyacrylamide gel electrophoresis. Electrophoresis 27: 3306–3321
3. Pyndiah S, Lasserre JP, Ménard A et al (2007) Two-dimensional blue native/SDS gel electrophoresis of multiprotein complexes from *Helicobacter pylori*. Mol Cell Proteomics 6: 193–206
4. Bernarde C, Lehours P, Lasserre JP et al (2010) Complexomics study of two *Helicobacter pylori* strains of two pathological origins: potential targets for vaccine development and new insight in bacteria metabolism. Mol Cell Proteomics 9:2796–2826
5. Lasserre JP, Nicaud JM, Pagot Y et al (2010) First complexomic study of alkane-binding protein complexes in the yeast *Yarrowia lipolytica*. Talanta 80:1576–1585
6. Lasserre JP, Sylvius L, Joubert-Caron R et al (2010) Organellar protein complexes of Caco-2 human cells analyzed by two-dimensional blue native/SDS-PAGE and mass spectrometry. J Proteome Res 9:5093–5107
7. Dresler J, Klimentova J, Stulik J (2011) Bacterial protein complexes investigation using blue native PAGE. Microbiol Res 166:47–62, Review
8. Liang J, Niu Q, Xu X et al (2009) Effective elimination of nucleic acids from bacterial protein samples for optimized blue native polyacrylamide gel electrophoresis. Electrophoresis 30:2454–2459
9. Krause F (2006) Detection and analysis of protein-protein interactions in organellar and prokaryotic proteomes by native gel electrophoresis: (membrane) protein complexes and supercomplexes. Electrophoresis 27:2759–2781
10. Schägger H, Cramer WA, von Jagow G (1994) Analysis of molecular masses and oligomeric states of protein complexes by blue native electrophoresis and isolation of membrane protein complexes by two-dimensional native electrophoresis. Anal Biochem 217:220–230
11. Wittig I, Beckhaus T, Wumaier Z et al (2010) Mass estimation of native proteins by blue-native electrophoresis: principles and practical hints. Mol Cell Proteomics 9:2149–2161
12. Marzoa J, Abel A, Sanchez S et al (2009) Analysis of outer membrane porin complexes of *Neisseria meningitidis* in wild-type and specific knock-out mutant strains. Proteomics 9: 648–656
13. Aufurth S, Schagger H, Muller V (2000) Identification of subunits a, b, and c1 from *Acetobacterium woodii* Na+-F1F0-ATPase. Subunits c1, c2, and c3 constitute a mixed c-oligomer. J Biol Chem 275:33297–33301
14. Tsirogianni E, Aivaliotis M, Papasotiriou DG et al (2006) Identification of inducible protein complexes in the phenol degrader *Pseudomonas* sp. strain phDV1 by blue native gel electrophoresis and mass spectrometry. Amino Acids 30: 63–72
15. Stenberg F, Chovanec P, Maslen SL et al (2005) Protein complexes of the *Escherichia coli* cell envelope. J Biol Chem 280:34409–34419
16. Kulajta C, Thumfart JO, Haid S et al (2006) Multi-step assembly pathway of the cbb3-type cytochrome c oxidase complex. J Mol Biol 355:989–1004

17. Schneider D, Pohl T, Walter J et al (2008) Assembly of the *Escherichia coli* NADH:-ubiquinone oxidoreductase (complex I). Biochim Biophys Acta 1777:735–739
18. Peters A, Kulajta C, Pawlik G et al (2008) Stability of the cbb3-type cytochrome oxidase requires specific CcoQ-CcoP interactions. J Bacteriol 190:5576–5586
19. Bessonneau P, Besson V, Collinson I et al (2002) The SecYEG preprotein translocation channel is a conformationally dynamic and dimeric structure. EMBO J 21:995–1003
20. Inacio JM, Correia IL, de Sa-Nogueira I (2008) Two distinct arabinofuranosidases contribute to arabino-oligosaccharide degradation in *Bacillus subtilis.* Microbiology 154: 2719–2729
21. Swanson KA, Taylor LD, Frank SD et al (2009) *Chlamydia trachomatis* polymorphic membrane protein D is an oligomeric autotransporter with a higher-order structure. Infect Immun 77:508–516
22. Wang ZJ, Xu XP, Fan KQ et al (2007) Sample preparation for two-dimensional blue native/SDS polyacrylamide gel electrophoresis in the identification of *Streptomyces coelicolor* cytoplasmic protein complexes. J Biochem Biophys Methods 70:565–572

Chapter 28

Immunoelectrophoresis: A Method with Many Faces

Gyorgy Csako

Abstract

Immunoelectrophoresis (IEP) was the first practical method that combined electrophoresis and immunoprecipitation for identifying and characterizing proteins within complex mixtures. Over the years, IEP has been extended to include a variety of techniques and, as a general name, has been applied to virtually any technique that involves electrophoresis and antigen–antibody precipitin reaction for proteins. Because of the diversity in technical details of different IEP versions, the method described here deals only with classic IEP. Although it requires some manual expertise, IEP is versatile, relatively easy to customize, and economical with no need for expensive instrumentation. Further, it can discern identity, partial identity, and nonidentity of the proteins. Any low-viscosity body fluid specimen or, possibly, culture fluid and tissue extract could be tested with IEP if proper antibodies are available. With these attributes, classic IEP remains a valuable tool for clinical diagnostic testing, purity checking of biochemical and pharmaceutical products, and research.

Key words: Countercurrent immunoelectrophoresis, Counter-immunoelectrophoresis, Crossed immunoelectrophoresis, Double immunodiffusion, Immunoglobulins, Immunoprecipitation, Laurell crossed immunoelectrophoresis (two-dimensional electroimmunodiffusion), Reversed rocket immunoelectrophoresis, Rocket immunoelectrophoresis, Starch gel

1. Introduction

In its classic form, first devised by Grabar and William (1–3) and then modified as a micromethod by Scheidegger (4, 5), immunoelectrophoresis (IEP) combines separation of protein antigens by electrophoresis and immunodiffusion against an antiserum. The proteins are first separated by electrophoresis in semisolid media (gels) then the specific proteins are visualized through an immunologic reaction, namely, antigen–antibody precipitin formation. Double immunodiffusion ensues after an antiserum prepared against the target proteins is applied into a trough cut into the gel

Biji T. Kurien and R. Hal Scofield (eds.), *Protein Electrophoresis: Methods and Protocols*, Methods in Molecular Biology, vol. 869, DOI 10.1007/978-1-61779-821-4_28, © Springer Science+Business Media, LLC 2012

parallel to the direction of electrophoretic separation and distinct elliptical precipitin arcs develop with the antisera for each protein antigen. Over the years, however, IEP has been extended to include a variety of techniques and, as a general name, has been applied to virtually any technique that involves electrophoresis and antigen–antibody precipitin reaction for separation, characterization, and, possibly, quantitation of proteins. In fact, the initial reports about the immunofixation (IFE) technique were misleadingly also called "quantitative IEP" (6) and "direct IEP" (7). "True" IEP techniques now include two versions of the one-dimensional (classic) IEP (1–5), counter(-current) or crossover (immuno)electrophoresis or electrosyneresis (8–10), crossed IEP ((two-dimensional) quantitative IEP, two-dimensional electroimmunodiffusion) (11–13), rocket-IEP or electroimmunoassay (14–17), fused rocket IEP (18), affinity IEP or crossed immuno-affinoelectrophoresis (19), line immunoelectrophoresis (20), crossed radioimmunoelectrophoresis (21), crossed-line immunoelectrophoresis (22), rocket-line immunoelectrophoresis (23), and numerous other modifications of these techniques (e.g., see ref. (24)). Even reverse version of some of these assays has been developed (e.g., reverse rocket IEP (25)). Because of their shared concept, these methods are all capable of identifying (and, depending on assay design, quantifying) antigen(s) (or antibodies in reversed versions) in complex specimens.

With much improved resolution and quantitation of proteins in biological fluids and biological extracts, the introduction of the IEP analysis had a great impact on protein chemistry and laboratory diagnostics. Interestingly, the history of classic IEP for human paraprotein immunotyping follows an opposite pattern to IFE. IFE was introduced for improved detection of protein polymorphisms and only years later was applied to the study of paraproteins as a clinical diagnostic technique, but it became the preferred test in this area. Classic IEP was developed primarily as a clinical diagnostic test for the detection and identification of paraproteins, but then it was gradually replaced by IFE, so that now it is only selectively used for that purpose. On the other hand, classic IEP is still widely employed for purity checks and identification of biochemical and pharmaceutical products (e.g., in major official reference publications for drug specifications (26, 27)), and research, further, with its many new versions, IEP reinvented itself for both clinical diagnostic and research applications. The following is the description of a classic IEP method for paraprotein immunotyping which can be readily modified for other diagnostic or research applications.

2. Materials

Prepare all solutions with distilled or deionized water and high-grade analytical reagents at room temperature (15–30°C). Except where stated otherwise, reagents and other materials are stored at room temperature. Unless obtained commercially, prepare and store all reagents and accessories used in these methods without adding sodium azide. Follow all waste disposal regulations for discarding reagents and disposables. As a safe laboratory practice, always wear clean gloves to minimize or eliminate exposure to infectious agents and hazardous chemicals.

2.1. Agarose Gel Plates for Protein Electrophoresis (See Notes 1 and 2)

1. Hydragel 7 and Hydragel 15/30 Protein(e) alkaline buffered precast 0.8% (w/v) agarose gels (Sebia, Norcross, GA, USA). These gels contain Tris–barbital buffer (pH 9.2 ± 0.1) and proprietary additives which are claimed to be "nonhazardous" at concentrations used. Store horizontally at room temperature or refrigerated (2–8°C) until stated expiration date.

2.2. Accessories for Preparing Agarose Gel Plates for Electrophoresis and Immunodiffusion (See Note 3)

1. Templates for positioning wells and troughs (Fig. 1a) (see Note 4).
2. "Cutting templates" for antiserum application (Fig. 1a) (see Notes 3 and 5).
3. Suction needle as a well cutter (well diameter of ~1 mm) (see Notes 3 and 6).

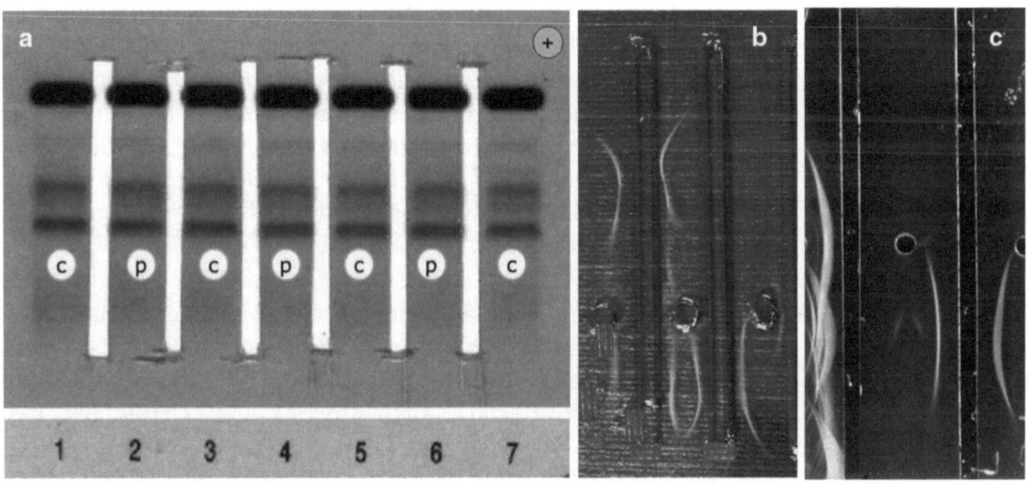

Fig. 1. Positioning template for wells and troughs, gel-cutting template for antiserum application, and photographs of wet agarose gel with precipitin arcs for IEP. (**a**) The "homemade" gel-cutting template for making troughs is shown on *top* of the "positioning" template (1:1 paper copy of amido black-stained human serum protein patterns after electrophoresis in agarose gel). Codes: *c* control sample positions, *p* patient sample positions. (**b**) Photograph of wet agarose gel that shows a developing human serum IEP pattern using "homemade" devices. (**c**) Photograph of wet agarose gel that shows part of a completed human serum IP pattern produced by a discontinued commercial kit.

4. Scalpel.
5. Glass Pasteur pipette connected by tubing to a filter flask (see Note 7).
6. Filter flask connected by tubing to a vacuum source (see Notes 3 and 5).
7. Leveling table.
8. Pipettes or syringes for delivering volumes between 0.5 and 50 μL.
9. Wet incubation chamber (see Note 8).

2.3. Apparatus and Accessories for Protein Electrophoresis in Agarose Gel

1. Hydrasys® semiautomated electrophoresis system (Sebia) (see Note 9).
2. Buffered, ready-to-use sponge strips for protein electrophoresis (Sebia). Each contains Tris–barbital buffer pH 9.2 ± 0.3, sodium azide, and proprietary additives which are claimed to be "nonhazardous" at the concentrations used. Store horizontally at room temperature or refrigerated until stated expiration date (see Note 10).
3. Precut, single-use filter papers (Sebia). Store in a dry place at room temperature or refrigerated.
4. Pipette for delivering 200 μL.
5. Bromphenol Blue sodium salt (MW 692; Sigma Aldrich Chemical Co., St. Louis, MO, USA): 0.05% (w/v) solution in a Tris–barbital buffer, pH 8.6 (first, prepare a 5× buffer in a measuring cylinder): Dissolve 0.24 g 5,5′-diethyl barbituric acid and 4.43 g Tris in distilled water to a final volume of 100 mL. Dilute by a factor of 5 before use. Store in a closed bottle for 6 months. Second, weigh 5 mg of the dye powder (3% difference due to the sodium content is negligible) and pour the dye into a 15-mL tube with a screw cap. Add 10 mL of the fivefold diluted stock buffer to the tube and cap it. Shake the tube well to dissolve the dye in the buffer. Store for up to 1 year (see Note 11).

2.4. Antibodies for Immunodiffusion (See Notes 12–14)

1. Polyclonal anti-human whole serum (Helena Laboratories, Beaumont, TX, USA) and monospecific rabbit antibodies to human IgG, IgA, IgM, and κ and λ (free and bound) light chains (all from Dako North America, Inc., Carpinteria, CA, USA).
2. Reconstitute lyophilized preparations according to the manufacturer's instructions. Store all antisera refrigerated until stated expiration date.

2.5. Solutions and Accessories for Protein Staining in Agarose Gel (See Note 15)

1. Amido black stain (Sebia). Reconstitute and store according to the manufacturer's instructions. The working solution is acidic (pH 2) and contains 0.4 g/dL amido black, 6.7% (v/v) ethylene glycol, and additives which are claimed to be "non-hazardous" at the concentrations used. Working solution in a closed container to prevent evaporation is stable for 1 month (see Note 16).
2. Destaining solution: distilled water.
3. Staining pans.
4. Paragon dryer (Beckman).

2.6. Specimens for IEP in Agarose Gel (See Note 17)

1. Human serum for illustration of the method (see Note 18).
2. Collect blood specimens with standard medical procedures. Store serum refrigerated for up to 1 week (preferably use within 48 h), freeze at −20°C for up to 3 months or at −70°C for up to 12 months.
3. Use serum samples straight or dilute with 0.85% saline when antigen excess is suspected (see Note 19).

2.7. Control (Reference) Specimens for IEP in Agarose Gel (See Note 20)

Normal human serum or serum pool for paraprotein testing. Store serum refrigerated for up to 1 week (preferably use within 48 h), freeze at −20°C for up to 3 months or at −70°C for up to 12 months.

3. Methods

Carry out all procedures at room temperature unless otherwise specified. For general guidance and safety issues consult the operating manual of the electrophoretic system used. Estimated total time needed to complete one gel with the electrophoretic and detection method described below is ~48 h.

3.1. Preparation of Wells and Troughs in Agarose Gel

1. Unpack a Hydragel plate selected for IEP (Hydragel 7 or 15/30 Protein(e)) and roll quickly and uniformly a thin filter paper onto the gel surface to absorb excess liquid. Remove and discard the paper immediately.
2. Place the gel plate over an appropriate template showing the position of wells for sample application and troughs for antiserum application (Fig. 1a) (see Note 4).
3. Carefully overlay the gel with the "cutting template" (Fig. 1a) (see Note 5).
4. Cut the desired number of troughs with a scalpel and aspirate the gel from the troughs by suction using a glass Pasteur pipette

connected to the side arm of a filter flask which, in turn, is connected to a vacuum line. Avoid damaging the walls of the trough. Clean the blade often by immersion into distilled water and wiping it dry on paper towels laid down on the benchtop. For safety, do NOT hold towels in your hand while wiping!

5. Remove the cutting template by making sure that the gel is not damaged (e.g., not lifted up from its plastic support) (see Note 21).
6. Guided by the template placed under the gel for assay layout, punch sample application wells and simultaneously aspirate the agarose plugs with the suction needle connected to the side arm of a filter flask which, in turn, is connected to a vacuum line. Wash suction needle between punching wells by aspirating distilled water (see Note 6).

3.2. Sample Preparation and Loading onto Precut Agarose Gel

1. Add 2 μL of 0.05% Bromphenol Blue solution to 18 μL of control specimen and briefly (1–2 s) vortex it. This results in a final dye concentration of 0.005% (w/v).
2. Apply 2 μL of control serum sample to each application well as follows: positions 1, 3, 5, and 7 for testing one unknown specimen on a 7 Protein(e) plate and positions 1, 3, 5, 7, 9, 11, 13, and 15 for testing two unknown specimens on a 15 Protein(e) plate (see Note 22).
3. Apply 2 μL of unknown serum sample to each application well as follows: positions 2, 4, and 6 for testing one unknown specimen on a 7 Protein(e) plate and positions 2, 4, 6, 12, and 14 for testing two unknown specimens on a 15 Protein(e) plate. Note that position 8 receives no sample; this position separates the two unknown specimen runs (see Note 22).

3.3. Protein Electrophoresis in Agarose Gel

Switch on the Hydrasys instrument and proceed according to the Hydragel 7 or 15/30 Protein(e) method for migration. Briefly,

1. Attach buffered strips as appropriate for protein run to the pins on the uplifted electrode carrier.
2. Place the 7 Protein(e) or15/30 Protein(e) plate preloaded with samples into the migration module by making sure that the sample application wells are toward the cathode.
3. Select migration program appropriate for the type of agarose plate.
4. Run electrophoresis under 10 W constant for serum proteins using Hydragel 7 Protein(e) plate (for one unknown IEP specimen) or 20 W constant using Hydragel 15/30 Protein(e) plate (for two unknown IEP specimens) at 20°C, controlled by Peltier effect, until 33 Vh have accumulated (~7 min).
5. Interrupt electrophoretic program after migration has been completed.

3.4. Immunodiffusion in Agarose Gel

1. After stopping electrophoresis, remove gel from the migration module and wipe excess solution from back of gel.
2. Verify proper migration by observing the position of the tracking dye, Bromphenol Blue: the second fastest migrating blue spot showing the position of albumin should be slightly above the anodic end of the troughs.
3. Gently blot the gel surface with a thin filter paper. Discard paper.
4. Place the plate on a leveling platform.
5. Slowly apply 15 μL of various antisera onto the anodic end of the appropriate troughs. Although antiserum usually spreads along the trough by capillary action, make sure that it has evenly spread along the entire length of the trough (see Note 23).
6. Place gel in level humidity incubation chamber.
7. Cover chamber and incubate gel for 18–24 h in a refrigerator (4°C) or until visible precipitin arcs develop at expected locations with special attention to the controls in the gel (Fig. 1b, c) (see Note 24).
8. After ~16 h of incubation or at the earliest convenient time, begin checking the progress of precipitin formation at every few hours. Visible precipitin arcs begin to appear after 6–8 h and the immunoprecipitation pattern is fully developed after 24 h in this system.

3.5. Processing of Agarose Gel for Staining (See Note 25)

1. At the end of incubation, remove gel from wet incubation chamber.
2. Stop immunodiffusion by washing the gel in ~75 mL of 0.85% saline on a rocking platform at a low speed for 1 h.
3. Continue washing with saline for 6–24 h. During this time, change saline 5–10 times to increase efficiency of washing (obviously, more frequently for a shorter wash time).
4. Soak gel in ~75 mL of distilled water with gentle shaking on a platform (or just by hands) to desalt for ~1 h. Change water 2–3 times during this time.
5. Wipe excess solution from back of gel and dry gel in a Paragon dryer at 60°C (5–10 min).

3.6. Amido Black Staining for Proteins (See Notes 26 and 27)

1. Stain the dried gel by immersing into the amido black stain for 3–5 min at room temperature under gentle agitation.
2. Destain gel with distilled water at room temperature under gentle agitation until the desired low background level is achieved.
3. Remove gel from water, wipe water from back of gel and dry gel in a Paragon dryer for 5–10 min at 60°C.

3.7. Interpretation of IEP Patterns (See Notes 28 and 29)

General comments:

1. Evaluate visually the gel by noting.
   (a) The presence of precipitin arc(s) at the expected location with each antiserum in the control (reference) sample.
   (b) The location and shape of specific precipitin arc(s) with each antiserum in the unknown sample in relationship to those in the control (reference) sample.
2. In case of paraprotein testing, distortion of an arc (broadening, bulging, or displacement) indicates paraprotein. Matching the location of distortions with the heavy chain-specific and light chain-specific antisera identifies the immunotype of paraprotein(s).
3. In case of paraprotein testing, if technical errors such as inappropriate or low-avidity antisera can be ruled out, then
   (a) Absence of heavy chain abnormality in the presence of light chain abnormality suggests the possibility of IgD or IgE paraprotein which requires further testing with specific antisera against these immunoglobulins (see Note 30).
   (b) Absence of light chain abnormality in the presence of heavy chain abnormality suggests the possibility of heavy chain disease.
   (c) Absence of abnormality in the routine IEP pattern but abnormal band in the gamma region on protein electrophoresis suggests the possibility of free light chain, IEP artifact (the so-called "umbrella effect"), or pseudoparaprotein (e.g., fibrinogen) (see Note 29).

Specific comments on IEP patterns produced with the manual method on human serum:

1. Even this minimally optimized IEP method shows abnormalities (distortion exemplified by broadening) at the cathodal end of the IgG and κ precipitin arcs with the patient serum Fig. 2c, indicating the presence of an IgG(κ) paraprotein. This paraprotein is even more apparent in the corresponding IFE pattern Fig. 2b.
2. IFE and IEP performed with optimized assays both show unequivocal presence of an IgG(κ) paraprotein. Note characteristic distortion (primarily broadening) of the precipitin arcs that correspond to the paraprotein (Fig. 3).
3. Since IEP assays do not routinely test for the presence of free light chains, testing with anti-free light chain antibodies may be indicated when either clinical suspicion arises or discordant routine IEP and protein electrophoretic results are obtained (no abnormality in the routine IEP pattern but abnormal band

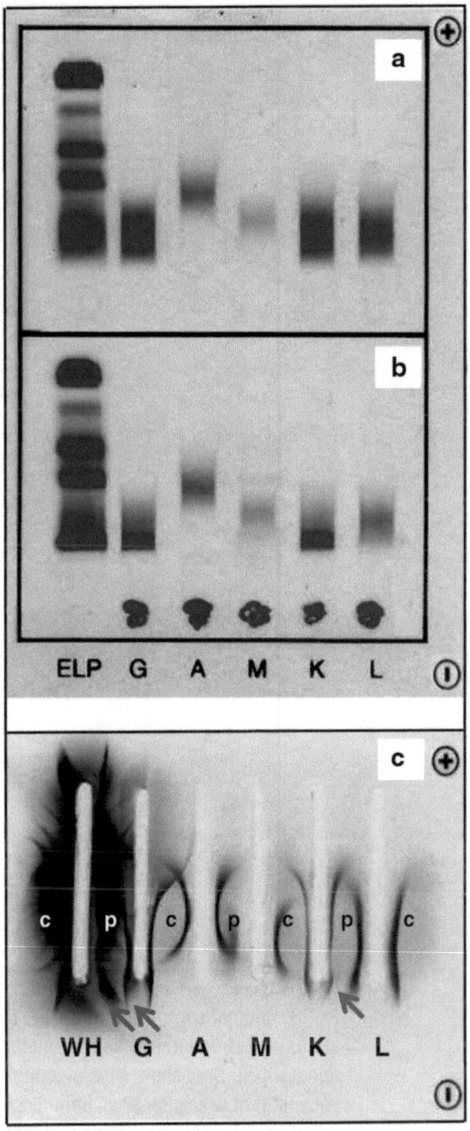

Fig. 2. Comparison of the manual IEP method with IFE obtained using a semiautomated commercial method for human serum paraprotein immunotyping in agarose gel. (**a**) Normal control human serum (**b**) human serum with IgG(κ) paraprotein, both analyzed with Hydragel 2 IF kit on Hydrasys® followed by staining with acid violet (Sebia Inc., Norcross, GA, USA), and (**c**) minimally optimized manual EIP method on a protein electrophoresis plate using a hand-cut IEP antiserum template for cutting the troughs prior to electrophoresis and manually applying various antisera after electrophoresis, followed by amido black staining. *Arrows* indicate abnormal distortion and broadening of precipitin arcs. Codes: *ELP* acid-fixed protein reference lanes; *G*, *A*, and *M* monospecific antibodies against human IgG, IgA, and IgM heavy chains; *K* and *L* monospecific antibodies against human κ and λ (free and bound) light chains; *WH* antibodies against whole human serum; *c* control serum [same as in (**a**)], *p* patient serum [same as in (**b**)].

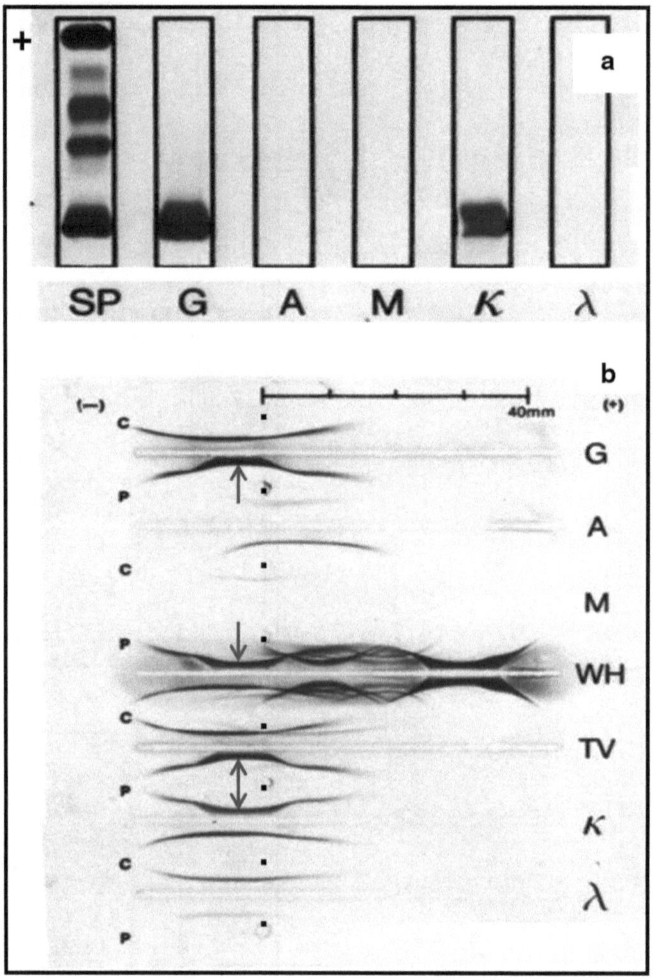

Fig. 3. Comparison of optimized IFE (TITAN Gel ImmunoFix, Helena) and IEP (TITAN Gel Immunoelectrophoresis, Helena) methods for immunotyping of a human serum paraprotein (IgG[κ]). *Dots* show sites of sample application, *arrows* indicate distortion (abnormal widening) of precipitin arcs. Immunoglobulin quantitation by radial immunodiffusion: IgG 3,524 mg/dL, IgA <40 mg/dL, and IgM 20 mg/dL. Codes: *SP* acid-fixed protein reference lane; *G*, *A*, and *M* monospecific antibodies against human IgG, IgA, and IgM heavy chains; *K* and *L* monospecific antibodies against human κ and λ (free and bound) light chains; *WH* polyclonal antibodies against whole human serum; *TV* trivalent antiserum against IgG, IgA, and IgM heavy chains; *c*, control (reference) serum; *p*, patient serum. (Courtesy of Helena Laboratories, Beaumont, TX, USA).

in the gamma region on protein electrophoresis). This figure shows a case where routine IEP panel suggested a normal pattern but testing with anti-free light chain antibodies revealed a weak but distinct band of free κ light chain in the patient's serum (Fig. 4).

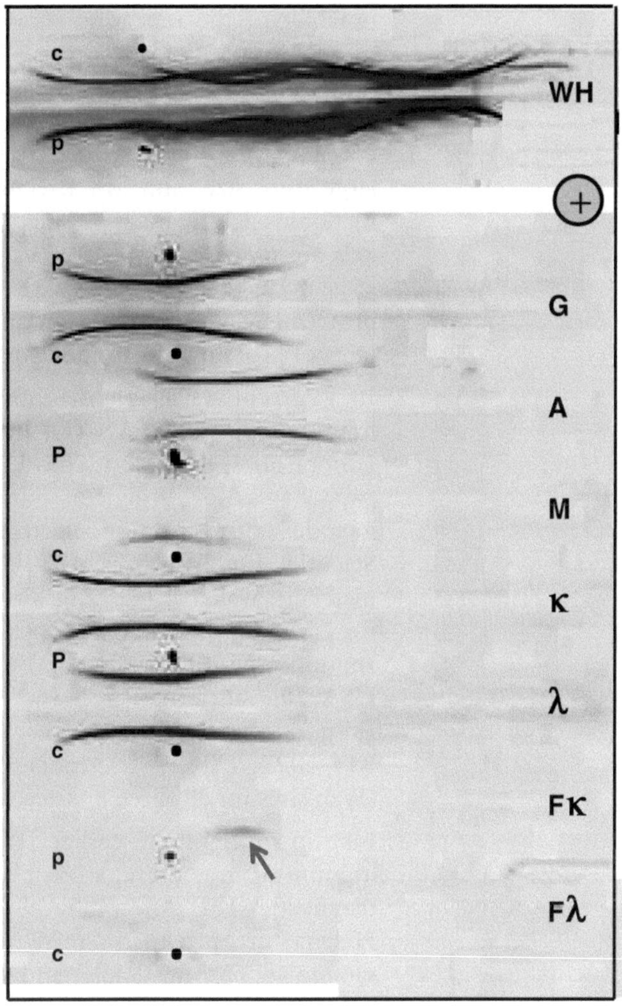

Fig. 4. Detection of serum-free κ light chain with IEP (TITAN Gel Immunoelectrophoresis, Helena). *Dots* show sites of sample application, *arrow* indicates free κ light chain. Codes: *WH* polyclonal antibodies against whole human serum; *G*, *A*, and *M* monospecific antibodies against human IgG, IgA, and IgM heavy chains; κ and λ monospecific antibodies against human κ and λ (free and bound) light chains; *F*κ and *F*λ monospecific antibodies against human free κ and free λ light chains. (Courtesy of Helena Laboratories, Beaumont, TX, USA).

4. Notes

1. Although agar and agarose gel have been used in the majority of classic IEP assays, applications have been also reported for starch, acrylamide, and cellulose acetate (Cellogel) gels, and even for dry cellulose acetate membrane (28). Agar, the original medium for IEP, consists of a mixture of agarose and agaropectin. Agarose, the main component of agar, is a linear

polymer of the repeating monomeric unit of a disaccharide (agarobiose). Agaropectin is a similar polymer but contains numerous acidic side groups and gels poorly. The neutral charge and relatively simple chemical composition of agarose coupled with the high gel strength, low electroendosmosis, large pore size, and low background staining of agarose gel made this gel the most popular medium for IEP. Most agarose gels are prepared as 0.7–1.5% slabs of ~1 mm thickness. The large pores of agarose gels allow free passage and separation of proteins, while providing an anchor for the immunoprecipitates of protein antigens and their specific antibodies.

2. In a currently available diagnostic kit for human paraprotein immunotyping (TITAN Gel Immunoelectrophoresis, Helena Laboratories, Beaumont, TX, USA) the IEP plates contain 1% (w/v) agarose. A general purpose commercial kit for IEP includes agar (Immunoelectrophoresis kit, Thermo Fisher Scientific Inc., Waltham, MA, USA), while two different commercially available teaching kits for IEP both supply agarose for preparing gel-coated glass slides (EDVO-Kit #272 Immunoelectrophoresis, EDVOTEK®, Inc., Rockville, MD, USA (kit also marketed by Thermo Fisher) and Manual Immunoelectrophoresis Teaching Kit, Catalog #IMI-KIT-1022, IMGENEX India Pvt Ltd, Bhubaneswar, India) (Web sites for all three kits accessed on July 3, 2011).
3. Like in most of the earlier IEP kits for human paraprotein immunotyping, the agarose plates in the currently available diagnostic IEP kit (TITAN Gel Immunoelectrophoresis, Helena) are provided "ready-to-use," i.e., precut for both sample application (wells) and subsequent antiserum application (troughs). A commercial IEP kit (Immunoelectrophoresis kit, Thermo Fisher) contains all the required materials to perform IEP, except for the electrophoresis system and power supply. Besides agar, glass slides, and several accessories, these materials include dies for two-well patterns, cutters for 1- and 2-mm troughs, suction needle, and gel knife. Along with agarose and other accessories, an IEP teaching kit by EDVOTEK® supplies well cutters (see Note 2).
4. Templates for positioning wells and troughs in 7 and 15 sample gels are 1:1 paper copies (preferably placed in a transparent plastic sleeve) of the respective gel plates stained with amido black after serum protein electrophoresis and marked for the position of wells and troughs (Fig. 1a).
5. Although the antiserum troughs (also called trenches and slits) may be cut before or after the electrophoresis, for convenience they are often cut before as was the case here. The "cutting templates" for 7 and 15 position gels used here were custom cut (~1-mm slits) with a scalpel from a previously exposed

X-ray film using the appropriate templates for well and trough positioning as a guide (Fig. 1a). A Pasteur pipette connected to the side arm of a filter flask, serving as a trap to receive pieces of gel and rinse water for protection of the vacuum line, was used to "clear" the troughs. Historically, troughs have been cut, in the simplest way, by means of single-edge razors, scalpels, or sharp (gel) knives using only "positioning templates." Most commonly, however, troughs were prepared by means of "gel cutters" which included blades (29). A simple trough cutter can be made with a pair of single-edge razor blades bolted together across metal spacers with machine screws (30). A U-shaped trough cutter device constructed from an aluminum profile and scalpel blade for making ~2-mm wide troughs was described in detail in an early book about IEP (29). In the "scoop method" a metal scoop was made of 24-gauge stainless steel bent in the form of a U along its long axis (31). A commercial IEP kit includes cutters for 1- and 2-mm troughs along with suction needle (Thermo Fisher), whereas an IEP teaching kit reports a gel "displacement" method for making troughs (EDVOTEK®). Most commonly, trough widths range from 1 to 2 mm (see Note 2).

6. A suction needle for making sample application wells in the gel can be "homemade" by cutting off the point of a hypodermic needle and sharpening the cut end on the inside surface. These "well cutters" are then connected to a filter flask which, in turn, is connected to a vacuum source. This suction needle may also be used to remove pieces of gel when cutting troughs (see Note 5). As a guide for choosing a hypodermic needle for gel punching, the nominal outer diameter is 1.1 mm for a 19-gauge, 1.5 mm for a 17-gauge, 2.1 mm for a 14-gauge, 3.0 mm for an 11-gauge, and 4.2 mm for an 8-gauge regular-wall needle. Gel punchers also can be made by cutting micropipette tips or using the back of these tips for making (large) sample application wells. Instead of using vacuum, the gel plugs can be removed with a toothpick or, in case of larger wells, with forceps. In an early book about IEP, a copper tube was used for cutting wells (29). A commercial IEP kit provides "dies for two-well patterns" and "suction needle" (Thermo Fisher), whereas an IEP teaching kit supplies "well cutters" (EDVOTEK®) for making sample application wells (see Note 3). Note that small sample volumes (up to ~5 μL) may be applied directly to the gel surface with a micropipette and this approach may be even preferred in case of very small volumes (0.5–2 μL). In some experiments, we also used this technique. However, this technique requires steady hands for precise sample application. In general, samples >5 μL applied as a droplet spread too widely on the gel surface and direct application is not

advisable (30). As another alternative to using wells, application templates made of thin plastic sheets either commercially or "handmade" in the laboratory may be used, especially for somewhat larger sample volumes (>2 μL). When directly applied to the surface, wait 3–5 min for the serum to absorb into the gel. The diameter of sample application wells in different applications varies from 1–2 to 8 mm, but most often probably is ~2 mm. Since the volume of these wells also depends on the thickness of the gel, sample volumes ranging from 1–2 to 30–40 μL can be loaded into these wells. However, most often only a few microliters (commonly 1–2 μL) of the sample are electrophoresed. Besides the size of the application wells, their position relative to the troughs also is an important factor to consider. In case of precast gels, this distance generally is defined by the manufacturer's plate design, though deviation from that is possible. In case of custom cast gels, the user can freely optimize the system for IEP. Although special needs allow for deviations, the distance between the rim of the application well and the edge of the nearest trough should be at least ~2 mm and <5 mm for optimal immunoprecipitation to occur.

7. This is used to aspirate pieces of gel from the troughs after cutting with scalpel.
8. Any level humidifying incubation chamber of proper size to accommodate the gels (e.g., a plastic container with wet paper towel at the bottom and tightly covered with a lid or plastic foil) is suitable for incubation. Use distilled water to saturate towel.
9. Electrophoresis equipments with a horizontal cooling plate are usually recommended for the electrophoresis. Currently, there is only one major IEP kit with dedicated instrumentation for human paraprotein immunotyping (TITAN Gel Immunoelectrophoresis on semiautomated SPIFE® systems, Helena).
10. These strips constitute the electrophoresis buffer reservoir and provide contact between the gel and electrodes.
11. Bromphenol Blue or Bromophenol Blue (3′,3″,5′,5″-Tetrabromophenolsulfonephthalein, Bromphenol Blue Sultone Form) is a light pink to purple crystalline powder (MW 670). It is used as a tracking dye to monitor the progress and completion of electrophoretic separation for proteins (and also serves as a size marker for nucleic acids) in agarose gel. It is also a pH indicator with a yellow color below pH 3 and blue-purple at more basic pH. Usually, a final concentration of 0.002–0.01% (w/v) Bromphenol Blue is recommended for tracking electrophoresis. While the dye is soluble in organic solvents such as alcohol, usually aqueous solutions are used.

The sodium salt of Bromphenol Blue is readily soluble in alkaline electrophoresis buffers and is a convenient way to dissolve it. If a stock solution is made in water, it should be slowly titrated with 1 N NaOH solution until the powder dissolves and the solution turns dark blue.

12. Because polyclonal antisera (and their IgG fraction) contain antibodies to multiple recognition sites (epitopes) on an antigen, the apparent affinity and avidity of the resultant antigen–antibody interactions is very high. This explains that high-affinity and high-avidity polyclonal antisera with optimized dilutions to avoid prozoning will produce maximal immunoprecipitation. These attributes, along with a relatively low price, make polyclonal antibodies, commonly produced in rabbit, sheep, goat, or donkey, the reagent of choice for most immunoprecipitation-based applications, including IEP. High antibody titer has the advantage of allowing for higher antiserum dilutions which, in turn, reduce the undesirable background protein "load" in the gel. Because of their generally low affinity, monoclonal antibodies are rarely suitable for IEP.
13. While anti-whole human serum has been the initial reference and continues to be a widely used reference for identifying specific protein components in IEPs, polyvalent antisera with well-defined specificities are increasingly used alternatives for human paraprotein testing. Trivalent (a mix of IgG, IgA, and IgM heavy chain-specific antibodies) and pentavalent (a mix of IgG, IgA, IgM, IgD, and IgE heavy chain-specific antibodies) are being used in a commercial kit (TITAN Gel Immunoelectrophoresis, Helena). The concept may be applied to other protein systems as well.
14. Inclusion of 3–5% of dextran (average MW 70,000) or polyethylene glycol (PEG, MW 6,000) in low-avidity antibody-containing agarose gel increased the detection sensitivity (i.e., immune complex formation) of two-dimensional IEP and electroimmunodiffusion at least up to fivefold (32, 33). Further, it was suggested that other high molecular weight polymers with good water solubility would have the same effect (32). Nevertheless, inclusion of these polymers in IEP has not gained wide acceptance.
15. After successful immunodiffusion, most precipitin arcs are visible to the naked eye in the wet agarose gel by holding the plate over a dark background close to a diffuse light source or using indirect illumination and these patterns can be documented by photography (Fig. 1b, c). However, the precipitin arcs almost always are stained with protein dyes in the dried gel for enhanced detection and for permanent records. Most commonly used stains in decreasing order of detection sensitivity include Coomassie Brilliant Blue R-250 > Acid Violet > Amido black.

Less commonly used or historic protein stains include Light green SF, Ponceau Red (only for cellulose acetate membrane), Nigrosin(e), Thiazine red, and Azocarmine. Although silver staining of the immunoprecipitates is possible in agarose and acrylamide gels and is more sensitive than staining with dyes, it is rarely used for IEP.

16. For convenience, a commercial amido black stain was used in these IEP runs. Alternative "homemade" amido black 10B staining solution: Prepare a saturated solution of the dye (~1 g/dL) in 10% acetic acid (v/v) in methanol by dissolving 1 g of amido black in 450 mL of methanol in a glass beaker, then adding 50 mL of glacial acetic acid. Mix well using a magnetic stirrer for at least 30 min, leave the solution at room temperature overnight, then filter. Store in a closed container for up to 2 months. For further information regarding amido black see Note 17 in Chapter 13 (on IFE) of this book.
17. In order to assess the standing of IEP methods in the literature, PubMed searches were performed on July 21, 2011. Using the search term "immunoelectrophoresis," a total of 27,727 articles were retrieved, of which 95% (26,403/27,727) were diagnosis related based on adding "diagnosis" as a second search term. These data indicate that (a) the classic IEP method and its various versions have been widely used in laboratories, and (b) IEP with its various versions was likely used as a diagnostic test in an overwhelming majority of the reports.
18. Any low-viscosity body fluid specimen, culture fluid, or tissue extract could be tested with classic IEP if proper antibodies are available. Human serum specimens with high viscosity due to the presence of large amounts of polymeric paraproteins like IgM can be pretreated with the reducing agent 2-mercaptoethanol for IEP. Prewarming of serum specimens containing cryoglobulins allowed immunotyping of these cryoglobulins with IEP only in less than one-third of the cases (34). Whether pretreatment of high-viscosity and cryogelating specimens with chaotropic agents would make them amenable to testing with IEP remains to be fully elucidated.
19. In general, IEP is much less sensitive to antigen excess than IFE. Consequently, human serum specimens initially are tested straight in IEP for paraproteins and diluted only when evidence for antigen excess (prozoning) emerges with incomplete precipitin reaction. Even in this case, an alternative approach includes retesting of the undiluted serum with doubling the amount of antiserum added to the trough immediately after electrophoresis or after the usual incubation period but before washing (TITAN Gel Immunoelectrophoresis method, Helena). Urine specimens are often preconcentrated for IEP

testing, but using both native and concentrated urine samples also is proposed in the above-mentioned diagnostic kit. It is noteworthy, however, that concentrating specimens may lead to artifacts, including protein losses that differ from sample to sample and may be substantial (up to 50%) with some concentrators.

20. Based on the presence, shape, intensity, and location of the corresponding precipitin arcs, control specimens are used to confirm the performance and correctness of antisera to human whole serum and to human immunoglobulin heavy and light chains when immunotyping for paraproteins. The controls also serve as a side-by-side reference for the evaluation of immunoprecipitin arcs developing with the unknown specimens. In case of purity checks, the presence or absence of contaminating proteins and degradation products of the protein of interest are evaluated against reference preparations and/or the "starting material."
21. The IEP antiserum application templates are reusable after washing with mild detergents and, when indicated, disinfection.
22. Sample volumes used in different IEPs vary widely from 0.5 to 40 μL.
23. It is desirable to fill up the troughs with the antisera. Depending on the volume of the troughs, antiserum volumes may vary from 10 μL to as much as 200 μL per trough.
24. Incubation time primarily depends on the type and concentration of the gel, the distance between the sample application well and the lateral wall of trough, and the incubation temperature. Room temperature incubations may be needed for up to 3 days in some gels. In order to shorten the time for immunoprecipitation, incubation at temperatures higher than room temperature (e.g., at 37°C) can be used up to ~45°C. Conversely, incubation in a refrigerator (2–8°C) may be needed to slow down immunodiffusion when overnight room temperature incubation would result in partial or total dissolution of precipitin arcs.
25. When whole antisera or large volume of antibody (>20 μL/mL of agarose gel) are used in IEP, the background staining will likely increase due to inadequate removal of unbound proteins. This can be overcome by including repeat press-dry cycles in preparation of the gel for subsequent staining. Press-dry can also markedly reduce the time for washing gels in the absence of protein "overload." For press-dry, multiple layers of filter papers and weight are added on top of the gels for periods of 10–15 min each between wash steps. The first filter paper

should be soaked with distilled water and, after completing the press step, should be removed last after soaking it again with distilled water. Regarding the amount of weight, at least 10 g/cm^2 pressure is suggested for optimal effect. (Bio-Pheresis® Horizontal Electrophoresis Cell (Cat. No. 170–2900) Instruction Manual, rev. 031193, Bio-Rad Laboratories, Hercules, CA, USA.) This corresponds to ~0.5 kg for the smaller 7-position and ~1 kg for the larger 15-position protein plate used in this work. A commercial IEP kit routinely involves press-dry steps for preparing the gel for staining (TITAN Gel Immunoelectrophoresis, Helena).

26. In addition to staining for proteins in immunoprecipitates "trapped" in the electrophoretic matrix, special stains or enzyme reactions can be applied to detect the nonprotein components of the antigens and enzyme antigens themselves. During electrophoresis in agar and agarose gels, the native structure and activities of the electrophoretically separated proteins are preserved. Further, precipitating antibody binds enzymes at antigenic sites and usually leaves active sites unmasked. The open structure of immunoprecipitates thus allows characterization of ligand binding and enzyme activities of the precipitated protein antigens. Examples include detection of lipids, calcium, iron, and esterase and oxidase activities. It is important to note, however, that azide is a potent "general" enzyme inhibitor, so that azide-containing antisera (which are common) may interfere with procedures that involve enzyme reactions. Using azide-free antisera obviously avoids this interference. The open structure of the immunoprecipitate in the agarose gel also allows additional binding of radioactively labeled antibodies to reveal specific proteins. Binding of ^{125}I-labeled IgE antibodies onto allergen-containing precipitates permitted identification of individual allergens in complex antigen extracts after autoradiography of the gel in crossed radio-IEP (CRIE) (21).

27. It was recently shown that electroimmunoprecipitated proteins can be extracted from agarose gel for identification (35). After excising the immunoprecipitates, the precipitates were extracted from the gel and dissociated. The protein components of the extracts were analyzed using peptide mass fingerprinting following sodium dodecyl sulfate-polyacrylamide gel electrophoresis (SDS-PAGE). In addition to characterizing antibody specificity, the method can also be used to identify antigens generating unknown precipitates in crossed IEP with polyspecific antisera, including human IgG–antigen complexes electroimmunoprecipitated by secondary antibodies. This method appears to be applicable to classic IEP as well.

28. Advantages of classic IEP
    (a) The primary advantage of classic IEP is that, in contrast to the alternative (immunotyping) method IFE, IEP is able to discern identity, partial identity, and nonidentity of the proteins under investigation (29, 30).
    (b) Classic IEP is also advantageous for assessing purity and integrity of various biochemical and pharmaceutical preparations (26, 27).
29. Disadvantages of classic IEP.
    (a) Classic IEP is very slow; it may take up to 3 days to complete a plate (18–48 h needed just for the diffusion step) versus IFE with a total of ~3 h completion time.
    (b) In contrast to IFE, classic IEP lacks sensitivity to detect low concentrations of monoclonal proteins or proteins in general.
    (c) In contrast to IFE, IEP has a relatively poor resolution and often is unable to distinguish two closely migrating protein bands. Further, unlike IFE, IEP has difficulty to detect gammopathies with multiple paraproteins (36); an increasingly common pattern due to bone marrow transplantation.
    (d) Unlike IFE, classic IEP exhibits the "umbrella effect" where polyclonal IgG molecules mask an IgM monoclonal gammopathy (36).
    (e) Although interpretation of IFE patterns can be also difficult, polyclonal increases, primarily IgG, in inflammatory conditions may be misdiagnosed by IEP as monoclonal gammopathy (36).
30. Routine IEP methods, including commercial kits for paraproteins, do not test initially for IgD and IgE heavy chains and free (unbound) κ and λ light chains. However, antibodies against these antigens are available commercially and a pentavalent antiserum against all five heavy chains is provided in the kit for diagnostic paraprotein immunotyping (TITAN Gel Immunoelectrophoresis, Helena).

Acknowledgments

The author wishes to thank Dr. Rita Ellerbrook of the Helena Laboratories, Beaumont, TX, USA for kindly providing illustrations to this article.

References

1. Grabar P, Williams CA Jr (1953) Méthode permettant l'étude conjugée des propriétés électrophorétiques et immunochimiques d'un mélange de proteins. Application au serum sanguine. [Method permitting the combined study of the electrophoretic and the immunochemical properties of protein mixtures; application to blood serum]. Biochim Biophys Acta 10:193–194 [French]
2. Martin E, Scheidegger JJ, Grabar P et al (1954) [Immuno-electrophoretic study of blood serum]. Bull Schweiz Akad Med Wiss 10: 193–198 [French]
3. Williams CA Jr, Grabar P (1955) [Immunoelectrophoretic method for the analysis of mixtures of antigenic substances]. Biochim Biophys Acta 17:67–74 [French]
4. Scheidegger JJ (1955) Une mict-méthode de l'immunoelectrophorése. Int Arch Allergy Appl Immunol 7:103–110 [French]
5. Scheidegger JJ, Roulet H (1955) [Practical application of the immuno-electrophoretic method; first results]. Praxis 44:73–76 [French]
6. Afonso E (1964) Quantitative immunoelectrophoresis of serum proteins. Clin Chim Acta 10:114–122
7. Wilson AT (1964) Direct immunoelectrophoresis. J Immunol 92:31–34
8. Bussard A (1959) [Description of a technic simultaneously combining electrophoresis and immunological precipitation in gel: electrosyneresis]. Biochim Biophys Acta 34:258–260 [French]
9. Vergani C (1971) Crossover electrophoresis for the rapid detection of serum hepatitis (Australia) antigen and antibody. J Clin Pathol 24:86–87
10. TiltonRC(1978)Counterimmunoelectrophoresis in biology and medicine. CRC Crit Rev Clin Lab Sci 9:347–365
11. Ressler N (1960) Two-dimensional electrophoresis of protein antigens with an antibody containing buffer. Clin Chim Acta 5:795–800
12. Laurell CB (1965) Antigen-antibody crossed electrophoresis. Anal Biochem 10:358–361
13. Clarke HG, Freeman T (1968) Quantitative immunoelectrophoresis of human serum proteins. Clin Sci 35:403–413
14. Laurell CB (1966) Quantitative estimation of proteins by electrophoresis in agarose gel containing antibodies. Anal Biochem 15:45–52
15. Laurell CB (1972) Electroimmuno assay. Scand J Clin Lab Invest Suppl 124:21–37
16. Weeke B (1969) [Quantitative immunoelectrophoresis. Rocket electrophoresis]. Ugeskr Laeger 131:1419–1423 [Danish]
17. Weeke B (1973) Rocket immunoelectrophoresis. Scand J Immunol Suppl 1:37–46
18. Svendsen PJ (1973) Fused rocket immunoelectrophoresis. Scand J Immunol Suppl 1:69–70
19. Bog-Hansen TC (1973) Crossed immuno-affinoelectrophoresis. An analytical method to predict the result of affinity chromatography. Anal Biochem 56:480–488
20. Kroll J (1973) Line immunoelectrophoresis. Scand J Immunol Suppl 1:61–67
21. Weeke B, Lowenstein H (1973) Allergens identified in crossed radioimmunoelectrophoresis. Scand J Immunol Suppl 1:149–153
22. Kroll J (1973) Crossed-line immunoelectrophoresis. Scand J Immunol Suppl 1:79–81
23. Kroll J (1973) Rocket-line immunoelectrophoresis. Scand J Immunol Suppl 1:83–87
24. Ohlenschläger G, Berger I (1974) [Tandem technic in Clarke, Freeman, Ressler and Laurell's 2 dimensional immunoelectrophoresis. Methods and criteria for use as a consequence of the problems of realization of reproducible separations by stability of electrical values]. Z Med Labortech 15:119–126 [German]
25. Axelsen NH, Svendsen PJ (1973) Reversed rocket immunoelectrophoresis. Scand J Immunol Suppl 1:155–157
26. United States Pharmacopeial Convention (2011) Quality of biotechnological products: stability testing of biotechnological/biological products. In: The United States Pharmacopeia–National Formulary (USP 34–NF 29 through First Supplement). http://www.uspnf.com. Accessed 26 July 2011
27. European Directorate for the Quality of Medicines and Health Care (2012) Immunochemical methods. In: ep703 – European pharmacopoeia online, 7th Edition 2012 (7.3). http://online6.edqm.eu/ep703/. Accessed 26 July 2011
28. Nerenberg ST, Ganger C, DeMarco L (1972) Multi-antisera applicator for use in solid-support immunoelectrophoresis on cellulose acetate. Clin Chem 18:1021–1023
29. Arquembourg PC, Salvaggio JE, Bickers JN (1970) Primer of immunoelectrophoresis with interpretation of pathologic human serum patterns. Ann Arbor-Humphrey Science Publishers, Inc, Ann Arbor, MI [S. Karger, Basel]
30. Williams CA (1971) Immunoelectrophoretic analysis. In: Williams CA, Chase MW (eds) Methods in immunology and immunochemistry, vol 3. Academic, New York and London, pp 234–287
31. Cawley LP, Schneider D, Eberhardt L et al (1965) A simple semi-automated method of

immunoelectrophoresis. Clin Chim Acta 12:105–110

32. Kostner G, Holasek A (1972) Influence of dextran and polyethylene glycol on sensitivity of two-dimensional immunoelectrophoresis and electroimmunodiffusion. Anal Biochem 46:680–683
33. Thoss K, Abendroth K (1974) [The effect of polyethylene glycol and dextran on the immunoelectrophoresis (author's transl)]. Z Immunitatsforsch Exp Klin Immunol 147:404–412 [German]
34. Musset L, Diemert MC, Taibi F et al (1992) Characterization of cryoglobulins by immunoblotting. Clin Chem 38:798–802
35. Beyer NH, Schou C, Houen G et al (2008) Extraction and identification of electroimmunoprecipitated proteins from agarose gels. J Immunol Methods 330:24–33
36. Keren DF (1994) High-resolution electrophoresis and immunofixation. Techniques and interpretation, 2nd edn. Butterworth-Heinemann, Boston

Chapter 29

Cell Surface Protein Biotinylation for SDS-PAGE Analysis

Giuliano Elia

Abstract

Cell surface proteins play a very important role in physiology and pathology and are receiving increased attention by the pharmaceutical industry as valuable targets for development of new therapeutics. However, owing to the very nature of this category of proteins, their comprehensive study remains an elusive task. A number of methods have been proposed to enrich and purify cell surface proteins. Among them, usage of biotinylating reagents and exploitation of the strong interaction between biotin and streptavidin for the purification of biotinylated proteins has rapidly gained in popularity and allowed some of the most significant progresses in quantitative proteomics. This chapter focuses on methods for cell surface biotinylation with commercially available reagents, capture by avidin-affinity chromatography and release of the biotinylated surface proteins for downstream analysis by electrophoretic techniques.

Key words: Biotin, Cell surface proteins, Electrophoresis, SDS-PAGE, Streptavidin

1. Introduction

Proteins expressed at the surface of cells are likely to represent about one-third of proteins encoded for by the human genome (1, 2). However, only a fraction of the cell surface proteome has been characterized so far (3, 4), as this protein category has proved to be pretty difficult to study. Proteins populating the cell surface carry out several functions, critical for the cell's biology, which include for instance establishment of cell–cell interactions, response to environmental stimuli via surface receptor-mediated activation of signaling cascades and the regulation of trafficking of ions and solutes in and out of the cell (3). Specific alterations of the cell surface protein makeup have been found to be associated with different pathologic states, including cancer (5), diabetes (6), and neurological disorders (7). As a consequence, in recent years, studies aiming at high-throughput identification, quantification, and characterization of the cell surface proteome have remarkably increased

Biji T. Kurien and R. Hal Scofield (eds.), *Protein Electrophoresis: Methods and Protocols*, Methods in Molecular Biology, vol. 869, DOI 10.1007/978-1-61779-821-4_29, © Springer Science+Business Media, LLC 2012

in number, with the scope of discovering and validating disease-specific protein signatures of diagnostic value or of unveiling new candidate targets for therapeutic intervention. Indeed, a number of biopharmaceuticals (mostly antibodies), which target cell surface proteins, have already entered the clinical arena for treatment of both solid tumors and lymphomas, as well as of other autoimmune diseases (8).

As said, investigating the cell surface proteome has proven difficult. This is due to different reasons including (1) heterogeneity of cell surface proteins (9), (2) hydrophobicity of most integral membrane proteins, which translates into poor solubility in aqueous media, (3) relatively low abundance, and (4) the relative lower frequency of tryptic cleavage sites in transmembrane hydrophobic domain of membrane proteins (10), which all have limited the applicability of the most widespread proteomic techniques (2D-PAGE, bottom-up mass spectrometry) to this category of proteins.

A number of different approaches have been proposed to specifically address the cell surface proteome. Protocols for purification of this category of proteins have principally exploited differences in density of subcellular fractions by ultracentrifugation (11, 12), hydrophobicity of the proteins (two-phase partitioning (13, 14) and sequential extraction (15, 16)), or affinity for the isolation of modified or non-modified cell surface proteins (antibody- or lectin-based approaches (17, 18)). However, all these methods have failed to produce highly pure cell surface proteomes. In more recent times, other techniques have been proposed to improve the enrichment of cell surface proteins and facilitate their study, both in vitro and in vivo. These include silica bead coating (19, 20), chemical capture of glycosylated cell surface proteins (21), and cell surface biotinylation (22–24).

This chapter discusses methods for cell surface biotinylation with commercially available reagents, capture by avidin-affinity chromatography and release of the biotinylated surface proteins (25) for downstream analysis by electrophoretic techniques.

Usage of a cell membrane-impermeable, cleavable reagent for biotinylation of primary amino groups in side chains of lysines (sulfosuccinimidyl-2-(biotinamido) ethyl-1,3-dithiopropionate; Sulfo-NHS-SS-biotin) will be illustrated in detail, as well as the downstream capture and release of the biotinylated proteins by means of streptavidin (SA) affinity chromatography. Protocols for the study of the purified proteins by SDS polyacrylamide gel electrophoresis (SDS-PAGE) and western blot with SA-conjugated detection enzymes will also be provided.

Sulfo-NHS-SS-biotin has a number of distinctive advantages that make it the ideal tool for labeling of cell surface proteins. First, the compound is polar because of the negative charge contributed by the sulfonate group and therefore is very soluble in aqueous buffers even at low temperatures. Second, the negative charge reduces uptake of the reagent by cells, thus limiting contamination due to labeling of intracellular proteins (see Subheading 4 below). Third, the long spacer arm present in this compound (about 10% longer than that of the corresponding long-chain Sulfo-NHS-LC-biotin) improves binding to avidin-based reagents or supports. Finally, the disulfide bridge contained in the linker can be cleaved by reducing agents for a facilitated release of proteins after capture on avidin-based supports.

Other reagents than the one described in this protocol (either commercially available or developed in the laboratory) can be used for the derivatization of different chemical functional groups in proteins with biotin (e.g., thiol groups in cysteine residues). These compounds, as well as methods for their usage in vitro, in vivo, and ex vivo, have been extensively discussed elsewhere (26, 27).

2. Materials

2.1. Biotinylation of Cell Surface Proteins

1. Cells growing in monolayer or as a suspension in T75 tissue culture flasks (Greiner Bio-One GmbH, Frickenhausen, Germany) in the appropriate cell culture medium.
2. Biohazard-grade, laminar flow sterile hood, equipped with vacuum flasks for aspiration and connected through an HEPA filter to a vacuum line (Thermo Herasafe KS Class II).
3. Phosphate Buffered Saline (PBS): 50 mM NaH_2PO_4, 50 mM Na_2HPO_4, 100 mM NaCl in MilliQ water, pH 7.4.
4. Sulfosuccinimidyl-2-(biotinamido)ethyl-1,3-dithiopropionate (Sulfo-NHS-SS-biotin, MW = 606.7) (Thermo Scientific, Rockford, IL).
5. Tris(hydroxymethyl) aminomethane hydrochloride (Tris, MW 157.6) dissolved in MilliQ water at 50 mM concentration, pH 7.4.
6. Oxidized glutathione (MW 612.6) (Sigma-Aldrich, St. Louis, MO, USA) dissolved in PBS at 1 mM concentration.

2.2. Capture and Release of Biotinylated Proteins from Streptavidin–Sepharose

1. Protein concentration measurement kit (Bio-Rad Protein Assay kit Bio-Rad, Hercules, CA).
2. Streptavidin Sepharose™ High Performance (GE Healthcare, Waukesha, WI).
3. Buffer A: 1% NP-40, 0.1% SDS in PBS.
4. Buffer B: 0.1% NP-40, 0.5 M NaCl.

5. Centrifugal filter units with microporous membrane (Ultrafree-MC 5.0 µm ®Durapore filter units; Millipore Co., Billerica, MA).
6. Elution solution: 6 M urea, 2 M thiourea, 30 mM D-biotin (Sigma), 2% SDS in PBS, pH 12.0.
7. Rotating mixer (Model SB3; Bibby Scientific Ltd, Stone, UK).
8. Tabletop refrigerated microcentrifuge (Mod. 5430R; Eppendorf Ag, Hamburg, Germany).
9. Heating block (e.g., Thermomixer Comfort, Eppendorf).

2.3. SDS-Polyacrylamide Gel Electrophoresis with Pre-cast Gels

1. XCell SureLock Mini Cell (Invitrogen, Carlsbad, CA).
2. NuPAGE Novex Bis-Tris 4–12% acrylamide gradient gels (Invitrogen). Store at 4°C.
3. 3-(*N*-morpholino)propanesulfonic acid (MOPS): 20× concentrated running buffer (Invitrogen). Store at room temperature.
4. Reducing loading buffer, fivefold concentrated: 235 mM SDS, 33.6% glycerol (v/v), 5% 2-mercaptoethanol (v/v), 0.67% bromophenol blue (w/v), and 210 mM Tris–HCl, pH 6.8. Store at room temperature.
5. Rainbow Molecular Weight markers (GE Healthcare).

2.4. Western Blotting of Biotinylated Proteins

1. XCell II Blot Module (Invitrogen).
2. 20× Transfer buffer: 500 mM Bicine, 500 mM Bis-Tris (free acid), 20 mM NaEDTA.
3. Methanol analytical grade.
4. Nitrocellulose membrane (Protran™ BA85; Schleicher and Schuell, Dassel, Germany) and blotting paper (Grade 3MMChr; Whatman, Florham Park, NJ).
5. Tris-buffered saline with Tween (TBS-T): Prepare 10× stock with 1.37 M NaCl, 27 mM KCl, 250 mM Tris–HCl, pH 7.4, 1% Tween-20. Dilute 100 mL with 900 mL MilliQ water before use.
6. Blocking buffer: 5% (w/v) nonfat dry milk in TBS-T.
7. Streptavidin conjugated to horse radish peroxidase (GE Healthcare, 2 mL).
8. Enhanced chemiluminescent (ECL) reagents (ECL Plus western Blotting Detection Reagents; GE Healthcare) and Bio-Max Light film (Kodak, Rochester, NY).
9. Saran Wrap.
10. X-ray film developing apparatus (e.g., Curix CP1000 automatic film processor; Agfa Gevaert, Mortsel, Belgium).

3. Methods

3.1. Biotinylation of Cell Surface Proteins

1. Cells growing as monolayers should be seeded at a density of 5×10^4 cells/cm^2, using 1 mL of the appropriate growth medium per 5 cm^2 of growing surface. Cells growing in suspension should be seeded at $2–5 \times 10^5$ cells/mL. Replace seeding medium with the same volume of fresh medium 24 h after seeding. Depending on the cell type, cells seeded as above should reach confluence 3–5 days after seeding (see Note 1). For cells growing as monolayers, fill a tray of appropriate size with crushed ice and place a metal surface (e.g., a second, smaller stainless steel tray) in direct contact with ice.
2. Remove the medium from the flasks by aspiration (or centrifugation in case of cells growing in suspension) and wash twice with 10 mL of ice-cold PBS (see Note 2). Remove each washing carefully, operating rapidly.
3. Add 15 mL of an ice-cold solution of Sulfo-NHS-SS-biotin in PBS (250 μg/mL; 4.1×10^{-4} M) to each flask and place them flat on the ice-cold (see Note 3) stainless steel tray (or resuspend the cell pellet in the same volume of PBS/biotinylating reagent and place tubes on ice). Allow to react for 10 min, gently rocking the tray on an orbital shaker or manually inverting the tubes a few times.
4. Quench the reaction by adding 10 mL of Tris–HCl solution per flask or per tube. Monolayer cells at this point are rounded up and can normally be detached by repeatedly pipetting the liquid contained in the flask over them. In case cells still held firmly to the plastic surface at the end of the labeling procedure, gently scrape them off the flask surface and collect them in a centrifuge tube.
5. In both cases, after removing cells from the flask, wash the surface once with additional 5 mL of oxidized glutathione solution (see Note 4) and add this washing to the collected cells. For cells in suspension, simply add 5 mL of oxidized glutathione solution to each tube.
6. Centrifuge cells in a refrigerated benchtop centrifuge, discard the supernatant, and wash two times with 10 mL of glutathione-containing PBS. Carefully discard the last wash and add the lysis buffer of choice to the cell pellet (see Note 5).

3.2. Capture and Release of Biotinylated Proteins from Streptavidin–Sepharose

1. Measure protein concentration on a fraction of the solubilized cell pellet with one of the many available commercial kits (e.g., the Bio-Rad Protein Assay kit), according to manufacturer's instructions and adjust to a final value of 0.2 mg/mL with PBS.

2. Transfer an aliquot of 500 μL to a fresh tube and add 72 μL of prewashed streptavidin–sepharose resin (see Note 6). After 2 h of incubation at room temperature with occasional mixing by manual inversion of the tubes, transfer the slurry to an Ultrafree centrifugal filter and spin down for 30 s at 16,100 × *g* in a tabletop centrifuge.
3. Remove or discard the filtrate, wash the resin three times with 400 μL of buffer A, followed by two additional washings with 400 μL of buffer B.
4. Quantitatively recover the washed resin from the filtering unit by means of two washing steps (each 50 μL) with PBS and transfer to a fresh tube.
5. Centrifuge the resin for 30 s at 16,100 × *g* in the tabletop centrifuge and remove the supernatant carefully, as the pellet can be very loose. Add 500 μL of the elution solution (see Note 7) to the pelleted resin, resuspend by flicking the tube gently with fingertips, and incubate for 15 min at room temperature.
6. Transfer the tube to a heating block set at 96°C and incubate for additional 15 min with occasional agitation.
7. Finally, transfer quantitatively the tube content to a new Ultrafree centrifugal filter, spin down for 30 s at 16,100 × *g* in the tabletop centrifuge, and collect and save the filtrate.

3.3. SDS-PAGE with Pre-cast Gels

1. SDS-PAGE is a powerful analytical technique, but if gels are to be casted in-house, it is time-consuming and sometimes reproducibility of results is not optimal. Usage of pre-cast gels is preferable and has become more affordable in recent times. This protocol describes the usage of the Invitrogen NuPAGE pre-cast gels, but gels from other brands (e.g., Bio-Rad, ChemBio) are equally suitable and lead to results of comparable good quality.
2. Remove one NuPAGE Novex Bis-Tris 4–12% acrylamide gel from its package, discard the preserving buffer, and gently take out the well-forming comb and the white tape strip at the bottom of the gel cassette. Carefully rinse the wells twice with MilliQ water, paying attention not to distort the well walls; drain the liquid each time by turning the gel upside down.
3. Assemble the gel in the XCell SureLock Mini Cell apparatus and seal the inner chamber (see Note 8).
4. Dilute 30 mL of the 20-fold concentrated MOPS running solution with MilliQ water to a final volume of 600 mL (see Note 9). Fill the inner chamber of the gel with enough buffer as to completely cover the gel wells. Pour the remaining running buffer in the outer chamber, making sure that the buffer level is such as to cover the slot in the bottom part of the gel cassette. The gel is now ready for loading.

5. Twenty-four microliters of the filtrate from the preceding section are mixed with 6 μL of fivefold concentrated loading buffer and heated at 95°C in a heating block for 5 min. After cooling and a quick spin to collect evaporated liquid, the sample is applied onto the wells of the SDS-PAGE gel.
6. Ten microliters of the Rainbow Molecular Weight Markers are also loaded in one of the outermost wells.
7. Electrophoretic separation is allowed to proceed at 110 mA, 180 V for 1 h or until the tracking dye front reaches the bottom of the gel.
8. Gels are then removed from the casting assembly and either processed for western blotting or fixed for 30 min and stained with Sypro Ruby (see Note 10).

3.4. Western Blotting for Biotinylated Proteins

1. Prepare 1 L of 1× NuPAGE Transfer Buffer by adding 50 mL 20× NuPAGE Transfer Buffer and 200 mL methanol to 750 mL of MilliQ water.
2. Gels obtained as described in the previous section are soaked for 10 min at RT in 1× NuPAGE Transfer Buffer with gentle agitation.
3. During this time, cut one piece of nitrocellulose membrane and several pieces of 3-MM filter paper in rectangles of 9.5 × 5 cm, the internal dimensions of the inner core of the XCell II Blotting Module. Soak the membrane, the filter paper, and the blotting pads (see Note 11) in 1× NuPAGE Transfer Buffer.
4. The blotting device is assembled as follows. Place two soaked blotting pads into the cathode(−) core of the blot module, followed by 4–5 sheets of soaked filter paper and by the soaked gel, flipped by 180° around its vertical axis (i.e., with the Rainbow markers position inverted with respect to the original loading sequence). Carefully place the presoaked nitrocellulose membrane on the gel, paying attention to remove any trapped air bubbles. Place another 4–5 sheets of soaked filter paper on top of the membrane and again two soaked blotting pads on top of the filter paper. Close with the anode (+) cover, squeezing firmly together the blot module and slide it into the guide rails of the lower buffer chamber.
5. Fill the blot module with 1× NuPAGE Transfer Buffer, the outer chamber with cold water for refrigeration, place the lid on the XCell II Blot Module unit and start the run. The power supply is set at 30 V constant, with maximum current limited to 220 mA, and the run is allowed to proceed for 2 h.
6. Once transfer is complete, open the blotting module, carefully remove the nitrocellulose, and transfer it to a clean vessel.

7. Wash twice at RT with blocking buffer (30 min each) on a rocking platform.
8. Dilute the streptavidin–horseradish peroxidase conjugate 1:1,000 with blocking buffer and pour 20 mL of this solution onto the membrane. Allow the biotin–streptavidin-binding reaction to proceed at room temperature for 45 min.
9. Remove the streptavidin–horseradish peroxidase solution and wash the membrane three times for 10 min each with 50 mL of TBS-T and once with PBS.
10. The remaining steps are performed in a dark room under safe light conditions. Place the membrane flat on a piece of Saran Wrap and mix the ECL reagents as specified by the manufacturer's instructions (100 μL/cm^2 of nitrocellulose membrane). Pour on the membrane and allow to react for 60 s.
11. After this time, drain away the excess of reagent, tap the membrane on clean Kim Wipes, and envelope it in clean Saran Wrap.
12. The membrane is exposed to X-ray film in an X-ray cassette for variable length of time, and films are developed in a developing machine.

4. Notes

1. In order to obtain sufficient starting material for downstream purification and analysis by SDS-PAGE, it is recommended to use ~0.5–1 × 10^8 cells. Normally, cells contained in two T-75 confluent flasks (per experimental condition) should represent an adequate amount to begin with. Only actively growing cell cultures, exhibiting normal growth curves should be used for labeling. Avoid using cells that grow sluggishly and/or cells that have been subcultured for an elevated number of doublings. The manipulations required to carry out the present protocol make it ideally suited to labeling of cells which grow as monolayers; however, it can also be applied to cells growing in suspension. In the latter case, all washing and medium/buffer replacement steps have to be performed by centrifugation of cells at 500 × *g* for 5 min at 4°C. Gentle centrifugation conditions are needed to prevent cell damage with consequent release of cytoplasmic proteins into the medium. Some cells in suspension have a tendency to form clumps; these should be gently but extensively disaggregated by pipetting prior to labeling, in order to ensure an even and effective access of reagents to the cell surface.

2. For cells growing as monolayers, care should be taken not to lose cells by accidentally detaching them from the plastic surface during experimental manipulations. Avoid pipetting cold buffer directly on the cells, direct the pipette jet toward the opposite flask surface, which is best done holding the flask with the top side oriented downward. Glass Pasteur pipettes connected to a vacuum device represents a fast and efficient way to remove and discard media and washes.

3. Maintaining cells at 0°C before and during the labeling procedure helps reducing the uptake of biotinylating reagents into the cell. Sulfonate group-containing reagents, which are supposed to be membrane-impermeable, are however able to cross the membrane barrier and to label intracellular proteins (28). A 4% paraformaldehyde short treatment has been shown to limit Sulfo-NHS-SS-biotin diffusion to the cytosolic compartment (28), but it is not entirely clear to which extent this impacts on further characterization of biotinylated proteins.

   Sulfo-NHS-based reagents are inherently prone to hydrolysis. Dissolve Sulfo-NHS-SS-biotin in PBS immediately before addition to cells. In experiments requiring labeling of very high number of samples, consider splitting the work among different operators to minimize artifacts due to partial hydrolysis of the reagent during the experimental time.

4. Oxidized glutathione is added to the solution to prevent untimely reduction of the disulfide bridge-containing cleavable linker in the biotinylating reagent. Oxidized glutathione should also be present during manipulations carried out to lyse cells and capture biotinylated proteins on avidin-based supports, prior to their release with reducing agents.

5. It is not possible to anticipate the precise degree of labeling that can be obtained by this protocol. Biotinylation of cell surface proteins is variable with the cell type used and also from protein to protein. The efficiency of reaction, estimated by western blot and avidin–horseradish peroxidase detection, using the conditions reported in this protocol is usually good; however, in case of poor outcome for a particular cell type, modification of reaction parameters such as pH or ionic strength might help improving the results (29).

6. Preservatives (e.g., ethanol, sodium azide) are normally included in commercial preparations of (strept)avidin-based affinity-chromatography resins. Slurries have to be carefully washed before use to completely remove these preservatives. Prewash the desired amount of resin (1 mL of slurry binds 300 nmol of biotin) twice with wash buffer A (1% NP-40, 0.1% SDS in PBS) and twice with PBS. After each washing step, pellet the resin by short centrifugation in a tabletop microcentrifuge.

As these pellets are extremely loose, particular care should be taken not to aspirate resin material while removing supernatants of each washing step.

7. D-Biotin is not soluble in water but can be dissolved in concentrated alkali. Prepare a stock 300 mM solution of D-biotin by adding 366.5 mg of the vitamin to 400 μL of 8 M NaOH and then bring to a final volume of 5 mL with MilliQ water.

8. The XCell SureLock Mini Cell apparatus can accommodate two NuPAGE gels at a time. If only one gel is needed, a plastic part of the same shape of a gel cassette (provided) must be used to close the opposite side of the inner core and form the inner buffer chamber.

9. Invitrogen proposes two different buffering agents for their NuPAGE system, MOPS and its structural analog 2-(*N*-morpholino)ethanesulfonic acid (MES). These zwitterionic buffering agents were developed in the 1960s for general use in biochemistry by Good et al. (30) to meet the criteria of a midrange *pKa*, maximum water solubility and minimum solubility in all other solvents, minimal salt effects, minimal change in *pKa* with temperature, chemical and enzymatic stability, minimal absorption in visible or UV range, and reasonable ease of synthesis. MOPS works better for separation of proteins with MW comprised between 200 and 14 kDa, while MES is more indicated for smaller size proteins (for 60–2.5 kDa). MOPS is very heat stable at the conditions used and can even be autoclaved. Concentrated solution might turn yellow with time, but this does not interfere with its buffering capacity. Straw-colored buffer is good but do not use darker buffer.

10. Sypro Ruby is a highly effective, though expensive, alternative to Silver staining for sensitive detection of proteins in SDS-PAGE experiments. Remove the gel from the pre-cast gel cassette, transfer it to a clean vessel, and soak it in fix solution (10% methanol, 7% acetic acid) for 30 min at RT. Stain the gel for 2 h (or overnight) with Sypro Ruby Protein Stain (Invitrogen; Cat. No. S12000). Destain by repeated changes of fix solution (30 min each) on an orbital shaker and finally rinse in MilliQ water. After destaining, gels are imaged using a UV Transilluminator and a suitable recording device (e.g., DIANA 2 imager, Raytest, Straubenhardt, Germany). The used Sypro Ruby solution can be recovered and reused two additional times without significant loss of activity.

11. For an efficient transfer of proteins to occur, a perfect contact between the gel and the nitrocellulose needs to be established. Air bubbles trapped at the interface between the two can generate a disruption of the buffer ion flow and prevent proteins from migrating out of the gel or seriously distort their profile.

Care should therefore be taken to thoroughly remove any macroscopically visible air bubbles. However, nitrocellulose should carefully be pre-wetted in order to also avoid air microbubble trapping in the bulk of the membrane itself. To effectively prevent microbubble formation, lay the nitrocellulose membrane in such a way as to make it float on the surface of the transfer buffer, without submerging it or pouring liquid on top of the membrane itself. Allow the buffer to be absorbed by capillarity through the membrane pores for some minutes, until tiny droplets appear on the upper surface (the membrane "sweats"). The membrane can then be wetted with excess buffer without problems.

Another major source of air bubbles is the blotting pads, especially when they have already been used several times. In our experience, wetting blotting pads with deionized water under pressure helps dislodging also very tiny air bubbles. Eliminate excess water by gently squeezing the pads and soak in 2–3 changes of 1× NuPAGE Transfer Buffer prior to assembling the blotting device.

Acknowledgements

G. E. is a Conway Fellow. Support received from the UCD Conway Mass Spectrometry Resource is gratefully acknowledged.

References

1. Wallin E, von Heijne G (1998) Genome-wide analysis of integral membrane proteins from eubacterial, archaean, and eukaryotic organisms. Protein Sci 7:1029–1038
2. Almen MS, Nordstrom KJ, Fredriksson R et al (2009) Mapping the human membrane proteome: a majority of the human membrane proteins can be classified according to function and evolutionary origin. BMC Biol 7:50
3. Wu CC, Yates JR 3rd (2003) The application of mass spectrometry to membrane proteomics. Nat Biotechnol 21:262–267
4. Macher BA, Yen TY (2007) Proteins at membrane surfaces—a review of approaches. Mol Biosyst 3:705–713
5. Dorsam RT, Gutkind JS (2007) G-protein-coupled receptors and cancer. Nat Rev Cancer 7:79–94
6. Rosengren AH, Jokubka R, Tojjar D et al (2010) Overexpression of alpha2A-adrenergic receptors contributes to type 2 diabetes. Science 327:217–220
7. Cooper EC, Jan LY (1999) Ion channel genes and human neurological disease: recent progress, prospects, and challenges. Proc Natl Acad Sci USA 96:4759–4766
8. Brekke OH, Sandlie I (2003) Therapeutic antibodies for human diseases at the dawn of the twenty-first century. Nat Rev Drug Discov 2:52–62
9. Tan S, Tan HT, Chung MC (2008) Membrane proteins and membrane proteomics. Proteomics 8:3924–3932
10. Zheng YZ, Foster LJ (2009) Biochemical and proteomic approaches for the study of membrane microdomains. J Proteomics 72: 12–22
11. Huber LA, Pfaller K, Vietor I (2003) Organelle proteomics: implications for subcellular fractionation in proteomics. Circ Res 92: 962–968
12. Castle JD (2003) Purification of organelles from mammalian cells. Curr Protoc Immunol Chapter 8, 8.1B.1–8.1B.57

13. Schindler J, Lewandrowski U, Sickmann A et al (2006) Proteomic analysis of brain plasma membranes isolated by affinity two-phase partitioning. Mol Cell Proteomics 5:390–400
14. Schindler J, Nothwang HG (2006) Aqueous polymer two-phase systems: effective tools for plasma membrane proteomics. Proteomics 6:5409–5417
15. McCarthy FM, Cooksey AM, Burgess SC (2009) Sequential detergent extraction prior to mass spectrometry analysis. Methods Mol Biol 528:110–118
16. McCarthy FM, Burgess SC, van den Berg BH et al (2005) Differential detergent fractionation for non-electrophoretic eukaryote cell proteomics. J Proteome Res 4:316–324
17. Lawson EL, Clifton JG, Huang F et al (2006) Use of magnetic beads with immobilized monoclonal antibodies for isolation of highly pure plasma membranes. Electrophoresis 27:2747–2758
18. Ghosh D, Krokhin O, Antonovici M et al (2004) Lectin affinity as an approach to the proteomic analysis of membrane glycoproteins. J Proteome Res 3:841–850
19. Robinson JM, Ackerman WE 4th, Tewari AK et al (2009) Isolation of highly enriched apical plasma membranes of the placental syncytiotrophoblast. Anal Biochem 387:87–94
20. Simonson AB, Schnitzer JE (2007) Vascular proteomic mapping in vivo. J Thromb Haemost 5(Suppl 1):183–187
21. Wollscheid B, Bausch-Fluck D, Henderson C et al (2009) Mass-spectrometric identification and relative quantification of N-linked cell surface glycoproteins. Nat Biotechnol 27:378–386
22. Scheurer SB, Roesli C, Neri D et al (2005) A comparison of different biotinylation reagents, tryptic digestion procedures, and mass spectrometric techniques for 2-D peptide mapping of membrane proteins. Proteomics 5: 3035–3039
23. Rybak JN, Ettorre A, Kaissling B et al (2005) In vivo protein biotinylation for identification of organ-specific antigens accessible from the vasculature. Nat Methods 2:291–298
24. Scheurer SB, Rybak JN, Roesli C et al (2005) Identification and relative quantification of membrane proteins by surface biotinylation and two-dimensional peptide mapping. Proteomics 5:2718–2728
25. Rybak JN, Scheurer SB, Neri D et al (2004) Purification of biotinylated proteins on streptavidin resin: a protocol for quantitative elution. Proteomics 4:2296–2299
26. Elia G (2010) Protein biotinylation. Curr Protoc Protein Sci, Chapter 3, 3.6.1–3.6.21
27. Elia G (2008) Biotinylation reagents for the study of cell surface proteins. Proteomics 8:4012–4024
28. Yu M-J, Pisitkun T, Wang G et al (2006) LC-MS/MS analysis of apical and basolateral plasma membranes of rat renal collecting duct cells. Mol Cell Proteomics 5: 2131–2145
29. Gottardi CJ, Dunbar LA, Caplan MJ (1995) Biotinylation and assessment of membrane polarity: caveats and methodological concerns. Am J Physiol 268:F285–F295
30. Good NE, Winget GD, Winter W et al (1966) Hydrogen ion buffers for biological research. Biochemistry 5:467–477

Chapter 30

A Multichannel Gel Electrophoresis and Continuous Fraction Collection Apparatus for High-Throughput Protein Separation and Characterization

Ming Dong, Megan Choi, Mark D. Biggin, and Jian Jin

Abstract

We developed a multichannel gel electrophoresis system that continuously collects fractions as protein bands migrate off the bottom of gel columns. The device uses several short linear gel columns, each of a different percent acrylamide, to achieve a separation power similar to that of a long gradient gel. A "Counter Free-Flow" elution technique allows continuous and simultaneous fraction collection from multiple channels at low cost. Using the system with SDS-PAGE, 300 μg samples of protein can be separated and eluted into 48–96 fractions over a mass range of 10–150 kDa in 2.5 h. Each eluted protein can be recovered at 50% efficiency or higher in ~500 μL. The system can also be used for native gel electrophoresis, but protein aggregation limits the loading capacity to about 50 μg per channel and reduces resolution. This system has the potential to be coupled with mass spectrometry to achieve high-throughput protein identification.

Key words: Continuous elution electrophoresis, Counter free-flow, High-throughput, Multichannel electrophoresis

1. Introduction

Proteomic characterization of complex protein samples by mass spectrometry requires pre-fractionation at both the protein level and peptide level. PAGE is a widely used method to fractionate proteins due to its high resolution. For example, SDS-PAGE is often used to process tagged pull-down samples (1). One limitation of all such gel-based methods is that it is laborious and time consuming to extract protein out of a gel. Some commercially available systems can elute protein from the bottom of an electrophoretic gel column. These systems, however, suffer from high sample dilution and low throughput.

Biji T. Kurien and R. Hal Scofield (eds.), *Protein Electrophoresis: Methods and Protocols*, Methods in Molecular Biology, vol. 869, DOI 10.1007/978-1-61779-821-4_30,

Table 1
Gel concentration and fractionation range

	Gel concentration (%)		MW range (kDa)
SDS single mode	12		<25
	10		25–100
	8		>100
Native single mode	4		>400
	6		150–350
	8		45–180
	Top gel (%)	**Lower gel (%)**	
SDS multimode	10	12	10–100
	8	12	10–150
	7	10	25–150
Native multimode	5	8	45–400

In order to separate and collect protein efficiently over a broad mass/size range with high throughput, we have developed a multichannel electrophoresis and continuous fraction collection apparatus (2). This system can be operated in two different modes. In the "single" mode, we use a single piece, long polyacrylamide gel, and the proteins are separated and eluted directly into the fraction collector. This mode can achieve effective separation over a finite mass range only; however, the range being chiefly determined by the concentration of polyacrylamide used (Table 1). In "multi" mode, we use two or three, short linear gel segments of different polyacrylamide concentrations (Fig. 1). We first stack these gels one on top of one another, with the lowest concentration gels uppermost (Fig. 1b). The sample is loaded on the very top. After an initial period of electrophoresis when faster-migrating, low-molecular-weight protein bands have traveled into the lower gel segments, the upper gel segments are removed from the stack and placed onto separate elution devices. Electrophoresis is then resumed simultaneously on each gel segment and protein bands are captured in the liquid phase using the fraction collector (Fig. 1c). This multimode provides an overall separation power similar to that of a traditional long gradient gel for proteins over a broad mass range and allows slow-migrating, larger-molecular-weight protein bands to be eluted in a realistic time.

A prototype 16-channel apparatus (Fig. 2) has been constructed and tested. It includes multiple gel boxes (each containing 4 channels of gel column), four elution units, and two fraction collectors.

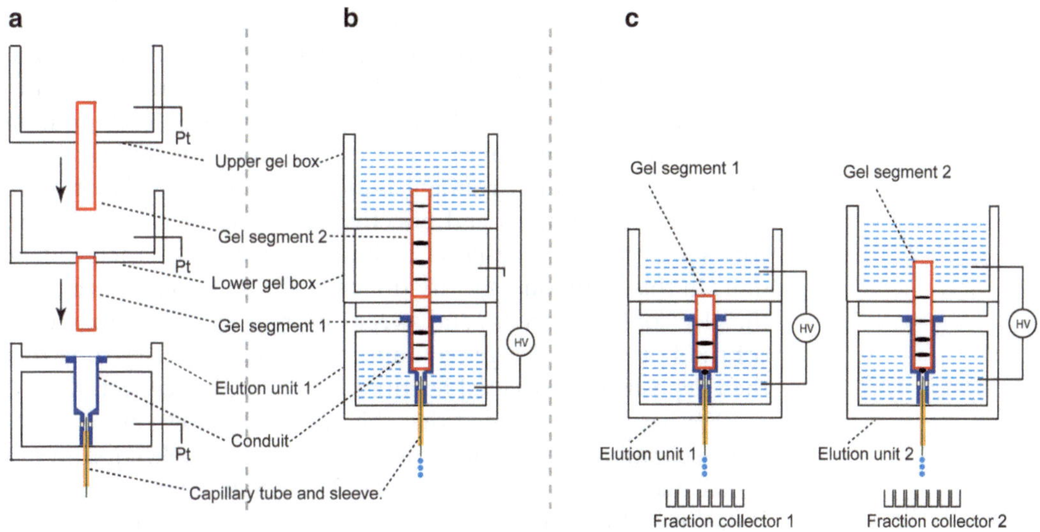

Fig. 1. System design and operation scheme for Counter Free-Flow electrophoresis. (**a**) Upper and lower gel boxes and an elution unit. Each gel box houses 4 gel columns (only one gel column is shown), and contains a cathode electrode and electrophoresis running buffer container. The upper gel (Gel segment 2) has a lower polyacrylamide concentration, while the lower column gel (Gel segment 1) has a higher polyacrylamide concentration. (**b**) The two gel boxes and the elution unit are stacked to form an approximation of a gradient gel. Gel segment 1 is inserted into a conduit in elution unit 1 and Gel segment 2 is stacked on top of this unit. Then, an initial phase of electrophoresis is used to separate proteins from a complex mixture loaded at the top of Gel segment 2. High-mobility protein bands migrate into Gel segment 1, whereas lower mobility proteins remain in Gel segment 2. (**c**) After separating (unstacking) the two gel columns, the buffer container associated with Gel segment 1 is filled, whereas Gel segment 2 is inserted into a conduit into elution unit 2 (see text and Fig. 2). Finally, electrophoresis is resumed on both gel columns to further separate and elute the protein bands captured in each gel segment. To stack more than two gel segments in a separation, additional gel boxes designed like Gel segment 1 can be included because they are constructed to allow "self-stacking." Reproduced from (1) with permission from Wiley-VCH Verlag GmbH & Co.

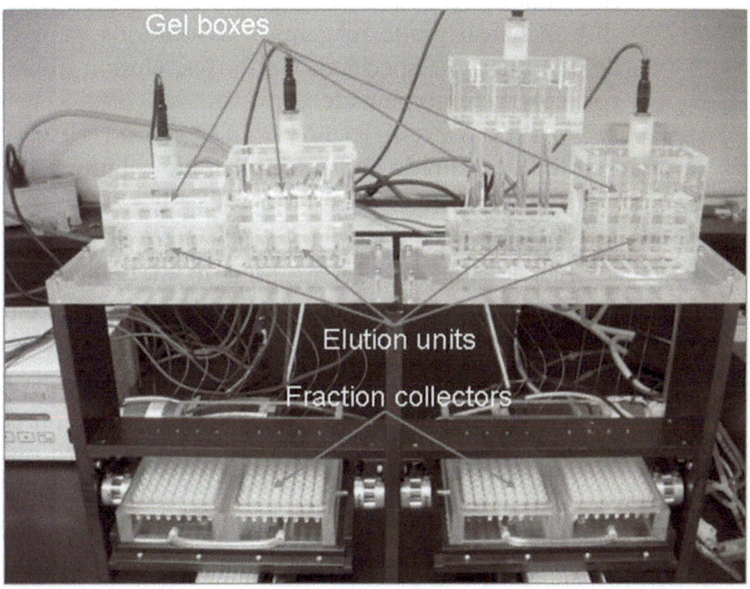

Fig. 2. Prototype 16-channel Counter Free-Flow electrophoresis and elution apparatus. It includes four gel boxes that each house four gel columns (*top*), four elution units (*middle*), and two motorized fraction collectors (*bottom*). Reproduced from (1) with permission from Wiley-VCH Verlag GmbH & Co.

Gel boxes were constructed for easy and quick "docking" and "undocking" of one box on the top of another. Each gel box also contains a cathode electrode and electrophoresis running buffer container. Gels are poured and polymerized inside each column (made out of a glass tube). Each elution unit contains an anode electrode and another running buffer container. The key feature that allows continuous and effective capturing of eluted protein bands is a "counter free-flow" technique employed in each elution unit. In this technique, gravity-driven buffer solution flows against the electrophoretic force and sweeps protein molecules migrating off the bottom of the gel column into a glass capillary tube that then deposits the collected fractions into a fraction collector.

We have demonstrated that this apparatus allows high resolution separation and fast and continuous fraction collection over a broad mass range (2). In a typical 2.5 h SDS counter free-flow PAGE run, for example, each sample can be separated and eluted into some 48–96 fractions over the mass range of ~10–150 kDa; sample recovery rate can be 50% or higher; up to ~300 μg of protein in 0.4 mL can be loaded onto each channel; and a purified protein band can be eluted within 400–600 μL. However, innovations can still be made to improve system performance. For example, current protocols for native PAGE gels on this system lead to lower resolution and loading capacity compared to SDS-PAGE gels.

2. Materials

All solutions were made using ultrapure water (18 MΩ/cm resistance) and analytical grade reagents. All solutions were stored at 4°C unless otherwise specified. Gel boxes, elution units, and fraction collector were homemade.

2.1. Gel Column Preparation

1. 30% acrylamide and bis-acrylamide solution: 29:1 ratio (Bio-Rad, Hercules, CA, USA).
2. Separation gel buffer: 1.5 M Tris–HCl, pH 8.8. Weigh 18.17 g of Tris base, dissolve it in about 90 mL of ultrapure water, adjust pH with HCl, and make up to 100 mL with ultrapure water. Autoclave stock solutions and store at room temperature.
3. Stacking gel buffer: 0.5 M Tris–HCl, pH 6.8. Weigh 6.06 g of Tris base, dissolve it in about 90 mL of ultrapure water, adjust pH with HCl, and make up to 100 mL with ultrapure water. Autoclave stock solutions and store at room temperature.
4. 12% separation gel solution: 20 mL each of 30% acrylamide and bis-acrylamide, 12.5 mL of separation gel buffer, and 17.5 mL of ultrapure water.

5. 2% separation gel solution: 4 mL of 30% acrylamide and bis-acrylamide solution, 15 mL of separation gel buffer, and 41 mL of ultrapure water.
6. 4% stacking gel solution: 8 mL of 30% acrylamide and bis-acrylamide solution, 15 mL of stacking gel buffer, and 37 mL of ultrapure water.
7. *N*, *N*, *N*, *N′*-tetramethylethylenediamine (TEMED) (Sigma-Aldrich, St. Louis, MO, USA).
8. 10% Ammonium persulfate (APS): Dissolve 0.1 g of APS in 10 mL of ultrapure water. Make 100 μL/tube aliquots, store in −20°C freezer. For each experiment, thaw and use sufficient aliquots and discard the unused solution.
9. Water-saturated butanol: Pour 25 mL of butanol and 25 mL of ultrapure water into a 50-mL centrifuge tube. Shake briefly and let it stand until layers form. Repeat three times. The top layer is water-saturated butanol.
10. Plastic film: Saran Wrap or Dura Seal.

2.2. Protein Sample Preparation

1. 6× SDS sample loading buffer: 150 mM Tris–HCl, 1.15 M glycine, 60% v/v glycerol, 18% v/v SDS, 0.06% w/v bromophenol blue. 0.91 g Tris base, 4.32 g glycine, 9 g SDS, 5.8 g DTT, 30 mg bromophenol blue, add 15 mL ultrapure water, 30 mL glycerol, mix well, and make up to 50 mL with ultrapure water.
2. 6× stacking gel sample loading buffer: 187.5 mM Tris–HCl (pH 6.8), 60% v/v glycerol, 0.06% w/v bromophenol blue. Dissolve 30 mg bromophenol blue in 10 mL ultrapure water, add 6.25 mL stacking gel buffer, 30 mL glycerol, and make up to 50 mL with ultrapure water.
3. Cell sonication buffer: 20 mM HEPES, pH 7.4, 100 mM NaCl, 12.5 mM $MgCl_2$, 0.1 mM EDTA, 20% glycerol, 1 mM PMSF.
4. Amicon Ultra-4 Centrifugal Filter Unit with Ultracel-3 membrane (Millipore, Billerica, MA, USA).
5. Disposable PD-10 Desalting Columns (GE Healthcare, Piscataway, NJ, USA).

2.3. Gel Running and Protein Collection

1. 10× gel running buffer: 0.25 M Tris–HCl, 1.92 M glycine. Weigh 121.1 g of Tris base and 576 g of glycine. Dissolve in 4 L of ultrapure water and store at RT.
2. Dilute 100 mL of 10× gel running buffer to 990 mL with water and add 10 mL of 10% SDS. Care should be taken to add SDS solution last, since it causes bubbles to form.
3. 1× native gel running buffer: 25 mM Tris–HCl, 192 mM glycine. Dilute 400 mL of 10× gel running buffer to 4 L.

4. 1× SDS gel running buffer: 25 mM Tris–HCl, 191.8 mM glycine, 0.2% SDS. Dilute 400 mL of 10× gel running buffer to 3.96 L, add 40 mL of 10% SDS, and mix well.
5. Collection Plate: 96-well plate.

2.4. Slab-Gel Analysis of Eluted Fractions

1. 6× native sample loading buffer: 150 mM Tris–HCl, 1.15 M glycine, 60% v/v glycerol, 0.03% w/v bromophenol blue.
2. Criterion Tris–HCl 4–15% polyacrylamide gel (Bio-Rad).
3. Criterion Tris–HCl 4–20% polyacrylamide gel (Bio-Rad).
4. SilverQuest™ Silver Staining Kit (Invitrogen, Carlsbad, CA, USA).

3. Methods (See Note 1)

All the steps were carried out at room temperature unless otherwise specified.

3.1. Gel Column Preparation

3.1.1. Single Mode SDS Gel (See Note 2)

1. Seal the bottom of gel glass tube with plastic film (see Note 3).
2. Choose suitable gel concentration based on fractionation range (Table 1), and mix ingredients based on Table 2 (see Note 4).
3. Pour 1,600 μL of separation gel solution into single mode gel columns (see Note 5).
4. Overlay with 200 μL of water-saturated butanol (see Note 6).
5. Allow the gel to set 30–60 min at room temperature (see Note 7).
6. Remove butanol, wash gel column twice with water, and absorb the last drop of water with paper towel (see Note 8).
7. Add 500 μL 1× SDS gel running buffer on the top of gel, and SDS gel is ready for use.

3.1.2. Single Mode Native Gel

1. Seal the gel glass tube with plastic film (see Note 3).
2. Chose suitable gel concentration based on fractionation range (Table 1) and mix ingredients based on Table 2 (see Note 4).

Table 2
Separation gel recipe (see Note 16)

	4%	5%	6%	7%	8%	10%	12%
12% Separation gel solution (mL)	2	3	4	5	6	8	10
2% Separation gel solution (mL)	8	7	6	5	4	2	0
10% APS (μL)	30	30	30	30	30	30	30
TEMED (μL)	10	10	10	10	10	10	10

3. Pour 1,600 μL of separation gel solution into single mode gel columns (see Note 5).
4. Overlay with 200 μL of water-saturated butanol.
5. Allow the gel to set 30–60 min at room temperature (see Note 7).
6. Remove butanol, wash gel column twice with water, and absorb the last drop of water with paper towel (see Note 8).
7. Pour 400 μL of 4% stacking gel solution on top (see Note 5).
8. Overlay with 200 μL of water-saturated butanol.
9. Allow the gel to set 30–60 min at room temperature (see Note 7).
10. Remove butanol, wash gel column twice with water, and absorb the last drop of water with paper towel (see Note 8).
11. Add 1× Native Gel Running buffer on the top of gel (see Note 9).

3.1.3. Multimode SDS-PAGE

1. Seal the gel glass tube with plastic film (see Note 3).
2. Chose suitable gel concentration based on fractionation range (Table 1) and mix ingredients based on Table 2 (see Note 4).
3. Pour 1,200 μL of separation gel solution into lower gel columns (see Note 5).
4. Overlay with 200 μL of water-saturated butanol.
5. Allow gels to set 30–60 min at room temperature (see Note 7).
6. Remove butanol, wash gel column twice with water, and absorb the last drop of water with paper towel (see Note 8).
7. Add 1× SDS Gel Running buffer on the top of gel (see Note 9).
8. Pour 1,200 μL separation gel solution into the top gel columns (see Note 10).
9. Overlay with 200 μL of water-saturated butanol.
10. Allow gels to set 30–60 min at room temperature (see Note 7).
11. Remove butanol, wash gel column twice with water, and absorb the last drop of water with paper towel (see Note 8).
12. Add 1× SDS Gel Running buffer on the top of gel (see Note 9).

3.1.4. Multimode Native Gel

1. Seal the gel glass tube with plastic film (see Note 3).
2. Choose suitable gel concentration based on Table 1 and mix ingredients according to Table 2 (see Note 4).
3. Pour 800 μL of separation gel solution into lower gel columns (see Note 5).
4. Overlay with 200 μL of water-saturated butanol.

5. Allow gels to set 30–60 min at room temperature (see Note 7).
6. Remove butanol, wash gel column twice with water, and absorb the last drop of water with a paper towel (see Note 8).
7. Add 500 μL of 1× native gel running buffer on top of the gel (see Note 10).
8. Pour 800 μL of separation gel solution into top gel columns (see Note 5).
9. Overlay with 200 μL of water-saturated butanol.
10. Allow gels to set 30–60 min at room temperature (see Note 7).
11. Remove butanol, wash gel column twice with water, and absorb the last drop of water with a paper towel (see Note 8).
12. Pour 400 μL of 4% stacking gel solution on top of top gel column (see Note 5).
13. Overlay with 200 μL of water-saturated butanol.
14. Allow the gel to set 30–60 min at room temperature (see Note 7).
15. Remove butanol, wash gel column twice with water, and absorb the last drop of water with paper towel (see Note 8).
16. Add 500 μL of 1× native gel running buffer on top of the gel (see Note 10).

3.2. Protein Sample Preparation

1. For native PAGE, add one-fifth volume of 6× native sample loading buffer to a crude or partially fractionated protein sample of 0.1–20 mg/mL, vortex, and briefly spin down (see Note 11).
2. For SDS-PAGE samples, add one-fifth volume of 6× SDS sample loading buffer to a crude or partially fractionated protein sample of 0.1–20 mg/mL, vortex, brief spin, heat at 95°C for 10 min to denature samples, and briefly spin down (see Note 11).

3.3. Electrophoresis and Protein Collection

Single mode is fast and effective if the application targets a specific protein band. Multimode is more efficient in separating a range of different molecular weight proteins and uses a smaller amount of sample, but it requires more control and parameter optimization for operation.

3.3.1. SDS- and Native PAGE Single Mode

1. Insert gel column into conduit and put conduit into elution units (see Fig. 1).
2. Fill buffer tank with SDS or native running buffer for SDS-PAGE or native PAGE, respectively.
3. Adjust the position of the capillary tube to 2 mm below the tapered section of the conduit (see Note 12).
4. Load protein sample on top of gel (see Note 11).

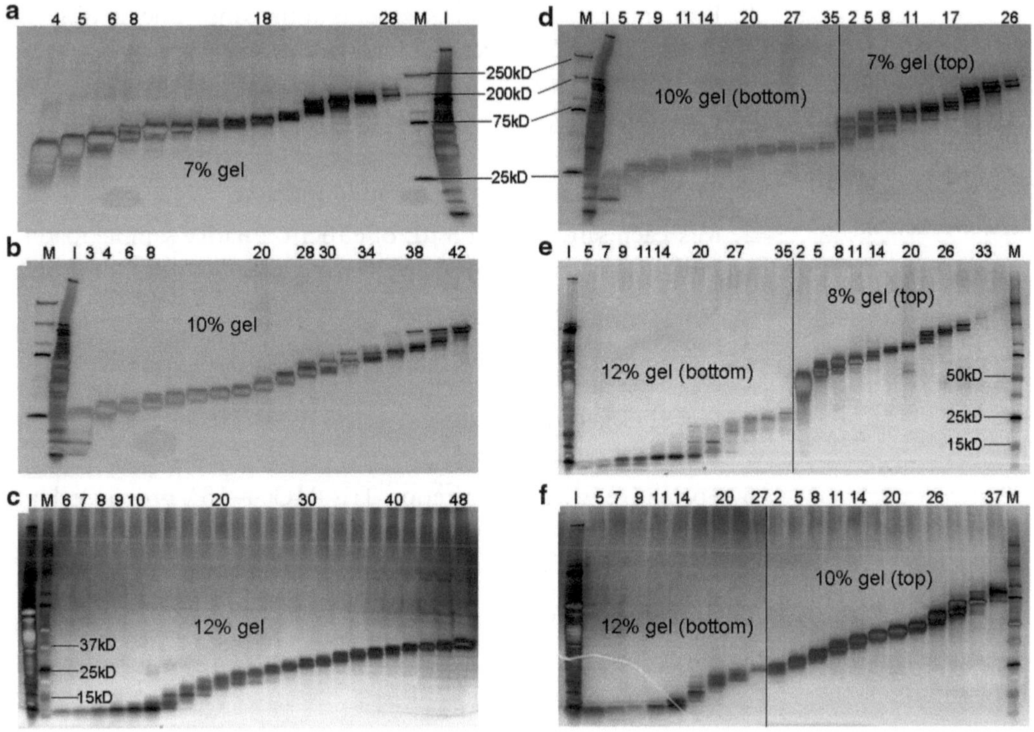

Fig. 3. Elution of SDS Counter Free-Flow PAGE gels. (**a–c**) Protein contents of collected fractions of a crude protein extract of *D. vulgaris* separated using the single mode and different gel concentrations. (**d–f**) Results obtained using two-gel-concentration multimode and different gel combinations. The concentrations in the gel columns used are given in each panel. The contents of each eluted fraction were visualized via slab SDS-PAGE and silver staining. The sample input lane is marked as "I." Protein markers are denoted as "M" and other lanes denoted according to eluted fraction number. Reproduced from (1) with permission from Wiley-VCH Verlag GmbH & Co.

5. Set electrical field to 20–30 V/cm and the flow rate of the elution buffer to 125 mL/min.
6. Collect ~200 μL fractions in a 96-well plate (Fig. 3).

3.3.2. SDS and Native PAGE Multimode

1. Stack multiple gel boxes together, insert gel column into conduit, and put conduit into elution units.
2. Fill buffer tank with SDS or native running buffer for SDS-PAGE or native PAGE, respectively (see Note 13).
3. Load protein sample on top of gel.
4. Set electrical field to 20–30 V/cm and the flow rate of elution buffer to 125 mL/min.
5. After running for about 30 min, unstack the gel boxes (see Note 14).
6. Fill the gaps above the lower gel columns with running buffer and turn on the electric field for 30–60 s (see Note 15).
7. Put the upper gel segments onto another elution unit and fill the buffer containers for each separated gel box.

8. Run upper and lower gel segments at the same time on different elution units. Set electrical field to 20–30 V/cm and flow rate to 125 mL/min.
9. Collect ~200 μL fractions in a 96-well plate.

3.4. Slab-Gel Analysis of Eluted Fractions

1. Sample 12.5 μL from each eluted fraction.
2. Mix each sample with 3 μL of either 6× native sample loading buffer for native gels or 6× SDS sample loading buffer for SDS-PAGE.
3. For SDS-PAGE only, denature sample at 95°C for 5 min before loading onto gel column.
4. For SDS gels, use Criterion Tris–HCl 4–20% precast gels and 1× SDS running buffer.
5. For native gels, use Criterion Tris–HCl 4–15% precast gel and 1× native running buffer.
6. Load samples.
7. Run electrophoresis at 200 V.
8. Silver stain the gel following manufacturer's instructions.

4. Notes

1. Our current gel column preparation method is suitable for small-scale laboratory operations. For large-scale and robust routine applications, we suggest that an industry approach must be adapted to improve gel quality and prepare gel columns in larger quantities with low cost. First, the structure of the gel box must be changed to allow quick and easy loading/unloading of individual gel columns, instead of the current batch of four. Second, a gel-pouring manifold must be constructed so that tens or even hundreds of individual gel columns can be poured at once. With adequate QC/QA, the quality of gel columns prepared would be greatly improved. In addition, the gel column tube should be made out of plastic to make them disposable and producible in large quantities with the possibility of having new features molded onto them. These changes can be realized with a small engineering effort.

   In addition, a gel column of smaller diameter should be used if an application involves a much smaller sample volume. A smaller diameter gel column would be beneficial since better-matched spot sizes between the eluting protein band and the capturing capillary would allow the use of a lower elution flow rate and/or higher electrophoresis speed. For small sample loading, alternatively, a sample loading well similar to that in a slab gel can be casted into the gel column. This alternative would eliminate the use of butanol, as the sample only migrates through the central core of the gel column during electrophoresis. Currently, the inner surface of our gel tube has not been treated (coated); therefore, there is a possibility of unwanted interactions when proteins come into contact with the wall. There was some evidence (particularly for the native gel electrophoresis) indicating that proteins were slipping through the interface between the gel and the inner glass wall, or were retained by the wall.

Another immediate benefit of using plastic tubes to house the gel columns is that they could be tapered, with a larger diameter at the upper sample loading end and a smaller one at the eluting end. This structure allows more samples to be loaded gradually by increasing the migration speed as a protein band travels into higher electric field zone. The most ideal configuration would be to have a smaller diameter at the end of the upper gel column feeding directly into the lower gel column to deposit proteins at the center of the lower gel column only.

2. Each gel box houses four gel columns, if not all the gel columns will be used, we pour gels into the unused tube to seal it.
3. Put the bottom of the glass tube on a 2″ × 2″ plastic film, then insert the tube and film into a 1.5-mL Eppendorf tube so that the bottom of the tube is covered by plastic film, which is fixed in place by the Eppendorf tube.
4. 2 and 12% separation gel solutions are made fresh every time, and APS and TEMED are added just before pouring the gel.
5. The gel column is about 4 cm long. As a rule of thumb, a 400 μL gel is about 1 cm long.
6. The butanol layer should be about 0.5 cm high.
7. Once the gel is set, a sharp optical discontinuity at the butanol/gel interface will be visible.
8. Residual butanol can reduce resolution of the protein bands and must be carefully removed. The butanol overlay should not be left on the gel longer than 2 h.
9. The gel can be stored at room temperature overnight.
10. The top and lower gel columns can be made in parallel.
11. For SDS-PAGE, the loading capacity is up to 300 μg and loading volume is 0.2–0.5 mL. For native PAGE, loading capacity is about 50 μg and loading volume is 0.2–0.5 mL.
12. If the capillary tube is set too close to the taper, the tip may become blocked by the gel if it slightly slides out or swells during electrophoresis.

13. Prior to stacking multiple gel boxes together, the space above each gel (in the hole on top of the lower gel boxes) should be filled with running buffer to prevent air bubbles from being trapped between the gel columns.
14. The time required before removing the upper gel segment needs to be determined by preliminary experiments and is typically 30 min.
15. This step is used to drive those protein molecules remaining in solution into the lower gel column.
16. SDS is not added in the gel column preparation step. For SDS-PAGE, SDS is added to the gel running buffer and will migrate into gels.

Acknowledgments

This work was part of ENIGMA, a Scientific Focus Area Program supported by the US Department of Energy, Office of Science, Office of Biological and Environmental Research, Genomics: GTL Foundational Science through contract DE-AC02-05CH11231 between Lawrence Berkeley National Laboratory and the US Department of Energy.

References

1. Butland G, Peregrín-Alvarez JM, Li J et al (2005) Interaction network containing conserved and essential protein complexes in Escherichia coli. Nature 433:531–537
2. Choi M, Nordmeyer R, Cornell E et al (2010) A multichannel gel electrophoresis and continuous fraction collection apparatus for high-throughput protein separation and characterization. Electrophoresis 31:440–447

Chapter 31

Gel-Absorption-Based Sample Preparation Method for Shotgun Analysis of Membrane Proteome

Xianchun Wang and Songping Liang

Abstract

Membrane proteins solubilized in a starting buffer containing high concentration of sodium dodecyl sulfate (SDS) are directly entrapped and immobilized into gel matrix when the membrane protein solution is absorbed by the vacuum-dried polyacrylamide gel. After the detergent and other salts are removed by washing, the proteins are subjected to in-gel digestion and the tryptic peptides are extracted and analyzed by CapLC-MS/MS. The newly developed method not only avoids protein loss and the adverse protein modifications during gel-embedment but also improves the subsequent in-gel digestion and the recovery of tryptic peptides, particularly the hydrophobic peptides. Thus, this method facilitates the identification of membrane proteins especially the integral membrane proteins.

Key words: Gel absorption, In-gel digestion, Membrane proteome, Mass spectrometry, Shotgun

1. Introduction

Membrane proteins as critical components of cellular structure and function play central roles such as signal transduction, cellular homeostasis, cell–cell recognition, and host defense (1, 2). A comprehensive analysis of membrane proteins would facilitate our understanding of their roles in cell life. However, the analysis of membrane proteins presents a special challenge due to their hydrophobic nature and low abundance, which seriously complicates their solubilization, separation, and identification (3, 4).

To enhance the solubility of membrane proteins, researchers have commonly used some organic solvents and detergents in their extraction buffer (3–11). Although these additives can improve the solubilization and extraction of membrane proteins, most of them

Biji T. Kurien and R. Hal Scofield (eds.), *Protein Electrophoresis: Methods and Protocols*, Methods in Molecular Biology, vol. 869, DOI 10.1007/978-1-61779-821-4_31, © Springer Science+Business Media, LLC 2012

reduce enzyme activity, interfere with high-performance liquid chromatography (HPLC) separation, and affect mass spectrometric analysis (8, 11, 12). On the other hand, although the conventional two-dimensional electrophoresis (2DE) has had some successes in the analysis of soluble proteins, this technique is unsuitable for the separation of membrane proteins, mainly due to protein aggregation during the isoelectric focusing (IEF) step (13). In view of this drawback, one-dimensional electrophoresis (SDS-PAGE) combined with mass spectrometry (1D-PAGE-MS) has been widely used for analysis of membrane proteins. In this protocol, a certain concentration of detergents can be used to enhance the solubility of membrane proteins during sample preparation and SDS-PAGE separation because the detergents, as well as other interfering substances, can be removed by washing, thus giving a clean environment for subsequent in-gel digestion and MS/MS analysis. However, problems such as protein insolubility still remain (14, 15).

For high-throughput analysis of membrane proteome, an alternative protocol named "tube-gel digestion" has been recently developed and applied (14–18). In this protocol, detergent-containing protein lysate was mixed with acrylamide/methylenebisacrylamide solution, and the proteins were incorporated into the polyacrylamide gel when the gel solution polymerized. Detergents were removed by washing prior to protein digestion and would not interfere with the subsequent analyses. However, this method unavoidably led to significant protein loss because there was usually a part of the protein solution excluded from the gel during gel polymerization. In addition, proteins incorporated into gel in this method have a higher risk of being chemically modified by reagents such as acrylamide (19, 20). Such limitations of the protocol will reduce the efficiency and confidence of protein identification and quantification. Therefore, its applicability to membrane proteome is limited, and there still exists a substantial need for more applicable methods to be explored and optimized.

Here we present a new method for gel-absorption-based sample preparation. In the method, the protein sample prepared using high concentration of sodium dodecyl sulfate (SDS) is directly and completely absorbed by vacuum-dried polyacrylamide gel, and then the interfering substances including SDS and some other salts are removed by in-gel washing steps while retaining the denatured proteins in the gel, thus offering a clean environment amenable to downstream buffer exchange, proteolytic digestion, digest recovery, etc. The newly developed protocol not only can more efficiently avoid protein loss and adverse chemical modifications but also improve the protein digestion and tryptic peptide recovery, thereby significantly facilitating the identification of membrane proteins, especially hydrophobic integral membrane proteins. In addition, it is worth mentioning that the volume of the used vacuum-dried gel can be conveniently adjusted according to the volume and quantity of a sample. Therefore, the method can be used for the preparation of protein samples varying in protein molecule size, sample volume, kind, and concentration of detergents and salts.

2. Materials

Prepare all solutions using ultrapure water obtained from a Milli-Q Plus purification system (Millipore, Bedford, MA) and reagents of highest grade available. Prepare and store all reagents at room temperature unless indicated otherwise. Diligently follow all waste disposal regulations when disposing waste materials.

1. Thirty percent acrylamide/Bis solution (29:1 acrylamide: Bis): Weigh 29.0 g of acrylamide monomer and 1.0 g Bis (cross-linker) and transfer to a 100 mL graduated cylinder containing about 40 mL of water. Use a glass rod to mix the solution until the reagents are completely solubilized. Make up to 100 mL with water and filter the solution through a filter paper. Store at 4°C in a bottle (see Note 1).
2. Ammonium persulfate (AP): 10% solution in water (see Note 2).
3. N, N, N′, N′-tetramethylenediamine (TEMED): 10% solution in water. Store at 4°C.
4. Ammonium bicarbonate (NH_4HCO_3): 25 mM solution in water.
5. SDS lysis buffer: 4% SDS in 25 mM NH_4HCO_3 (see Note 3).
6. Dithiothreitol (DTT): 50 mM DTT in 25 mM NH_4HCO_3 (see Note 4).
7. Iodoacetamide (IAA): 40 mM IAA in 25 mM NH_4HCO_3 (see Note 5).
8. Sucrose solutions: prepared in weight/weight (W/W) percent concentration (see Note 6).
9. Homogenization buffer: 50 mM 4-(2-hydroxyethyl)-1-piperazineethanesulfonic acid (HEPES), 1 mM $CaCl_2$, and 0.1 mM phenylmethylsulfonyl fluoride (PMSF), pH 7.4.
10. Automated Agilent 1,200 LC system coupled with an ion trap mass spectrometer (Bruker Daltonics, Bremen, Germany).
11. C18 pre-column (500 μm i.d., 3.5 cm long, Bruker).
12. C18 reverse phase capillary column (PepMap, 180 μm i.d., 15 cm long, LC-Packings).
13. Data Analysis™ 3.4 software (Bruker Daltonics).
14. Chemstation B01 (Agilent Technologies, Waldbronn, Germany).

15. EsquireControl™ 6.0 software (Bruker Daltonics).
16. Mascot software 2.0 (Matrixscience, London, UK).

3. Methods

Carry out all the procedures at room temperature unless otherwise specified. Obtain appropriate approval for carrying out animal experiments from Institutional Animal Care and Use Committee.

3.1. Enrichment of Rat Liver Plasma Membrane

1. Sacrifice the rats, excise liver, and remove liver blood. Cut the liver into pieces and homogenize them in ice-cold homogenization buffer.
2. Centrifuge the homogenate at 1,500 × *g* for 10 min at 4°C. Recover the resulting supernatant and centrifuge it again at 2,400 × *g* (Beckman, SW-28 rotor) for 30 min at 4°C. Recover the pellet.
3. Mix the pellets with 44% sucrose, over the top of which layer 42.3% sucrose carefully. After centrifugation at 90,000 × *g* for 2.5 h at 4°C, the crude plasma membranes at the top of 42.3% sucrose were collected.
4. For further enrichment, mix the obtained membrane fraction with 50% sucrose and add 30% sucrose to the mixture until the final concentration of sucrose solution was 44%, over which layer 42.8, 42.3, 41.8, 41.0, 39.0, and 37.0% sucrose sequentially.
5. After centrifugation at 100,000 × *g* for 6 h at 4°C, collect the plasma membranes floating in the upper part of the sucrose solution column, wash the sample three times with homogenization buffer (see Note 7), and then store it at −80°C until further analysis.

3.2. Proteins Extraction and Their Reduction and Alkylation

1. Extract the proteins from plasma membranes with 4% SDS/25 mM NH_4HCO_3 (see Note 8).
2. Reduce the proteins with 50 mM DTT at 50°C for 1 h and alkylate them with 40 mM IAA at room temperature in the dark for 45 min.

3.3. Incorporation of Membrane Proteins into Polyacrylamide Gel and In-Gel Digestion

1. For incorporation of the proteins, an 80 μL polyacrylamide gel was firstly prepared (see Note 9). Mix 26.7 μL of acrylamide/bisacrylamide (30%, w/v, 29:1) solution, 48.5 μL of 25 mM NH_4HCO_3, 4 μL of 1% AP, and 0.8 μL of 10% TEMED in an Eppendorf tube.

2. After polymerization, wash the gel thoroughly with Milli-Q water (see Note 10), followed by dehydration in 100% acetonitrile (ACN) and lyophilization in a Speed Vac.
3. Add 20 μL of the membrane protein solution (100 μg of protein) into the tube to make the proteins absorbed by the dried gel at 4°C.
4. Cut the gel lumps into small pieces and wash thoroughly with 25 mM NH_4HCO_3 and 50% ACN/25 mM NH_4HCO_3 sequentially. Dehydrate the gel pieces in 100% ACN and then dry them in a Speed Vac.
5. Add trypsin (1:100 w/w enzyme-to-protein ratio) prepared in 25 mM NH_4HCO_3 (see Note 11) to the gel pieces. Incubate the gels overnight at 37°C.
6. Extract the tryptic peptides from the gel pieces twice, each by adding 40 μL 67% ACN containing 5% formic acid with sonication. Combine the supernatants and then concentrate the sample solution in a Speed Vac. Store it at −20°C until CapLC-MS/MS analysis.

3.4. CapLC-MS/MS Analysis

1. Analyze the tryptic digests prepared as above by an automated Agilent 1200 LC system coupled with an ion trap mass spectrometry, equipped with an autosampler and a C18 reverse phase capillary column (PepMap, 180 μm i.d., 15 cm long, LC-Packings), using the following eluting solvents: solvent A (98% H_2O, 1.9% ACN, and 0.1% formic acid) and solvent B (95% ACN, 4.9% H_2O, and 0.1% formic acid).
2. Before separation on the capillary column, pre-concentrate the sample on a C18 pre-column (500 μm i.d., 3.5 cm long, Bruker) that is connected in the front of the capillary column using solvent A at a flow rate of 20 μL/min (see Note 12).
3. When the sample is eluted from the pre-column and further separated on the C18 PepMap capillary column, set the flow rate at 3 μL/min and the column temperature at 25°C. Use a gradient from 5 to 31% B in 137 min and then 31 to 55% B in 10 min, followed by 80% B for 10 min, then by 5% B for 10 min (see Note 13).
4. Online direct the peptides eluted from the capillary column into the ion source of the mass spectrometry. Control the LC-MS system using Chemstation B01 and EsquireControl™ 6.0 softwares. Set the main instrumental parameters as follows: nebulizer pressure, 10 psi; drying gas flow rate, 5 L/min; drying gas temperature, 300°C; capillary voltage, 4,000 V; and the full MS scan mode, standard enhanced (m/z 350–1,600 Da). Select the seven most abundant ions (see Note 14) in each MS scan for collision-induced dissociation (CID) analysis, using data-dependent MS/MS mode over the m/z range of 100–2,000 Da.

3.5. Data Processing and Bioinformatics Analysis

1. Process the obtained raw mass spectrometry data and create Mascot compatible mgf files using Data Analysis™ 3.4 software with the following main parameters: compounds (autoMS) threshold, 10,000; number of compounds, 300; and retention time windows, 1.0 min.
2. For protein identification, perform database searches using Mascot software 2.0 against the international protein index (IPI) rat combined forward/reverse database (vision 3.56, http://www.ebi.ac.uk/IPI, 79768 protein sequences). Set the search parameters as follows: enzyme, trypsin; allowance of up to one missed cleavage; peptide mass tolerance, 1.2 Da and MS/MS mass tolerance, 0.6 Da; fixed modification, carbamidomethylation (C); and variable modification, oxidation (at Met) (see Note 15).
3. Identify proteins generally on the basis of two or more peptides whose ion scores exceed the threshold, $P < 0.05$, which indicates identification at the 95% confidence level. If a protein is identified by a single peptide, inspect manually its spectrum systematically to confirm or cancel the Mascot suggestion. For a protein to be confirmed, its spectrum should be checked and interpreted in detail (see Note 16).
4. Retrieve the theoretical molecular weight (MW) and isoelectric point (p*I*) of the identified proteins from Mascot output files.
5. Carry out the mapping of putative transmembrane domains (TMDs) for identified proteins using the TMHMM 2.0 program based on transmembrane hidden Markov model (http://www.cbs.dtu.dk/services/TMHMM) by submitting the FASTA files (21).
6. Calculate the grand average of hydropathy (GRAVY) values for identified proteins using the ProtParam program, available at http://tw.expasy.org/tools/protparam.html.
7. Predict the information on the subcellular location and function of identified proteins by using gene ontology (GO) component and function terms, respectively. The text-based annotation files are available for download from GO database ftp site at http://geneontology.org/pub/go (22).

   By using this method, a total of 190 nonredundant membrane proteins were identified from a rat liver membrane-enriched fraction in triplicate experiments. The identified membrane proteins were distributed in the ranges of MW 6.5–538 kDa, pI 4.37–10.72, and GRAVY values −0.86 to +0.80. One hundred and twenty-five (65.8%) of the 190 identified membrane proteins were transmembrane proteins with 1–15 TMDs (23).

4. Notes

1. The acrylamide solution can be stored at 4°C for 1 month. Acrylamide monomer is a neurotoxin. Wear a mask when weighing acrylamide. Wear gloves even while handling polymerized gels.
2. It is best to prepare fresh AP solution each time.
3. SDS precipitates at 4°C. The SDS solution needs to be warmed prior to use.
4. It is best to prepare fresh solution each time.
5. It is best to prepare fresh solution each time.
6. The concentration of sucrose solution is expressed in percentage by weight.
7. Centrifuge the sample solution at 100,000×*g* for 30 min at 4°C to retrieve the plasma membranes each time.
8. About 2% final concentration.
9. The volume of vacuum-dried gel is adjusted according to the amount of a protein sample to be analyzed, guaranteeing that all the protein solution is completely absorbed by the gel.
10. Thorough washing guarantees that some residual reagents such as acrylamide in the polymerized gel do not react with amino acid residues in proteins, such as His, Lys, Tyr, Cys, or the N terminus (19).
11. It is helpful to add a certain concentration of organic solvent or enzyme-friendly detergent such as 10% ACN or 0.1% sodium deoxycholate (SDS) (18).
12. The residual detergents and other interfering substances, if any, in the sample are removed prior to mass spectrometric analysis.
13. The eluting gradient can be adjusted according to the type of samples.
14. The number of the ions chosen can be set at five or so.
15. The search parameters can be modified as desired.
16. Proteins identification and characterization should adhere to the Paris guidelines (http://www.mcponline.org/misc/ParisReport_Final.dtl).

Acknowledgments

This work was supported by grants from Special Program for Key Basic Research of the Ministry of Science (2004CCA00300), National High Technology Research and Development Program or "863 Program" of China (2006AA02Z141), and National Basic Research Program or "973 Program" of China (2010CB529800, 2007CB914203).

References

1. Josic D, Clifton JG (2007) Mammalian plasma membrane proteomics. Proteomics 7:3010–3029
2. Speers AE, Wu CC (2007) Proteomics of integral membrane proteins–theory and application. Chem Rev 107:3687–3714
3. Blonder J, Goshe MB, Moore RJ et al (2002) Enrichment of integral membrane proteins for proteomic analysis using liquid chromatography-tandem mass spectrometry. J Proteome Res 1:351–360
4. Lin Y, Chen Y, Yang X et al (2009) Proteome analysis of a single zebrafish embryo using three different digestion strategies coupled with liquid chromatography–tandem mass spectrometry. Anal Biochem 394:177–185
5. Ruth MC, Old WM, Emrick MA et al (2006) Analysis of membrane proteins from human chronic myelogenous leukemia cells: comparison of extraction methods for multidimensional LC-MS/MS. J Proteome Res 5:709–719
6. Quach TT, Li N, Richards DP et al (2003) Development and applications of in-gel CNBr/tryptic digestion combined with mass spectrometry for the analysis of membrane proteins. J Proteome Res 2:543–552
7. Mitra SK, Gantt JA, Ruby JF et al (2007) Membrane proteomic analysis of Arabidopsis thaliana using alternative solubilization techniques. J Proteome Res 6:1933–1950
8. Zhou J, Lin Y, Deng X et al (2008) Development and application of a two-phase, on-membrane digestion method in the analysis of membrane proteome. J Proteome Res 7:1778–1783
9. McCarthy FM, Burgess SC, van den Berg BHJ et al (2005) Differential detergent fractionation for non-electrophoretic eukaryote cell proteomics. J Proteome Res 4:316–324
10. Nagaraj N, Lu A, Mann M et al (2008) Detergent-based but gel-free method allows identification of several hundred membrane proteins in single LC-MS runs. J Proteome Res 7:5028–5032
11. Zhang N, Chen R, Young N et al (2007) Comparison of SDS- and methanol-assisted protein solubilization and digestion methods for Escherichia coli membrane proteome analysis by 2-D LC-MS/MS. Proteomics 7:484–493
12. Zhang N, Li L (2004) Effects of common surfactants on protein digestion and matrix-assisted laser desorption/ionization mass spectrometric analysis of the digested peptides using two-layer sample preparation. Rapid Commun Mass Spectrom 18:889–896
13. Santoni V, Molloy M, Rabilloud T (2000) Membrane proteins and proteomics: un amour impossible. Electrophoresis 21:1054–1070
14. Lu X, Zhu H (2005) Tube-gel digestion: a novel proteomic approach for high throughput analysis of membrane proteins. Mol Cell Proteomics 4:1948–1958
15. Cao R, He Q, Zhou J et al (2008) High-throughput analysis of rat liver plasma membrane proteome by a nonelectrophoretic in-gel tryptic digestion coupled with mass spectrometry identification. J Proteome Res 7:535–545
16. Han CL, Chien CW, Chen WC et al (2008) A multiplexed quantitative strategy for membrane proteomics: opportunities for mining therapeutic targets for autosomal-dominant polycystic kidney disease. Mol Cell Proteomics 7:1983–1997
17. Yu H, Wakim B, Li M et al (2007) Quantifying raft proteins in neonatal mouse brain by 'tube-gel' protein digestion label-free shotgun proteomics. Proteome Sci 5:17
18. Zhou J, Huang S, Bi D et al (2009) Analysis of integral membrane proteins by heat gel-embedment combined with improved in-gel digestion system. Electrophoresis 30:4109–4117
19. Righetti PG, Gelfi C, Bosisio AB (1981) Polymerization kinetics of polyacrylamide gels. III. Effect of catalysts. Electrophoresis 2:291–295
20. Liu T, Martin AM, Sinai AP et al (2008) Three-layer sandwich gel electrophoresis: a method of salt removal and protein concentration in proteome analysis. J Proteome Res 7:4256–4265
21. Krogh A, Larsson B, von Heijne G et al (2001) Predicting transmembrane protein topology with a hidden markov model: application to complete genomes. J Mol Biol 305:567–580
22. Ashburner M, Ball CA, Blake JA et al (2000) Gene ontology: tool for the unification of biology. The Gene Ontology Consortium. Nat Genet 25:25–29
23. Zhou J, Xiong J, Li J et al (2010) Gel absorption-based sample preparation for the analysis of membrane proteome by mass spectrometry. Anal Biochem 404(2):204–210

Chapter 32

Interactions of Hemoglobin in Live Red Blood Cells Measured by the Electrophoresis Release Test

Yan Su, Lijun Gao, and Wenbin Qin

Abstract

Electrophoresis release test (ERT) is the starch–agarose mixed gel electrophoresis of live red blood cells (RBCs). Mixed gel electrophoresis used to be one of the classic methods to isolate proteins, and in our laboratory, this technique is usually performed to isolate hemoglobins. Recently, combined with sodium dodecyl sulfate polyacrylamide gel electrophoresis (SDS-PAGE) and liquid chromatography coupled with tandem mass spectrometry (LC/MS/MS), ERT has been used to study the interactions between hemoglobin and other proteins in live RBCs.

Key words: Red blood cell, Hemoglobin, Interaction, Electrophoresis, Release test

1. Introduction

It is believed that biological processes are carried out through precise protein–protein interactions (1, 2). A better understanding of these interactions is crucial for elucidating the structural/functional relation of proteins, investigating their roles in the development of associated disease, and determining potential drug targets for clinical applications (3, 4). Now, increasing number of approaches, such as yeast two-hybrid systems, tandem affinity purification (TAP), protein chip, co-immunoprecipitation, and glutathione-S-transferase (GST) pull-down methods, have been developed to detect protein–protein interactions (5). However, functional protein–protein interactions are dynamic processes, and many are maintained by non-covalent bonds. Therefore, the detection of interactions in live cells remains difficult. Red blood cell (RBC) is the simplest human cell, and the loss of internal organelles becomes the great limitation to study its protein–protein interaction by the usual detection approaches such as TAP and GST pull-down in

Biji T. Kurien and R. Hal Scofield (eds.), *Protein Electrophoresis: Methods and Protocols*, Methods in Molecular Biology, vol. 869, DOI 10.1007/978-1-61779-821-4_32, © Springer Science+Business Media, LLC 2012

the RBC (4). The electrophoresis release test (ERT) is a newly developed experiment to study the interactions between hemoglobin (Hb) and other proteins in live RBCs (4). During ERT, live RBCs can be added directly onto the gel, and the electric current perforates the membrane instantaneously (6). The protein complexes in live RBCs can be thus released directly to the electric field, and the different electrophoresis behaviors of RBC proteins can be directly compared with hemolysate, in which the protein–protein interactions would have been damaged during preparation, especially interactions mediated by membrane proteins (4). Thus, in vivo interactions in the RBC can be identified by ERT.

2. Materials

Prepare all solutions using ultrapure water (prepared by purifying deionized water to attain a sensitivity of 18 MΩ cm at 25°C) and analytical grade reagents. Prepare and store all reagents at room temperature (unless indicated otherwise). Diligently follow all waste disposal regulations when disposing waste materials.

2.1. Apparatus

1. Horizontal electrophoresis cell: DYY-9B, Beijing Liuyi instrument factory, Beijing, China.
2. Electrophoresis power supply: DYY-2C, Beijing Liuyi instrument factory, Beijing, China.
3. Glass plate: 17 × 10 cm or 17 × 17 cm; about 2–3 mm thick (see Note 1).
4. High-speed centrifuge: TGL-16G, Shanghai Anting Instrument Factory, Shanghai, China.
5. Vortex mixer.
6. Sprit level.
7. Whatman® 3-MM filter paper.

2.2. Fresh Anticoagulated Blood

1. Venous blood samples were anticoagulated with ethylenediaminetetra acetic acid (EDTA) (see Note 2), stored at 4°C, and generally analyzed within 24 h.

2.3. RBCs and Hemolysate Preparation Components

1. Physiological saline: Dissolve 0.9 g NaCl in 100 mL water. Store at 4°C.
2. Carbon tetrachloride (CCl_4).

2.4. Starch–Agarose Mixed Gel Electrophoresis Components

1. Gel preparation buffer (1 × TEB pH 8.6): Weigh 5.1 g Tris, 0.5 g EDTA, and 0.4 g boric acid; transfer them to a cylinder; and add water to a volume of 900 mL. Mix and adjust pH to 8.6 with 1 N HCl and then make up the volume to 1 L with water.

2. Electrophoresis buffer (boric acid buffer, pH 9.3): Weigh 18.55 g boric acid and 4.0 g NaOH, transfer them to a cylinder, and add water to a volume of 900 mL. Mix and adjust pH to 9.3 with 1 N NaOH and then make up the volume to 1 L with water.
3. Potato starch ($C_6H_{10}O_5$): 162, molecular weight (see Note 3).
4. Agarose (see Note 4).

2.5. Staining Solution Components

1. Ponceau red staining solution: Dissolve 1.0 g Ponceau (see Note 5), 50 mL glacial acetic acid, and 20 mL glycerol (see Note 6) in 1 L water.
2. Benzidine staining solution: 0.6 g benzidine (see Note 7), 25 mL glacial acetic acid, and 10 mL glycerol, add water to 500 mL, then heat the solution in 75°C water bath for 1 h until the benzidine is dissolved completely. Sodium nitroprusside and 30% hydrogen peroxide (H_2O_2) should be added to this solution before use (see Note 8).
3. Rinsing solution: 50 mL glacial acetic acid and 20 mL glycerin, add water to 1 L.

3. Methods

3.1. Preparation of RBCs Suspension

1. Centrifuge the anticoagulated blood at 700×*g* for 5 min and aspirate the upper plasma (see Note 9).
2. Transfer 200 μL of the lower RBCs to 1.5-mL Eppendorf tube (see Note 10) and then add 1 mL physiological saline to the Eppendorf tube, mixed gently.
3. Centrifuge at 700×*g* for 4 min and aspirate the supernatant.
4. Repeat this washing operation four to five times until the supernatant becomes colorless and transparent.
5. Finally, aspirate the supernatant and dilute the RBCs 1:1 with saline to prepare RBC suspension.

3.2. Preparation of Hemolysate

1. Add 200 μL CCl_4 to the well-prepared RBC suspension.
2. Mix on the vortex mixer for at least 5 min.
3. Centrifuge at 10,000×*g* for 10 min.
4. Transfer the upper red hemolysate carefully to a new 1.5-mL Eppendorf tube (see Note 11) and store at 4°C for later use.

3.3. Preparation of the Starch–Agarose Mixed Gel

1. Clean a 17×10-cm glass and then place it onto the horizontal desk top (see Note 12).

2. Weigh 1.0 g potato starch and 0.15 g agarose and transfer them to a 250-mL triangle flask, then add 50 mL 1 × TEB (see Note 13), mix together.
3. Heat the triangle flask on the electric stove repeatedly (see Note 14) until the starch and agarose are dissolved completely (see Note 15). Cool the gel solution at room temperature to about 60°C (see Note 16).
4. Pour the gel solution onto the horizontal glass plate (see Note 17), then solidify at room temperature for at least 20 min (see Note 18).

3.4. Sample Loading

1. Cut the 3-MM filter paper to 0.1 × 1-cm-small pieces (see Note 19).
2. Add 5 μL RBC suspension to the 0.1 × 1-cm-small piece of 3-MM filter paper (see Note 20).
3. Clamp the filter papers with tweezers and insert (see Note 21) them one by one into the gel which is 1.5–2 cm away from one end of the gel (see Note 22), the intervals between every two samples being about 0.5–1 cm. (Fig. 1).

3.5. RBCs Release Electrophoresis

1. Add 500 mL electrophoresis buffer (boric acid buffer, pH 9.3) to each side of the electrophoresis cell.
2. Immerse proper size of the gauze into the electrophoresis buffer.

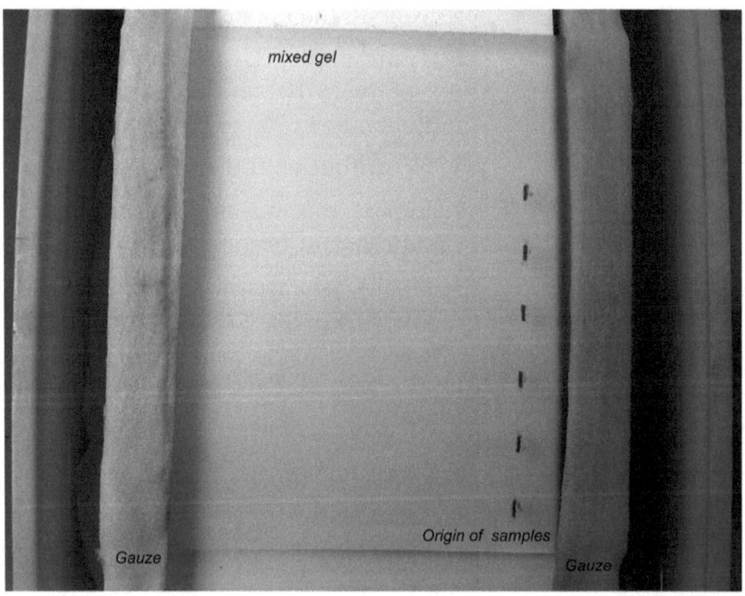

Fig. 1. The methods of sample loading and electrophoresis. The origin of samples is 1.5–2 cm away from one end of the gel, and the interval between every two samples is about 0.5–1 cm. On each side, 4–6 layers of gauze are put on the gel. The origin should be placed on the negative side.

3. Put the gel into the horizontal electrophoresis cell, the origin of sample should be placed on the negative side.
4. Put 4–6 layers of gauze over each end of the gel (see Note 23).
5. Electrophoresis is performed in borate buffer at constant voltage (6 V/cm) for 2–3 h (see Note 24).
6. After the red hemoglobin band moves to the other side, turn off the power.

3.6. Staining and Rinsing

1. First, directly observe the red bands on the gel with naked eyes (see Note 25) and take pictures or scan the gel to keep the results.
2. Put the gel into Ponceau red staining solution for at least 4 h.
3. Rinse the gel with rinsing solution repeatedly until the background is clean.
4. Dry the gel in the air or in the oven.
5. Put the gel into benzidine staining solution for about 20 min, until the Hb bands are stained blue-black.
6. Take out the gel and wash it with water and then dry it in the air.
7. Take pictures or scan the gel to keep the results.

3.7. Analysis of Results

All of protein can be stained red by Ponceau red, but only hemoglobin can be stained black by benzidine specifically. Compare the electrophoresis behavior of Hbs in RBC and hemolysate group (see Note 26). If there are some differences, proteins can be recovered and enriched from the bands of starch–agarose mixed gel (see Note 27) and further isolated by SDS-PAGE. The important band can be cut out and detected by LC/MS/MS to disclose the protein components. This will help us to find the important protein which has interaction with Hbs in live RBCs.

4. Notes

1. The size of the glass is determined both by the size of the electrophoresis tank and the requirements of the experiment.
2. Venous blood can be anticoagulated by a variety of anticoagulants, such as EDTA, heparin, and sodium citrate, but hemolytic blood should not be used here.
3. The results obtained are not that good when fresh potato starch is used, compared to that obtained with starch stored for several years before use. Potato starch can also be replaced by lotus root starch and corn starch sold in the market.

4. In the past, the mixed gel was made up of starch and agar, but now agar has been substituted by agarose which can increase the electrophoretic resolution.
5. The dissolution progress of Ponceau is slow, so the 10× stock solution should be prepared and stored at room temperature. Before use, it should be 1:10 diluted with water.
6. Proper concentration of glycerol can avoid gel cracking after it is dry, but if too much, it will reduce the gel's ability to dry. Also, if the results need not be kept for a long time, glycerol can be omitted in the staining or rinsing solutions.
7. Benzidine is highly carcinogenic; it could also be substituted by leuco base of malachite green which is noncarcinogenic, but its staining specificity is lower than benzidine.
8. 1 g sodium nitroprusside and 1.4 mL 30% H_2O_2 should be added to 500 mL well-prepared benzidine staining solution before use.
9. Normal ratio of blood cells to plasma in the whole blood is about 1:1.
10. When RBCs are transferred, the pipette tip must be placed under the liquid surface at least 2–3 min to avoid the contamination of white blood cells and platelets.
11. After centrifugation, three layers of the solution can be visible: the lower layer is CCl_4, the middle layer is membrane fragments, and the upper layer is hemolysate. Pipette out the upper supernatant carefully and avoid touching the middle layer.
12. Level must be placed on the horizontal desktop. If the glass is not horizontal, the gel is easy to leak from the edge of glass when gel solution is poured on the glass.
13. The concentrations of potato starch and agarose are 2% and 0.3%, respectively. Different size of glasses need different volume of 1 × TEB; the average volume is about 3.2 mL/cm^2.
14. When the triangle flask is heating on the electric stove, the flask must be shaken slowly to avoid the starch and agarose attach to the bottom of flask. In addition, when the gel solution begins boiling, overflow must be avoided at any time. When the solution is boiling, the flask should be removed from the electric stove immediately and cooled at room temperature for 1–2 min, and then continue to heat, and so forth, until the gel is completely dissolved.
15. When the solution bubbles become larger during heating, the starch/agarose should have been well dissolved.
16. The proper temperature to pour the gel is 60°C. If the temperature is too high, the gel solution is easy to leak from the edge of glass; if too low, the gel solution will become solid

during pouring, which will cause incomplete and uneven surface of gel.

17. Pour the gel solution at the center of glass slowly and constantly. The gel solution will stop flowing at the edge of glass because of the tension of edge.
18. If the room temperature is high, it will take a longer time for the gel to get solid; if the gel is not used immediately, please put it in the 4°C refrigerator.
19. The length of filter paper is determined by the volume and number of samples, generally the size of a 17 × 10 cm small plate can load up to eight samples (5 μL/sample).
20. If the size of filter paper is 0.1 × 2 cm, it can load 10–15 μL samples.
21. A 3-MM filter paper has certain degree of hardness and can be inserted into the gel directly. Also, the comb of proper size and thickness can be used to groove on the gel firstly and then insert the filter paper into them.
22. Keep each sample located at the same level as far as possible.
23. When putting the gauze on the gel, the action must be gentle to avoid torn the gel. The gauze should closely contact with the gel surface, and the edge of gauze should be neat and be parallel with the sample origin.
24. We call this kind of electrophoresis primary release electrophoresis of RBCs; there are also point release electrophoresis, ladder release electrophoresis, and bidirectional ladder release electrophoresis.
    (a) Point release electrophoresis of RBCs: Electrophoresis is performed in borate buffer at (6 V/cm) for 1 h, then pause for 5 min and run for another 1 h; at this time, a Hb band released from the origin will appear between HbA_2 and carbonic anhydrase (CA) (Fig. 2). Changing the time of the first and second electrophoresis can change the location of the released Hb band.
    (b) Ladder release electrophoresis of RBCs: Electrophoresis is performed in borate buffer at (6 V/cm) for 1 h, then pause for 5 min and run for 15 min by turns. It takes about 3–4 h for the entire electrophoresis (Fig. 3).
    (c) Bidirectional ladder release electrophoresis of RBC: Two vertical-direction electrophoresis runs are performed at 6 V/cm. The first directional electrophoresis is run for 15 min with a break of 15 min alternately. After the first-direction electrophoresis, the direction of the electric field is altered vertically, and electrophoresis continued for about 1 h (Fig. 4).

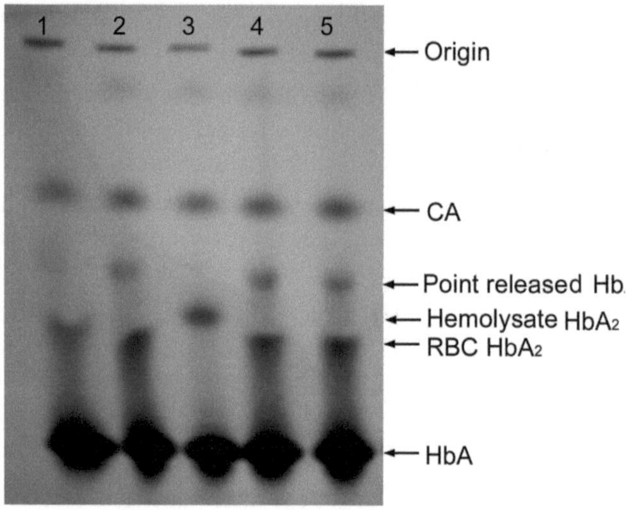

Fig. 2. Point release electrophoresis of RBCs. Samples (*1*), (*2*), (*4*), and (*5*) are RBCs of different patients; sample (*3*) is hemolysate. All of the RBC samples have different degree of released Hbs which locate between HbA_2 and CA, but hemolysate sample does not have released Hb band, and its electrophoretic speed of HbA_2 is slower than that of RBCs obviously.

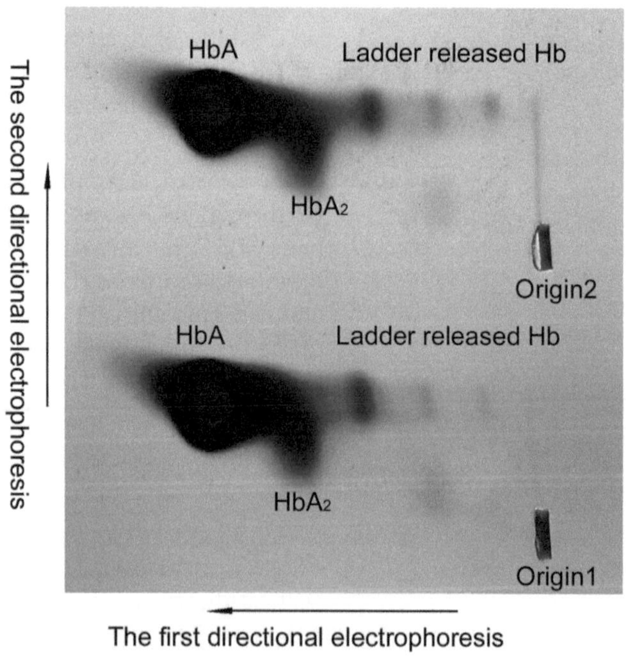

Fig. 3. Ladder release electrophoresis of RBCs. Samples 1–6 are RBCs from different patients; the ladder released Hbs of samples 3 and 4 are significantly more than the other samples.

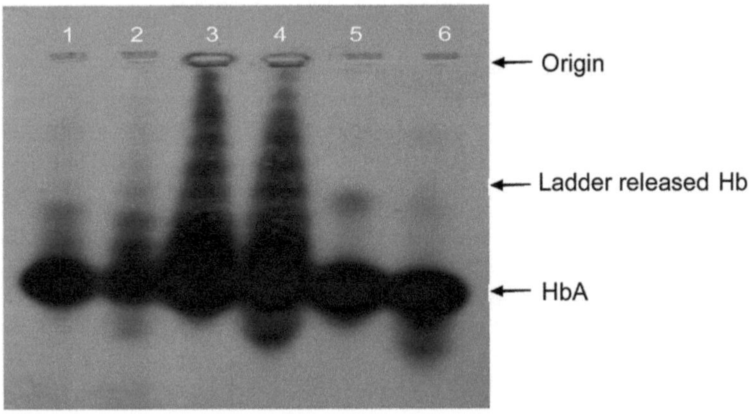

Fig. 4. Bidirectional ladder release electrophoresis of RBC. Samples (*1*) and (*2*) are RBCs from different patients, and the contents of ladder-released Hb are different from each other.

25. Hb is red, so we can see the main Hb bands directly without staining.
26. RBCs come from different patients; different treatments and different physiological and pathological conditions can be analyzed by this method.
27. When recovering and enriching the protein from the starch–agarose mixed gel, the gel should not be stained by any dye after starch–agarose mixed gel electrophoresis. The red band of HbA, HbA_2, and the gel located between HbA and HbA_2 (HbA–HbA_2) of RBC and hemolysate can be cut out separately and frozen at −80°C for at least 30 min. Before use, the gels should be taken out and thawed at room temperature. After centrifuging at 10,000×*g* for 10 min, the supernatants can be pipetted out into new Eppendorf tubes and dried in a vacuum freeze drier for about 6–8 h.

Acknowledgments

This work was supported by grants from the Key Science and Technology Research Project of the Ministry of Education (210039), and Natural Science Foundation of China (81160214), the Major Projects of Higher Education Scientific Research in the Inner Mongolia Autonomous Region (NJ09157), Natural Science Foundation of Inner Mongolia (2010BS1101), and Inner Mongolia Health Department Fund. We also especially acknowledge all of the people who donated their blood samples for our research.

References

1. Kim KK, Kim HB (2009) Protein interaction network related to helicobacter pylori infection response. World J Gastroenterol 15:4518–4528
2. Bu D, Zhao Y, Cai L, Xue H, Zhu X, Lu H, Zhang J, Sun S, Ling L, Zhang N, Li G, Chen R (2003) Topological structure analysis of the protein-protein interaction network in budding yeast. Nucleic Acids Res 31:2443–2450
3. Hase T, Tanaka H, Suzuki Y et al (2009) Structure of protein interaction networks and their implications on drug design. PLoS Comput Biol 5:e1000550
4. Su Y, Gao L, Ma Q et al (2010) Interactions of hemoglobin in live red blood cells measured by the electrophoresis release test. Electrophoresis 31:2913–2920
5. Drewes G, Bouwmeester T (2003) Global approaches to protein-protein interactions. Curr Opin Cell Biol 15:199–205
6. Su Y, Shao G, Gao L et al (2009) RBC electrophoresis with discontinuous power supply: a newly established hemoglobin release test. Electrophoresis 30:3041–3043

Chapter 33

Extraction of Proteins from Gels: A Brief Review

Biji T. Kurien and R. Hal Scofield

Abstract

Gel electrophoresis is an important methodology employed for protein analysis. It is often necessary to elute and recover proteins separated by sodium dodecyl sulfate polyacrylamide gel electrophoresis (SDS-PAGE). The procedure involves localizing the protein of interest on the gel following SDS-PAGE, eluting the protein from the gel, removing SDS from the eluted sample, and finally renaturing the protein (enzymes, for example) for subsequent analysis. Proteins are extracted from gels by several methods. These include dissolution of the gel matrix, passive diffusion, and electrophoretic elution. Proteins eluted from gels have been used successfully in a variety of downstream applications, including protein chemistry, proteolytic cleavage, determination of amino acid composition, polypeptide identification by trypsin digestion and matrix-assisted laser desorption ionization-time of flight mass spectroscopy, as antigens for antibody production, identifying a polypeptide corresponding to an enzyme activity and other purposes. Protein yields ranging from nanogram levels to 100 μg have been obtained.

Key words: Elution, SDS-PAGE, Passive diffusion, Electroelution, Renaturation

1. Introduction

To recover a specific protein of interest from a gel, it has to be liberated out of the gel matrix that imprisons them. In principle, one has to just simply reverse the way these molecules entered the gel. But for all practical purposes, this process is easily beset with problems, especially when working with large protein complexes (1). Omitting the gel matrix and performing preparative electrophoresis in solution is an alternative to the arduous process of recovering proteins from gels (2). However, free-flow electrophoresis is even more challenging with respect to the technical requirements compared to the traditional method of gel electrophoresis followed by protein elution.

Biji T. Kurien and R. Hal Scofield (eds.), *Protein Electrophoresis: Methods and Protocols*, Methods in Molecular Biology, vol. 869, DOI 10.1007/978-1-61779-821-4_33, © Springer Science+Business Media, LLC 2012

Proteins recovered from gels have been used in protein chemistry, proteolytic cleavage, determination of amino acid composition, polypeptide identification by trypsin digestion, and matrix-assisted laser desorption ionization-time of flight mass spectroscopy, as antigens for antibody production, identifying a polypeptide corresponding to an enzyme activity and other purposes. Protein yields ranging from nanogram levels to 100 μg have been obtained. Denatured enzymes obtained in this manner are often amenable to renaturation and regain enzymic activity (3).

The procedure basically involves (a) gel electrophoresis, (b) localizing the protein of interest on the gel, (c) protein elution from the gel, (d) SDS removal from the eluted sample, and (e) protein renaturation for subsequent analysis (3).

In this chapter, we report the various methods that have been employed to elute proteins from gels.

2. Elution Procedures

2.1. Diffusion-Mediated Elution of Proteins from Gels

This method, known as elution by diffusion (4) or passive elution, is the simplest method to release protein molecules from gel matrix and can be accomplished by adding water (or buffer) to a piece of gel containing the protein of interest. Trapped protein molecules, generally proteins below 60 kDa (5), diffuse out of the gel matrix into the surrounding medium if one waits sufficiently long and can be separated from the gel by centrifugation or filtration. Contaminants such as acrylamide monomers or remainders of the polymerization initiators which also diffuse out of the matrix may necessitate additional purification procedures (such as ion exchange chromatography). In addition, this method does not work well with protein complexes in acrylamide gels (6), unless harsh detergents like sodium dodecyl sulfate (SDS) is added to the elution process (7). Complete elution by passive diffusion has been achieved by crushing a gel slice with a small Teflon pestle in an elution buffer containing 0.1% SDS and incubating the crushed gel fragments for 4 h (for a 36 kDa protein) and 16–24 h (for a 150 kDa protein) on a rotator (3).

2.1.1. SDS Removal and Concentration

SDS removal following elution has been effectively achieved by acetone precipitation. This procedure also concentrates the protein (3).

2.2. Protein Recovery by Dissolving Gels

The easiest way to elute proteins from a gel matrix would be to dissolve this matrix. However, the ability of acrylamide to form stable and chemically inert matrix becomes a disadvantage in this context. Very harsh conditions are required to recover proteins from acrylamide gels containing the standard cross-linker

N,N′-methylene bisacrylamide. Typically, a gel slice is incubated in 30% hydrogen peroxide at 50°C for this purpose. This results in irreversible damage to all proteins. Other cross-linkers, such as *N,N′*-diallyltartardiamide (gel slice that has this cross-linker needs to be dissolved with 2% periodic acid), *N,N′*-(1,2-dihydroxyethylene) bisacrylamide (protein recovery requires harsh conditions), and ethylene diacrylate (gel slice needs to be dissolved with 1 M sodium hydroxide) have been used. However, these methods have serious problems as well (as noted above). Cleavable disulfide cross-linkers have been used that require milder dissolving conditions. However, the cleavable bond is unstable under alkaline conditions (1).

2.3. Electroelution Devices

Several methods have been described to elute proteins out of gels by the use of an electric field (reviewed in (6)). The simplest method requires no special device, except for a dialysis tubing and a flat-bed electrophoresis chamber. The proteins are trapped by the large surface of the tubing as they migrate electrophoretically out of the gel. The nonspecific adsorption by the dialysis tube may be problematic.

Other electroelution methods include devices like the vertical-type eluter, horizontal-type electroeluter, and bridge-type electroeluters and methods such as reversed electrophoresis using a discontinuous conductivity gradient, elution by employing a steady-state stacking buffer system, and continuous elution devices. These methods have been reviewed nicely by Seelert and Krause (1).

As described, several interesting methods have been employed to elute proteins out of gels. An interesting method to elute proteins out of gels has been described in detail in this book (see Chapter 35). The method is simple, convenient, and easy to perform.

References

1. Seelert H, Krause F (2008) Preparative isolation of protein complexes and other bioparticles by elution from polyacrylamide gels. Electrophoresis 29:2617–2636, Review
2. Ressler N (1958) Application of direct photographic photometry to preparative electrophoresis. Nature 182:463–464
3. Burgess RR (2009) Elution of proteins from gels. Methods Enzymol 463:565–572, Review
4. Lewis UJ, Clark MO (1963) Preparative methods for disk electrophoresis with special reference to the isolation of pituitary hormones. Anal Biochem 6:303–315
5. Jacobs E, Clad A (1986) Electroelution of fixed and stained membrane proteins from preparative sodium dodecyl sulfate-polyacrylamide gels into a membrane trap. Anal Biochem 154:583–589
6. Shoji M, Kato M, Hashizume S (1995) Electrophoretic recovery of proteins from polyacrylamide gel. J Chromatogr A 698:145–162
7. May LT, Santhanam U, Sehgal PB (1991) On the multimeric nature of natural human interleukin-6. J Biol Chem 266:9950–9955

Chapter 34

A New Algorithm for the Simulation of SDS 2D-PAGE Datasets

Emilio Marengo and Elisa Robotti

Abstract

This chapter describes a new software for the generation of simulated realistic sodium dodecyl sulfate two-dimensional polyacrylamide gel electrophoresis (SDS 2D-PAGE) images. In order to choose the simulation strategy to provide realistic 2D-PAGE maps the statistical characteristics of such images were taken into account, such as the distributions of sizes, intensities, and volumes of the spots. Also, the low reproducibility typical of replicated SDS 2D-PAGE maps of the same sample was simulated. This approach can be used to generate simulated datasets useful in the development and performance evaluation of new classification and/or image analysis algorithms applied to two-dimensional electrophoresis datasets, given the usually small number of experimental replications available.

Key words: Software, Simulation, 2D-PAGE maps, Volume variation, Spot displacement

1. Introduction

Sodium dodecyl sulfate two-dimensional polyacrylamide gel electrophoresis (SDS 2D-PAGE) (1) is an analytical separation method that permits to characterize the protein content of biological samples through a two-step procedure. In the first step, the proteins are separated on the basis of their isoelectric points, whereas in the second they are separated on the basis of their masses. The result of each separation is a gel electrophoretic map where the proteins, after the application of a proper coloring procedure, are represented by spots, dispersed on the maps, easily recognizable also by a visual inspection. For the successive treatment and analysis, the map is usually scanned by a densitometer or a fluorescence laser scanner (2) and the images obtained are successively subjected to further analysis with the purpose of recognizing the proteins (spots) present. The analysis of sets of proteomic SDS 2D-PAGE images is normally performed on a comparative basis,

Biji T. Kurien and R. Hal Scofield (eds.), *Protein Electrophoresis: Methods and Protocols*, Methods in Molecular Biology, vol. 869, DOI 10.1007/978-1-61779-821-4_34,

by searching the spots common to all maps, which correspond to proteins present in all samples, while the proteins differently expressed (present/not present or characterized by a relevant variation of their content) contain information on the differences existing between the different classes of samples. This permits, for example, to achieve information on the effect of a therapy, the mechanism of evolution of an illness, or more simply for diagnostic/prognostic purposes. The standard image analysis procedure performed using commercial software is developed following some steps common to almost all the commercial software devoted to SDS 2D-PAGE analysis:

- Automated spot detection
- Spot filtering and editing
- Background correction
- Creation of the reference gel
- Gel matching

After completion of these steps each original 2D map is transformed into a so-called "spot volume dataset," where every protein is characterized by the total optical density, measured by the densitometer or the fluorescence laser scanner, of the corresponding spot present on the map. A volume dataset of several samples is then a matrix of volume values, where each sample is described by the volume of the spots present. Obviously, the five steps performed through the commercial software, which go from the recognition of the spots to the matching of the corresponding proteins present in different maps, are crucial to the final result of the analysis. This final analysis can be accomplished by means of several methods and algorithms that perform the classification of the maps either using the volume datasets or directly working on the unfolded images through, for example, some global or local image descriptors by means of a plethora of different statistical tools (3–14). In this perspective, for example, the authors have recently proposed a new method for the automatic classification of gel electrophoretic maps that involves the decomposition of the images via Legendre and Zernike moments (15, 16). All the strategies for image dataset analysis present in the literature involve usually algorithms like principal component analysis (PCA) (17–20), partial least squares discriminant analysis (PLS-DA) (21–23), or linear discriminant analysis (LDA) (24, 25). The evaluation of the performances of the classification as well as of the matching of the different SDS 2D-PAGE maps are tasks strictly related to the complexity and to the variability of the gel electrophoretic images. The intrinsic variability of sets of different maps obtained starting from the same biological sample, namely replications of the two-dimensional analytical procedure, is mainly related to changes of the number of spots, modifications of their shape and position (15, 26), variations of local image intensity that can be related to variations of protein concentration and presence of spurious or noise spots. Also, the background can be fairly different among replicated maps obtained from the same sample. Many variables can influence the experimental reproducibility of the separation, like, for example, the sample pretreatment procedure, stainer time exposure, operator ability, structure of the polyacrylamide gel, condition of the buffer solution. If two replications of the same separation are observed, it is very common to detect some differences between the two images. An important question is how the algorithms for automatic image analysis and image classification will respond to these differences.

There is another problem that heavily affects the use of classification algorithms to two-dimensional proteomic datasets, namely the limited numbers of available images. The sample preparation and the separation procedure are expensive and time consuming, so that frequently the number of available maps is rather small, often in the range 4–10. Unfortunately, this collides with the necessity of the several images that are needed for training and testing classification algorithms, especially if the classification model is requested to show good prediction performances, which is definitely mandatory, for example, for diagnostic/prognostic applications. During the last few years some authors have started to use simulated data containing images for testing their algorithms, especially in military and civil vehicle identification (5, 6, 27).

For what regards proteomic two-dimensional maps, numerical simulations and modeling have been applied since the 1980s to investigate the mechanisms of separation involved in 2D gel electrophoresis (28–32). Simulations in this field usually involve the development of theoretical models miming the effect of voltages and a large number of carrier ampholytes on the final 2D separation (33–38). Other recent studies on 2D gel electrophoresis regard instead the decoding of 2D maps by the auto-covariance function and the statistical model of peak overlapping (SMO) (39, 40). In these last two approaches, applied by Pietrogrande et al., the 2D auto-covariance function (2D-ACVF) is computed on experimental digitized maps, while the SMO is applied to the spot coordinates on the final map. These approaches are used to estimate features of the electrophoretic separation (i.e., number of proteins present in the sample), to provide information on the separation performance, to identify spot trains related to post-translational modifications, and to evaluate spot overlapping.

However, in proteomics very few has been done for the simulation of 2D maps focused on the achievement of a set of replicated artificial maps starting from few experimental replications: simulations in the field of proteomics are mostly devoted to understand the mechanisms underlying the separation. Here, the considerations drawn from the statistical analysis of several SDS 2D-PAGE datasets available were the basis for the simulation of realistic proteomic maps (41). The final aim is to develop an algorithm that is able to provide a series of replicated simulated maps on the basis of a few experimental 2D maps available: the sets of artificial maps can therefore be used for testing 2D maps image analysis tools and classification tools.

2. Materials

2.1. Experimental SDS 2D-PAGE Maps Database

1. Take into consideration several datasets of gel electrophoresis maps (21, 42, 43) to draw some indications on how to produce realistic simulated maps: a total of 50 maps belonging to 8 classes. Each class contains four or five replications of the same sample.
2. Perform all separations on maps that are approximately square shaped using 17-cm long pH gradient strips (pH 3–10).

2.2. Software

1. Analyze all the maps using PDQuest (BioRad, Hercules, CA), Melanie (Swiss Institute of Bioinformatic, Lausanne, Switzerland) (44–46), or other dedicated software.
2. Export in Matlab (R2007b, The Mathworks, Natick, MA) the data extracted using image analysis software (the algorithm in its present form is developed in Matlab).
3. All graphical representations presented here were performed by Matlab, Statistica version 7.1 (Statsoft Inc., Tulsa, OK) and Excel 2007 (Microsoft Corporation, Redmond, WA).

3. Methods

3.1. Experimental SDS 2D-PAGE Maps Database

1. For all separations, use 17-cm long pH gradient strips (pH 3–10), rehydrated for 8–10 h with a solubilizing solution (21, 42, 43). Carry out isoelectric focusing (IEF) with an increasing voltage gradient, the range of *time·voltage* applied being 70,000–75,000 Vh for each strip ($T=20°C$).
2. Equilibrate the IPG (immobilized pH gradient) strips by rocking for 24–26 min in a solution containing urea, SDS, glycerol, and Tris–HCl.
3. In the second dimension, use acrylamide–bisacrylamide gradient gels (average dimension 19×18 cm).
4. Perform the electrophoretic run according to the following current profile: 2 mA (1–2 h), 5 mA (2–3 h), 10 mA (20 h), and 15–20 mA until end of the run.

5. Perform staining by Coomassie Brilliant Blue or SYPRO Ruby.
6. Perform acquisition by the VersaDoc (Biorad) system. The resolution of the images ranges from 200 to 500 dpi with a dynamic range of acquisition that varies from 0–3.0 to 0–3.4 OD (optical density).
7. Analyze the maps (here 50 maps are used) to identify, for each spot present, its position, size (diameter measured along pH and mass separations), volume, and maximum intensity.
8. The final database contains approximately a total of 20,000 spots, each described by seven variables: x_1 (position along pH separation), x_2 (position along mass separation), dx_1 and dx_2 (size of the spot along the two dimensions, in millimeters), I (intensity, in absorbance units), V (volume, in absorbance units), and A (area, in squared millimeters) (see Note 1).
9. The positions (x_1 and x_2) and diameters (dx_1 and dx_2) of the spots are not scaled, whereas their volumes, intensities, and areas are normalized using maximum scaling normalization.
10. According to literature, approximate each spot by a Gaussian function (2, 26).

3.2. Strategy for the Simulation of Realistic Gel Electrophoresis Maps

1. The pseudocode flowchart of the proposed algorithm is reported in Fig. 1. In this representation, the crucial choices to be taken by the user and the most important steps performed by the program are reported.
2. The approach can be summarized in two main blocks: (1) generation of the first synthetic, i.e. simulated, map (FSM, namely the progenitor) and (2) generation of replications of the progenitor map which is used as a reference map. In both the blocks, the statistical characteristics of the 50 maps are used as a guide to produce realistic results: in the first block of the software, in order to obtain a progenitor map as similar as possible to a real map; in the second block, in order to produce replications that can correctly simulate the low reproducibility of SDS 2D-PAGE maps (see Note 2).

3.3. Generation of the Reference Map

1. All the steps that lead to the generation of the progenitor map are included in the pseudocode in the routine named "Generation of the First Synthetic Map" (FSM). This map represents the progenitor of all the maps belonging to this class.
2. First, choose the input parameters that define the type of map to be simulated, namely the number (N) of spots that the algorithm must produce on the simulated map, the size of the gels (usually 200×200 or 100×100 mm), and the image resolution (usually from 127 to 508 dpi) (see Note 3).

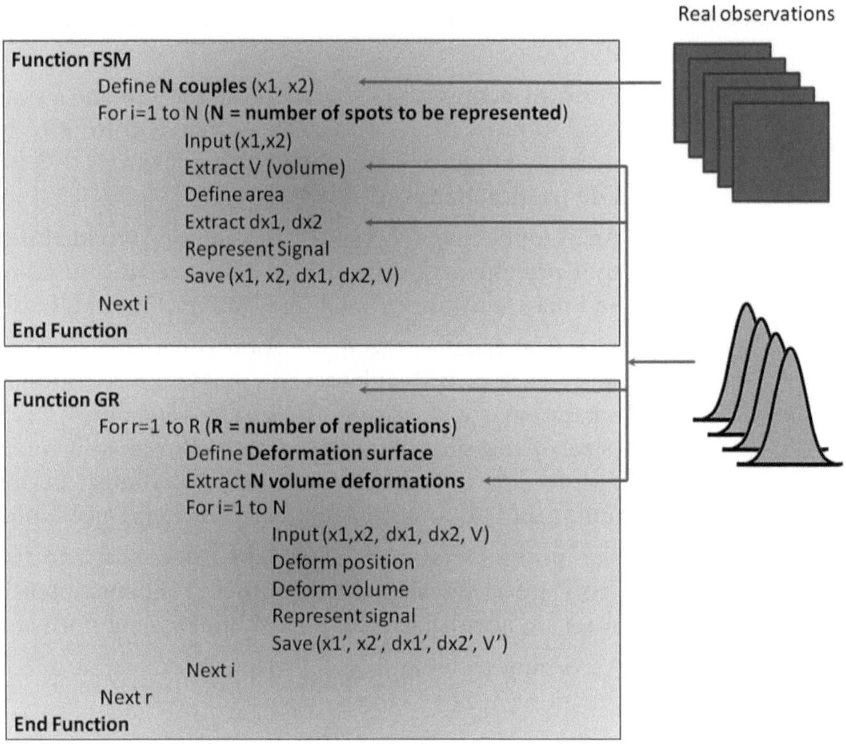

Fig. 1. Pseudocode of the algorithm used for the generation of simulated 2D-PAGE maps.

3. A simple way of generating realistic dispersions of the spots is to randomly extract them from a database of spots like the one we obtained by collecting the 20,000 experimental spots previously described. Once the position of the *N* spots has been defined, the next step consists in assigning them an intensity and a proper shape (area). This is a crucial phase in order to achieve realistic images. For this purpose, perform a statistical analysis, devoted to study the population characterizing intensities and areas of the experimental spots collected (in the present case we used the maps collected by our group along the years in previous researches). The two-dimensional distributions of the two diameters (dx_1, dx_2), and of peak areas and intensities, obtained from the database of experimental maps are reported in Fig. 2. The two-dimensional distribution of areas and intensities (Fig. 2a) shows a high abundance of points in the low left zone of the graph, namely small areas and small intensities. A few spots present average areas and intensity values. There is no data in the upper left and lower right corners of the graph, showing that spots with a large area and small intensity or small area and large intensity are not present in real maps. In general, the majority of the spots occupy a well-defined region that is probably related to the focusing efficiency of the separation procedure, as well as to

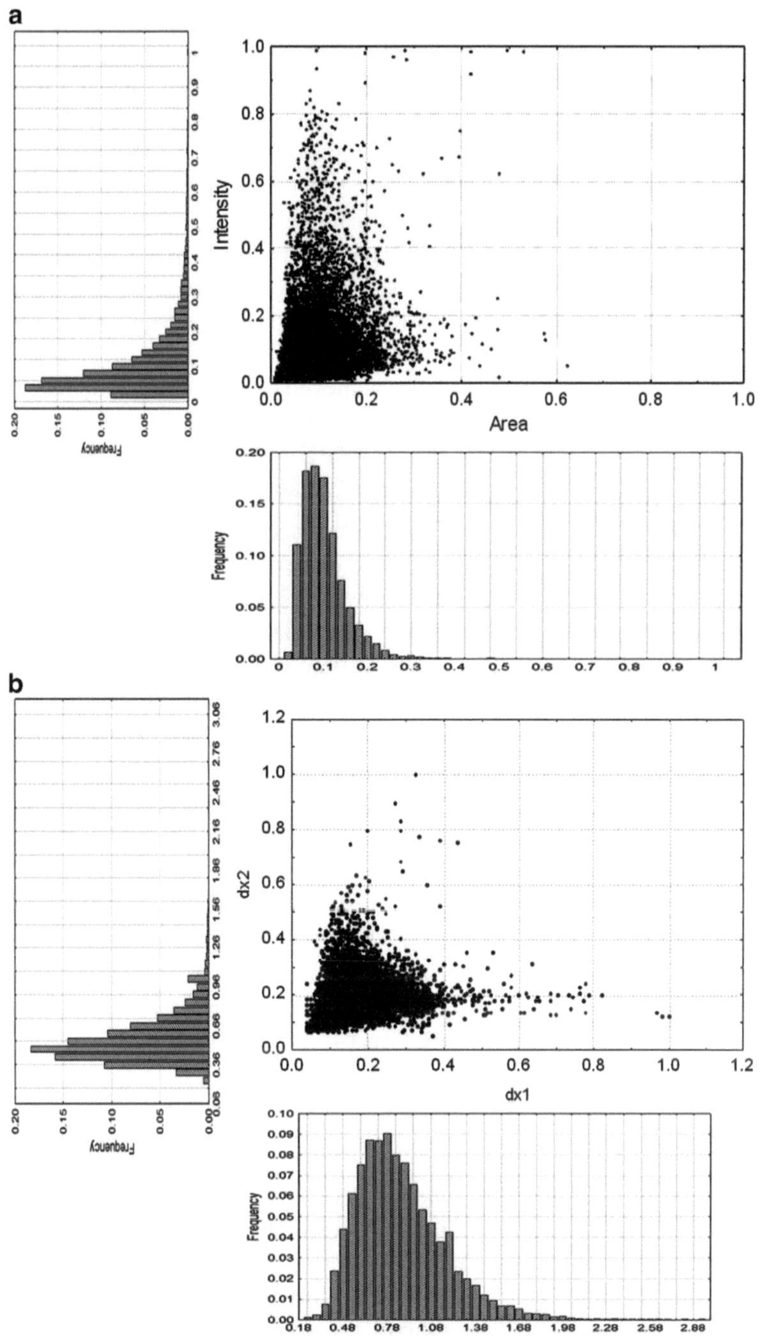

Fig. 2. (**a**) Scatter plot of the observed area and intensity: marginal distributions are represented close to the scatter plot. (**b**) Scatter plot of the observed diameters dx_1 and dx_2: marginal distributions are represented close to the scatter plot.

the concentration of the corresponding protein present in the sample (2, 26). There is generally an obvious relationship between the area of the spot and its volume that states that the volume can increase because of the spot area and/or of the spot intensity: when the volume increases because of an increase in the area, there is sometimes a phenomenon of saturation of the signal measured by the densitometer. In this case, typically only observed for large spots, there is a loss of the Gaussian shape of the signal which is truncated at the maximum possible value at the center of the spot. In the other case, typical of the majority of the spots and in particular of the medium and small size spots, the expected Gaussian shape is instead preserved. Given the saturation effect is due to an overload effect of the sample unto the strips, it was decided to extract the intensities from the population that does not show the saturation effect. In fact, if several saturated spots are present in a map, their elimination can be easily achieved experimentally by reducing the amount of sample deposited at the beginning of the separation procedure. Thus, the simulated map will not contain spots whose shape is not basically Gaussian.

4. For what concerns the size of the spots along the two electrophoretic dimensions, the marginal distribution of $\mathrm{d}x_1$ and $\mathrm{d}x_2$ are shown in Fig. 2b. As it can be observed, the two distributions are similar, both are right tailed but this tendency is less relevant than in the case of intensity and area. Values of intensities and areas are then randomly extracted from the observed populations. In this case, there are two possible approaches for the choice of the positions: (a) extract them from the database of observed values, using a simple random number generator (RNG); (b) obtain a random selection from the two distributions, using the rejection method (47, 48). In the present version of the software the first approach is chosen, but the second is viable and easy to implement.

5. Once position, size, area, and intensity of the spots are defined, calculate their volume by the bivariate Gaussian shape approximation (49) that leads to an elliptic shape spot, using the following equation:

$$f(x) = I\mathrm{e}^{\left(-1/2(\mathbf{x}-\mathbf{C})^{\mathrm{T}} S^{-1}(\mathbf{x}-\mathbf{C})\right)} \tag{1}$$

where vector $\mathbf{x}$ identifies the point of the image where the intensity must be evaluated ($\mathbf{x} = (x_1, x_2)$); $\mathbf{C}$ contains the coordinates of the center of the spot ($\mathbf{C} = (x_{c1}, x_{c2})$); I is the extracted intensity; and $\mathbf{S}$ contains information on the size and orientation of the axes of the elliptic spot: $\mathbf{S}$ is a diagonal matrix as the two electrophoretic dimensions can be normally considered to be independent. The values of $\mathbf{S}$ elements can be obtained by imposing that the Gaussian function (Eq. 1) respects the boundary constraint b of the intensity:

$$\begin{cases} b = I\mathrm{e}^{\wedge(-1/2(\mathrm{d}x_1/2)^2 S_{x1}^{-1})} \\ b = I\mathrm{e}^{\wedge(-1/2(\mathrm{d}x_2/2)^2 S_{x2}^{-1})}, \end{cases} \tag{2}$$

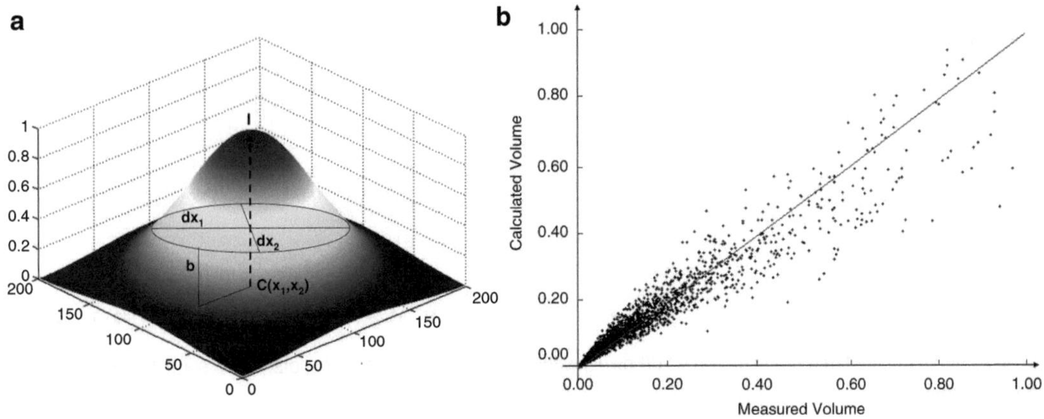

Fig. 3. Example of a Gaussian signal (**a**) with the parameters used in the algorithm. Plot of observed vs. calculated volumes (**b**). Reprinted from ref. (41) with permission from the American Chemical Society.

where S_{x1} and S_{x2} are the diagonal elements of **S**. The boundary intensity b is selected by the user and corresponds to the intensity on the visual edge of the spot, in fact from a physical point of view the spot has an asymptotic behavior that never reaches 0 (given the Gaussian approximation), so it must be cut in agreement with the densitometer capability to register an intensity signal different from the blank (Fig. 3a) (see Note 4).

The two values of **S** can then be obtained by the following equation:

$$S_x = -(d_x/2)^2 / (2\ln(b/I)), \tag{3}$$

where S_x is either S_{x1} or S_{x2} and d_x is either d_{x1} or d_{x2}. The volume of the spot is then

$$V = I\int_{x_1}\int_{x_2} \mathrm{e}^{\left(-1/2(\mathbf{x}-\mathbf{C})^{\mathrm{T}} S^{-1}(\mathbf{x}-\mathbf{C})\right)} = 2\pi I\sqrt{S}. \tag{4}$$

6. The volumes simulated with this approach show a good fit when compared to the ones present in the database (Fig. 3b) demonstrating that the model employed is satisfactory. Figure 3b shows that, for some large observed volumes, the calculated volume is smaller than expected: in these cases, a possible saturation effect could be present. This relationship could therefore be used for the identification of the presence of saturated spots. Generate a realistic map by applying the described procedure: this can be used as the starting point for generating a population of replications. Figure 4 shows a series of simulated maps with an increasing number of spots, obtained following the procedure described previously. As can be noticed they are very realistic.

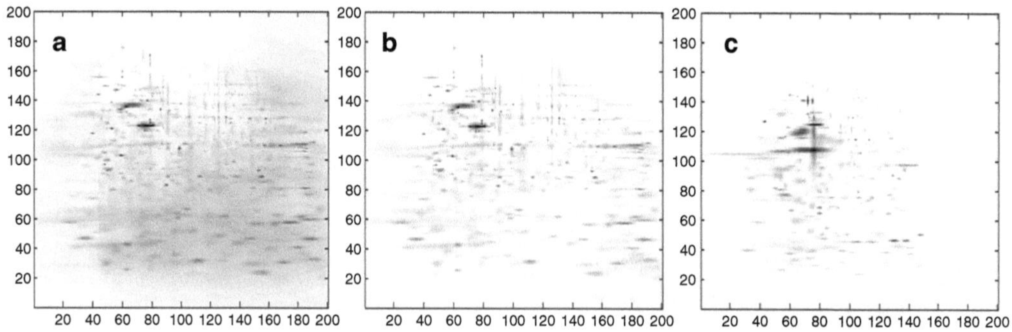

Fig. 4. Example of simulated maps of 200 × 200 pixels with an increasing number of spots: 1,119 spots (**a**), 560 spots (**b**), and 266 spots (**c**).

3.4. Generation of Replications

1. Use the same database of experimental spots (here the database of 20,000 spots is used) as a basis for the study and simulation of the distortions of volumes and positions in a series of replicated SDS 2D-PAGE maps of the same sample. The database contains the spots obtained in several replicated analyses of the same sample so that information on the reproducibility of the two-dimensional separation technique is surely present. The analysis of the experimental data shows that the principal aspects to be taken into consideration are the variability of position and volume of the spots. The former is probably due to several uncontrolled or badly controlled factors affecting the separation, such as non-homogenous electrical field, elongations of the gels, bad alignments, etc. All these factors can produce shifts of the spots (see Fig. 5a). The latter aspect is probably connected with the variability of the sample load unto the strips, the staining procedure and also the same factors that may affect the position of the spots.
2. Deepening the analysis of the position changes, take into consideration that the spots present in the maps analyzed in previous researches had been automatically matched by the algorithms implemented in some of the most used commercial software. The statistical analysis of the historical data suggests the possibility that some spots were not correctly matched by the automatic algorithm. In fact, the majority of the spot displacements are shown to belong to a normal distribution which can be expected as an effect of the central limit theorem, but there are some outliers that do not belong to this type of distribution. These spots can be easily identified by means of normal probability plots (18): eliminate the spots that diverge from normality from the study as they are probably caused by a mismatching problem of the automatic matching algorithm. Figure 6a shows the distribution obtained after the elimination of the outliers and confirms that the spot shifts are normally distributed.

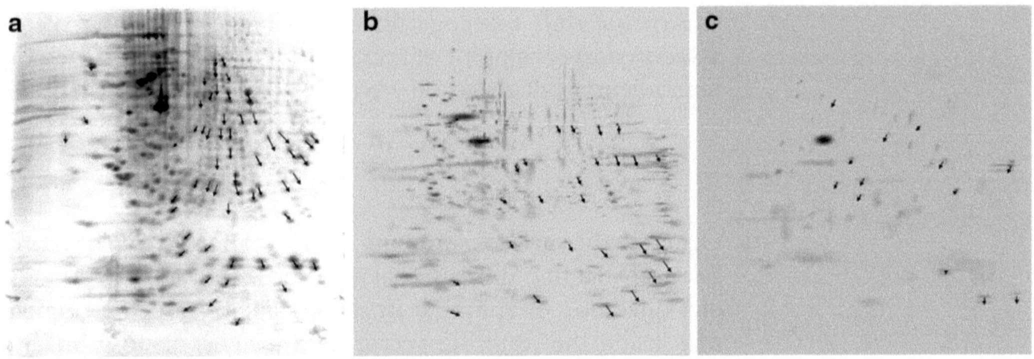

Fig. 5. SDS 2D-PAGE real separation: two replicated maps superimposed (**a**). Some displacements of the same proteins are presented using vectors. Four SDS 2D-PAGE simulated separations: two superimposed images deformed using a vector field obtained from a spherical surface (**b**); the same two simulated maps deformed with a vector field obtained with a spherical surface of double radius (**c**).

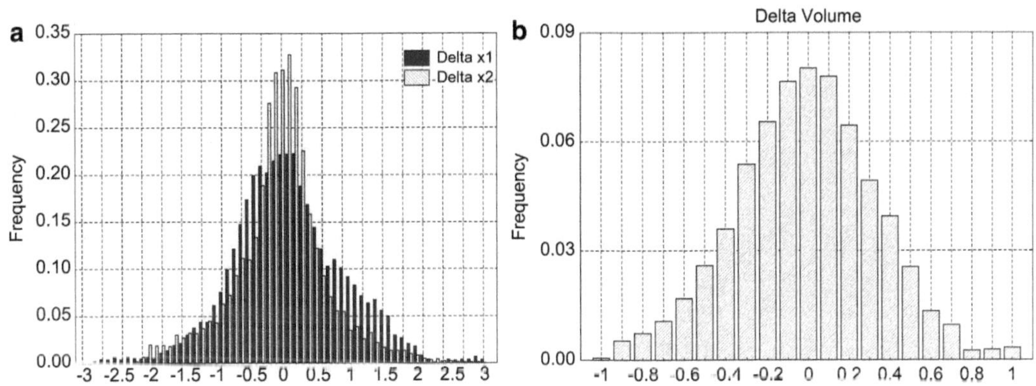

Fig. 6. (**a**) Distributions of the observed differences with respect to the spot position along x_1 and x_2, expressed in millimeters; (**b**) distribution of the observed relative difference with respect to the volumes.

3. Another important remark can be drawn by considering the position changes that characterize a given map: it can be observed that the shifts are not random (Fig. 5a). The displacement of spots close in the map shows a certain correlation for what concerns their direction and size (Fig. 5a). The deformation seems to present a sort of "organization" so that spots close on the same map show displacements similar in direction and size. It is very rare to observe shifts of spots close on the map opposite in size and direction and a possible cause could be again a mismatching problem. A random shift generated that does not take these indications into consideration could accidentally invert the positions of some spots if the shift is sufficiently large. For a correct simulation of the shifts of the spots a different procedure is employed that maintains the local correlation showed by the displacements: a shift vector field is introduced characterized by correlated direction and

size of the shift topologically close on the map. The vector associated to each point of these fields is obtained as the countergradient $\vec{v}$ calculated on a concave or convex surface:

$$\vec{v} = -\nabla f\left(\frac{\partial f}{\partial x_1}, \frac{\partial f}{\partial x_2}\right). \quad (5)$$

4. Generate the surfaces summing a similar effect of two or more, relatively flat two-dimensional functions that can be, for example, quadratic or spherical in shape. Then, warp the original map using the resulting vector field (only spot positions are warped, the form and the intensity are preserved), as shown in Fig. 7a–c. Extract the center and the curvature of the functions used for the generation of the surface from a uniform population of possible values which can be defined by the user. Use an RNG to select the quadratic function parameters, so that every replicated map contains different deformations with respect to the others.
5. The curvature of the surface that generates the displacement vector field controls and determines the deformation that is applied to every map as shown in Fig. 7a–c, where spherical surfaces of different size and internal and external positions of the center of the function are compared. In Fig. 5b and 5c, two examples of two matched maps obtained using different spherical surfaces are represented. As it can be noticed the result is quite realistic, very similar to a real situation reported in Fig. 5a.

   The procedure that performs the deformation of the spots position is indicated in the pseudocode (Fig. 1) as "Position Deformation".
6. The second deformation effect affects the spot volumes. Figure 6b shows the histogram of the relative variation of volumes of matched spots in replicated maps. The distribution is approximately normal around zero and with a standard deviation of 0.3. Using a normal distribution with the same parameters it is possible to generate a population of volume deformations similar to the experimental one. Changing the parameters of this distribution it is possible to increase or decrease the variations of the volumes in the artificial images. No local correlation of the volumes is taken into consideration since it is not observed in real cases.
7. Then, shift every spot present in the reference map according to "Position Deformation" and change its volume according to the function "Deform Volume" that modifies the volume through a multiplicative factor that is extracted from the proper normal population previously described.

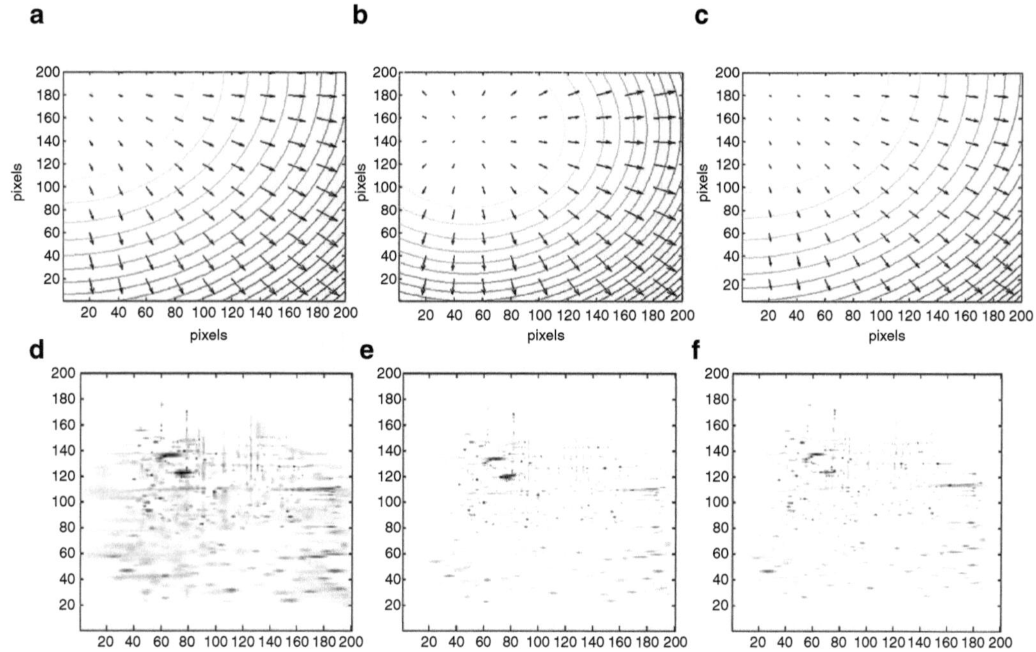

Fig. 7. Three examples of vector fields associated to spherical surfaces used for the deformations. (**a**) Center of the sphere positioned outside the map. (**b**) Center of the sphere positioned inside the map. (**c**) Center of the sphere as in (**a**) and curvature more accentuated with respect to fields (**a**) and (**b**). Three artificial maps with deformation of spot volumes: original map with 0% of deformation (**d**); the same spots as in map (**d**) with a deformation of 10% (**e**); the same spots as in map (**d**) with a deformation of 30% (**f**).

8. When the new volume is defined for all spots present on the simulated map, it is possible to calculate the new spot intensities and diameters using the following equation:

$$V' = I' 2\pi \sqrt{s'_{x1} s'_{x2}}, \tag{6}$$

where s'_{x1} and s'_{x2} are the new diagonal elements of $\mathbf{S}$, and I' and V' are the modified intensity and volume, respectively. The new parameters can be estimated using the equations presented in group 8 under the hypothesis that the spots uniformly increase or reduce their size:

$$\begin{aligned} V' &= Ik2\pi\sqrt{ks_{x1}ks_{x2}} = Ik^2 2\pi\sqrt{s_{x1}s_{x2}} \\ I' &= kI \\ s'_{x1} &= ks_{x1} \\ s'_{x2} &= ks_{x2} \end{aligned} \tag{7}$$

where k is a multiplicative coefficient. This procedure permits the extraction of plausible values of area and intensity that respect the volume distribution and preserve the shape of the original spot. Three examples of maps with modified spot volumes are reported in Fig. 7d–f. The first map is the reference

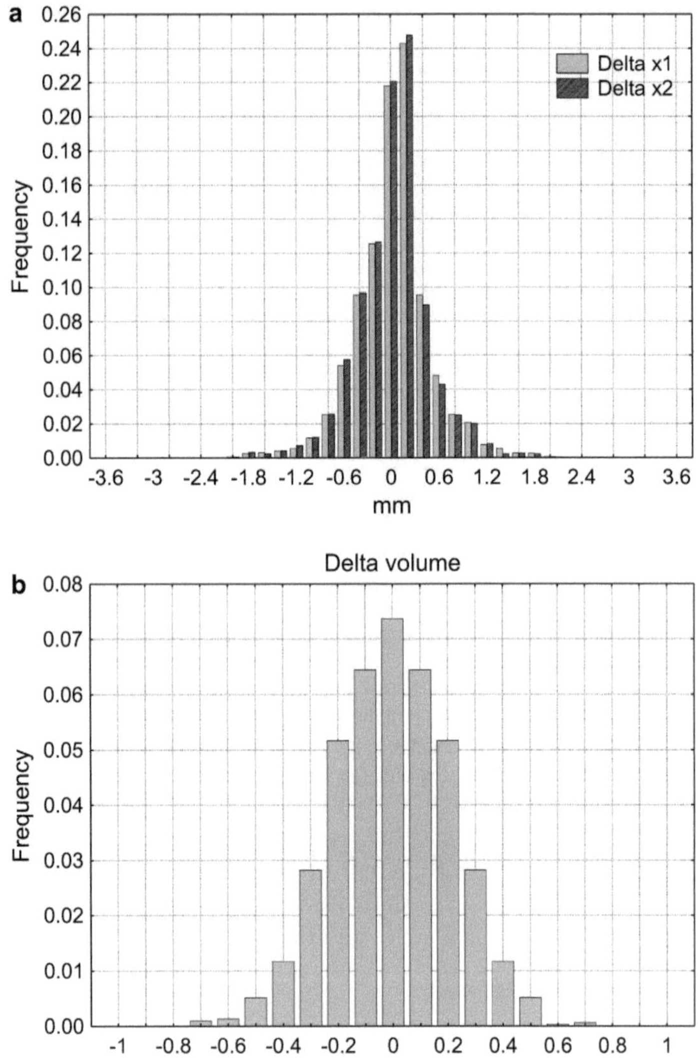

Fig. 8. Distribution of the differences relative to 30 simulated maps: distributions of the differences with respect to the spot position along x_1 and x_2 (the differences are expressed in millimeters) (**a**) and distribution of the observed relative difference with respect to the volumes (**b**).

map while the other images contain spots that present volume variations of 10 and 30%. Some spots have changed their size and intensity according to Eq. 8. This effect is more evident on observing the third figure (*f*), where many spots present different intensities and base areas but preserve their original shape. Figure 7f presents a level of deformation of the spot volumes that is very similar to the one observed in real maps.

9. Figure 8a shows the distribution of the displacements of position of the corresponding spots for a dataset of 30 simulated maps. The simulated maps are 400 × 400 pixels and each pixel

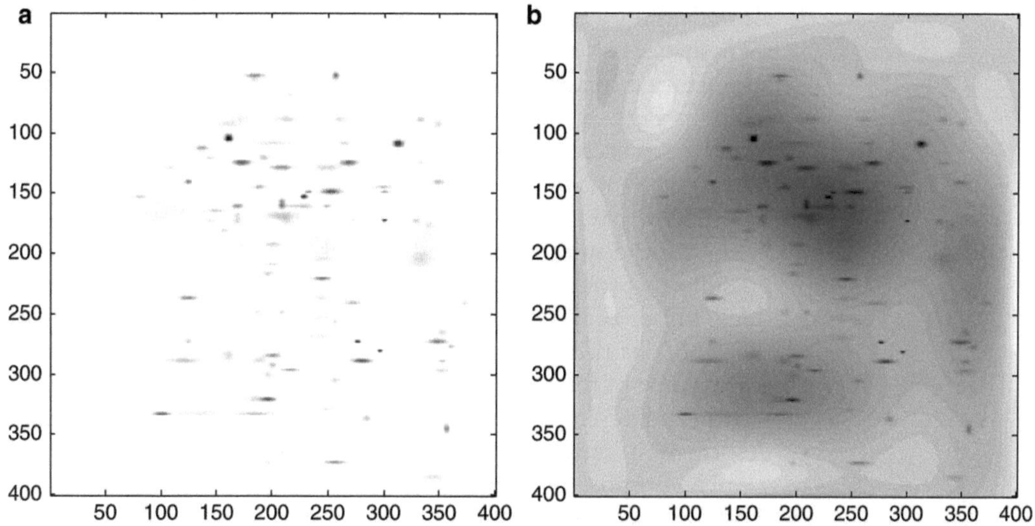

Fig. 9. Two examples of virtual maps: a virtual map without background (**a**) and a virtual map added to a background generated using a weighted spline (**b**).

is approximately 50 μm. Figure 8b shows the distribution of the relative variation of the volume of the spots in the same dataset. As it can be noticed, the distributions presented in Fig. 8, calculated on the virtual maps, are similar to the distributions calculated from real data (Fig. 6) (see Note 5).

3.5. Background

Once the virtual set of maps is generated, in order to make the simulation even more realistic, a background noise can be added. A simple way for generating a background is to use a weighted B-spline algorithm (50). The procedure can be described as follows:

1. Define a matrix **W** ($n_1 \times n_2$) of random values in the range (0, 1) where 1 indicates a large background noise and 0 indicates no background. These values are the seeds for the generation of the overall background noise.
2. Create the background **M** by fitting the seeds using a cubic spline. Once the background is created it can be added to the simulated proteomic map.

Figures 9a and 9b show, respectively, a virtual map before and after the addition of a background surface **M**. The resulting image is a simple sum of the virtual map and the artificial background (see Note 6).

3.6. Final Considerations

Using the software described in this chapter it is possible to obtain very realistic images that simulate experimental SDS 2D-PAGE maps. Figures 5a and 5b show an experimental and an artificial image. It can be noted that the artificial image is very similar to the real one except for the background that in real images could be cleared using a spline (51) technique. The position of the spots, their shape, volume, and intensity are extremely similar in the real and artificial images. The superimposition of the SDS 2D-PAGE maps in Figs. 5b and 5c (some displacements of the spots are indicated as vectors) highlights how the deformation introduced in the artificial images is similar to the one observed in real cases. Varying the parameters related to the image deformations (vector field of the displacements and rate of volume deformation) it is possible to produce any desired simulation with different reproducibility of the SDS 2D-PAGE technique. Varying the database that contains the reference values for the position, intensity, area, and volume of the spots it is possible to reproduce cases characterized by a different spot overlapping (52) and then apply the image deformation steps.

The proposed strategy is to be intended as a guideline for the representation of sets of maps that undergo changes similar to the real separations: this strategy can be therefore a good approach for the generation of artificial sets of data for developing and testing new image analysis techniques and/or multivariate statistical classification algorithms. The algorithm presented here can be useful in different fields of research where well-defined sets of artificial 2D maps are needed: in image recognition, for testing the automatic procedures developed; in diagnostics, for testing the number of replications required to provide reliable classification models; in bioinformatics, for evaluating algorithms for the correction of 2D maps (in terms of the artifacts introduced in the maps by background removal or warping procedures); and for testing manual or automatic matching procedures. Good virtual maps can also be a precious help in all cases where a fast and economic evaluation of new algorithms and software is needed. All the variables used for the generation of the virtual maps (following the proposed pseudocode) can be manipulated and all the features of the final datasets of images can be controlled and measured. In conclusion, the proposed strategy offers the opportunity of generating realistic SDS 2D-PAGE separations using a simple sequence of steps. The proposed pseudocode is quite simple and its implementation using commercial programming languages is easy.

4. Notes

1. It is obviously possible to use an alternative database, but this would require to perform the same statistical analysis that is performed in the present case, which permits to obtain realistic simulated maps and is the basis for a correct consideration of the statistical features of the typical SDS 2D-PAGE proteomic datasets.

2. Instead of using the procedure of generation of the progenitor map, this can be directly input by the user, and only the replication module can be used, if necessary.
3. In spite of the great and always increasing power of computers, image datasets are usually rather big and limitations to their size and resolution can still be a problem. Resolution of the scanner and size of the images can be very important in image analysis and this is the reason they were included as parameters of the simulation.
4. The maps presented in this chapter were obtained by using a *b*-value equal to 1% of the maximum intensity (I_{max}) observed in the map.
5. The experimental maps present some other anomalies, like the simulation of artifacts, background, spurious signals, etc., which could be also simulated, but the two effects considered here are surely the most important causes of misclassification, together with background variations.
6. In a more sophisticated way, it is possible to define a threshold intensity that mimes saturated spots. Pixel values of the resulting images that exceed the threshold (cut-off level) are set to the threshold value.

Acknowledgements

The authors gratefully acknowledge the collaboration of Prof. Pier Giorgio Righetti (Polytechnic of Milan, Italy) and Dr. Daniela Cecconi (University of Verona, Italy) who provided the proteomic datasets used in this study.

References

1. O'Farrel PH (1975) High resolution two dimensional electrophoresis of proteins. J Biol Chem 250:4007–4021
2. Westermeier R, Naven T (2002) Proteomics in Practice. Wiley, Freidburg
3. Bhanu B (1986) Automatic target recognition: state of the art survey. IEEE Trans Aerosp Electron Syst 22:364–379
4. Duda RO, Hart PH (1970) Pattern classification and scene analysis. Wiley, New York
5. Taniguchi M, Mokuno Y, Kintaba K et al (1999) Correlation filter design for classification of road sign by multiple optical correlators. Proc SPIE 3804:140–147
6. Song X, Wyrwicz AM (2009) Unsupervised spatiotemporal fMRI data analysis using support vector machines. Neuroimage 47: 204–212
7. Golnabi H, Asadpour A (2007) Design and application of industrial machine vision systems. Robot Comput Integr Manuf 23: 630–637
8. Foran DJ, Comaniciu D, Meer P et al (2000) Computer-assisted discrimination among malignant lymphomas and leukemia using immunophenotyping, intelligent image repositories, and telemicroscopy. IEEE Trans Inf Technol Biomed 4:265–273
9. Antoine J, Bouyoucef K, Vanderhheyst P et al (1995) Target detection and recognition using two dimensional continuous isotropic and anisotropic wavelets. Proceedings of automatic

object recognition V. SPIE symposium on optical engineering/aerospace sensing and dual use photonics, Orlando, FL
10. Der S, Chellappa R (1997) Probe-based automatic target recognition in infrared imagery. IEEE Trans Image Process 6:92–102
11. Kramer A, Perschbacer D, Johnston R et al (1993) Relational template matching algorithm for FLIR automatic target recognition. Proc SPIE 1957:29–37
12. Færgestad EM, Rye M, Walczak B et al (2007) Pixel-based analysis of multiple images for the identification of changes: a novel approach applied to unravel proteome patters of 2-D electrophoresis gel images. Proteomics 7:3450–3461
13. Daszykowski M, Stanimirova I, Bodzon-Kulakowska A et al (2007) Start-to-end processing of two dimensional gel electrophoretic images. J Chromatogr A 1158:306–317
14. Mukundan R, Ramakrishnan KR (1998) Moment functions in image analysis. World Scientific, London
15. Marengo E, Robotti E, Bobba M et al (2008) A new method of comparing 2D-PAGE maps based on the computation of Zernike moments and multivariate statistical tools. Anal Bioanal Chem 391:1163–1173
16. Marengo E, Bobba M, Liparota MC et al (2005) Use of Legendre moments for the fast comparison of two-dimensional polyacrylamide gel electrophoresis maps images. J Chromatogr A 1096:86–91
17. Massart DL, Vanderginste BGM, Buydens LMC et al (1998) Handbook of chemometrics and qualimetrics: part A. Elsevier, Amsterdam
18. Massart DL, Vanderginste BGM, Deming SM et al (1988) Chemometrics: a textbook. Elsevier, Amsterdam
19. Heidema AG, Thissen U, Boer JMA et al (2009) The association of 83 plasma proteins with CHD mortality BMI, HDL-, and total-cholesterol in men: applying multivariate statistics to identify proteins with prognostic value and biological relevance. J Proteome Res 8:2640–2649
20. Rodriguez-Pineiro AM, Rodriguez-Berrocal FJ, de la Cadena MP (2007) Improvements in the search for potential biomarkers by proteomics: application of principal component and discriminant analyses for two-dimensional maps evaluation. J Chromatogr B 849: 251–260
21. Marengo E, Robotti E, Bobba M et al (2008) Application of partial least squares—discriminant analysis and variable selection procedures: a 2D-PAGE proteomic-based study. Anal Bioanal Chem 390:1327–1342
22. Martens H, Naes T (1989) Multivariate calibration. Wiley, London
23. Karp NA, Griffin JL, Lilley KL (2005) Application of partial least squares discriminant analysis to two-dimensional difference gel studies in expression proteomics. Proteomics 5: 81–90
24. Eisenbeis RA, Avery RB (1972) Discriminant analysis and classification procedures: theory and applications. D.C. Heath and Company, Lexington, MA
25. Klecka WR (1980) Discriminant analysis. Sage, Beverly Hills
26. Hamdan M, Righetti PG (2005) Proteomics today. Wiley, Hoboken, NJ
27. Javidi B (2002) Image recognition and classification. Marcel Dekker Inc, New York
28. Thormann W, Mosher RA, Bier M (1986) Experimental and theoretical dynamics of isoelectric focusing: elucidation of a general separation mechanism. J Chromatogr 351:17–29
29. Mosher RA, Thormann W, Bier M (1988) Experimental and theoretical dynamics of isoelectric focusing. II. Elucidation of the impact of the electrode assembly. J Chromatogr 436:191–204
30. Mosher RA, Dewey D, Thormann W et al (1989) Computer simulation and experimental validation of the electrophoretic behavior of proteins. Anal Chem 61:362–366
31. Mosher RA, Thormann W (1990) Experimental and theoretical dynamics of isoelectric focusing: IV. Cathodic anodic and symmetrical drifts of the pH gradient. Electrophoresis 11: 717–723
32. Mosher RA, Thormann W (2002) High-resolution computer simulation of the dynamics of isoelectric focusing using carrier ampholytes: the post-separation stabilizing phase revisited. Electrophoresis 23:1803–1814
33. Thormann W, Huang TM, Pawliszyn J et al (2004) High-resolution computer simulation of the dynamics of isoelectric focusing of proteins. Electrophoresis 25:324–337
34. Thormann W, Mosher RA (2006) High-resolution computer simulation of the dynamics of isoelectric focusing using carrier ampholytes: focusing with concurrent electrophoretic mobilization is an isotachophoretic process. Electrophoresis 27:968–983
35. Mosher RA, Bier M, Righetti PG (1986) Computer simulation of immobilised pH gradients at acidic and alkaline extremes: a quest for extended pH intervals. Electrophoresis 7:59–66
36. Tonani C, Faupel M, Righetti PG (1991) Isoelectric membrane simulator: a computational

approach for isoelectric immobiline membranes. Electrophoresis 12:631–636

37. Tonani C, Righetti PG (1991) Immobilized pH gradient simulator: an additional step in pH gradient engineering: I. Linear pH gradients. Electrophoresis 12:1011–1021
38. Lam HT, Josserand J, Lion N et al (2007) Modeling the isoelectric focusing of peptides in an OFFGEL multicompartment cell. J Proteome Res 6:1666–1676
39. Pietrogrande MC, Marchetti N, Tosi A et al (2005) Decoding two-dimensional polyacrylamide gel electrophoresis complex maps by autocovariance function: a simplified approach useful for proteomics. Electrophoresis 26:2739–2748
40. Pietrogrande MC, Marchetti N, Dondi F et al (2006) Decoding 2D-PAGE complex maps: relevance to proteomics. J Chromatogr B 833:51–62
41. Marengo E, Demartini M, Robotti E et al (2009) A new algorithm for the simulation of sodium dodecil sulfate two-dimensional polyacrylamide gel electrophoresis data sets. J Proteome Res 9:1864–1872
42. Marengo E, Robotti E, Bobba M et al (2006) Multivariate statistical tools applied to the characterization of the proteomic profiles of two human lymphoma cell lines by two-dimensional gel electrophoresis. Electrophoresis 27: 484–494
43. Marengo E, Robotti E, Cecconi D et al (2004) Identification of the regulatory proteins in human pancreatic cancers treated with Trichostatin A by 2D-PAGE maps and multivariate statistical analysis. Anal Bioanal Chem 379:992–1003
44. Garrels JI (1989) The Quest system for quantitative analysis of two-dimensional gels. J Biol Chem 264:5269–5282
45. PdQuest User Guide, version 6 (1998) Bio Rad Laboratories, Hercules, CA
46. Appel RD (1997) Melanie II—a third-generation software package for analysis of two-dimensional electrophoresis images: I. Features and user interface. Electrophoresis 18: 2724–2734
47. Press WH, Flannery BP, Teukolsky SA et al (1999) Numerical Recipes in C. Cambridge University Press, London
48. Knuth DE (1981) Seminumerical algorithms, vol 2, 2nd edn, The art of computer programming. Addison-Wesley, Upper Saddle River, NJ
49. Stuart A, Ord K (1994) Kendall's advanced theory of statistics distribution theory. Oxford University Press, London
50. Eilers PHC, Currie ID, Durban M (2006) Fast and compact smoothing on large multidimensional grids. Comput Stat Data Anal 50:71–76
51. Kaczmarek K, Walczak B, de Jong S et al (2005) Baseline reduction in two dimensional gel electrophoresis images. Acta Chromatogr 15:82–95
52. Pietrogrande MC, Marchetti N, Dondi F et al (2002) Spot overlapping in two-dimensional polyacrylamide gel electrophoresis separations: a statistical study of complex protein maps. Electrophoresis 23:282–291

Chapter 35

Isolation of Proteins from Polyacrylamide Gels

Irene Michalk, Stefanie Koristka, Claudia Arndt, and Michael Bachmann

Abstract

Minute amounts of proteins are required for immunization of mice for the development of antibodies including monoclonal antibodies. Here, we describe a rapid procedure for the isolation of proteins from polyacrylamide gels after sodium dodecyl sulfate polyacrylamide gel electrophoresis in sufficient amounts for immunization of animals.

Key words: Polyacrylamide gels, Isolation of proteins

1. Introduction

For the development of polyclonal or monoclonal antibodies (mAbs), enriched or better homogenously purified antigens are required. The purer the antigen the easier to establish a polyclonal serum or mAbs. Nowadays, recombinantly expressed proteins are useful starting materials for immunizations of animals. Frequently, proteins are recombinantly expressed in *Escherichia coli* as His-tagged fusion proteins which can efficiently be purified using an affinity chromatography step on Nickel NTA columns. Although proteins purified thus are highly enriched, they may still be contaminated with traces of other bacterial antigens, degradation, or truncated expression products. The immune system is by far more sensitive than any staining procedure and can therefore respond easily against such contaminants. Antigens for immunizations can be further enriched by separation on and elution from SDS-polyacrylamide gels.

Here, we describe a rapid and simple method for the isolation of proteins for immunization of mice and the preparation of mAbs based on sodium dodecyl sulfate polyacrylamide gel electrophoresis (SDS-PAGE).

Biji T. Kurien and R. Hal Scofield (eds.), *Protein Electrophoresis: Methods and Protocols*, Methods in Molecular Biology, vol. 869,
DOI 10.1007/978-1-61779-821-4_35, © Springer Science+Business Media, LLC 2012

2. Materials

1. Mortar and pestle.
2. Plastic columns: e.g., Mobicol tubes (MoBiTec, Göttingen, Germany).
3. Phosphate-buffered saline (PBS).
4. Distilled water.
5. SDS-polyacrylamide gels (see Chapter 22).
6. SDS-sample buffer.
7. Coomassie Brilliant Blue for staining and destaining solutions (see Chapter 40).
8. Nitrocellulose for immunoblotting.
9. Anti-his mAb and anti-mouse antibodies conjugated with alkaline phosphatase.

3. Methods

3.1. SDS-PAGE

For immunization of a mouse about 30–50 μg of antigen is sufficient per shot. Therefore, the antigen of interest has to be separated on a preparative polyacrylamide gel as schematically shown in Fig. 1a. In dependence on the molecular weight and the epitope of the antigen, the size and percentage of acrylamide in the gel and also the procedure has to be adapted. However, adaptations are easy and not critical. Both, native as well as denaturing gel electrophoresis conditions can be applied (see Chapters 22 and 40). The aim of the presented experiment was to separate the components of a protein complex consisting of a 50 and a 29 kDa protein.

1. For this experiment use a 10% discontinuous SDS-polyacrylamide gel.
2. Dissolve the sample containing about 200 μg protein mixture in SDS-sample buffer and denature by heating for 3 min to 95°C.

3.2. Staining of the Gels and Identification of the Protein of Interest

1. Stain the gels with Coomassie Brilliant Blue (see Chapter 40) for the identification of the protein of interest.
2. Destain the gels with methanol and acetic acid (see Chapter 40).
3. Destain the gel in distilled water (see Note 1) for several times (at least 5 × 100 mL for 1 h each) to remove methanol and acetic acid out of the gel. In the case of the presented work, an unstained gel was used.

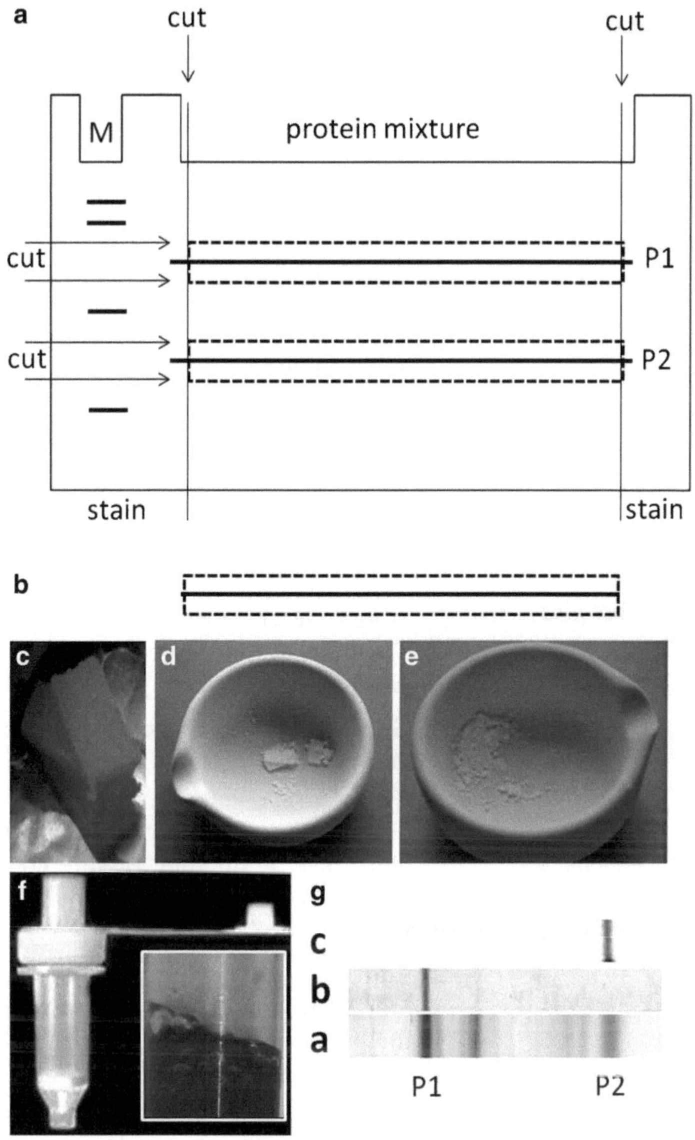

Fig. 1. Example of a protein isolation procedure. (**a**) For separation of the protein complex containing several His-tagged proteins [*see* also (**g**), *lane a*] we prepared a preparative 10% polyacrylamide gel. (**b**) The unstained gel was cut according to the expected position based on prestained molecular weight markers as indicated which resulted in two gel samples containing the respective protein. (**c**) The gel samples were frozen at −80°C over night. (**d**) After transfer of the gel (**c**) to a mortar and (**d**, **e**) the gel was ground to a fine powder. The powder was transferred into a suitable plastic column [e.g., mobicol tube, (**f**)]. [(**f**), *inset*] Elution buffer (here for visualization blue SDS-PAGE sample solution was used) was added dropwise until re-swelling of the gel. The protein samples were then eluted by adding further elution buffer. (**g**) The purity and elution efficiency was then checked by SDS-PAGE and immunoblotting. The two proteins P1 (50 kDa) and P2 (29 kDa) present in the starting material (*lane a*) were sufficiently separated and purified by the gel purification step (*lanes b, c*). The proteins were detected by an anti-His antibody followed by an anti-mouse antibody conjugated with alkaline phosphatase. *M* marker proteins.

3.3. Cutting Out the Protein Gel Region

1. Identify the region containing the protein of interest. For this purpose, cut small lanes at either side (and if necessary also in the middle of the gel) and stained as schematically shown in Fig. 1a. Alternatively, the positions of the proteins can be roughly estimated relative to the positions of the molecular weight markers.
2. Based on the identified positions of the proteins of interest, cut suitable gel fragments from the gel with a scalpel or razor plate (Fig. 1a, b). As we did not stain the gel before cutting, we cut relatively wide (5–7 mm) gel regions out of the gel where we expected to run the respective protein relative to the position of prestained marker proteins.
3. Stain the remaining gel with Coomassie Brilliant Blue to see that the proteins were correctly removed from the gel (data not shown).

3.4. Grinding of the Polyacrylamide Gel to a Fine Powder (See Note 2)

1. Freeze the isolated gel fragments containing the proteins overnight at −80°C.
2. Transfer each frozen gel (Fig. 1c) into a mortar (Fig. 1d). The mortar and pestle was cooled down overnight at −20°C (see Note 2).
3. Grind the gel to a fine powder using the pestle (Fig. 1d, e). Perform the grinding on dry ice.

3.5. Elution of the Proteins from the Gel Powder

1. Transfer the resulting gel powder (Fig. 1e) containing the protein into a suitable plastic column, e.g., Mobicol tube (MoBiTec, Göttingen, Germany; http://www.mobitec.com; Fig. 1f).
2. Add SDS-PAGE sample buffer dropwise containing bromophenol blue (see Note 3) after transfer of the powder into the column. The powder swells, reforming a blue gel in the column (see Fig. 1f, insert).
3. Elute the protein by passing further SDS-PAGE sample solution (2×, 500 μL) through the column comparably to a gel filtration (see Note 4). Using Mobicol tubes the elution procedure can be facilitated either by the pressure of a syringe or by centrifugation of the column in an Eppendorf tube using an Eppendorf centrifuge (5 min, 10,000 × *g*). The collected eluates contained the respective protein (Notes 5).
4. After gel isolation confirm the separation of the purified proteins by a further SDS-PAGE (Fig. 1g).
5. Aliquots of samples separated on 10% SDS-polyacrylamide gels were transferred to nictrocellulose membrane. As both antigens were tagged with 6xHis, the isolated antigens were detected with a murine monoclonal antibody directed to the

His-tag and an anti-mouse antibody conjugated with alkaline phosphatase. Figure 1g (lane a) shows the starting material. As shown in Fig. 1g, the gel purification procedure resulted in an efficient separation of the 50 kDa (Fig. 1g, lane b) and the 29 kDa protein (Fig. g, lane c).

4. Notes

1. All solutions should be prepared in water that has a resistance greater than 18 MΩ and total organic content of less than five parts per billion. This standard is referred to as "water" in this text.
2. *Please cool down the mortar and pestle overnight in a freezer to at least –20°C. Handle with care and wear safety goggles! Mortar and pestle can break under these conditions especially if they were not cooled down. This can lead to severe injuries!*
3. For documentation we used sample buffer solution (containing bromophenol blue) for elution of the gel. For recovery of protein useful for immunization of animals use PBS instead. The amount of buffer requested for the formation of the gel column depends on the amount of gel powder. Add your selected buffer dropwise until a clear gel with little, if any, supernatant has formed in the column.
4. Usually, two to three times the gel volume in the column are sufficient for elution (depends on the molecular weight of the protein and the pore size of the gel). If necessary, the buffer volume has to be increased up to ten times of the gel volume.
5. After gel elution the samples should be dialyzed (e.g., against PBS) prior to immunization of animals.

Acknowledgments

Our special thanks for excellent technical support to Christin Gräfe, Livia Schulze, and Barbara Uteß (Institute Immunology, Dresden, Germany).

Chapter 36

Gel Drying Methods

Slava Stamova, Irene Michalk, Holger Bartsch, and Michael Bachmann

Abstract

For some instances, protein gels need to be dried after SDS-PAGE, for example, if autoradiography should be performed from radioactive-labeled proteins after their separation on SDS-polyacrylamide gels. Another reason may be to simply store the gel in the laboratory book. Aside from expensive commercial solutions, especially for storage of the dried gel in the lab book, the simple and cheap drying protocol here presented may be sufficient.

Key words: Polyacrylamide gels, Proteins, Gel drying

1. Introduction

After separation of metabolically labeled proteins by SDS-PAGE, prior to autoradiography, a gel drying step is required to enhance the sensitivity and the sharpness of the bands. Especially for this purpose, one of the commercially available vacuum heat dryers in combination with a vacuum pump will certainly fulfill this job. However, such an equipment is not necessary if, for example, a simple Coomassie Brilliant Blue-stained nonradioactive gel needs to be dried in order to just store it in the laboratory book. Here we describe a simple and cheap procedure for drying of polyacrylamide gels as previously described for example by Krishnan (1). For this purpose, the gel will be mounted between two cellophane sheets which are kept together by two acrylic frames as schematically shown in Fig. 1.

Biji T. Kurien and R. Hal Scofield (eds.), *Protein Electrophoresis: Methods and Protocols*, Methods in Molecular Biology, vol. 869, DOI 10.1007/978-1-61779-821-4_36, © Springer Science+Business Media, LLC 2012

1A

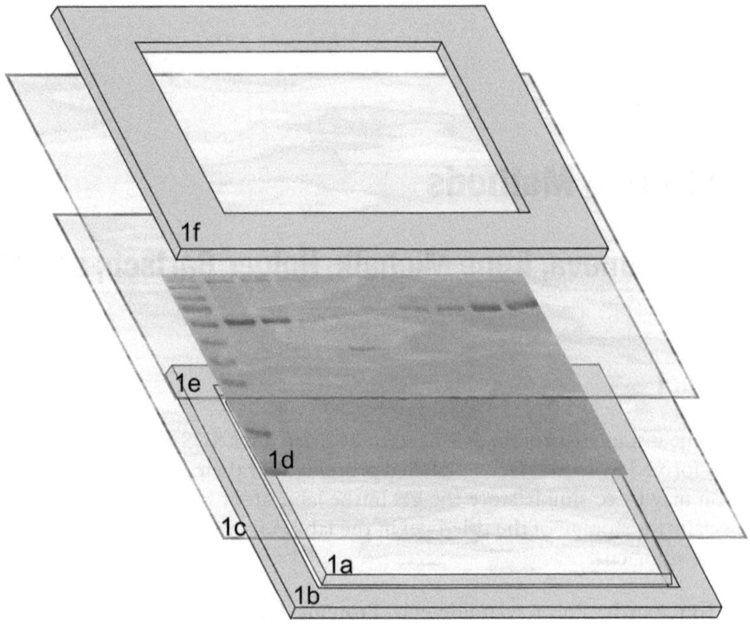

1B

Fig. 1. Gel drying apparatus. Schematic drawing of the components (**A**) or the mounted apparatus (**B**).

2. Materials

1. Equilibration solution: Add 250 mL ethanol and 3 mL glycerol (see Note 1) to a 1-L measuring cylinder. Add deionized water to 1 L (see Note 2).
2. Cellophane membrane.
3. Binder clips (6×).
4. Two acrylic frames and optionally a plate (glass or acryl) fitting (same thickness) inside one of the frames (see Note 3).

3. Methods

1. For equilibration, place the gel into about 100 mL of equilibration solution for 1 h (see Note 4).
2. Place the glass or acrylic plate (Fig. 1A, a) inside of the lower acrylic frame (Fig. 1A, b).
3. Wet two cellophane sheets that are larger than the outer frame.
4. One of the cellophane sheets (Fig. 1A, c) is spread over the first lower acrylic frame (Fig. 1, b).
5. Pour 5–10 mL of equilibration solution onto the membrane.
6. Place the gel (Fig. 1A, d) onto the cellophane membrane (Fig. 1A, c).
7. The second cellophane membrane (Fig. 1A, e) is now carefully placed over the gel and the first membrane.
8. Air bubbles and the surplus of equilibration solution are removed by carefully rolling a pipette over the membrane/gel sandwich.
9. Place the second frame on top of the sandwich (Fig. 1A, f) and fix it with the binder clips (see Note 5).
10. Remove the inner glass or acrylic plate from the final sandwich (Fig. 1B).
11. Store the gel/cellophane sandwich upside down until the gel is dry.
12. Cut it with a sharp knife out of the frame and store it.

4. Notes

1. Depending on the size and percentage of the gel, the glycerol content may have to be increased for some gels.
2. Unless stated otherwise, all solutions should be prepared in deionized water that has a resistance greater than 18 Mohm and total organic content of less than five parts per billion. This standard is referred to as "deionized water" in this text.
3. The size of the acrylic frames and the inner plate depends on the respective gel: For common slab gels, an outer measurement of 30 cm × 28 cm and an inner measurement of 22 cm × 20 cm may work. The inner plate must fit inside of the lower frame and must have the same thickness as this frame. If the inner measurements of the upper frame is about 10%

smaller (e.g., 20 cm × 18 cm), this can facilitate the mounting of the sandwich.

4. In dependence on the size and percentage of the gel for some gels, the equilibration time has to be prolonged.
5. Do not move the inner plate up or down until the gel sandwich is fixed between the frames.

Acknowledgments

Special thanks for excellent technical support to Christin Gräfe. Livia Schulze und Barbara Uteß (Inst. Immunology, Dresden, Germany).

Reference

1. Krishnan M, Nguyen HT (1990) Drying acrylamide slab gels for fluorography without using gel drier and vacuum pump. Anal Biochem 187:51–53

Chapter 37

Long-Term, Buffer-Less, Wet Gel Storage in Non-sealed Polyethylene Bags

Biji T. Kurien and R. Hal Scofield

Abstract

Electrophoresed gels are normally fixed, stained, destained, and dried. Drying is normally carried out with commercial gel dryers or by drying between two cellophane sheets held together by two acrylic frames. Here, we report that stained and destained gels (7.5, 10, or 15%, denaturing or native gels; 0.4 or 1.5 mm in thickness) could be stored wet, unsealed, and without any storage buffer for several months at room temperature within flexible polyethylene bags without significant shrinking or protein diffusion. The gel remains hydrated because of the *de facto* sealing achieved by the polyethylene sheets (PS) adhering air-tightly to the gel on either sides. The microsaturated environment generated by the thin film of water molecules trapped between the gel and the PS, along with the nonporous nature of the PS, apparently protects the gel from cracking as well as shrinking significantly. The intensity of stained proteins increased during storage probably from the slight gel shrinkage observed. Wet gel storage is useful (a) when low abundance protein spots from multiple two-dimensional electrophoresis gels have to be excised for in-gel tryptic digestion or electroelution and (b) for wet gel autoradiography. In addition to avoiding dryer contamination and saving drying time, these bags prevent the moist gel from sticking to X-ray film. Such storage could also prove useful for electrophoretic transfer of fixed and stained gels.

Key words: SDS-PAGE, Coomassie blue, Protein staining, Gel storage

1. Introduction

Sodium dodecyl sulfate-polyacrylamide gel electrophoresis (SDS-PAGE) (1) and two-dimensional gel electrophoresis (2) have served as an invaluable analytical tool in basic and applied research. Gels are traditionally dried following electrophoresis, for purposes of scanning and/or documentation. A variety of drying methods have been used for this purpose, including the use of commercial gel dryers (3) and drying between two cellophane sheets held

Biji T. Kurien and R. Hal Scofield (eds.), *Protein Electrophoresis: Methods and Protocols*, Methods in Molecular Biology, vol. 869, DOI 10.1007/978-1-61779-821-4_37, © Springer Science+Business Media, LLC 2012

together by two acrylic frames. We observed that we could store stained and destained gels (0.4 or 1.5 mm in thickness, denaturing or native, 7.5, 10, or 15%) wet, unsealed, and without any storage buffer for several months at room temperature within flexible polyethylene bags without significant shrinking or protein diffusion.

Gels were obtained from various experiments to study the efficacy of polyethylene bags to serve as a long-term storage medium. First, isoelectric focusing (IEF) of human granulocyte membrane proteins was carried out in a tube gel in the first dimension followed by SDS-PAGE (10%, 1.5 mm thickness) in the second dimension. The gel was immunoblotted non-electrophoretically immediately following SDS-PAGE (4). After transfer, the gel was stained, destained, and stored in a polyethylene bag. Secondly, increasing concentrations of bovine serum albumin (BSA) was electrophoresed (7.5 and 10% SDS-PAGE gels, 1.5 mm thick), stained, destained, and stored in polyethylene bags. Thirdly, HeLa cell extract was electrophoresed in a 15% SDS-PAGE gel (1.5 mm thick), stained, destained, and stored. Finally, flatbed native IEF (0.4-mm gel thickness) of rabbit anti-spectrin was carried out (5). After focusing was complete, the gel, still on the gel support film, was stained, destained, and stored.

Figure 1a shows a two-dimensional electrophoresis (2-DE) gel after being placed in a polyethylene bag. The same gel, after storage for over 2 months in the polyethylene bag at room temperature is shown in Fig. 1b. Similarly, Figs. 2a, b show a SDS-PAGE gel (10%) containing BSA before and after storage in a polyethylene bag for over a month, respectively. Similarly, storage of a 7.5% SDS-PAGE gel containing BSA and a 15% SDS-PAGE gel with HeLa cell extract analyzed on it in a polyethylene bag was also successful (data not shown). In addition, an electrophoresed native IEF gel (5) was also stored successfully in a polyethylene bag for over 2 months (data not shown). Several additional 10% SDS-PAGE gels were also stored similarly for at least 3 months and some more than 6 months. The gels did not crack or display any significant change from its original form. The patterns of the protein bands did not change over this period, even though there was a slight shrinkage of the gel and increase in intensity of stained protein bands.

We also found that wet gel autoradiography with gels in polyethylene bags prevents the X-ray film from getting wet and thus prevents the film from sticking to the gel, unlike performing this with the radioactive gel wrapped in clear plastic wrap (unpublished observation).

SDS-PAGE gels (precast) are sold stored in buffer inside sealed plastic bags or aluminum foil. Storage buffer is essential to provide a saturated atmosphere to prevent gel cracking (dehydration) or swelling (hydration). When gels are stored in polyethylene bags, the polyethylene sheets trap water molecules between themselves and the gels on either side and thus provide a microsaturated

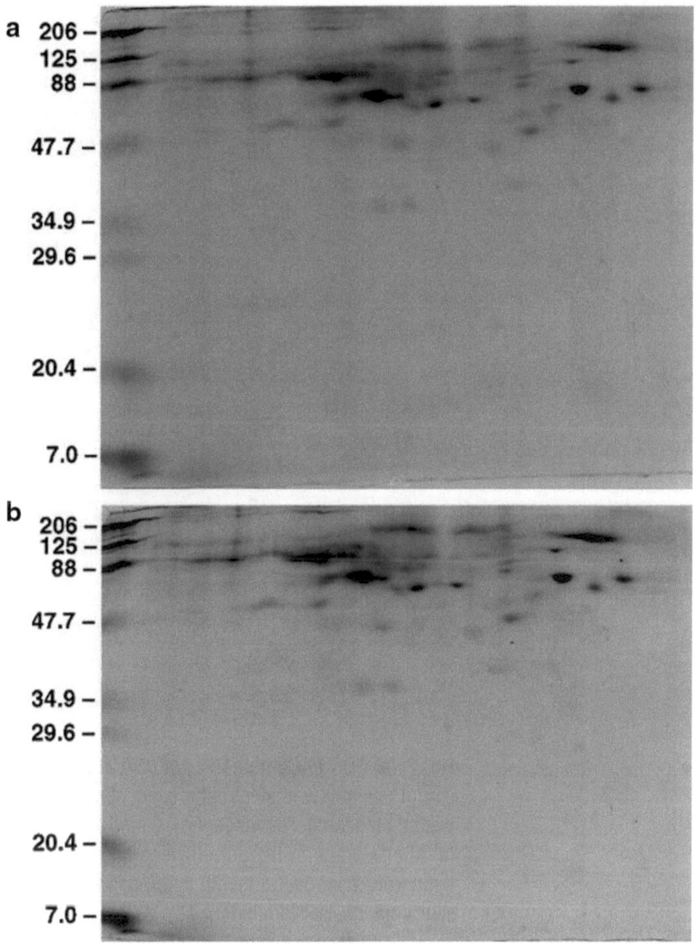

Fig. 1. (**a**) Two-dimensional electrophoresis gel (1.5 mm thick) containing separated granulocyte membrane proteins at day 1 of storage in a polyethylene bag. Granulocyte membrane proteins were first separated on the basis of charge in the first dimension and on the basis of size in the second dimension. (**b**) The gel shown in (**a**) after storage in the polyethylene bag for over 2 months. The gel with the separated granulocyte membrane proteins was stored for over 2 months in the bag at room temperature. Reproduced from (6) with permission from Elsevier Limited.

environment (6). The sheets adhere airtightly to either sides of the gel to provide a *de facto* sealing. Furthermore, the sheets stick to themselves in the regions around the gel where liquid droplets from the gel drip into the bag (when the gel is placed in the bag). The polyethylene sheets do not appear to be porous and therefore are able to maintain this microsaturated atmosphere for several months. We observed an increase in protein stain intensity in the gels stored in the polyethylene bags. This is more than likely due to the slight shrinkage observed (6).

Even though all these gels were stored at room temperature, we observed that they could be stored at 4°C without significant

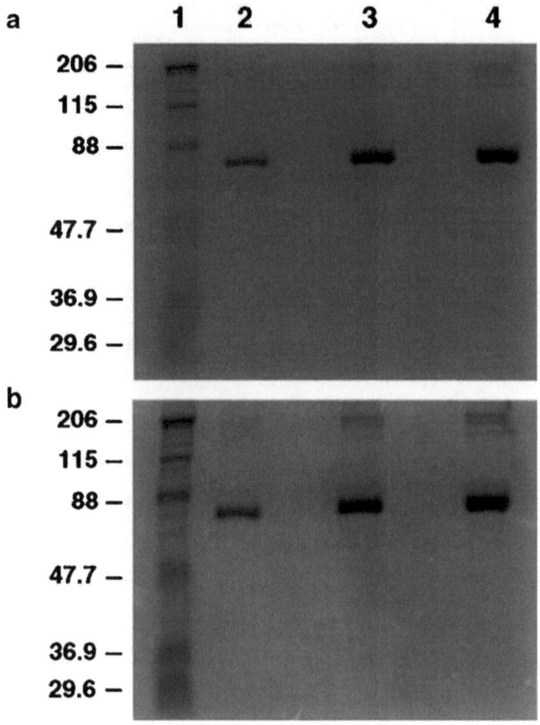

Fig. 2. (**a**) SDS-PAGE gel (10%) of BSA at day 1 of storage in a polyethylene bag. Bovine serum albumin was electrophoresed, stained, destained, and stored. (*Lane 1*) molecular weight standards. (*Lanes 2–4*) 3, 6, and 12 mg of BSA, respectively. (**b**) The gel shown in (**a**) after storage in the polyethylene bag for over a month. The gel was stored at room temperature for over a month in the polyethylene bag. Reproduced from (6) with permission from Elsevier Limited.

shrinkage for over a month. However, there was significant shrinkage (without cracking) when stored at –20°C for over a month (data not shown).

Wet gel autoradiography with these bags can help avoid possible contamination of gel dryers, reduce exposure to radioisotopes, and save time/effort (by not having to dry the gel). Storing gels in this manner also enables electrophoretic transfer of fixed and stained gels at a future date, if necessary (7). Dried gels, of course, can be rehydrated and used. However, we observed that filter paper (used for drying) remnants remain stuck to the gel after rehydration, which necessitates painstaking effort to remove them to prevent the gel from tearing.

Storage of gels in polyethylene bags permits the investigator to compare gels for reproducibility of protein spots and also to excise protein spots when enough number of gels has been collected (especially for protein of low abundance). Various kinds of gels, thin and thick gels, gels of different concentrations, and also denaturing and native gels can be stored in polyethylene bags.

2. Materials

Prepare all solutions using ultrapure water (prepared by purifying deionized water, to attain a sensitivity of 18 MΩ cm at 25°C) and analytical grade reagents. Prepare and store all reagents at room temperature (unless indicated otherwise). Diligently follow all waste disposal regulations when disposing waste materials. We do not add sodium azide to reagents.

1. SDS-PAGE gels (7, 10 and 15%; 1.5 mm thick) (1).
2. IEF tube gels (2).
3. Flatbed native IEF gel (5).
4. Flexible polyethylene bag (Fisher Scientific Company, Springfield, NJ).
5. Granulocyte membranes (8).
6. HeLa cell extract.
7. Coomassie Brilliant Blue (CBB) stain: 0.05% (CBB) in 25% methanol/10% acetic acid (see Note 1).
8. Destaining solution: 25% methanol/10% acetic acid.

3. Methods

All procedures are carried out at room temperature unless otherwise specified.

3.1. Two Dimensional Gel Electrophoresis

1. Carry out IEF of granulocyte membranes in a tube gel in the first dimension and SDS-PAGE on a large gel for the second dimension (1, 3).

3.2. SDS-PAGE

1. Electrophorese HeLa cell extract on 15% mini-gel SDS-PAGE as well as BSA on 10 and 7.5% mini-gels (1).

3.3. Flatbed Native Isoelectric Focusing

1. Electrofocus rabbit anti-spectrin on flatbed native IEF (0.4-mm gel thickness) (4).

3.4. Gel Staining

1. Stain gels with 0.05% Coomassie Brilliant Blue (see Note 2).
2. Rinse with water (see Note 3).
3. Destain gel with destaining solution (see Note 4). Add a few sheets of Kimwipes to absorb the released Coomassie stain (see Note 5).

3.5. Gel Storage in Polyethylene Bags

1. Incubate the destained gels in 50% methanol containing 5% glycerol for 30 min.
2. For a large, regular SDS-PAGE gel, follow the following protocol. Cut a polyethylene bag (originally open at one end) along one of its sides. Lift the top flap of the bag and place the SDS-PAGE gel on the lower flap. Lower the top flap slowly on top of the gel, such that no air bubbles are trapped (see Note 6).
3. Store the polyethylene bag at room temperature without heat sealing.
4. For a mini SDS-PAGE gel as well as the smaller native IEF gel, insert the gel into the bag through the open end of the bag without cutting the side of the bag. Store the gels at room temperature without sealing.

4. Notes

1. Weigh the required amount of the dye and dissolve it in methanol first. Filter the dissolved CBB through a circular Whatman 1-M filter paper folded into a cone and fitted inside a glass funnel. For a quicker way to filter the CBB solution, layer a circular 1-M Whatman filter paper (125-mm diameter circles) inside a Buchner funnel (Coors, USA; 135 mm inner diameter). The filter paper should cover all the holes in the funnel and lay flat inside the funnel. Attach the Buchner funnel to a sidearm fitted conical flask that is attached to the house vacuum through the sidearm. Pour the CBB solution into the Buchner funnel and apply vacuum. If filter paper gets clogged, replace with a fresh one and filter. Collect the filtered CBB. Adjust volume with water, methanol, and acetic acid to obtain a 0.05% CBB in 25% methanol/10% acetic acid.
2. Gels can be stained faster by heating the CBB staining solution. Place the gel in a container with the staining solution, with its lid in place. The container can be heated in a shaking water bath maintained at 65°C for 20 min or in a microwave for a few minutes (9).
3. Rinse the CBB off the gel with few milliliter of ultrapure water prior to adding destaining solution. Do not discard the wash into the sink. Save the rinse in a bottle for proper disposal.
4. Gels can be destained faster by heating the destaining solution. Place the Coomassie stained gel in a container with the destaining solution, with its lid in place. The container can be heated in a shaking water bath maintained at 65°C for 20 min or in a microwave for a few minutes (9).

5. Kimwipes bind the released CBB very strongly. The Kimwipes do not release the bound CBB even after they were incubated in water for over 5 weeks or in high salt solutions (10). The Kimwipes with the bound CBB can thus be discarded as solid waste. This serves as an environmentally safe way to discard CBB and enable the destaining solution to be recycled (if the gel did not contain radioisotopes).
6. Lower the top flap slowly on the gel, starting from one side of the gel, to cover the entire gel. The thin film of water on either side of the gel enables the polyethylene sheets to seal airtightly and thus provide a microsaturated atmosphere over both sides of the gel.

References

1. Laemmli UK (1970) Cleavage of structural proteins during the assembly of the head of bacteriophage T4. Nature 227:680–685
2. O'Farrell PH (1975) High resolution two-dimensional electrophoresis of proteins. J Biol Chem 250:4007–4021
3. Sambrook J, Fritsch EF, Maniatis T (1989) Molecular cloning, 2nd edn. Cold Spring Harbor Laboratory Press, Cold Spring Harbor, NY
4. Kurien BT, Scofield RH (1997) Multiple immunoblots after non-electrophoretic bidirectional transfer of a single SDS-PAGE gel with multiple antigens. J Immunol Methods 205:91–94
5. Kurien BT, Scofield RH (2009) Affinity immunoblotting for analysis of antibody clonotype distribution in a lupus patient developing anti-Ro 60 over time. Methods Mol Biol 536:45–53
6. Kurien BT, Scofield RH (1998) Long-term wet storage of sodium dodecyl sulfate-polyacrylamide gels in polyethylene bags. Anal Biochem 261: 116–118
7. Phelps DS (1984) Electrophoretic transfer of proteins from fixed and stained gels. Anal Biochem 141:409–412
8. Kurien BT, Newland J, Paczkowski C et al (2000) Association of neutropenia in systemic lupus erythematosus (SLE) with anti-Ro and binding of an immunologically cross-reactive neutrophil membrane antigen. Clin Exp Immunol 120:209–217
9. Kurien BT, Scofield RH (1998) Heat mediated quick coomassie blue protein staining and destaining of SDS-PAGE gels. Indian J Biochem Biophys 35:385–389
10. Dorri Y, Kurien BT (2010) Environmentally safe removal/disposal of coomassie brilliant blue from gel destain and used gel stain. Anal Biochem 404:193–196

Chapter 38

A Brief Review of Other Notable Electrophoretic Methods

James D. Fesmire

Abstract

The use of electrophoretic methodologies for use in qualitative and preparative purposes are commonly used in biological research and have been well established as an integral analytical tool for a long time in most research laboratories. Listed here are some of the more specialized innovations that have been developed in recent times for special purposes of study. These include micropreparative isoelectric focusing in liquid suspension, accelerated protease digestion by SDS-PAGE, two-dimensional SDS-PAGE for membrane protein resolution, carbon nanotube-modified page for resolution of complement C3, electrophoretic resolution of ultra-acidic proteomes in acidic media, and two-dimensional immunoelectrophoresis of pre-beta/alpha lipoprotein A-I in agarose. All these methods are briefly reviewed in this chapter.

Key words: Micropreparative IEF, Controllable pore expansion, Cross-linker cleavage, Doubled SDS-PAGE carbon nanotube-assisted PAGE

1. Micropreparative Isoelectric Focusing Protein Separation in a Suspended Drop

This methodology (1) offers the resolving power of isoelectric focusing combined with the potential advantage of analysis of samples in very low concentrations (0.6 μg) contained in volumes as low as 12 μL.

Electrophoresis is performed by suspending the microvolume of sample between the anode and cathode in open air. The electrophoresis procedure is completed when the buffer suspension is evaporated to the extent that it can no longer maintain adhesion connecting both electrodes and is therefore separated. The proteins are recovered from each electrode separately according to their migrational positioning.

The mild electrophoretic conditions of 5 V or less for the procedure generally protect the sample integrity, therefore permitting recoveries of the intact, separated proteins of 50% or more.

Biji T. Kurien and R. Hal Scofield (eds.), *Protein Electrophoresis: Methods and Protocols*, Methods in Molecular Biology, vol. 869, DOI 10.1007/978-1-61779-821-4_38, © Springer Science+Business Media, LLC 2012

The procedure is comparably fast requiring 1 h or less running time, depending on migrational rates of the sample proteins in question.

The method is relatively simple requiring little or no sample preparation for complex equipment. Samples may be applied in an intact or denatured state for resolution.

2. Accelerated Protease Digestion and MS Recovery by Modified SDS-PAGE Media

The authors (2) present a novel and effective innovation to the study of proteomic research.

The procedure consists of adjusting the pore size of the bis-acrylamide PAGE gel to permit increased access to the peptides of study and thereby increase the peptide numbers as much as 15–45% as determined by mass spectroscopy.

This increase is achieved by the addition of the alkaline hydrolysis reagent, ethylene-glycol-diacrylate (EDA), with the bis-acrylamide which cleaves the cross-linkages permitting greater access of the proteases to the peptides of interest.

Acrylamide pore size is controlled by means of the proportions of the additions in accordance with the requirements of the peptides in question.

3. Two-Dimensional Bicine-Doubled SDS-PAGE Method for Enhanced Membrane Protein Resolution

This method (3) consists of a two-dimensional bis-tricine (bicine) acrylamide media in both dimensions using a modified bicine buffer as an effective approach to resolve the difficulties of insolubility in membrane protein chemistry.

To further enhance resolution, acrylamide was increased from 10 to a 12% gradient in bicine buffer from the first to the second dimension.

Large-format one-dimensional gels were run at 24 mA/gel for approximately 7 h. Gel thickness ranged from 0.75 mm for small gels to 1.0 mm for the large-format gels.

The authors demonstrated a 151% increase in spot count number applying this method in a bicine buffer system to membrane proteins from the anaerobic bacterium *Clostridium thermocellum*.

4. Carbon Nanotube-Modified Polyacrylamide Gel Electrophoresis Method for Enhanced Resolution of Complement C3 Proteins

A modified form of carbon nanotube technology combined with polyacrylamide gel electrophoresis was developed in this study (4) for the purpose of separation and identification of proteins in small concentration related to liver dysfunction.

Due to the inherently high-retention properties owing to the structure and composition of the single-walled, water-soluble carbon nanotube, a carboxylated form in presence of Triton X-100 was developed for use in combination with acrylamide separating gels in the polyacrylamide gel electrophoresis methodology to be applied toward resolving the proteins complement C3 and haptoglobin as may be diagnostic in hepatic cirrhosis patients.

Theory suggests that the adsorption kinetics is improved by means of the addition of carboxylated nanotubes with the proteins of study.

This combined with the driving electrophoretic forces results in a much greater resolving power.

Therefore, this novel modification may be of great advantage for use as a discerning diagnostic tool for a variety of disorders as compared with normal responses in human serum.

5. Electrophoretic Resolution of Ultra-Acidic Proteomes in Acidic Acrylamide Media

This innovative methodology was successfully developed for the purpose of resolving and identifying proteomes in hyperphosphorylated forms which characteristically result in developing an isoelectric point of 2.5 or less (5).

This acidic nature categorically presents a restrictive obstacle for study using traditional two-dimensional electrophoretic systems. One protein of interest of this type is the tau (*t*) protein, which has been identified as being related to the neurodegenerative Alzheimer's disease.

Authors of the method provide a suitable modification here to circumvent the problem by devising a compatible acidic gel system of low pH (2.5–5.0).

The method consists of first dimension being comprised of an anodic acidic gel with up to 5 mg per lane. The lanes were cut, transferred, and separated by SDS-PAGE as the second dimension.

Proteins are visualized using Coomassie G-250 as stain.

6. Two-Dimensional Electrophoresis

This modified immunological technique, originally developed by Laurel (6), provides a powerfully discerning immunological tool for the purpose of examination of antibody specificity and or homogeneity of antigens.

The method presented here is a modified form presented by Weeke (7) in that the 4 × 4-in. gels are poured separately, but at the same event to create the entire gel.

First, 1.5 mL of 2% Indubiose (agarose) without antibody is poured into a 4 × 4-in. glass frame separated by a 1.5-mm thick u-frame and 0.004" clear Cronar hydrophilic photography film, to which the gel will bond permanently.

This casting is held together with binder clips which also provide a vertical support stand during the curing process.

As soon as the bottom gel is set, but still warm, the second layer containing the antibody in 12.5 mL of agarose is poured into the casting. Pouring both layers while hot, but separately creates a more consistent matrix at the boundary which permits the protein to smoothly migrate into the gel containing the antisera.

The casting is stood at room temperature until cured sufficiently for use or can be sealed and stored overnight at 4° centigrade with no appreciable diffusion.

Samples may be applied to wells or troughs for the first dimension as determined by the assay requirements in 2% agarose, with veronal buffer, pH 8.6, and 80–100 V DC measured at 4.5 in. across the gel, using cotton wicks producing 50–60 milliamps; the gel resolves the protein bands in approximately 45 min–1 h for human high-density lipoproteins.

The plate is rotated 90° and electrophoresed at the same rate for 3–4 h for the second dimension. Upon completion of the electrophoresis, the gel is washed in saline overnight, or 2 h minimum with stirring.

The washed plate is blotted and dried with filter paper, then stained as desired with Coomassie Blue protein stain.

Areas measured under the individual completed peaks are directly proportional to the concentration of the protein in question in one- and two-dimensional assays, making this methodology a powerful analytical tool.

The two-dimensional pattern shown in Fig. 1 is presented to represent expressions of the lipoprotein A-I as separate peaks represented by the pre-beta A-I as the slower peak and the alpha A-I as the faster moving, larger peak. The migration is from left to right.

This gel also contains antisera to the lipoprotein A-II and is shown as the precipitin line closely associated with the A-I in the normal sample but quite slower in migration in the hypertriglyceridemic sample, with the pre-beta A-II peak slower than pre-beta A-I.

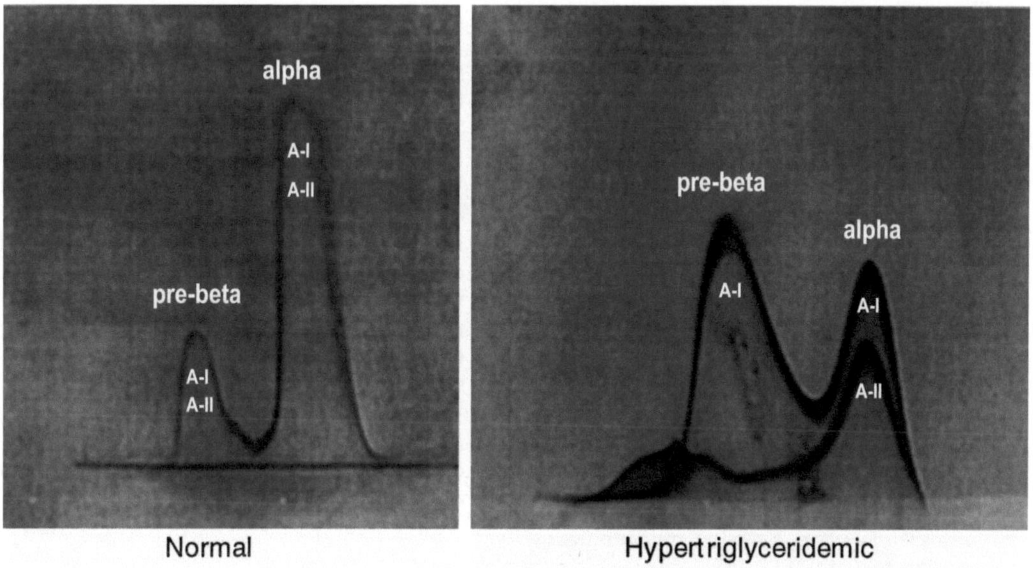

Fig. 1. Second dimension agarose patterns of apolipoprotein A-I and apolipoprotein A-II expressions of normal and hyperlipemic sera. Immunological continuity is shown to represent concentrations of particles positioned within the pre-beta and alpha regions for both the A-I and A-II lipoproteins, both of which migrate in advance of the beta position. Areas in favor of the alpha position peaks for A-I and A-II reflect lipid transport efficiency.

Overall, the hypertriglyceridemic sample shows the A-I distribution to be greatly in favor of the pre-beta position and therefore much slower in migration of the A-I peaks.

This resulting areas of the lipoprotein A-I peaks are interpreted and presented here to be a reflection of the effects of the enzyme LCAT (lecithin cholesterol acetyl transferase) activity on the reverse cholesterol transport system in accordance with (8).

The status of metabolic HDL cholesterol transport efficiency is expressed as a function of A-I clearance to the higher density region (alpha) in HDL. The distribution of the measured areas is presented as reflective of the status of risk for HDL particle clearance.

Pre-beta A-I is shown to be a positive risk factor in this assay, significantly correlating with LDL-cholesterol and total serum cholesterol:

$$\text{The distribution ratio} = \frac{\text{Alpha A-I mm}^2}{\text{Prebeta A-I mm}^2} = (\text{AIDR})$$

A favorable percent clearance to the alpha region (AIDR) = 85% or higher.

References

1. Egatz-Gomez A, Thormann W (2011) Micropreparative isoelectric focusing protein separation in a suspended drop. Electrophoresis 32:1433–1437
2. Bornemann S, Rietschel B, Baltruschat S et al (2010) A novel polyacrylamide gel system for proteomic use offering controllable pore expansion by crosslinker cleavage. Electrophoresis 31:585–592
3. Williams TI, Combs JC, Thakur AP et al (2006) A novel Bicine running buffer system for doubled sodium dodecyl sulfate - polyacrylamide gel electrophoresis of membrane proteins. Electrophoresis 27:2984–2995
4. Jiang F, Wang Y, Hu X et al (2010) Carbon nanotubes-assisted polyacrylamide gel electrophoresis for enhanced separation of human serum proteins and application in liverish diagnosis. J Sep Sci 33:3393–3399
5. Hempel K, Rosen R, Becher D et al (2009) Analysis of ultra acidic proteins by the use of anodic acidic gels. Anal Biochem 385:208–214
6. Laurel CB (1965) Antigen-antibody crossed electrophoresis. Anal Biochem 10:358
7. Weeke B (1973) Crossed immunoelectrophoresis. In: Axelsen NH, Krøll J, Weeke B (eds) Quantitative immunoelectrophoresis. Universitetsforlaget, Oslo
8. Fielding PE, Miida T, Fielding CJ (1991) Metabolism of low-density lipoprotein free cholesterol by human plasma lecithin-cholesterol acyltransferase. Biochemistry 30:8551–8557

Chapter 39

Protein Stains and Applications

Ranjini K. Sundaram, Natarajan Balasubramaniyan, and Pazhani Sundaram

Abstract

Staining of proteins separated on gels provides the basis for determination of the critical properties of these biopolymers, such as their molecular weight and/or charge. Detection of proteins on gels and blots require stains. These stains vary in sensitivity, ease of use, color, stability, versatility, and specificity. This review discusses different stains and applications with details on how to use the advantages and disadvantages of each stain. It also compiles some important points to be considered in imaging and evaluation. Commonly used colorimetric and fluorescent dyes for general protein staining, and posttranslational modification-specific detection methods are also discussed.

Key words: SDS-PAGE, 2-DE, Coomassie Brilliant Blue, Silver nitrate, Zinc staining, Fluorescent stains, Pre-electrophoresis staining

1. Introduction

The most important hurdle that any scientist faces on a daily basis is obtaining proteins of high purity and decent yield. The solution to check purity issues can be tackled by running a denaturing sodium dodecyl sulfate (SDS)-polyacrylamide gel. This process separates proteins from a mixture based on size. Detection of separated proteins is carried out by colorimetric or fluorescent staining. Protein posttranslational modifications such as phosphorylation and glycosylation can be reliably determined with several fluorescence-based protocols.

Here, we focus on protein detection after separation by electrophoresis, highlighting on most popular methods for detection of both total protein and most frequently encountered protein posttranslational modifications such as phosphorylation and glycosylation (1). A typical visual example of showing protein purification

Biji T. Kurien and R. Hal Scofield (eds.), *Protein Electrophoresis: Methods and Protocols*, Methods in Molecular Biology, vol. 869, DOI 10.1007/978-1-61779-821-4_39, © Springer Science+Business Media, LLC 2012

from a mix involves showing fewer number of proteins with each step along with enrichment of the protein of interest (2–5).

One-dimensional SDS-PAGE protein analysis is the most common method to separate and detect proteins but 2D gel electrophoresis format is preferred for analysis of complex protein mixtures. The common factor in both these methods is the use of colorimetric and fluorescent stains for total protein detection for further downstream applications (6–9). A detection method that accurately assesses the total protein profile is generally sufficient for a general purification protocol. Protein detection in SDS-PAGE can be done by noncovalent postelectrophoresis staining with organic dyes or by metal deposition techniques or by labeling samples by covalent modification with a fluorescent dye prior to running gels. Posttranslational modifications can be measured by specialized staining formulations and protocols that target the modified functional groups from a background of total protein.

1.1. Universal Practice

Common protocols of all gel-staining methods require good laboratory practices. Gel staining is generally accomplished in covered polycarbonate or polypropylene dishes. Dishes should be cleaned with 70–100% ethanol or methanol followed by a water rinse. Gels are incubated on a rocking platform or an orbital shaker or at a moderate speed of 40–70 rpm and should float freely in solution. Always remove the stacking gel and trim the gel at the bottom by at least 1 mm. Pure deionized or double distilled water should be used when required for all solutions.

1.2. Instrumentation for Visualizing Staining

Most purification detection and documentation requirements now can be accomplished with flatbed scanners or with camera-mounted visible-light box stations. For fluorescent applications, excitation and emission setting are frequently identified by the wavelength settings. Simple transilluminators with light-emitting diode or other inexpensive filtered monochromatic light sources, paired with an orange-filtering emission cover, offer inexpensive, safe modes of detection of many of the UV-excitable dyes.

2. Total Protein Detection

2.1. Colorimetric Total Protein Stains

Coomassie Brilliant Blue (CBB) is the most frequently used total protein gel stain. This stain can be detected simply by visual inspection and is relatively easy to use.

2.1.1. Coomassie Brilliant Blue Staining

CBB R-250 and the dimethyl derivative CBB G-250 are disulfonated triphenylmethane dyes that stain protein bands bright blue. The dyes bind through electrostatic interaction with protonated basic amino acids (lysine, arginine, and histidine) and by hydrophobic associations with aromatic residues of separated proteins. The dye does not bind to the polyacrylamide with high affinity but does infiltrate the gel matrix and bind with low affinity which can be removed by destaining. However, when in colloidal preparations, destaining step is not necessary. Coomassie stains are endpoint stains, which are reversible and do not interfere with subsequent mass spectrometry of excised protein bands.

Reagents and procedure: Staining with CBB R-250 has been done with 0.025–0.10% (w/v dye) in an acidic alcohol formulation [30–50% (v/v) methanol (or, less frequently, ethanol)) with 7–10% (v/v) acetic acid solution. The dye is dissolved, and the solution is then filtered through Whatman #1 paper to produce a stable formulation. Fixation and staining are typically done in 10 gel volumes of solution. For the most sensitive and consistent results, the gel is fixed and stained in the same acidic alcohol solution; under these conditions, the gel shrinks. Destaining is in 7% acetic acid, minus the alcohol to return the gel to full size. Destaining generally requires several solution changes and can be accelerated by addition of paper or foam adsorbant.

Advantages and disadvantages: It is very easy and most commonly used stain. Staining and destaining requires more time and reagents.

2.1.2. Colloidal Coomassie Staining

Thorough analyses of methods have resulted in colloidal formulations such that the stain does not effectively enter the gel matrix while binding specifically to protein bands, allowing low-background staining with reliable quantitation and increased sensitivity relative to the standard CBB R-250 staining method above.

Reagents and procedure: For colloidal staining, CBB G-250 is preferred to CBB R-250. An improved, widely used formulation is prepared by dissolving 10% (w/v) ammonium sulfate in 2% (w/v) phosphoric acid followed by addition of CBB G-250 to 0.1% (w/v) from a 5% (w/v) aqueous stock solution; this is combined with 20% (v/v) methanol prior to staining (10). The addition of methanol may increase the proportion of monodispersed dye, resulting in more rapid staining and more intense bands, but with some background; increasing ammonium sulfate concentrations drives the dye toward the colloidal state. Another, more latest formulation used greater dye and phosphoric acid concentrations in a final formulation of 0.12% (w/v) dye, 10% (w/v) ammonium sulfate, 10% (w/v) phosphoric acid, and 20% (v/v) methanol in a single stable staining solution, prepared by sequentially combining desired amounts of phosphoric acid, ammonium sulfate, and powdered dye in an aqueous solution to 80% of the desired volume, then adding methanol to the final volume (11). Colloidal staining solutions should not be filtered. Following electrophoresis the gel is washed in water to remove most of the SDS placed in the staining solution for few minutes to overnight. Protein bands are visible within the first few minutes, and background is initially clear, and eventually weakly blue. Following staining the gel is washed with water to reduce background. Several of the commercially available formulations include simple microwave-based protocols to accelerate staining (and destaining) with CBB G-250 colloidal formulations.

2.1.3. Silver Staining

Silver staining continues to be the alternative colorimetric method for increased detection sensitivity over Coomassie staining. It is regarded as the standard by which all other "ultrasensitive" staining methods are judged. Silver staining may set the standard of rigor for ultimate detection sensitivity but quantitation is complex in nature. There are commercially available silver staining kits that provide working protocols and reagent mixtures from the literature that allow reproducible results. Silver staining principles, methods, and protocols have been reviewed and summarized (5, 12, 13).

Reagents and procedure: Fixer reagent contains 50% methanol, 12% acetic acid, and 0.05% formalin. Wash solution contains 35% ethanol (96%). Sensitizing solution contains 0.02% sodium thiosulfate. Silver nitrate solution contains 0.2% silver nitrate and 0.076% formalin. Developer solution contains 6% sodium carbonate, 0.05% formalin, and 0.0004% sodium thiosulfate. Stop solution contains 50% methanol and 12% acetic acid. After separation, fix gel for 2 h or overnight followed by three washes for 20 min each. Sensitize gel for 2 min followed by three washes in water for 5 min each. Stain gel in silver nitrate solution for 20 min followed by two washes in water. Develop gel by addition of developer and stop staining by leaving gel for 5 min in stop solution. Gels can be stored at 4°C in 1% acetic acid.

Advantages and disadvantages: Silver staining is the most sensitive colorimetric method for detecting total protein. Silver staining protocols require several steps that are affected by reagent quality as well as incubation times and thickness of the gel. An advantage of commercially available silver staining kits is that the formulations and protocols are optimized and consistently manufactured, helping to minimize the effects of minor differences in day-to-day use.

2.1.4. Zinc Staining

Zinc staining is unlike all other staining methods in that it is a negative stain. Instead of staining the proteins, this procedure stains all areas of the polyacrylamide gel in which there are no proteins. Zinc ions complex with imidazole, which precipitates in the gel matrix except where SDS-saturated protein occur. The milky-white precipitate renders the background opaque while the protein bands remain clear.

Reagents and procedure: 0.2 M imidazole, 0.1% SDS, and 0.3 M zinc sulfate. Following electrophoresis, the gel is incubated for 15 min in 0.2 M imidazole, 0.1% SDS. The imidazole solution is discarded and the gel is developed by incubation in 0.3 M zinc sulfate for 30–45 s. The developer is discarded and the gel is washed several times, with water, 1 min per wash. The gel is stored in 0.5% (w/v) sodium carbonate.

Advantages and disadvantages: Sensitive and detects 0.25 ng of protein per band in a mini gel. The entire protocol is complete in 15 min. Reversible and completely possible to erase the stain from the gel to easily recover proteins or proceed with other methods. Over development can be problematic, but can be corrected by incubation in 100 mM glycine to dissolve excess zinc imidazolate. Though the transparent protein bands against the opaque background do not provide the striking colorimetric contrast of Coomassie or silver staining, imaging can be done with the gel illuminated against a black background (14, 15).

2.2. Fluorescent Total Protein Stains

Fluorescent protein gel staining, when combined with an appropriate imager, is a sensitive and quantitative approach for protein analysis. Fluorescent gel stains also offer advantages such as ease of use, sample stability, and safety. Several dyes have been reported for fluorescence staining of proteins either before or after electrophoresis. Fluorescent staining methods can combine detection sensitivity that rivals silver staining with workflow advantages similar to Coomassie or zinc ion staining, and offer linear quantitation ranges 10–100-fold greater than the colorimetric methods. Detection relies on instrumentation requiring a monochromatic excitation light source, selective optical filter, excitation light, and a detection mode. For many fluorescent stains, detection can also be by visual inspection, but with reduced sensitivity in comparison to photographic or instrumentation methods. Photobleaching to some extent can be expected. Fluorescent stains fall into two general categories: (a) fluorogenic stains that show significant fluorescent enhancement corresponding to localization with protein bands and (b) intrinsically fluorescent stains that bind selectively to protein bands and do not bind to the gel matrix. SYPRO™ Orange, Red, Ruby, and Tangerine protein gel stains are one-step fluorescent stains optimal for rapid and efficient staining of one-dimensional (1D) protein gels. These stains provide sensitivity that is equivalent to the silver staining method for 1D gels and enable protein detection that is not affected by the presence of nucleic acids and lipopolysaccharides.

2.2.1. SYPRO Orange, SYPRO Red, and SYPRO Tangerine Protein Gel Stains

SYPRO Red and SYPRO Orange protein gel stains are described by Steinberg et al. (16) and are available commercially as proprietary dyes in 10 mM stock solutions in dimethylsulfoxide (DMSO).

Reagents and procedure: To prepare a staining solution, the dye is diluted 5,000-fold to 2 mM in dilute acetic acid (7%, v/v). Following SDS-PAGE, the gel is placed in staining solution. The gel is washed briefly with water. Staining can be supervised periodically by viewing on a UV light box. The gel is washed with two changes of water for 2–3 min per wash before acquiring the gel image.

Advantages and disadvantages: Staining of proteins is rapid, with a majority of the staining time being required for reducing the background that occurs in the course of the first 30–45 min of incubation. Destaining is easy. Highly sensitive and signal is photostable.

2.2.2. SYPRO Ruby Gel Stain

SYPRO Ruby is a ruthenium metal chelate that binds to the basic amino acids in proteins. In contrast to other SYPRO dyes, there is no intercalation into SDS micelles; binding is via direct electrostatic interaction with basic amino acid residues by a mechanism similar to colloidal CBB. The staining procedure is quite simple and allows high-throughput and large-scale proteomic applications. Excitation is achieved either with UV light or with laser scanners.

Reagents and procedure: The gel is fixed in 50% (v/v) methanol, 10% (v/v) acetic acid; fixation can be 30 min to overnight following which the gel is stained with SYPRO ruby stain. The gel is briefly destained with 10% (v/v) methanol, 7% (v/v) acetic acid. Excitation can be with UV light or with blue light sources; the protein bands appear orange-red to the eye.

Advantages and disadvantages: Various qualities like intensity, stability, detection sensitivity, and ease of use has led to SYPRO Ruby becoming the comparison standard against subsequently developed protein gel stains.

2.2.3. Nile Red Protein Gel Stain

Nile red (also known as Nile blue oxazone) is a phenoxazone dye that shows strong fluorescence enhancement upon transition from aqueous to hydrophobic environments such as SDS micelles or protein–SDS complexes. Nile red does not interact significantly with SDS monomers. This property has been utilized to develop a rapid, nonfixative total protein staining method for SDS gels.

Reagents and procedure: Nile red is diluted from a stock solution (0.4 mg/mL in DMSO) 200-fold into water to 2 mg/mL, and a tenfold volume excess (e.g., 50 mL staining solution for a 5 mL mini gel) is added to a gel with immediate and thorough agitation. The dye precipitates quickly in water such that staining is optimal in 2–5 min and does not improve thereafter. Following staining, the gels are briefly washed with water. Excitation can be with UV light or with green light sources; protein bands appear pale red.

Advantages and disadvantages: Electrophoresis should be run in nonstandard conditions with running buffer SDS at 0.05% (w/v), below the detergent's critical micelle concentration, rather than 0.1% (w/v) SDS typically used in 1D and 2D SDS-PAGE. Detection sensitivity is similar to that obtained with Coomassie stains. Because there is no fixation, Nile red-stained gels can be subsequently electroblotted with good transfer efficiency (17, 18). High fluorescent background and photostability can be problems encountered with this technique.

2.2.4. Epicocconone Stain

Bell and Karuso (19) isolated a compound from the fungus *Epicoccum nigrum* useful as a fluorescent stain for polyacrylamide gels. Epicocconone is an azaphilone that reacts with primary amines and NH_3 almost instantaneously to produce red fluorescent compounds. Formerly named Lightning Fast, it is now distributed by GE Health Care under the trade name Deep Purple Total Protein Stain.

Reagents and procedure: Fixation solution contains 15% (v/v) ethanol and 1% (w/v) citric acid (approx. pH 2.3) in water. Staining solution contains 100 mM sodium borate, pH 10.5–10.8 in water. Add 1 part Deep Purple to 200 parts borate buffer. Washing solution is 15% (v/v) ethanol in water. Acidification reagent is the same as fixation solution [15% (v/v) ethanol; 1% citric acid, pH 2.3 in water]. Fix gels in the fixation solution for a minimum of 1 h with gentle rocking. Remove the gels from the fixation solution and place into the staining solution with gentle rocking. Remove the gels from the staining solution and wash the gels by gentle rocking in the washing solution for 30 min. Remove the gels from the washing solution and acidify by placing them in the acidification solution and rock gently for 30 min. (This step may be repeated or extended up to overnight to reduce background staining.)

Advantages and disadvantages: Rapid, ultrahigh sensitivity fluorescent total protein stain suitable for both protein gels and blots. Low background fluorescence and no "speckling." This stain is highly compatible with mass spectrometry, Edman sequencing, and Ettan™ DIGE system. It is also environmentally friendly and free from heavy metals, allowing safe disposal after use. Works well with both UVA transilluminators and industry standard fluorescence scanners.

3. Stains That Detect Posttranslational Modifications

3.1. Phosphorylation

Phosphorylation is a reversible modification widely used by cells for activation/inhibition of specific pathways. Phosphorylation of residues in proteins results in a shift of the protein p*I* to more acidic values. In some cases (multiple), phosphorylation results in a shift in *Mr* with band thickening and even band doubling in SDS-PAGE. The Pro-Q® Diamond phosphoprotein gel stain provides a method for selectively staining phosphoproteins in polyacrylamide gels. It is ideal for the identification of kinase targets in signal transduction pathways and for phosphoproteomic studies (20–22). This proprietary fluorescent stain allows direct, in-gel detection of phosphate groups attached to tyrosine, serine, or threonine residues, without the need for antibodies or radioisotopes (20, 23). The stain can be used with standard SDS-polyacrylamide gels or with 2D gels (24–26). The simple and reliable staining protocol delivers results in as little as 4–5 h. Pro-Q® Diamond stain is fully compatible with mass spectrometry, allowing meaningful analysis of the phosphorylation state of entire proteomes.

Reagents and procedures: Pro-Q® Diamond phosphoprotein gel stain is supplied ready to use as a stand-alone reagent by Molecular Probes. Fix solution contains a 50% methanol and 10% acetic acid. Destain solution contains 20% acetonitrile, 50 mM sodium acetate, pH 4.0. A delipidated and desalted sample is essential for adequate separation of the proteins by electrophoresis and subsequent staining by Pro-Q® Diamond phosphoprotein gel stain. Following separation, immerse the gel in ~100 mL of fixing solution and incubate at room temperature with gentle agitation for at least 30 min. Repeat the fixation step once more to ensure that all of the SDS is washed out of the gel. Gels can be left in the fix solution overnight. Following this, incubate the gel in ~100 mL of ultrapure water with gentle agitation for 10 min. It is important that the gel be completely immersed in water in order to remove all of the methanol and acetic acid from the gel. Residual methanol or acetic acid will interfere with Pro-Q® Diamond phosphoprotein staining. Repeat this step twice for a total of three washes. Stain the gel by incubating the gel in a volume of Pro-Q® Diamond phosphoprotein gel stain equivalent to ten times the volume of the gel with gentle agitation in the dark for 60–90 min. Destaining is done by incubating the gel in 80–100 mL of destain solution with gentle agitation for 30 min at room temperature, protected from light. Repeat this procedure two more times. The optimal total destaining time is about 1.5 h. Wash twice with ultrapure water at room temperature for 5 min per wash. If the background is high or irregular, the gel may be left in the second wash for 20–30 min and reimaged. Heating the water washes and the stain in a microwave oven greatly shortens the required time for staining and destaining. Stained gels are best visualized using excitation at 532–560 nm, such as that provided with a visible-light laser-based or xenon arc lamp-based gel-scanning instrument.

3.2. Glycosylation

The most classical procedure for glycoprotein detection is periodate acid-Schiff (PAS) stain. This method specifically detects glycosylated proteins having sialic acid and other oxidizable carbohydrate groups. A kit from Pierce provides three essential reagents and a complete protocol to specifically stain glycosylated proteins (glycoproteins) that have been separated by polyacrylamide gel electrophoresis (PAGE). A gel or membrane containing separated proteins is treated with a periodate solution, which oxidizes *cis*-diol sugar groups in glycoproteins. The resulting aldehyde groups are detected through the formation of Schiff base bonds with a reagent that produces magenta bands.

Reagents and procedure: Fix the gel with 50% methanol for 30 min followed by three washes for 10 min in 3% acetic acid. To use the Pierce Glycoprotein Staining Kit, simply treat the gel or membrane containing separated proteins treated with the provided periodate-based Oxidation Reagent. This treatment oxidizes *cis*-diol sugar groups in glycoproteins. The resulting aldehyde groups are reacted with the Pierce Glycoprotein Stain Reagent to form Schiff base bonds that result in the production of magenta bands of the stain. Traditionally called the PAS reagent method, this staining technique is made easier by the convenient packaging of essential reagents in one kit with easy-to-follow instructions.

Advantages and disadvantages: Magenta bands specific for glycoproteins appear against a faint pink or colorless gel background. Detection limit depends on the extent of glycosylation and protein size and abundance. Effective for use with polyacrylamide mini gels or nitrocellulose membranes. More sensitive stains have substituted the use of this stain.

Pro-Q® Emerald 300 Glycoprotein Gel Stain from Molecular Probes® provides a powerful method for differentially staining glycosylated and nonglycosylated proteins in the same gel. The Pro-Q® Emerald 300 glycoprotein stain reacts with periodate-oxidized carbohydrate groups, creating a bright green fluorescent signal on glycoproteins. Using this stain, it is possible to detect as little as 0.5 ng of glycoprotein per band, depending upon the nature and the degree of glycosylation, making it about 50-fold more sensitive than the standard periodic acid-Schiff base method using acidic fuchsin dye. Pro-Q® Emerald 300 glycoprotein stain also provides easier and much more reliable glycoprotein detection than mobility shift assays, which only detect glycoproteins susceptible to specific deglycosylating enzymes. The green fluorescent signal from Pro-Q® Emerald 300 stain can be visualized with 300 nm UV illumination.

Reagents and procedure: Add 6 mL of DMF to the vial containing the Pro-Q® Emerald 300 reagent and mix gently and thoroughly to dissolve the contents. Store the stock solution at −20°C. Fix solution is a solution of 50% methanol and 5% acetic acid in dH_2O. The fix solution should be prepared fresh for each experiment. Wash solution contains 3% glacial acetic acid in dH_2O. Oxidizing solution contains periodic acid dissolved in 250 mL of 3% acetic acid. Perform SDS-PAGE, following which the gel is fixed in fixing solution by incubation at room temperature with gentle agitation. Wash the gel with gentle agitation for 10–20 min twice. Incubate the gel in 25 mL of oxidizing solution with gentle agitation for 30 min followed by three more washes. Prepare fresh Pro-Q® Emerald 300 staining solution and place gel in stain for 90 min. The signal can be seen after about 30 min and maximum sensitivity is reached at about 120 min. Wash the gel at room temperature for

15 min. Repeat this wash once for a total of two washes. Do not leave the gel in wash solution for more than 2 h, as the staining will start to decrease. Stained glycoproteins can be visualized using a 300-nm UV transilluminator.

Advantages and disadvantages: Highly sensitive stain and most popular choice for proteomics.

4. Common Staining Agents on Blots

Western blotting is the transfer of proteins from the SDS-PAGE gel to a solid supporting membrane. There are three different supports commonly in use: nitrocellulose, nylon, and polyvinylidene fluoride. The choice of membrane depends on the protein being transferred, as different proteins may bind more efficiently to one membrane in comparison with another. Blots are stained to check the status of transfer (electroblotting). Ponceau Red, Coomassie BB R-250, Amido Black, Indian ink, silver stain, SYPRO Ruby, and AuroDye are commonly used to stain blots. Here we discuss Ponceau S and Amido Black stains as they are the most commonly used stains.

Ponceau S staining solution is used for the detection of proteins on cellulose acetate, PVDF, and nitrocellulose membranes. Microgram quantities of transferred protein can be detected with a clear background and red protein bands. This staining technique is reversible to allow further immunological detection. The limit of detection for this stain is 250 ng. Ponceau S is a negative stain which binds to the positively charged amino groups of the protein. It also binds noncovalently to nonpolar regions in the protein. Ponceau S is not suitable for use with nylon membranes.

Reagents and procedure: Ponceau S 0.1% in 1% (v/v) acetic acid. After electrophoresis, immerse the blotted membrane in a sufficient amount of Ponceau S staining solution and stain for 5 min. After staining, immerse the membrane in an aqueous solution containing 5% acetic acid (v/v) for 5 min, change the aqueous solution, and immerse the membrane for another 5 min. Transfer the membrane into water for two washes of 5 min each. Remove the membrane and block as normal.

Amido Black, also known as Naphthalene Blue-Black or Naphthalene Black 12B, is a protein dye that stains the proteins in a blue-black color.

Reagents and protocol: Prepared as 0.1% (w/v) solution of Napthol Blue Black in 10% (v/v) methanol, 2% acetic acid. Proteins are transferred from the 2D gels onto nitrocellulose membranes according to standard methods for electroblotting. The blot is immersed in an Amido Black solution while shaking for 3–6 s, followed by immediate removal of the stain. Longer contact with the stain would interfere with the destaining procedure. Destaining is carried out for 3–5 min by rinsing with running deionized water. The spots then become visible against a faint blue background. The localization of the spots can be marked either by photography or careful encircling with indelible ink without damaging the membrane.

5. Pre-electrophoresis Staining

Proteins are stained prior to electrophoresis by attaching a (usually fluorescent) label in a defined chemical way. If dyes with similar structure but different spectral properties are available, combination of these fluorophores in one gel is possible, thus allowing differential imaging (DIGE).

5.1. Modified Lysine Residues

One of the very first fluorophores used to stain proteins either before or after the electrophoretic separation is MDPF(2-methoxy-2,4-diphenyl-3(2H)-furanone). It reacts irreversibly with the ε-amino-group of lysine residues and protein N-terminal groups, giving a very stable and sensitive fluorescent signal. It can be applied prior to electrophoresis (SDS-PAGE) (27) as well as after the IEF step or after 2-DE (28). Today, cyanine-based dyes have found wide use as the commercially available CyDyes. There are three dyes (Cy2, Cy3, and Cy5) that covalently label lysine residues of proteins via an amide linkage. Although structurally similar, they are spectrally quite distinct (29). With three dyes of this kind, a multiplexed system is possible, allowing differential images (DIGE) to be created, as first described by Uenlue et al. (30). Up to three samples, each labeled with one of these fluorophores, may be mixed and separated electrophoretically in only one run. Subsequent scanning at appropriate wavelengths generates an image for each sample, all run under identical conditions, without interference of run-to-run differences. The general idea behind DIGE is that with multiplexing fewer gels are needed to enable reliable evaluation of a given sample set (29). Besides a lower number of gels and replicates necessary for statistical analysis, this new method is supposed to allow smaller differences to be accurately detected and quantified with statistical confidence. DIGE should exert its full strength in settings comprising highly similar but not identical biological conditions (31). This includes comparison of treated/untreated or healthy/pathological states, and more multifaceted experiments like time courses and dose–response experiments (31). Some other recent applications include the study of pattern changes induced by treatment, e.g., phosphorylation (32), or for disease markers. Post-staining with other dyes is recommended prior to spot picking from minimally labeled DIGE gels. Reproducible results were reported for SYPRO Ruby (33). DIGE technology requires specialized equipment and evaluation software. For imaging, specific fluorescence scanners are necessary, usually point scanners with laser light sources with different wavelength and a wide dynamic range. Dedicated algorithms have been developed, similar to those from microarrays, to eliminate dye influences (34–36). Differentially expressed proteins are extracted by means of statistical tests, such as *t*-test, ANOVA, or cluster analysis (31–37).

5.2. Modified Cysteine Residues

Cysteine is a less prevalent amino acid; it is chemically modified through a reaction with its thiol group. The compound 3,3′,5,5′-Tetramethylbenzidine (TMB) has been used to fluorescently label sulfhydryl groups in protein solutions prior to electrophoresis. Small changes in p*I* and *Mr* have been found for TMB-labeled proteins as well as variations in relative spot intensities. More recently, a very sensitive type of cyanine dyes has been developed that also allows multiplexing, although with only two fluorophores. The dyes CyDye DIGE Fluor Cy3 and CyDye DIGE Fluor Cy5 contain maleimide reactive groups that covalently bind to the cysteine residues on proteins via a thioether linkage. The general mass shift is about 0.7 kDa, the dyes have the same spectral qualities as minimal CyDyes (38). This saturation labeling reaction has to be carefully optimized for each type of sample to ensure that all accessible cysteine residues contained within a protein are labeled ("saturation labeling"). Excess dye may give side reactions with lysines. These dyes offer increased sensitivity over their minimal-labeling counterparts (the method is also propagated as "scarce sample labeling"). Most applications described are in the field of cancer research, in search of potential biomarkers, using samples obtained by laser capture microdissection (39, 40). Being very sensitive, saturation labeling can be performed on protein lysates obtained from as little as 1,000 microdissected cells (41). It has also been described for comparing protein expression profiles of human hepatocellular carcinoma cell lines with primary culture hepatocytes.

FlaSHpro dyes are another set of cysteine-specific fluorescent dyes that react with the cysteine residues on proteins via their maleimide group.

6. Conclusion

This chapter reviews different types of stains for proteins used in different applications. It summarizes procedures, reagent preparations, and advantages of each stain. It also compiles some important points to be considered in imaging and evaluation. Although there are a large number of different staining methods, the patterns obtained may vary depending on the detection mechanisms of the applied methods. Hence, it is necessary to choose the staining method depending on the downstream applications planned. Some points to be attentive of when selecting a stain for a particular experiment are composition of the proteins of interest (presence/absence of single amino acids), availability of sample, and post-staining application of the gel/protein.

References

1. Steinberg TH (2009) Protein gel staining methods: an introduction and overview. Methods Enzymol 463:541–63
2. Dunbar BS, Kimura H, Timmons TM (1990) Protein analysis using high resolution two-dimensional polyacrylamide gel electrophoresis. In: Deutscher MP (ed) Methods in enzymology, vol 182. Academic, San Diego, CA, pp 441–459
3. Garfin DE (1990) One-dimensional gel electrophoresis. In: Deutscher MP (ed) Methods in enzymology, vol 182. Academic, San Diego, CA, pp 425–441
4. Garfin DE (1990) Isoelectric focusing. In: Deutscher MP (ed) Methods in enzymology, vol 182. Academic, San Diego, CA, pp 459–478
5. Merril CR (1990) Gel-staining techniques. In: Deutscher MP (ed) Methods in enzymology, vol 182. Academic, San Diego, CA, pp 477–488, xxixþ894pp
6. Miller I, Crawford J, Gianazza E (2006) Protein stains for proteomic applications: which, when, why? Proteomics 6:5385–5408
7. Patton WF (2000) A thousand points of light: the application of fluorescence detection technologies to two-dimensional gel electrophoresis and proteomics. Electrophoresis 21:1123–1144
8. Patton WF (2002) Detection technologies in proteome analysis. J Chromatogr B 771:3–31
9. Smejkal GB (2004) The Coomassie chronicles: past, present and future perspectives in polyacrylamide gel staining. Expert Rev Proteomics 1:381–387
10. Neuhoff V, Arold N, Taube D, Ehrhardt W (1988) Improved staining in polyacrylamide gels including isoelectric focusing gels with clear background at nanogram sensitivity using Coomassie Brilliant Blue G-250 and R-250. Electrophoresis 9:255–262
11. Candiano G, Bruschi M, Musante L et al (2004) Blue silver: a very sensitive colloidal Coomassie G-250 staining for proteome analysis. Electrophoresis 25:1327–1333
12. Rabilloud T (1990) Mechanisms of protein silver staining in polyacrylamide gels: a 10-year synthesis. Electrophoresis 11:785–794
13. Poland J, Rabilloud T, Sinha P (2005) Silver staining of 2-D gels. In: Walker JM (ed) The proteomics protocol handbook. Humana Press, Totowa, NJ, pp 215–222
14. Fernandez-Patron C (2005) Zn2þ reverse staining technique. In: Walker JM (ed) The proteomics protocol handbook. Humana Press, Totowa, NJ, pp 215–222
15. Fernandez-Patron C, Castellanos-Serra L, Hardy E et al (1998) Understanding the mechanism of the zinc-ion stains of biomacromolecules in electrophoresis gels: generalization of the reverse-staining technique. Electrophoresis 19:2398–2406
16. Steinberg TH, Hart CR, Patton WF (2005) Rapid, Sensitive Detection of Proteins in Minigels with Fluorescent Dyes. Coomassie Fluor Orange, SYPRO Orange, SYPRO Red, SYPRO Tangerine, Protein Gel Stains. In: Walker JM (ed), The proteomics handbook. Humana Press, Totowa, NJ, pp 215–222
17. Daban JR (2001) Fluorescent labeling of proteins with Nile red and 2-methoxy-2, 4-diphenyl-3(2H)-furanone: physicochemical basis and application to the rapid staining of sodium dodecyl sulfate polyacrylamide gels and Western blots. Electrophoresis 22:874–880
18. Daban JR, Bartolome S, Samso M (1991) Use of the hydrophobic probe Nile red for the fluorescent staining of protein bands in sodium dodecyl sulfate-polyacrylamide gels. Anal Biochem 199:169–174
19. Bell PJ, Karuso P (2003) Epicocconone, a novel fluorescent compound from the fungus epicoccumnigrum. J Am Chem Soc 125: 9304–9305
20. Schulenberg B, Goodman TN, Aggeler R et al (2004) Characterization of dynamic and steady-state protein phosphorylation using a fluorescent phosphoprotein gel stain and mass spectrometry. Electrophoresis 15:2526–2532
21. Murray J, Marusich MF, Capaldi RA et al (2004) Focused proteomics: monoclonal antibody-based isolation of the oxidative phosphorylation machinery and detection of phosphoproteins

using a fluorescent phosphoprotein gel stain. Electrophoresis 15:2520–2525

22. Schulenberg B, Aggeler R, Beechem JM et al (2003) Analysis of steady-state protein phosphorylation in mitochondria using a novel fluorescent phosphosensor dye. J Biol Chem 278:27251-5
23. Steinberg TH, Agnew BJ, Gee KR et al (2003) Global quantitative phosphoprotein analysis using Multiplexed Proteomics technology. Proteomics 3:1128–44
24. Schulenberg B, Patton WF (2004) Combining microscale solution-phase isoelectric focusing with Multiplexed Proteomics dye staining to analyze protein post-translational modifications. Electrophoresis 25:2539–44
25. Hayduk EJ, Choe LH, Lee KH (2004) A two-dimensional electrophoresis map of Chinese hamster ovary cell proteins based on fluorescence staining. Electrophoresis 25:2545–56
26. Ge Y, Rajkumar L, Guzman RC et al (2004) Multiplexed fluorescence detection of phosphorylation, glycosylation, and total protein in the proteomic analysis of breast cancer refractoriness. Proteomics 4:346
27. Barger BO, White RC, Pace JL et al (1976) Estimation of molecular weight by polyacrylamide gel electrophoresis using heat stable fluorophors. Anal Biochem 70:327–335
28. Jackson P, Urwin VE, Mackay CD (1988) Rapid imaging, using a cooled charge-coupled-device, of fluorescent two-dimensional polyacrylamide gels produced by labelling proteins in the first-dimensional isoelectric focusing gel with the fluorophore 2-methoxy-2,4-diphenyl-3(2H)furanone. Electrophoresis 9:330–339
29. Tonge R, Shaw J, Middleton B et al (2001) Validation and development of fluorescence two-dimensional differential gel electrophoresis proteomics technology. Proteomics 1:377–396
30. Uenlue M, Morgan ME, Minden JS (1997) Difference gel electrophoresis: a single gel method for detecting changes in protein extracts. Electrophoresis 18:2071–2077
31. Sitek B, Scheibe B, Jung K, Schramm A, Stuehler K (2006) Difference gel electrophoresis (DIGE): the next generation of two-dimensional gel electrophoresis for clinical research. In: Marcus K, Stuehler K, van Hall A, Hamacher M et al. (eds) Proteomics in drug research. Wiley-VCH, Weinheim, pp 33–55
32. Raggiaschi R, Lorenzetto C, Diodato E et al (2006) Detection of phosphorylation patterns in rat cortical neurons by combining phosphatase treatment and DIGE technology. Proteomics 2006(6):748–756
33. Gharbi S, Gaffney P, Yang A et al (2001) Evaluation of two-dimensional differential gel electrophoresis for proteomic expression analysis of a model breast cancer cell system. Mol Cell Proteomics 1:91–98
34. Huber W, von Heydebreck A, Sultmann H et al (2002) Variance stabilization applied to microarray data calibration and to the quantification of differential expression. Bioinformatics 18(Suppl 1):S96–104
35. Karp NA, Kreil DP, Lilley KS (2004) Determining a significant change in protein expression with DeCyder during a pair-wise comparison using two-dimensional difference gel electrophoresis. Proteomics 4:1421–1432
36. Kreil DP, Karp NA, Lilley KS (2004) DNA microarray normalization methods can remove bias from differential protein expression analysis of 2D difference gel electrophoresis results. Bioinformatics 20:2026–2034
37. Fodor IK, Nelson DO, Alegria-Hartman M et al (2005) Statistical challenges in the analysis of two-dimensional difference gel electrophoresis experiments using DeCyder. Bioinformatics 21:3733–3740
38. Somiari RI, Russell S, Somiari SB, Sullivan AG et al (2005) Differential In-Gel Electrophoresis in a High-Throughput Environment. In: Walker JM (ed) The proteomics handbook. Humana Press, Totowa, NJ, pp 223–237.
39. Greengauz-Roberts O, Stoppler H, Nomura S et al (2005) Saturation labeling with cysteine-reactive cyanine fluorescent dyes provides increased sensitivity for protein expression profiling of laser-microdissected clinical specimens. Proteomics 5:1746–1757
40. Wilson KE, Marouga R, Prime JE et al (2005) Comparative proteomic analysis using samples obtained with laser microdissection and saturation dye labeling. Proteomics 5: 3851–3858
41. Sitek B, Luttges J, Marcus K et al (2005) Application of fluorescence difference gel electrophoresis saturation labelling for the analysis of microdissected precursor lesions of pancreatic ductal adenocarcinoma. Proteomics 5: 2665–2679

Chapter 40

Coomassie-Brilliant Blue Staining of Polyacrylamide Gels

Claudia Arndt, Stefanie Koristka, Anja Feldmann, Holger Bartsch, and Michael Bachmann

Abstract

Over the past, a series of staining procedures for proteins were published. The most commonly used staining dye for proteins is still Coomassie-Brilliant Blue. The major reason is Coomassie-Brilliant Blue staining is simple, fast, and sensitive. As Coomassie-Brilliant Blue is almost insoluble in water, a series of procedures including colloidal aqueous procedures were described.

Key words: Coomassie-Brilliant Blue, Polyacrylamide gels, Proteins

1. Introduction

The name Coomassie originates from the name of the town Coomassie, nowadays known as Kumasi, in Ghana. The dye was originally invented for the staining of textiles. However, in 1963, it was described for the staining of proteins after separation by electrophoresis (1). Coomassie-Brilliant Blue can not only be used for staining of proteins in gels. It is also known as Bradford reagent for quantitative photometric estimation of protein concentrations in solution. Moreover, upon binding to protein, Coomassie-Brilliant Blue will transfer an overall negative charge to the protein which allows the separation of proteins using polyacrylamide gel electrophoresis under native conditions, a technique that is termed Blue Native PAGE (2). The most commonly used two dye forms which only slightly differ in their chemical structure are Coomassie-Brilliant Blue R-250 or G-250: The R stands for the more red, the G for the more green color of the respective dye. In its anionic form, Coomassie forms a stable blue complex with proteins. Over the past years, a series of procedures were developed and described

Biji T. Kurien and R. Hal Scofield (eds.), *Protein Electrophoresis: Methods and Protocols*, Methods in Molecular Biology, vol. 869, DOI 10.1007/978-1-61779-821-4_40, © Springer Science+Business Media, LLC 2012

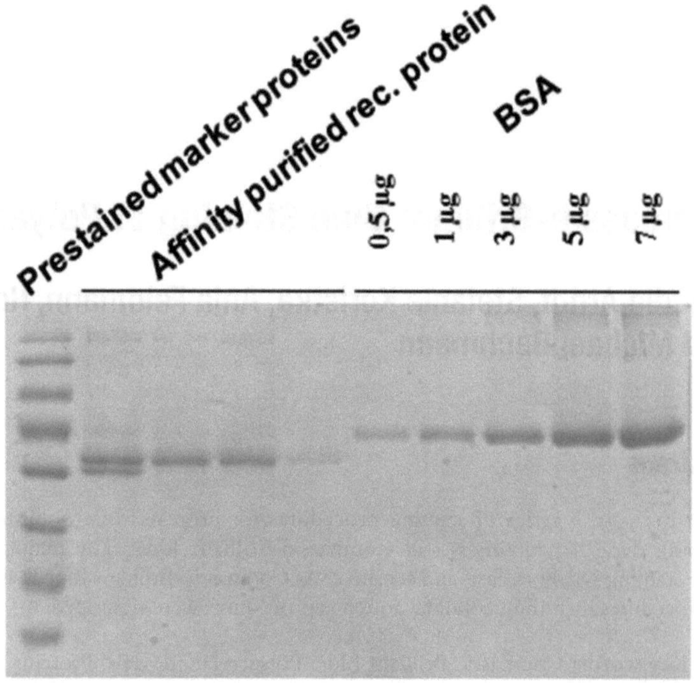

Fig. 1. Example of a 1D gel stained with Coomassie-Brilliant Blue according to the standard staining protocol (see Subheading 3.1).

to facilitate and speed up the staining of proteins with Coomassie-Brilliant Blue after gel electrophoresis. As Coomassie-Brilliant Blue is insoluble in water, the staining is usually based on either acidic solutions of the dye in methanol or colloidal forms for aqueous procedures (e.g., (3, 4)). The former versions require a staining/destaining procedure. Here we include two Coomassie-Brilliant Blue staining procedures according to a standard staining/destaining protocol and a colloidal Coomassie-Brilliant Blue staining protocol. According to the literature, the colloidal staining procedure is more sensitive. As shown in Figs. 1 and 2 for visualization of the selected proteins, here, we achieved comparable sensitivities with both staining protocols.

2. Materials

2.1. Materials for a Standard Coomassie Staining Protocol

1. Coomassie-Brilliant Blue R-250 (or G-250).
2. Staining solution:
   (a) Dissolve 2.5 g of Coomassie-Brilliant Blue in 450 mL methanol and stir overnight.
   (b) Add 450 mL of deionized water (see Note 1).

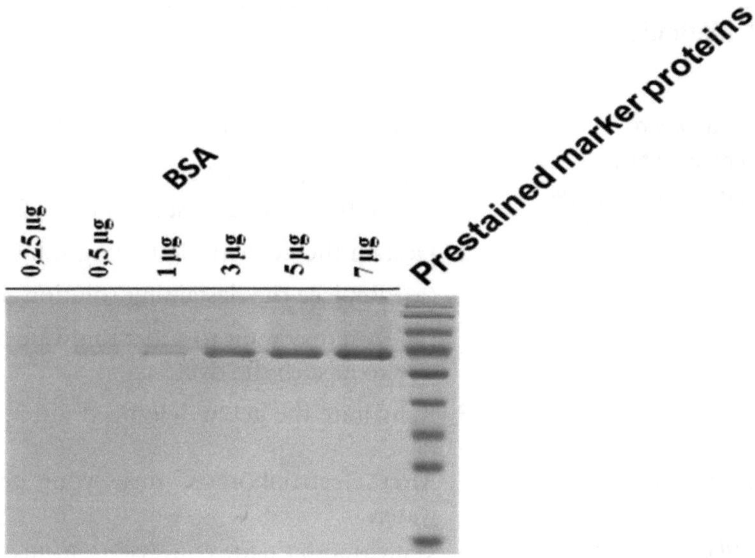

Fig. 2. Example of a 1D gel stained with colloidal Coomassie-Brilliant Blue staining procedure according to protocol (see Subheading 3.2).

(c) Add 100 mL of glacial acetic acid.

(d) If necessary, filter the solution using Whatman filter paper (see Note 2).

3. Destaining solution (see Note 3):

   (a) 450 mL methanol.

   (b) 450 mL deionized water.

   (c) 100 mL glacial acid.

2.2. Materials for a Colloidal Coomassie Staining Protocol

1. Coomassie-Brilliant Blue G-250.
2. Staining solution:

   (a) Suspend 1 g of Coomassie-Brilliant Blue in 20 mL of deionized water.

   (b) Dissolve separately 100 g of ammonium sulfate in about 600 mL of deionized water.

   (c) Add the prepared 20 mL [see Subheading 2.2, step 2(a)] of Coomassie-Brilliant Blue solution to the ammonium sulfate solution (when completely dissolved).

   (d) Then add 30 mL of orthophosphoric acid, followed by 200 mL of ethanol.

   (e) Add water to make up volume to 1,000 mL (see Note 1).

3. Destaining solution: deionized water.

3. Methods

3.1. Staining According to a Standard Coomassie Staining Protocol

After electrophoresis, rinse your gel twice with deionized water.

1. Stain the gel with Coomassie-Brilliant Blue staining solution (see Subheading 2.1, step 2) for 1–2 h (see Note 4).
2. Destain the gel with destaining solution:
   (a) Replace the destaining solution several times.
   (b) You may want to add a Kimwipe to the destaining solution to adsorb the dye.
3. Rehydrate the gel in deionized water.

3.2. Staining According to a Colloidal Coomassie Staining Protocol

1. After electrophoresis, rinse your gel twice with deionized water.
2. Stain the gel with Coomassie-Brilliant Blue staining solution (see Subheading 2.2, step 2) for 1–2 h (see Note 4).
3. The bands should become visible without destaining. However, if necessary, you can destain the gel with water.

4. Notes

1. Unless stated otherwise, all solutions should be prepared in deionized water that has a resistance greater than 18 MΩ and total organic content of less than five parts per billion. This standard is referred to as "deionized water" in this text.
2. It will take some time to filter the Coomassie Blue stain.
3. Destaining solution is actually staining solution minus Coomassie-Brilliant Blue.
4. In order to increase the sensitivity, it may be necessary to extend the staining time over night.

Acknowledgments

Special thanks for excellent technical support to Christin Gräfe, Livia Schulze, and Barbara Uteß (Inst. Immunology, Dresden, Germany).

References

1. de St F, Groth S, Webster RG, Datyner A (1963) Two new staining procedures for quantitative estimation of proteins on electrophoretic strips. Biochim Biophys Acta 71:377–391
2. Wittig I, Braun HP, Schägger H (2006) Blue native PAGE. Nat Protoc 1:418–428
3. Neuhoff V, Arold N, Taube D et al (1988) Improved staining of proteins in polyacrylamide gels including isoelectric focusing gels with clear background at nanogram sensitivity using Coomassie Brilliant Blue G-250 and R-250. Electrophoresis 9:255–262
4. Kang D, Gho YS, Suh M et al (2002) Highly sensitive and fast protein detection with coomassie brilliant blue in sodium dodecyl sulfate-polyacrylamide gel electrophoresis. Bull Korean Chem Soc 11:1511–1512

Chapter 41

Accelerated Coomassie Blue Staining and Destaining of SDS-PAGE Gels with Application of Heat

Biji T. Kurien and R. Hal Scofield

Abstract

Coomassie Brilliant Blue is commonly used for the detection of proteins in sodium dodecyl sulfate–polyacrylamide gel electrophoresis, owing to its reliability and simplicity. Here, we report dramatically decreased protein staining and destaining time, as well as significantly increased detection sensitivity with the application of enhanced heat. The staining time was 5 min at 55, 62.5, or 70°C for a 1.5-mm gel, while it took 45, 45, and 20 min, respectively, for destaining. The staining time could be reduced to 1 min for a 0.8 mm gel stained at 65°C, to 2 min at 60°C and 5 min at 55°C. The destaining of proteins analyzed on a 0.8 mm gel could be accomplished in 8, 15, and 20 min at 65, 60, and 55°C, respectively. Application of heat, thus, enables proteins to be stained and destained rapidly, as well as enhancing detection sensitivity.

Key words: SDS-PAGE, Coomassie Brilliant Blue, Heat, Destaining

1. Introduction

Sodium dodecyl sulfate–polyacrylamide gel electrophoresis (SDS-PAGE) (1) is an invaluable technique employed to separate proteins on the basis of size and is a vital part of almost every laboratory involved in proteomics. Detection of proteins on SDS-PAGE gels has become a rapidly evolving technology and has led to the development of a variety of colorimetric protein stains as well as several new fluorescent stains.

Numerous colorimetric staining methods for proteins have been described, including staining with Coomassie Brilliant Blue (CBB), Coomassie and Bismarck brown mixture, amido black (2), and silver (2–5). Proteoglycans and glycosaminoglycans, stained weakly by traditional protein stains, have been detected with combined Alcian

Biji T. Kurien and R. Hal Scofield (eds.), *Protein Electrophoresis: Methods and Protocols*, Methods in Molecular Biology, vol. 869, DOI 10.1007/978-1-61779-821-4_41,

blue-silver stain (6). In addition to colorimetric staining methods, several fluorescent stains have become available.

Fluorescent staining methods for proteins include staining with Nile Red, SYPRO Red, SYPRO Orange, SYPRO Tangerine, Coomassie Fluor Orange, SYPRO Ruby, Epicocconone (Deep Purple, Lightning Fast protein gel stain), fluorescein derivatives, Krypton, Krypton infrared, Flamingo LUCY stain (3, 7), and Alta [0.8% Crocein scarlet (brilliant Crocein) and 0.2% Rhodamine B] (8). Phosphoproteins have been detected using Pro-Q Diamond and Phos-tag (3). Glycoproteins have been detected on gels using acid fuchsin dye, Pro-Q Emerald 300, Pro-Q Emerald 488 glycoprotein gel stain kits, and azide–alkyne "click chemistry" reagents (3). A nontoxic fluorescent stain derived from the food spice turmeric has been reported recently (9).

Protein labeling with radioactive isotopes, even though the most sensitive method to detect proteins on gels, requires special equipment and very complicated handling procedures (5). Also, there is the issue of health and safety concerns associated with this procedure.

CBB protein stain has been used more commonly than other staining procedures, owing to its simplicity and reliability. Here, we show the use of heat for the rapid staining and destaining of proteins on SDS-PAGE with CBB, as well as for enhancing detection sensitivity.

Figure 1 depicts gels stained and destained at 55, 62.5 and 70°C using a hot water bath. A gel stained for 5 min at 55°C and destained for 45 min at the same temperature is shown in Fig. 1a. Figure 1b displays a gel stained at 62.5°C for 5 min and destained at 62.5°C for 45 min. Shown in Fig. 1c is a gel stained at 70°C for 5 min and destained at 70°C for 20 min. As seen in Fig. 1, the protein molecular weight standards myosin (200 kDa), phosphorylase B (97 kDa), BSA (68 kDa), ovalbumin (43 kDa), carbonic anhydrase (29 kDa), β-lactoglobulin (18 kDa), and lysozyme (14 kDa) stained significantly better at 62.5°C and 70°C compared to 55°C (at 2,500 and 500 ng level). Lower amounts (100 and 50 ng) of the marker, however, stained only faintly at both 62.5 and 70°C. Myosin (2,500 ng), on the other hand, did not get stained at 55°C (5 min). Similar results were observed when the amount of phospholipase B was reduced to 1,000 ng (Fig. 1a, lane 2) and bovine serum albumin (BSA) to 500 ng (Fig. 1a, lane 3). Destaining of the gel could be carried out in half the time at 70°C compared to the time taken to destain at 55 and 62.5°C. The intensity of the protein bands (Fig. 1) obtained following densitometric analysis is given as absolute numbers in Table 1. The increased sensitivity of staining at 70°C, compared to 62.5 and 55°C, is clearly evident from Table 1.

Figure 2 presents the Coomassie stained and destained HeLa cell lysate proteins following separation on an SDS-PAGE gel.

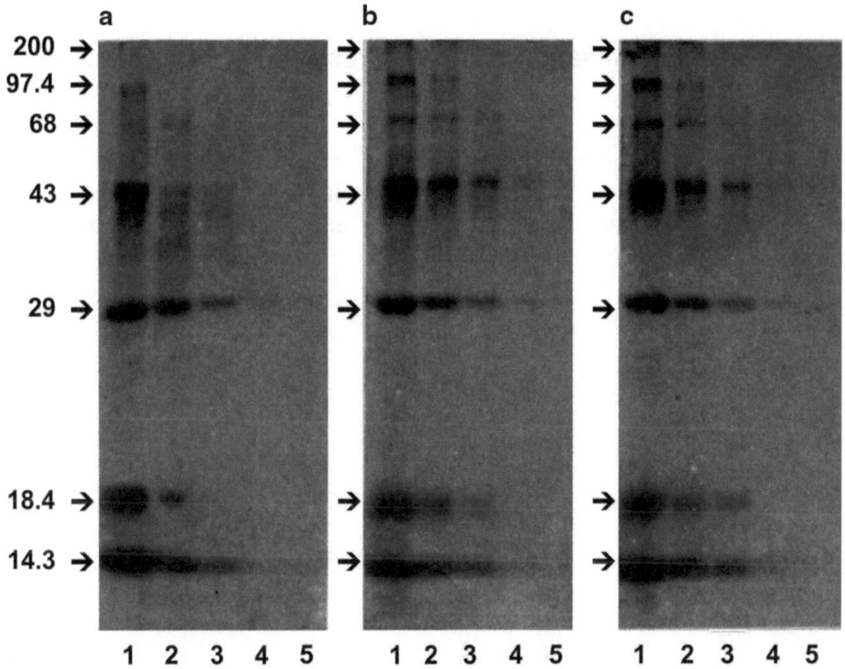

Fig. 1. High molecular weight unstained protein markers on a 10% SDS-PAGE gel (1.5 mm) stained and destained at (**a**) 55°C, (**b**) 62.5°C, and (**c**) 70°C. The protein markers are in the following order from the top of the gel–myosin H chain (200 kDa), phosphorylase B (97.4 kDa), BSA (68 kDa), ovalbumin (43 kDa), carbonic anhydrase (29 kDa), β-lactoglobulin (18.4 kDa), and lysozyme (14.4 kDa). The markers were stained with CBB at 55, 62.5 and 70°C. Lanes 1, 2, 3, 4, and 5 in each gel contains 2,500, 1,000, 500, 100, and 50 ng, respectively, of protein in each lane. Reproduced from ref. 2 with permission from National Institute of Science Communication and Information Resources, New Delhi, India.

Staining and destaining at 70°C yielded the best staining pattern, compared to that observed at 62.5 and 55°C. Figure 3 shows the staining of bovine serum albumin analyzed on a SDS-PAGE gel (0.8 mm thickness) at 55, 60 and 65°C. The protein could be stained in 5, 2, and 1 min at 55, 60 and 65°C, respectively, while it took 20, 15, and 8 min for destaining (at 55, 60 and 65°C, respectively).

The enhanced heat used did not affect the integrity of the gel in any way. We have used temperatures of 80°C for staining and destaining without any adverse effect to the gel (however, the protein bands appeared to be a little fuzzy when stained at this high temperature—data not shown). Enhanced temperatures appear to make the gel more permeable to staining and destaining solutions, as evident from the short time required for staining and destaining. Wu and Welsh have reported the combination of 0.13% Coomassie and high heat (55°C) for staining SDS-PAGE protein gels in 30 min (10). A heating step for staining has been suggested earlier as well (11).

Thus, CBB staining and destaining of proteins on SDS-PAGE gels will be useful to detect proteins that are available in small amounts rapidly and in a simple way.

Table 1
The intensity of the protein bands of Fig. 1 represented in terms of numbers

Protein	Myosin H	Phosphorylase	BSA	Ovalbumin	Carbonic anhydrase	Lactoglobulin	Lysozyme
Fig. 1a							
1	–	97	92	201	162	140	177
2	–	–	86	92	160	95	147
3	–	–	–	48	111	–	114
4	–	–	–	–	43	–	–
Fig. 1b							
1	98	138	122	192	147	143	174
2	68	75	76	164	136	117	140
3	–	–	55	56	96	72	115
4	–	–	–	39	50	–	–
Fig. 1c							
1	132	164	169	216	177	141	194
2	86	78	95	165	123	107	162
3	–	–	61	63	89	72	126
4	–	–	–	45	51	–	–

The brightness/area of each band in the scanned picture of the gel was found out with a NIH Image 1.6 PPC software on a Macintosh computer, and the numbers are presented. No numbers are shown when a particular band is absent. Figure 1a–c represents the gels that were stained and destained at 55, 62.5 and 70°C, respectively. The numbers 1, 2, 3, and 4 in each case refer to the lane number in each figure
Reproduced from ref. 2 with permission from National Institute of Science Communication and Information Resources, New Delhi, India

2. Materials

Prepare all solutions using ultrapure water (prepared by purifying deionized water, to attain a sensitivity of 18 MΩ cm at 25°C) and analytical grade reagents. Prepare and store all reagents at room temperature (unless indicated otherwise). Diligently follow all waste disposal regulations when disposing waste materials. We do not add sodium azide to reagents.

1. 10% SDS-PAGE gels (1.5 or 0.8 mm thick).
2. Coomassie Brilliant Blue (CBB) stain: 0.25% CBB (see Note 1) in 25% isopropanol/10% acetic acid (see Note 2).
3. Destaining solution: 25% isopropanol/10% acetic acid (see Note 2).
4. Heated, shaking water bath.
5. Unstained molecular weight protein standards.
6. Prestained molecular weight protein standards.
7. HeLa cell extract.

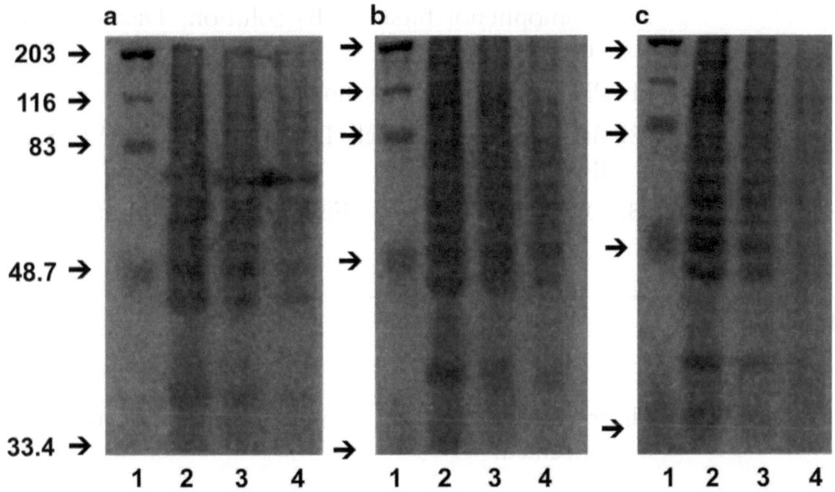

Fig. 2. SDS-PAGE of HeLa cell lysate on a 10% gel (1.5 mm) stained and destained at (**a**) 55°C, (**b**) 62.5°C, and (**c**) 70°C. *Lane 1* in each figure shows prestained molecular weight standards. *Lanes 2–4* in each figure show 10, 5, and 2.5 μL of the HeLa cell extract. Reproduced from ref. 2 with permission from National Institute of Science Communication and Information Resources, New Delhi, India.

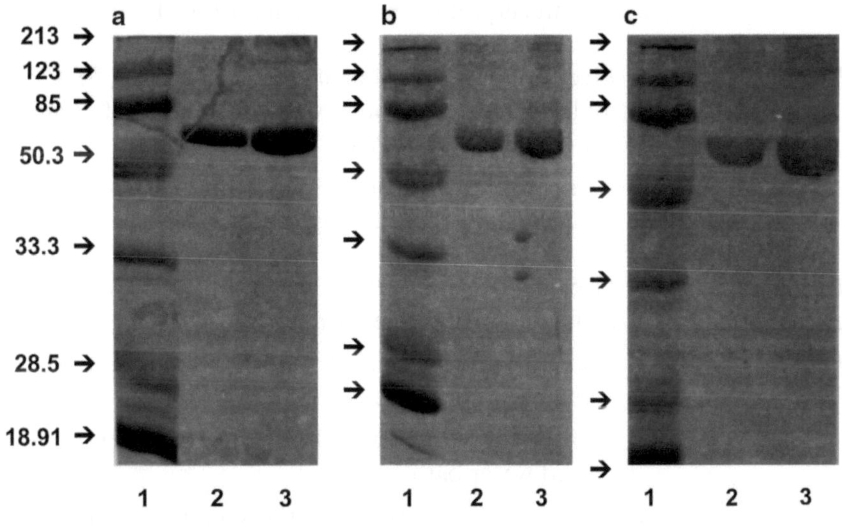

Fig. 3. Analysis of bovine serum albumin on a 10% SDS-PAGE gel (0.8 mm) stained and destained at (**a**) 55°C, (**b**) 60°C, and (**c**) 65°C. *Lane 1* in each gel show broad-range molecular weight standards, 10 μg of BSA and 15 μg of BSA, respectively. Reproduced from ref. 2 with permission from National Institute of Science Communication and Information Resources, New Delhi, India.

8. SDS-PAGE running buffer: 0.025 M Tris, 0.192 M glycine, 0.1% SDS, pH 8.3 (see Note 3).
9. SDS lysis buffer (5×): 0.3 M Tris–HCl (pH 6.8), 10% SDS, 25% β-mercaptoethanol, 0.1% bromophenol blue, 45% glycerol. Leave one aliquot at 4°C for current use and store remaining aliquots at −20°C (see Note 4).

10. Bromophenol blue (BPB) solution: Dissolve 0.1 g BPB in 100 mL water.
11. Phosphate buffered saline (PBS), pH 7.4.
12. Branson Sonifier Cell Disruptor 185 (VWR Scientific Co., Boston, MA, USA).
13. FB300 power supply (Fisher Scientific, Houston, TX, USA).

3. Methods

All procedures are carried out at room temperature unless otherwise specified.

3.1. Preparation of HeLa Cell Extract

1. Harvest freshly cultured HeLa cells by centrifuging at 800 × *g* and wash twice with PBS.
2. Lyse cells by sonication in PBS buffer using a Branson sonicator (setting 4) and centrifuge at 10,000 × *g* for 10 min at 4°C (see Note 5).
3. Use an aliquot of the supernatant for SDS-PAGE.

3.2. SDS-PAGE

1. Electrophorese aliquots of HeLa cell extract, BSA, or unstained molecular weight protein standards on 10% (0.8 mm or 1.5 mm gel thickness) SDS-PAGE.
2. Extricate the gel from between the glass plates carefully and proceed to the gel staining step.

3.3. Gel Staining with Heated Coomassie and Destaining with Heated Destain

1. Place 100 mL of CBB stain in a plastic container. Close the container tightly (see Note 6).
2. Take 250 mL of destain in another plastic container. Close the container tightly (see Note 6).
3. Preheat the contents of both the containers at 55°C in the heated water bath.
4. Add gel (1.5 mm thick) to the staining solution and incubate for 5 min.
5. Rinse gel with warm tap water (see Note 7).
6. Transfer the stained gel to the container containing the destaining solution kept at 55°C.
7. Destain for 45 min (see Note 8).
8. Transfer gel to a container containing ultrapure water and destain the gel for 10–30 min at room temperature with gentle shaking (see Note 9).

9. Impregnate the destained gel in 50% methanol containing 5% glycerol (see Note 10) for 10–20 min at room temperature with shaking.
10. Dry gel using gel drying apparatus or between two sheets of cellophane held together by two acrylic frames (see Note 11).
11. Repeat steps 1–10 for staining and destaining the 1.5-mm-thick gels at 60 and 70°C (see Note 12).
12. Carry out steps 1–10 for staining and destaining 0.8 mm SDS-PAGE gels at 55, 60, or 65°C. Carry out CBB staining for 5, 2, and 1 min for gels stained at 55, 60, and 65°C, respectively. In addition, carry out destaining for 20, 15, and 8 min for gels stained at 55, 60, and 65°C, respectively (see Note 13).

4. Notes

1. Weigh the required amount of the CBB dye and dissolve it in methanol first. Add water to make up the volume. Filter the dissolved dye through a circular Whatman 1M filter paper folded into a cone and fitted inside a glass funnel. For a quicker way to filter the Coomassie solution, layer a circular 1M Whatman filter paper (125-mm-diameter circles) inside a Buchner funnel (Coors, USA; 135-mm inner diameter). The filter paper should cover all the holes in the funnel and fit within the walls of the funnel. Attach the Buchner funnel to a side-arm-fitted conical flask and attach the conical flask to the house vacuum through the side arm. Pour the Coomassie solution into the Buchner funnel and apply vacuum. If filter paper gets clogged, replace with a fresh one and filter. Collect the filtered Coomassie. Adjust volume with water, methanol, and acetic acid to obtain a 0.05% Coomassie staining solution in 25% methanol/10% acetic acid.
2. Methanol can also be used, instead of isopropanol.
3. Simple method of preparing running buffer: Prepare 10× native buffer (0.25 M Tris, 1.92 M glycine). Weigh 30.3 g Tris and 144 g glycine, mix, and make it to 1 L with water. Dilute 100 mL of 10× native buffer to 990 mL with water and add 10 mL of 10% SDS. Care should be taken to add SDS solution last, since it makes bubbles.
4. SDS precipitates at 4°C. Therefore, the lysis buffer needs to be warmed prior to use.
5. Cool microcentrifuge tube, containing HeLa cells suspended in PBS, on ice first. Clean sonicator probe with ethanol first

and then with distilled water. Wipe the probe dry with Kimwipes. Sonicate for 10 s, with the tube held inside a 100-mL beaker containing ice, and let it cool on ice. Repeat this step three more times. Make sure that the probe does not touch the bottom of the tube during sonication, to avoid the probe from puncturing a hole in the tube. Sonicators generate high-frequency sound waves in the 20,000-Hz range, outside our normal range of hearing. These sound waves can cause hearing damage, and therefore, laboratory personnel should wear sound mufflers when sonication is in process. (http://www.labmanager.com/?articles.view/articleNo/1103/article/Sonicator-Safety).

6. The container needs to be covered tightly to prevent evaporation of acetic acid and to prevent acetic acid fumes from contaminating the laboratory.
7. Rinse the CBB off the gel with few mL of water prior to adding destaining solution. Do not discard the wash into the sink. Save the rinse in a waste bottle for proper disposal. Alternatively, adding a few Kimwipes into the waste container will help adsorb the CBB. Kimwipes bind the released CBB very strongly and does not release the bound CBB even after they are incubated in water for over 5 weeks or in 1 M salt solutions for a week. Thus, the Kimwipes with the bound CBB can be discarded as solid waste. CBB removal with Kimwipes can thus serve as an environmentally safe way to discard the dye. This procedure also helps to recycle the destaining solution, provided the solution is nonradioactive (12).
8. Add a few Kimwipes to soak up the released CBB (12).
9. Destaining with just water helps clear the background blue color without destaining the stained protein bands.
10. This helps prevent the gel from cracking during drying.
11. Alternatively, the wet gel can be scanned using a regular scanner to generate picture for analysis. The gel can also be stored in polyethylene bags without buffer for several months and scanned when convenient (13).
12. Destaining at 70°C needs to be carried out for only 20 min, since the higher temperature helps destain faster.
13. Lower heating temperatures (60 and 65°C were employed instead of 62.5 and 70°C) and higher concentration of CBB (0.5% instead of 0.25%) were used for staining since the gel was only 0.8 mm in size instead of 1.5 mm. Also, incubation time for destaining was reduced for these gels owing to the same reason.

References

1. Laemmli UK (1970) Cleavage of structural proteins during the assembly of the head of bacteriophage T4. Nature 227:680–685
2. Kurien BT, Scofield RH (1998) Heat mediated quick Coomassie blue protein staining and destaining of SDS-PAGE gels. Indian J Biochem Biophys 35:385–389
3. Steinberg TH (2009) Protein gel staining methods: an introduction and overview. Methods Enzymol 463:541–563, review
4. Jin LT, Hwang SY, Yoo GS et al (2006) A mass spectrometry compatible silver staining method for protein incorporating a new silver sensitizer in sodium dodecyl sulfate-polyacrylamide electrophoresis gels. Proteomics 6:2334–2337
5. Hwang SY, Yoo GS, Choi JK (2004) Sensitive silver staining of protein in sodium dodecyl sulfate-polyacrylamide gels using an azo dye, calconcarboxylic acid, as a silver-ion sensitizer. Electrophoresis 25:2494–2500
6. Møller HJ, Heinegård D, Poulsen JH (1993) Combined alcian blue and silver staining of subnanogram quantities of proteoglycans and glycosaminoglycans in sodium dodecyl sulfate-polyacrylamide gels. Anal Biochem 209: 169–175
7. Mackintosh JA, Choi HY, Bae SH et al (2003) A fluorescent natural product for ultra sensitive detection of proteins in one-dimensional and two-dimensional gel electrophoresis. Proteomics 3:2273–2288
8. Pal JK, Godbole D, Sharma K (2004) Staining of proteins on SDS polyacrylamide gels and on nitrocellulose membranes by Alta, a colour used as a cosmetic. J Biochem Biophys Methods 61: 339–347
9. Kurien BT, D'Souza A, Scofield RH (2010) Heat-solubilized curry spice curcumin inhibits antibody–antigen interaction in in vitro studies: a possible therapy to alleviate autoimmune disorders. Mol Nutr Food Res 54: 1202–1209
10. Wu W, Welsh MJ (1996) Rapid Coomassie blue staining and destaining of polyacrylamide gels. Biotechniques 20:386–388
11. Allen G (1989) Sequencing of proteins and peptides. In: Burdon RH, van Knippenberg PH (eds) Laboratory techniques. Elsevier, New York, pp 112–113
12. Dorri Y, Kurien BT (2010) Environmentally safe removal/disposal of Coomassie Brilliant Blue from gel destain and used gel stain. Anal Biochem 404:193–196
13. Kurien BT, Scofield RH (1998) Long-term wet storage of sodium dodecyl sulfate-polyacrylamide gels in polyethylene bags. Anal Biochem 261:116–118

Chapter 42

Silver Staining Techniques of Polyacrylamide Gels

Holger Bartsch, Claudia Arndt, Stefanie Koristka, Marc Cartellieri, and Michael Bachmann

Abstract

Although the main application for polyacrylamide gels is the separation and subsequent blotting of proteins for immunodetection, there are tasks that need staining of proteins in the polyacrylamide gel. Several different staining techniques exist for protein staining in SDS gels that differ in their sensitivity, their expenditure of time, and other aspects. Still, silver staining is the most sensitive and reliable staining technique. Because this technique was developed in the 1970s, a huge number of variations exist. Therefore, we will provide herein three methods, which are robust and easy to perform.

Key words: Silver staining, Polyacrylamide gels, Proteins

1. Introduction

Although a large portion of protein SDS gels will be used for blotting and detection of proteins on the blot itself, sometimes, it is sufficient or required to stain the proteins directly in the sodium dodecyl sulfate (SDS) gel, for example, after two-dimensional (2-D) electrophoresis when proteins have to be eluted for further analysis like matrix-assisted laser desorption ionization-time-of-flight (MALDI-ToF) mass spectrometry.

Several different staining techniques exist for protein staining in SDS gels that differ in their sensitivity, their expenditure of time, and other aspects (Table 1). So, the most sensitive method is still the silver staining technique (about 10–100 times more sensitive than various Coomassie Blue staining techniques). Although some drawbacks exist, silver staining technique is the method of choice when very low amounts of protein have to be detected on electrophoresis gels. The main problems concerning silver staining are that blotting after staining is not possible and quantification is

Biji T. Kurien and R. Hal Scofield (eds.), *Protein Electrophoresis: Methods and Protocols*, Methods in Molecular Biology, vol. 869, DOI 10.1007/978-1-61779-821-4_42, © Springer Science+Business Media, LLC 2012

Table 1
Comparison of different staining techniques used for detection of proteins in polyacrylamide gels

Method	Sensitivity (ng/band)	Duration (h)	Remarks
Silver staining	5–30	1–24	a
Silver staining (w/o aldehyde)	5–30	4–24	b
Eosin	10	0.5	b
SYPRO Orange	30–50	1	b
Coomassie	200–400	3–4	b
Nitro blue tetrazolium	200–400	0.3	b

Remarks: *a* not suitable for blotting and MS, *b* suitable for blotting and MS

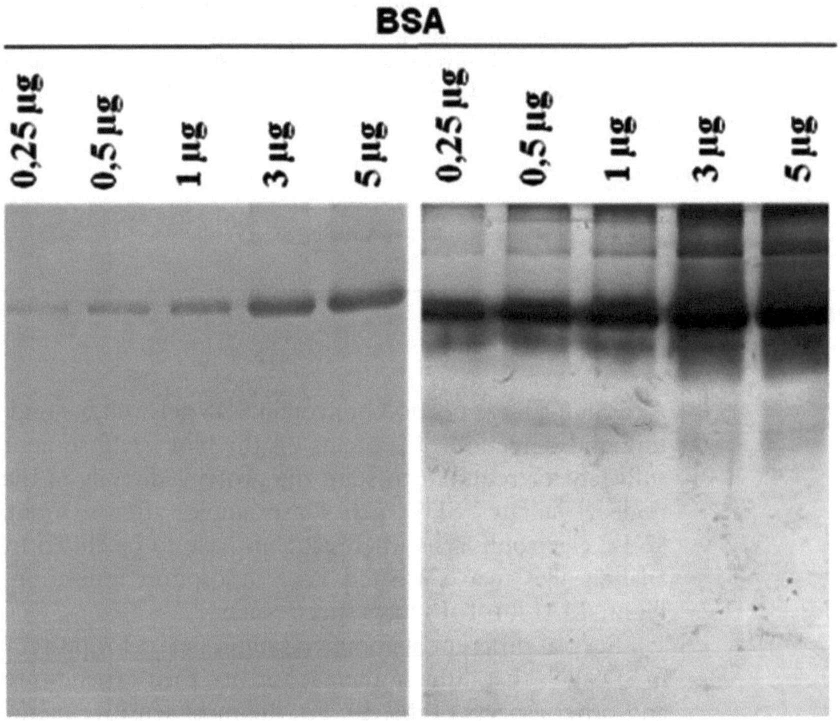

Fig. 1. Example of the same one-dimensional gel either stained with Coomassie Brilliant Blue (*left panel*) or with silver (*right panel*).

only possible for low amounts of protein (between 100 and 150 ng) (see ref. 1). Many variations of the traditional protocol of Switzer et al. (2) exist (see refs. 3, 4). Here we provide the reader with three variations that are robust and sensitive and useful for staining of proteins, but might also be used for staining of nucleic acids in polyacrylamide gels. Figure 1 gives an example of a silver-stained polyacrylamide gel.

2. Materials

2.1. Pretreatment of the SDS Gel

1. Fixation solution I: 40% EtOH, 10% acetic acid in bidest. water (see Note 1).
2. Fixation solution II (6): 50% methanol, 12% acetic acid, 0.05% formaldehyde; sensitizing solution: 0.02% sodium thiosulfate in bidest. water.
3. Fixation/sensitizing (5):
   (a) Solution IA: 0.05% glutaraldehyde, 0.01% formalin (buffered formaldehyde, 37%), 40% EtOH (see Notes 2 and 3).
   (b) Solution IB: 0.02% sodium thiosulfate in bidest. water.
4. Fixation/sensitizing solution II ((4), *modified*): 30% EtOH (in 0.4 M sodium acetate buffer, pH 6.0), 0.3% sodium thiosulfate, 0.5% glutaraldehyde [alternative protocol using methanol instead EtOH [(6), *modified*)].

2.2. Silver Staining (Method 1) (see Ref. 4, Modified)

1. Staining solution: 0.1% silver nitrate, 0.05% formaldehyde in bidest. water.
2. Develop solution: 3% sodium carbonate, 0.025% formaldehyde in bidest. water.
3. Stop solution: 1% glycine in bidest. water.

2.3. Silver Staining (Method 2) (see Ref. 5, Modified)

1. Staining solution: 0.1% silver nitrate in bidest. water.
2. Wash solution: 40% EtOH in bidest. water.
3. Develop solution: 2.5% sodium carbonate, 0.04% formalin in bidest. water.
4. Stop solution: 5% acetic acid in bidest. water.

2.4. Silver Staining (Method 3) (see Ref. 6, Modified)

1. Staining solution: 0.2% silver nitrate, 0.076% formaldehyde in bidest. water.
2. Wash solution: 35% methanol in bidest. water.
3. Develop solution: 3.6% sodium carbonate, 0.05% formaldehyde, 0.0004% sodium thiosulfate in bidest. water.
4. Stop solution: 50% methanol, 12% acetic acid.

3. Methods (see Notes 4–7)

3.1. Pretreatment of the SDS Gel and Silver Staining Procedure

3.1.1. Method 1

1. Fix the gel in fixation solution I for 30 min (see Note 8).
2. Wash with bidest. water for 5 min.
3. Treat your gel with fixation/sensitizing solution II for 60 min.
4. Wash the gel six times with bidest. water for 10 min each.

5. Perform the silver reaction using staining solution for 30 min.
6. Wash three times in bidest. water for 20 s each.
7. Develop your staining by incubating in develop solution for 3–5 min (see Note 9).
8. Shortly wash the gel once in bidest water.
9. Stop the staining by incubating the gel in stop solution for 5 min (see Note 10).
10. Wash the gel three times in bidest. water for 10 min each.

3.1.2. Method 2

1. Fix your gel in fixation solution I for 10 min.
2. Wash with bidest. water for 10 min.
3. Treat your gel with fixation/sensitizing solution IA for 5 min.
4. Wash the gel in wash solution for 20 min. Add a second washing step for another 20 min using bidest. water.
5. Sensitize the gel by incubating it in fixation/sensitizing solution IB for one min.
6. Wash the gel twice in bidest. water for one min each.
7. Perform the silver reaction using staining solution for 20 min.
8. Wash four times in bidest. water for one min each.
9. Develop your staining by incubating in develop solution. The first incubation is until the solution becomes yellowish (takes approximately 1 min). Then immediately pour off the solution and replace it with fresh develop solution. Incubate until you get the desired level of staining intensity (4–15 min) (see Note 9).
10. Stop the staining by incubating the gel in stop solution for 5 min (see Note 10).

3.1.3. Method 3

1. Fix the gel in fixation solution II for 2 h.
2. Wash with wash solution three times 20 min each
3. Sensitize the gel by incubating with fixation/sensitizing solution IB (5) for 2 min.
4. Wash the gel three times with bidest. water for 5 min each.
5. Perform the silver reaction using staining solution for 60 min.
6. Wash twice in bidest. water for 1 min each.
7. Develop your staining by incubating in develop solution. Incubate until you get the desired level of staining intensity (see Note 9).
8. Stop the staining by incubating the gel in stop solution for 5 min. (see Note 10).

4. Notes

1. Unless stated otherwise, all solutions should be prepared in water that has a resistance greater than 18 M ohm and total organic content of less than five parts per billion. This standard is referred to as "water" in this text.
2. Always prepare all buffers containing any aldehydes fresh shortly before usage!
3. If the stained proteins will be used in downstream techniques like mass spectroscopy, you should use solutions without glutaraldehyde or formaldehyde. You might use formaldehyde-containing buffers, like in the protocol of Swain and Ross (5), when you proceed with your further analysis as soon as possible, or you can also try to use the staining protocols without any aldehyde.
4. We have provided three slightly different protocols taken from Li et al. (6), Swain and Ross (5), and Heukeshoven and Dernick (4) that perform well in most standard conditions. Nevertheless, in your special case, you might try all protocols in order to identify the one which works best for your conditions.
5. Prior to gel staining the proteins must be fixed. This is done to denature the proteins in order to prevent their diffusion with time. Problematic are basic proteins as they cannot be fixed with acetic acid, when the gel is stained with silver.
6. Depending on the further protocol you want to proceed after the staining process, you might need to modify your staining protocol. Always use the fixation, which is essential. Sensitizing and washing will reduce the background and will enhance detection. If you need to characterize your proteins by MS, you have to omit glutaraldehyde (and, to a lesser extent, also formaldehyde) in all buffers used in the procedures. Although this will reduce the sensitivity, it is the only way to use the stained proteins in downstream MS application (note: you might use formaldehyde-containing buffers, like in the protocol of Swain et al. (5), when you proceed with your further analysis as soon as possible, or you can also try to use the staining protocols without any aldehyde).
7. It is advisable not to pre-stain gels that will be subjected to autoradiography, as it will cause quenching (this is more a problem with silver staining than with Coomassie staining). Gels are soaked in an autoradiography reagent for 30 min at room temperature if using 35S or 14C labeled proteins, whereas gels containing 32P-labeled proteins do not require such a treatment.

8. Never touch the gel with bare fingers. Always wear gloves or use forceps.
9. For nearly comparable staining results in repeating experiments, you should always use identical developing times.
10. Alternately, you can use an EDTA solution (40 mM) as stop solution. This will chelate residual silver ions.

Acknowledgments

Special thanks for excellent technical support to Christin Gräfe, Livia Schulze, und Barbara Uteß (Inst. Immunology, Dresden, Germany).

References

1. Patton WF (2000) A thousand points of light: The application of fluorescence detection technologies to two-dimensional gel electrophoresis and proteomics. Electrophoresis 21: 1123–1144
2. Switzer RC 3rd, Merril CR, Shifrin S (1979) A highly sensitive silver stain for detecting proteins and peptides in polyacrylamide gels. Anal Biochem 98:231–237
3. Blum H, Beier H, Gross HJ (1987) Improved silver staining of plant proteins, RNA and DNA in polyacrylamide gels. Electrophoresis 8:93–99
4. Heukeshoven J, Dernick R (1988) Improved silver staining procedure for fast staining in PhastSystem Development Unit. I. Staining of sodium dodecyl sulfate gels. Electrophoresis 9:28–32
5. Swain M, Ross NW (1995) A silver stain protocol for proteins yielding high resolution and transparent background in sodium dodecyl sulfate-polyacrylamide gels. Electrophoresis 16:948–951
6. Li ZB, Flint PW, Boluyt MO (2005) Evaluation of several two-dimensional gel electrophoresis techniques in cardiac proteomics. Electrophoresis 26:3572–3585

Chapter 43

Revisit of Imidazole–Zinc Reverse Stain for Protein Polyacrylamide Gel Electrophoresis

Han-Min Chen

Abstract

Imidazole–zinc reverse stain (ZN stain) is known for its high sensitivity, ease of use, and cost-effective feature. ZN stain is compatible to many experiments of which those are proteomics-related in particular. Here, we describe the ZN staining procedures and the subsequent procedures incorporated in detail, along with the improvements of setup in aspects of visualization and documentation for post-processing ZN-stained gel images.

Key words: Imidazole–zinc reverse stain, SDS-PAGE, Native PAGE, Electro-blotting, Electro-elution, Mass spectrometry

1. Introduction

Imidazole–zinc reverse stain (ZN stain) that utilizes zinc and imidazolate ions for protein visualization in polyacrylamide gels was originally developed and introduced in the 1990s (1). This stain method is based on the selective precipitation of imidazolate–zinc complex in gel matrix except for the portions (2, 3) where proteins or other macromolecules, such as DNA (4, 5) or lipopolysaccharides (6, 7), stand. ZN staining offers many advantages over the other approaches utilized for life science research, particularly proteomics of which researches simultaneously investigate vast proteins. ZN staining is highly sensitive for protein visualization by detecting as little as 1 ng of protein that is equal to detection limit of silver staining (8–10). Moreover, performing ZN staining procedures are remarkably simple in comparison with other available staining methods in aspects of preparation and performance. Additionally, the process of ZN staining is extraordinarily short, which generally takes less than 15 min to complete (8, 9, 11).

Biji T. Kurien and R. Hal Scofield (eds.), *Protein Electrophoresis: Methods and Protocols*, Methods in Molecular Biology, vol. 869, DOI 10.1007/978-1-61779-821-4_43, © Springer Science+Business Media, LLC 2012

Finally, the staining is compatible to subsequent applications and experiments, such as mass spectrometry (9, 10, 12–14), Edman sequencing (2, 15, 16), electro-elution (2, 4, 9, 15), and membrane blotting techniques (8, 9). In proteomics, ZN staining is considered as an alternative approach to restain the fluorescently stained two-dimensional electrophoresis (2-DE) gels for the manual pick of minor protein targets by research groups that do not access synchronized spot pickers on the fluorescent scanners.

Recently, improvements in gel image with transparent bands, sharp spots, and a white background have broadened the applicability of ZN stain. Previously, ZN-stained gel was visualized by placing above a dark background, and visualization of gel images is limited with bare eyes. Recently, a Snell's law-based backlit light plate setup has been introduced and demonstrated to significantly improve the visualization of ZN-stained gel (17). Additionally, previous researches in which cameras were utilized to photograph the ZN-stained gels against a dark background have experienced uneven documented gel images and unacceptable contrasting quality for quantitative analysis. Recently, the transparency scanning procedure for documenting the ZN-stained gels was proposed (10). With this new documentation setup, high-quality contrast images of ZN-stained gels can be acquired. High background present in the ZN-stained gel images usually compromises the results of quantitative analysis of protein gel image. Recently, a gel image tuning procedure has been proposed to increase the visualization of gel images with higher background but not to jeopardize the quantitation results (18). Such a tuning procedure is considered as a solution for the long-existing dilemma of high background occurring in ZN-stained gel image analysis.

Given aforementioned fascinating features along with the economical character, ZN staining is of use for researches related to life science. Here, we detailed the staining procedure and described cautions that are essential while performing ZN staining. We also described the relevant preparation procedures for the ZN compatible applications.

2. Materials

1. 10× Staining solution 1: 2 M imidazole, 0.1% (w/v) SDS. Weigh 136.2 g imidazole and 1 g SDS and prepare a 1 L solution. Store at room temperature.
2. 10× Staining solution 2: 1 M zinc chloride. Weigh 136.3 g zinc chloride and prepare a 1 L solution. Adjust pH to < pH 6.5 with HCl. Store at room temperature.
3. SDS electrophoresis buffer: 25 mM Tris, 192 mM glycine, 0.1% SDS, pH 8.3. Weigh 3.03 g Tris, 14.4 g glycine, 1 g SDS

and prepare a 1 L solution at pH 8.3. Store at room temperature.

4. Native electrophoresis buffer: 25 mM Tris, 192 mM glycine, pH 8.3. Weigh 3.03 g Tris and 14.4 g glycine and prepare a 1 L solution at pH 8.3. Store at room temperature.
5. Acidic destaining solution: 10% acetic acid. Weigh 0.1 L acetic acid and prepare a 1 L solution. Store at room temperature.
6. Black plastic staining boxes.
7. Backlit light plate (GP1002-1, Gel lighting plate, Visual Protein, Taipei, Taiwan).
8. Flatbed scanner with transparency unit (V750, Epson, Nagano, Japan).

3. Methods

3.1. PAGE

1. Carry out SDS-PAGE according to Laemmli's protocol.
2. Conduct native PAGE according to Laemmli's protocol except that use of SDS is avoided in the gel and electrophoresis solutions.

3.2. Gel Staining Procedure

1. After gel electrophoresis, remove gels from electrophoresis units and transfer gels into stain boxes with dark background or plastic black staining box (see Note 1).
2. Briefly rinse the gel with distilled water.
3. Add sufficient amount of SDS electrophoresis buffer into the staining box and gently agitate (50–75 rpm) for 5–10 min. This is a tuning procedure to increase the quality of ZN stain (see Notes 2, 3).
4. Add sufficient amount of staining solution 1 into the stain box and agitate gently for 10–15 min (see Notes 4–7).
5. Discard staining solution 1. Wash the gel with large quantity of distilled water two times (see Note 8).
6. Add adequate amount of staining solution 2 into the staining box. *Warning: Do not pour staining solution 2 directly onto the gel surface.* Vigorously rock the gel, to develop the staining result, by hands. The background of gel should turn white, and protein spots and bands can be visualized between 10 and 30 s (see Notes 9, 10).
7. Discard staining solution 2. Wash the developed reverse-stained gels immediately by adding large quantity of distiller water three times. An example of ZN-stained gels is shown in Fig. 1a; as low as 1–2-ng protein band can be seen by naked eyes (see Note 11).

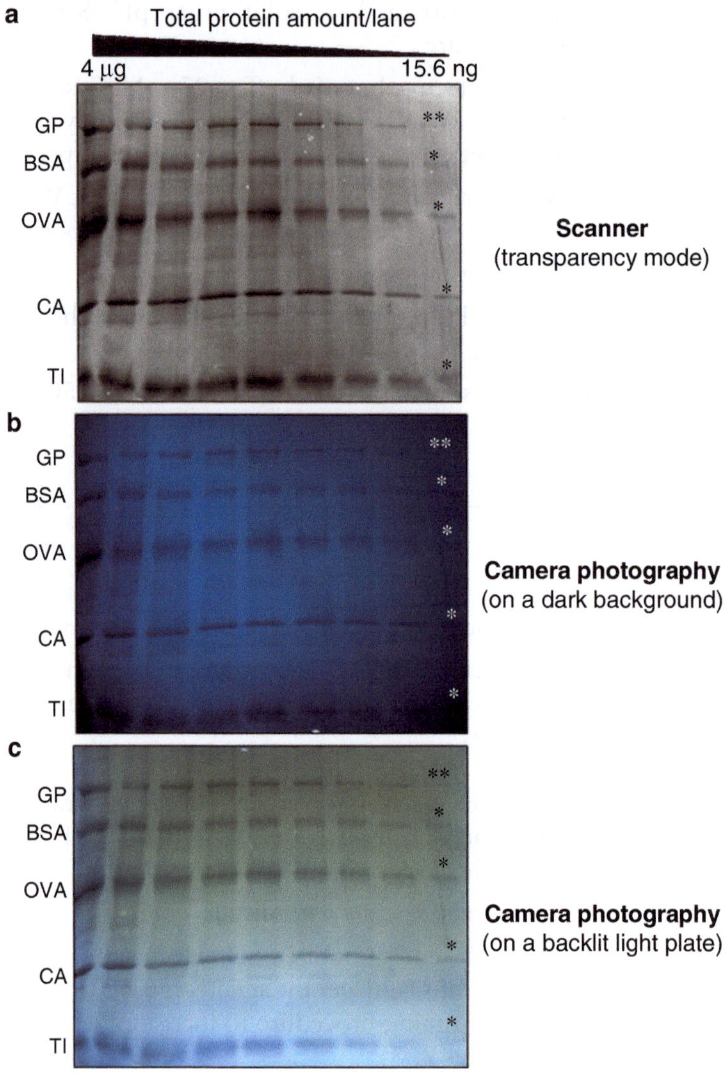

Fig. 1. Images of the ZN-stained gel documented by different methods. 12.5% SDS-PAGE gel that separated a serially diluted standard protein mixture (low molecular weight protein markers, GE Healthcare) from 4 μg to 15.6 ng was used as the evaluation materials. The gel image was (**a**) documented by scanner with the transparency mode, (**b**) photographed against a dark background, or (**c**) photographed on a backlit light plate. Asterisks indicate the position of least visible band of each standard protein on gels or gel images. The lowest amount protein detected is 1.8 ng of glycogen phosphorylase b (indicated by a *double asterisk*). Abbreviations used for standard protein mixture are as follows: glycogen phosphorylase b, *GP*; bovine serum albumin, *BSA*; ovalbumin, *OVA*; carbonic anhydrase, *CA*; and trypsin inhibitor, *TI*.

8. The developed gels are now ready for subsequent analysis or can be stored by wrapping between two transparency sheets for storage at 4°C. No significant protein degradation or diffusion after several months occurs.

3.3. Gel Visualization

1. Place the reverse-stained gels on a dark background; the location of protein spots or bands in the reverse-stained gel can be spotted by bare eyes (Fig. 1b).

2. The contrast of the reverse-stained gel can be significantly enhanced when using the backlit light plate. The white background of gels may glow, and protein spots or bands are seen as dark signals (Fig. 1c).

3.4. Gel Image Documentation

1. Scan reverse-stained gel using the flatbed scanner with the transparency unit. Acquire the negative gel images under transparency mode. A positive gel image will be documented accordingly (see Note 12).
2. When using CCD camera to document gel images, a black background has to be placed underneath the gels.

3.5. Destain/Restain and Fixation of Gels

1. If the staining results are unsatisfactory, the reverse-stained gel can be destained with SDS electrophoresis buffer. The white background can be completely removed in minutes after soaking the gels in the SDS electrophoresis buffer with agitation. The destained gels can be immediately restained by adding staining solution 1.
2. The reverse-stained gel can also be destained by incubating gels in10% acetic acid. The white stained background will vanish completely in less than 1 min. This fast destaining procedure also fixes proteins.
3. Fixation of proteins in gel can be achieved by extending the incubation of 10% acetic acid to 30 min (see Note 13).
4. To restain the acidic destained or fixed gels, an extensive washing procedure is required. At least, washing gels with distilled water three times (each of 10 min) may be sufficient to remove any residual acetic acid. The SDS electrophoresis buffer tuning procedure is also mandatory before restaining the gels (see Note 14).

3.6. Prepare ZN-Stained Gels for Mass Spectrometry (MS) Sample Preparation

1. Visualize the ZN-stained gels on the backlit light plate. Excise protein spots or bands and transfer them into individual microcentrifuge tubes.
2. If acidic fixation procedure is not performed for the ZN-stained gels, fix the excised gel spots or bands in 10% acetic acid solution for 30 min (see Note 15).
3. Wash the fixed gel spots or bands by vigorously agitating in 1-mL distilled water for at least three times.
4. The washed gel spots or bands can now be processed by using the standard protocol for MS sample preparation (19).
5. After vacuum drying, the washed gel spots or bands can be stored at −80°C for at least 3 months.

3.7. Prepare ZN-Stained Gels for Electro-Blotting and Electro-Elution

1. For electro-blotting: Regarding western blot experiments, the ZN stain can be used to monitor the electrophoresis results prior to performing the electro-blotting procedure. Just incubate the ZN-stained gels in the native electrophoresis

buffer (without SDS) for 10–15 min to eliminate the white background before performing the regular electro-blotting procedure (see Notes 16, 17).

2. For electro-elution: The ZN reverse stain is suitable for visualizing the location of protein targets in preparative electrophoresis. Just excise the protein targets from the ZN-stained gels and perform electro-elution. No destaining or acidic fixation procedure is required (see Note 17).

3.8. Restain by Other Protein Staining Methods

1. The ZN-stained gels can be restained by other protein staining methods, such as Coomassie Brilliant Blue stain, silver stain, and various kinds of fluorescent protein stains.
2. Prior to performing subsequent protein stain methods, destain and fix the ZN-stained gels by 10% acetic acid solution.
3. If activity staining needs to be performed, destain the ZN-stained gels by using the native electrophoresis buffer (without SDS).
4. ZN stain is very suitable for recovering protein targets from fluorescently stained protein gels, which is not easily seen by the eyes (see Note 18). ZN stain can be also used for manually recovering protein target in difference in gel electrophoresis (DIGE). Just perform Zn staining procedure as described after documenting the fluorescent gel images (see Note 19).

4. Notes

1. Performing the reverse gel stain procedure on a one gel per stain box basis.
2. The pH of gels might be changed during the prolonged electrophoresis procedure. This may affect the result of ZN stain. It is found that including the gel tuning procedure by using the SDS electrophoresis, buffer may prevent the above problem and increase the contrast of ZN stain. When native PAGE gels are being stained, tune the gels by incubating with SDS-PAGE electrophoresis buffer for 10 min.
3. Before adding staining solution 1, there is no need to wash the tuned gel with distilled water.
4. Pre-warming the staining solutions 1 and 2 and distilled water to 25°C may be required when the room temperature is extremely low. Use of chilled solutions or distilled water may affect the staining result.
5. Staining solutions have to fully cover the gel in a staining box. The volume of staining solutions has to be enough to cover the gel (typically 50 mL for 1 minigel [10 cm × 7 cm × 0.75 mm]).

6. The longer the gels are incubated in solution 1, the whiter and earlier background staining will occur during the gel development.
7. The larger the gel thickness or acrylamide concentration, the longer the equilibration step. A prolonged incubation of staining solution 1 (up to 20 min) will be needed for gels thicker than 1.5 mm.
8. Incomplete removal of residual staining solution 1 will result in uneven staining background.
9. Pour solution 2 into the corner of the staining box. Do not pour solution 2 directly onto the surface of gels; it will cause uneven staining results.
10. In any case, the gel development procedure should be longer than 1 min; it will cause over-staining of gels. Our experience shows it is better to stop the development of the ZN stain before the background is really turned white. The background of the ZN-stained gel may even become whiter when stopping the development procedure.
11. The sensitivity of ZN stain may depend on the formulation of staining solution. Some commercial ZN stain kits may comprise a higher staining sensitivity than the original recipe. Alternatively, the employed visualization and documentation methods may affect the quality of ZN-stained gel images as well as the sensitivity.
12. For the purpose of gel image analysis, gel images are recommended to be saved at 16-bit grayscale format with at a least 300-dpi resolution.
13. Prolonged acetic acid fixation of protein gels is not recommended because it may cause modification, e.g., acetylation and acid hydrolysis of proteins. Our experience shows that 30 min incubation in 10% acidic acid solution is sufficient to fix proteins in gels.
14. Satisfactory restaining results can be obtained for the destained ZN gels by less than two times. Whiter background will be observed when a gel is repeatedly restained for more than two times.
15. Although some reports have suggested using the unfixed ZN protein gel spots or bands for MS sample preparation (9), our experience shows that the acidic protein fixation for ZN-stained gel spots or bands may increase the yield of peptides.
16. The SDS electrophoresis buffer should not be used under this circumstance.
17. Acidic fixation of the ZN-stained gel should not be performed when subsequent electro-blotting or electro-elution is on demand. Acidic fixation will decrease the efficiency and recovery of proteins during electro-blotting and electro-elution, respectively.

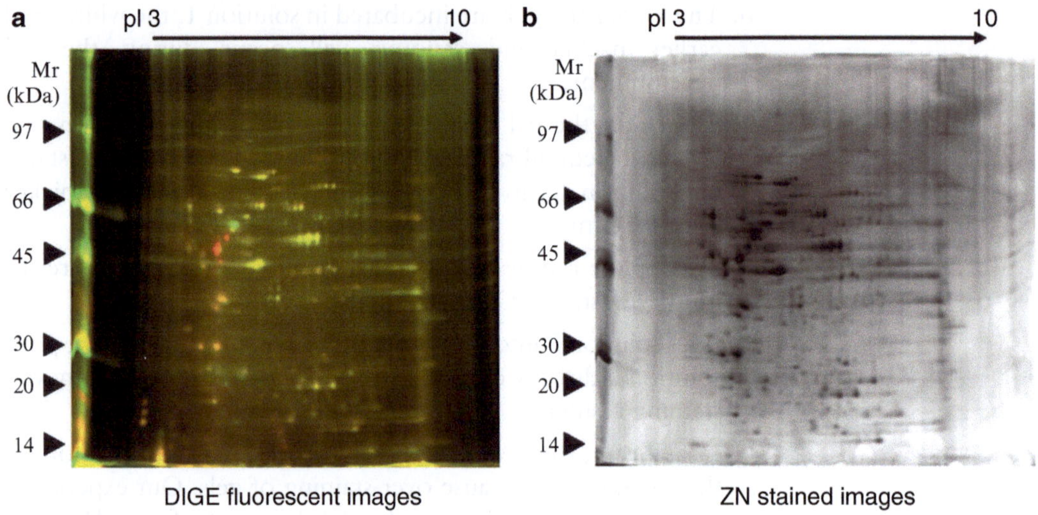

Fig. 2. The fluorescent and corresponding ZN-stained gel images of a DIGE experiment. Fifty-mg proteins extracted from the chemical hypoxia-treated (5 mM $NiCl_2$ for 8 h) human teratocarcinoma NT2 cells and the control sample were labeled by Cy5 and Cy3 dyes (CyDye DIGE Fluor minimal dye, GE healthcare), respectively, and simultaneously analyzed by 2-DE using 13-cm pH 3–10 linear IPG strip and 12.5% SDS-PAGE. (**a**) The fluorescent gel image pairs were documented using laser gel scanner (Typhoon, Trio+, GE healthcare) with excitation/emission wavelength as follows: Cy3 image: ex/em, 532/580 nm; Cy5 image: ex/em, 633/670 nm. (**b**) The same gel was stained by ZN stain and documented using scanner under transparency mode. A comparable number of protein spots can be detected.

18. Some versions of ZN stains are more sensitive than fluorescent gel staining methods such as SYPRO Ruby™ stain (10). Therefore, ZN stain can be used for manual recovery of minor protein targets detected in the fluorescently stained protein gel images.
19. In our experience, most Cy2-, Cy3-, and Cy5-labeled protein targets that appear in the DIGE results can be visualized by ZN stain. A comparison between the Cy2, Cy3, and Cy5 fluorescently stained and the corresponding restained ZN-stained 2-D gel images is shown in Fig. 2. Comparable number of protein spots can be observed by the eyes.

Acknowledgements

This work was supported by the grants from Catholic Fu-Jen University (9991A02-109931041001-2) and National Science Council of the Republic of China (NSC 99-2627-E-030-001). We thank Dr. Kuo Cheng-Yi for linguistic edition and Mr. Tsai Tzung-Lin and Dr. Wu Szu-Yu for providing gel images.

References

1. Fernandez-Patron C, Castellanos-Serra L (1990) Abstract booklet. In: Eight international conference on methods on protein sequence analysis, Kiruna, Sweden, July 1–6 1990
2. Fernandez-Patron C, Castellanos-Serra L, Rodriguez P (1992) Reverse staining of sodium dodecyl sulfate polyacrylamide gels by imidazole-zinc salts: sensitive detection of unmodified proteins. Biotechniques 12:564–573
3. Ferreras M, Gavilanes JG, Garcia-Segura JM (1993) A permanent Zn2+ reverse staining method for the detection and quantification of proteins in polyacrylamide gels. AnalBiochem 213:206–212
4. Hardy E, Pupo E, Casalvilla R et al (1996) Negative staining with zinc-imidazole of gel electrophoresis-separated nucleic acids. Electrophoresis 17:1537–1541
5. Hardy E, Sosa AE, Pupo E et al (1996) Zinc-imidazole positive: a new method for DNA detection after electrophoresis on agarose gels not interfering with DNA biological integrity. Electrophoresis 17:26–29
6. Hardy E, Pupo E, Castellanos-Serra L et al (1997) Sensitive reverse staining of bacterial lipopolysaccharides on polyacrylamide gels by using zinc and imidazole salts. Anal Biochem 244:28–32
7. Pupo E, Lopez CM, Alonso M et al (2000) High-efficiency passive elution of bacterial lipopolysaccharides from polyacrylamide gels. Electrophoresis 21:526–530
8. Castellanos-Serra L, Hardy E (2001) Detection of biomolecules in electrophoresis gels with salts of imidazole and zinc II: a decade of research. Electrophoresis 22:864–873
9. Hardy E, Castellanos-Serra LR (2004) "Reverse-staining" of biomolecules in electrophoresis gels: analytical and micropreparative applications. Anal Biochem 328:1–13
10. Lin CY, Wang V, Shui HA et al (2009) A comprehensive evaluation of imidazole-zinc reverse stain for current proteomic researches. Proteomics 9:696–709
11. Pupo E, Lopez CM, Alonso M et al (2000) High-efficiency passive elution of bacterial lipopolysaccharides from polyacrylamide gels. Electrophoresis 21:526–530
12. Scheler C, Lamer S, Pan Z et al (1998) Peptide mass fingerprint sequence coverage from differently stained proteins on two-dimensional electrophoresis patterns by matrix assisted laser desorption/ionization-mass spectrometry (MALDI-MS). Electrophoresis 19: 918–927
13. Castellanos-Serra L, Proenza W, Huerta V et al (1999) Proteome analysis of polyacrylamide gel-separated proteins visualized by reversible negative staining using imidazole-zinc salts. Electrophoresis 20:732–737
14. Gulin S, Pupo E, Schweda EK et al (2003) Linking mass spectrometry and slab-polyacrylamide gel electrophoresis by passive elution of lipopolysaccharides from reverse-stained gels: analysis of gel-purified lipopolysaccharides from Haemophilus influenzae strain Rd. Anal Chem 75:4918–4924
15. Fernandez-Patron C, Calero M, Collazo PR et al (1995) Protein reverse staining: high-efficiency microanalysis of unmodified proteins detected on electrophoresis gels. Anal Biochem 224:203–211
16. Fernandez-Patron C, Castellanos-Serra L, Hardy E et al (1998) Understanding the mechanism of the zinc-ion stains of biomacromolecules in electrophoresis gels: generalization of the reverse-staining technique. Electrophoresis 19:2398–2406
17. Lin CY, Huang HM, Chen HM (2006) Use of backlit light plate to enhance visualization of imidazole-zinc reverse stained gels. Biotechniques 41:560, 562, 564
18. Wu HC, Yen CC, Tsui WH et al (2010) A red line not to cross: evaluating the limitation and properness of gel image tuning procedures. Anal Biochem 396:42–50
19. Westermeir R, Naven T (2002) Proteomics in practice. Wiley-VCH Verlag-GmbH, Weinheim

Chapter 44

Counterion Dye Staining of Proteins in One- and Two-Dimensional Sodium Dodecyl Sulfate-Polyacrylamide Gel Electrophoresis and Tryptic Gel Digestion of Stained Protein for Mass Spectrometry

Wei-Tao Cong, Xu Wang, Sun-Young Hwang, Li-Tai Jin, and Jung-Kap Choi

Abstract

A fast and matrix-assisted laser desorption/ionization mass spectrometry compatible protein staining method in one- and two-dimensional sodium dodecyl sulfate-polyacrylamide gel electrophoresis is described. It is based on the counterion dye staining method that employs oppositely charged two dyes, zincon and ethyl violet, to form an ion-pair complex. The protocol, including fixing, staining, and quick washing steps, can be completed in 1–1.5 h, depending upon gel thickness. It has the sensitivity comparable to the colloidal Coomassie Brilliant Blue G stain using phosphoric acid as a component of staining solution (4–8 ng). The counterion dye stain does not induce protein modifications that complicate interpretation of peptide mapping data from mass spectrometry. Considering the speed, sensitivity, and compatibility with mass spectrometry, the counterion dye stain may be more practical than any other dye-based protein stains for routine proteomic researches.

Key words: Protein stain, Proteomics, 1-D and 2-DE, Counterion dye stain, MALDI-MS

1. Introduction

In proteomic researches, the ability to analyze and identify protein from polyacrylamide gels at high sensitivity is important (1). Most commonly, protein samples separated by 1- or 2-D SDS-PAGE are visualized by various staining methods using visible organic dye, fluorescence, and silver stains (2). Of these, sensitive visible organic dye stain is still the most popular for preparative or further analytical purposes since protein can be visualized directly without special equipment and subsequently excised from the band or spot easily.

Biji T. Kurien and R. Hal Scofield (eds.), *Protein Electrophoresis: Methods and Protocols*, Methods in Molecular Biology, vol. 869, DOI 10.1007/978-1-61779-821-4_44, © Springer Science+Business Media, LLC 2012

Particularly for matrix-assisted laser desorption/ionization time-of-flight mass spectrometry (MALDI-TOF-MS) analysis, it is important that the staining method used does not interfere with MS methods and detection limit should be usually low nanogram range of protein (3–5). Many organic visible dye stains are available. However, only a few of these are compatible with protein digestion and mass spectrometric analysis. Among them, colloidal Coomassie Brilliant Blue G (CBBG) stain with ethanol (EtOH) and phosphoric acid as components of staining solution has been used for this purpose (6). It has the highest sensitivity among visible organic dye stains (7). However, the CBBG stain still has several disadvantages in troublesome multisteps and very long staining time (more than 24 h) (6).

Therefore, we developed a visible counterion dye staining method that employs zincon (ZC) and ethyl violet (EV) in EtOH/acetic acid staining solution (see Note 1; ref. 8). The sensitivity of this method is comparable to that of colloidal CBBG stain, about 4–8 ng per protein band (see Note 2). Furthermore, another interesting feature of the staining protocol described here is the applicability to the identification of proteins by MALDI-TOF-MS. In this report, we compared the EZ stain to colloidal CBBG stain with respect to sensitivity in 1- and 2-D SDS-PAGE and compatibility in MALDI-TOF-MS analysis.

2. Materials

2.1. SDS-PAGE

1. Separating buffer (4×): 1.5 M Tris–HCl, pH 8.7, and 0.4% SDS. Store at room temperature.
2. Stacking buffer (4×): 0.5 M Tris–HCl, pH 6.8, and 0.4% SDS. Store at room temperature.
3. Rehydration buffer: 8 M urea, 2% (w/v) CHAPS, 2% IPG buffer, and 0.002% bromophenol blue. Aliquot and store at −20°C.
4. Monomer stock solution: 30% acrylamide and 0.8% *N*,*N*′-methylene-bisacrylamide (see Note 3).
5. Ammonium persulfate: Prepare 10% solution in water, aliquot, and store at −20°C.
6. Running buffer: 25 mM Tris, 192 mM glycine, and 0.1% SDS. Store at room temperature.
7. Loading buffer: 60 mM Tris–HCl, pH 6.8, 25% glycerol, 2% SDS, 1% DTT, and 0.002% bromophenol blue. Aliquot and store at −20°C.
8. Molecular weight marker proteins: Add 2.5 mL of sample buffer to 3 mg of SDS6H2 protein standard (Sigma-Aldrich,

St. Louis, MO, USA), mix by inversion, and then mix again using a vortex mixer till complete solubilization. Aliquot and store at –20°C.

9. *Escherichia coli* (*E. coli*) BL 21 total cell proteins: Harvest *E. coli* by centrifugation at 5,000*g* for 10 min at 4°C, sonicate the cells using ultrasonic homogenizer (Sonopuls HD 2200, Bandelin electronic, Germany) in buffer containing 50 mM Tris, pH 7.5, 1 mM EDTA, and 0.4 mM PMSF 5 times for 1 min (1 g cell/10 mL), and centrifuge at 20,000*g* for 20 min at 4°C. Determine the protein amount by Bradford's method using the Bio-Rad protein assay kit (Bio-Rad Lab., Hercules, CA, USA) (9). For 1-D SDS-PAGE, dissolve the extract of *E. coli* BL 21 cells in a buffer containing 60 mM Tris, pH 6.8, 25% glycerol, 2% SDS, 2% β-mercaptoethanol, and 0.002% bromophenol blue. Aliquot and store at –20°C. Prior to electrophoresis, the protein samples are heated at 100°C for 5 min in a boiling water bath and then cooled to room temperature. For 2-D SDS-PAGE, solubilize the extract of *E. coli* BL 21 cells in IPG rehydration buffer with the addition of 2.5 M DTT to the final concentration of 250 mM just prior to IEF.
10. Equilibration solution I: 1% DTT, 50 mM Tris–HCl, pH 8.8, 6 M urea, 30% glycerol, 2% SDS, and 0.002% bromophenol blue.
11. Equilibration solution II: 2.5% iodoacetamide, 50 mM Tris–HCl, pH 8.8, 6 M urea, 30% glycerol, 2% SDS, and 0.002% bromophenol blue.
12. Agarose sealing solution: Prepare 0.5% agarose, 0.002% bromophenol blue solution in running buffer, and microwave to dissolve.
13. Strip holder (Amersham Ettan IPGphor Strip Holder).

2.2. Protein Staining

1. All working solutions should be freshly prepared with distilled water and clean glassware or plastic ware, and overnight storage is not recommended. All steps are carried out at room temperature with shaking.
2. Fixing solution: 40% v/v EtOH and 10% v/v acetic acid solution.
3. Staining solution: 0.004% w/v ZC, 0.003% w/v EV in 24% EtOH, and 7% acetic acid solution (see Notes 4 and 5, Fig. 1).

2.3. MALDI-MS Sample Preparation

1. Reduction solution: 40 μL 10 mM DTT in 100 mM ammonium bicarbonate (pH 7.8). Store at room temperature.
2. Alkylation solution: 55 mM iodoacetamide (fresh) in 100 mM ammonium bicarbonate (pH 7.8). Avoid light and store at room temperature.

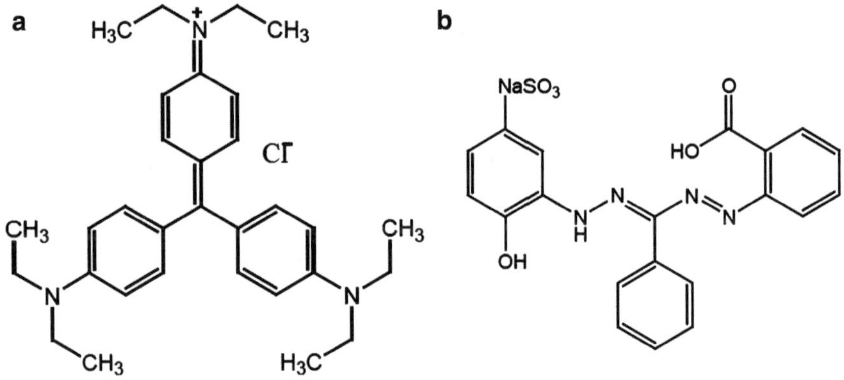

Fig. 1. The structure of ethyl violet (**a**) and zincon (**b**).

3. In-gel digestion solution: 12.5 ng/μL trypsin in chilled 5 mM $CaCl_2$ and 80% acetonitrile (ACN) (v/v). Store in ice bath.
4. Matrix solution: 10 mg/mL α-cyano-4-hydroxycinnamic acid in 100 mM ammonium bicarbonate and 60% ACN (v/v). Store at room temperature.
5. Calmix 2 peptide standard solution: 2.0 μM angiotensin, 2.0 μM ACTH 1–17 and 1.5 μM ACTH 18–39 solution in water. Aliquot and store at −70°C.
6. PerSeptive Biosystems MALDI-TOF Voyager-DE STR mass spectrometer (PerSeptive Biosystems, Framingham, MA, USA).

3. Methods

3.1. 1-D SDS-PAGE

1. The electrophoresis device is a Mini-PROTEAN 3 cell gel system. Clean the glass plate and rinse extensively with distilled water. They can be kept clean until use in a plastic rack.
2. Prepare a 0.75-mm-thick, 12% polyacrylamide gel by mixing 7.5 mL of 4× separating buffer, with 12 mL monomer stock solution, 10.5 mL distilled water, 150 μL ammonium persulfate solution, and 30 μL TEMED. Pour the gel, leaving space for a stacking gel, and overlay with distilled water. The gel should polymerize within 30 min. Pour off the water.
3. Prepare the stacking gel by mixing 3.0 mL of 4× stacking buffer, with 1.8 mL monomer stock solution, 7.1 mL distilled water, 60 μL ammonium persulfate solution, and 12 μL *N,N,N′,N′*-tetramethylethylenediamine (TEMED). Pour the gel and insert the comb. The stacking gel should polymerize within 30 min.

4. Remove the comb and wash the wells with running buffer, then fix the 1-D gels onto the gel unit, and add the running buffer to the upper and lower chambers of gel unit.
5. Dilute the protein samples with loading buffer containing 60 mM Tris, pH 6.8, 25% glycerol, 2% SDS, 1% DTT, and 0.002% bromophenol blue, load 5 μL of each sample onto gel lane at 2–1,000 ng (molecular weight marker proteins) and 8–4,000 ng (*E. coli* total cell proteins), respectively.
6. Complete the assembly of the gel unit and run at a constant current of 22 mA per slab gel. The electrophoresis should complete within 50 min.

3.2. 2-D SDS-PAGE

1. Wipe out strip holder with paper towels to clean the holder.
2. Mix 80 μg of *E. coli* total cell lysate with rehydration buffer to make a total volume of 125 μL.
3. Add rehydration buffer and sample mixture to the strip holder boat, spreading the mixture evenly along the grove.
4. Transfer the 7-cm immobilized pH 4–7 gradient strip (Pharmacia's pH 4–7 Strip) to the boat by using tweezers to hold the strip at one end so as not to disturb the gel. Be sure to have the positive end of the strip at the pointed end of the strip holder boat.
5. Add 300 μL of Dry Strip cover fluid into each end of strip holder boat until meet in the middle wetting entire strip and then cover with the lid.
6. Run rehydration and focusing protocol on the Ettan IPGphor II. For example, 7-cm strips:

Step 1	30 V	360 Vh
Step 2	100 V	100 Vh
Step 3	500 V	500 Vh
Step 4	1,000 V	1,000 Vh
Step 5	2,000 V	2,000 Vh
Step 6	4,000 V	4,000 Vh
Step 7	8,000 V	24,000 Vh

7. Once the program has ended and the maximum voltage has been reached, remove strips from the boat with tweezers and transfer to a Falcon tube containing 5 mL equilibration solution I. Make sure gel side of strip is facing up and overlaid with solution. Place on a rocker for 15 min. Carefully pour solution out of tubes.

8. Add 5 mL equilibration solution II. Make sure gel side of strip is facing up and overlaid with solution. Place on a rocker for 15 min.
9. Make a 12% SDS polyacrylamide gel without the upper stacking gel, leaving about 10 mm of room from the top of the glass plate.
10. Rinse top of gel with running buffer and transfer the equilibrated gel strip onto the gel with tweezers keeping the positive end to right, with the gel side facing outside.
11. Overlay gel well with 1 mL of preheated agarose sealing solution. Using a pair of forceps, carefully press the strip down onto the gel surface. Eliminate any bubbles between the gel surface and the strip.
12. Load the 2-D gels with immobilized pH gradient strips sealed in place into the electrophoresis tank by sliding gel plate sets between the rubber gaskets. Dip the gel plate into the SDS running buffer to lubricate prior to inserting them between the gaskets. Make sure gaskets are not folded and that they form a smooth seal along the entire length of each set of gel plates.
13. Turn on power supply and run the gel at a constant current of 20 mA per slab gel with a running buffer. The electrophoresis should complete within 1.5 h.

3.3. Protein Staining

1. Fixation: After electrophoresis, fix the gels (0.75-mm thickness, 8 × 10 cm) in 150 mL (400 mL for 2-DE) of fixing solution (40% v/v EtOH, 10% v/v acetic acid solution) for 15–30 min, and overnight soaking is recommended (see Note 6).
2. Staining: Soak the gel in 50 mL (200 mL for 2-DE) of 0.004% w/v ZC, 0.003% w/v EV (see Note 7) in 24% v/v EtOH, and 7% v/v acetic acid staining solution for 15–60 min (see Notes 8 and 9).
3. Destaining: Wash the gel in 50 mL (200 mL for 2-DE) of fixing solution for 30 s to remove the dyes on the gel surface and then soak the gel in distilled water for 5 min to improve band intensity (see Figs. 2 and 3).

3.4. Excision, Destaining, and Washing for MALDI-MS

1. Wash the gel piece twice for 10 min with deionized water and excise the gel band as closely as possible with a clean scalpel (see Notes 10–12).
2. Cut the gel band into approx. 1 × 1-mm pieces and transfer them into a 1.5-mL Eppendorf tube (see Note 13).
3. Destain the gel pieces with 1 mL 30% EtOH (v/v) and 10% acetic acid (v/v) on a platform shaker for gentle agitation for at least 1 h.

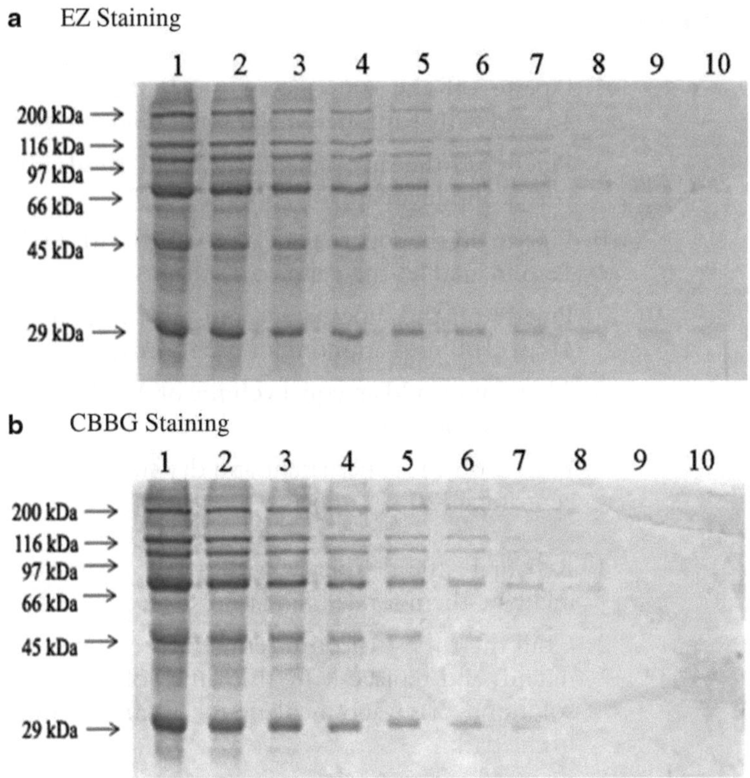

Fig. 2. Comparison of sensitivity of (**a**) EZ stain with (**b**) colloidal CBBG stain in 1-D SDS-PAGE. Conditions: 0.75-mm-thick polyacrylamide gels, 12% SDS-PAGE. Samples: myosin (200 kDa), phosphorylase b (97 kDa), BSA (66 kDa), ovalbumin from chicken (45 kDa), bovine carbonic anhydrase (29 kDa), and peroxiredoxin I from *Homo sapiens* (22 kDa). Twofold serial dilutions of proteins loaded onto the gels (from left to right): *lane* (1) 1,000; (2) 500; (3) 250; (4) 125; (5) 62; (6) 31; (7) 15; (8) 8; (9) 4; and (10) 2 ng per band, respectively.

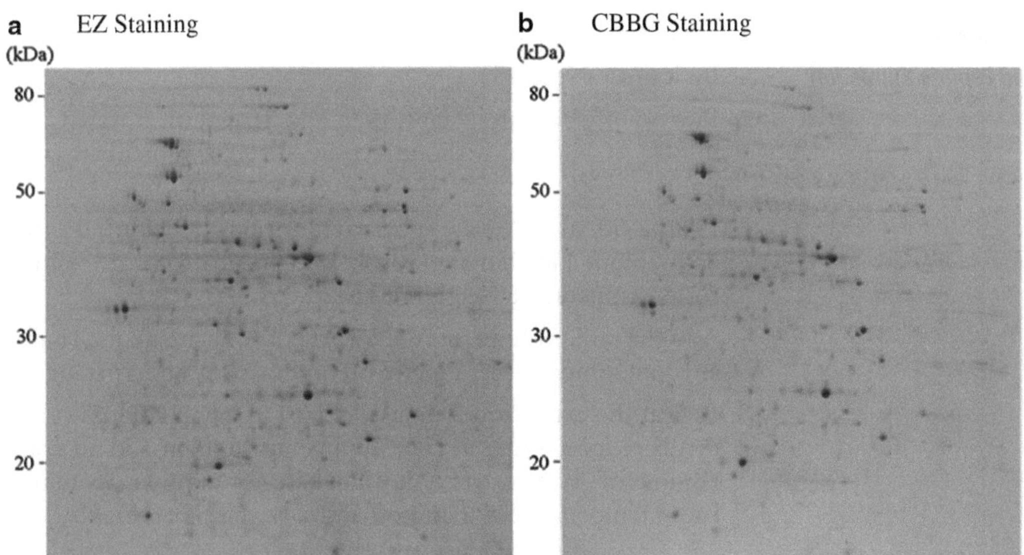

Fig. 3. Comparison of protein staining methods, (**a**) EZ and (**b**) colloidal CBBG stain, on *E. coli* BL 21 proteins in 2-D SDS-PAGE. Electrophoresis was performed in the first dimension using a linear 4–7 pH gradient, 7-cm IPG gel strip, and in the second dimension by a 1-mm-thick, 12% polyacrylamide gel. Gels are oriented from the *left*, pH 4 to the *right*, and pH 7 and high molecular mass proteins are from the *top*. Total amount of protein loaded per each gel was 80 μg, respectively.

4. Pipette off the supernatant, wash the gel pieces twice for 15 min with 1 mL deionized water, and vortex several times.
5. Pipette off the supernatant, then add 40 μL 50% ACN (v/v); vortex and incubate for 15 min.
6. Pipette off the supernatant, then add 40 μL ACN, incubate for 15 min, and let the gel pieces shrink.
7. Pipette off the supernatant and rehydrate the gel pieces with 40 μL 100 mM ammonium bicarbonate (pH 7.8).
8. After 5 min, add an equal volume of ACN and then vortex and incubate for 15 min.
9. Pipette off the supernatant and dry the gel pieces in a vacuum centrifuge (20–30 min).

3.5. Reduction and Alkylation

1. Rehydrate the gel pieces with 40 μL reduction solution and incubate them for 45 min at 56°C (see Note 14).
2. Chill the tubes to room temperature, pipette off the supernatant, and replace it by the same volume of fresh alkylation solution; vortex and incubate for 30 min at room temperature in the dark.
3. Pipette off the supernatant and wash the gel pieces with 40 μL 100 mM ammonium bicarbonate (pH 7.8).
4. After 5 min, add an equal volume of ACN and then vortex and incubate for 15 min.
5. Pipette off the supernatant and dry the gel pieces in a vacuum centrifuge (20–30 min).

3.6. In-Gel Digestion and Peptide Extraction

1. Add 40 μL in-gel digestion solution, then incubate for 45 min on ice (see Note 15).
2. Pipette off the supernatant, then add 10–20 μL 5 mM $CaCl_2$ in 80% ACN (v/v) without trypsin; vortex and incubate overnight at 37°C.
3. Pipette off and save the supernatant, then add 20 μL 25 mM ammonium bicarbonate; vortex and incubate for 15 min at room temperature (see Note 16).
4. Add the same volume of ACN; vortex and incubate for 15 min at room temperature. Pipette off and save the supernatant.
5. Repeat the extraction 2 times: Add 20 μL 5% HCOOH (v/v), incubate for 15 min at room temperature, then add an equal volume of ACN; vortex and incubate for another 15 min at room temperature. Pipette off and save the supernatants.

6. Pool all extracts, then add 2.8 μL 100 mM DTT; vortex and dry them in a vacuum centrifuge.

3.7. ZipTip Elution

1. Add 15 μL 0.1% TFA (v/v); vortex and incubate for 30 min.
2. Prior to using the ZipTip, perform a washing step with 5 × 10 μL of 50% ACN (v/v), followed by an equilibration step with 5 × 10 μL 0.1% TFA (v/v), using the maximum volume setting at a 10-μL pipette.
3. Dispense 10 μL sample solution into the tip, aspirate, and dispense up to 10 cycles for maximum binding of complex mixtures.
4. After loading the sample, perform a washing step with 5 × 10 μL 0.1% TFA (v/v).
5. Dispense 3 μL of 50% ACN (v/v) into the tip, aspirate, and dispense eluant through ZipTip at least three times without introducing air.

3.8. Matrix and MALDI-MS

1. Dilute Calmix 2 peptide standard solution into different concentrations in matrix solution.
2. Mix 1 μL of sample and 1 μL of diluted Calmix 2 peptide standard solution, spin down and spot 1 μL on the MALDI plate, and allow to dry in the ambient air.
3. Analyze the samples on a PerSeptive Biosystems MALDI-TOF Voyager-DE STR mass spectrometer in the delayed extraction and ACTH reflector mode. ACTH reflector mode was set in 20 kV of accelerating voltage, 65% grid voltage, 1.12 of mirror voltage ratio, and 150 ns of extraction delay time.
4. Masses were internally calibrated with standard peptides: Angiotensin ([M+H]+=1296.6853), ACTH 1–17 ([M+H]+= 2093.0867), and ACTH 18–39 ([M+H]+=2465.1989).

3.9. Database Searching

Peptide mass fingerprinting was conducted with the database search tool MS-Fit in the program ProteinProspector (version 5.7.3), available at http://prospector.ucsf.edu. A number of restrictions are applied to the initial search based on the species of the identified proteins: species = *E. coli* (β-galactosidase), *Oryctolagus cuniculus* (phosphorylase b), *Bos taurus* (BSA, carbonic anhydrase), *Gallus gallus* (ovalbumin); p*I* range = 4–7; mass range = 20–120 kDa; with a minimum for four peptides to match and a maximum of one missed cleavage; allow a mass tolerance of 50 ppm. Top candidate proteins identified by MS-Fit with a minimum MOWSE score of 1×10^5 and a minimum sequence coverage of 25% are selected as acceptable results (see Table 1).

Table 1
Summary of MALDI-TOF-MS data for five standard proteins from EZ and colloidal CBBG-stained gel

Protein species	ng/band		125 ng	62 ng	30 ng	15 ng	8 ng	4 ng
Phosphorylase b	Mass matched	CBBG	38%	27%	21%	21%	11%	10%
	Sequence coverage	EZ	35%	27%	32%	10%	10%	3%
	Data Tol ppm	CBBG	49%	43%	29%	27%	24%	14%
		EZ	50%	45%	44%	21%	18%	8%
		CBBG	32.8	36.5	39.3	25.9	41.0	42.1
		EZ	24.3	27.1	29.3	32.6	41.2	48.7
BSA	Mass matched	CBBG	35%	19%	19%	25%	18%	5%
	Sequence coverage	EZ	30%	40%	29%	21%	7%	7%
	Data Tol ppm	CBBG	42%	28%	32%	35%	29%	10%
		EZ	49%	46%	46%	39%	24%	26%
		CBBG	30.8	33.6	23.3	30.1	39.1	20.3
		EZ	21.8	18.7	23.3	32.5	44.2	37.3
Ovalbumin	Mass matched	CBBG	27%	23%	12%	12%	14%	7%
	Sequence coverage	EZ	27%	18%	12%	9%	20%	6%
	Data Tol ppm	CBBG	55%	49%	44%	45%	45%	29%
		EZ	57%	51%	44%	44%	48%	30%
		CBBG	28.3	18.3	20.2	30.6	23.4	22.7
		EZ	20.6	26.9	18.4	14.2	27.6	33.0

Carbonic anhydrase	Mass matched	CBBG	13%	9%	11%	9%	8%	5%
	Sequence coverage	EZ	7%	7%	11%	11%	8%	7%
	Data Tol ppm	CBBG	48%	58%	48%	45%	35%	29%
		EZ	47%	41%	41%	41%	50%	23%
		CBBG	19.2	32.5	37.7	21.0	14.9	50.6
		EZ	31.4	27.9	28.2	22.0	41.6	33.8
Peroxiredoxin	Mass matched	CBBG	25%	14%	18%	14%	11%	7%
	Sequence coverage	EZ	26%	21%	12%	16%	7%	9%
	Data Tol ppm	CBBG	71%	70%	52%	42%	42%	32%
		EZ	67%	71%	59%	66%	49%	46%
		CBBG	24.9	18.0	23.8	34.4	39.2	27.9
		EZ	22.4	14.0	28.7	31.4	28.8	29.3

4. Notes

1. The protein staining mechanism of ZC should be the same as that of CBB dyes. It has been suggested that in CBB stains, the dyes bind to proteins primarily by electrostatic interaction between the sulfonate group of the dye and protonated amino groups of proteins. In addition, hydrophobic interaction, van der Waals forces, and hydrogen bonding also contribute to the binding of the dye to proteins (8).
2. The linear dynamic ranges of the amount of proteins stained with EZ were phosphorylase b (8–1,000 ng, correlation coefficient, 0.997), BSA (8–500 ng, 0.983), OVA (8–1,000 ng, 0.996), CA (4–1,000 ng, 0.991), and Prx1 (4–1,000 ng, 0.987), while the ranges with colloidal CBBG were phosphorylase b (8–500 ng, 0.987), BSA (30–500 ng, 0.985), OVA (8–500 ng, 0.992), CA (15–500 ng, 0.981), and Prx1 (8–500 ng, 0.983), respectively. In general, EZ stain showed wider linear dynamic ranges than colloidal CBBG stain since the slopes of band intensity in colloidal CBBG stain were much steeper than those of EZ stain.
3. Acrylamide is a neurotoxin when unpolymerized, and so care should be taken by wearing gloves at all stages.
4. EZ staining solution was prepared to be 0.004% ZC/0.003% EV by diluting the stock dye solutions of 0.4% ZC (in methanol) and 0.3% EV (in EtOH) with 24% EtOH/7% acetic acid.
5. The working staining solution should be prepared fresh, just before use to prevent precipitation because of the loss of EtOH.
6. For quick staining, fixing time was 15 min. For 1-mm-thick mini gel, fixing time should be at least 1 h. The fixing procedure should be done first to obtain the best result.
7. At the higher concentrations of ZC than 0.004%, band intensity decreased due to a strong background stain. At the higher concentrations of EV than 0.003%, EV tended to precipitate with ZC easily, and at the lower EV concentrations than 0.003%, the band-enhancing effect was much reduced.
8. 30-min staining is enough to see 10 ng of protein band. Overnight staining is not recommended to avoid a possible dye precipitation due to the evaporation of EtOH.
9. EZ counterion stain is an end-point staining method. Maximum staining intensity is reached in an hour, and it remains at a steady state as long as no precipitation occurs due to the evaporation of EtOH. Even after the precipitation, there is a very little change in sensitivity because it induces an overall destaining effect on both the protein band and gel matrix showing clear background. However, for a routine experiment, 1 h staining is recommended.

10. Gel staining and preparation of peptides must be performed with labware that has never been in contact with nonfat milk, BSA, or any other protein-blocking agent to prevent carryover contamination.
11. Always use non-latex gloves when handling samples; keratin and latex proteins are potential sources of contamination.
12. Never reuse any solutions; abundant proteins will partially leach out and contaminate subsequent samples.
13. Optionally, excise a blank gel piece of the same size as a control.
14. Reduction is still recommended if reducing sample and running buffer were used in the PAGE operation.
15. This digestion protocol is compatible with the following electrophoretic parameters: 1-/2-D SDS-PAGE; gel thickness: 0.5–1 mm; acrylamide concentration: 7.5–18%; sample buffer: Laemmli, 0.1% SDS.
16. After peptide extraction, mass spectrometry should be performed as soon as possible.

Acknowledgments

This work was supported by Mid-career Researcher Program through NRF grant funded by the MEST (No. 2010-0000309).

References

1. Steinberg TH (2009) Protein gel staining methods: an introduction and overview. Methods Enzymol 463:541–563
2. Patton WF (2002) Detection technologies in proteome analysis. J Chromatography B 771:3–31
3. Ball MS, Karuso P (2007) Mass spectral compatibility of four proteomics stains. J Proteome Res 11:4313–4320
4. Lin JF, Chen QX, Tian HY et al (2008) Stain efficiency and MALDI-TOF MS compatibility of seven visible staining procedures. Anal Bioanal Chem 7:1765–1773
5. Lauber WM, Carroll JA, Dufield DR et al (2001) Mass spectrometry compatibility of two-dimensional gel protein stains. Electrophoresis 22:906–918
6. Anderson NL, Esquer-Blasc OR, Richardson F et al (1996) The effects of peroxisome proliferators on protein abundances in mouse liver. Toxicol Appl Pharmacol 137:75–89
7. Neuhoff V, Stamm R, Eibl H (1985) Clear background and highly sensitive protein staining with Coomassie Blue dyes in polyacrylamide gels: A systematic analysis. Electrophoresis 6:427–448
8. Choi JK, Chae HZ, Hwang SY et al (2004) Fast visible dye staining of proteins in one- and two-dimensional sodium dodecyl sulfate-polyacrylamide gels compatible with matrix-assisted laser desorption/ionization-mass spectrometry. Electropho-resis 25:1136–1141
9. Bradford MM (1976) A rapid and sensitive method for the quantitation of microgram quantities of protein utilizing the principle of protein-dye binding. Anal Biochem 72: 248–254

Chapter 45

TEMED-Enhanced Photoluminescent Imaging of Human Serum Proteins by Quantum Dots After PAGE

Na Na and Jin Ouyang

Abstract

Polyacrylamide gel electrophoresis (PAGE) has been one of the most powerful and widely used separation techniques for complex biological samples, whose traditional detection methods include organic dye or silver staining. As a simple, convenient, and ultrasensitive detection of proteins for PAGE, a novel enhanced photoluminescent (PL) imaging method was developed. Thioglycolic acid (TGA)-capped CdTe quantum dots (QDs) and the enhanced reagent of tetramethylethylenediamine (TEMED) are introduced, achieving the direct detection of various proteins in native 1-DE, 2-DE, and SDS gels. Here, we describe the general protocol of TEMED-enhanced PL imaging by QDs, including materials, practical procedures, as well as some notes.

Key words: CdTe Quantum dots, TEMED, Photoluminescent imaging, Polyacrylamide gel electrophoresis, Human serum proteins

1. Introduction

Proteomic characterization of human serum promises to be a powerful diagnostic tool for early disease detection, clinical diagnosis and therapy, as well as defining the onset, progression, and prognosis of these pathological processes (1). The native physicochemical properties and biological functions of serum analytes are normally detected by polyacrylamide gel electrophoresis (PAGE), which is considered to be one of the most powerful and widely used separation techniques for complex biological samples. However, traditional and prevalent protein detection method of CBB staining meets with drawbacks of low sensitivity and tedious staining procedure. Therefore, simple, convenient, and ultrasensitive detection methods for human serum proteins (especially for

Biji T. Kurien and R. Hal Scofield (eds.), *Protein Electrophoresis: Methods and Protocols*, Methods in Molecular Biology, vol. 869, DOI 10.1007/978-1-61779-821-4_45, © Springer Science+Business Media, LLC 2012

low-abundance proteins) after PAGE are dramatically in great demand. In the past decade, quantum dots (QDs) have attracted much attention because of potential applications in biological markers by combining advantages of ultrasensitive photoluminescence (PL), high photobleaching threshold, good chemical stability, and readily tunable spectral properties (2–4). Additionally, some molecules with electronegative atoms or groups, such as amines (N–) and thiols (S–), have been reported for enhancing PL intensity by the coordination to the surface of QDs (5–7). Tetramethylethylenediamine (TEMED), a simple electron-donating ligand, has also been reported for the synthesis of QDs with higher fluorescence efficiency due to the coordination of N– (8–11). Therefore, it has potential for the enhancement of PL intensity of QDs, which might be effective in PL imaging (12).

Thus, an enhanced PL imaging method for the detection of proteins in human serum with the addition of TEMED was developed. Thioglycolic acid (TGA)-capped CdTe was applied for staining gels after PAGE in alkaline conditions. Some low-abundance proteins can be easily identified using this method, but cannot be detected by CBB-R250 staining or direct PL imaging. As demonstrated, this method can be used after proteins are separated by native 1-DE and native 2-DE, as well as SDS gels.

2. Materials

All solutions were prepared using ultrapure water (prepared by purifying deionized water to attain a sensitivity of 18 MΩ cm at 25°C) from a Millipore Simplicity 185 water purification system (Millipore, Milford, MA). All reagents are analytical grade agents. Prepare and store all reagents at room temperature (unless indicated otherwise). Diligently follow all waste disposal regulations when disposing waste materials.

2.1. PAGE Compounds

1. Gel stock solution (30%, *m/v*): Weigh 29.2 g acrylamide (see Note 1) and 0.8 g *N,N′*-methylenebisacrylamide (Bis) into a glass beaker. Dissolve to 100 mL with water and then filter through 0.45-μm Corning filter (see Note 2). Store at 4°C in a bottle wrapped with aluminum foil (see Note 3).
2. Stacking gel buffer (0.5 M Tris–HCl, pH 6.8): Weigh 6.05 g tris(hydroxymethyl)aminomethane (Tris) into a glass beaker. Dissolve to 100 mL with water and then set pH to 6.8 with concentrated HCl. Store at 4°C.
3. Separating gel buffer (1.5 M Tris–HCl, pH 8.8): Weigh 18.15 g Tris into a glass beaker. Dissolve to 100 mL with water and then set pH to 8.8 with concentrated HCl. Store at 4 °C.

4. Ammonium persulfate solution (10%): Dissolve 0.10 g of ammonium persulfate to 1 mL with water (see Note 4).
5. *N*,*N*,*N*′,*N*′-Tetramethylethylenediamine (TEMED). Store at 4°C.
6. Ampholytes (pH 4–6): Serva (Heidelberg/New York).
7. 3′,3″,5′,5″-Tetrabromophenolsulfonphthalein (Bromophenol Blue) solution (0.02%, v/v): Dissolve 0.02 g Bromophenol Blue to 100 mL with water.
8. Glycerine buffer (20%, v/v): Dissolve 2 mL of glycerol to 10 mL with water.
9. Running buffer: Dissolve 3.0 g Tris and 14.4 g glycine to 1,000 mL with water, and adjust pH to 8.3 with concentrated HCl.
10. Cathode electrode buffer in IEF (0.1 M NaOH): Dissolve 4 g of NaOH to 1,000 mL with water.
11. Anode electrode buffer in IEF (0.03 M H_3PO_4): Dissolve 1.8 mL of H_3PO_4 to 1,000 mL with water.
12. Coomassie Brilliant Blue staining solution (0.1 g/100 mL): Dissolve 0.1 g CBB R-250 into the solution composed of 35 mL methanol, 10 mL acetic acid, and 55 mL water.
13. Destaining solution: Dissolve 70 mL acetic acid, 100 mL ethanol with 830 mL water.

2.2. QDs Compounds

1. Tellurium (powder) (Beijing Fine Chemical Factory, Beijing, China).
2. Sodium borohydride ($NaBH_4$) (Beijing Fine Chemical Factory).
3. Cadmium chloride hydrate ($CdCl_2 \cdot 2.5H_2O$) (Beijing Fine Chemical Factory).
4. Thioglycolic acid (TGA).
5. 1,2-Ethylenediamine (EDA).
6. *N*,*N*,*N*′,*N*′,*N*″-Pentamethyldiethylenetriamine (PMDTA).
7. Nitrogen gas (99.99%).

2.3. Protein Samples

1. Molecular-weight markers for proteins (Amersham Pharmacia Biotech, USA): The protein mixture (250 μg/vial) contains the following proteins: thyroglobulin, 76 μg, 669 kDa; ferritin, 50 μg, 440 kDa; catalase, 36 μg, 232 kDa; lactate dehydrogenase, 48 μg, 140 kDa; albumin, 40 μg, 66 kDa.
2. Serum samples: After careful settlement for 2 h, the supernatant of fresh blood samples was extracted as serum with a mini-sample collector and then centrifuged at 2,500 rpm (RCF: 1100 g) thrice (each time 10 min). Store at −20°C.

3. Methods

Carry out all procedures at room temperature unless otherwise specified.

3.1. Protein Electrophoresis

3.1.1. Native One-Dimensional Electrophoresis (1-DE)

1. Sample buffer preparation: Dissolve 105 μL Bromophenol Blue solution (0.02% w/v) and 140 μL glycerol (20% v/v) with 70 μL of water.
2. Sample preparation: Mix 35 μL serum with 315 μL sample buffer.
3. Separating gel preparation (7.5%, *m/v*): Mix 4 mL of gel stock solution (30%, *m/v*), 4 mL of separating gel buffer, 150 μL ammonium persulfate solution (10%), 15 μL TEMED, and 8 mL water in a 50-mL flask. Then cast gel within a gel cassette of 83 mm × 75 mm × 1.5 mm. Allow space for stacking gel and gently overlay with water. Polymerize for 15 min.
4. Stacking gel preparation (4.0%, *m/v*): Aliquot 0.7 mL of gel stock solution (30%, *m/v*), 1.25 mL of stacking gel buffer, 40 μL ammonium persulfate solution (10%), and 4 μL TEMED and make up volume to 5 mL with water. Then cast gel over separating gel and insert a 10-well gel comb immediately without introducing air. Polymerize for about 30 min.
5. Load 15 μL of sample for each channel.
6. Perform electrophoresis in a vertical discontinuous gel system composed of a DYCZ-24D vertical electrophoresis tank and a DYY-6C electrophoresis instrument. Supply a voltage of 120 V in the stacking gels for 0.5 h and then 100 V for 3 h when the sample enters the separating gel.

3.1.2. Native Two-Dimensional Electrophoresis (2-DE)

1. Sample buffer preparation: Dissolve 165 μL Bromophenol Blue solution (0.02% w/v) and 110 μL glycerol (20% v/v) into 110 μL water.
2. Sample preparation: Dissolve 55 μL serum into 385 μL sample buffer for 2-DE.
3. IEF gel preparation: Mix 0.45 mL of gel stock solution, 0.135 mL carrier ampholytes, 11.5 μL ammonium persulfate (10% *m/v*), and 2.5 μL TEMED. Dilute this to 2.25 mL with water. Then cast gel solution within a gel tube (ϕ1.0 mm × 170 mm) to polymerize for 3–6 h.
4. Separating gel preparation (7.5%, *m/v*): Same as 1-DE.
5. Load 100 μL of the prepared serum sample onto the IEF gel.
6. Electrophoresis was performed in a vertical discontinuous gel system composed of a DYY-6B electrophoresis instrument and DYCZ-21 vertical electrophoresis tank. Serum proteins were isoelectrically focused at 200 V for 30 min and 400 V for 14 h.

Electrophoresis for the second-dimensional gel was run at 100 V for 3 h, using the same buffer as in 1-DE.

3.1.3. SDS 1-DE

The procedures were similar to those in native 1-DE (see Subheading 3.1.1) except for the use of SDS and dithiothreitol (DTT):

1. Sample buffer: 60 mM of Tris–HCl, pH 6.8, 25% glycerol, 2% SDS, 0.015 g/mL DTT, and 0.1% Bromophenol Blue.
2. Separating gel preparation (8.0%, *m*/*v*): Mix 4 mL of gel stock solution (30%, *m*/*v*), 3.8 mL of separating gel buffer, 150 μL ammonium persulfate solution (10%), 10 μL TEMED, 150 μL of SDS solution (10%), and 6.9 mL water in a 50-mL flask. Then cast gel within a gel cassette of 83 mm × 75 mm × 1.5 mm. Allow space for stacking the gel and gently overlay with water. Polymerize for 15 min.
3. Stacking gel preparation (5.0%, *m*/*v*): Mix 0.83 mL of gel stock solution (30%, *m*/*v*), 0.63 mL of stacking gel buffer (0.5 M, pH 6.8), 50 μL ammonium persulfate solution (10%), 10 μL TEMED, and 50 μL of SDS solution (10%) into 3.4 mL water. Then cast gel over separating gel and insert a 10-well gel comb immediately without introducing air. Polymerize for 30 min.
4. Running buffer: 0.1% SDS in the running buffer as used in native 1-DE.

3.2. Preparation of QDs

The TGA-stabilized CdTe QDs were synthesized by the hydrothermal method (13, 14).

3.2.1. NaHTe Preparation

1. Add 126.7 mg of tellurium powder and 10 mL water into a three-necked flask (25 mL), keep on stirring the solution.
2. Degas with nitrogen for 20 min.
3. Add 80 mg of $NaBH_4$ into the flask (see Note 5).
4. Keep on degassing with nitrogen and stir the solution at room temperature until the color changes from violet to colorless (about 4 h). During the reaction, a small outlet connected to the flask was kept open to release the produced hydrogen.

3.2.2. QDs Preparation

1. Dissolve 0.4567 g of $CdCl_2 \cdot 2.5H_2O$ by 200 mL water into a three-necked flask.
2. Add 418 μL TGA and keep on stirring.
3. Adjust pH to 9.2 by NaOH.
4. Degas with nitrogen for 30 min.
5. Add the fresh prepared NaHTe solution immediately (see Note 6) and keep on vigorous stirring for 20 min. Then obtain the solution with the color in yellow.
6. Heat the solution at 80°C for 3.5 h. Therefore, obtain the QDs with the average diameter of 3.0 nm (Fig. 1).

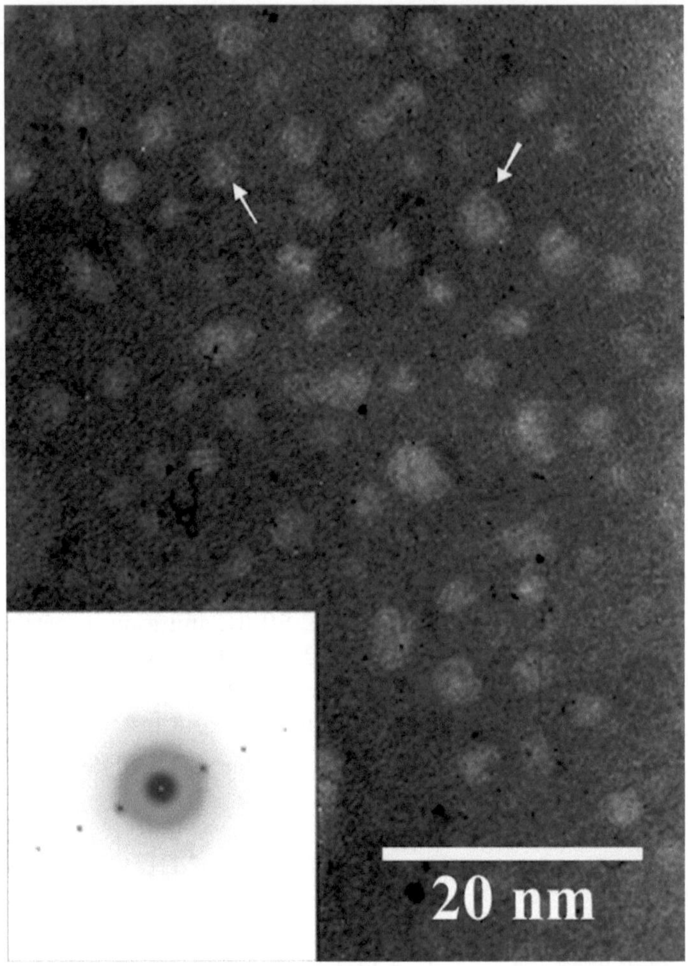

Fig. 1. TEM image of thioglycolic acid-stabilized CdTe QDs (the insert of FT-ED pattern) (reproduced from (15) with permission from ACS).

3.3. Detection After Electrophoresis

3.3.1. CBB R-250 Staining

Stain the gel in CBB R-250 (0.1 g/100 mL) solution for 2 h. Then wash and submerge the gel into destaining solution for 10 h.

3.3.2. QD-Based PL Staining

1. QDs staining solution (0.27 mM): Add 540 μL of QDs (see Note 7) to 10 mL of NaOH solution (2.1 mol/L) (see Note 8).
2. Wash the gel three times by water (see Note 9).
3. Soak the gel into the QDs staining solution to obtain the QDs–protein complex (see Note 10), for 1 h (see Note 11).

3.3.3. QD-Based TEMED-Enhanced PL Staining

1. Staining solution: Add 540 μL of QDs (see Note 7), 27 μL (see Note 12) of TEMED (see Note 13) to 10 mL of NaOH solution (2.1 mol/L) (see Note 8).

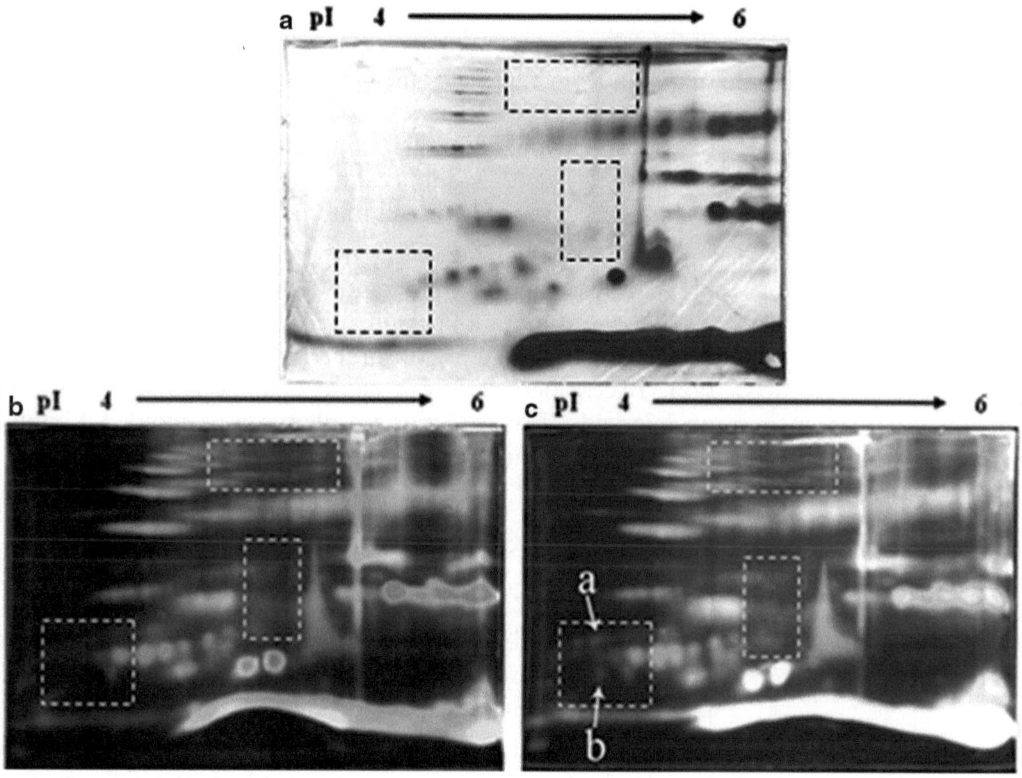

Fig. 2. Comparison of (**a**) CBB-R250 staining, (**b**) direct PL imaging detection, and (**c**) TEMED-enhanced PL imaging detection of human serum after native 2-DE. Time of UVA exposure was 1 s (reproduced from (15) with permission from ACS).

2. Wash the gel three times by water to avoid the interference of electrophoresis buffer (see Note 9).
3. Soak the gel into the QDs staining solution for 1 h (see Note 11).

3.4. Imaging Record

1. For CBB-R250 staining, scan the gel and transfer the results to the computer.
2. For PL and TEMED-enhanced PL imaging, insert the gel into a bioimaging system (UVP EC3 Imaging System, UVP, Inc.). At the excitation wavelength of 365 nm (see Note 14), the protein images of 2-DE (Fig. 2) (see Note 15), 1-DE (see Note 16), and SDS 1-DE (see Note 17) can be obtained.

4. Notes

1. Acrylamide is a suspected human carcinogen, severe neurotoxin, and causes irritation of the eyes, skin (is readily absorbed), and respiratory tract. Wear a mask when weighing acrylamide.

2. The presence of some impurities will influence the polymerization of the gel.
3. Acrylamide is thermally unstable, and thus, the acrylamide solution can be stored at 4°C for 1 month. It easily hydrolyzes to acrylic acid and ammonia. Polymerization may be caused by exposure to heat, UV light, oxidizers, or peroxides. Therefore, it should be wrapped with aluminum foil.
4. Ammonium persulfate is very hygroscopic, which makes it to break down almost immediately once dissolved in water. Therefore, ammonium persulfate solutions should be prepared fresh daily. Persulfate is consumed in the polymerization reaction. Excess persulfate can cause oxidation of proteins and nucleic acids. This oxidation problem can be avoided if inhibitor-free gel-forming reagents are used, and ammonium persulfate is used at the recommended levels.
5. $NaBH_4$ is hygroscopic and the solid reagent hydrolyses if exposed to moist air. It should therefore be kept in a desiccator and the weighing performed quickly.
6. NaHTe is made by reducing tellurium with $NaBH_4$. It is very sensitive to air oxidation and therefore should be freshly prepared and added to the flask quickly.
7. Although the higher QDs concentration is beneficial for achieving the higher PL intensity, excessively high concentration leads to the decrease of PL. This might be due to the aggregation of QDs in alkaline condition, which results in the less effective interaction between QDs and proteins.
8. Concentration of NaOH solutions is higher than 1 mol/L, as proteins cannot be PL imaged in gentle alkaline conditions. Although the high NaOH concentration is beneficial for imaging, excessive NaOH leads to PL decrease, which is due to aggregation of QDs in this condition.
9. The residual running buffer will change the composition of the QDs staining solution, which leads to the decrease of the staining ability. Therefore, the gel should be cleaned carefully before PL staining.
10. With proteins, PL intensity increases with increasing protein concentrations, and a red shift occurs in PL spectra comparing with the QDs solution. The added protein molecules can be used as a PL strengthening agent to QDs, which is due to its stabilization to QDs by attaching at the defect site on the QDs surface (16, 17). Thus, protein imaging can be obtained on the gel.
11. Although the highest PL intensity is obtained after 0.5 h, the high background will decrease the resolution. In addition, the precipitation of QDs and PL quenching will be lead when the time is longer than 2 h. Therefore, 1 h is appropriate for the staining.

12. PL emission intensity increases at a certain range of TEMED concentration. The redundant TEMED will decrease PL intensity, which is generated from the competition of redundant TEMED and protein to integrate with QDs.
13. TEMED is a negative reagent, whose PL enhancing ability is ascribed to the electronegativity of N atom. Other negative reagents, such as 1,2-ethylenediamine (EDA) and N,N,N',N',N''-pentamethyldiethylenetriamine (PMDTA), also have the PL enhancing ability. But their ability is much weaker than TEMED, which is due to the lower electronegativity or stereohindrance effect.
14. The excitation spectrum of CdTe QDs is in the range of 260–400 nm, confirming the feasibility of adopting the ultraviolet lamp of 365 nm for the PL imaging.
15. Comparing with CBB R-250 staining, proteins are easily detected by PL staining. In particular, the highest PL intensity of proteins is obtained by TEMED-enhanced PL staining, and some new proteins can be observed.
16. There is a linear increase of PL intensity with the increase of protein concentration. As demonstrated, the sensitivity of TEMED-enhanced PL imaging is higher than the other two methods.
17. Compared with native gels, proteins in SDS gels are more likely to bind onto the surfaces of positive charged QDs. This binding is so advantageous in PL imaging, that the addition of electronegative reagents such as TEMED cannot improve the PL intensity.

Acknowledgments

This work was supported by the National Nature Science Foundation of China (20975016, 21005007, 91027034) and National Grant of Basic Research Program of China (2011CB915504).

References

1. Anderson NL, Anderson NG (2002) The human plasma proteome—History, character, and diagnostic prospects. Mol Cell Proteomics 1:845–867
2. Bruchez M, Moronne M, Gin P, Weiss S, Alivisatos AP (1998) Semiconductor nanocrystals as fluorescent biological labels. Science 281:2013–2016
3. Medintz IL, Uyeda HT, Goldman ER, Mattoussi H (2005) Quantum dot bioconjugates for imaging, labelling and sensing. Nat Mater 4:435–446
4. Gill R, Zayats M, Willner I (2008) Semiconductor quantum dots for bioanalysis. Angew Chem Int Ed 47:7602–7625
5. Duncan TV, Polanco MAM, Kim Y, Park SJ (2009) Improving the quantum yields of semiconductor quantum dots through photoenhancement assisted by reducing agents. J Phys Chem C 113:7561–7566

6. Rene-Boisneuf L, Scaiano JC (2008) Sensitivity versus stability: making quantum dots more luminescent by sulfur photocuring without compromising sensor response. Chem Mater 20:6638–6642
7. Park C, Yoon TH (2010) L-Cysteine-induced photoluminescence enhancement of CdSe/ZnSe quantum dots in aqueous solution. Colloid Surface B 75:472–477
8. DeGroot MW, Taylor NJ, Corrigan JF (2003) Zinc chalcogenolate complexes as capping agents in the synthesis of ternary II-II '-VI nanoclusters: Structure and photophysical properties of [(N, N′-tmeda)(5)Zn5Cd11Se13(SePh)(6)(thf)(2)]. J Am Chem Soc 125:864–865
9. Cusack J, Drew MGB, Spalding TR (2004) Syntheses and spectroscopy of diamine complexes of Zn(II) and Cd(II) ethylxanthates and the molecular structures of [M(S2COEt)(2) TMEDA]: formation of US nanoparticles from [Cd(S2COEt)(2)] and [Cd(S2COEt)(2) TMEDA]. Polyhedron 23:2315–2321
10. Kedarnath G, Kumbhare LB, Jain VK, Phadnis PP, Nethaji M (2006) Group 12 metal monoselenocarboxylates: synthesis, characterization, structure and their transformation to metal selenide (MSe; M = Zn, Cd, Hg) nanoparticles. Dalton Trans 22:2714–2718
11. Jun YW, Koo JE, Cheon J (2000) One-step synthesis of size tuned zinc selenide quantum dots via a temperature controlled molecular precursor approach. Chem Comm 14:1243–1244
12. Na N, Liu L, Taes YEC, Zhang CL, Huang BR, Liu YL, Ma L, Ouyang J (2010) Direct CdTe quantum-dot-based fluorescence imaging of human serum proteins. Small 6: 1589–1592
13. Wang Q, Kuo Y, Wang Y, Shin G, Ruengruglikit C, Huang Q (2006) Luminescent properties of water-soluble denatured bovine serum albumin-coated CdTe quantum dots. J Phys Chem B 110:16860–16866
14. Zhang H, Zhou Z, Yang B, Gao M (2003) The influence of carboxyl groups on the photoluminescence of mercaptocarboxylic acid-stabilized CdTe nanoparticles. J Phys Chem B 107: 8–13
15. Huang CG, Na N, Huang LY, He DC, Ouyang J (2010) TEMED enhanced photoluminescent imaging detection of proteins in human serum using quantum dots after PAGE. J Proteome Res 9:5574–5581
16. Ge SG, Zhang CC, Zhu YN, Yu JH, Zhang SS (2010) BSA activated CdTe quantum dot nanosensor for antimony ion detection. Analyst 135:111–115
17. Idowu M, Lamprecht E, Nyokong T (2008) Interaction of water-soluble thiol capped CdTe quantum dots and bovine serum albumin. J Photochem Photobiol A 198:7–12

Chapter 46

Microwave-Assisted Protein Staining, Destaining, and In-Gel/In-Solution Digestion of Proteins

Jennie R. Lill and Victor J. Nesatyy

Abstract

Rapid evolution of state-of-the-art proteomic analyses has encompassed development of high-throughput analytical instrumentation and bioinformatic tools. However, recently, there has been a particular emphasis on increasing the throughput of sample preparation, which has become one of the rate-limiting steps in protein characterization workflows. Researchers have been investigating alternative methods to conventional convection oven incubations to try and reduce sample preparation time for protein characterization. Several protocols have appeared in the literature, which employ microwave irradiation as a tool for the preparation of biological samples for subsequent characterization by a variety of analytical techniques. In this chapter, techniques for microwave-assisted protein staining, destaining, and digestion are described. In general, the application of microwave-assisted technologies resulted in the drastic reduction of overall sample preparation time, though discrepancies in the reproducibility of several published digestion protocols still remain to be clarified.

Key words: Microwave, SDS gels, Protein staining, Destaining, Digestion

1. Introduction

Proteomic analysis has undergone a complete revolution during the past decade (1). New technologies allowing higher-throughput, increased sensitivity, and improved characterization of biomolecules have rapidly emerged and established themselves as firm competitors to previously assumed workflows (2). This evolution has been due in part to breakthroughs in technology development including the introduction of lower-flow rate and more sensitive chromatography systems (3), faster scanning, and more sensitive and accurate mass spectrometric instrumentation (4), and this evolution in chromatographic and mass spectrometric technologies has

Biji T. Kurien and R. Hal Scofield (eds.), *Protein Electrophoresis: Methods and Protocols*, Methods in Molecular Biology, vol. 869, DOI 10.1007/978-1-61779-821-4_46, © Springer Science+Business Media, LLC 2012

also allowed for more confident characterization, validation, and quantitation of proteomic samples. Bioinformatic power for database searching and sophisticated informatic workflows has increased the speed and confidence in which one may annotate proteins and posttranslational modifications (PTMs) (5). Consequently, sample preparation has gradually become the limiting factor in terms of throughput for many protein characterization protocols. To address the issue of time-consuming sample preparation, several methodologies have been deployed, enabling a drastic reduction in previously lengthy protein visualization and enzymatic incubation protocols (6, 7). Microwave-assisted methods are just one protocol being explored as a method to increase proteolysis efficiency, and while this field is growing, other alternative energies such as ultrasonic vibrations (via a sono-reactor or ultrasonic probe) or pressure are also being explored as options to increase proteolytic catalysis (8, 9), demonstrating a continuing trend to push the limitations of sample preparation.

Microwave-assisted energies have gained the most momentum out of the energies explored, and these have spanned across a wide range of disciplines including decomposition processing for hydrolysis of proteins and peptides (10); the preparation of plant, fish, and soil samples prior to trace metal determination by atomic spectroscopy (11); polymer technology (12); drug-release targeting (13); and inorganic and solid-state synthesis (14).

Directly related to bottom-up proteomic workflows for protein characterization, microwave irradiation has proven an invaluable method to dramatically decrease the time required for Coomassie Blue staining, silver staining, and destaining of SDS-PAGE samples (6). Not only does microwave assistance decrease sample preparation time by allowing the rapid visualization of gels, but in many examples, it improves sensitivity and therefore allows the researcher to characterize low-level samples which would otherwise be below the limit of detection. In addition to microwave assistance being applicable to improving the efficiency of staining and destaining gels, it has also demonstrated utility in the silver staining of histological samples too (15).

Over the past 9 years or so, microwave-assisted tryptic proteolysis has been described in multiple studies and adopted as a standard protocol in many laboratories (16). Accelerated proteolytic cleavage of proteins under controlled microwave conditions (i.e., set temperature, pressure, and power) in a scientific monomode microwave system was first described by Pramanik et al. (17). It was demonstrated that microwave-assisted digestions could be achieved using the proteolytic enzymes endoproteinase Lys-C (Lys-C) and trypsin. Figure 1 below demonstrates a clear example of a protein being digested more efficiently with trypsin in the presence of microwave irradiation.

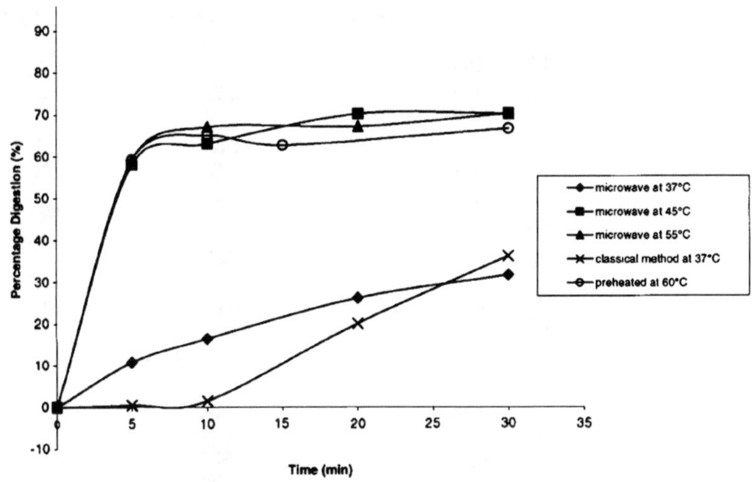

Fig. 1. Plots showing the rate of trypsin digestion of IFN-α-2b in the first 30 min under microwave irradiation and classic conditions. Reproduced with permission from Protein Science.

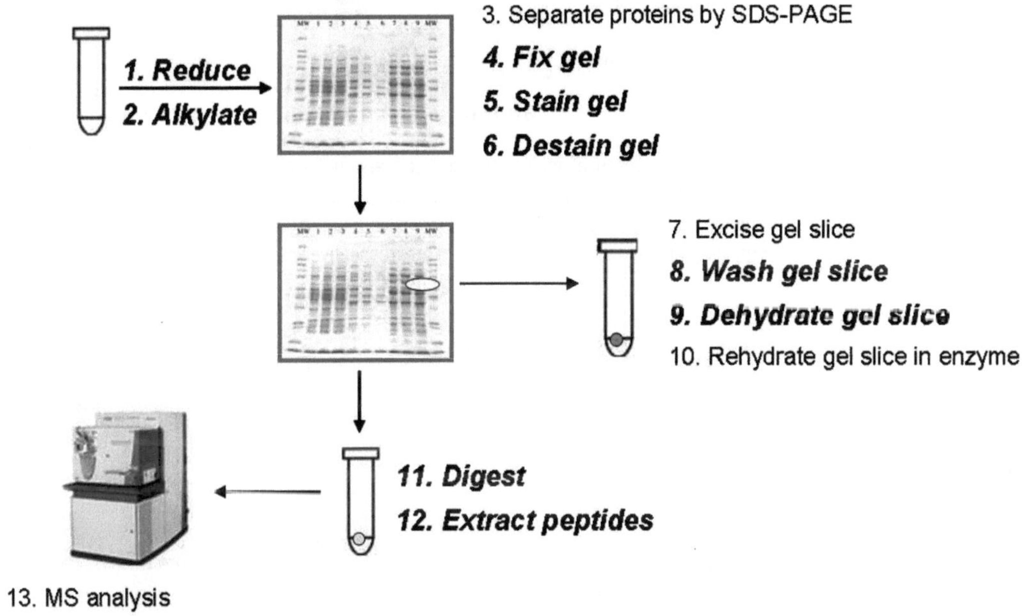

Fig. 2. Typical protocol for the analysis of proteins using microwave-assisted workflows. Steps that can be drastically reduced by the introduction of microwave assistance are highlighted in ***bold italic***.

This chapter describes the application of microwave irradiation as a valuable tool in proteomic research, allowing for higher-throughput sample processing in bottom-up proteomic workflows. Figure 2 dissects a typical proteomic workflow and highlights parts of the protocol which can benefit significantly from microwave assistance, while Figs. 3 and 4 (see Subheadings 3.1 and 3.2) highlight application of this approach in practice.

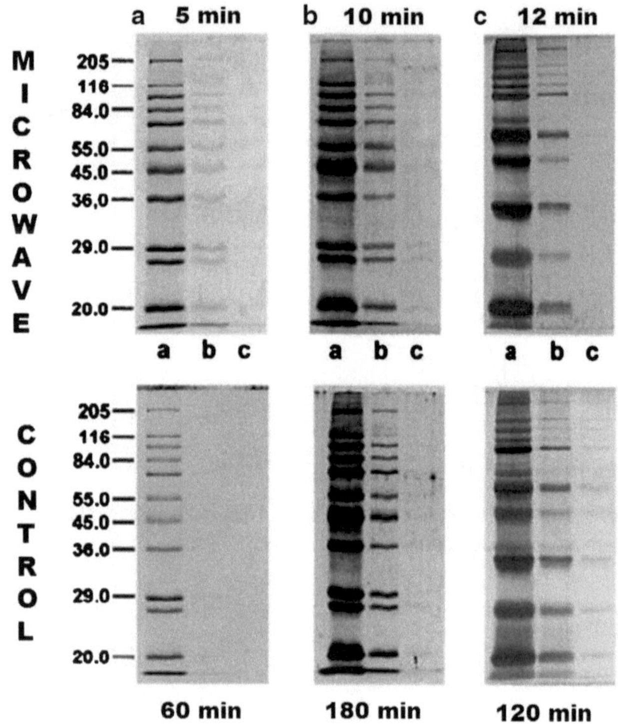

Fig. 3. Comparison of microwave and conventional staining methods for visualization of protein standards run on 0.75 mm thick SDS-PAGE minigels. Numbers on the left indicate molecular mass in kiloDaltons. (**a**) Coomassie Blue, (**b**) SYPRO® Ruby stain, and (**c**) Silver stain compatible with MS. Lanes (**a–c**) are, respectively, 10, 1, and 0.1 μg loads of protein.

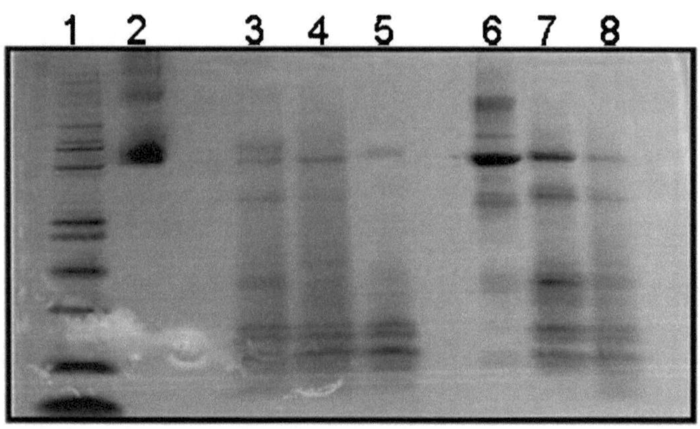

Fig. 4. SDS-PAGE analysis of BSA after tryptic digestion at 37°C in both the water bath and the microwave. *Lane 1*, mark 12 molecular weight standard; *Lane 2*, start material at 66 kDa; *Lane 3*, microwave-assisted digestion 5 min; *Lane 4* microwave-assisted digestion 10 min; *Lane 5*, microwave-assisted digestion 30 min; *Lane 6*, water bath digestion 5 min; *Lane 7*, water bath digestions 10 min; and *Lane 8*, water bath digestion 30 min.

2. Materials

All solutions should be prepared using ultrapure water and analytical grade reagents. It is always preferable to use freshly prepared solutions. When disposing waste material, please follow all waste disposal regulations.

2.1. Microwave-Assisted Protein Staining and Destaining

2.1.1. Coomassie Blue Staining

1. Fixing solution: 10% acetic acid/40% methanol. For a 1 L fixing solution, add 100 mL of glacial acetic acid to a 1-L graduated cylinder or a glass beaker containing 200 mL deionized water. Add 400 mL of methanol. Make up to 1 L with 300 mL of water. Mix thoroughly.
2. Coomassie Blue preparation: 0.25% Coomassie Blue, 7.5% acetic acid, and 50% ethanol. For a 1 L staining solution, dissolve 2.5 g Coomassie Blue in 250 mL water. Slowly add 75 mL of glacial acetic acid. Add 500 mL of ethanol. Make up to 1 L with water.
3. Destaining solution: For a 1 L destaining solution add 100 mL of glacial acetic acid to a 1-L graduated cylinder or a glass beaker containing 200 mL deionized water. Add 400 mL of methanol. Make up to 1 L with 300 mL of water. Mix thoroughly.

2.1.2. SYPRO® Ruby Staining

1. Fixing solution: 7% acetic acid/10% methanol. For a 1 L fixing solution, add 70 mL of the glacial acetic acid to a 1-L graduated cylinder or a glass beaker containing 200 mL deionized water. Add 100 mL of methanol. Made up to 1 L with 630 mL of water. Mix thoroughly.
2. SYPRO® Ruby protein gel stain (commercial preparation from Bio-Rad, Hercules, CA).
3. Destaining solution: 10% acetic acid/10% methanol. For a 1 L fixing solution, add 100 mL of the glacial acetic acid to a 1-L graduated cylinder or a glass beaker containing 200 mL deionized water. Add 100 mL of methanol. Make up to 1 L with 600 mL of water. Mix thoroughly.

2.1.3. Silver Staining (Compatible with MS Analysis)

1. Fixing solution: 10% acetic acid/50% methanol. For a 1 L fixing solution, add 100 mL of glacial acetic acid to a 1-L graduated cylinder or a glass beaker containing 200 mL deionized water. Add 500 mL of methanol. Make up to 1 L with 200 mL of water. Mix thoroughly.
2. Incubation solution: 50% methanol. For a 1 L incubation solution, add 500 mL of deionized water to a 1-L graduated cylinder or a glass beaker. Add 500 mL of methanol. Mix thoroughly.
3. Sensitizing solution: 0.02% sodium thiosulfate. For a 400-mL solution, weigh 0.08 g of sodium thiosulfate, transfer to a

graduated cylinder or glass beaker, make up to 400 mL with water, and mix thoroughly.

4. Impregnating solution: 0.2% silver nitrate. For a 400 mL solution, weigh 0.8 g of silver nitrate, transfer to a graduated cylinder or glass beaker, make up to a 400 mL with water, and mix thoroughly.
5. Developing solution: 3% sodium carbonate, 0.025% formalin. For a 400 mL developing solution, weigh 12 g of sodium carbonate, transfer to a cylinder or glass beaker, add 200 mL of water, mix thoroughly, add 100 μL of formalin (35% formaldehyde), make up to a 400 mL with water, and mix thoroughly.
6. Stopping solution: 1.4% EDTA disodium salt. For a 250 mL stopping solution, weigh 3.65 g of EDTA disodium salt, make up to a 250 mL with water, and mix thoroughly.

2.2. Microwave-Assisted In-Gel Digestion

1. Ammonium bicarbonate (aqueous): 25 mM solution. For a 50 mL solution, weigh 100 mg of ammonium bicarbonate, transfer to a graduated cylinder or glass beaker, add 50 mL of water, and mix thoroughly.
2. Trypsin digestion solution: Recommended Promega sequence-grade modified porcine trypsin. To a 20 μg vial of Promega Sequencing Grade trypsin, add 1 mL of 25 mM ammonium bicarbonate.
3. Ammonium bicarbonate (in 50% acetonitrile): For a 50-mL solution, weigh 100 mg of ammonium bicarbonate, transfer to a graduated cylinder or glass beaker, add 25 mL of water, and mix thoroughly. Add 25 mL of HPLC grade acetonitrile and mix thoroughly.
4. Extracting solution: 50% acetonitrile/5% tri-fluoroacetic acid (may substitute with formic or acetic acid). For a 50-mL extracting solution, add 2.5 mL of concentrated 100% acid to the cylinder or glass beaker containing 22.5 mL of water, mix thoroughly, add 25 mL of acetonitrile, and mix thoroughly.

2.3. Microwave-Assisted In-Solution Digestion

1. Ammonium bicarbonate (aqueous): 25 mM solution. For a 50 mL solution, weigh 100 mg of ammonium bicarbonate, transfer to a graduated cylinder or glass beaker, add 50 mL of water, and mix thoroughly.
2. Trypsin digestion solution: Recommended Promega sequence-grade modified porcine trypsin. To a 20-μg vial of Promega Sequencing Grade trypsin, add 1 mL of 25 mM ammonium bicarbonate.
3. Dithiothreitol (DTT) solution: Make a 15.4-mg/mL solution of DTT (this equates to a 100-mM stock solution). Dilute your sample 1/10 with this sample which will make the final concentration 10 mM DTT.

4. Iodoacetamide (IAA) solution: Make 18.4 mg/mL of IAA (this equates to a 100 mM stock solution). Dilute sample ¼ with this solution to make the final concentration 25 mM IAA.

3. Methods

All protocols unless specified otherwise should be carried out at room temperature, preferably under the fume hood, paying utmost attention to safety (see Notes 1–4). Also, please note that keratin contamination can be problematic in bottom-up proteomic analyses; therefore, please try and reduce the presence of keratin by keeping all solvents filtered and the microwave oven clean at all times and work in an environment which minimizes contamination (see Note 5). For all steps described below, the gel was placed in a glass dish and covered with a minimum of 100 mL of the appropriate solution at room temperature.

3.1. Microwave-Assisted Protein Staining and Destaining

During staining, it is recommended to manipulate each gel individually in a glass container, using a minimum of 100 mL of each staining solution. The size of the glass staining containers should be larger than the size of the stained gels (see Note 6). The microwave-assisted staining protocols described below were optimized for the solid-state Kenmore model # 87780 (Sears, Canada) microwave oven with the frequency of 2,450 MHz and maximum output power of 700 W.

3.1.1. Coomassie Blue Staining

1. Place the gel in a glass container with 100 mL of fixation solution into a domestic microwave oven. Turn on at full power for 30 s. Gently agitate container inside microwave to distribute the heat evenly and continue irradiating for an additional 30 s. Drain the fixation solution from the dish (beware of noxious vapors and high temperatures), holding the gel in the dish with a plastic holder. This step could be repeated for improvement in clearer gel background (see Note 7).
2. Add 100 mL of the Coomassie Blue staining solution to the container with the gel. Place it in the microwave for two 30 s bursts as described above. Drain the Coomassie Blue staining solution from the gel container into an appropriate waste container, holding the gel with plastic holder (see Note 8). This step could be repeated for the improvement in signal intensity. Please note that Coomassie Blue solution may be reused.
3. Add 100 mL of the destaining solution into the container with the gel. Place it in the microwave for 30 s, wait 30 s, and then repeat. Drain the destaining solution from the container into an appropriate waste disposal container, holding the gel with

either tweezers or a plastic holder. This step could be repeated to obtain a clearer background (see Note 9). The gel is now ready for scanning or spot/band removal for further mass spectrometry analysis.

3.1.2. SYPRO® Ruby Staining

1. Place the gel in a glass container with 100 ml of fixation solution into domestic microwave. Turn on at full power for 30 s. Gently mix container inside microwave and continue for additional 30 s. Drain the fixation solution from the dish (beware of the vapors and temperature), holding the gel in the dish with a plastic holder. This step may be repeated for improvement in signal intensity and clear gel background.
2. Add 100 mL of the SYPRO® Ruby staining solution to the container with the gel. Place it in the microwave for 2×30 s. Drain the SYPRO® Ruby staining solution from the container into an appropriate container, holding the gel with plastic holder. This step may be repeated for the improvement in signal intensity.
3. Add 100 mL of the destaining solution to the container with the gel. Place it in the microwave for 2×30 s. Drain the destaining solution from the container into an appropriate container, holding the gel with plastic holder. This step may be repeated to obtain a clearer gel background. The gel is now ready for scanning (is recommended to do as soon as possible).

3.1.3. Silver Staining

It is known that silver staining is susceptible to interference from a wide variety of contaminants. To avoid artifacts during gel handling, it is recommended to wear powder-free gloves at all times. Also, please avoid placing pressure on the gels during solution transfers, which could be a source of additional interferences. Make sure to have all solutions available and easily accessible to avoid any delay in executing the protocol.

1. Place the gel in a glass container with 100 ml of fixation solution into a domestic microwave oven. Turn on at full power for 30 s. Gently agitate container inside microwave and continue for an additional 30 s. Drain the fixation solution from the dish (beware of the noxious vapors and high temperature), holding the gel in the dish with a plastic holder.
2. Add another 100 mL of incubation solution and place in the microwave for 2×30 s. Discard the incubation solution.
3. Add 100 mL of rinsing water to the container and place in the microwave for a 2×30 s. Drain the water and repeat the procedure twice more.
4. Add 100 mL of sensitizing solution to the container. Place it in the microwave for 2×30 s. Drain the sensitizing solution from the container, holding the gel with plastic holder.

5. Add 100 mL of rinsing water to the container and place in the microwave for a 2×30 s. Drain the water from the container, holding the gel in the dish with a plastic holder.
6. Add 100 mL of silver nitrate solution to the container. Place it in the microwave for 2×30 s. Drain the silver nitrate solution from the container into an appropriate waste container, holding the gel with plastic holder.
7. Add 100 mL of rinsing water to the container and place in the microwave for a 2×30 s. Drain the water from the container, holding the gel in the dish with a plastic holder.
8. Add 100 mL of developing solution to the container and place in the microwave for 2×30 s. Drain the developing solution, holding the gel in the dish with a plastic holder, and repeat the procedure one more time. It is always useful to watch the gels individually after each microwave burst, while determining the proper time for the development. Typically, it is assessed by the appearance of dark colored protein spots, while the gel itself remains slightly yellow. Time requirements for proper development vary with each individual gel. Be careful to avoid overdevelopment.
9. Add 100 mL of the solution stopping development and place in the microwave oven for 30 s. Drain the solution from the container, holding the gel in the dish with a plastic holder.
10. Add 100 mL of the rinsing water and place in the microwave for 30 s. Drain the water from the container, holding the gel in the dish with a plastic holder. The gel is now ready for scanning or spot removal for further mass spectrometry analysis. It is recommended to scan gels as soon as possible.

3.2. Microwave-Assisted In-Gel Digestion

1. Excise gel band and cut into 4–8 pieces.
2. Wash gel pieces in 25 μL of 25 mM aqueous ammonium bicarbonate followed by 25 μL of 25 mM ammonium bicarbonate in 50% acetonitrile.
3. Dehydrate gel band in 100% acetonitrile, then rehydrate with trypsin solution in 25 mM ammonium bicarbonate (25 μL or enough volume to rehydrate the gel pieces). For best results, allow gel to rehydrate on ice for 30 min prior to incubation.
4. Incubate in the microwave at 5 W, 50°C for 15 min.
5. Extract from the gel with 30 μL of extracting solution (50% acetonitrile/5% TFA) followed by 30 μL of 100% acetonitrile. If gel bands are not completely dehydrated (white and crisp) after this step, repeat.
6. Reduce volume of organic solvent (preferably by speed-vac) and reconstitute in appropriate buffer for assay (see Note 10).

3.3. Microwave-Assisted In-Solution Digestion

1. Samples may be reduced and alkylated prior to digestion if required. This may be performed by adding 10 mM final concentration of DTT into the sample and microwaving at 60°C for 1 min followed by alkylation with either 25 mM iodoacetamide or Ni-isopropyl-iodoacetic acid at 37°C and a further 2 min in DTT (to neutralize the activity of any excess DTT).
2. For tryptic digestion, dilute trypsin at recommended concentration (1:100 enzyme:substrate) in 25 mM ammonium bicarbonate to sample.
3. Add vials/Eppendorf tubes into holder, place into the cavity, and program the microwave for 30 min at 5 W at 55°C with a 2-min ramp time (see Note 11). Domestic microwave oven can be employed; however, conditions may vary (see Note 12). If employing a domestic microwave oven, it is suggested to place a beaker of cold water within the cavity and replace every 5 min.
4. Test an aliquot of sample using either MALDI-TOF MS or SDS-PAGE to check completion of digestion (see Note 13).
5. As trypsin will degrade rapidly at higher temperatures, if digestion is incomplete add further aliquot of trypsin and repeat, or for future experiment, add small percentage (<20%) of organic solvent such as acetonitrile or mass spectrometry-friendly surfactant (e.g., Rapigest™).

4. Notes

1. It is strongly advised to wear appropriate protective laboratory clothing, breathing mask, and safety glasses at all times.
2. To minimize the risk of microwave exposure, the microwave oven should be turned off and unplugged from the power after each use. In the case of microwave oven malfunction that may result in continued radiation generation even after it is switched off, unplugging the unit from the power should prevent any potential accident from such failure.
3. Take care in handling acids and possible acid vapors during and after microwave irradiation. It is recommended to allow the sample vial to cool inside the microwave oven prior to removal and then to open the cooled sample vial in a well-vented bench area, such as inside a fume hood. To minimize the risk of the exposure from the hot vapors altogether, it is recommended to place the microwave system under a fume hood.
4. Dilution of concentrated acid solution should always be done in a fume hood. For safety reasons, especially for concentrated acids, it is important to pour acid into water and not vice versa.
5. The contamination of gel slices with keratins from the skin, accumulated during gel handling, may prevent identification

of low-abundant proteins by mass spectrometry. Make sure to wash all surfaces that may come in contact with the gel by ethanol and then water. Always wear gloves. Remember to use a clean glass surface for the excision of the gel band. Use new razor blade for a new gel. Avoid contact of the gel slices with the surface that may have been in contact with yours or someone else's hands.

6. To minimize spilling, ensure easy handling, and good mixing, it is recommended to use glass containers with 5–10-cm wall height.
7. Microwave-assisted processing of the thicker 1.5-mm gels may require slight modifications in the published protocols.
8. Remember, superheating of solvents may occur and may be much hotter than anticipated!
9. To increase destaining efficiency, a lint-free tissue can be added. Wearing gloves (to minimize keratin contamination), tie the tissue into a knot and place in destain solution next to the gel.
10. For useful overview of in-gel enzymatic digestions in general, please refer to ref. 19.
11. For a safety review on handling microwave irradiation in the lab please refer to ref. 18.
12. For microwave-assisted in-solution digestion, it is recommended to use an industrial microwave such as the Discover unit by CEM Corporation (http://www.cem.com).
13. It is recommended to thoroughly test newly published microwave-assisted digestion protocols and contact publication authors in case inconsistencies between published results and experimental data are observed.

Acknowledgements

Victor J. Nesatyy is grateful to Prof. Yury Tsybin for his support, encouragement, and creative atmosphere during writing of this manuscript.

References

1. Ahrens C, Brunner E, Qeli E et al (2010) Generating and navigating proteome maps using mass spectrometry. Nat Rev Mol Cell Biol 11:789–801
2. Hüttenhain R, Malmström J, Picotti P et al (2009) Perspectives of targeted mass spectrometry for protein biomarker verification. Curr Opin Chem Biol 13:518–525
3. Horvatovich P, Hoekman B, Govorukhina N et al (2010) Multidimensional chromatography coupled to mass spectrometry in analysing complex proteomics samples. J Sep Sci 33:1421–1437
4. Chalkley R (2010) Instrumentation for LC-MS/MS in proteomics. Methods Mol Biol 658:47–60

5. Jacob R (2010) Bioinformatics for LC-MS/MS-based proteomics. Methods Mol Biol 658:61–91
6. Nesatyy V, Dacanay A, Kelly J et al (2002) Microwave-assisted protein staining: mass spectrometry compatible methods for rapid protein visualisation. Rapid Commun Mass Spectrom 16:272–280
7. Juan H, Chang S, Huang H et al (2005) A new application of microwave technology to proteomics. Proteomics 5:840–842
8. López-Ferrer D, Capelo J, Vázquez J (2005) Ultra fast trypsin digestion of proteins by high intensity focused ultrasound. J Proteome Res 4:1569–1574
9. Lopez-Ferrer D, Petritis K, Hixson K et al (2008) Application of pressurized solvents for ultrafast trypsin hydrolysis in proteomics: proteomics on the fly. J Proteome Res 7:3276–3281
10. Chen S, Chiou S, Chy Y et al (1987) Rapid hydrolysis of proteins and peptides by means of microwave technology and its application to amino acid analysis. Int J Peptide Protein Res 30:572–576
11. Aysola P, Anderson P, Langford C (1987) Wet washing in biological samples in a microwave oven under pressure using poly (tetrafluoroethylene) vessels. Anal Chem 59:1582–1583
12. Watanabe S, Hayama K, Park KH et al (1993) New microwave-assisted rapid synthesis of polyamides from nylon salts. Macromol Chem Rapid Commun 14:481–484
13. Miyazaki S, Yokouchi C, Takada M (1989) External control of drug release IV. Controlled release of 5-fluorouracil from a hydrophilic polymer matrix by microwave irradiation. Chem Pharm Bull Jpn 37:208–210
14. Caddick S (1995) Microwave assisted organic reactions. Tetrahedron 38:10403–10432
15. Churukian C, Kazee A, Lapham LW et al (1992) Microwave modification of bielschowsky silver impregnation method for diagnosis of Alzheimer's disease. J Histotechnol 15:299–302
16. Capelo J, Carreira R, Diniz M et al (2009) Overview on modern approaches to speed up protein identification workflows relying on enzymatic cleavage and mass spectrometry-based techniques. Anal Chim Acta 650:151–159
17. Pramanik BN, Mirza UA, Ing Y et al (2002) Microwave-enhanced enzyme reaction for protein mapping by mass spectrometry: a new approach to protein digestion in minutes. Protein Sci 11:2676–2687
18. Lew A, Krutzik P, Hart M et al (2002) Increasing rates of reaction microwave-assisted organic synthesis for combinatorial chemistry. J Comb Chem 4:95–105
19. Granvogl B, Plöscher M, Eichacker L (2007) Sample preparation by in-gel digestion for mass spectrometry-based proteomics. Anal Bioanal Chem 389:991–1002

Chapter 47

Electrophoretic Separation and Detection of Metalloproteins by X-Ray Fluorescence Mapping

Tripti Khare, Yasmin Chishti, and Lydia A. Finney

Abstract

All living systems depend on metalloproteins. Yet, while tools for the separation and identification of apo-proteins are well developed, those enabling identification and quantitation of individual metalloproteins within complex mixtures are still nascent. Here, we describe the electrophoretic separation of a mixture of carbonic anhydrase, ceruloplasmin, urease, and hemoglobin using native 2D gel electrophoresis and X-ray fluorescence mapping—an approach we have developed to be broadly applicable, not require specialized equipment for sample preparation, and likely to be extensible in the future.

Key words: X-ray fluorescence, Metalloprotein, Imaging, Electrophoresis, Metal, Iron, Copper, Zinc, Nickel

1. Introduction

The detection of metals in bulk protein samples has become routine in recent years and has greatly aided our understanding of bioinorganic chemistry and the physiology of metalloproteins in the cell. However, to advance from understanding the chemistry of these proteins in vitro to observing and understanding their behavior in vivo, we need methods that enable us to detect and measure the metal ion component of metalloproteins isolated from nature. This means looking at cell lysates and extracts, at samples where rather than one purified protein, we are observing dozens of proteins in the same sample, each of which may be binding some of the zinc, copper, or iron present in total mass.

Here, we describe an approach to this that pairs gel electrophoresis with synchrotron X-ray fluorescence mapping. Similar

Biji T. Kurien and R. Hal Scofield (eds.), *Protein Electrophoresis: Methods and Protocols*, Methods in Molecular Biology, vol. 869,
DOI 10.1007/978-1-61779-821-4_47, © Springer Science+Business Media, LLC 2012

procedures are being used by others; notably, electrophoretic separation is being combined with LA-ICP-MS in some cases to accomplish similar aims with some success (1–7)—also, synchrotron X-ray fluorescence has been combined in some cases with gel electrophoresis, often where the gels have been imaged with linear scans through areas that stain positive for protein (8–18). In this approach, we describe separations that are sensitive to metalloproteins as coordination complexes (see Note 1), instrumentation that is freely and publicly available through a general user proposal system, mapping that produces 2D images of the electrophoresis gels that can intuitively be compared to gels stained for protein, and methods that allow further examination of the electrochemical state of the metal through micro-X-ray absorption near edge spectroscopy (XANES) (19). This method allows one to separate and detect metalloproteins using 2D gel electrophoresis, with simultaneous, quantitative, detection of elemental content for sulfur through zinc.

2. Materials

Prepare all solutions using ultrapure water (with a sensitivity of 18 MΩ at room temperature), and analytical grade reagents. Prepare and store all reagents at room temperature unless otherwise indicated. Diligently follow all waste disposal regulations when disposing waste materials. All gel tanks, cassettes, spatulas, scrapers, etc., should be rinsed with ultrapure water and allowed to dry before use.

2.1. Native Isoelectric Focusing Components

1. Zoom IPGRunner mini-Cell, cassette, sealing stickers, wicking strips, and nonlinear pH 3–10 strip.
2. Power supply. Must be able to run at high voltage (2,000 V) at low amperage (<1 mA). This typically requires a specialized source; many electrophoresis power supplies cannot operate at these parameters.
3. Rotating platform for gentle rocking of samples.
4. Rehydration buffer: 2% CHAPS, 0.2% Zoom Carrier Ampholyte (pH 3–10), 0.001% (w/v) bromophenol blue. To prepare 25 mL, add 0.5 g CHAPS, 50 μL Carrier Ampholyte, and 0.5 mL of a 0.1% (w/v) solution of bromophenol blue to a 50 mL Falcon tube. Add ultrapure water to a final volume of 25 mL.
5. Protein dilution buffer (see Note 2): 0.5 M Tris–Cl, pH 6.8. To prepare 250 mL, add 19.7 g Tris-base to an Erlenmeyer flask. Dissolve with about 200 mL ultrapure water. Adjust pH with trace-element grade HCl (see Note 3). Dilute to 250 mL final volume with ultrapure water. Store in a glass or plastic bottle.

6. Protein samples: bovine ceruloplasmin (Sigma-Aldrich, Saint Louis, MO), bovine carbonic anhydrase (isoform II) (Sigma-Aldrich, Saint Louis, MO), bovine hemoglobin (Sigma-Aldrich, Saint Louis, MO), and jack bean urease (Sigma-Aldrich, Saint Louis, MO) were obtained as lyophilized stocks or solutions.
7. Equilibration buffer: 0.375 M Tris–HCl, 20% glycerol (v/v), and 0.001% (w/v) bromophenol blue, pH 6.8. To prepare 25 mL, add 18 mL of 0.5 M Tris–Cl, pH 6.8 buffer, 2 mL glycerol, and 0.5 mL of a 0.1% (w/v) bromophenol blue stock to a 50 mL Falcon tube. Dilute to 25 mL with ultrapure water.

2.2. Native Polyacrylamide Gel Electrophoresis Components

1. Separating gels: 4–15% Tris–Cl gels (Bio-Rad) were used. Store at 4°C.
2. Running buffer: Tris-glycine buffer (25 mM Tris, 192 mM glycine, pH 8.3) was prepared from a 10× stock obtained from Bio-Rad. Store stock at 4°C.
3. IEF strip loading buffer: Prepare a 0.5% (w/v) agarose solution in 0.5 M Tris–Cl, pH 6.8 buffer, stored in a 50-mL Falcon tube. To prepare, add 0.3 g agarose to 30 mL of 0.5 M Tris–Cl, pH 6.8 stock buffer.
4. Prestained molecular weight marker (Precision Plus Protein Kaleidoscope, Bio-Rad).

2.3. Electroblotting Components

1. PVDF membranes (Immobilon, Millipore—see Note 4).
2. Blotting buffer: same as running buffer above, do not add methanol.
3. Plastic tweezers.
4. One large (9″ × 13″), shallow plastic bin, and two small (5″ × 7″) shallow plastic bins.
5. Blotting Roller (Invitrogen Corporation, Carlsbad, CA).
6. Precut blotting papers.
7. Methanol for activating the PVDF.
8. Bio-Rad mini-protean wet transfer unit, including plastic sandwich, frozen ice pack, and two foam pieces.
9. Magnetic stir bar and stirrer.

2.4. X-Ray Imaging

1. Arrange in advance for use of public resources at an appropriate synchrotron facility. To use dedicated instrumentation facilitated by the X-ray Science Division of the Advanced Photon Source (APS), beamtime is awarded on a competitive basis through a general user proposal system (for more information, see Note 5).
2. Small plastic container for sample transport.

3. Methods

Throughout this protocol, gloves should certainly be worn, but the gel and the membranes should not be touched by hand, even with gloved hands, whenever possible. If the membrane is touched with hands or gloves, the XRF image will show fingerprints of zinc and other metals. Likewise, the membrane should not come into contact with any metal implements. Everything should be minimally manipulated with plastic, preferably Teflon, tweezers.

3.1. Isoelectric Focusing

1. Prepare stocks of each protein by dissolving the lyophilized stocks to a concentration of approximately 10 mg/mL (w/v) with protein dilution buffer. Store these stocks at 4°C and keep on ice while in use.
2. Prepare the sample by adding 100 μg of each protein to an Eppendorf tube (about 10 μL of each stock) and then enough sample rehydration buffer to reach a final volume of 155 μL (see Note 6).
3. Fill a lane of the cassette with the sample. Then, gently slide the Zoom strip into the lane, gel-side up. Seal the ends of the cassette with stickers. Rehydrate overnight.
4. Remove the stickers and plastic cover of cassette and replace them with the wicks. Add 600 μL of water to each wick and assemble unit as described in the Invitrogen IPGRunner manual.
5. Run the focusing step using a low-current power supply and a step voltage protocol (200 V for 20 min, 450 V for 15 min, 750 V for 30 min).
6. Turn off the power supply and disconnect the cables. Disassemble the cassette and place the strip, gel-side up, in a 15-mL Falcon tube containing about 5 mL of sample equilibration buffer, on a rotating platform for about 15 min.

3.2. Native-PAGE

1. While the strip is equilibrating, melt the 0.5% agarose solution by loosening the top of the tube and placing it into a beaker of boiling water on a hot plate (see Note 7). Turn off the hot plate and let the hot water melt the agarose while the strip equilibrates.
2. Once the strip is equilibrated, and the agarose is just slightly warmer than room temperature but still liquid, setup the gel for native-PAGE. Use tweezers to take the IPG strip from the falcon tube and place it on the edge of the gel. Use scissors to cut off the extra plastic from the ends of the strip. Then, gently nudge the strip into the well of the gel using the tweezers or a pipette tip. The goal is to avoid any air bubbles between the

strip and gel (see Note 8). Once the strip is in the well and in contact with the gel, gently fill the rest of the well with the agarose solution to seal it in place. Fill the molecular weight marker lane with about 10 μL of pre-stained marker.

3. Electrophorese the samples at 100 V for about 1 h or until the lowest molecular weight marker of the standard reaches the bottom of the gel. Meanwhile, refrigerate 1 L of running buffer for the blotting step (see Note 9).

3.3. Transfer

1. Prepare the large shallow bin with about 300 mL of the refrigerated transfer buffer. In one of the small bins, activate the PVDF with methanol by dipping it for about 30 s, then equilibrate it with a small amount of transfer buffer. The PVDF will float—you can just flip it over once or twice to make sure both sides contact liquid. Place a small amount of the transfer buffer into the third bin.
2. Crack open the gel cassette with a small plastic scraper or spatula. Using this tool, coax the gel to gently fall into the small bin of buffer.
3. In the large bin of buffer, begin to assemble the transfer cassette, which is assembled as in a standard Western blot. Place the dark side of the sandwich down into the buffer. Then, layer the foam pad, blotting paper, gel, PVDF, and blotting paper. Using the roller (see Note 10), and wearing gloves, with the sandwich completely submerged in buffer, gently roll over the sandwich to squeeze any air bubbles that might be between the PVDF and gel out. Keeping the sandwich submerged, add the last layer of foam and press the top down into the buffer to close the cassette. Latch it. Transfer it into the transfer unit and add the buffer from the bin.
4. Add the ice pack (see Note 11), and stir bar, and then pour cold transfer buffer into the unit until it just reaches the bottom of the lip of the ice pack. With the stir bar spinning, run the transfer at 0.35 A for 1 h.
5. Following the transfer, disassemble the sandwich using plastic tweezers. Carefully, with the PVDF immersed in buffer if necessary, ensure that all pieces of gel are removed from the membrane (see Notes 12 and 13). Allow the membrane to rest on blotting paper and to completely dry.

3.4. Imaging

1. Using tweezers, place the membrane in a sealed, clean plastic dish, such as a petri dish or tissue culture plate.
2. Transport the membrane to the APS high-throughput imaging facility, currently at beamline 8-BM-B or an equivalent facility (arrangement for beamtime in advance is necessary; contact the responsible beamline scientist for more information, see Note 5).

3. Tape the membrane to an acrylic mount provided at APS, being sure that the tape does not contact any parts of interest on the membrane and that the membrane does not contact anything dusty (to facilitate this, place a large, clean Kimwipe or sheet of blotting paper to cover your working surface). Keep covered, for example, with the top of a petri dish, if work is stopped.
4. Mount the acrylic holder to the linear motion sample stages of the Metalloproteomics instrument, which are mounted at an approximately 45° angle to the X-ray beam. Close the sample chamber door and flush the sample chamber with helium for 30 min at 5 L/min. Once, flushed, turn down the flow of helium 300 mL/min.
5. Using the Metalloproteomics instrument, comprised of a 3×10^{10} ph/s beam of 10 keV X-rays, passing through a 0.5-mm pinhole to excite the sample, and a SII Vortex single-element energy dispersive detector run by dxp software at a distance of approximately 3 mm from the sample and approximately a 90° angle to the X-ray beam, raster-scan the sample. Scan the sample in 0.5-mm steps, with 1 s dwell-time per step. A typical scan of a mini-gel, approximately 60 by 80 mm, takes about 8 h.
6. Using NBS1832 and NBS1833 thin film standards mounted at the same sample-to-detector distance as above, with a dwell-time of 180 s, collect spectra of both films to use in fitting and quantification.
7. Using MAPS software (see Note 14), create a fit of the spectrum collected at each point of the raster-scan image, pixel-by-pixel, to Gaussians at characteristic X-ray emission energies, normalizing to incident X-ray intensity and correcting for background. As shown in Fig. 1, an image of the blot is generated for each element. Apply the spectra of the NBS standards using the software to quantitate the image of each element.

4. Notes

1. In considering this method, one should be aware that metals may be lost from proteins to some extent during electrophoresis—they are, after all, coordination complexes under an applied voltage. Given this, one must be aware of the caveat that not all metalloproteins in a given sample may be observed by this method. False negatives are possible.
2. It is important that the sample buffer be low salt. If a high salt buffer is used for the sample, the IPG strip may ignite. The Zoom IPGRunner Manual has further information on this topic.

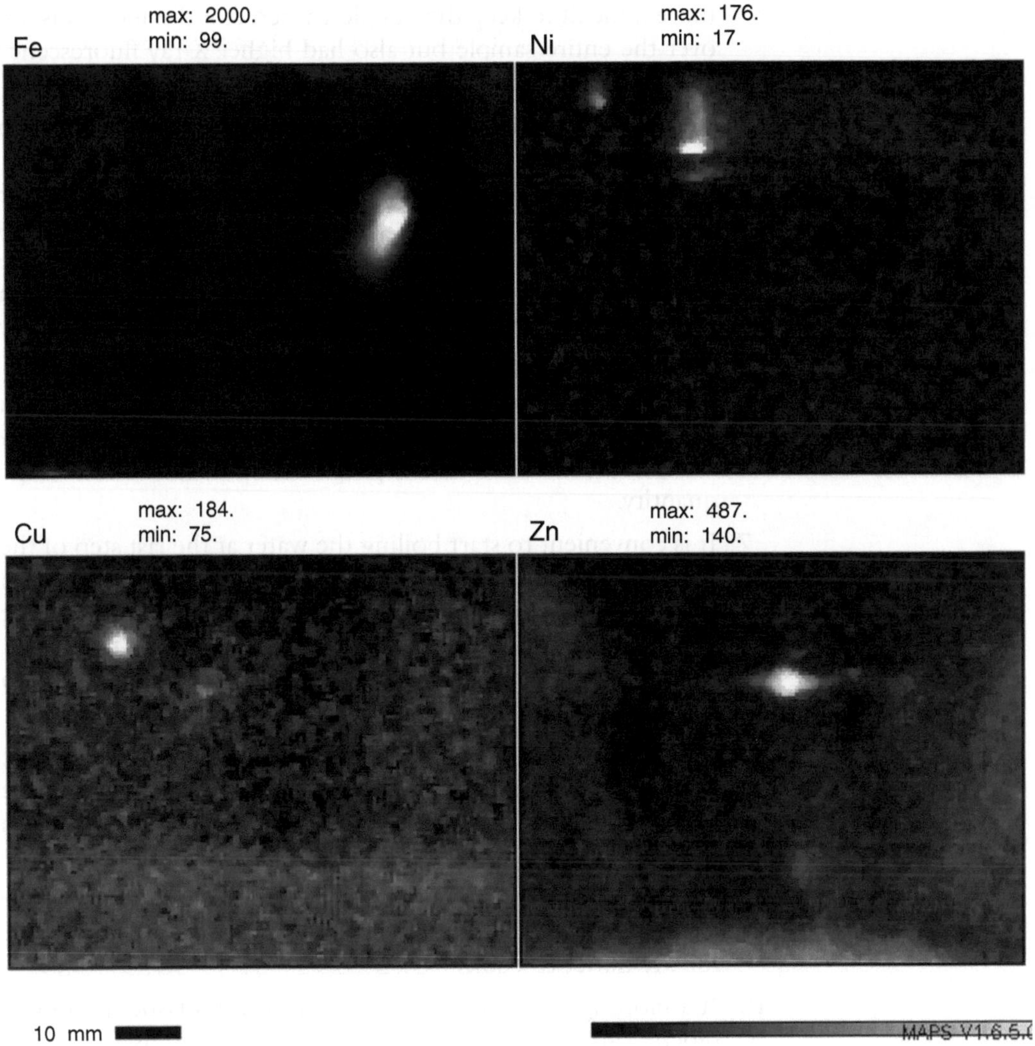

Fig. 1. Map of metals in separated proteins. Each panel represents the fitted distribution of iron in hemoglobin, nickel in urease, copper in ceruloplasmin, and zinc in carbonic anhydrase, respectively, as marked. Images of over a dozen elements, from S through Zn in the periodic table, were simultaneously collected. The thresholds shown are in counts/s.

3. Concentrated hydrochloric acid dissolves contaminant metals present in plastics and glass into solution. Use trace-element grade hydrochloric acid, and to avoid contamination of stock solution, pour out small quantities for each use, rather than inserting a pipette into the stock bottle.
4. The Immobilon PVDF from Millipore is a membrane that we have found to consistently have low X-ray fluorescence background. Other PVDF we examined sometimes had quite high background—sources should be evaluated prior to use. Nitrocellulose we examined not only was less rigid, making it

more difficult to keep the sample-to-detector distance constant over the entire sample but also had higher X-ray fluorescence background.

5. Information on the APS and the general user proposal system may be found at http://www.aps.anl.gov/Users/Scientific_Access/General_User/. Instrumentation is available for use free of charge for nonproprietary work on a competitive basis through the general user proposal system, and awarded time includes the support of staff scientists. The instrumentation currently at beamline 8-BM-B of the APS was developed by the authors for use by the user community.
6. The amount of protein loaded should be optimized for each sample type. In this case, since the proteins are commercially available, for demonstration purposes, we chose a rather large quantity.
7. It is convenient to start boiling the water at the last step of the ramped voltage protocol for the IEF, if you plan to proceed immediately. If you turn off the hot plate when you put the agarose tube in the beaker, there is sufficient heat to melt it, and it should cool down to room temperature by the time you are ready to start the electrophoresis.
8. The gels come with a small amount of buffer present in the wells. Remove most, but not quite all, of this buffer to leave a little bit of liquid in the well helps keep any air bubbles from forming between the strip and gel.
9. If you start refrigeration of the buffer early in the day, for example, while the IEF is running, it will ensure sufficient time for the buffer to cool.
10. If a blotting roller is not available, a 10-mL plastic disposable pipette can be cut into about a 4 in. section and used instead.
11. It is important that the PVDF remains flat. If gel adheres to the PVDF following transfer, do not scrape the surface of the PVDF with the tweezers. Instead, try to gently remove the gel rinsing with buffer.
12. If the samples become warm, either during the electrophoresis or during the transfer, we find that you will probably not detect any metals during imaging. They may be lost through denaturation.
13. If you use expired gels, they will often either stick to the PVDF during transfer, making it curl as it dries, or they will simply shatter.
14. Fitting the data using MAPS software is supported by APS staff—assistance is available either to learn to fit the data yourself or to have it processed on your behalf. MAPS software is freely distributed.

Acknowledgments

This work was supported by the DOE Office of Science under contract DE-AC02-06CH11357. The authors thank Ken Kemner and Carol Giometti for their support and for initiating efforts in this area at ANL and also thank Stefan Vogt for helpful discussions regarding imaging and for facilitating our efforts at APS.

References

1. Becker JS, Zoriy MV, Pickhardta C et al (2005) Copper, zinc, phosphorus and sulfur distribution in thin section of rat brain tissues measured by laser ablation inductively coupled plasma mass spectrometry: possibility for small-size tumor analysis. J Anal At Spectrom 20:912–917
2. Chery CC, Gunther D, Cornelis R et al (2003) Detection of metals in proteins by means of polyacrylamide gel electrophoresis and laser ablation-inductively coupled plasma-mass spectrometry: Application to selenium. Electrophoresis 24:3305–3313
3. Ma R, McLeod CW, Tomlinson K et al (2004) Speciation of protein-bound trace elements by gel electrophoresis and atomic spectrometry. Electrophoresis 25:2469–2477
4. Marshall PS, Leavens B, Heudi O et al (2004) Liquid chromatography coupled with inductively coupled plasma mass spectrometry in the pharmaceutical industry: selected examples. J Chromatogr A 1056:3–12
5. Neilsen JL, Abildtrup A, Christensen J et al (1998) Laser ablation inductively coupled plasma-mass spectrometry in combination with gel electrophoresis: a new strategy for speciation of metal binding serum proteins. Spectrochimica Acta Part B-Atomic Spectroscopy 53:339–345
6. Raab A, Pioselli B, Munro C et al (2009) Evaluation of gel electrophoresis conditions for the separation of metal-tagged proteins with subsequent laser ablation ICP-MS detection. Electrophoresis 30:303–314
7. Cvetkovic A, Menon AL, Thorgersen MP et al (2010) Microbial metalloproteomes are largely uncharacterized. Nature 466:779–782
8. Chevreux S, Roudeau S, Fraysse A et al (2009) Multimodal analysis of metals in copper-zinc superoxide dismutase isoforms separated on electrophoresis gels. Biochimie 91:1324–1327
9. Dong YX, Gao YX, Chen CY et al (2006) Analysis of trace elements in protein bands with synchrotron radiation X-ray fluorescence after electrophoresis separation. Chin J Anal Chem 34:443–446
10. Gao YX, Chen CY, Chai ZF et al (2002) Detection of metalloproteins in human liver cytosol by synchrotron radiation X-ray fluorescence combined with gel filtration chromatography and isoelectric focusing separation. Analyst 127:1700–1704
11. Garcia JS, da Silva GA, Arruda MAZ et al (2007) Application of Kohonen neural network to exploratory analyses of synchroton radiation x-ray fluorescence measurements of sunflower metalloproteins. X-Ray Spectrom 36:122–129
12. Garcia JS, de Magalhaes CS, Arruda MAZ (2006) Trends in metal-binding and metallo protein analysis. Talanta 69:1–15
13. Kemner KM, Kelly SD, O'Loughlin EJ et al (2005) XRF and XAFS analysis of electrophoretically isolated nondenatured proteins. Phys Scr T115:940–942
14. Kuhbacher M, Weseloh G, Thomzig A et al (2005) Analysis and localization of metal and metalloid-containing proteins by synchrotron radiation x-ray fluorescence spectrometry. X-Ray Spectrom 34:112–117
15. Liu YB, Gao YX, Chen CY et al (2005) Quantitative analysis of iron in protein bands existing in polyacrylamide gel matrix with synchrotron radiation x-ray fluorescence. High Energy Phys Nuc Phys – Chin Ed 29:61–64
16. Sussulini A, Garcia JS, Mesko MF et al (2007) Evaluation of soybean seed protein extraction focusing on metalloprotein analysis. Microchim Acta 158:173–180
17. Verbi FM, Arruda SC, Rodriguez AP et al (2005) Metal-binding proteins scanning and determination by combining gel electrophoresis,

synchrotron radiation X-ray fluorescence and atomic spectrometry. J Biochem Biophys Methods 62:97–109

18. Weseloh G, Kuhbacher M, Bertelsmann H et al (2004) Analysis of metal-containing proteins by gel electrophoresis and synchrotron radiation X-ray fluorescence. J Radioanal Nucl Chem 259:473–477

19. Finney L, Chishti Y, Khare T et al (2010) Imaging metals in proteins by combining electrophoresis with rapid x-ray fluorescence mapping. ACS Chem Biol 5:577–587

Chapter 48

Fluorescent Staining of Gels

Engelbert Buxbaum

Abstract

Certain transition metal complexes show intensive fluorescence when bound to proteins. They can be used to stain gels after electrophoresis with a sensitivity approaching that of silver staining, but in a much simpler and more reproducible procedure. Stains can be prepared easily and at a fraction of the cost of commercially available reagents. Hydrophobic dyes can be used to stain gels without fixing; they do not interfere with later blotting or electro-elution.

Key words: SDS-PAGE, Fluorescent gel staining, Ruthenium(II) tris(bathophenantroline disulfonate), RuBPS, Nile red

1. Introduction

Since the introduction of polyacrylamide gel electrophoresis (PAGE) by Ornstein (1) and Davis (2), gels were stained with dyes, of which Coomassie Brilliant Blue R250 (CBB-R250) is the most well known (3). This allows spots of a few nanograms of proteins to be detected. If higher sensitivity was required, silver staining (4, 5) was used instead, a complicated, time-consuming, and somewhat finicky technique. However, it has a tenfold lower detection limit than CBB-R250. For a more thorough discussion of the theory of electrophoresis, see (6).

Some fluorescent stains bind to the protein-SDS-complex, the hydrophobic environment increases their fluorescence. The first stain used in this manner was Nile red, 9-diethylamino-5H-benzo(α)phenoxazine-5-one (7). Proprietary fluorescent stains (Epicocconone, Sypro Red, Orange, and Tangerine) have slightly higher sensitivities. Such dyes are used without removal of SDS,

Biji T. Kurien and R. Hal Scofield (eds.), *Protein Electrophoresis: Methods and Protocols*, Methods in Molecular Biology, vol. 869, DOI 10.1007/978-1-61779-821-4_48, © Springer Science+Business Media, LLC 2012

that is, either without fixing at all or after fixing in 7.5% acetic acid. Even higher sensitivities can be achieved if the protein is fluorescently labeled before electrophoresis (CyDyes). Gel staining has been reviewed recently by Miller et al. (8).

Sensitivity and simplicity would make fluorescent gel staining the method of choice. However, the commercial dyes are forbiddingly expensive. In addition, their composition is a trade secret. If they were ever withdrawn from the market, how are experiments to be reproduced? For these reasons, the publication of a cheap reagent, ruthenium(II) tris(bathophenantroline disulfonate) (RuBPS, "Dye hard"), that can be produced easily in the lab, was a milestone (9, 10). Several modifications of this original procedure exist; the interested reader is referred to www.ruthenium.ag.vu/rubps.dmc.html. The dye binds preferentially to basic amino acid side chains (Arg, Lys, and His). It is used with gels fixed in the usual manner with alcohol and acetic acid, as residual detergent would interfere with staining.

2. Materials

All reagents should be of analytical quality and can be obtained from the usual suspects. Potassium aquapentachlororuthenate(III) can be bought from Alfa Aesar, Ward Hill, MA. Electrophoresis was performed in a MightySmall SE250 system (Hoefer, Holliston, MA).

2.1. SDS-PAGE

The following recipe is for SDS-PAGE gels according to Laemmli (11), but other systems might work as well:

1. Running buffer (10× stock): 30 g Tris (250 mM), 144 g glycine (1.92 M) and 10 g SDS, make to 1 L with water. Stable at room temperature for several months.
2. Separating gel buffer: Dissolve 90.9 g Tris (1.5 M) and 2 g SDS (0.4%) in about 400 mL water, adjusted to pH 8.8 with 6 M HCl, and then made to 500 mL with water. Stable at room temperature for several months.
3. Stacking gel buffer: Dissolve 60.6 g Tris (1 M) and 4 g SDS (0.8%) in about 400 mL of water and adjust to pH 6.8 with 6 M HCl. Add 500 μL 20% phenol red (optional, see Note 1) and make to 500 mL with water. Stable at room temperature for several months.
4. Sample buffer (5×): Mix 1 g SDS (10%), 3 mL 1 M Tris–HCl pH 6.8 (300 mM), 1 mL 10% bromophenol blue, 2 mL

β-mercaptoethanol (20%), and 4 mL glycerol (40%). Aliquot into 1 mL portions, keep one at 4°C, and freeze the rest at –20°C (see Note 2).

5. Acrylamide/bisacrylamide: 40% solutions of these reagents are commercially available, both 37:1 and 19:1 are required (see Note 3). Store refrigerated over an anion exchanger to remove any acrylic acid that may form.
6. Ammonium peroxodisulfate (APS): 10% in water. Make fresh every day and keep on ice.
7. *n*-Butanol, water-saturated: Add some water to 100 mL *n*-butanol and shake. There should be some undissolved water left. Allow for phase separation and then use the upper, organic phase. Stable at room temperature.
8. Fixative: 300 mL *i*-propanol (30%), 100 mL glacial acetic acid (10%), and water to 1 L.

2.2. Nile Red Staining

1. Nile red stock solution: 0.4 mg/mL in DMSO. Stable for months at room temperature if protected from light.

3. Methods

3.1. Preparation of RuBPS

The synthesis (9) is performed most conveniently under reflux. If such an apparatus is not available, work in an Erlenmeyer flask on a stirring hotplate and replace the evaporated water from time to time. During the procedure, the solution may foam; use a 50-mL vessel:

1. Dissolve 0.2 g $K_2Cl_5Ru \times H_2O$ (0.5 mmol) in 20 mL boiling water, the solution will be deep purple.
2. Add 0.9 g (1.5 mmol) bathophenantroline disulfonate and gently boil for 20 min. The solution will now be deep green.
3. Add 4.59 g (25 mmol) sodium ascorbate and boil for another 20 min to reduce the Ru(III) to Ru(II). The solution will turn orange-brown.
4. Cool the solution down and neutralize it with NaOH to pH 7. Make the volume to 25 mL, giving a 20 mM stock solution. Store refrigerated.

3.2. Casting SDS-PAGE Gels

1. Depending on the molecular mass of the proteins of interest, gels with different pore sizes are used. For a wider range, use gradient gels (5–15% I found the most useful). The recipe is for

ten mini gels with 1-mm spacers. Mix the reagents according to the following table:

	5%	7.5%	10%	12.5%	15%	20%	25%	30%	Stack (4%)
Stacking gel buffer (mL)	–	–	–	–	–	–	–	–	1.25
Separating gel buffer (mL)	7.5	7.5	7.5	7.5	7.5	7.5	7.5	7.5	–
40% Acrylamide 37:1 (mL)	3.8	5.6	7.5	9.4	11.3	15.0	18.8	22.5	–
40% Acrylamide 19:1 (mL)	–	–	–	–	–	–	–	–	1.0
Water (mL)	to 30	to 30	to 30	to 30	to 30	to 30	to 30	0	to 10

2. Thoroughly degas the mixture either by exposure to vacuum or in an ultrasonic water bath. Any residual oxygen may interfere with polymerization.
3. Add 15 μL TEMED and 45 μL APS to the separating gel, mixing gently so as not to shake air into the solution. Pour into the gel chamber.
4. Overlay each gel with 0.5 mL water-saturated butanol and allow to polymerize. This will take about 10 min. The butanol ensures that the gel will have a nice, flat surface. The phase border between gel and butanol will become more prominent when polymerization is complete.
5. Pour of the butanol.
6. Add 5 μL TEMED and 30 μL APS to the stacking gel mixture, mix gently, and pour over the separating gel. Insert the combs quickly and allow the gel to polymerize. If phenol red was included in the buffer, the color will deepen during the process.
7. Keep the gels in the fridge overnight so that reactive intermediates of the polymerization process may break down. This is particularly important if proteins are to be characterized by Edman degradation or mass spectrometry after electrophoresis.
8. Gels may be stored in the fridge for up to 2 weeks, wrapped in moist tissue, and enclosed in an air-tight plastic box.

3.3. Running the Gel

1. Remove the combs and mount the gel in the electrophoresis apparatus.
2. Dilute the running buffer and fill into the apparatus; 200 mL of buffer is required for the Hoefer mini gel system.
3. Use a tuberculin syringe to rinse all wells with running buffer, make sure there are no air bubbles left. This process is greatly aided by the phenol red in the stacking gel, which makes the wells easy to see (13).
4. Load your sample into the wells. All wells of a gel should receive the same sample volume so that they later run evenly. Loading is most comfortably achieved with a Hamilton syringe (25 μL, with flat needle).
5. Close the apparatus and connect to a power supply for cooling. Mini gels are run with 10 mA constant current per gel, and once the front has reached the separating gel, current may be increased to 20 mA per gel. Voltage will increase from about 50 to 200 V during the run.
6. Allow to run until the bromophenol blue marker dye has just reached the bottom of the gel. Disconnect from the power supply, discard the running buffer, and open the gel cassettes. Place the gel into fixative (at least 20 volumes) on an orbital shaker for at least an hour. Fixing may also be done overnight, if that is more convenient.

3.4. Staining Gels with RuBPS

The following recipe is from ref. (10), I found it not only more convenient but also slightly more sensitive than the original method from (9). Staining is performed in scrupulously clean plastic dishes with lid, which can be obtained in a household store:

1. Add 1 μL RuBPS stock to 20 mL of fixative (1 μM final concentration) and stain the gel in this solution for 3 h to overnight in the dark.
2. Wash twice for 10 min in water.
3. The gel can now be placed on a 302-nm UV box (see Note 4) or in a 488-nm laser scanner for documentation. Protein bands will fluoresce orange. For photography, use a strong yellow filter to improve contrast.
4. Detection limit is a few nanogram of protein per band, about 1/10 that of CBB-R250.

3.5. Staining with Nile Red

This method was first published in ref. (7) and modified in ref. (12). For optimal results, the concentration of SDS in the gel and in the running buffer should be reduced to 0.05% rather than the usual 0.1%. The gel is not fixed, dye binding occurs to

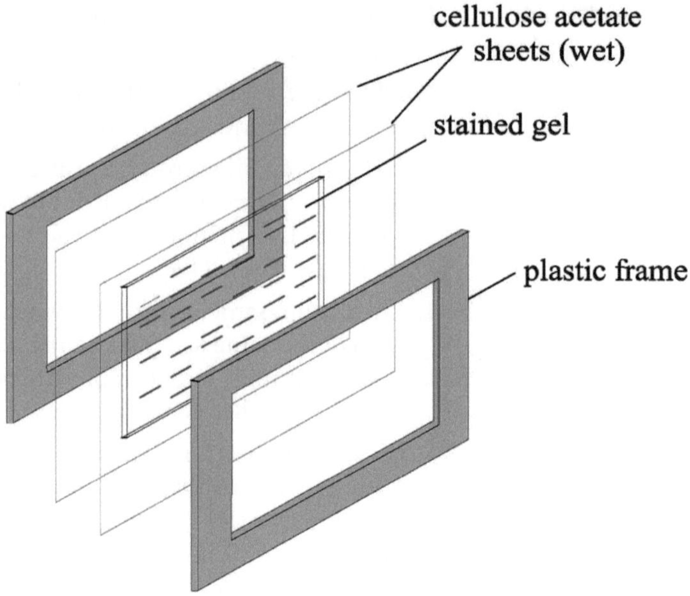

Fig. 1. Drying gels between cellulose acetate sheets. The assembly is held together with binder clips. The figure is taken from ref. 6.

the SDS/protein complex. The method can therefore be used before blotting or electro-elution. See Notes 4 and 5 for safety issues:

1. Add 0.5 mL of the Nile red stock solution to 50 mL bidistilled water in a plastic dish under vigorous agitation for 5 s.
2. Add the gel immediately and stain for 5 min in the dark on an orbital shaker. Longer staining will not improve the result, as Nile red precipitates rapidly from aqueous solutions.
3. Wash the gel for 10 s with bidistilled water.
4. Place on a transilluminator (302 nm) or into a fluorimetric scanner (590/630 nm) to observe the red fluorescence. Photography is improved by an orange and a yellow filter in series.

3.6. Drying Gels

Gels stained with RuBPS may be dried and then stored in a lab book. Incubate the gels with 1% glycerol in water, then place them between two sheets of wet cellophane (can be obtained in household stores or from BioRad) in a plastic frame (see Fig. 1). The assembly is held together with binder clips (two per side, i.e., eight in total). Let it air-dry over night. As the cellophane dries, it shrinks. This results in a flat, wrinkle-free, transparent gel, from which interesting bands can be cut out and rehydrated later, if need arises. Gels stained with Nile red and the proprietary fluorescent dyes cannot be dried in this manner.

4. Notes

1. The difference in refractive index between the stacking gel and the running buffer is low, making sample loading difficult for those with less than 20/20 vision. Addition of phenol red helps. Some of the dye is polymerized into the stacking gel network, the reminder runs with the front, slightly faster than bromophenol blue (13).
2. β-Mercaptoethanol in the sample buffer may be replaced with dithiothreitol (Cleland's reagent, DTT, HS-CH_2–$(CHOH)_2$–CH_2–SH), which has a more reducing redox potential (−332 mV at pH 7.0 and −366 mV at pH 8.1 versus about 260 mV at pH 7.0 for β-mercaptoethanol) (14, 15) and a less offensive odor. However, it is more expensive. Use at 100 mM.
3. Acrylamide is a neurotoxin that can penetrate even the intact skin. Gloves should be worn whenever this reagent is handled. Gloves should be washed free of powder, as powder residues on the gel may later bind dye. Although polyacrylamide is nontoxic, gels should be handled with gloves until thoroughly washed free of residual acrylamide (i.e., after fixing and staining). Ready-made solutions of acrylamide/bisacrylamide are commercially available. That way, no dust can develop during weighing of the dry reagents. If you want to make your own solutions, add the required amount of water and bisacrylamide directly to a freshly opened container of acrylamide, assuming the manufacturer has sold you the correct weight. Gel casting stands sometimes leak. Work in a plastic tea tray to contain any spillage.
4. UV light is dangerous to skin and eyes. Minimize exposure by wearing an UV-blocking full face shield and by utilizing the UV-shield that transilluminators come with.
5. DMSO is a flammable solvent, which can penetrate intact skin and facilitate the entry of dissolved compounds into the body. Wear impenetrable gloves when handling such solutions.

Acknowledgments

This work was supported by funding from Ross University.

References

1. Ornstein L (1964) Disc electrophoresis – I. Background and theory. Ann N Y Acad Sci 121:321–349, http://www.pipeline.com/~lenornst/DiscEle1.pdf
2. Davis BJ (1964) Disc electrophoresis – II. Method and application to human serum protein. Ann N Y Acad Sci 121:404–427, http://www.pipeline.com/~lenornst/DiscEle2.pdf
3. Fazekas de St. Groth S, Webster RG, Datyner A (1963) Two new staining procedures for quantitative estimation of proteins on electrophoretic strips. Biochim Biophys Acta 71:377–391. doi:10.1016/0006-3002(63)91092-8
4. Heukeshoven J, Dernick R (1988) Improved silver staining procedure for fast staining in phastsystem development unit. I. staining of sodium dodecyl sulfate gels. Electrophoresis 9:28–32. doi:10.1002/elps.1150090106
5. Merril CR, Goldman D, Sedman SA, Ebert MH (1981) Ultrasensitive stain for proteins in polyacrylamide gels shows regional variation in cerebrospinal fluid proteins. Science 211:1437–1438. doi:10.1126/science.6162199
6. Buxbaum E (2010) Biophysical chemistry of proteins: an introduction to laboratory methods. Springer, New York. ISBN ISBN 978-1-4419-7250-7
7. Daban J-R, Bartolomé S, Samsó M (1991) Use of the hydrophobic probe Nile red for the fluorescent staining of protein bands in sodium dodecyl sulfate-polyacrylamide gels. Anal Biochem 199:169–174. doi:10.1016/0003-2697(91)90085-8
8. Miller I, Crawford J, Gianazza E (2006) Protein stains for proteomics application: which, when, why? Electrophoresis 6:5385–5408. doi:10.1002/pmic.200600323
9. Rabilloud T, Strub J-M, Luche S, Dorsselaer AV, Lunardi J (2001) A comparison between Sypro ruby and ruthenium II tris(bathophenantroline disulfonate) as fluorescent stains for protein detection in gels. Proteomics 1:699–704. doi:10.1002/elps.1150130190
10. Pluder F, Beck-Sickinger AG (2007) One-step procedure for staining of proteins with ruthenium II tris(bathophenanthroline disulfonate). Anal Biochem 361:299–301. doi:10.1016/j.ab.2006.11.003
11. Laemmli UK (1970) Cleavage of structural proteins during the assembly of the head of bacteriophage T4. Nature 227:680–685, http://bioinfcpcri.org/misc/nature.pdf
12. Alba FJ, Bermudez A, Bartolome S, Daban J-R (1996) Detection of five nanograms of protein by two-minute nile red staining of unfixed SDS gels. Biotechniques 21:625–626, http://www.biotechniques.com/multimedia/archive/00009/96214bm12_9719a.pdf
13. Smith I, Cromie R, Stainsby K (1988) Seeing gel wells well. Anal Biochem 169:370–371. doi:10.1016/0003-2697(88)90297-7
14. Cleland WW (1964) Dithiothreitol, a new protective reagent for SH groups. Biochemistry 3:480–482. doi:10.1021/bi00892a002
15. Crans DC, Zhang B, Gaidamauskas E, Keramidas AD, Willsky GR, Roberts CR (2010) Is vanadate reduced by thiols under biological conditions? changing the redox potential of V(V)/V(IV) by complexation in aqueous solution. Inorg Chem 49:4245–4256. doi:10.1021/ic100080k

Chapter 49

A Single-Step Simultaneous Protein Staining Procedure for Polyacrylamide Gels and Nitrocellulose Membranes by Alta During Western Blot Analysis

Jayanta K. Pal, Sunil K. Berwal, and Rupali N. Soni

Summary

A simple method for staining of proteins simultaneously on sodium dodecyl sulfate (SDS) polyacrylamide gels and nitrocellulose membranes by Alta during western blot analysis is described. A 5% solution of Alta, a commercially available cosmetic preparation, is added in the upper tank buffer during electrophoresis. On completion of electrophoresis, the gel is washed in distilled water and viewed on a white light plate and a transilluminator to photograph the protein profiles. The gel is processed for western blot transfer of proteins onto a nitrocellulose membrane, and upon completion, the protein profiles on the membrane are viewed and photographed as stated above. The membrane can then be processed for immunostaining as per the standard procedure. Thus, the staining procedure using Alta is simple, rapid (without any need of destaining), and cost-effective.

Key words: SDS polyacrylamide gels, Nitrocellulose membrane, Western blotting, Alta, Protein stain

1. Introduction

Sodium dodecyl sulfate (SDS) polyacrylamide gel electrophoresis (1) and western blot analysis (2) have become a part and parcel of research in life sciences. Considering this reality, there has been a continuous effort towards developing better versions of these procedures. Among others, detection of proteins in SDS polyacrylamide gels and also on nitrocellulose membranes, both of which are important in the western blot analysis, has attracted much attention of researchers involved in improvement of these techniques (3–5). Along this line, a number of protein staining procedures are available in the literature for staining proteins independently on polyacrylamide gels and on nitrocellulose/PVDF membranes (6–11). The commonly used stains for gels and membranes are Coomassie Brilliant Blue (R-/G-250) and Ponceau Red S, respectively.

Biji T. Kurien and R. Hal Scofield (eds.), *Protein Electrophoresis: Methods and Protocols*, Methods in Molecular Biology, vol. 869, DOI 10.1007/978-1-61779-821-4_49, © Springer Science+Business Media, LLC 2012

However, no single-step staining procedure is available by which staining of both the gel and the nitrocellulose membrane, two successive components of the western blot, can be accomplished. Thus, there is a need to develop such a single-step staining method using a suitable stain to avoid the existing elaborate and dual-step staining–destaining procedures for western blot analysis.

Towards such an approach, a procedure was developed using a commercially available Red stain called the Alta. Alta, an azo dye, contains Crocein scarlet, which makes it red, and Rhodamine B, a fluorescence compound (4, 12). Interestingly, Alta is traditionally used as a cosmetic by Asian women for decorating their feet, particularly during festivities. Initially, we used this stain for staining of proteins on polyacrylamide gels along with Coomassie Brilliant Blue R-250, in parallel. It worked favorably, and the results were comparable with those obtained by Coomassie Brilliant Blue (4). Considering the presence of Rhodamine B in Alta, we tried observing the gels on a transilluminator, and as expected, the signals on protein bands appeared more intense than that obtained by observing the gel on a white light plate. This signal was manyfold more than the signal obtained by conventional Coomassie Blue stain. In the next step, we tried staining the nitrocellulose membranes with Alta, and we obtained satisfactory result as obtained for the gels. The intensity was much higher when the membrane was viewed on a transilluminator. These results prompted us to develop a procedure by which a single-step staining would be adequate for both the gel and the blot without the traditional elaborate procedures. Among various approaches taken, the final procedure of applying the stain directly in the upper tank buffer appeared most useful. This combined approach resulting in a single-step and simple staining procedure using Alta has been already published (4). In this chapter, the details of the procedure, protocol, and specific notes for the benefit of the users are described.

2. Materials

All chemicals and reagents to be used should be of molecular biology grade. Prepare all solutions and reagents in high-quality distilled water and store them at appropriate temperatures and conditions.

2.1. Protein Extraction and Estimation

1. Protein source: *Escherichia coli* cells expressing a histidine-tagged recombinant protein of 50 kDa.
2. Lysis buffer: 50 mM Tris–HCl, pH 8.0, 2 mM EDTA, and 100 mM NaCl. Store at 4°C.
3. Bovine serum albumin standard (BSA): 10 mg/mL. Store at −80°C as small aliquots.

4. Protein assay reagent: 5× Bradford reagent (13). Dissolve 50 mg of Coomassie Brilliant Blue G-250 in 25 mL of ethanol. Add 50 mL of orthophosphoric acid drop-wise with continuous stirring. Make the volume to 100 mL by adding deionized water drop-wise and store it in a dark bottle at 4°C.

2.2. SDS-PAGE

1. Gel apparatus (Hoefer Inc., Holliston, MA, USA/Techno Source, Mumbai, India).
2. Constant current/voltage power supply (Bio-Rad, Hercules, CA, USA/Techno Source, Mumbai, India).
3. Gel solution: Dissolve 29.2 g acrylamide and 0.8 g bis-acrylamide in distilled water to a final volume of 100 mL (see Note 1).
4. Stacking gel and sample buffer: 0.5 M Tris–HCl, pH 6.8. Dissolve 6.05 g Tris base in 90 mL distilled water, adjust the pH to 6.8 by adding 1 N HCl, make up the volume to 100 mL, and filter sterilize and store at 4°C.
5. Separating gel buffer: 1.5 M Tris–HCl, pH 8.8. Dissolve 18.17 g Tris base in 90 mL distilled water, adjust the pH to 8.8 by adding 1 N HCl, make up the volume to 100 mL, and filter sterilize and store at 4°C.
6. 10% SDS: Dissolve 10 g SDS in 100 mL distilled water, filter sterilize, and store at room temperature (see Note 2).
7. TEMED.
8. 10% Ammonium persulfate: Dissolve 100 mg in 1 mL of distilled water (see Note 3).
9. Tank buffer: Tris–glycine (0.025, 0.192 M, respectively), pH 8.3, 0.1% SDS.
10. Sample buffer (2x): Mix 2.5 mL 0.5 M Tris–HCl, pH 6.8, 4 mL of 10% SDS, 2 mL glycerol, 1.0 mL β-mercaptoethanol, 0.05% bromophenol blue, and 0.5 mL distilled water. Store the solution in small aliquots at −20°C, except one aliquot at room temperature for regular use.
11. Molecular weight marker proteins (Bangalore Genei, Bengaluru, Karnataka, India).
12. Coomassie Brilliant Blue (CBB) R-250 stain: Prepare 0.125% CBB R-250 in methanol, acetic acid, and distilled water (5:1:4).
13. Alta stain: Alta liquid formulation (commercially available as a cosmetic solution) or a solution made by mixing 0.4% Crocein scarlet and 0.1% Rhodamine B in 50% methanol and 10% acetic acid in distilled water.
14. Gel doc system (Bio-Rad, Hercules, CA, USA).

2.3. Western Blot Analysis

1. Wet transfer apparatus (Hoefer Inc., Holliston, MA, USA)/ semidry transfer apparatus (Bio-Rad, Hercules, CA, USA).
2. Nitrocellulose/PVDF membranes (Millipore Billerica, MA, USA).
3. Transfer buffer: Tris–glycine (0.025, 0.192 M, respectively), pH 8.3 containing 20% methanol.
4. Tris-buffered saline (TBS): Dissolve 6.05 g Tris and 8.76 g NaCl in 900 mL distilled water, adjust the pH to 7.5 by adding 1 N HCl, and make up the volume to 1 L.
5. TBST: TBS containing 0.1% Tween 20.
6. Antibodies: Rabbit anti-histidine polyclonal antibody (Santa Cruz Biotechnology, Santa Cruz, CA, USA) and goat anti-rabbit IgG-peroxidase (Sigma-Aldrich, St. Louis, MO, USA).
7. Blocking reagent and enzyme substrates: BM Chemiluminescence western blotting kit (Roche, Mannheim, Germany).

3. Methods

3.1. Protein Extraction and Estimation

1. Harvest *E. coli* cells by centrifugation at 16,500 g for 1 min at 4°C and resuspend in Lysis buffer.
2. Add freshly prepared lysozyme (final concentration 0.25 mg/mL) and mix well by inverting the contents 4–5 times.
3. Incubate the mixture at the room temperature for 1 h.
4. Add Triton X-100 (0.2%, v/v), $MgCl_2$ (10 mM), and DNase (1 μg/mL).
5. Incubate the mixture for 1 h at room temperature and centrifuge at 16,500 g for 5 min at 4°C.
6. Collect the supernatant and store at −70°C till further use.
7. Estimate protein concentration in the supernatants by the method of Bradford (13) using BSA as the standard protein.

3.2. SDS-PAGE (1)

1. Preparation of separating gel: Prepare a 10% separating gel by mixing the following solutions: 3.98 mL distilled water, 2.5 mL 1.5 M Tris–HCl, pH 8.8, 3.3 mL gel solution, 100 μL 10% SDS, 7.5 μL TEMED, and 75 μL 10% APS, and by layering the mixed solution in the gel plate assembly.
2. Preparation of stacking gel: Prepare a 4% stacking gel by mixing 3.05 mL distilled water, 1.25 mL 0.5 M Tris–HCl, pH 6.8, 650 μL gel solution, 50 μL 10% SDS, 5 μL TEMED, and 37.5 μL 10% APS, and by layering the mixed solution on top of the polymerized separating gel.

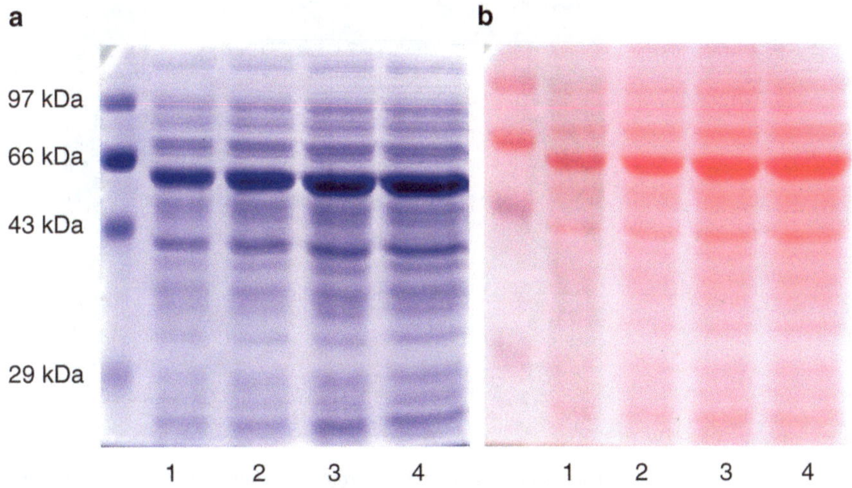

Fig. 1. Photographs of the protein profiles (*E. coli* extract containing an overexpressed histidine-tagged recombinant protein of 50 kDa) on gels stained with either Coomassie Brilliant Blue R-250 (**a**) or Alta (**b**). Lanes in both the gels contain different quantities (μg) of proteins: *Lanes 1*, *2*, *3*, and *4* contain 50, 60, 75, and 90 μg, respectively. Molecular weight marker proteins (97–29 kDa) are loaded on the left most lanes in both the gels.

3. Preparation of protein samples: Prepare protein samples of definite concentrations by mixing with (2x) sample buffer and heating on a boiling water bath for 2–4 min. Centrifuge (1 min) the treated samples prior to their use for loading onto the gel.
4. Electrophoresis: Load protein samples in the range of 15 90 μg of total proteins and the molecular weight marker proteins [phosphorylase b (97 kDa), bovine serum albumin (67 kDa), ovalbumin (45 kDa), and carbonic anhydrase (29 kDa)], and subject them to electrophoresis at 25 mA (constant current) till the bromophenol blue reaches the end of the gel.
5. Staining: Stain the proteins in the gel with either in 0.1% Coomassie Brilliant Blue R-250 in methanol/acetic acid/water (5:1:4) for 2 h followed by destaining in the same solution without the stain or in Alta mixed with methanol and acetic acid (4:5:1) for 2 h followed by destaining in 7% acetic acid (see Note 4).
6. Photograph the gels on a white plate in a gel doc (Fig. 1).
7. For western blot analysis, when both the gel and the membrane are to be simultaneously stained, add 5% (v/v) Alta (see Note 5) or a mixture of 0.4% Crocein scarlet and 0.1% Rhodamine B (prepared in 50% methanol and 10% acetic acid in distilled water) directly into the upper tank buffer (Fig. 2). In this procedure, the protein bands get stained in situ during the electrophoresis. The detailed staining procedure is schematically presented in Fig. 3.

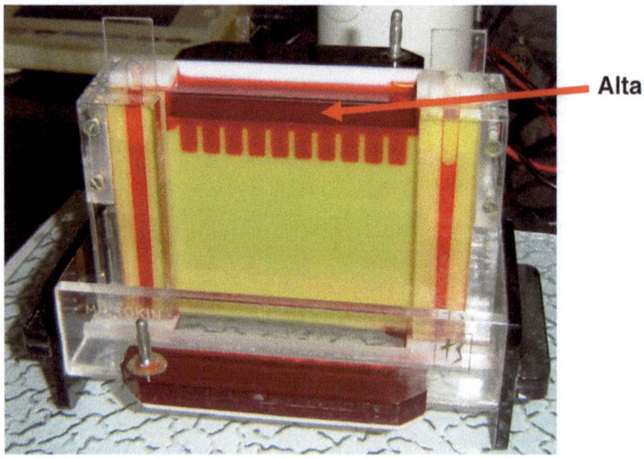

Fig. 2. Photograph of a gel in the electrophoresis setup with Alta added into the upper tank buffer.

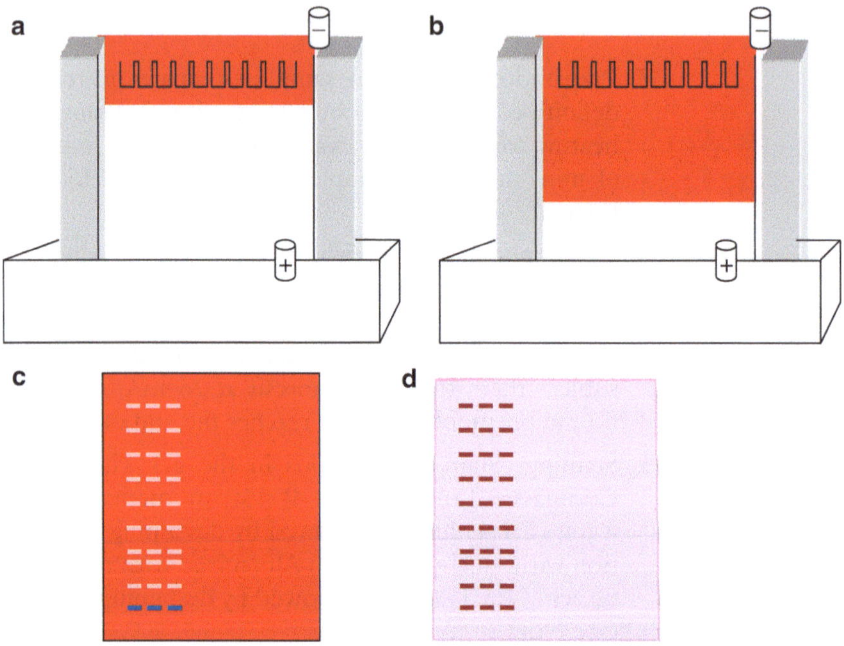

Fig. 3. A schematic representation of the staining procedure during electrophoresis and western blot. (**a**) Alta added in the upper tank buffer before beginning the electrophoresis. (**b**) The gel along with the protein bands gets stained red as the dye front moves ahead in the gel during electrophoresis. (**c**) The protein bands on the gel after completion of electrophoresis as observed under UV. (**d**) The protein bands as observed under UV on the nitrocellulose membrane after their electrotransfer from the gel.

3.3. Western Blot Analysis

1. After completion of electrophoresis, transfer the proteins from the gel onto a nitrocellulose membrane in a wet transfer system (Hoefer Inc., Holliston, MA, USA) at 70 mA overnight or in a semidry electrotransfer system (Bio-Rad, Hercules, CA, USA) at a constant voltage of 15 V for 40 min (see Note 6) according to Towbin et al. (2).
2. Following completion of transfer, view and photograph the protein profiles on the nitrocellulose membrane (blot) on both a white plate and an UV transilluminator in a gel doc system (Fig. 4a, b).
3. Process the blot for immunoreaction with the antibodies. In brief, saturate the blot with 1x blocking reagent for 1 h at room temperature, wash it 3 times with 1x TBST for 5 min each.

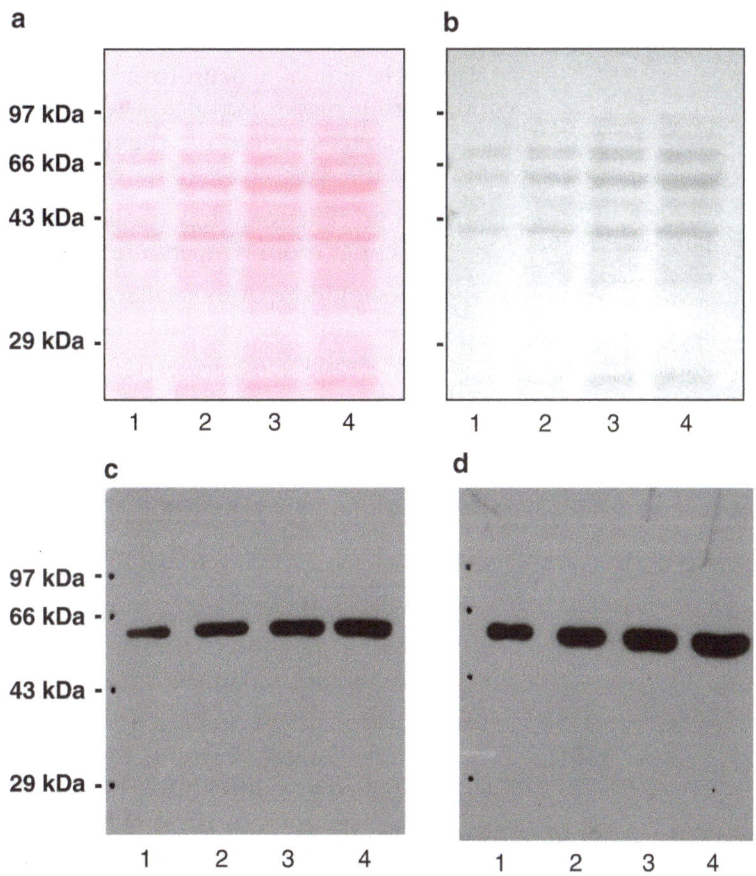

Fig. 4. Photographs of the protein profiles (*E. coli* extract containing an overexpressed histidine-tagged recombinant protein of 50 kDa) on the nitrocellulose membrane after electroblotting as taken on a (**a**) white light plate and on an (**b**) UV transilluminator and of Western blots (**c** and **d**) detected by anti-histidine antibody. *Lanes 1–4* in each profile contain the same extract with 15, 25, 40, and 50 μg of proteins. Western blots were done by using (**c**) the conventional method or by using (**d**) Alta in the upper tank buffer.

4. Incubate the blot with the primary antibody (rabbit anti-histidine polyclonal antibody), wash it with 1x TBST (5 min, 3 times), and then incubate with the secondary antibody conjugate, i.e., goat anti-rabbit IgG-peroxidase. The dilution of antibodies and the period of incubation should be as per the recommended protocol of the manufacturer.
5. Develop signal on the blot using BM chemiluminescence kit using H_2O_2 and luminol as the substrates (see Note 7) and capture the signal on an X-ray film (Fig. 4c, d). Figure 4c, d represent Western blot signals of a 50 kDa his-tagged recombinant protein using routine technique and Alta-staining method, respectively.

4. Notes

1. Acrylamide is a neurotoxin. Take appropriate measures like wearing masks and gloves while weighing and making solutions, etc. One can use the ready-made gel solutions to minimize the risks.
2. SDS precipitates at low temperature. Thus, SDS solution is to be kept at room temperature.
3. APS should be of molecular biology grade and absolutely dry (it is highly hygroscopic). The solution should preferably be made fresh before use.
4. For sharpening of the bands (desirable for photography), the gel may briefly be incubated with methanol:acetic acid:water (5:1:4) following destaining in 7% acetic acid.
5. Depending on the formulation available, the concentration of Alta to be used may vary from 2 to 5%. In case the Alta solution (ready-made) is not available or is not of proper quality, use a mixture of 0.4% Crocein scarlet and 0.1% Rhodamine B (prepared in 50% methanol and 10% acetic acid in distilled water) as an alternative.
6. If a wet transfer system is chosen, use the following parameters: 70 mA (constant current) for 16 h at 4°C (on top of a magnetic stirrer to uniformly dissipate the heat generated). Please note that, in case of semidry transfer, the large molecular weight proteins transfer poorly.
7. One can develop the blot by using either chemiluminescence detection method or colorimetric method. In the latter, the protocol to be followed would vary, particularly with reference to the antibody dilution and period of incubation with blocking reagent and antibodies.

Acknowledgments

This work is supported by a research grant from the Department of Atomic Energy, Government of India, Bhabha Atomic Research Centre, Mumbai (BRNS Grant number 2007/37166/BRNS), and a grant from the Department of Science and Technology (DST), Government of India under the DST PURSE Programme. We wish to thank Dr. Jayant Londhe for his assistance in initiating some preliminary experiments.

References

1. Laemmli UK (1970) Cleavage of structural proteins during the assembly of the head of bacteriophage T4. Nature 227:680–685
2. Towbin H, Staehelin T, Gordon J (1979) Electrophoretic transfer of proteins from polyacrylamide gels to nitrocellulose sheets: procedure and some applications. Proc Natl Acad Sci USA 76:4350–4354
3. Pal JK, Sarkar A, Katoch B (2001) Detergent-mediated destaining of Coomassie brilliant blue-stained SDS polyacrylamide gels. Ind J Exptl Biol 39:95–97
4. Pal JK, Godbole D, Sharma K (2004) Staining of proteins on SDS polyacrylamide gels and on nitrocellulose membranes by Alta, a colour used as a cosmetic. J Biochem Biophys Methods 61:339–347
5. Kurien BT, Scofield RH (1998) Heat mediated quick Coomassie blue protein staining and destaining of SDS PAGE gels. Ind J Biochem Biophys 35:385–389
6. Merril CR, Goldman D, Sedman SA et al (1981) Ultrasensitive stain for proteins in polyacrylamide gels shows regional variation in cerebrospinal fluid proteins. Science 211:1437–1438
7. Zehr BD, Savin TJ, Hall RE (1989) A one-step low background coomassie staining procedure for polyacrylamide gels. Anal Biochem 182: 157–159
8. Wu W, Welsh MJ (1996) Rapid coomassie blue staining and destaining of polyacrylamide gels. BioTechniques 20:386–388
9. Wu M, Stockley PG, Martin WJ II (2002) An improved western blotting technique effectively reduces background. Electrophoresis 23: 2373–2376
10. Antharavally BS, Carter B, Bell PA et al (2004) A high-affinity reversible protein stain for western blots. Anal Biochem 329:276–280
11. Steinberg TH (2009) Protein gel staining methods: an introduction and overview. Methods Enzymol 463:541–563
12. Bajaj AK, Pandey RK, Misra K et al (1998) Contact depigmentation caused by an azo dye in Alta. Contact Dermatisis 38:189–193
13. Bradford MM (1976) A rapid and sensitive method for quantitation of microgram quantities of protein utilizing the principle of protein-dye binding. Anal Biochem 72: 248–254

Chapter 50

Detection of Glycoproteins in Polyacrylamide Gels Using Pro-Q Emerald 300 Dye, a Fluorescent Periodate Schiff-Base Stain

Padmaja Mehta-D'souza

Abstract

Pro-Q Emerald 300 glycoprotein stain generates a bright-green fluorescent signal upon reacting with periodic acid-oxidized carbohydrate groups on proteins. With this dye, it is possible to detect proteins directly in the gel without the need to transfer them to a membrane. This dye is more sensitive than the standard periodic acid Schiff's base which uses acidic fuchsin dye.

Key words: SDS-PAGE, Glycoprotein staining, P-selectin glycoprotein ligand-1

1. Introduction

Proteins expressed on the membranes of eukaryotic cells are often posttranslationally modified by the addition of oligosaccharides. Protein glycosylation plays a critical role in many physiological processes, including cell adhesion, cell migration, tissue differentiation, and inflammation (1). An alteration in the glycosylation profile of a surface molecule is often an indicator of a pathological condition (2). Hence, there has always been an interest in developing efficient methods for the detection of glycoproteins.

Two methods have been widely used for detecting glycoproteins in gels or on membrane blots: the non-covalent binding of lectins to specific oligosaccharides and the oxidation of carbohydrate groups followed by conjugation with a chromogenic or tagged substrate. Staining with lectins and detection of specific tags often requires a subsequent enzyme amplification system whereas the conventional method involving the oxidation of carbohydrate groups can be challenging and lacking in sensitivity.

Biji T. Kurien and R. Hal Scofield (eds.), *Protein Electrophoresis: Methods and Protocols*, Methods in Molecular Biology, vol. 869, DOI 10.1007/978-1-61779-821-4_50, © Springer Science+Business Media, LLC 2012

Pro-Q Emerald 300 dye is a fluorescent hydrazide that can be used to detect glycoproteins directly in gels or on blots. It also relies upon the oxidation of carbohydrate groups but is a much simpler and more sensitive method for glycoprotein detection. Depending on the glycoprotein and the degree of glycosylation, as little as 0.5 ng of glycoprotein can be detected per band in a polyacrylamide gel (3).

P-selectin glycoprotein ligand-1 (PSGL-1) is expressed as a dimer on the surface of certain leukocyte subsets and plays an important role in inflammation by binding to P-selectin expressed on endothelial cells (4). PSGL-1 is present in very small amounts on the cell surface. This makes it difficult to obtain purified protein in large quantities. We have generated a recombinant, soluble form of PSGL-1 in the CHO-K1 cell line for structure-function studies. The recombinant protein was purified by affinity chromatography over an HPC4 antibody column (5) and its purity checked by sodium dodecyl sulfate polyacrylamide gel electrophoresis (SDS-PAGE) followed by silver staining. We stained the gel with Pro-Q Emerald 300 and could detect small quantities of PSGL-1 at the appropriate molecular weight.

2. Materials

2.1. P-Selectin Glycoprotein Ligand-1

1. P-selectin glycoprotein ligand-1 was purified by HPC4 affinity chromatography from conditioned medium of transfected CHO-K1 cells.

2.2. SDS-Polyacrylamide Gel Electrophoresis (SDS-PAGE)

1. 30% Acrylamide/bis solution (37.5:1 with 2.6% C) (see Note 1).
2. Separating gel buffer (3×): 1 M Tris–HCl, pH 8.8. Store at 4°C.
3. Stacking gel buffer (5×): 1 M Tris–HCl, pH 6.8. Store at 4°C.
4. Gel loading buffer: 62.5 mM Tris–HCl, pH 6.8, 5% β-mercaptoethanol, 10% glycerol, and 2% SDS. A 3× version of the gel loading buffer was also made by reducing the amount of water used to make up the final volume (see Note 2).
5. 10% SDS solution (exercise care when using this chemical in the dry form). Store at room temperature.
6. *N,N,N,N′* tetramethylethylenediamine (TEMED): Make a 10% solution in water. This should be made fresh every time to ensure good polymerization of gels.
7. Ammmonium persulfate (APS): Make a 10% solution in water. This should also be made fresh every time to ensure good polymerization (see Note 3).
8. Running buffer for SDS-PAGE: 25 mM Tris, 190 mM glycine, and 0.2% (w/v) SDS. This buffer was made as a 10× solution and diluted with water before use. Store at room temperature.

9. β-Mercaptoethanol.
10. CandyCane glycoprotein molecular weight standards (Molecular Probes/Invitrogen, Carlsbad, CA, USA).
11. Minigel apparatus (Atto, Tokyo, Japan).
12. Syngene G-Box (Imgen technologies, Alexandria, VA, USA).

2.3. Glycoprotein Staining

Pro-Q Emerald 300 Glycoprotein Gel and Blot Stain Kit (Molecular Probes) containing:

1. Pro-Q Emerald 300 stock solution: Add 6 mL DMF to Pro-Q Emerald 300 reagent. Store at −2°C (see Note 4).
2. Fixing solution: 50% Methanol and 5% acetic acid.
3. Wash solution: 3% Glacial acetic acid.
4. Oxidizing solution: Add 250 mL of 3% acetic acid to the bottle containing periodic acid (see Note 5).

3. Methods

3.1. P-Selectin Glycoprotein Ligand-1 (Sample for SDS-PAGE)

1. Purify P-selectin glycoprotein ligand-1 by HPC4 affinity chromatography from conditioned medium of transfected CHO-Lec1 cells.
2. Concentrate the purified protein.
3. Dialyze the purified protein and store at −20°C until further use.

3.2. SDS-Polyacrylamide Gel Electrophoresis (SDS-PAGE)

1. Carry out SDS-PAGE using a minigel apparatus.
2. Electrophorese the proteins in a buffer containing 2-mercaptoethanol, essentially as described by Laemmli (6), using a discontinuous SDS-polyacrylamide gel system with a 10% separating gel.

3.3. Glycoprotein Staining

1. Once the electrophoresis is complete, immerse the gel in 100 mL of fixing solution and place on a shaker at room temperature for 45 min. Discard the fixing solution and repeat the fixation step once more (see Note 6).
2. Immerse the gel in 100 mL of wash solution and place on a rocker for 20 min. Repeat the wash step once more (see Note 6).
3. Immerse the gel in 25 mL of the oxidizing solution and incubate on a rocker with gentle agitation for 30 min (see Note 7).
4. Wash the gel 3 times by immersing it in 100 mL of the wash solution and placing it on the rocker for 20 min (see Note 8).
5. Prepare Pro-Q Emerald 300 staining solution by adding 500 μL of the Pro-Q 300 Emerald Stock Solution (*prepared in* Subheading 2.3, step 1) above to 25 mL of the staining buffer provided with the kit (see Note 9).

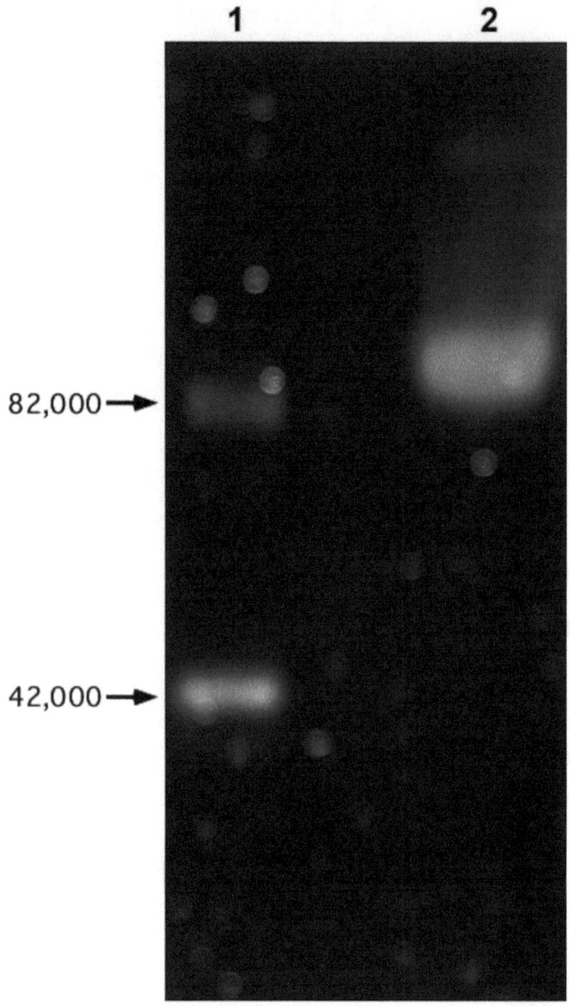

Fig. 1. Electrophoresis of soluble PSGL-1 expressed in CHO-K1- cells. After SDS-PAGE, the gel was stained using the Pro-Q Emerald 300 glycoprotein stain kit (Molecular Probes). *Lane 1* shows the CandyCane molecular weight standards with only the glycosylated proteins stained with the dye. *Lane 2* shows intense staining of the heavily glycosylated PSGL-1.

6. Stain the gel for 90 min by immersing it in 25 mL of the Pro-Q Emerald 300 Staining Solution (*prepared in* Subheading 3.3, step 5). Gently agitate the gel on a rocker and keep in the dark by covering the container with aluminum foil (see Note 10).
7. Wash the stained gel twice by immersing it in 100 mL of wash solution and incubating it at room temperature for 20 min (see Note 11).
8. Visualize the glycoprotein bands on the gel (see Fig. 1) on a Syngene G-Box by choosing the fluorescence detection protocol specifically set for Pro-Q Emerald 300 dye (using UV transillumination at 300 nm).

4. Notes

1. Acrylamide is a neurotoxin, so great care should be taken while handling this chemical. Use gloves and avoid breathing fumes.
2. When running samples under nonreducing conditions, the β-mercaptoethanol should be omitted from the buffer.
3. For the sake of convenience, this can be made in a larger volume and then stored in single use (200 μL) aliquots at −20°C.
4. Ensure that the contents of the vial are dissolved thoroughly by gentle and constant mixing.
5. Ensure complete mixing. Store at 4°C after use. Use gloves when handling the gel during all subsequent steps.
6. Use at least 200 mL fixing and wash solutions for an 8 × 10-cm gel. These two steps help remove SDS from the gel and are essential to ensure good staining as SDS can interfere with staining. Perform fixing and washing with the recommended volumes for the recommended time.
7. Gels larger than 8 × 10 cm will require a larger volume of oxidizing solution and should be incubated for a longer time.
8. This wash step is important to remove the periodate, which can also interfere with the staining. Use recommended volumes and incubation times.
9. Gels larger than 8 × 10 cm will require a larger volume of staining solution and should be incubated for a longer time.
10. Ensure that the gel is protected from light during the incubation with staining solution. The gel may be incubated with staining solution for up to 120 min, but an overnight incubation is not required. Depending on the amount of protein loaded in the gel and the degree of glycosylation of the individual proteins, bands may be visible even within 20 min.
11. If it is noticed that the background on the gel is very high when imaging, then the gel should be washed two more times. Longer than necessary incubation with the staining solution and/or incorrect dilution of the staining solution may produce a high background.

Acknowledgments

The author thanks Dr. Charles Esmon for the HPC4 antibody and Dr. Rodger P. McEver for purified soluble PSGL-1.

References

1. Spiro R (2002) Protein glycosylation: nature, distribution, enzymatic formation, and disease implications of glycopeptide bonds. Glycobiology 12:43R–56R
2. Wandall HH, Blixt O, Tarp MA, Pedersen JW, Bennett EP, Mandel U, Ragupathi G, Livingston PO, Hollingsworth MA, Taylor-Papadimitriou J, Burchell J, Clausen H (2010) Cancer biomarkers defined by autoantibody signatures to aberrant O-glycopeptide epitopes. Cancer Res 70:1306–1313
3. Molecular Probes: Product Information 2007
4. McEver RP (2002) Selectins: Lectins that initiate cell adhesion under flow. Curr Opin Cell Biol 14:581–586
5. Stearns DJ, Kurosawa S, Sims PJ, Esmon NL, Esmon CT (1988) The interaction of a Ca^{2+}-dependent monoclonal antibody with the protein C activation peptide region. Evidence for obligatory Ca^{2+} binding to both antigen and antibody. J Biol Chem 263:826–832
6. Laemmli UK (1970) Cleavage of structural proteins during the assembly of the head of bacteriophage T4. Nature 227:680–685

Chapter 51

Spicy SDS-PAGE Gels: Curcumin/Turmeric as an Environment-Friendly Protein Stain

Biji T. Kurien, Yaser Dorri, and R. Hal Scofield

Abstract

Gel proteins are commonly stained with calorimetric/fluorescent dyes. Here, we demonstrate that heat-solubilized curcumin can serve as a nontoxic and environment-friendly fluorescent/colorimetric reversible protein stain. Curcumin, the yellow pigment found in the rhizomes of the perennial herb *Curcuma longa* (turmeric), is insoluble in aqueous solvents. However, heat (100°C) solubilization in water renders 1.5% of curcumin soluble. Curcumin solubilized by ethanol or alkali is ineffective in staining proteins. Heat solubilized curry spice turmeric stains proteins similarly. Staining is achieved in 30 min, with a sensitivity almost equaling that of Coomassie Brilliant Blue (CBB). Destaining is not required, and excess curcumin/turmeric can be discarded into the sink. Binding of proteins by silver inhibits curcumin binding, suggesting similarity of protein binding by silver and curcumin. It costs $1.5–2.0 to stain a mini-gel with curcumin, while turmeric costs less than 0.005 cent. CBB staining/destaining costs about 2 cents. However, CBB is toxic and its use necessitates specialized disposal efforts. Curcumin/turmeric, thus, can serve as an ideal nontoxic protein stain.

Key words: Curcumin, Turmeric, SDS-PAGE, Environment-friendly protein stain, Coomassie Brilliant Blue, Silver nitrate

1. Introduction

Sodium dodecyl sulfate polyacrylamide gel electrophoresis (SDS-PAGE), native PAGE, and two-dimensional gel electrophoresis are essential methodologies employed by various laboratories involved in proteomic work. The usefulness of these gel-based methods has led to the development of a variety of colorimetric/fluorescent protein stains.

An appropriate staining method is very important for quantitative assessment of expression levels of proteins in biological samples.

Biji T. Kurien and R. Hal Scofield (eds.), *Protein Electrophoresis: Methods and Protocols*, Methods in Molecular Biology, vol. 869, DOI 10.1007/978-1-61779-821-4_51, © Springer Science+Business Media, LLC 2012

The following criteria needs to be considered in determining the right detection method: (a) the detection limit should be as low as possible with a high signal to noise ratio, (b) the method should also have a wide linear relationship between the quantity of protein and the staining intensity, (c) it should be easy and fast to perform, (d) the procedure should be nontoxic, environment-friendly, and (e) most importantly, the procedure should not be too expensive (1). A variety of staining methodologies are available to detect proteins analyzed on gels.

1.1. Colorimetric Protein Stains

Numerous colorimetric staining methods for proteins have been described, including staining with Coomassie Brilliant Blue (CBB), Coomassie and Bismarck brown mixture, amido black (2), and silver (2–5). Proteoglycans and glycosaminoglycans, stained weakly by traditional protein stains, have been detected with combined Alcian blue-silver stain (6).

1.2. Fluorescent Protein Stains

Flourescent staining methods for proteins include staining with Nile Red, SYPRO Red, SYPRO Orange, SYPRO Tangerine, Coomassie Fluor Orange, SYPRO Ruby, Epicocconone (Deep Purple, Lighting Fast protein gel stain), fluorescein derivatives, Krypton, Krypton infrared, Flamingo LUCY stain (1), and Alta [0.8% Crocein scarlet (brilliant Crocein) and 0.2% Rhodamine B) (7). Phosphoproteins have been detected using Pro-Q Diamond and Phos-tag (3). Glycoproteins have been detected on gels using acid fuchsin dye, Pro-Q Emerald 300 and Pro-Q Emerald 488 glycoprotein gel stain kits, and azide–alkyne "click chemistry" reagents (3).

1.3. Radiolabeling

Protein labeling with radioactive isotopes is the most sensitive method to detect proteins on gels. However, the procedure requires special equipment and very complicated handling procedures (5). In addition, there is the issue of health, security, and safety concerns associated with using radioisotopes.

1.4. Reverse (Negative) Staining Procedures

Zn^+ reverse staining of proteins is a non-fixative procedure that does not involve the use of organic dyes. This procedure stains the areas of the gel in which there are no proteins and does not stain proteins (8).

1.5. Curcumin/Turmeric as a Colorimetric as well as Fluorescent Protein Stain

Of the numerous protein stains available for gel protein detection, none of them can be really be classified as nontoxic. Here, we describe a nontoxic nutraceutical that can serve as a colorimetric as well as fluorescent protein stain for protein detection, possessing a staining sensitivity almost equal to that of Coomassie.

Curcumin is a naturally occurring "nutraceutical," derived from curry spice turmeric (*Curcuma longa*). The yellow polyphenolic pigment curcumin is the most active component in turmeric and has a long history of use in Indian herbal medicine and in the traditional diet of Asian countries. Curcumin has been found to be nontoxic up to 8 g/day and has been studied in prostate, esophagus, lung, breast, and oral cancers. It acts as an antioxidant as well as an anti-inflammatory phytochemical. With its ability to selectively induce apoptosis in cancer cells in vitro, it has been touted for it antitumor potential. Curcumin has been shown to be effective when used in in vitro experiments involving breast cancer cell lines. In addition, curcumin has been demonstrated to modulate multiple sclerosis and Alzheimer's disease, lower cholesterol, suppress diabetes, enhance wound healing, and block HIV replication. In addition, curcumin has been reported to inhibit inflammatory cytokine production, tumorigenesis, metastasis, platelet aggregation, cataract formation, inflammatory bowel disease, and myocardial infarction (9–13).

Lack of solubility in water and low bioavailability have been the bane in using curcumin. However, we have shown that it was possible to increase curcumin's solubility 12-fold with the use of heat. In addition, we have found that the heat-solubilized curcumin is stable and biologically active after heat treatment (14–17). We have used the heat-solubilized curcumin to demonstrate its binding to proteins (18) mainly to support the possible mechanism of curcumin mediated inhibition of antigen–antibody interaction.

Here, we show in detail the ability of curcumin or turmeric to stain proteins. The yellow pigment, curcumin, fluoresces at 446–549 nm when irradiated with ultraviolet light (excitation 355 nm), which enables curcumin-bound proteins to be visualized and recorded with gel-documenting systems. In addition, proteins bound by curcumin/turmeric can be directly visualized by the naked eye. This enables gels to be scanned by a regular scanner for documenting results.

Figure 1 shows decreasing amounts of unstained protein markers (2 μg/protein marker to 100 ng/protein marker) analyzed on SDS-PAGE and stained with CBB (Fig. 1a) or curcumin (Fig. 1b). Regular staining/destaining with CBB takes four hours or more.The staining process with curcumin can be completed in 30 min, and destaining is unnecessary. Staining with curcumin for two hours (Fig. 1c) does not make a significant difference compared to staining for 30 min. It can be seen that curcumin faintly stained the protein at about 45,000 molecular weight at 100 ng level (Fig. 1c, Lane 4), almost similar to CBB staining (Fig. 1a, Lane 4). The staining of proteins at 500 ng, 1 μg, and 2 μg levels with both CBB and curcumin was similar (Fig. 1a, b).

Curcumin is insoluble in water at room temperature. By solubilizing curcumin in water at 100°C, we were able to increase curcumin solubility 12-fold over curcumin solubilized in water at room temperature. However, the bulk of the curcumin (98.5%) remains insoluble.

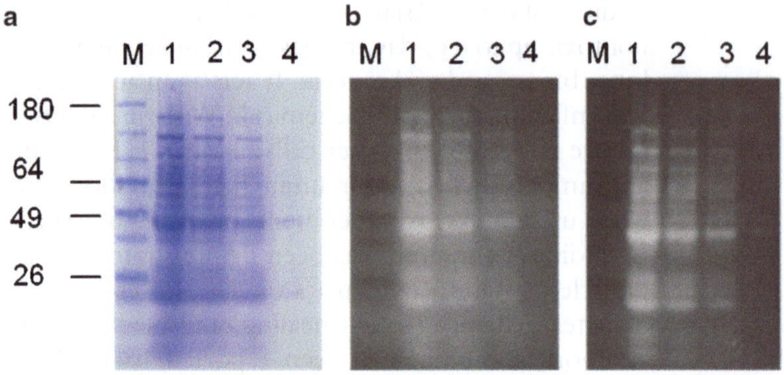

Fig. 1. Staining of unstained protein molecular weight standards by heat-solubilized curcumin and Coomassie Brilliant Blue (CBB). (**a**) Stained by CBB, (**b**) stained by heat-solubilized curcumin for 30 min, and (**c**) stained by heat-solubilized curcumin for 2 h. *Lane 1*—2 μg/protein marker, *lane 2*—1 μg/protein marker, *lane 3*—500 ng/protein marker, and *lane 4*—100 ng/protein marker.

Curcumin, however, is soluble in ethanol, sodium hydroxide, and dimethylsulfoxide. Curcumin was solubilized in these solvents at 5 mg/mL in order to study their protein staining efficacy compared to staining with heat-solubilized curcumin (5 mg/mL water). Figure 2 shows that heat solubilized curcumin-stained bovine serum albumin (2, 0.1, and 0.5 μg) efficiently (Fig. 2a). Curcumin solubilized in ethanol did not stain as well as heat-solubilized curcumin (Fig. 2b), whereas curcumin solubilized in dimethylsulfoxide (DMSO) or alkali did not stain. Curcumin is soluble in DMSO, but when diluted in water, it precipitates, thus rendering it useless to stain proteins.

Since curcumin is derived from turmeric, we attempted to stain proteins with turmeric as well. Various proteins, including glycoproteins, were electrophoresed on SDS-PAGE and stained with heat-solubilized curcumin or turmeric and CBB to compare the efficacy of staining between these stains. As seen in Fig. 3, IgM, Fc fragment, interleukin 13 receptor, mouse serum, and bovine serum albumin were stained efficiently by all three stains (Figs. 3a, b, c). One of the advantages of staining proteins with curcumin or turmeric is that detection sensitivity can be fine tuned with the UVP detector using changes in exposure time (compare Fig. 3b vs. 3c). Proteins stained by heat-solubilized curcumin (Fig. 3e) or turmeric (Fig. 3f) could also be observed by the naked eye, without the use of ultraviolet light. However, the detection sensitivity was lower compared to UV light detection.

In order to understand the possible mechanism by which curcumin binds to proteins, we carried out several experiments. First, 4 μg/lane of bovine serum albumin (BSA) was analyzed on SDS-PAGE and stained with curcumin (Fig. 4a) or stained with CBB (Fig. 4b). Then, a third set of BSA (4 μg/lane) was first stained with curcumin for 30 min and then immediately stained with CBB for another 30 min (Fig. 4c). The Coomassie dye bound to the

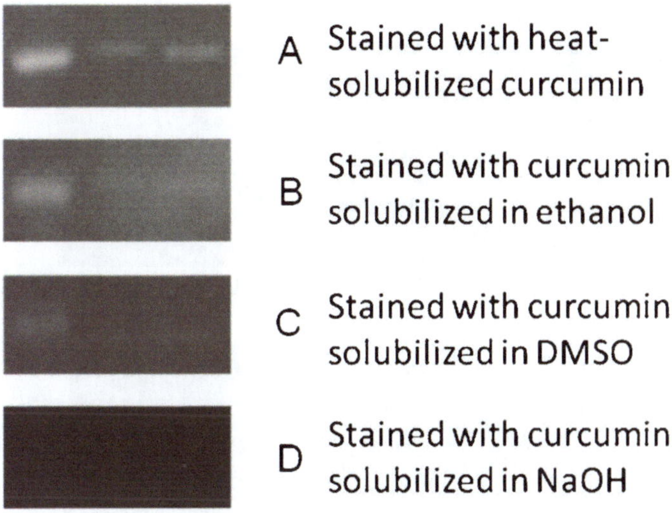

Fig. 2. Comparison of protein staining efficacy of (**a**) curcumin solubilized in water with heat, (**b**) curcumin solubilized in absolute ethanol, (**c**) curcumin solubilized in DMSO (and diluted with water to obtain a 0.1% DMSO solution), and (**d**) curcumin solubilized in 0.5 N NaOH.

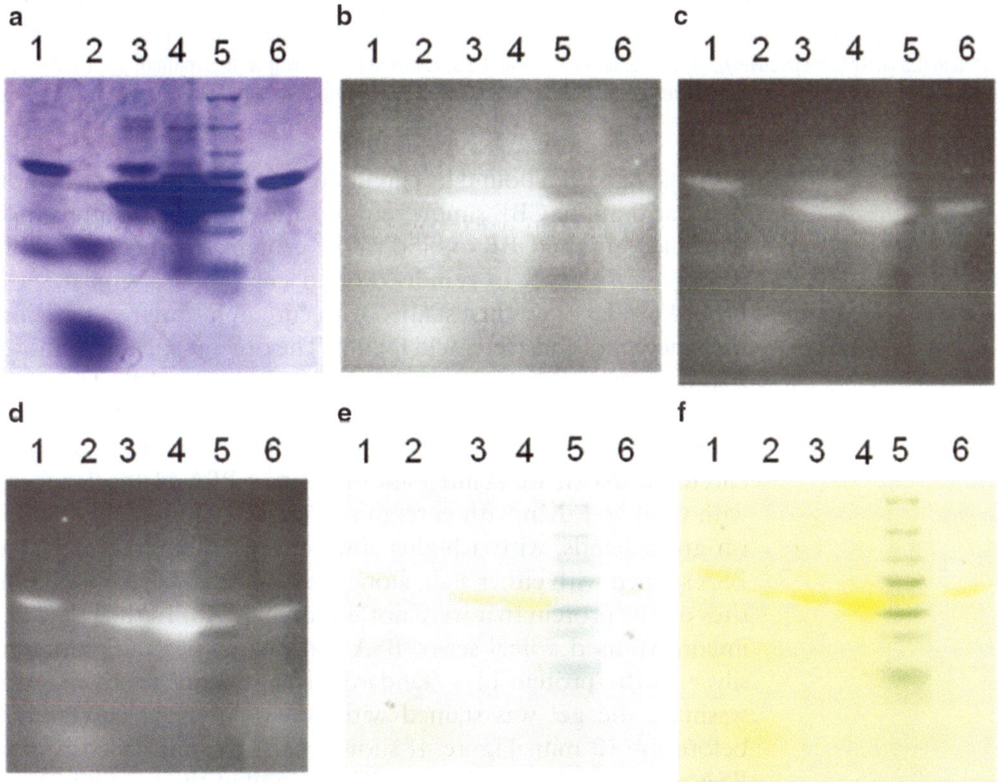

Fig. 3. Staining of various proteins by CBB and heat-solubilized curcumin or turmeric. IgM, Fc fragment, interleukin 13 receptor, mouse serum, and bovine serum albumin stained with (**a**) CBB, (**b**) heat-solubilized curcumin (longer exposure), (**c**) heat-solubilized curcumin (shorter exposure), (**d**) heat-solubilized turmeric, (**e**) stained with heat solubilized curcumin and visualized without ultraviolet light, (**f**) stained with heat-solubilized turmeric and visualized without ultraviolet light.

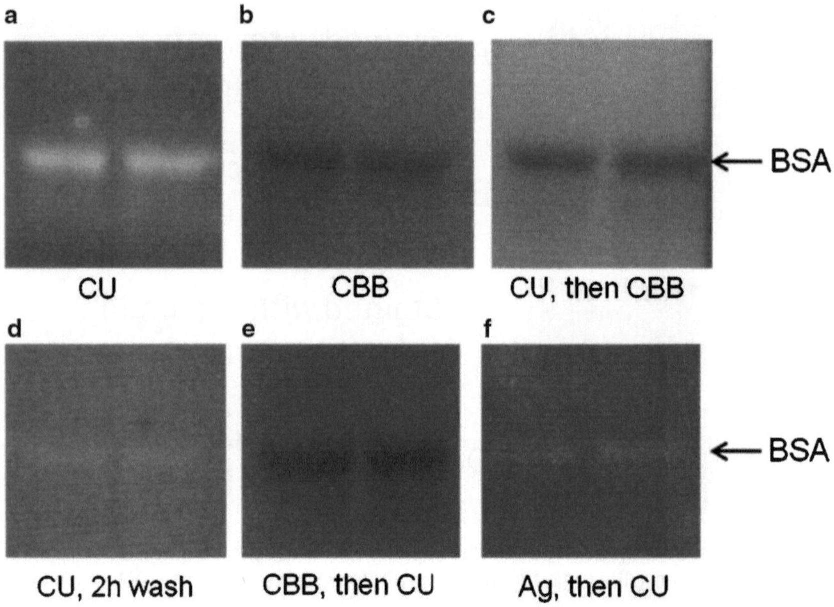

Fig. 4. Possible mechanism of curcumin (CU) binding to proteins. (Please note: Heat-solubilized curcumin was used to stain BSA, whenever curcumin is mentioned.) (**a**) Curcumin binding to BSA; (**b**) Coomassie (CBB) binding to BSA; (**c**) BSA stained first by heat-solubilized curcumin, followed by staining with CBB; (**d**) curcumin-stained BSA washed for 2 h with methanol/acetic acid (25%:10%, respectively); (**e**) BSA stained first with CBB, followed by staining with curcumin; and (**f**) BSA incubated first with silver (using standard silver staining protocol), followed by staining with curcumin.

BSA that had been bound by curcumin. It is possible that methanol/acetic acid in the CBB simply removes the curcumin from the protein, allowing the CBB to bind to the protein. To check the ability of organic solvents to remove bound curcumin, a similar set of BSA (4 μg/lane) was first stained with curcumin and then destained with methanol and acetic acid for 2 h. The organic solvents removed a significant amount of the curcumin in 2 h (Fig. 4d). Therefore, the binding of CBB to BSA following curcumin binding to proteins must have occurred mainly as a result of the removal of bound curcumin. Next, we stained another set of BSA (4 μg/lane) first with CBB and then with curcumin (Fig. 4e). This resulted in bluish-green bands, with a higher intensity of staining compared to BSA stained with either stain alone, showing that curcumin bound sites on the protein that were not bound by CBB. In the last experiment, we used a final set of BSA and carried out the binding of silver to the protein in a standard silver staining protocol. After washing, the gel was stained with heat-solubilized curcumin as before for 30 min. Figure 4f shows that curcumin failed to bind BSA, showing that the bound silver inhibited the binding of curcumin to the protein.

Thus, we have a method of staining proteins fluorescently or colorimetrically using heat-solubilized curcumin/turmeric. Curcumin is the active principle of the food spice turmeric, used in a variety of foods (especially in curry), and has been shown to be well tolerated in humans up to 8 g/day. Therefore, heat-solubilized curcumin is perhaps the only protein stain in the market that is truly nontoxic and does not involve the use of toxic solvents. Thus, used stain can be discarded in the sink after use. The staining sensitivity is almost equal to that of CBB. The binding of curcumin to proteins appear similar to the binding of silver to proteins.

2. Materials

Prepare all solutions using ultrapure water (prepared by purifying deionized water to attain a sensitivity of 18 MΩ cm at 25°C) and analytical grade reagents. Prepare and store all reagents at room temperature (unless indicated otherwise). Diligently follow all waste disposal regulations when disposing waste materials. We do not add sodium azide to reagents.

1. Curcumin (1,7-bis[4-hydroxy-3-methoxyphenyl]-1,6-heptadiene-3,5-dione; Indian saffron; turmeric yellow) >90% (Cayman Chemical Company, Ann Arbor, MI, USA) (see Note 1).
2. Turmeric (local Indian grocery store).
3. Coomassie Brilliant Blue (CBB) stock: Make a stock of 0.5% CBB in 25% methanol (see Note 2) and 10% acetic acid (see Note 3). Dilute the stock 10 times with 25% methanol and 10% acetic acid to obtain 0.05% working solution of Coomassie at the time of gel staining.
4. Precast (10%, 4–20% gradient) SDS-PAGE gels (Bio-Rad, Hercules, CA, USA).
5. IgM, Fc fragment (Sigma-Aldrich. St. Louis, MO, USA).
6. Interleukin 13 receptor (Sigma).
7. BALB/c mouse serum (see Note 4).
8. SDS-PAGE running buffer: 0.025 M Tris, pH 8.3, 0.192 M glycine, and 0.1% SDS (see Note 5).
9. SDS lysis buffer (5×): 0.3 M Tris–HCl (pH 6.8), 10% SDS, 25% β-mercaptoethanol, 0.1% bromophenol blue, and 45% glycerol. Leave one aliquot at 4°C for current use and store remaining aliquots at −20°C (see Note 6).
10. Bromophenol blue (BPB) solution: Dissolve 0.1 g BPB in 100 mL water.
11. Phosphate-buffered saline (PBS), pH 7.4.
12. UVP BioDoc-It™ system, Upland, CA, USA.
13. Branson Sonifier Cell Disruptor 185 (VWR Scientific Co., Boston, MA, USA).

14. FB300 power supply (Fisher Scientific, Houston, TX, USA).
15. Corning PC-351 magnetic stirrer.

3. Methods

All procedures are carried out at room temperature unless otherwise specified.

3.1. Preparation of HeLa Cell Extract

1. Harvest freshly cultured HeLa cells by centrifuging at 800 × g and wash twice with PBS.
2. Lyse cells by sonication (see Note 7) in PBS buffer using a Branson sonicator (setting 4) and centrifuge at 10, 000 × *g* for 10 min (see Note 8).
3. Use an aliquot of the supernatant for SDS-PAGE.

3.2. Heat Solubilization of Curcumin or Turmeric

1. Weigh curcumin or turmeric and place in 50 mL blue-capped centrifuge tube.
2. Add hot (about 90°C) distilled water into tube to obtain a 5 mg/mL solution of curcumin or turmeric. Mix to suspend curcumin or turmeric in the water (see Note 9).
3. Heat tube contents for 10 min in a boiling water bath. Mix once every 2 min.
4. Centrifuge at 1,800 × *g* for 20 min at room temperature using a bench top centrifuge.
5. Transfer supernatant to a fresh 50-mL blue-capped centrifuge tube and repeat step 4.
6. Transfer supernatant to a fresh 50-mL blue-capped centrifuge tube. Use this clear supernatant for staining gels (see Note 10).

3.3. SDS-PAGE

1. Electrophorese (mini-gel electrophoresis) prestained protein marker, unstained protein marker, bovine serum albumin, HeLa cell extract, Fc fragment (IgM), or interleukin 13 receptor on 10% or 4–20% precast SDS-PAGE gels according to Laemmli's procedure (19).
2. Run separate gels for staining with CBB, curcumin, turmeric, or silver nitrate.

3.4. Staining SDS-PAGE Gel Proteins with Heat-Solubilized Curcumin or Turmeric

1. Carefully remove gel from gel cassette.
2. Fix gel with 25% methanol and 10% acetic acid for 10–20 min (see Note 11).

3. Rinse gel with distilled water to remove all traces of the fixative (see Note 12). Discard water from the container completely.
4. Add 25 mL of heat-solubilized curcumin or turmeric to the gel and incubate at room temperature for a minimum of 30 min with mild shaking.
5. Visualize and document (see Figs. 1 and 3) stained gel with ultraviolet light using an UVP BioDoc-It™ system (see Note 13).

3.5. Staining SDS-PAGE Gel Proteins with Curcumin Solubilized in Ethanol

1. Dissolve curcumin in absolute ethanol (see Note 14) to obtain a 5 mg/mL solution.
2. Carry out steps 1–3 (see Subheading 3.4).
3. Add 25 mL of curcumin dissolved in ethanol and stain for 30 min at room temperature with gentle shaking.
4. Visualize and document (see Fig. 2) stained gel with ultraviolet light using an UVP BioDoc-It™ system (see Note 13).

3.6. Staining SDS-PAGE Gel Proteins with Curcumin Solubilized in DMSO

1. Dissolve curcumin in DMSO. Dilute with water to obtain a final concentration of 0.1% DMSO and a 5 mg/mL solution curcumin solution (see Note 15).
2. Carry out steps 1–3 (see Subheading 3.4)
3. Add 25 mL of curcumin dissolved in 0.1% DMSO and stain for 30 min at room temperature with gentle shaking.
4. Visualize and document (see Fig. 2) stained gel with ultraviolet light using an UVP BioDoc-It™ system.

3.7. Staining SDS-PAGE Gel Proteins with Curcumin Solubilized in 0.5 N Sodium Hydroxide

1. Dissolve curcumin in 0.5 N sodium hydroxide to obtain a 5 mg/mL solution (see Note 16).
2. Carry out steps 1–3 (see Subheading 3.4).
3. Add 25 mL of curcumin dissolved in alkali and stain for 30 min at room temperature with gentle shaking.
4. Visualize and document (see Fig. 2) stained gel with ultraviolet light using an UVP BioDoc-It™ system.

4. Notes

1. Commercial curcumin contains pure curcumin (77%), demethoxycurcumin (17%), and bisdemethoxycurcumin (3%). Curcumin, demethoxycurcumin, and bisdemethoxycurcumin differ from each other structurally by the presence and position of a methoxy group.

2. Weigh the required amount of the dye and dissolve it in methanol first. Filter the dissolved CBB through a circular Whatman 1 M filter paper folded into a cone and fitted inside a glass funnel. For a quicker way to filter the CBB solution, layer a circular 1 M Whatman filter paper (125-mm diameter circles) inside a Buchner funnel (Coors, USA; 135 mm inner diameter). The filter paper should cover all the holes in the funnel and lay flat inside the funnel. Attach the Buchner funnel to a side-arm-fitted conical flask that is attached to the house-vacuum through the side-arm. Pour the CBB solution into the Buchner funnel and apply vacuum. If filter paper gets clogged, replace with a fresh one and filter. Collect the filtered CBB. Adjust volume with water, methanol, and acetic acid to obtain a 0.05% CBB in 25% methanol/10% acetic acid.
3. Add acid to water and not vice versa.
4. Mice were purchased from Jackson Laboratories, Bar Harbor, ME, USA. The animals were housed at the Laboratory Animal Resource Facility, OMRF, OKC, OK 73104. Mice studies were approved by the Institutional Animal Care and Use Committee.
5. Simple method of preparing running buffer: Prepare 10× native buffer (0.25 M Tris, 1.92 M glycine). Weigh 30.3 g Tris and 144 g glycine, mix, and make it to 1 L with water. Dilute 100 mL of 10× native buffer to 990 mL with water and add 10 mL of 10% SDS. Care should be taken to add SDS solution last since it makes bubbles.
6. SDS precipitates at 4°C. Therefore, the lysis buffer needs to be warmed prior to use.
7. Sonicators generate high-frequency sound waves in the 20,000 Hz range, outside our normal range of hearing. These sound waves can cause hearing damage, and therefore, laboratory personnel should wear sound mufflers when sonication is in process. (http://www.labmanager.com/?articles.view/articleNo/1103/article/Sonicator-Safety).
8. Cool microcentrifuge tube, containing HeLa cells suspended in PBS, on ice first. Clean sonicator probe with ethanol first and then with distilled water. Wipe the probe dry with Kimwipe. Sonicate for 10 s and let cool on ice. Repeat this step 3 more times. Make sure that the probe does not touch the bottom of the tube when sonication is in progress to avoid the probe from puncturing a hole in the tube.
9. Heat distilled water in a clean 1-L glass beaker. When temperature reaches about 90°C, transfer required volume with a disposable pipette. Take extra caution when handling hot water. Bring the water to boil and use this for heating the tube containing curcumin or turmeric in the next step.

10. Most of the curcumin/turmeric will remain insoluble. A 5 mg/mL curcumin solution (solubilized at room temperature with water) yielded 0.6 μg/mL of soluble curcumin. Heat treatment increases the solubility 12-fold to yield 7.4 μg/mL of soluble curcumin (14).
11. We use a yellow pipette tip (10–200 μL) box cover to fix the gel (as well as to stain the gel). This container helps to reduce the volume of fixative or stain needed.
12. Curcumin dissolved in water is very sensitive to pH changes. Therefore, it is important to wash fixative away with water prior to the addition of curcumin solution.
13. There is no need to rinse the curcumin or turmeric from the gel prior to visualization and documentation with the UVP gel doc system.
14. Curcumin at 5 mg/mL dissolves completely in 200-proof ethanol, yielding a yellow-colored solution.
15. Curcumin dissolves initially in DMSO. However, curcumin precipitates when the stock is diluted with water to obtain 0.1% final concentration of DMSO.
16. Curcumin at 5 mg/mL dissolves completely in 0.5 N sodium hydroxide, yielding a dark red solution.

References

1. D'souza A, Scofield RH (2009) Protein stains to detect antigen on membranes. Methods Mol Biol 536:433–440
2. Kurien BT, Scofield RH (1998) Heat mediated quick Coomassie blue protein staining and destaining of SDS-PAGE gels. Indian J Biochem Biophys 35:385–389
3. Steinberg TH (2009) Protein gel staining methods: an introduction and overview. Methods Enzymol 463:541–563, Review
4. Jin LT, Hwang SY, Yoo GS et al (2006) A mass spectrometry compatible silver Staining method for protein incorporating a new silver sensitizer in sodium dodecyl sulfate-polyacrylamide electrophoresis gels. Proteomics 6:2334–2337
5. Jin LT, Hwang SY, Yoo GS et al (2004) Sensitive silver staining of protein in sodium dodecyl sulfate-polyacrylamide gels using an azo dye, calconcarboxylic acid, as a silver-ion sensitizer. Electrophoresis 25:2494–2500
6. Møller HJ, Heinegård D, Poulsen JH (1993) Combined alcian blue and silver staining of subnanogram quantities of proteoglycans and glycosaminoglycans in sodium dodecyl sulfate-polyacrylamide gels. Anal Biochem 209:169–175
7. Pal JK, Godbole D, Sharma K (2004) Staining of proteins on SDS Polyacrylamide gels and on nitrocellulose membranes by Alta, a colour used as a cosmetic. J Biochem Biophys Methods 61:339–347
8. Lin CY, Wang V, Shui HA et al (2009) A comprehensive evaluation of imidazole-zinc reverse stain for current proteomic researches. Proteomics 9:696–709
9. Aggarwal BB, Kumar A, Aggarwal MS et al (2004) Curcumin derived from turmeric (*Curcuma longa*): a spice for all seasons. In: Bagchi D, Preuss HG (eds) Phytochemicals in cancer chemoprevention. CRC, Boca Raton, FL, pp 349–387
10. Aggarwal BB, Sung B (2009) Pharmacological basis for the role of curcumin in chronic diseases: an age-old spice with modern targets. Trends Pharmacol Sci 30:85–94
11. Bright JJ (2007) Curcumin and autoimmune disease. Adv Exp Med Biology 595:425–451, Review
12. Rowe DL, Ozbay T, O'Regan RM et al (2009) Modulation of the BRCA1 protein and induction of apoptosis in triple negative breast cancer cell lines by the polyphenolic. Breast Cancer 3:61–75

13. Anand P, Sundaram C, Jhurani S et al (2008) Curcumin and cancer: an "old-age" disease with an "age-old" solution. Cancer Lett 267:133–164, Review
14. Kurien BT, Singh A, Matsumoto H et al (2007) Improving the solubility and pharmacological efficacy of curcumin by heat treatment. Assay Drug Dev Technol 5:567–576
15. Kurien BT, Scofield RH (2009) Curry spice curcumin and prostate cancer. Mol Nutr Food Res 53:939–940
16. Kurien BT (2009) Comment on Curcumin attenuates acrylamide-induced cytotoxicity and genotoxicity in HepG2 cells by ROS scavenging. J Agric Food Chem 57:5644–5646
17. Kurien BT, Scofield RH (2009) Bubbling hookah smoke through heat-solubilized curcumin/turmeric and incorporation of the curry spice as an additive or filter in cigarettes to minimize tobacco smoke-related toxicants. Med Hypotheses 73:462–463
18. Kurien BT, D'Souza A, Scofield RH (2010) Heat-solubilized curry spice curcumin inhibits antibody–antigen interaction in in vitro studies: A possible therapy to alleviate autoimmune disorders. Mol Nutr Food Res 54:1202–1209
19. Laemmli UK (1970) Cleavage of structural proteins during the assembly of the head of bacteriophage T4. Nature 227:680–685

Chapter 52

In-Gel Staining of Proteins in Native Polyacrylamide Gel Electrophoresis Using Meso-Tetrakis(4-Sulfonatophenyl) Porphyrin*

K. Divakar, G. Nandhini Devi, and Pennathur Gautam

Abstract

Protein identification in polyacrylamide gel electrophoresis (PAGE) requires post-electrophoretic steps like fixing, staining, and destaining of the gel, which are time-consuming and cumbersome. A new method for direct visualization of protein bands in PAGE has been developed using meso-tetrakis(4-sulfonatophenyl)porphyrin (TPPS) as a dye without the need for any post-electrophoretic steps; thus, separation and recovery of enzymes become much easier for further analysis. Activity staining was carried out to show that the biochemical activity of the enzymes was preserved after electrophoresis.

Key words: TPPS, PAGE, Staining, In-gel

1. Introduction

Development in the field of proteomics relies on efficient strategies for the detection of proteins (1). A highly sensitive protein detection method is essential for the growth of proteomics. Meso-tetrakis(4-sulfonatophenyl)porphyrin (TPPS) is a water-soluble porphyrin possessing negative charges on four sulfonic acid groups. Porphyrin derivatives form complexes with biopolymers (proteins and nucleic acids) non-covalently, bound by forces which are electronic, ionic, and hydrophobic nature (2). TPPS binds to surface positive residues of the protein and confers a negative charge without denaturing the protein. Proteins with acidic or basic isoelectric points are conferred a net negative charge and migrate to the anode.

*This work is dedicated to the memory of Mr. Riki Krishnan.

Biji T. Kurien and R. Hal Scofield (eds.), *Protein Electrophoresis: Methods and Protocols*, Methods in Molecular Biology, vol. 869, DOI 10.1007/978-1-61779-821-4_52,

In recent years, there has been a growing interest in the use of porphyrins and related compounds as tools for cancer detection and photosensitizers in photodynamic therapy for cancer (3). Derivatives of protoporphyrinate have been used to detect protein–porphyrin interaction using gel electrophoresis (4).

Coomassie Blue stain, silver stain, and fluorescent stain are the commonly used stains for protein detection in electrophoresis. Coomassie Blue staining is popular and easy method for detecting proteins, but it needs post-electrophoretic processing. Silver staining is a more sensitive method used to detect proteins even in lower concentrations, but it is more time-consuming. Fluorescent staining is a popular alternative to traditional staining procedures mainly because it is more sensitive and easier to process than Coomassie staining. The limitations of labeling proteins by fluorescent tags are that both protein and dye be present in higher concentrations for the reaction to proceed (5), although exceptions have been reported (6). A method has been developed for in-gel staining of protein, without affecting the bioactivity. It is easy to perform and also aids in efficient recovery of protein for further characterization.

2. Materials

2.1. Sulfonation of Porphyrin

Prepare sulfonated porphyrin by method described by Song et al. (7).

2.2. Polyacrylamide Gel

1. Monomer solution: Dissolve 29.2 g acrylamide and 0.8 g bisacrylamide in 100 mL of double distilled water (see Note 1).
2. Separating gel buffer: 1.5 M Tris–HCl, pH 8.8 (see Note 2).
3. Stacking gel buffer: 1.5 M Tris–HCl, pH 6.8 (see Note 2).
4. Running gel buffer (RGB): Prepare RGB by dissolving 3.1 g Tris base and 14.4 g glycine in 1.0 L of double distilled H_2O. pH of the buffer was 8.3.
5. Ammonium persulfate (APS): 100 mg APS in 1 mL double distilled H_2O (see Note 3).
6. Sodium dodecyl sulfate (SDS): 10 g SDS in 100 mL dd.H_2O.
7. *N*, *N*, *N*, *N*′-tetramethylethylenediamine (TEMED).
8. 2× Sample solubilizing buffer: 0.5 mL meso-tetrakis(4-sulfonatophenyl)porphyrin (TPPS), 200 μL of 0.5 M Tris–HCl, pH 6.8, and 500 μL glycerol. Make up to 2 mL with water. Store at 4°C (8).

9. Filter all gel buffers and acrylamide monomer solution through Whatman # 1 paper and store at 4°C in brown bottles.

2.3. Activity Staining

1. Activation buffer: 25 mM Tris–HCl buffer (pH 7.5).
2. Substrate solution for protease: Prepare 1% (w/v) gelatin solution by dissolving 0.5 g gelatin in 50 mL double distilled water (see Note 4).
3. Substrate-agar emulsion for esterases: 1.3% (w/v) agar, 1% (v/v) tributyrin, 25 mM Tris–HCl, pH 8.0, 10 mM $CaCl_2$, and 0.1% (v/v) Triton X-100 (see Note 5).

3. Methods

1. For 8% w/v resolving gel, mix 4.0 mL of monomer solution, 3.8 mL separating gel buffer, 7.0 mL water, 75 μL APS, and 5 μL TEMED and pour the gel mix into gel cassette to fill up to 3/4 volume of gel cassette (see Note 6). Overlay gently with water or 2-butanol. Allow gel mix to polymerize (see Note 7).
2. For 4% w/v stacking gel, mix 1.0 mL of acrylamide solution, 1.5 mL stacking gel buffer, 60 μL 10% SDS, 3.4 mL water, 50 μL APS, and 3 μL TEMED and pour into gel cassette over resolving gel and insert the comb to make wells (see Note 8).
3. Mix 20 μL of protein sample (3 μg) (see Note 10) with equal volume of sample buffer freshly before use. Incubate the mixture at 4°C for 30 min before loading onto the gel (see Note 9).
4. Perform electrophoresis at a constant current of at 60 V at 22°C for 4 h. Stop electrophoresis when the dye reaches 1 cm from the bottom of the gel (see Note 11).
5. After electrophoresis, remove the gel from electrophoresis setup and visualize the protein bands in the gel under UV light (Fig. 1) (see Note 12).

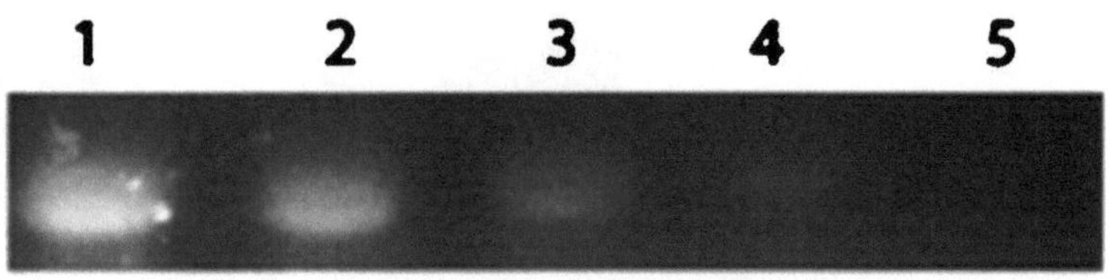

Fig. 1. Polyacrylamide gel electrophoresis of BSA. Fluorescence of protein–TPPS complex observed under UV light. Samples contained constant amounts of protein (3 mg/mL) and various concentrations of TPPS; *lane 1*, 2.5; *lane 2*, 2; *lane 3*, 1.5; *lane 4*, 1; and *lane 5*, 0.5 mg/mL. Reproduced from Analytical sciences (8).

3.1. Activity Staining for Hydrolases

For activity staining, equilibrate the gel in the activation buffer for 30 min at 25°C.

3.1.1. Zymogram for Esterases

1. For esterase zymogram, weigh 630 mg agar powder and dissolve agar in 50 mL H_2O. Add 250 μL tributyrin and emulsify by adding 50 μL Triton X-100. Then add 1.25 mL of 1 M Tris–HCl, pH 8.0 (at 25°C), and 0.5 mL of 1 M $CaCl_2$. Mix well to make homogenous substrate-agar solution (see ref. 9).
2. Pour half the volume of substrate-agar solution to clean petri dish and allow it to solidify. Overlay the gel on the solidified substrate-agar then pour remaining solution over it to sandwich the gel (see Note 13).
3. Incubate the plate overnight at 37°C until a clear zone is observed (Fig. 2a).

3.1.2. Zymogram for Proteases

1. Transfer the gel lane (loaded with protease) to a container filled with equilibration buffer containing 1% (w/v) gelatin for 1 h at 37°C in order to allow the substrate to diffuse into the gel (see Note 14).
2. Then transfer the gel to clean petri dish and incubate at 37°C for 1 h for digestion of the gelatin by the active protease (see ref. 10).

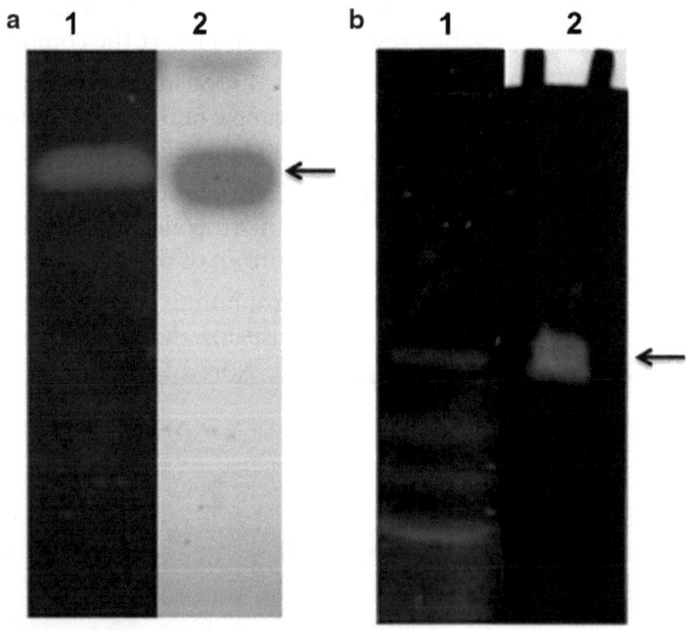

Fig. 2. Native polyacrylamide gel electrophoresis. (**a**) *Lane 1*, esterase labeled with TPPS visualized under UV light (3 μg). *Lane 2*, zymogram for TPPS-labeled esterase (3 μg). (**b**) Partially purified protease. *Lane 1*, sample labeled with TPPS visualized under UV light (20 μg total protein). *Lane 2*, zymogram for TPPS-labeled sample (20 μg total protein). Arrow indicates respective enzyme activity. Reproduced from Analytical sciences (8).

3. After incubation, stain the gel with CBB R-250 for 1 h and then destain in destaining solution for 1–2 h.
4. Clear zone of hydrolysis around the protein band shows protease activity in the gel (Fig. 2b).

4. Notes

1. Acrylamide is a neurotoxin and must be handled with appropriate safety precautions. Wear hand gloves and face mask to avoid exposure.
2. Dissolve 181.7 g tris(hydroxymethyl)aminomethane in double distilled water and adjust the pH by adding HCl with constant mixing, maintain constant temperature at 25°C.
3. APS should be prepared fresh because it is hygroscopic, and the accumulation of water in ammonium persulfate results in a rapid loss of reactivity.
4. Gelatin will not readily dissolve in water. Add gelatin powder to water and boil it for 2–3 min to dissolve it completely.
5. Dissolve agar completely in water by boiling it for 3–5 min. Then add tributyrin and emulsify it by adding Triton X-100 followed by vigorous mixing.
6. Excess TEMED can increase pH of the buffer; in addition, it causes oxidation of sample proteins.
7. Once the gel polymerizes, then three layers will be visible, i.e., polymerized gel, water layer, and 2-butanol layer at the top.
8. Insert the comb immediately after pouring the gel mix and avoid formation of air bubbles between the comb. Air bubbles leads to improper/uneven gel shape.
9. Care should be taken while handling enzymes because prolonged incubation may alter enzyme activity.
10. The protein content was determined by the Bradford method with bovine serum albumin as standard (11).
11. Allow tracking dye to run into anode buffer because tracking dye will have unbound TPPS which interfere while visualizing protein bands under UV light.
12. Wear UV-resistant goggles while viewing proteins under UV light.
13. Temperature of the substrate-agar solution poured above the gel should be ambient, high temperature may denature the enzyme activity.
14. This allows the gelatin to diffuse into the gel.

Acknowledgments

We thank Department of Biotechnology and Department of Science and Technology, Government of India, for their continued support to our laboratory. K. Divakar thanks Council of Scientific and Industrial Research, Government of India, for research fellowship.

References

1. Krause F (2006) Detection and analysis of protein-protein interactions in organellar and prokaryotic proteomes by native gel electrophoresis: (Membrane) protein complexes and supercomplexes. Electrophoresis 27:2759–2781
2. Lang K, Mosinger J, Wagnerova DM (2004) Photophysical properties of porphyrinoid sensitizers non-covalently bound to host molecules; models for photodynamic therapy. Coordination. Chem Rev 248:321–350
3. Berg K, Selbo PK, Weyergang A et al (2005) Porphyrin-related photosensitizers for cancer imaging and therapeutic applications. J Microsc 218:133–147
4. Kowalska A, Kwiek P, Milosz E et al (2003) Application of electrophoretic methods for detection of protein-porphyrin complexes. Acta Biochim Pol 50:1155–1163
5. Stoyanov AV, Ahmadzadeh H, Krylov SN (2002) Heterogeneity of protein labeling with a fluorogenic reagent, 3-(2-furoyl)quinoline-2-carboxaldehyde. J Chromatogr B 780: 283–287
6. Hu S, Jiang J, Cook LM et al (2002) Capillary sodium dodecyl sulphate-DALT electrophoresis with laser-induced fluorescence detection for size-based analysis of proteins in human colon cancer cells. Electrophoresis 23: 3136–3142
7. Song R, Robert A, Bernadou J et al (1998) Sulfonated and acetamidosulfonylated tetraarylporphyrins as biomimetic oxidation catalysts under aqueous conditions. Inorg Chimi Acta 272:228–234
8. Divakar K, Sujatha V, Barath S et al (2011) In-gel staining of proteins in native poly acryl amide gel electrophoresis using tetrakis(4-sulfonato phenyl) porphyrin (TPPS). Anal Sci 27: 101–103
9. Saminathan M, Muthukumaresan KT, Rengarajan S et al (2008) Blue native electrophoresis study on lipases. Anal Biochem 377: 270–271
10. Divakar K, Priya JDA, Gautam P (2010) Purification and characterization of thermostable organic solvent-stable protease from *Aeromonas veronii* PG01. J Mol Cat B Enz 66: 311–318
11. Bradford M (1976) A rapid and sensitive method for quantitation of microgram quantities of protein utilising the principle of protein-dye binding. Anal Biochem 72:248–254

Chapter 53

Rapid Detection of Proteins in Polyacrylamide Electrophoresis Gels with Direct Red 81 and Amido Black

David Choveaux, Robert G.E. Krause, and J.P. Dean Goldring

Abstract

Proteins separated by SDS-polyacrylamide gel electrophoresis need to be stained with organic dyes to be visualized and to enable comparisons to be made between the intensity of protein bands to observe and determine differences in protein concentration. The standard protein staining is with Coomassie Blue R-250. Coomassie staining takes 1 h to complete. Direct Red 81 and Amido Black stain proteins within 10 min. This chapter describes Direct Red 81 and Amido Black staining in comparison to staining with Coomassie Blue R-250.

Key words: Protein staining, Amido Black, Coomassie Blue, Direct Red 81

1. Introduction

Proteins in SDS-PAGE gels can be stained with Coomassie Blue R-250, silver, Amido Black, Fast Green, Congo Red, Bismarck Brown, Direct Red 81, or other organic dyes (1–4). The most popular stain is Coomassie Blue, whilst the most sensitive and involved protocol is with the silver stain. The Coomassie Blue stain takes up to 60 min to stain protein (4). We have developed staining with Direct Red 81 and found that both Direct Red 81 and Amido Black dyes stain proteins within 10 min in a 1-mm thick electrophoresis gel (4). Protein bands prominently stained with Direct Red 81 are visible immediately upon addition of the destain solution. The Direct Red 81 staining technique is much more rapid than the Coomassie Blue R-250 method.

Biji T. Kurien and R. Hal Scofield (eds.), *Protein Electrophoresis: Methods and Protocols*, Methods in Molecular Biology, vol. 869, DOI 10.1007/978-1-61779-821-4_53, © Springer Science+Business Media, LLC 2012

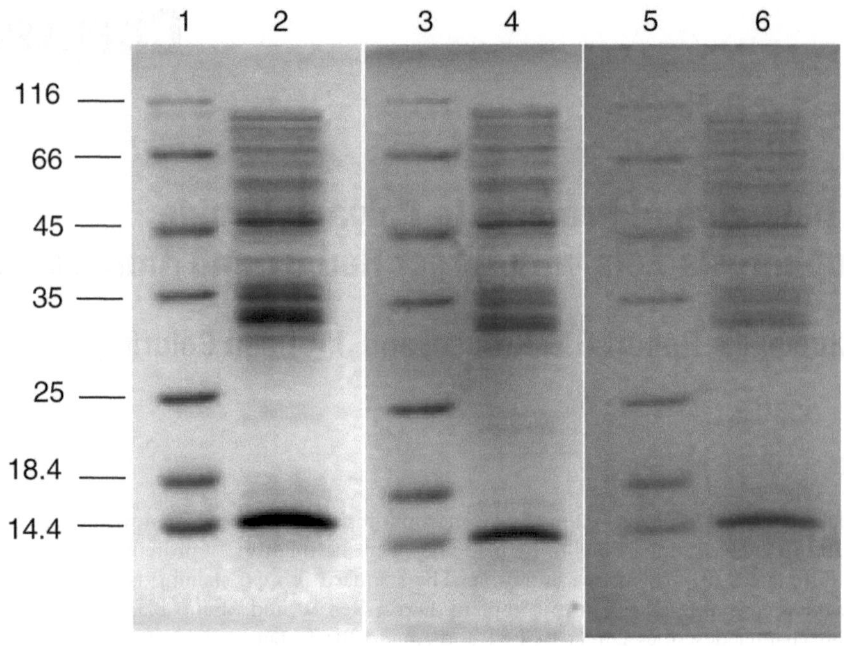

Fig. 1. Staining of molecular mass standards and *E. coli* lysate with Coomassie Blue R-250, Amido Black, and Direct Red 81. Proteins were resolved in a 12.5% reducing SDS-polyacrylamide gel and stained (left to right) for 60 min with Coomassie Blue (lanes 1 and 2) and 10 min with Amido Black (lanes 3 and 4) and Direct Red 81 (lanes 5 and 6). Lanes 1, 3, and 5, molecular mass standards, β-galactosidase, bovine serum albumin, ovalbumin, lactate dehydrogenase, REase Bsp981, β-lactoglobulin, and lysozyme. *Lanes* 2, 4, and 6, an *E. coli* BL21 cell lysate.

1.1. Staining Proteins with Direct Red 81

Direct Red 81 rapidly enters a gel, and the staining of proteins is complete within 10 min of applying the dye (4). Protein bands are visible upon addition of destain, though removal of most residual staining of the gel takes between 2 and 3 h (see Notes 1, 2, 7, and 8). Direct Red 81 staining of an *Escherichia coli* lysate and molecular mass standards is shown in Fig. 1.

1.2. Staining Proteins with Amido Black

Amido Black stains proteins in an electrophoresis gel within 10 min (3, 4). Protein bands take nearly an hour to become visible after the addition of the destaining solution, and it takes up to 3 h to remove most of the residual stain from the gel (see Notes 3, 4, 7, and 8). Amido Black staining of an *E. coli* lysate and molecular mass standards is shown in Fig. 1.

1.3. Staining Proteins with Coomassie Blue R-250

Coomassie Blue R-250 staining of proteins in an electrophoresis gel at room temperature increases until saturation is reached at 60 min (4). The more prominently stained protein bands are visible upon the addition of destain, and removal of most residual stain from the gel takes between 2 and 3 h (see Notes 5–8).

2. Materials

2.1. Staining Protein with Direct Red 81

1. Direct Red 81 (Sigma-Aldrich, St Louis, USA).
2. Water (40% v/v), methanol (50% v/v), and acetic acid (10% v/v).
3. Molecular mass standards: β-galactosidase, bovine serum albumin, ovalbumin, lactate dehydrogenase, REase Bsp981, β-lactoglobulin, and lysozyme.

2.2. Staining Proteins with Amido Black

1. Amido Black (Sigma-Aldrich, St Louis, USA).
2. Water (40% v/v), methanol (50% v/v), and acetic acid (10% v/v).

2.3. Staining Proteins with Coomassie Blue R-250

1. Coomassie Blue R-250 (Sigma-Aldrich, St Louis, USA).
2. Water (40% v/v), methanol (50% v/v), and acetic acid (10% v/v).

3. Methods

3.1. Direct Red 81 Staining

1. After electrophoresis, place polyacrylamide gels in the staining solution containing 1.5 μmol/mL (0.1014 g/100 mL) Direct Red 81 in 50% (v/v) methanol, 10% (v/v) glacial acetic acid, and 40% (v/v) water for 10 min. Gloves should be worn to avoid staining skin keratin.
2. After staining, destain the gel with a 50% (v/v) methanol, 10% (v/v) glacial acetic acid, and 40% (v/v) water solution until the background is clear. The destain can be changed after 1 h to increase the rate of removal of residual stain from the gel.

3.2. Amido Black Staining

1. After electrophoresis, place polyacrylamide gels in the staining solution containing 1.5 μmol/mL (0.0924 g/100 mL) Amido Black in 50% (v/v) methanol, 10% (v/v) glacial acetic acid, and 40% (v/v) water for 10 min. Gloves should be worn to avoid staining of skin keratin.
2. After staining, destain the gel with a 50% (v/v) methanol, 10% (v/v) glacial acetic acid, and 40% (v/v) water solution until the background is clear. The destain can be changed after 1 h to increase the rate of removal of residual stain from the gel.

3.3. Coomassie Blue R-250 Staining

1. After electrophoresis, place polyacrylamide gels in the staining solution containing 1.5 μmol/mL (0.124 g/100 mL) Coomassie Blue R-250 in 50% methanol, 10% glacial acetic acid, and 40% water for 1 h. Gloves should be worn to avoid staining skin keratin.

2. After staining, destain the gel with a 50% (v/v) methanol, 10% (v/v) glacial acetic acid and 40% (v/v) water solution until the background is clear. The destain can be changed after 1 h to increase the rate of removal of residual stain from the gel.

4. Notes

1. Advantages of Direct Red 81 staining: (a) Proteins are stained within 10 min. (b) Proteins can be seen 10 min after electrophoresis and immediately upon addition of the destain solution. (c) The reagents are easy to make up and are stable. (d) Proteins are stained much faster than with Coomassie Blue R-250.
2. Disadvantages of Direct Red 81 staining: (a) Staining is less sensitive than staining with Coomassie Blue. 0.3 μg of protein is readily seen after Coomassie Blue staining and is just detectible with Direct Red 81. (b) The red stain of the proteins appears less intense to the naked eye than the staining intensity of Amido Black or Coomassie Blue but has a higher intensity when determined with image processing software.
3. Advantages of Amido Black staining: (a) Proteins are stained within 10 min. (b) The reagents are easily made up and stable for many months. (c) The staining is as intense as Coomassie Blue staining and more intense than with Direct Red 81 to the naked eye.
4. Disadvantages of Amido Black staining: (a) The background gel staining is very intense, and proteins are not readily visible until 1 h of destaining. (b) Destaining is much slower than with Coomassie Blue or Direct Red 81.
5. Advantages of Coomassie Blue staining: (a) The technique is well established and the most popular method to stain proteins in electrophoresis gels. (b) The reagents are easily made and are stable for many months. (c) The staining and destaining protocols are simple to follow.
6. Disadvantages with Coomassie Blue staining: (a) The dye stains containers and skin. (b) It takes up to 1 h to complete staining of proteins at room temperature (4). (c) Staining is slower than with Amido Black and Direct Red 81.
7. Comparison of staining with Direct Red 81, Amido Black, and Coomassie Blue staining: (a) Proteins take 1 h to stain with Coomassie Blue and 10 min with Amido Black and Direct Red 81. (b) Prominent protein bands can be seen immediately upon addition of the destaining solution with both Direct Red 81 and Coomassie Blue but not with Amido Black. (c) One

hour after addition of destain solution, most proteins are visible with all three stains. It should be noted that there are still high residual levels of stain within the gel with all three stains. (d) Removing the background stain to a minimum requires destaining (with changes of destain solution each hour) for between 2 and 3 h for Coomassie Blue and Direct Red 81 and up to 3 h for Amido Black. (e) The intensity of staining of the lightly stained proteins, e.g., those between the 18 and 25 kDa markers on Fig. 1, decreased with prolonged incubation in the destain solution with all three dyes.

8. When taking a picture of the stained protein in a gel, it should be noted that gel documentation systems employ cameras or scanners with differing resolution. Additionally, some cameras and scanners use an algorithm to produce the black and white image. The algorithms differ for each software package. As a result, the computed intensity of the stained protein band may differ depending on the image capture equipment and software that is used. Comparing band intensity between images taken with different equipment and software is not valid unless a calibration is performed.

References

1. Wirth PJ, Romano A (1995) Staining methods in gel-electrophoresis, including the use of multiple detection methods. J Chromatogr A 698: 123–143
2. Merril CR (1990) Gel-staining techniques. Methods Enzymol 182:477–488
3. Schaffner W, Weissman C (1973) Rapid, sensitive, and specific method for determination of protein in dilute-solution. Anal Biochem 56: 502–514
4. Achilonu I, Goldring JPD (2010) Direct red 81 and amido black stain proteins in polyacrylamide electrophoresis gels within 10 min. Anal Biochem 400:139–141

Chapter 54

Zymographic Assay of Oxidases Using Peroxidase or Hemin Entrapped in Polyacrylamide Gel

Mircea Alexandru Mateescu, Carmen Calinescu, Tien Canh Le, Rodolfo Federico, and Bruno Mondovi

Abstract

This chapter describes a zymographic assay of oxidases which is based on a coupled peroxidase or hemin reaction. The enzymatic activity of oxidases (i.e., diamine oxidase/DAO, glucose oxidase, galactose oxidase) can be directly monitored on polyacrylamide gels containing horseradish peroxidase or hemin, in the presence of their specific substrates and *ortho*-phenylenediamine (OPDA), an oxidizable chromogen. In the presence of hydrogen peroxide, OPDA is oxidized to azo-aniline, which led to well-defined yellow-brown bands on gels, with intensities corresponding to the enzymatic activity of oxidases.

Key words: Zymographic assay, Entrapped peroxidase, Entrapped hemin, Polyacrylamide gel electrophoresis, Oxidases, *ortho*-Phenylenediamine

1. Introduction

Gel electrophoresis of proteins is a well-known technique, commonly used for analyzing proteins. In non-denaturing (native) polyacrylamide gel electrophoresis, the electrophoretic mobility is dependent on various factors such as the charge, the size, and the conformation of protein. Generally, in non-denaturing systems, enzymes often retain their biological activity (1), which can be detected on gels in the presence of their specific substrates, the products of the enzymatic reactions being converted into detectable final products on gels. This type of assay requires more than one substrate. The buffer selection is important because the charge on a native protein is pH dependent. Short migration is generally observed into the gel when the pH of the buffer is near the isoelectric point of the protein. In nonreducing sodium dodecyl sulfate

Biji T. Kurien and R. Hal Scofield (eds.), *Protein Electrophoresis: Methods and Protocols*, Methods in Molecular Biology, vol. 869, DOI 10.1007/978-1-61779-821-4_54, © Springer Science+Business Media, LLC 2012

polyacrylamide gel electrophoresis (SDS-PAGE), the proteins are negatively charged, and their electrophoretic mobility depends more on their hydrodynamic size. In the same time, the biological activity of the protein can also be affected by the presence of SDS, i.e., diminution or activation of enzymatic activity (2–4). However, the presence of SDS seems to afford a better resolution of the separation comparatively to native PAGE (2).

Hydrogen peroxide (H_2O_2) is the product of almost all oxidases that use oxygen as an acceptor, and a coupled peroxidase reaction is widely used to detect enzymes producing H_2O_2. Generally, the chromogenic peroxidase reaction is used as the last step in enzyme-linked detection systems. For this zymographic assay, it is of interest to mention that, to our knowledge, hemin, a nonenzymatic molecule, was for the first time used instead of peroxidase to detect enzymes producing H_2O_2 in zymographic assays. The development of potential enzyme mimics is one of the interesting trends in analytical biochemistry. Hemin is an iron protoporphyrin IX, a constituent of the active site of many hemoproteins, able to catalyze the reactions of many chromogenic agents with H_2O_2 (5). It is known that the peroxidase is one of the most widely used enzymatic reactions. However, when compared to hemin, it is expensive and relatively unstable. In addition, some preparations of peroxidase containing catalase as impurities can reduce the sensibility of the test, and such contamination cannot occur with hemin.

By using polyacrylamide gel electrophoresis, the protein pattern (staining with Coomassie Brilliant Blue) and the enzymatic activity (visualization of a zymographic pattern) can be monitored simultaneously. To be relevant, a protein band (stained with Coomassie Blue) must correspond to the enzyme of interest and must maintain an enzymatic activity on zymographic gels. The oxidase of interest can first be migrated by electrophoresis, and then, its enzymatic activity can be visualized via a coupled peroxidase or hemin reaction (Figs. 1–4). Several specific staining methods for detection of oxidase activities on gels, in the presence of peroxidase solution, have been reported (6, 7). In this chapter, we describe a zymographic method to detect the enzymatic activities of some oxidases on polyacrylamide (PAA) gels using the peroxidase or hemin immobilized in these gels, and OPDA, with the formation of a stable product, azo-aniline. This product is easily detected on gels, and its intensity is related to the amount of released H_2O_2 resulting from enzymatic activity of oxidases (2).

$$\text{Substrate for oxidase} + O_2 \longrightarrow \mathbf{H_2O_2}$$

$$\downarrow \text{Peroxidase/hemin} \mid \text{OPDA}$$

Azo-aniline
(yellow-brown compound)

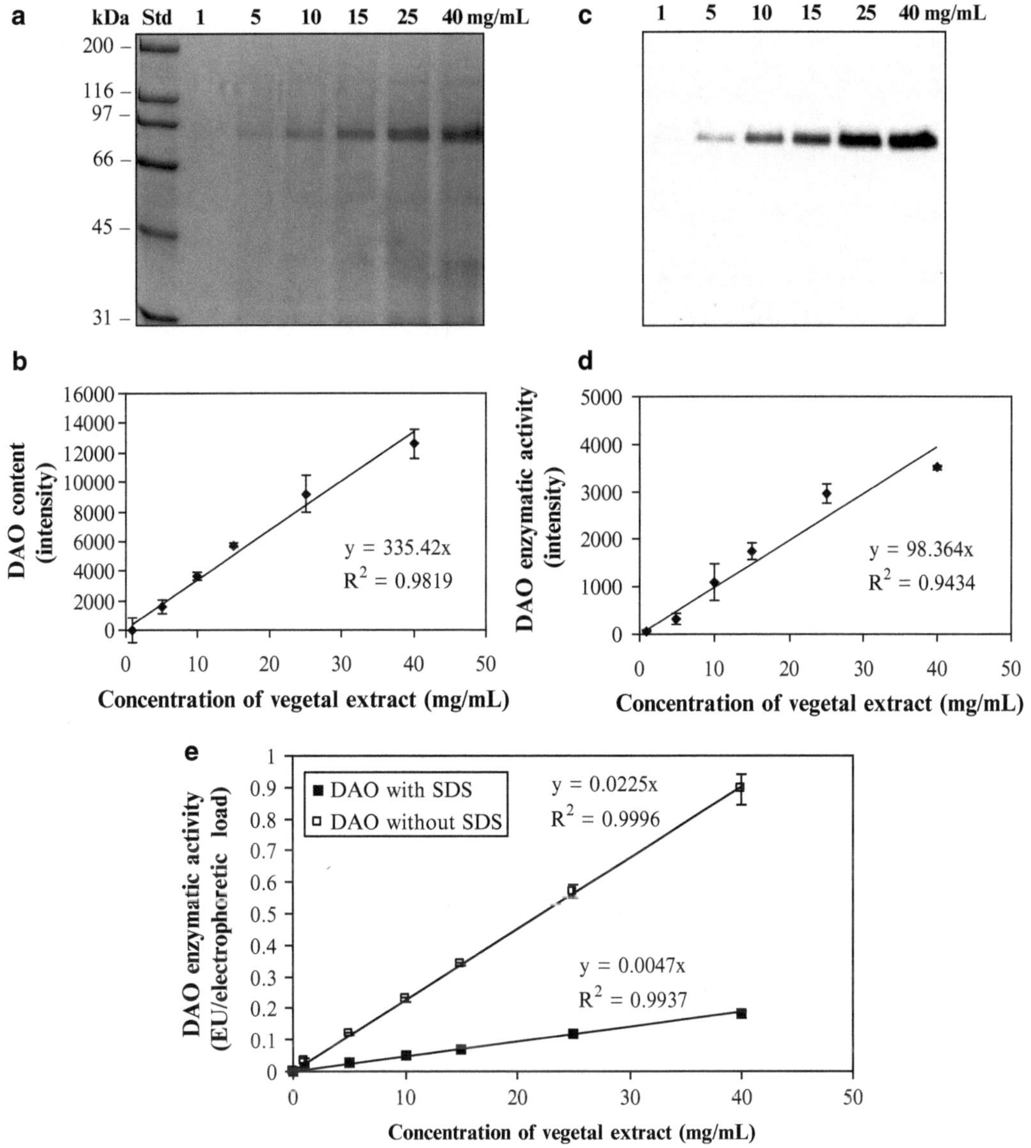

Fig. 1. Electrophoretic pattern of vegetal extract from *L. sativus* seedlings. *Coomassie staining* (**a**) on SDS-PAGE (8% PAA resolving gel without peroxidase) of different concentrations of DAO vegetal extract (1, 5, 10, 15, 25, 40 mg powder/mL), treated (1:1, v/v) with nonreducing SDS loading buffer (30 μL/well). The gels were stained with Coomassie Brilliant Blue G-250. *Image analysis* (**b**) of DAO bands intensity with the Quantity One program ($n=3$). *Zymographic DAO enzymatic activities* (**c**) of different concentrations of DAO vegetal extract (1, 5, 10, 15, 25, 40 mg powder/mL), treated (1:1, v/v) with nonreducing SDS loading buffer (30 μL/well). Oxidase activity was directly monitored on 8% PAA gels (containing entrapped peroxidase) after the electrophoretic run, in the presence of putrescine and *ortho*-phenylenediamine. *Image analysis* (**d**) of DAO bands intensity (enzymatic activity) with the Quantity One program. *Enzymatic activities of DAO* (**e**) on gel (EU/electrophoretic load of 30 μL), spectrophotometrically determined, in function of the vegetal extract concentrations. Protein content of vegetal extract: 15 ± 1.4 μg/mL (1 mg/mL), 132 ± 8 μg/mL (5 mg/mL), 302 ± 13 μg/mL (10 mg/mL), 435 ± 17 μg/mL (15 mg/mL), 719 ± 33 μg/mL (25 mg/mL), and 1,091 ± 27 μg/mL (40 mg/mL) ($n=3$). Reproduced from (2) with permission from Springer. One enzymatic unit of DAO is defined as the amount of enzyme catalyzing the oxidation of 1.0 μmol of putrescine per 10 min at pH 7.4, at 37°C.

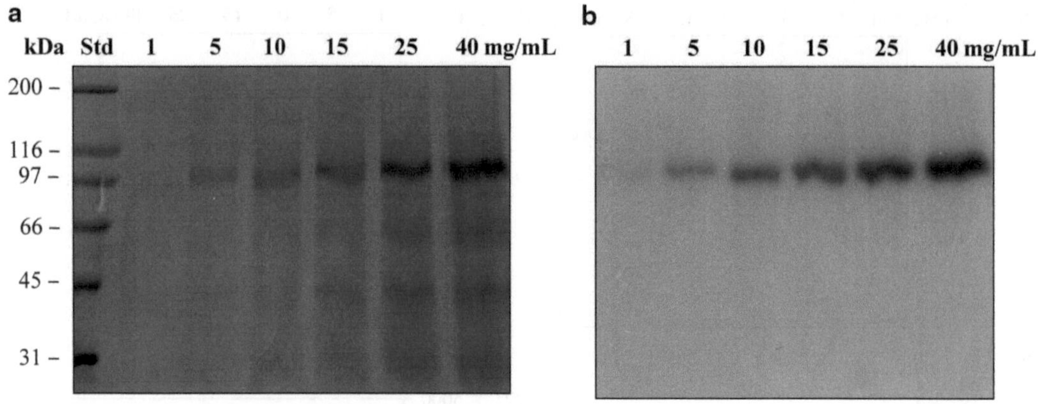

Fig. 2. Electrophoretic pattern of vegetal extract from *L. sativus* seedlings on polyacrylamide gels containing entrapped hemin. *Coomassie staining* (**a**) and *zymographic DAO enzymatic activities* (**b**). Different concentrations of DAO vegetal extract (1, 5, 10, 15, 25, 40 mg powder/mL PBS) were mixed (1:1, v/v) with nonreducing SDS loading buffer, and 30 μL of treated sample was deposited in every well. The DAO protein profile was monitored on 8% PAA resolving gel containing entrapped hemin by Coomassie staining. The oxidase activity was also directly monitored on 8% PAA resolving gels containing entrapped hemin in the presence of putrescine and *ortho*-phenylenediamine. The protein content and the enzymatic activities of vegetal extract were the same as in Fig. 1 for the mentioned concentrations.

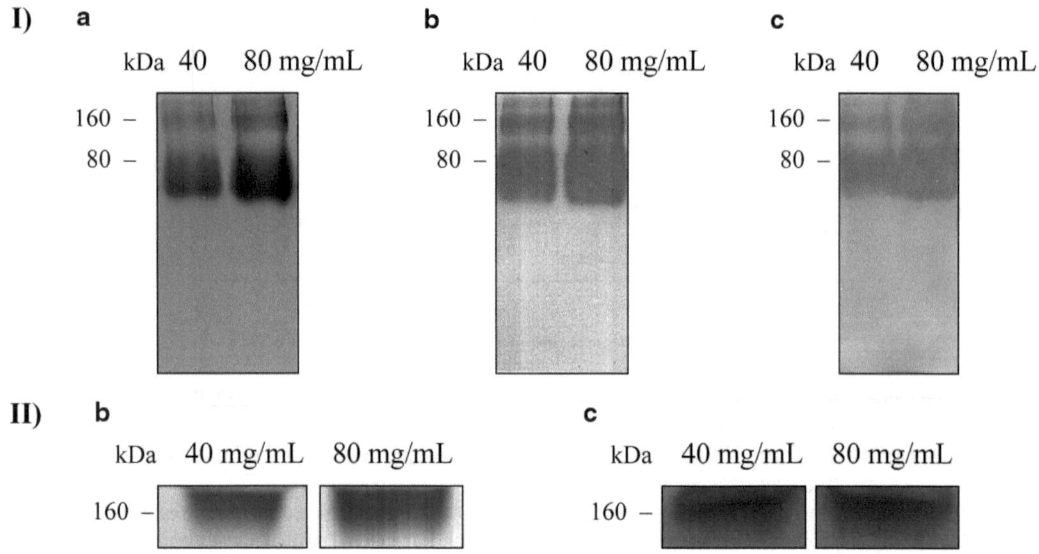

Fig. 3. Electrophoretic pattern of glucose oxidase. *Coomassie staining* (**a**) and *zymography* on PAA gels containing entrapped peroxidase (**b**) or hemin (**c**). Two different concentrations of glucose oxidase (40 and 80 mg powder/mL sodium acetate buffer) were mixed (1:1, v/v) with nonreducing SDS loading buffer, and 30 μL of treated sample was deposited in every well. The nonreducing SDS loading buffer contained Tris–HCl, pH 6.8 (I) or Tris–HCl, pH 5.5 (II), SDS, glycerol, and bromophenol blue, and the SDS electrophoresis buffer used had a pH of 8.6 (I) or 5.5 (II). The oxidase activity was directly monitored on 8% (I) or 6% (II) PAA resolving gels containing entrapped peroxidase/hemin in the presence of D-(+)-glucose and *ortho*-phenylenediamine. Glucose oxidase: 5,800 units/g solid. One enzymatic unit oxidizes 1.0 μmol of β-D-glucose to D-gluconolactone and H_2O_2 per min at pH 5.1 and 35°C.

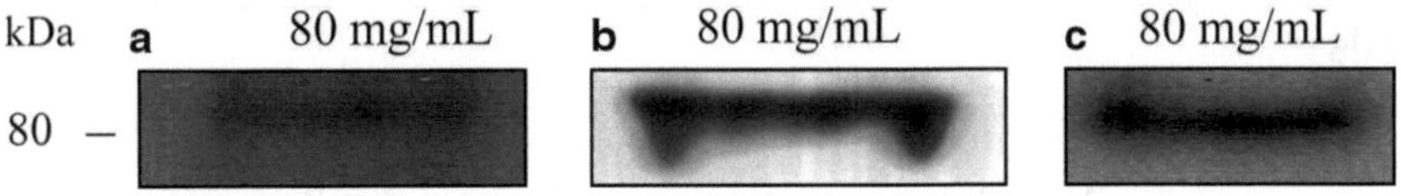

Fig. 4. Electrophoretic pattern of galactose oxidase. *Coomassie staining* (**a**) and *zymography* on 6% PAA gels containing entrapped peroxidase (**b**) or hemin (**a**, **c**). The galactose oxidase sample (80 mg powder/mL sodium acetate buffer) was centrifuged (1000 × *g*, 2 min), and the supernatant was mixed (1:1, v/v) with nonreducing SDS loading buffer. Then, 30 μL of the treated sample was deposited in every well. The galactose oxidase was directly monitored on PAA resolving gels containing entrapped peroxidase/hemin in the presence of D-(+)-galactose and *ortho*-phenylenediamine. Galactose oxidase: 3,685 units/g solid. One enzymatic unit produces a ΔA_{425} of 1.0 per min at pH 6.0 at 25°C, in a peroxidase and *ortho*-toluidine system.

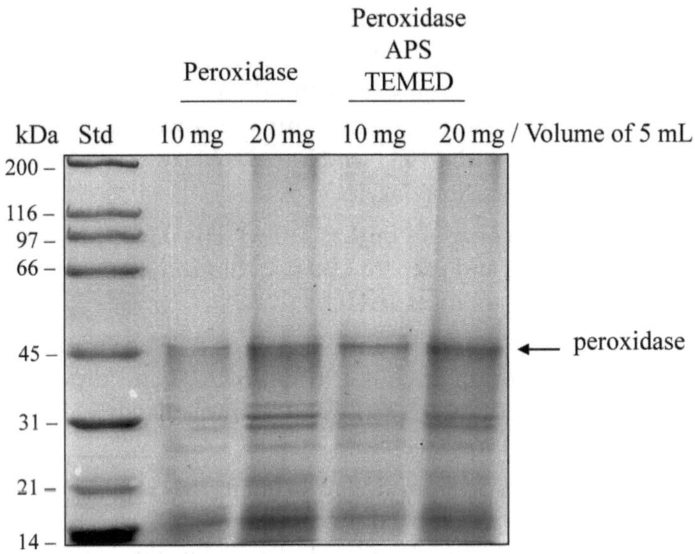

Fig. 5. SDS-PAGE of horseradish peroxidase in the absence and the presence of ammonium persulfate (APS) and *N,N,N′,N′*-tetramethyl-ethylenediamine (TEMED). The peroxidase (type I from horseradish, 96 purpurogallin units/mg solid) was incubated in the presence/absence of APS and TEMED for 2 h at room temperature. Every mixture sample containing peroxidase (10 mg powder/5 mL and 20 mg powder/5 mL) was mixed (1:1, v/v) with nonreducing SDS loading buffer (containing Tris–HCl, pH 6.8, SDS, glycerol, bromophenol blue) and then deposited on a 8% PAA gel (30 μL/well). After electrophoresis (nonreducing SDS-PAGE), the gel was stained with Coomassie Brilliant Blue G-250. No considerable difference between control and treated peroxidase was found.

The advantage of this zymographic assay is the immobilization of peroxidase or of hemin by entrapment into the PAA gel. Often, immobilization may cause changes in the structure of the enzyme and even steric impediments, depending on how the enzyme links to the support. In our case, under conditions of PAA radical polymerization, neither change of size nor degradation of peroxidase is observed (Fig. 5). For instance, the entrapped enzyme exerts its normal activity and accomplishes the chromogenic reaction of OPDA with H_2O_2. In addition, using PAA gel as support to entrap enzyme is a benefit because PAA is inert and nonionic, easily allowing the diffusion of reactants into gel immobilized enzyme.

2. Materials

All solutions were prepared using distilled water and analytical grade reagents. The reagents were kept at room temperature, unless indicated otherwise.

2.1. Sodium Dodecyl Sulfate Polyacrylamide Gel Electrophoresis (Nonreducing Conditions): Components

1. Stock acrylamide solution: 30% acrylamide/*N,N′*-methylene-*bis*-acrylamide solution (29.2 g acrylamide and 0.8 g *N,N′*-methylene-*bis*-acrylamide). Weigh 29.2 g of acrylamide monomer and 0.8 g *N,N′*-methylene-*bis*-acrylamide (cross-linker) (see Note 1). Make up to 100 mL with distilled water and filter. The solution is stable at 4°C for a few months (see Note 2).
2. Separating gel buffer: 1.5 M Tris–HCl, pH 8.8. Weigh 18.18 g Tris and transfer to 40 mL of water in a glass beaker. Add some more water and mix until the powder is totally dissolved. Adjust pH with HCl (see Note 3). Make up to 100 mL with distilled water. Store at 4°C.
3. Stacking gel buffer: 1.0 M Tris–HCl, pH 6.8. Weigh 12.12 g Tris and prepare 100 mL solution, as described in step 2. Store the solution at 4°C.
4. 10% Ammonium persulfate (APS) solution. Prepare 1 mL of solution (weigh 100 mg of APS and dissolve in 1 mL of distilled water). Store at 4°C immediately before use or freeze the solution (see Note 4).
5. 10% Sodium dodecyl sulfate (SDS). Prepare 100 mL of solution: Weigh 10 g of SDS powder (electrophoresis grade) and dissolve in 40–50 mL distilled water. Make up to 100 mL with water (see Note 5).
6. *N,N,N′,N′*-tetramethyl-ethylenediamine (TEMED). Store at 4°C.
7. Sample buffer/nonreducing SDS loading buffer (2×).

   Mix the following:

   (a) 1.2 mL Tris–HCl (1.0 M), pH 6.8

   (b) 4 mL SDS 10%

   (c) 2 mL glycerol

   (d) 1 mL bromophenol blue 0.04%, stock (see Note 6).

   Make up volume to 10 mL with distilled water (see Note 7).
8. Electrophoresis running buffer (SDS-PAGE): 0.025 M Tris base, 0.192 M glycine, and 0.1% SDS. Dissolve 3.0 g Tris base, 14.4 g glycine, and 1 g SDS in 800 mL distilled water. Adjust the pH to 8.8 and make up to 1 L with distilled water (see Note 8a).
9. Solution of fixation-methanol:acetic acid:water (50:10:40, v/v/v): Mix 500 mL methanol, 100 mL glacial acetic acid, and 400 mL distilled water to obtain 1 L of solution.

10. Protein staining solution-methanol:acetic acid:water (40:10:50, v/v/v) containing 0.5% Coomassie Brilliant Blue G-250: Weigh 1.25 g Coomassie Brilliant Blue G-250 and dissolve it in 100 mL methanol first. Then add 25 mL glacial acetic acid and 125 mL distilled water (final volume 250 mL) (see Note 9).
11. Destaining solution-methanol:acetic acid:water (20:5:75, v/v/v): Mix 200 mL methanol, 50 mL glacial acetic acid, and 750 mL distilled water to obtain 1 L of solution.
12. Molecular weight (MW) protein standards-Broad Range (Bio-Rad, Hercules, CA). Store at −20°C.

2.2. Non-denaturing Polyacrylamide Gel Electrophoresis (Native PAGE): Components

1. Stock acrylamide solution-30% acrylamide/*N*,*N*′-methylene-*bis*-acrylamide solution.
2. Separating gel buffer: 1.5 M Tris–HCl, pH 8.8.
3. Stacking gel buffer: 1.0 M Tris–HCl, pH 6.8.
4. 10% APS solution.
5. TEMED.
6. Sample buffer (2×).
7. Electrophoresis running buffer (native PAGE).
8. Solution of fixation: methanol:acetic acid:water (50:10:40, v/v/v).
9. Protein staining solution: methanol:acetic acid:water (40:10:50, v/v/v) containing 0.5% Coomassie Brilliant Blue G-250.
10. Destaining solution: methanol:acetic acid:water (20:5:75, v/v/v).
11. Kit for molecular weights 14,000–500,000. Non-denaturing PAGE (Sigma-Aldrich Chemical Company, St. Louis, MO, USA).

The preparation and the storage of some components used in non-denaturing PAGE (i.e., stock acrylamide solution, separating gel buffer, stacking gel buffer, APS solution, solution of fixation, protein staining solution, destaining solution) are the same as that described in Subheading 2.1.

In this section, only the components that are different are described.

Sample buffer (2×). Mix the following:

(a) 1.2 mL Tris–HCl (1.0 M), pH 6.8
(b) 2 mL glycerol
(c) 1 mL bromophenol blue 0.04%, stock (see Note 6). Make up to 10 mL with distilled water (see Note 7). *Electrophoresis running buffer (native PAGE)*: 0.025 M Tris base and 0.192 M glycine. Dissolve 3.0 g Tris base and 14.4 g glycine in 800 mL distilled water. Adjust the pH to 8.8 and make up to 1 L with distilled water (see Note 8b).

2.3. Enzymatic Activity of Oxidases: Components

1. Solutions of oxidases: Every oxidase has to be solubilized in its appropriate buffer (see Note 10). For example, diamine oxidase (DAO) from *Lathyrus sativus* seedlings, prepared from grass pea *Lathyrus sativus* seedlings (8), can be solubilized in a phosphate buffer solution (PBS, 10 mM, pH 7.4); glucose oxidase from *Aspergillus niger* and galactose oxidase from *Dactylium dendroides* (Sigma-Aldrich) can be solubilized in a sodium acetate buffer (50 mM, pH 5.5), etc.
2. Peroxidase solution (type I from horseradish, Sigma-Aldrich): 1 mg solid/mL (96 purpurogallin units/mg solid). Weigh 10 mg powder of peroxidase and dissolve it into 1 mL of cold distilled water. Before use, make a dilution 1:10 in distilled water (see Note 11). Store at 4°C.
3. Hemin solution (BioXtra, from porcine, Sigma-Aldrich, 1 mg solid/mL): Weigh 10 mg powder of hemin and dissolve into 1 mL of cold 1.5 M Tris–HCl buffer (pH 8.8). Before use, make a dilution 1:10 in the Tris–HCl buffer, pH 8.8 (see Note 11). Store at 4°C.
4. Solutions of substrates for oxidases (30 mM): Weigh the appropriate quantity of a specific substrate. For every specific substrate of oxidases, make up to 12.5 mL with the appropriate buffer in a 50-mL conical flask (i.e., for DAO, weigh 60.4 mg of putrescine dihydrochloride/12.5 mL buffer). The solutions of substrate can be kept at room temperature or at 4°C before utilization depending of their stability (see Note 12).
5. Solution of oxidizable dye-17 mM OPDA (Sigma-Aldrich): Weigh 38.47 mg OPDA and dissolve it into 12.5 mL of buffer, the same of that used at the preparation of substrate solutions. Store the solution at 4°C before utilization (see Note 12).

3. Methods

All the procedures indicated below can be carried out at room temperature, unless indicated otherwise.

3.1. Sodium Dodecyl Sulfate Polyacrylamide Gel Electrophoresis

This protocol is written for use with electrophoresis systems having a separate casting stand, such as Bio-Rad Protean II:

1. Before casting the gel, clean the internal surfaces of the gel plates, dry, and then assemble the plates and spacers to form a vertical gel mold cassette, using a flat surface to align the bottom of the spacers with the bottom edge of the glass (see Note 13).
2. Insert the cassette into the casting stand and make a leak test of the cassette by filling it with distilled water. If alignment is incorrect, it will leak (see Note 14).

3. Preparation of 8% separating gel containing peroxidase (5 mL separating gel solution containing 19.2 purpurogallin units of peroxidase) or hemin (5 mL separating gel solution containing 1 mL hemin, 1 mg/mL): Mix 1.3 mL of distilled water, 1 mL of peroxidase solution (1 mg/mL) or 1 mL of hemin solution (1 mg/mL), 1.3 mL of acrylamide mixture (30%), 1.3 mL resolving buffer (1.5 M Tris–HCl, pH 8.8), 50 μL SDS solution (10%), 50 μL APS solution (10%), and 3 μL TEMED in a 50-mL conical flask (see Note 15).
4. Preparation of 8% separating gel containing no entrapped peroxidase or hemin: Mix the same ingredients as indicated in step 3, by replacing 1 mL of peroxidase or hemin solution with 1 mL of distilled water (see Note 16).
5. To cast a vertical separating gel, fill the cassette with the separating gel solution high enough to allow for 1 cm of stacking gel beneath the teeth of the well-forming comb (see Note 17).
6. Overlay the gel solution with a thin layer (3 mm) of isobutanol/isobutanol:distilled water/distilled water. Allow to polymerize for 1 h.
7. Once the polymerization of separating gel is complete, remove the isobutanol/isobutanol:distilled water/distilled water gel layer (see Note 18).
8. Preparation of stacking gel (3 mL) containing no peroxidase or hemin. Mix 2.1 mL of distilled water, 500 μL of acrylamide mixture (30%), 380 μL stacking buffer (1.0 M Tris HCl, pH 6.8), 30 μL APS solution (10%), 30 μL SDS solution (10%), and 3 μL TEMED in a 50-mL conical flask (see Note 15).
9. Fill the cassette to the top with solution for the stacking gel. Rapidly, insert a well-forming comb into this solution and allow to polymerize (see Note 19).
10. Remove the well-forming comb after polymerization of the stacking gel (see Note 20). Transfer the gel cassette from the casting stand to the electrophoresis reservoir by fixing it into the electrophoresis reservoir.
11. Fill the electrophoresis reservoir with the electrophoresis running buffer, ensuring that the buffer fully fills the sample loading wells (see Note 21). Check that there are no leaks.
12. Preparation of the samples containing the oxidase of interest: Mix 1:1 (v/v) the oxidase samples (i.e., DAO samples from *Lathyrus sativus* vegetal extract, glucose oxidase samples from *Aspergillus niger*, or galactose oxidase from *Dactylium dendroides*) with the sample buffer, containing Tris–HCl, glycerol, bromophenol blue, and SDS. The mixtures do not contain *beta*-mercaptoethanol. Do not heat the mixtures.

13. Preparation of molecular weight (MW) protein standards: Dilute the standards 1:20 in SDS reducing sample buffer (see Note 22), heat for 5 min at 95°C, then cool and load 10 μL/well to the PAA gel, as indicated in instruction sheet from Bio-Rad.
14. Load 30 μL of the treated samples into the wells (see Note 23). Each sample has to be run on two PAA gels: one gel with entrapped peroxidase or hemin for zymography (detection of enzymatic activity) and another with or without peroxidase or hemin for protein pattern (Coomassie staining).
15. Connect the electrodes to the power supply and start the electrophoresis (room temperature, 120 V) until the bromophenol blue dye reaches the bottom of the gel (see Note 24a, b). In the presence of SDS, as the proteins are negatively charged, the separation will proceed toward the positive electrode (anode).
16. When the bromophenol blue dye reaches the bottom of the gel, the electrophoresis is completed. Stop the electrophoresis system, dismantle the apparatus, and open the gel plates. Remove the stacking gels and keep one *separating gel with or without peroxidase/hemin* for fixing, Coomassie staining/destaining, and one *separating gel containing entrapped peroxidase/hemin* for detection of enzymatic activity.

3.2. Native Polyacrylamide Gel Electrophoresis

The procedure for native PAGE is similar with that of SDS-PAGE. Therefore, the description of similar steps will not be repeated. Hereto, we described only the separating and stacking gels preparation and some other aspects that are different:

1. Preparation of 8% separating gel containing entrapped peroxidase (5 mL separating gel solution containing 19.2 purpurogallin units of peroxidase) or entrapped hemin (5 mL separating gel solution containing 1 mL hemin, 1 mg/mL). Mix 1.35 mL of distilled water, 1 mL of peroxidase or hemin solution (1 mg/mL), 1.3 mL of acrylamide mixture (30%), 1.3 mL resolving buffer (1.5 M Tris–HCl, pH 8.8), 50 μL APS solution (10%), and 3 μL TEMED in a 50-mL conical flask (see Note 15).
2. Preparation of stacking gel (3 mL) containing no peroxidase or hemin. Mix 2.13 mL of distilled water, 500 μL of acrylamide mixture (30%), 380 μL stacking buffer (1.0 M Tris–HCl, pH 6.8), 30 μL APS solution (10%), and 3 μL TEMED in a 50-mL conical flask (see Note 15).
3. Fill the electrophoresis reservoir with the electrophoresis running buffer without SDS, ensuring that the buffer fully fills the sample loading wells. Check that there are no leaks.
4. Preparation of the samples containing the oxidase enzyme: Mix 1:1 (v/v) the oxidase samples with the sample buffer,

containing Tris–HCl, glycerol, and bromophenol blue. The mixtures do not contain SDS and *beta*-mercaptoethanol. Do not heat the mixtures.

5. Preparation of molecular weight protein standards: Reconstitute the contents of each vial of protein marker kit with 1 mL of water, with exception of urease (which is reconstituted with 0.5 mL of water and 0.5 mL of glycerol). Keep at -20°C. Immediately before use, thaw an aliquot of each prepared protein marker and dilute with an equal volume of sample buffer, then load on gel (see Note 25).

3.3. Coomassie Blue Protein Staining

1. After the electrophoresis is finished, incubate the gels into the fixation solution (methanol:acetic acid:water, 50:10:40) for 30 min under mild agitation.
2. Stain the gels with 0.5% Coomassie Brilliant Blue G-250 (in a methanol:acetic acid:water solution, 40:10:50) for 1 h under mild agitation.
3. Destain the gels with methanol:acetic acid:water solution (20:5:75) until visible protein bands appear on gels.

3.4. Detection of Oxidase Enzymatic Activity Bands on PAA Gels

1. At the end of electrophoresis, rinse rapidly the gels containing entrapped peroxidase or hemin with distilled water (see Note 26).
2. Place the gels into the oxidase substrate solution (i.e., 30 mM putrescine for plant DAO, 30 mM D-(+)-glucose for glucose oxidase). After a few seconds, add an equal volume of a 17 mM OPDA solution.
3. Incubate the gels, containing entrapped peroxidase or hemin and electrophoretically separated oxidase samples, into the two solutions (substrate and oxidizable dye) for minimum 1 h, under dark conditions and weak agitation (see Note 27).
4. The oxidase enzymatic activity is detected on PAA zymographic gels in function of newly produced H_2O_2 (substrate of peroxidase) and in the presence of OPDA (co-substrate of peroxidase).

4. Notes

1. Acrylamide monomer is a *neurotoxin*. Take care when handling the powder by wearing gloves and avoiding breathing the acrylamide dust (wear an appropriate face mask).
2. The acrylamide solution is usually kept at 4°C in brown bottles or in bottles wrapped with aluminum foil. In our laboratory, we use an acrylamide solution which is prepared every month or more frequently.

3. A solution of HCl (6 M) can be used to adjust the pH of the solution. It should be added in small volumes to avoid a sudden drop in pH below the required value.
4. The solution of APS can be prepared each time before utilization or can also be frozen (−20°C). We did not observe significant differences between the two cases.
5. The SDS solution has to be kept at room temperature. At low temperature, the SDS comes out of solution. In this situation, the bottle containing the SDS can be simply warmed up in a water bath.
6. Bromophenol blue stock solution: Dissolve 40 mg bromophenol blue powder in 100 mL distilled water. This solution stock can be kept at room temperature.
7. Samples already in solution have to be diluted accordingly with 2× sample buffer to obtain a final solution that is 1× sample buffer. Solid samples can be dissolved directly in 1× sample buffer. The sample buffer (nonreducing SDS loading buffer) may contain Tris–HCl of different pH (i.e., for glucose oxidase, a Tris–HCl with a pH of 5.5 was used).

8a. A simple way is to prepare a solution 10× of electrophoresis running buffer (0.25 M Tris base, 1.92 M glycine, and 1% SDS). Weigh 30 g Tris base, 144 g glycine, and 10 g SDS and add the ingredients in the same order to 800 mL distilled water. Then, make up to 1 L with water. Care should be taken when working with this solution as, because of the presence of SDS, it makes bubbles. Before use, dilute 10× electrophoresis running buffer with distilled water (final pH 8.8). At pH 8.8 (the pH of the separating gel), most proteins have a negative charge, and they will run to the anode (positive electrode). Differently, the basic proteins will migrate in the opposite direction and will not be detected on the gels, because they are lost. In this case, different conditions that are not described here have to be used (i.e., migration under acidic conditions). In some cases, the pH of the electrophoresis running buffer can be different than 8.8 (i.e., pH 5.5 for glucose oxidase, to confer more stability, as it is known that under denaturing SDS conditions, the subunits of glucose oxidase dissociate accompanied by the loss of cofactor FAD (9)).

8b. A solution 10× of electrophoresis running buffer containing 0.25 M Tris base and 1.92 M glycine (without SDS) can also be prepared as described in Note 8a.

9. Because the dye solubility can be a problem, it is indicated to dissolve the Coomassie Brilliant Blue in the methanol first, then to add the acid and water. The solution can also be filtered. The same solution can be used several times, but better results can be obtained with fresh solutions.

10. If it is necessary, the enzyme solution can be filtered or centrifuged before use (i.e., galactose oxidase from *Dactylium dendroides* is not completely soluble in sodium acetate buffer, pH 5.5, and the solution has to be centrifuged at 1000 × *g* for 2 min).
11. It is better to prepare peroxidase/hemin fresh solutions immediately before use. During the preparation of resolving gels, 1 mL of horseradish peroxidase or hemin (1 mg/mL) has to be added to the gel solutions prior to polymerization for a final volume of 5 mL gel (i.e., 19.2 purpurogallin units of peroxidase/5 mL of resolving gel). It is recommended for either entrapped peroxidase or hemin to have a high degree of purity.
12. In some cases, the solutions of substrate have to be prepared some time before their utilization (i.e., the solution of glucose, the substrate of glucose oxidase, has to stand 2 h at 25°C to allow mutarotation). Differently, the solution of OPDA (dihydrochloride salt form) can be prepared immediately before use and kept in the dark at 4°C. The gels have to be completely immersed into 12.5 mL of oxidase substrate and completed with 12.5 mL of OPDA (a final volume of 25 mL to well cover one gel).
13. The clips holding the plates and spacers together cannot be removed until the electrophoresis is complete.
14. Most leakage occurs because the spacers are not completely flush with the edges of the glass at the bottom. The cassette has to be made leakproof by realignment. If this does not work, try to grease the spacers with a nonconductive grease (i.e., Celloseal). Check the edges of electrophoresis plates since chipped edges may cause the cassette to leak.
15. It is better to degas the acrylamide solution (containing no initiator), because gas bubbles can be sometimes a problem during the polymerization. The degassing of acrylamide solution can take a few minutes. The initiator of polymerization that has been omitted can then be added, and the final mixture can be introduced into cassette. The percentage of the separating gel has to be chosen in function of the size of the oxidase of interest, maintaining the same concentration of peroxidase or hemin/resolving gel (see Note 11). For the oxidases with high or very high molecular weight, the migration on PAA gels could be difficult. Generally, gels in the range of 5% should separate proteins with MW > 150,000, and gels of 5–10% should separate proteins with MW between 20,000 and 150,000, whereas gels of 10–15% should separate proteins with MW between 10,000 and 80,000.
16. As the protein migration patterns (Coomassie staining) were the same on gels containing or not entrapped peroxidase or hemin, gels containing entrapped peroxidase or hemin will be used only for the evaluation of enzymatic activity of oxidases.

17. After adding the separating gel solution until it reaches 1 cm from the bottom of the well-forming comb, an excess solution can remain in your flask. Keep this remaining solution in order to see when polymerization is done.
18. When polymerized, a change in refractive index between the polymerized gel and the isobutanol/isobutanol:water/water layer will be observed. The surface of polymerized gel has to be well washed with distilled water to completely remove any trace of isobutanol that could interfere with the polymerization of stacking gel. The resolving gel containing hemin has to be kept longer for polymerization than the resolving gel containing peroxidase.
19. A well-forming comb has to exclude enough air to permit polymerization all the way to the top of the cassette. Refractive index changes around the comb indicate that the gel is polymerized.
20. After removing the well-forming comb from the stacking gel, rinse the wells with distilled water to completely remove any non-polymerized acrylamide solution.
21. When determining enzymatic activities, use the non-denaturing conditions of electrophoresis (gels and electrophoresis running buffer prepared without SDS).
22. Prepare the SDS reducing sample buffer immediately before use by adding 25 μL *beta*-mercaptoethanol to 475 μL stock sample buffer (4.8 mL distilled water, 1.2 mL 0.5 M Tris–HCl pH 6.8, 1.0 mL glycerol, 2.0 mL 10% SDS, 0.5 mL 0.1% bromophenol blue).
23. Because of the presence of glycerol in the sample buffer, the density of the sample will increase, causing it to sink to the bottom of the well. A volume of 10 μL to 20 μL/well is more appropriate for most gels.

24a. Maintenance of pH 6.7–6.8 is important for proteins to stack properly. If the sample is not buffered to pH 6.8 by addition of sample buffer alone, then the sample has to be neutralized first. A clear indication that a sample is acidic is that the solution turns from blue to yellow when adding the sample buffer.

24b. If the proteins are separated only in the top of the gel, the electrophoresis has to be carried out for more than 1 h.

25. Concerning the molecular weight protein standards, we suggest to use the kit for non-denaturing PAGE: molecular weights 14,000–500,000 (Sigma-Aldrich). The sample buffer used to dilute each protein marker should be prepared as described by Sigma-Aldrich: combine one volume Tris buffer (dilute 1 M Tris–HCl 20-fold with water to yield 50 mM

Tris–HCl buffer), one volume glycerol, and one volume of 0.04% (w/v) bromophenol blue solution.

26. It is important to rinse the gels before the detection of enzymatic activity to eliminate all traces of SDS and other components that remain from electrophoresis run.
27. The temperature of incubation is chosen in function of the studied oxidase, to afford the optimal conditions (i.e., for plant DAO, the optimal temperature for its enzymatic activity is 37°C).

References

1. Walker JM (2002) Nondenaturing polyacrylamide gel electrophoresis of proteins. In: Walker JM (ed) The protein protocols handbook, 2nd edn. Humana Press, New Jersey
2. Calinescu C, Federico R, Mondovi B et al (2010) Zymographic assay of plant diamine oxidase on entrapped peroxidase polyacrylamide gel electrophoresis. A study of stability to proteolysis. Anal Bioanal Chem 396:1281–1290
3. Moore BM, Flurkey WH (1990) Sodium dodecyl sulfate activation of a plant polyphenoloxidase. J Biol Chem 265:4982–4988
4. Schrock HL, Gennis RB (1977) High affinity lipid binding sites on the peripheral membrane enzyme pyruvate oxidase. J Biol Chem 252:5990–5995
5. Brusova Z, Magner E (2009) Kinetics of oxidation of hydrogen peroxide at hemin-modified electrodes in nonaqueous solvents. Bioelectrochemistry 76:63–69
6. Houen G, Leonardsen L (1992) A specific peroxidase-coupled activity stain for diamine oxidases. Anal Biochem 204:296–299
7. Lee MH, Chuang MT, Hou WC (2002) Activity staining of plasma amine oxidase after polyacrylamide gel electrophoresis and its application to natural inhibitor screening. Electrophoresis 23:2369–2372
8. Federico R, Cona A, Caliceti P et al (2006) Histaminase PEGylation: preparation and characterization of a new bioconjugate for therapeutic application. J Control Release 115:168–174
9. Jones MN, Manley P, Wilkinson A (1982) The dissociation of glucose oxidase by sodium *n*-dodecyl sulphate. Biochem J 203:285–291

Chapter 55

Coomassie Brilliant Blue Removal/Disposal from Gel Destain and Used Gel Stain in an Environment-Friendly Manner

Biji T. Kurien and Yaser Dorri

Abstract

Toxic reagents are employed to destain Coomassie Brilliant Blue (CBB) stained gels. We tested the efficacy of various paper adsorbents in adsorbing CBB released from gels during destaining. Kimwipes were the most efficient, followed by Teri towels, multifold towels, and Whatman (numbers 1 and 3) filter papers. Three Kimwipes added during destaining of a CBB-stained mini-gel helped adsorb the released dye. Thus, stain removal with Kimwipes helps reduce destain use and organic waste accumulation, enables recycling of nonradioactive destaining solution, and is 7.5-fold cheaper than an available method for CBB disposal. Next, we used Kimwipes to deplete the dye from a used CBB staining solution awaiting proper disposal by our Institutional Safety Office. Seventy-five Kimwipes successfully helped remove the dye from a 0.05% CBB staining solution in 5 to 10 min. The blue-colored Kimwipes did not release the stain even when squeezed dry after incubation in various salts, water, or acid solutions for five weeks. The CBB removed thus can be simply disposed as solid waste and will not leach out from solid landfills. Kimwipes, thus, enables CBB disposal in an environmentally friendly manner and allows recycling of destaining solution.

Key words: SDS-PAGE, 2-Dimensional gel electrophoresis, Coomassie Brilliant Blue, Gel destaining, Kimwipes, Environment-friendly destaining

1. Introduction

Laboratories involved in proteomics commonly employ sodium dodecyl sulfate polyacrylamide gel electrophoresis (SDS-PAGE) (1) and two-dimensional gel electrophoresis (2-DE) (2) to analyze proteins. Proteins separated by SDS-PAGE/2-DE are usually visualized by Coomassie Brilliant Blue (CBB) stain due to its simplicity and reliability (3). Methanol (4, 5) or isopropanol (6), along with acetic acid fixative, is commonly used to destain CBB-stained gels. As the gel loses its excess stain, the destaining solution becomes

Biji T. Kurien and R. Hal Scofield (eds.), *Protein Electrophoresis: Methods and Protocols*, Methods in Molecular Biology, vol. 869, DOI 10.1007/978-1-61779-821-4_55, © Springer Science+Business Media, LLC 2012

colored and normally needs to be replaced with fresh destaining solution multiple times for efficient destaining. CBB has been reported to be toxic (7) and requires expensive special treatment for disposal. Similarly, acetic acid and methanol are also toxic and cannot be disposed into the sink. The fact that these toxic reagents can ultimately get into drinking water supplies poses an environmental hazard. Disposal of small quantities of 10% acetic acid is allowed, following neutralization to pH 7 (http://www.scholarchemistry.com/msds/Acetic_Acid_10pct.pdf).

Special waste disposal procedures are recommended for even buffers with only 10% methanol, which are thought to be hazardous (8). Procedures to dispose of these reagents are expensive and time-consuming.

We have used a paper product (9) to safely eliminate CBB from the destain solution in a simple fashion that allows acetic acid and methanol to be recycled. This work resulted from the observation that the inclusion of two multifold towels in the destaining solution helped to adsorb most of the stain. Subsequently, we compared the efficiency of stain removal of two multifold towels with equivalent weights of Kimwipes, Whatman numbers 1 and 3 filter papers, and Teri towels. Kimwipes were found to be the most efficient, followed by Teri towels, multifold towels, Whatman number 1, and Whatman number 3 in decreasing order.

Charcoal cartridges have been employed for removing anionic detergents and CBB (10). In addition, activated charcoal has been employed to adsorb stain. But, this procedure (http://www.p2000.umich.edu/chemical_waste/cw4.htm) seems cumbersome compared to our procedure. In addition, there is the added disadvantage of particulate release when using activated charcoal (11). Cozap, a spongelike material, is packed in a Mylar envelope and sold commercially for CBB stain removal by adsorption. It (http://www.bioexpress.com/index.html? wscdet_show=000000000100114000114050) costs about $3.12 per piece, with each piece reportedly capable of destaining 10 mini-gels at the cost of about 30 cents/gel. (http://www.the-scientist.com/article/display/17674/). However, it costs slightly less than 4 cents/gel to destain gels using Kimwipes. Thus, destaining with Kimwipes is 7.5-fold cheaper compared to Cozap, as well as being easier and cleaner.

Then, we also investigated the ability of Kimwipes to remove the dye from a used CBB staining solution. Laboratories routinely collect used CBB stain as organic waste, following gel staining, for proper disposal by the Institutional Safety Office. A licensed professional waste disposal service is to be contacted, according to regulations, to dispose such material. Disposal regulations stipulate (in the United States and Europe) that this combustible material may be burned in a chemical incinerator equipped with an afterburner and a scrubber following all federal, state, and local environmental rules (http://www.teknova.com/v/vspfiles/files/

MSDS/C1040.pdf). Such methods are time-consuming and expensive. Kimwipes were found to efficiently remove nearly all of the dye from the used stain in 5–10 min in a manner similar to the removal of CBB from the destaining solution. The CBB removed thus remains tightly bound to the Kimwipes and does not leach out in 1 M salts, water, or acidic solutions even after lengthy incubation. This procedure allows the disposal of CBB as solid waste and thus helps to keep the environment safer, with little or no chance for toxic reagents and chemicals to contaminate drinking water reservoirs. Removal of CBB from nonradioactive destaining solution in this fashion permits it to be recycled for further use.

2. Materials

Prepare all solutions using ultrapure water (prepared by purifying deionized water, to attain a sensitivity of 18 MΩ cm at 25°C) and analytical grade reagents. Prepare and store all reagents at room temperature (unless indicated otherwise). Diligently follow all waste disposal regulations when disposing waste materials. We do not add sodium azide to reagents.

1. 10% SDS-PAGE gels (1-mm thickness).
2. CBB stain stock: 0.25% CBB in water (see Note 1).
3. CBB stain (0.05%): Dilute 0.25% CBB stain with water to obtain 0.05% stain.
4. CBB stain: 0.25% CBB in 25% methanol/10% acetic acid (see Note 2).
5. Destaining solution: 25% methanol/10% acetic acid.
6. Prestained molecular weight protein standards.
7. HeLa cell extract.
8. SDS-PAGE running buffer: 0.025 M Tris, 0.192 M glycine, 0.1% SDS, pH 8.3 (see Note 3).
9. SDS lysis buffer (5×): 0.3 M Tris–HCl (pH 6.8), 10% SDS, 25% β-mercaptoethanol, 0.1% bromophenol blue, 45% glycerol. Leave one aliquot at 4°C for current use and store remaining aliquots at −20°C (see Note 4).
10. Bromophenol blue (BPB) solution: Dissolve 0.1 g BPB in 100 mL water.
11. Phosphate buffered saline (PBS), pH 7.4.
12. Branson Sonifier Cell Disruptor 185 (VWR Scientific Co., Boston, MA, USA).
13. FB300 power supply (Fisher Scientific, Houston, TX, USA).
14. Kimwipes (Kimberly–Clarke, Roswell, GA, USA).

15. Tradition multifold towels (Kimberly–Clarke).
16. Whatman filter paper (numbers 1 and 3).
17. Teri towels (see Note 5).
18. Fisher Scientific balance (XE series, model XL100).

3. Methods

All procedures are carried out at room temperature unless otherwise specified.

3.1. Determining CBB Adsorption by Different Paper Adsorbents

1. Take 250 mL of water in each of five identical plastic containers.
2. Add 5 mL of 0.25% CBB solution (dissolved in water) to each container and mix well. Save an aliquot (1 mL) from each container to measure zero time optical density of the diluted CBB solution at 560 nm
3. Weigh two Tradition multifold towels in a Scientific balance. Weigh equal weights (see Note 6) of Kimwipes (~10.5 counts), Teri towels (approximately half), and Whatman filter papers (numbers 1 and 3).
4. Add each paper product to one of the five containers and incubate with gentle shaking at room temperature for 90 min (see Note 7).
5. Aliquots (1 mL) were taken from each container at periodic intervals and finally at the end of 90 min (see Table 1).
6. Read optical density of the saved samples at 560 nm using a Spectronic Genesys 5 Spectrophotometer (see Table 1).

3.2. Destaining CBB-Stained SDS-PAGE Mini-Gel with Kimwipes

1. Electrophorese aliquots of HeLa cell extract (see Note 8) and bovine serum albumin on 10% (1 mm thick) SDS-PAGE gel.
2. Stain with CBB stain in a plastic container (see Note 9). Remove the used CBB stain into a container employed for collecting used CBB.
3. Rinse the CBB off the gel with few mL of ultrapure water prior to adding destaining solution (see Note 10).
4. Insert the CBB-stained gel between the folds of three Kimwipes (see Fig. 1 and Note 11) and place the gel-Kimwipe sandwich into another plastic container.
5. Add 200 mL of destaining solution into the container containing the gel-Kimwipe sandwich. Close the container tightly and destain the gel at room temperature with gentle shaking (see Note 12). The Kimwipes will readily adsorb all the stain released by the gel (see Fig. 1).

Table 1
CBB adsorption from a solution containing excess CBB by various paper products. The CBB was dissolved in ultrapure water only. Since there was no gel involved, we did not use either acetic acid or methanol. The residual amount of stain remaining after incubation with the different paper products at the different timings was measured at 560 nm. The optical density of the stain prior to the addition of paper adsorbents is shown at zero time

Optical density at 560 nm[a]

Product	0 min	10 min	30 min	90 min
3 M Whatman	1.75	1.62	1.61	1.6
1 M Whatman	1.75	1.55	1.55	1.53
Multifold towels	1.77	1.26	1.06	0.84
Teri towels	1.76	0.6	0.39	0.25
Kimwipes	1.67	0.36	0.114	0.04

Reproduced from (9) with permission from Elsevier Limited, Oxford, OX5 1 GB, UK
[a]Values are means of three determinations

3.3. Effect of Prolonged Incubation of CBB-Stained Kimwipes in Water and Salt Solutions

1. Prepare CBB-stained Kimwipes by incubating fresh Kimwipes in 0.05% CBB (see Note 1).
2. Rinse the intensely blue-colored Kimwipes in water to remove any nonadherent color.
3. Incubate CBB-stained Kimwipes in water for 5 weeks at room temperature.
4. Incubate CBB-stained Kimwipes also in 1 M sodium bicarbon ate, 1 M sodium chloride, or 1 M sodium phosphate (dibasic anhydrous) for 5 weeks at room temperature.
5. Wring the Kimwipes dry at the end of experimental period (see Note 13).

3.4. CBB Removal from an Experimental Gel-Staining Solution with Kimwipes

1. Pour 250 mL of 0.05% CBB stain (dissolved in water) into a 1-L beaker.
2. Add 15 Kimwipes into the solution in the beaker, one at a time in quick succession.
3. Swirl the Kimwipes in the solution with a glass rod for a minute or less.
4. Remove the colored Kimwipes from the beaker and squeeze dry. Repeat this procedure four more times, each time with 15 fresh Kimwipes (see Notes 13–15 and Fig. 2).

Fig. 1. Destaining of CBB-stained gel using Kimwipes and destain solution. (**a**) The CBB-stained gel is positioned over one-half of Kimwipes (3 sheets) in destaining solution. (**b**) The gel is covered completely with the other half of Kimwipes. (**c**) Stained gel inserted between Kimwipes prior to destaining. (**d**) Kimwipes with CBB adsorbed from the gel. Reproduced from (9) with permission from Elsevier Limited, Oxford, OX5 1 GB, UK.

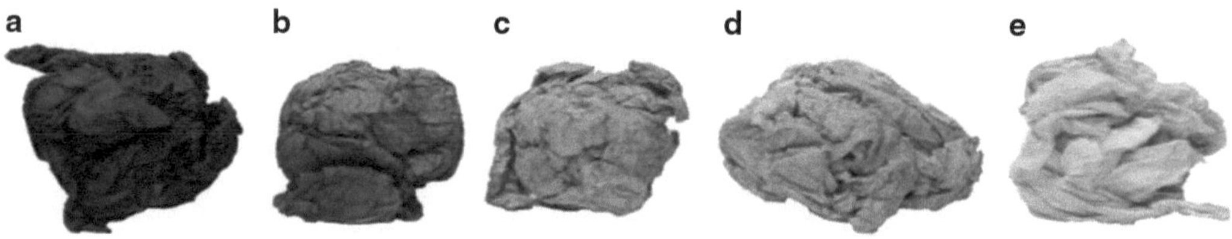

Fig. 2. Rapid CBB removal from gel stain using Kimwipes. Kimwipes can be employed to deplete CBB from used gel stain earmarked as organic waste for proper disposal by the Institutional Safety Office. Kimwipes (15 numbers used each time) were dipped into a 1-L glass beaker containing 250 mL of 0.05% CBB dissolved in water. Each time, they were removed immediately with a glass rod and wrung dry (**a**–**e**). The majority of the stain was adsorbed by the first set of Kimwipes (**a**). Kimwipes shown in B-E can be saved for adsorbing more CBB. Reproduced from (9) with permission from Elsevier Limited, Oxford, OX5 1 GB, UK.

4. Notes

1. Prepare CBB without methanol or acetic acid since this solution is to be used to test the adsorption capacity of various paper products such as Kimwipes, Teri towels, multifold paper towels, and 1- and 3-M Whatman filter papers. Filter (0.22-μm cellulose acetate membrane filter) the reagent to remove any insoluble CBB particles. This solution is not to be used for staining gels.
2. Weigh the required amount of the dye and dissolve it in methanol first. Filter the dissolved CBB through a circular Whatman 1-M filter paper folded into a cone and fitted inside a glass funnel. For a quicker way to filter the CBB solution, layer a circular 1-M Whatman filter paper (125-mm-diameter circles) inside a Buchner funnel (Coors, USA; 135-mm inner diameter). The filter paper should cover all the holes in the funnel and lay flat inside the funnel. Attach the Buchner funnel to a side-arm fitted conical flask that is attached to the house vacuum through the side-arm. Pour the CBB solution into the Buchner funnel and apply vacuum. If filter paper gets clogged, replace with a fresh one and filter. Collect the filtered CBB. Adjust volume with water, methanol, and acetic acid to obtain a 0.05% Coomassie in 25% methanol/10% acetic acid. Make sure to add acetic acid to water and not vice-versa.
3. Simple method of preparing SDS-PAGE running buffer: Prepare 10× native buffer (0.25 M Tris, 1.92 M glycine). Weigh 30.3 g Tris and 144 g glycine, mix and make it to 1 L with water. Do not adjust pH. Dilute 100 mL of 10× native buffer to 990 mL with water in a graduated cylinder and add 10 mL of 10% SDS. Care should be taken to add SDS solution last since it makes bubbles. Mix the contents of the graduated cylinder using a magnetic stirrer or manually by hand (cap the graduated cylinder with Parafilm and mix by inverting several times).
4. SDS precipitates at 4°C. Therefore, the lysis buffer needs to be warmed prior to use.
5. Kimwipes, Tradition multifold paper towels, Whatman paper, and Teri towels can be purchased from Fisher Scientific Co., Dallas, TX, USA.
6. Weigh enough Kimwipes to equal the weight of two Tradition multifold towels. Weigh the other paper products similarly.
7. Kimwipes (about 10) was found to be the most efficient, adsorbing almost the entire stain in 90 min. It took up more than three-fourths of the stain in about 10 min (see Table 1). An increase in surface area (e.g., that provided by ~10 Kimwipes) combined with the good adsorption property of

Kimwipes appears to be the reason for the efficient adsorption of the CBB stain. The order of adsorbing efficiency was Kimwipes > Teri towels > multifold towels > Whatman No. 1 > Whatman No. 3. Whatman filters did not take up any significant stain and also left behind some turbidity (particularly Whatman No. 1 filter paper) resulting from apparent onset of disintegration of the filter paper with increase in incubation time.

8. Preparation of HeLa cell extract: Harvest freshly cultured HeLa cells by centrifuging at 800 × *g* and wash twice with PBS. Lyse cells by sonication in PBS buffer using a Branson sonicator (setting 4) and centrifuge at 10, 000 × *g* for 10 min. Use an aliquot of the supernatant for SDS-PAGE. Sonicators generate high-frequency sound waves in the 20,000 Hz range, outside our normal range of hearing. These sound waves can cause hearing damage, and therefore, laboratory personnel should wear sound mufflers when sonication is in process (http://www.labmanager.com/?articles.view/articleNo/1103/article/Sonicator-Safety). Cool microcentrifuge tube, containing HeLa cells suspended in PBS, on ice first. Clean sonicator probe with ethanol first and then with distilled water. Wipe the probe dry with Kimwipe. Sonicate for 10 s and let cool on ice. Repeat this step three more times. Make sure that the probe does not touch the bottom of the tube when sonication is in progress to avoid the probe from puncturing a hole in the tube.
9. The gel can be heated at 65°C in a water bath for 20 min. Heating has been shown to accelerate protein staining and increasing the sensitivity of staining (12). If controlled heating temperatures are not essential, the plastic container with the CBB stain and the gel can be heated using a microwave. However, care needs to be taken to not overheat the container in the microwave. Also, the container needs to be covered tightly to prevent evaporation of acetic acid and to prevent acetic acid fumes from contaminating the laboratory.
10. Do not discard the wash into the sink. Save the rinse in a bottle for proper disposal (see Subsection 3.4).
11. It is not necessary to place the gel within the folds of the Kimwipes to enable adsorption of the dye. Kimwipes can be moistened in the destaining solution and placed as a crumpled ball in one corner of the container for CBB removal. However, inserting the gel between the folds of the Kimwipes enables the adsorption of the CBB as it is released from the gel.
12. The use of heat has been shown to accelerate the destaining process. It takes approximately 15–30 min to destain with the application of heat (12).

13. The Kimwipes will not leach any of the bound Kimwipe into water or any of the salts used. The CBB remains tightly bound to the Kimwipes even after lengthy incubation in salts or water, at various pH values used in these experiments (pH 2.4 with CBB destaining solution, pH 7.95 with sodium bicarbonate, and pH 8.8 with sodium phosphate [dibasic anhydrous]). Thus, CBB will not leach out from the Kimwipes dumped at solid waste landfills even under conditions of high salt concentration or highly acidic conditions.
14. Thus, the CBB-stained Kimwipes can be discarded as solid waste and will not release the bound CBB during rains or in situations involving high salts.
15. Almost all the dye from a 0.05% CBB staining solution could be removed with 75 Kimwipes (see Fig. 2) in just 5–10 min. Interestingly, the Kimwipes adsorb the stain quickly when there is excess CBB dye present, as seen with the 0.05% staining solution. Once the dye level decreases, the stain adsorbed by the Kimwipes quickly is reduced. This can be seen in Fig. 2 where the intensity of color decreases from Fig. 2a–e. Therefore, when using dilute CBB solutions, as obtained with gel destaining, it is better to leave the Kimwipes in the container for a longer time (the duration of gel destaining).

Acknowledgments

This work was supported in part by an Oklahoma Center for the Advancement of Science and Technology grant to B.T.K. We thank Hal Scofield (Oklahoma Medical Research Foundation [OMRF]) for his support of this work. B.T.K. first started using multifold paper towels during the early 1990s to remove CBB from gel destain in Hiroyuki Matsumoto's laboratory. We also thank Justin Simmons, safety officer at OMRF, for his input regarding proper disposal of CBB in the institutional setting.

References

1. Laemmli UK (1970) Cleavage of structural proteins during the assembly of the head of bacteriophage T4. Nature 227:680–685
2. Dorri Y, Kurien BT, Scofield RH (2009) A simpler and faster version of two-dimensional gel electrophoresis using vertical, mini SDS-PAGE apparatus. Iran J Chem Chem Eng 28:51–56
3. Schaffner W, Weissman C (1973) A rapid, sensitive, and specific method for the determination of protein in dilute solution. Anal Biochem 56:502–514
4. Merrill CR (1990) Gel-staining techniques. In: Deutscher MP (ed.). Methods in Enzymology, vol. 182. Guide to Protein Purification. Academic Press, San Diego, pp 477–488

5. Sambrook J, Fritsch EF, and Maniatis T (1989). Detection and analysis of proteins expressed from cloned genes. In: Sambrook J, Fritsch EF and Maniatis J (eds.). Molecular cloning: a laboratory manual, vol. 18. Cold Spring Harbor Laboratory Press, Cold Spring Harbor, NY, pp. 1819–1885
6. Fairbanks G, Steck TL, Wallach DFH (1971) Electrophoretic analysis of the major polypeptides of the human erythrocyte membrane. Biochemistry 10:2606–2617
7. Hoffman JI, Guz A (1961) Toxicity of Coomassie blue. Am Heart J 61:665–666
8. Villanueva MA (2008) Electrotransfer of proteins in an environmentally friendly methanol-free transfer buffer. Anal Biochem 373:377–339
9. Dorri Y, Kurien BT (2010) Environmentally safe removal/disposal of Coomassie Brilliant Blue from gel destain and used gel stain. Anal Biochem 404:193–196
10. Duhamel RC, Meezan E, Brendel K (1981) A charcoal cartridge for the removal of anionic detergent and electrophoresis stains. J Biochem Biophys Methods 4:73–80
11. Chandy T, Sharma CP (1998) Activated charcoal microcapsules and their applications. J Biomater Appl 13:128–157
12. Kurien BT, Scofield RH (1998) Heat mediated quick Coomassie blue protein staining and destaining of SDS-PAGE gels. Indian J Biochem Biophys 35:385–389

Chapter 56

A Brief Review of Other Notable Protein Detection Methods on Acrylamide Gels

Biji T. Kurien and R. Hal Scofield

Abstract

Several methods have been described to stain proteins analyzed on acrylamide gels. These include ultrasensitive protein detection in one-dimensional and two-dimensional gel electrophoresis using a fluorescent product from the fungus *Epicoccum nigrum*; a fluorescence-based Coomassie Blue protein staining; visualization of proteins in acrylamide gels using ultraviolet illumination; fluorescence visualization of proteins in sodium dodecyl sulfate-polyacrylamide gels using environmentally benign, nonfixative, saline solution; and increasing the sensitivity four- to sixfold for detecting trace proteins in dye or silver stained polyacrylamide gels using polyethylene glycol 6000. All these methods are reviewed briefly in this chapter.

Key words: Fluorophore, Epicocconone, Visible SYPRO, SDS-PAGE, Blue Silver staining

1. Ultrasensitive Protein Detection in One-Dimensional and Two-Dimensional Gel Electrophoresis Using a Fluorescent Product from the Fungus *Epicoccum nigrum*

Several protein detection methods have been devised for detection of proteins in protein gel electrophoresis. Gel protein staining with colloidal Coomassie Brilliant Blue (CBB) G-250, the best of the CBB family of stains, can detect proteins down to 6–8 ng while staining with the soluble CBB R-250 can only detect proteins down to about 50 ng and requires the additional step of destaining. Silver staining of proteins separated by sodium dodecyl sulfate-polyacrylamide gel electrophoresis (SDS-PAGE) or two-dimensional PAGE has helped attain higher sensitivity of detection in the low-nanogram quantities of protein. Fluorescence-based detection techniques offer higher sensitivities, are safer than radioactive detection techniques, and allow for multiple labeling and fluorescence resonance energy transfer-based applications and greater stability, both prior to and after an assay (1).

Biji T. Kurien and R. Hal Scofield (eds.), *Protein Electrophoresis: Methods and Protocols*, Methods in Molecular Biology, vol. 869, DOI 10.1007/978-1-61779-821-4_56, © Springer Science+Business Media, LLC 2012

Mackintosh et al. (1) have developed an ultrasensitive fluorescent stain "Lightning Fast" that can detect proteins down to less than 100 pg in one-dimensional gels. The stain contains the fluorophore epicocconone, derived from the fungus *E. nigrum*, which interacts noncovalently with SDS and protein. Proteins stained with "Lightning Fast" can be excited optimally by near-ultraviolet light (about 395 nm) or with visible light (about 520 nm). Sources used in image analysis systems such as ultraviolet A (about 365 nm) and ultraviolet B (about 302 nm) transilluminators, xenon-arc lamps, 488-nm and 457-nm argon-ion lasers, 473-nm and 532-nm neodymium: yttrium aluminum garnet (Nd:YAG) solid-state lasers, 543-nm helium-neon lasers, and emerging violet, blue, and green diode lasers can excite the stain. Maximum fluorescence emission of the dye is at approximately 610 nm. The new stain was found to be considerably more sensitive than SYPRO Ruby, SYPRO Orange, Silver, and Coomassie Brilliant Blue G-250. Staining can be accomplished in 3.5 h and is compatible with peptide mass fingerprinting using matrix-assisted laser desorption ionization-mass spectrometry and Edman sequencing (1).

2. A Fluorescence-Based Coomassie Blue Protein Staining

The SYPRO Ruby family of fluorescent dyes has high sensitivity and broad linear dynamic ranges and is compatible with mass spectrometry and other downstream protein identification (2, 3). However, proteins stained by these fluorescent protein stains cannot be visualized by the naked eye. Specialized equipment, such as an ultraviolet transillumination system, is required for protein detection. This greatly limits manual cutting of stained spots (4). Proteins stained by Coomassie Brilliant Blue (CBB) can be visualized by the naked eye with relatively low sensitivity, compared to fluorescent protocols.

Yang et al. (5) developed the visible "SYPRO" staining protocol for detecting proteins following sodium dodecyl sulfate-polyacrylamide gel electrophoresis (SDS-PAGE). The authors sensitized protein gels with SYPRO Ruby and then stained the proteins using the Coomassie Brilliant Blue G-250 protocol (Blue Silver) (6). This procedure that combined the two protocols had similar or better linearity than staining with only SYPRO Ruby or Blue Silver, respectively. Also, this method was more sensitive than Blue Silver staining, simpler than staining with SYPRO Ruby, and compatible with mass spectrometry analysis (5).

3. Visualization of Proteins in Acrylamide Gels Using Ultraviolet Illumination

Kazmin et al. (7) describe a fast, one-step method for quick detection of proteins following sodium dodecyl sulfate-polyacrylamide gel electrophoresis (SDS-PAGE). The method depends on ultraviolet-light-induced reaction of tryptophan in the presence of trichloroacetic acid, which results in fluorescent products emitting in the visible region of the spectrum. The authors show, using fluorescence spectra, the similarity of this reaction to the light-dependent reaction of chloroform with tryptophan (8, 9). The ultraviolet light generated by standard transilluminators (used for detecting ethidium bromide-stained DNA) is enough to drive the required reaction to produce visible light in the protein bands that can be visualized and recorded (7).

Proteins in polyacrylamide gels are visualized rapidly by soaking in trichloroacetic acid or chloroform followed by ultraviolet light illumination. In less than 20 min, protein bands can be rapidly identified on a gel using this method. The authors showed that this method is less sensitive than staining with Coomassie Blue, visualizing protein bands with as little as 1 μg of protein (a protein with typical tryptophan percentages) on thin acrylamide gels.

4. Fluorescence Visualization of Proteins in Sodium Dodecyl Sulfate-Polyacrylamide Gels Using Environmentally Benign, Nonfixative Saline

Protein staining often requires the use of toxic solvents such as methanol, ethanol, acetic acid, trichloroacetic acid, perchloric acid or phosphoric acid, glutaraldehyde, dichromate, formaldehyde, and ammonium hydroxide (10). Steinberg et al. report the use of SYPRO Tangerine stain that does not use any toxic reagent as the solvent (10). SYPRO Tangerine stock is diluted in phosphate buffered saline or simply 150 mM sodium chloride and used to stain protein gels in a one-step procedure (no need to destain). Proteins stained in this manner can be excited using ultraviolet (about 300 nm) or visible light (about 490 nm). SYPRO Tangerine's fluorescence emission maximum is about 640 nm. The dye binds noncovalently to proteins, mediated by SDS and by hydrophobic amino acid residues to a lesser extent. The stain's sensitivity is about 4–10 ng per protein band, similar to that obtained with SYPRO Red and SYPRO Orange. Colloidal Coomassie Blue staining and rapid silver staining procedures also achieve similar staining sensitivity. The advantages of using SYPRO Tangerine staining are (a) the stained proteins are not fixed and hence are amenable to easy elution procedures, (b) stained proteins can be used in zymography assay if SDS does not inactivate the protein of interest, (c) the stained gels can be used for western blotting, and (d) the staining procedure is environmentally benign (10).

5. A Four- to Sixfold Increase in Sensitivity for Detecting Trace Proteins in Dye or Silver Stained Polyacrylamide Gels

Palumbo and Tecce (11) demonstrate the possibility of increasing or regulating the sensitivity of detecting trace proteins in gels stained by different procedures. The method is based on the use of concentrated polyethylene glycol 6000 solutions, which results in a very noticeable, reversible, and uniform size reduction of gels. Gel size reduction depends mainly on the polyethylene glycol 6000 concentration and on the acrylamide content of gels. The authors showed, by employing an appropriate polyethylene glycol 6000 concentration, the possibility of regulating the final size and then the sensitivity as required. Palumbo and Tecce (11) were able to increase the sensitivity of detection by a factor of 4–6 or more based on the conditions described above. Thus, it may be possible to eliminate the need for the cumbersome and frequently impossible pre-concentration of samples by reducing the volume of the gel in this manner. Coomassie Blue and Silver stained gels were also treated in a similar manner with a concomitant increase in sensitivity. In addition, this method enables gels with radiolabeled materials to be autoradiographed in shorter times, without significant loss of resolution. A gel that is reduced in size by the use of polyethylene glycol 6000 can be restored to its original size with ease by just soaking it in the destaining solution. Alternatively, it can be converted to a transparent glass-like plate by using isopropyl alcohol washing (11).

References

1. Mackintosh JA, Choi HY, Bae SH et al (2003) A fluorescent natural product for ultra sensitive detection of proteins in one-dimensional and two-dimensional gel electrophoresis. Proteomics 3:2273–2288
2. Chevalier F, Rofidal V, Rossignol M (2007) Visible and fluorescent staining of two-dimensional gels. Methods Mol Biol 355:145–156
3. Harris LR, Churchward MA, Butt RH et al (2007) Assessing detection methods for gel-based proteomic analyses. J Proteome Res 6:1418–1425
4. Ball MS, Karuso P (2007) Mass spectral compatibility of four proteomics stains. J Proteome Res 6:4313–4320
5. Yang Y, Wang J, Bu D et al (2011) A fluorescence-based Coomassie Blue protocol for two-dimensional gel-based proteomics. Biotechnol Lett 33:119–121
6. Candiano G, Bruschi M, Musante L et al (2004) Blue Silver: a very sensitive colloidal Coomassie G-250 staining for proteome analysis. Electrophoresis 25:1327–1333
7. Kazmin D, Edwards RA, Turner RJ et al (2002) Visualization of proteins in acrylamide gels using ultraviolet illumination. Anal Biochem 301:91–96
8. Voropei AV, Chernitskii YeA, Konev SV et al (1992) Chloroform dependent photoproducts of tryptophan. Biophysics 37:743–745
9. Pinchuk SV, Vorobei AV (1993) Spectral characteristics of mechanisms of forming "chloroform-dependent" tryptophan photoproducts. J Appl Spectrosc 59:711–715
10. Steinberg TH, Lauber WM, Berggren K et al (2000) Fluorescence detection of proteins in sodium dodecyl sulfate-polyacrylamide gels using environmentally benign, nonfixative, saline solution. Electrophoresis 21:497–508
11. Palumbo G, Tecce MF (1983) A four- to sixfold enhancement in sensitivity for detecting trace proteins in dye or silver stained polyacrylamide gels. Anal Biochem 134: 254–258

Chapter 57

Ultrasensitive Protein Detection and Imaging: Comparison of Lumitein™, ProteoSilver™, SYPRO® Ruby, and Coomassie® Brilliant Blue Gel Stains

Guang Ren, Carl K. Okerberg, and Suresh T. Mathews

Abstract

Following sodium dodecyl sulfate polyacrylamide gel electrophoresis, proteins can be visualized by various methods of detection and imaging. Traditional methods of protein gel detection and imaging have been improved and expanded through technological advancement. Today, the detection of proteins, resolved on gels, can be accomplished with a variety of stains with various sensitivities. Digital cameras used in the imaging of protein gels are not only more sensitive than their film precursors, but they can be used in combination with imaging software that offers a host of useful applications. Here we describe the UVP BioImaging System in combination with LabWorks Image and Acquisition software to provide a comparison of four different protein gel stains: Lumitein™, ProteoSilver™, SYPRO® Ruby, and Coomassie® Brilliant Blue. We demonstrate that the detection sensitivity limit appears to be between 100 and 500 ng/protein band of protein with Coomassie® Brilliant Blue, 10–50 ng/protein band with Lumitein™ and SYPRO® Ruby, and as little as 5 ng/protein band with the ProteoSilver™ stain.

Key words: SDS-PAGE, Gel stain, Coomassie, Silver stain, Lumitein, SYPRO Ruby, LabWorks, CCD, UV transilluminator

1. Introduction

1.1. Protein Gel Staining

Protein gel staining is a standard procedure for detecting protein samples after separation by sodium dodecyl sulfate polyacrylamide gel electrophoresis (SDS-PAGE). There are several kinds of stains and staining procedures, all of which have their own individual advantages and disadvantages.

Coomassie® Brilliant Blue is perhaps the most traditional protein gel dye, having been first used to stain protein samples following electrophoretic separation in a polyacrylamide gel in 1965 (1). Coomassie® Brilliant Blue is a triphenylmethane dye that, upon

Biji T. Kurien and R. Hal Scofield (eds.), *Protein Electrophoresis: Methods and Protocols*, Methods in Molecular Biology, vol. 869,
DOI 10.1007/978-1-61779-821-4_57, © Springer Science+Business Media, LLC 2012

formation of a protein–dye complex, is stabilized, producing a bright blue color wherever it is bound to proteins within a gel. Coomassie® can detect protein samples of as little as 0.3–1 μg/protein band (2). Thus, Coomassie® staining may not offer the sensitivity to detect extremely minute quantities of protein samples.

Silver staining, first described in 1979, is more sensitive than Coomassie® (3). The ProteoSilver™ stain utilizes the compound silver nitrate, which under weakly acidic or neutral pH conditions binds to various chemical groups (e.g., sulfhydryl and carboxyl moieties) in proteins (4). The stain is sensitive enough to detect 2–5 ng of protein. Some disadvantages of silver staining are (a) that it is more time-consuming than Coomassie®; however, rapid silver staining procedures are currently available, and (b) that it is easy to cause a deep background and decrease detection sensitivity. But with the improvement of sensitizers and destain solutions, newer methodology helps decrease background staining and increase detection sensitivity (5–7).

Fluorescent stains, such as SYPRO® Ruby and Lumitein™, provide important alternatives to traditional gel stains because of their ability to rapidly detect small quantities of protein. SYPRO® Ruby protein stain provides a highly sensitive method for detecting proteins from SDS-PAGE gels (including glycoproteins, phosphoproteins, and lipoproteins) (8). SYPRO® Ruby stain binds to amino acids and the peptide backbone of the separated protein, resulting in increased protein–protein stability than silver stain (9, 10). While earlier procedures using SYPRO® Ruby required an overnight incubation, newer rapid staining procedures are currently available that can match the 90-min staining procedure of Lumitein™. Lumitein™ is a luminescent dye that is used for detecting proteins in polyacrylamide gels. Not only are both of these dyes incredibly user-friendly, requiring less hands-on time than most other stains, but these two fluorescent stains also have a linear quantitation range of over three orders of magnitude and can detect as little as 1 ng of protein or less.

Coomassie® and ProteoSilver™-stained gels can be imaged using a simple light box and camera. SYPRO® Ruby and Lumitein™ fluorescent stains are compatible with an UV transilluminator. SYPRO® Ruby stain has two excitation maxima, one at ~280 nm and one at ~450 nm, and an emission maximum near 610 nm. This excitation spectrum is compatible with a common 300-nm UV transilluminator, a blue-light transilluminator, or a laser scanner. Lumitein™ has the same UV excitation and emission maximum as SYPRO® Ruby stain and, therefore, is compatible with a simple UV transilluminator or a high-end laser scanner. Other fluorescent stains for detecting proteins in polyacrylamide gels include Coomassie® Fluor™ Orange, Pro-Q® Diamond, Pro-Q® Emerald, and SYPRO® Orange.

1.2. Protein Gel Imaging

Cooled charge coupled device (CCD) cameras have become increasingly popular over their film precursors. CCD cameras are more sensitive and offer a dynamic range of 3–4 orders of magnitude. This generates a high-quality image compared with the limited linear dynamic range of film, which is capable of 1.5 orders of magnitude. Cooling CCD cameras to −20°C increases exposure times to over 5 min without generating a grainy picture, and cooling to −40°C can integrate exposures for an hour without significant loss of image quality due to "hot pixels." Additionally, the available imaging software has served to ease the frustrations associated with the process of analyzing stained gels. Imaging software can be used for a variety of image enhancements and analyses including estimating molecular weights, calculating band densities, and correcting for nonuniform protein migration that can occur during electrophoresis.

In this chapter, we describe the methodology for protein gel image analysis using the UVP BioImaging System and LabWorks Image Acquisition and Analysis software. We have compared two fluorescent protein stains Lumitein™ and SYPRO® Ruby with traditional Coomassie® and silver staining methods. Using Coomassie® Brilliant Blue, the detection sensitivity limit appears to be between 100 ng and 500 ng/protein band. The detection sensitivity of Lumitein™ and SYPRO® Ruby was 10–50 ng/protein band, and the ProteoSilver™ stain could detect as little as 5 ng/protein band (Table 1 and Fig. 1).

2. Materials

2.1. Protein Gel Imaging: Lumitein™

1. Lumitein™ 1× staining solution: 1% Lumitein™ (Biotium Inc., Hayward, CA, USA) 100×, 58% ultrapure water, 28% methanol, 13% acetic acid.
2. Gel destaining solution: 30% methanol, 15% acetic acid, and 55% ultrapure water.
3. UVP BioImaging System with LabWorks Image Acquisition and Analysis software Version 4.6 (UVP LLC, Upland, CA, USA).
4. Shaker.

2.2. Protein Gel Imaging: SYPRO® Ruby

1. SYPRO® Ruby protein gel stain solution (Invitrogen, Carlsbad, CA, USA).
2. Fixing solution: 50% methanol, 7% acetic acid, 43% ultrapure water.
3. Wash solution: 10% methanol, 7% acetic acid, 83% ultrapure water.
4. UVP BioImaging System with LabWorks Image Acquisition and Analysis software.
5. Shaker.

Table 1
Bovine serum albumin (BSA: 500, 100, 50, 10, 5, and 1 ng) was separated by SDS-PAGE and stained with Lumitein™, ProteoSilver™ staining, SYPRO Ruby®, and Coomassie® stains

	Detection sensitivity					
Protein stain	**500 ng**	**100 ng**	**50 ng**	**10 ng**	**5 ng**	**1 ng**
Lumitein™	1	0.6	0.3	0.1	–	–
ProteoSilver™	1	0.8	0.6	0.4	0.3	–
SYPRO Ruby®	1	0.4	0.4	0.1	–	–
Coomassie®	1	–	–	–	–	–

Protein gel images were captured on a UVP BioImaging System and quantitated with LabWorks Image Acquisition and Analysis software (UVP LLC, Upland, CA). Relative detection sensitivity of various stains is shown. Silver staining was the most sensitive detecting as little as 5 ng BSA. Lumitein and SYPRO Ruby stains were able to detect 10 ng of protein

2.3. Protein Gel Imaging: ProteoSilver™

1. Fixing solution: 50% ethanol, 10% acetic acid, 40% ultrapure water.
2. 30% Ethanol solution: 30% ethanol, 70% ultrapure water.
3. Sensitizer solution: 1% ProteoSilver™ Sensitizer (Sigma, St Louis, MO), 99% ultrapure water. This solution should be used within 2 h of preparation.
4. Silver solution: 1% ProteoSilver™ Silver solution (Sigma, St Louis, MO, USA), 99% ultrapure water.
5. Developer solution: 5% ProteoSilver™ Developer 1 (Sigma, St. Louis, MO), 0.1% ProteoSilver Developer 2 (Sigma), 95% ultrapure water. The developer solution should be prepared less than 20 min before use.
6. Stop solution: Supplied by the manufacturer.
7. UVP BioImaging System with LabWorks Image Acquisition and Analysis software.
8. Shaker.

2.4. Protein Gel Imaging: Coomassie® Brilliant Blue R

1. Coomassie® solution: 10% acetic acid, 25% isopropanol, 65% ultrapure water, 1% w/v Coomassie® Brilliant Blue R.
2. Fixing and destaining solutions (these are the same solutions): 5% acetic acid, 10% isopropanol, 85% ultrapure water.
3. UVP BioImaging System with LabWorks Image Acquisition and Analysis software.
4. Shaker.

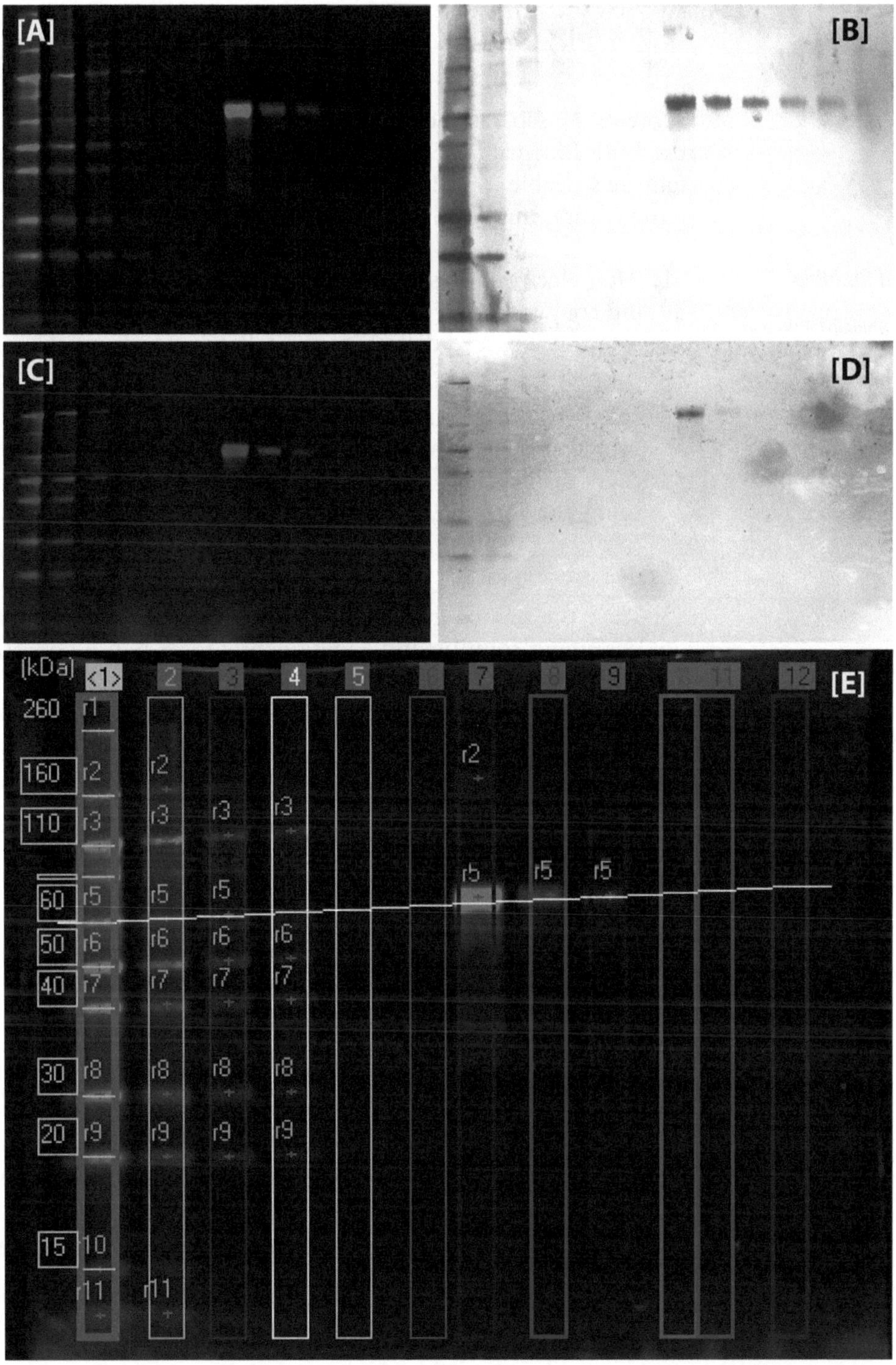

Fig. 1. Comparison of visible imaging using UVP BioImaging System. Lumitein™ (**a**), ProteoSilver™ (**b**), SYPRO® Ruby (**c**), and Coomassie® (**d**) protein gel stains. Proteins [1–6 contain decreasing loads of Novex Sharp Unstained Protein Standard Range from 5 μL of original ladder (lane 1) to 20 μL of 200-fold dilution (lane 6); 7–12: 500, 100, 50, 10, 5, and 1 ng BSA, respectively] were separated by SDS-PAGE and stained by appropriate staining kits. Images were captured on a UVP BioImaging System and analyzed with LabWorks Image Acquisition and Analysis software. A slant line was used to correct unequal protein migration in order to accurately estimate the molecular weight in a Lumitein™-stained gel (**e**).

3. Methods

Proteins are separated on SDS-PAGE, as described earlier (2), and the UVP BioImaging System is used to image all four stained protein gels (Table 1 and Fig. 1). All steps are performed at room temperature, unless otherwise noted (see Note 1).

3.1. Gel Staining

3.1.1. Lumitein™ Protein Gel Stain

1. After electrophoresis, place the gel directly in a clean gel staining container. It is not necessary to incubate the gel in a fixing solution. Proceed directly to the following steps.
2. Add 80 mL Lumitein 1× staining solution into the container. Incubate the gel in the container by agitating gently on an orbital shaker for at least 90 min. The staining solution must cover the gel (the volume of staining solution can be calculated using the formula, $V\,(\text{mL}) = 80\ \text{mL} \times (S/64)$, where S is the size of the gel in cm^2). Insufficient staining solution may cause dye precipitation.
3. Place the stained gel in a clean container. Add 100 mL of the destaining solution and agitate gently for 5 min on an orbital shaker.
4. Decant the destaining solution, add at least 100 mL of ultrapure water, and agitate gently for at least another 5 min before imaging the gel.

3.1.2. ProteoSilver™ Silver Stain Kit

1. After electrophoresis, stain the gels using the ProteoSilver™ Silver Staining Kit. Place the gel into a clean tray with 100 mL of fixing solution for 40 min or overnight.
2. Decant the fixing solution and wash the gel for 10 min with 100 mL of 30% ethanol solution.
3. Decant the ethanol solution and wash the gel for 10 min with 200 mL of ultrapure water.
4. Decant the water and incubate the gel for 10 min with 100 mL of the sensitizer solution.
5. Decant the sensitizer solution and wash the gel twice, each time for 10 min with 200 mL of ultrapure water.
6. Decant the water and equilibrate the gel for 10 min with 100 mL of the Silver solution.
7. Decant the Silver solution and wash the gel for 1.5 min with 200 mL of ultrapure water.
8. Develop the gel with 100 mL of developer solution for 3–7 min depending on the staining density. Typically, 7 min is sufficient for mini gels. Background staining will occur on overdeveloped gels.

9. Add 5 mL of ProteoSilver™ Stop solution directly to the developer solution and incubate for 5 min to stop the developing reaction.
10. Decant the mixture of developer and stop solution and wash the gel for 15 min with 200 mL of ultrapure water. Store the gel in fresh, ultrapure water.

3.1.3. SYPRO® Ruby Protein Gel Stain

1. After electrophoresis, place the gel into a clean container, add 100 mL fixing solution, and gently agitate on an orbital shaker for 30 min. Repeat once with fresh fixing solution.
2. Add 60 mL SYPRO® Ruby gel stain and then gently agitate on an orbital shaker overnight.
3. Transfer the gel to a clean container and wash in 100 mL of wash solution for 30 min. Before imaging, rinse the gel in ultrapure water for 4 min at least twice to prevent possible corrosive damage to the imager.

3.1.4. Coomassie® Brilliant Blue R Protein Gel Stain

1. After electrophoresis, place the gel into a clean container, add 100 mL fixing solution, and gently agitate on an orbital shaker for 1 h (see Note 2).
2. Decant the fixing solution, add 80 mL of Coomassie® stain solution, and then gently agitate on an orbital shaker for 2 h.
3. Transfer the gel to a clean container and wash it in 100 mL wash solution for 30 min. Repeat 3 times.
4. Switch to fresh wash solution and gently agitate on an orbital shaker overnight with some floating sponges to soak up the dye as it is released from the gel.

3.2. Image Acquisition

1. Turn on the power of the CCD camera 15 min before image acquisition to allow it to fully cool to –28°C.
2. Open UVP BioImaging System dark box, push the white light board up, remove the UV shield, and set the UV wavelength to 302 nm (see Note 3).
3. UVP BioImaging System dark box's Overhead Light setting should be turned off. Change the settings according to the particular stain used, as detailed below:
   (a) Lumitein™: Set the emission filter to ethidium bromide and the transillumination controls to UV safety.
   (b) ProteoSilver™: Set the emission filter to clear and the transillumination controls to white light on.
   (c) SYPRO® Ruby: Set the emission filter to ethidium bromide and the transillumination controls to UV safety.
   (d) Coomassie® Brilliant Blue: Set the emission filter to Coomassie® Blue and the transillumination controls to white light on.

4. To focus the UVP BioImaging System, place the UVP fluorescent sheet provided with the imager inside the dark box onto the UV transilluminator.
5. Click the Video/Digital capture icon on the toolbar, which opens the Live Workspace Preview. Now focus the image by manually adjusting the focus ring on the camera. Close the Live Workspace Preview upon completion.
6. Position the gel inside the dark box in the middle of the UV transilluminator and then close the dark box.
7. Keep the power on, open the Acquire menu, and choose the Video/Digital option to preview the gel image.
8. It may be useful to take a series of images to determine the best exposure. Click Multiframe/Timed under the Acquire menu, and then choose Sequential Integration in the Integration Tab of the dialog box. Click Sequential Settings and select Regular Intervals. Then input the number of images and total time into the Sequential Integration dialog box. Click OK. It is also possible to capture the image at irregular intervals using the Sequential Integration dialog box.
9. Click Capture in the Camera toolbar. This begins the capture process, during which images are sequentially added to the Desktop as they are captured.
10. When the capture is complete, an image of appropriate band intensity can be chosen and saved. The whole sequential multi-image can be saved if it is desired to choose the appropriate image at a later time.
11. Alternatively, a single image can be captured for a defined exposure time using the Video/Digital tab.

3.3. Image Enhancement

Before enhancing an image, it is important to keep in mind that the UVP BioImaging System captures images as 12-bit or 16-bit files, which may not be compatible with other imaging software and cannot be imported into Microsoft™ Office. In order to do these things, an image needs to be converted to an 8-bit file. If 12-bit or 16-bit files are compatible with the intended use, proceed to Step 3:

1. Highlight the image and then click Edit and choose "Convert to."
2. Next, click Gray Scale 8 and then click Convert. This converts the image to an 8-bit file.
3. Click Display Range under the Edit menu.
4. In Display Range, slide the green and blue bars from the right or the left side of the graph inward until the image's band densities are satisfying. Clicking the Z_+ button will allow for greater detail enhancement.

5. Click Contrast Enhancement under the Edit menu to further adjust the visual qualities of your image including the brightness, contrast, and gamma settings. The Contrast Enhancement dialog box can be used to reshape the curve that describes the image's intensity scale (see Note 4).
6. Click Process menu for sharpening or star field subtraction (see Notes 5 and 6).
7. Click Pseudo-Color under the Edit menu to colorize the image if desired. In the Pseudo-Color dialog box, click the color spread and for:
   (a) Lumitein™: Choose SYPRO® Orange for an orange background.
   (b) ProteoSilver™: Choose Silver for a silver background.
   (c) SYPRO® Ruby: Choose SYPRO® Red for a red background.
   (d) Coomassie® Brilliant Blue: Choose Coomassie® for a blue background.
8. After enhancement is complete, click Contrast Enhancement again and then click Apply to make the changes in the image permanent (see Note 7).

3.4. Estimation of Molecular Weights and Quantitation of Band Densities

1. Click Show Toolbar under the 1-D gel menu.
2. Click Lanes and LabWorks will automatically detect all of the lanes on the gel (see Note 8).
3. Open the Bands dialog box in the 1-D gel toolbar and change the Min. Band Height (%) parameter if it is desired to increase or decrease the band identifying sensitivity. Alternatively, bands can be added to the image manually by clicking Add Bands (see Note 9).
4. Click M. W. Standard in the 1-D gel toolbar. To specify the marker used in the gel, click New and then enter gel-specific information about the protein markers used, the units, the type of "genetic material" (choose protein), and the diffusion (the default is logarithmic). Click Add and then input the molecular weight of the highest molecular weight protein marker. Click Add again and repeat for all of the other molecular weights in decreasing order until all of them have been added (see Note 10).
5. LabWorks assumes that the marker is in the first lane of the gel. If this is not the case or there are more lanes containing the marker, the Select/Unselect button in the M.W. Standard dialog box can be used to manually select which lanes contain the marker.
6. Next, click Auto-Locate in the M. W. Standard dialog box. The software will automatically detect bands and estimate their molecular weights (see Note 11).

7. Based on the intensity of the protein marker bands, some molecular weights may not be accurately labeled. To manually correct this, click Locate in the M. W. Standard dialog box and then click Detach All. The band indicators can now be moved proportionately to each other. If desired, it is possible to move them individually. Clicking on an indicator locks it into place (a rectangle will appear around the indicator number), and it will not move with the others. Once the position of all markers is confirmed, click OK (see Notes 12 and 13).
8. Click the Slant button in the 1-D gel toolbar. Click Auto. Slant Lines and LabWorks will automatically construct yellow slant lines as long as at least two marker lanes were specified as in Step 5. Slant lines can be added manually by clicking on Add Slant. This will bring up an additional dialog box that requires the user to match bands with identical molecular weights (see Note 14).
9. To obtain results for gel analysis, click Results on the 1-D gel toolbar. Under the Show menu, select Both to show the Amount or IOD and Molecular Weight together. The results can be displayed in terms of the percentage, relative abundance, or mass units of the bands. It is also possible to represent them in Absolute Integrated Optical Density (IOD) or Maximum IOD (see Note 15).
10. Click File and then DDE to Excel to export the results to a Microsoft™ Excel spreadsheet.

4. Notes

1. There are many variables that can influence staining sensitivity, so in order to obtain the best result, researchers should be sure to use clean glassware, Milli-Q water, and quality reagents.
2. One hour of fixing will ensure that the dye is dissolved completely.
3. The UVP BioImaging System has a built-in safety mechanism that turns the UV light off whenever the door to the imager is opened. The user may not realize that the UV lights are on until the image is being captured. To make sure that all lights are turned off, a "safe lab practice" may be to turn off the power to the system completely.
4. The curve that describes the intensity scale, found in the Contrast Enhancement dialog box, can be a useful tool. LabWorks provides several curve types including 1/4 Tone, which can be used to modify the curve using 5 control points.
5. To refocus a blurry image, or to enhance the detail of an image, open the Process menu and click Sharpen. However, this will not correct an overly blurred image.

6. Star field Subtraction is another image filter tool that can be used to remove hot spots from an image. Keep in mind that all image enhancements should only be applied to improve image clarity, not for quantitative analysis. Filters can be removed under the Edit Undo menu.
7. It may be a good idea to have a reserve copy of the original image before clicking Apply to make Contrast Enhancement, because this cannot be undone.
8. Sometimes the software will detect unnecessary or invalid lanes. In order to delete a lane, click Delete Lane in the Lane dialog box and specify the unwanted lane.
9. By default, LabWorks assumes that the well origin line is straight and runs across the top of the gel image. If the gel image is not horizontal, the well origins should be defined using the Well Origin option before doing any band definition operations.
10. Below the image, the Lane Profile dialog box can be used to plot lane information on a line graph. In this profile, every peak represents the molecular weight value for the particular band from the selected or manually created standard. The area under the peaks is used to calculate the mass of each band by right clicking in the Lane Profile window to display the band information. If desired, the sensitivity of band detection can be increased by decreasing the area of the square brackets. The band information can be copied as a chart to Windows™ clipboard or converted to an Excel spreadsheet. If LabWorks displays a DDE error message, check to make sure your computer's Search Path is connected to the Excel directory.
11. To display band information directly on the image, use the Band Options dialog box. Using this tool, band numbers, molecular weight, or the protein amount can be displayed. Band numbering patterns can be controlled by choosing between rows of equal molecular weight, rows of equal Rf, or sequential numbering.
12. Background option can be used to correct backgrounds that are too dark. Using this option, background definitions can be changed without altering the image and compromising the results. Changes are automatically reflected in lane profile, molecular weight, and mass calculations.
13. If software-generated band definitions do not match up correctly with separated protein bands (representing the molecular weight markers) on gel, this can be corrected using the Options dialog box by choosing Manual positioning. Click Locate button and then proceed to select the bands manually. Choose Match Band for Band to designate which bands should be matched together. Alternatively, allow LabWorks to

compensate for additional or missing bands. This can be done by selecting Allow one or two missing bands.

14. Sometimes, it may be a good idea to load the outer lanes of a gel with markers so that a slant line can be constructed between them. In this way, the user can optimize the software's ability to estimate the protein sample's molecular weight. It is important to remember that if SDS-PAGE is performed too quickly, the proteins in the middle lanes will migrate faster than those in the outer lanes. In this case, spreading out the markers, as described above, will not help with slant line construction.
15. In Results, the Amounts option will display the absolute mass determined for each band, which can be useful for analysis.

Acknowledgment

This work was supported by a USDA Hatch award (ALA043-1-08033).

References

1. Meyer TS, Lambert BL (1965) Use of Coomassie brilliant blue R250 for the electrophoresis of microgram quantities of parotid saliva proteins on acrylamide-gel strips. Biochim Biophys Acta 107:144–145
2. Sasse J, Gallaghe SR (2003) Staining proteins in gels. Curr Protoc Mol Biol Chapter 10: Unit 10.6
3. Shapiro AL, Vinuela E, Maizel JV Jr (1967) Molecular weight estimation of polypeptide chains by electrophoresis in SDS-polyacrylamide gels. Biochem Biophys Res Commun 28: 815–820
4. Coligan JE, Dunn BM, Ploegh HL et al (1995) Current Protoc Prot Sci Chapter 1: Unit 10.5.11
5. Rabilloud T, Vuillard L, Gilly C et al (1994) Silver -staining of proteins in polyacrylamide gels: a general overview. Cell Mol Biol 40:57–75
6. Blum H, Beier H, Gross HJ (1987) Improved silver staining of plant proteins, RNA and DNA in polyacrylamide gels. Electrophoresis 8:93–99
7. Rabilloud T, Carpentier G, Tarroux P (1988) Improvement and simplification of low-background silver staining of proteins by using sodium dithionite. Electrophoresis 9:288–291
8. Berggren K, Chernokalskaya E, Steinberg TH et al (2000) Background-free, high sensitivity staining of proteins in one- and two-dimensional sodium dodecyl sulfate-polyacrylamide gels using a luminescent ruthenium complex. Electrophoresis 21:2509–2521
9. Lopez MF, Berggren K, Chernokalskaya E et al (2000) A comparison of silver stain and SYPRO Ruby Protein Gel Stain with respect to protein detection in two-dimensional gels and identification by peptide mass profiling. Electrophoresis 21:3672–3683
10. Smejkal GB, Robinson MH, Lazarev A (2004) Comparison of fluorescent stains: relative photostability and differential staining of proteins in two-dimensional gels. Electrophoresis 25: 2511–2519

Chapter 58

Common Artifacts and Mistakes Made in Electrophoresis

Biji T. Kurien and R. Hal Scofield

Abstract

Proteases that act at room temperature upon proteins in the sample buffer prior to heating, cleavage of the Asp–Pro bond upon prolonged heating of proteins at high temperatures, contamination of sample or sample buffer with keratin, leaching of chemicals from disposable plasticware, contamination of urea with ammonium cyanate are some subtle artifacts that can have significant deleterious effects on carefully planned and executed experiments. In addition, researchers are culpable of committing mistakes with respect to (a) calculating the cross-linking factor of a gel, (b) polymerization temperature and time for a polyacrylamide gel, (c) inducing aggregates in samples for electrophoresis, (d) titrating the running buffer in electrophoresis, (e) proper sample preparation, (f) amount of protein to be loaded on a gel, (g) sample buffer-to-protein ratios, (h) incompletely removing phosphate buffered saline from cells prior to cell lysis and (i) overfocusing of IPG strip in two-dimensional gel electrophoresis. Taking proper heed to all these factors can greatly help generate perfect experimental results.

Key words: Artifacts, Common mistakes in electrophoresis, Aggregates, Keratin, Asp–Pro bond, Proteases, Ammonium cyanate, Cross-linking factor

1. Introduction

Polyacrylamide gel electrophoresis (PAGE) is an invaluable technique for investigating the protein repertoire of a cell in health and disease. Westermeier (1) and Burgess (2) have recently reported regarding frequently made mistakes in electrophoresis and important but little known artifacts in protein biochemistry. Erroneous protocols abound, in spite of widespread information available via text books, instrument manuals, and online tutorials (1). Sometimes, subtle and obscure artifacts can significantly affect the outcomes of carefully performed experiments (2). This chapter details important artifacts and commonly made mistakes that can derail an otherwise perfectly executed experiment.

Biji T. Kurien and R. Hal Scofield (eds.), *Protein Electrophoresis: Methods and Protocols*, Methods in Molecular Biology, vol. 869, DOI 10.1007/978-1-61779-821-4_58, © Springer Science+Business Media, LLC 2012

2. Artifacts in Gel Electrophoresis

2.1. Sample Preparation for SDS-PAGE

Burgess (2) describes three ways in which one may find contaminants in a purified or nearly purified protein in sodium dodecyl sulfate (SDS) PAGE.

2.1.1. Proteases

Protein samples are normally added to sample buffer, containing SDS, β-mercaptoethanol or dithiothreitol, sucrose, or glycerol and heated at 95–100°C for 5 min. The heating is carried out to enable better denaturation and reduction of the proteases and thus bring about its inactivation (3). Gradually, it was recognized that the heating had to be carried out as soon as the samples were diluted in the sample buffer. This is because SDS, even though it easily unfolds most proteins, does not unfold some proteases. Consequently, the proteases will have time to digest the proteins of interest in the sample buffer if the heating step is not carried out immediately. One can design an experiment to find out if proteases are causing multiple bands to appear in SDS-PAGE analysis of a purified protein or a nearly pure protein. One way would be to add the protein of interest to two portions of the sample buffer. Mix well and heat one portion immediately. Leave the other at room temperature for 2–4 h and then heat. Analyze both samples on SDS-PAGE and check for degradation of the protein in the sample not heated immediately. As little as 1 pg protease in a protein sample has been shown to bring about major degradation, if the sample is added to the lysis buffer and not heated immediately (2).

2.1.2. Cleavage of Asp–Pro Bond at 100°C

The aspartic acid–proline bond is the most susceptible bond for cleavage by heat or acidic conditions (4). The bond is cleaved by heating at high temperatures for too long. Heating at 75°C for 5 min instead of 95–100°C for 5 min has been found to avoid D–P bond cleavage, as well as completely inactivating proteases (2). However, it has to be also noted that several proteins have been found to be stable at 100°C for several hours (5).

2.1.3. Contamination of SDS Lysis Buffer or Sample with Keratin

Keratin runs as a heterogeneous cluster of contaminating bands around 55–65 kDa on reducing SDS gels (with β-mercaptoethanol) and near the top of the gel on nonreducing SDS gels (without β-mercaptoethanol). Keratin occurs in skin, dander, etc. Commonly, the lysis buffer itself gets contaminated with skin contact or by a flake of dandruff. This is normally a minor contaminant and can be visualized mainly in silver stained gels and occasionally on Coomassie Blue stained gels. Contamination of lysis buffer with keratin can be ruled out by running sample buffer alone without the addition of proteins to a lane in the gel. The buffer should be remade if keratin bands are observed in this lane. Aliquoting lysis buffer and storing them at –80°C are practiced by many laboratories.

An aliquot can be thawed and used within a day or two. Keratin has been observed occasionally in western blots as a result of contamination of antigen used to prepare a polyclonal antibody and reacting with keratin transferred from SDS gel to membrane (2).

2.2. Leaching of Chemicals from Plasticware

Certain chemicals leach out from common laboratory plasticware (disposable) into standard aqueous buffers. In some cases, this can have significant effects on the results of experiments. Chemicals like oleamide are employed as lubricating agents in the molding process. Other chemicals, such as certain cationic biocides, are used to help prevent bacterial colonization of the plastic surface. Most of this material can be removed by washing the plasticware in water or even better with methanol or DMSO (2, 6).

2.3. Contamination of Urea with Cyanate

Urea is used widely as a denaturing agent for proteins. However, urea solutions contain significant amounts of ammonium cyanate. Ammonium cyanate is in chemical equilibrium with urea [$(H_2N)_2C{=}O$ in equilibrium with $NH_4^+ + NCO^-$)]. Isocyanic acid (H–N=C=O) has been shown to react with lysine's ε-amino group and the amino terminus as well as with arginine and cysteine to a lesser extent to form a carbamylated protein (7). Carbamylation can interfere with enzyme function in some cases, alter charge, and block certain protease cleavage reactions. In addition, as measured by mass spectrometry, it can add 43 Da per carbamylation event to the mass of the protein. To remove these contaminant ions (to diminish or prevent carbamylation), one can treat a urea solution with a mixed bed resin (Bio-Rad AG 501-X8). However, since this is a chemical equilibrium, the ammonium cyanate level builds up again to levels in the 0.5–3 mM range in an 8 M urea solution, with the potential to reach up to 20 mM (8). Chemical scavengers like ethylenediamine, glycylglycine, or glycinamide (in the 5–25 mM range) have been shown to reduce cyanate to less than 0.1 mM in 8 M urea, Tris–HCl (pH 8) (8). Burgess (2) suggests that the best general practice, if urea is to be used, is to replace some of the NaCl in the buffer with some ammonium salt (like 25–50 mM ammonium chloride) to push the equilibrium back by the common ion effect toward less cyanate. Lower temperature and acidic conditions have been shown to slow the reaction of isocyanic acid with amino groups in proteins. This effect can be also minimized by restricting the time of protein exposure to the urea solution to the shortest time possible (2).

3. Commonly Made Mistakes in Electrophoresis

3.1. Sample Preparation

Some commonly made mistakes (9) during sample preparation include employing an incorrect sample buffer-to-protein ratio, failure to remove insoluble material, and overloading and underloading of protein. In order to prevent inadequate sample buffer-to-protein ratios, overloading, and underloading of samples, the protein concentration of the sample should be determined using a standard protein assay (see Chapter 3).

Distorted, poorly resolved bands in the overloaded lane and distorted protein bands in adjacent lanes result from loading too much protein. On the contrary, underloading leads to lack of detection of minor protein bands and even makes major bands too faint for photographic reproduction. One should load purified protein in the 0.5–4.0 μg range (depending on well size and gel thickness) and from 40 to 60 μg for crude samples if a Coomassie Blue stain is to be used for gel staining. Less protein per sample is required when silver staining method is employed, since it is about 100-fold more sensitive than Coomassie staining (9). Enough sample buffer should be used in order to maintain an excess of SDS. Most proteins bind SDS in a constant mass ratio of 1.4 μg SDS per 1.0 μg protein. However, Hames (10) recommends a ratio of 3:1.

Certain proteins like histones and membrane proteins may need addition of 6–8 M urea or a nonionic detergent such as Triton X-100 since they may not completely dissolve by heating in SDS sample buffer alone (11, 12). The insoluble material should be removed by a 2-min centrifugation (at 17, 000 × *g*) following heat treatment in SDS lysis buffer. Streaking within the gel will happen if precipitated insoluble material is not removed. If the supernatant is not loaded immediately, it can be stored at 4°C overnight or frozen at −20°C for longer periods. However, the stored samples need to be warmed briefly at 37°C to redissolve the SDS and centrifuge once again to remove insoluble material before loading (9).

Pre-treatment methods such as lyophilization, spin concentration, dialysis against concentrated polyethylene glycol (PEG), and excess solvent absorption by exposing dialysis bag with sample to dry PEG, Aquacide or Sephadex® (gel filtration media) can be employed to concentrate samples that are too dilute for analysis. Prior to addition of SDS sample buffer, samples concentrated by the above said methods may be dialyzed against 50 mM Tris–HCl, pH 6.8 to get rid of low molecular weight contaminants. Trichloroacetic acid or acetone can be used to precipitate and concentrate proteins from dilute samples and acidic samples as well as get rid of contaminants such as potassium, guanidine hydrochloride, or ionic detergents from samples (1, 13).

Extremely viscous samples such as crude cell extracts owe their viscosity to the high concentration of unsheared nucleic acids. Treatment of samples with Benzonase® Nuclease (recombinant endonuclease) prior to addition of sample buffer can help eliminate the viscosity. This endonuclease lacks proteolytic activity and completely degrades all forms of DNA and RNA. Vigorous vortexing of the heated sample or physical shearing of the nucleic acids through sonication can also help reduce viscosity (9).

3.2. Miscalculating Cross-Linking Factor of a Polyacrylamide Gel

Two factors control the pore size of a polyacrylamide gel: (a) the total concentration of acrylamide T and (b) the degree of cross-linking C:

$$T = \frac{(a+b)\times 100}{V}[\%], \quad C = \frac{b\times 100}{a+b}[\%],$$

a = mass of acrylamide in g, b = mass of methylenebisacrylamide in g, and V = volume in mL.

However, sometimes, one mistakenly assumes that the given total concentration of acrylamide T is the percentage of acrylamide per volume and C (the cross-linking factor) is the percentage of methylenebisacrylamide per volume. This will lead to too much amounts of cross-linker in the gel. This will result in gels that are opaque, brittle, and highly hydrophobic. The correct method is to prepare the solutions as per the equation noted above. Alternatively, it would be better to use commercially available acrylamide/methylenebisacrylamide stock solutions that are ready to use (1).

3.3. Temperature and Time of Polymerization for a Polyacrylamide Gel

It takes about 30 min to 1 h for the polymerization of acrylamide and methylenebisacrylamide to occur. The matrix formation is unfinished at this time point, however. Polymerization continues, in a silent manner for several hours for the complete matrix formation. In general, complete polymerization is brought about efficiently, only when it is carried out at room temperature (1). However, Haeberle (14) has described a high-temperature SDS-PAGE (running gels at 70°C) that can be run in as little as 5 min for a mini gel. An added advantage of this method is the greatly accelerated rate of polyacrylamide cross-linking. According to this paper, the gel can be transferred to the 70°C buffer as soon as it becomes rigid enough to be removed from the casting stand, and by the time the samples are loaded, the cross-linking is almost complete (14).

Polyacrylamide gels are polymerized in the refrigerator or cold room in some laboratories, or the gels are used one or a few hours following the initiation of polymerization. Using incompletely polymerized gels can lead to disturbances during the protein separation, especially on a native gel. In addition, high background noise will be a problem in downstream mass spectrometry analysis as a consequence of the incompletely polymerized acrylamide mono- and oligomers. The correct way is to allow the gel to polymerize overnight at room temperature. The gel can be stored in a refrigerator, if necessary, afterwards (1).

3.4. Protein Aggregates in SDS Samples

After boiling, the sample is often directly loaded on the SDS gel after it has cooled down. Consequently, the reductant often becomes partly oxidized, a part of the cysteines remain unprotected and results in back-folding and the creation of inter-polypeptide aggregates. Back-folding leads to blurred zones and sometimes the formation of double bands. While some aggregates precipitate in the sample well (too big to enter the gel), some induce artifactual zones in the high molecular weight area. Two or three horizontal lines across the entire gel in the molecular weight range between 40 and 60 kDa are formed by an excess of reductant. The right procedure is to allow the sample to cool to approximately 60°C (after boiling) and then add iodoacetamide [10 μL of a 20% (w/v) aqueous iodoacetamide solution to 100 μL sample] and incubate the sample for 30 min at room temperature. Sharper bands are obtained with this procedure and artifacts (double bands, additional high molecular weight bands, precipitate in the sample well, and lines across the gel) are abolished (1).

3.5. Titrating Running Buffer in SDS-PAGE

The discontinuous buffer system based on Läemmli's method (15) consists of the stacking and resolving gel buffers that are composed of Tris and chloride buffers with defined pH values (with or without SDS). The running buffer has SDS, Tris, and glycine. The discontinuity between the chloride (acting as the leading ion) and the glycine (acting as the trailing ion) permits a controlled slow protein entry into the polyacrylamide gel and the stacking effect, resulting in very sharp and well-resolved zones.

However, some protocols call for adjusting the pH of the Tris–glycine buffer to a pH of 8.3 or 8.4. Doing so brings about a high load of chloride in the upper buffer chamber, and this causes the following effects: (a) The protein separation will take much longer than expected, with some zones remaining poorly resolved. (b) Since chloride has a very high electrophoretic mobility compared to all other ions, only chloride ions will migrate toward the anode until there is no chloride remaining in the upper buffer chamber. Only after the depletion of chloride in the upper chamber will the protein ions start to migrate. Therefore, the electrophoretic run will take several hours to overnight in a large format chamber. (c) Finally, the stacking will not be effective as in a perfect discontinuous buffer. The right procedure is to use only Tris base, glycine, and SDS for the running buffer. The buffer should not be titrated even when pH is above 9 (1).

3.6. Overfocusing of IPG Strips in 2-DE Should Be Avoided

Two-dimensional electrophoresis (2-DE) gels ideally show well-defined pattern of spots corresponding to all the individual proteins. However, the spots are sometimes blurred in certain areas. Vertical or horizontal streaks are seen instead of round spots. One reason for this is overfocusing. Of late, isoelectric focusing (the first dimension) mostly involves the use of polyacrylamide strips containing immobilized pH gradients. Since the pH gradient is fixed to the polyacrylamide matrix, it cannot go away owing to long exposure to an electric field. However, other disturbances in 2-DE maps can be brought about by overfocusing (1).

Some proteins are unstable, when they are at their isoelectric point for a certain time. It has been seen that some basic proteins are especially vulnerable to hydrolysis at their isoelectric point pH. This can bring about horizontal streaking from certain protein spots. Consequently, the focusing time applied to basic IPG strips should be limited to the minimum (1).

The addition of thiourea to urea, for protein extraction and IPG rehydration (16), results in increased solubility of hydrophobic proteins. As a result, there is a marked increase in protein spots in the 2-DE pattern compared with using urea alone. However, a strong vertical streak develops around pH 5.5 area in some cases. The protein spots along the acidic side of this streak appear more blurred than the spot pattern in the remaining areas of the gel. This phenomenon is not observed with all batches of thiourea. However, the phenomenon correlates also with the Volt hour load placed on an IPG strip. The amount of Volt hours placed on an IPG strip should be reduced if such an effect is observed (1).

3.7. PBS Must Be Removed Completely from Cells Prior to Cell Lysis

Cells derived from cell culture have to be washed efficiently with an isoosmotic solution, such as phosphate-buffered saline (PBS), prior to cell lysis to obtain antigens for high-resolution two-dimensional electrophoresis. Sometimes, the PBS solution is not removed completely from the cells. Consequently, the salt contamination of the sample causes many and long horizontal streaks in the 2-DE electrophoresis pattern.

To correct this, the PBS must be removed from the cells with a pipette completely after the final wash. Care must be taken to see that not even a tiny droplet of PBS is left behind. To ensure this, it would be good to have one to three additional washes with 250 mM sucrose and 10 mM Tris–HCl (pH 7.0) (1).

References

1. Westermeier R (2007) Frequently made mistakes in electrophoresis. Practical Proteomics 1/2007. Wiley-VCH Verlag GmbH & Co. KGaA, Weinheim. www.practical.proteomics-journal.com
2. Burgess RR (2009) Important but little known (or forgotten) artifacts in protein biochemistry. Methods Enzymol 463:813–820, Review
3. Pringle JR (1975) Methods for avoiding proteolytic artefacts in studies of enzymes and other proteins from yeasts. Methods Cell Biol 12:149–184
4. Volkin DB, Mach H, Middaugh CR (1995) Degradative covalent reactions important to protein stability. Methods Mol Biol 40:35–63
5. Deutsch DG (1976) Effect of prolonged 100°C heat treatment in SDS upon peptide bond cleavage. Anal Biochem 71:300–303
6. McDonald G, Hudson A, Dunn S et al (2008) Bioactive contaminants leach from disposable laboratory plasticware. Science 322:917
7. Stark GR (1965) Reactions of cyanate with functional groups of proteins: Reaction with amino and carboxyl groups. Biochemistry 4: 1030–1036
8. Lin M-F, Williams C, Murray MV et al (2004) Ion chromatographic quantification of cyanate in urea solutions: Estimation of the efficiency of cyanate scavengers for use in recombinant protein manufacturing. J Chromatogr B 803:353–362

9. Grabski A, Burgess RR (2001) Preparation of protein samples for SDS-polyacrylamide gel electrophoresis: Procedures and tips. inNovations 13:10–12
10. Hames BD (1990) In: Hames BD, Rickwood D (eds) One-dimensional polyacrylamide gel electrophoresis. In: Gel Electrophoresis of Proteins. Oxford University Press, New York, pp. 1–147
11. Franklin SG, Zweidler A (1977) Non-allelic variants of histones 2a, 2b and 3 in mammals. Nature 266:273–275
12. Siu CH, Lerner RA, Loomis WF (1977) Rapid accumulation and disappearance of plasma membrane proteins during development of wild-type and mutant strains of Dictyostelium discoideum. J Mol Biol 116: 469–488
13. Marshak DR, Kadonaga JT, Burgess RR et al (1996) Strategies for Protein Purification and Characterization. Cold Spring Harbor Laboratory Press, New York
14. Haeberle JR (1997) High-temperature sodium dodecyl sulfate polyacrylamide gel electrophoresis. *BioTechniques* 23:638–640
15. Läemmli UK (1970) Cleavage of structural proteins during the assembly of the head of bacteriophage T4. Nature 227:680–685
16. Rabilloud T (1998) Use of thiourea to increase the solubility of membrane proteins in two-dimensional electrophoresis. Electrophoresis 19:758–760

Index

Biji T. Kurien and R. Hal Scofield (eds.), *Protein Electrophoresis: Methods and Protocols*, Methods in Molecular Biology, vol. 869, DOI 10.1007/978-1-61779-821-4, © Springer Science+Business Media, LLC 2012

C

D

E

H

I

Q

R

S

MIX
Papier aus verantwortungsvollen Quellen
Paper from responsible sources
FSC® C105338

If you have any concerns about our products,
you can contact us on
ProductSafety@springernature.com

In case Publisher is established outside the EU,
the EU authorized representative is:
Springer Nature Customer Service Center GmbH
Europaplatz 3, 69115 Heidelberg, Germany

Printed by Libri Plureos GmbH
in Hamburg, Germany